Annals of Mathematics Studies

Number 159

Moments, Monodromy, and Perversity

A Diophantine Perspective

Nicholas M. Katz

PRINCETON UNIVERSITY PRESS

PRINCETON AND OXFORD

2005

Published by Princeton University Press, 41 William Street,
Princeton, New Jersey 08540
In the United Kingdom: Princeton University Press,
3 Market Place, Woodstock, Oxfordshire OX20 1SY

Library of Congress Cataloging-in-Publication Data
Katz, Nicholas M., 1943-
Moments, monodromy, and perversity : a diophantine perspective /
by Nicholas M. Katz.
p. cm. — (Annals of mathematics studies ; v. 159)
Includes bibliographic references and index.
ISBN-13: 978-0-691-12329-5 (acid-free paper)—
ISBN-13: 978-0-691-12330-1 (pbk. : acid-free paper)
ISBN-10: 0-691-12329-2 (acid-free paper)—
ISBN-10: 0-691-12330-6 (pbk. : acid-free paper)
1. Geometric group theory. 2. Diophantine analysis.
I. Title. II. Annals of mathematics studies ; no. 159.
QA183.K38 2005
512'.2—dc22 2004062828

British Library Cataloging-in-Publication Data is available

The publishers would like to acknowledge the author of this
volume for providing the camera-ready copy from which this book was printed.

Printed on acid-free paper. ∞

pup.princeton.edu

Printed in the United States of America

1 3 5 7 9 10 8 6 4 2

Contents

Introduction

It is now some thirty years since Deligne first proved his general equidistribution theorem [De-Weil II, Ka-GKM, Ka-Sar-RMFEM], thus establishing the fundamental result governing the statistical properties of suitably "pure" algebro-geometric families of character sums over finite fields (and of their associated L-functions). Roughly speaking, Deligne showed that any such family obeys a "generalized Sato-Tate law", and that figuring out which generalized Sato-Tate law applies to a given family amounts essentially to computing a certain complex semisimple (not necessarily connected) algebraic group, the "geometric monodromy group" attached to that family.

In our earlier books [Ka-GKM], [Ka-ESDE], and [Ka-TLFM], computations of geometric monodromy groups were carried out either directly on an open curve as parameter space, or by restriction to a well-chosen open curve in the parameter space. Once on an open curve, our main tool was to compute, when possible, the local monodromy at each of the missing points. This local monodromy information told us that our sought-after semisimple group contained specific sorts of elements, or specific sorts of subgroups. We typically also had a modicum of global information, e.g., we might have known that the sought-after group was an irreducible subgroup of GL(N), or of the orthogonal group O(N), or of the symplectic group Sp(N). It was often then possible either to decide either exactly which group we had, or to show that our group was on a very short list of possibilities, and then to distinguish among those possibilities by some ad hoc argument.

In this book, we introduce new techniques, which are resolutely global in nature. They are sufficiently powerful that we can sometimes prove that a geometric monodromy group is, say, the symplectic group Sp(N), without knowing the value of N; cf. Theorem 3.1.2 for an instance of this. The price we pay is that these new techniques apply only to families which depend on very many parameters, and thus our work here is nearly disjoint from our earlier "local monodromy" methods of analyzing one-parameter families. However, it is not entirely disjoint, because the new techniques will often leave us knowing, say, that our group is either SO(N) or O(N), but not knowing which. In such cases, we sometimes prove that the group is in fact O(N) by restricting to a suitable curve in the parameter space and then proving that the local monodromy at a particular missing point of this curve is a reflection: since SO(N) contains no reflections, we must have O(N).

Our work is based on two vital ingredients, neither of which yet existed at the time of Deligne's original work on equidistribution. The first of these ingredients is the theory of perverse sheaves, pioneered by Goresky and MacPherson in the topological setting, and then brilliantly transposed to algebraic geometry by Beilinson, Bernstein, Deligne, and Gabber. The second is Larsen's Alternative, discovered by Larsen ten odd years ago, which very nearly characterizes classical groups by their fourth moments.

This book has two goals, one "applied" and one "theoretical". The applied goal is to calculate the geometric monodromy groups attached to some quite specific universal families of (L-functions attached to) character sums over finite fields. The theoretical goal is to develop general techniques, based on combining a diophantine analysis of perverse sheaves and their higher moments with Larsen's Alternative and other group-theoretic results, which can be used to achieve the applied goal, and which are of interest in their own right.

Let us begin by describing some of the universal families we have in mind. Grosso modo, they are of three sorts:

families of additive character sums,

families of multiplicative character sums, and

Weierstrass (and other) families of L-functions of elliptic curves over function fields in one variable.

In the additive character case, we fix a finite field k and a nontrivial $\mathbb{C}$-valued additive character ψ of k. For any finite extension E/k, we denote by ψ_E the additive character of E defined by

$$\psi_E(x) := \psi(\mathrm{Trace}_{E/k}(x)).$$

Fix a pair of integers $n \ge 1$ and $e \ge 3$. We denote by $\mathcal{P}(n,e)(E)$ the space of polynomials over E in n variables of degree $\le e$. We are concerned with the families of sums, parameterized by f in $\mathcal{P}(n,e)(E)$, given by

$$\mathrm{Sum}(E, f, \psi) := \Sigma_{x_1, \ldots, x_n \text{ in } E}\, \psi_E(f(x_1, \ldots, x_n)).$$

It turns out that these sums are, up to sign, the local traces of a perverse sheaf, say $M(n, e, \psi)$, on $\mathcal{P}(n, e)/k$. On some dense open set, say $U(n, e, \psi)$ of $\mathcal{P}(n, e)/k$, this perverse sheaf is a [shift and a Tate twist of a] single lisse sheaf, say $\mathcal{M}(n, e, \psi)$, which is pure of weight zero. [When the degree e is prime to char(k), we can take $U(n, e, \psi)$ to be the open set $\mathcal{D}(n, e)$ consisting of "Deligne polynomials" of degree e in n variables, those whose leading forms define smooth, degree e hypersurfaces in $\mathbb{P}^{n-1}$.] It is the geometric monodromy of this lisse sheaf $\mathcal{M}(n, e, \psi)$ on $U(n, e, \psi)$ which we wish to calculate.

In the multiplicative character case, we fix a finite field k and a nontrivial $\mathbb{C}$-valued multiplicative character χ of $k^\times$, extended to all of k by $\chi(0) := 0$. For any finite extension E/k, we denote by χ_E the multiplicative character of $E^\times$ defined by

$$\chi_E(x) := \chi(\mathrm{Norm}_{E/k}(x)),$$

again extended to all of E by $\chi_E(0) := 0$. Fix a pair of integers $n \ge 1$ and $e \ge 3$. We are concerned with the families of sums, parameterized by f in $\mathcal{P}(n,e)(E)$, given by

$$\mathrm{Sum}(E, f, \chi) := \Sigma_{x_1, ..., x_n \text{ in } E}\ \chi_E(f(x_1, ..., x_n)).$$

It turns out that these sums are, up to sign, the local traces of a perverse sheaf, say $M(n, e, \chi)$, on $\mathcal{P}(n, e)/k$. On some dense open set, say $U(n, e, \chi)$ of $\mathcal{P}(n, e)/k$, this perverse sheaf is a [shift and a Tate twist of a] single lisse sheaf, say $\mathcal{M}(n, e, \chi)$. [When the degree e is prime to char(k), we can take $U(n, e, \chi)$ to be the open set $\mathcal{SD}(n, e)$ consisting of "strong Deligne polynomials" of degree e in n variables, those which define smooth hypersurfaces in $\mathbb{A}^n$ and whose leading forms define smooth, degree e hypersurfaces in $\mathbb{P}^{n-1}$.] When χ^e is nontrivial, $\mathcal{M}(n, e, \chi)$ is pure of weight zero, and it is the geometric monodromy of $\mathcal{M}(n, e, \chi)$ we wish to calculate. When χ^e is trivial, $\mathcal{M}(n, e, \chi)$ is mixed of weight ≤ 0, and it is the weight zero quotient of $\mathcal{M}(n, e, \chi)$ whose geometric monodromy we wish to calculate.

In the simplest instance of Weierstrass families of L-functions of elliptic curves, we fix a finite field k of characteristic $p \ge 5$. We denote by χ_2 the quadratic character of $k^\times$. Fix a pair of integers $d_2 \ge 3$ and $d_3 \ge 3$. For each finite extension E/k, we have the product space $(\mathcal{P}(1,d_2)\times\mathcal{P}(1,d_3))(E)$ of pairs $(g_2(t), g_3(t))$ of one-variable polynomials over E of degrees at most d_2 and d_3 respectively. We are concerned with the families of sums, parameterized by (g_2, g_3) in $(\mathcal{P}(1,d_2)\times\mathcal{P}(1,d_3))(E)$,

$$\mathrm{Sum}(E, g_2, g_3) := \Sigma_{x, t \text{ in } k}\ \chi_{2,E}(4x^3 - g_2(t)x - g_3(t)).$$

It turns out that these sums are, up to sign, the local traces of a perverse sheaf, say $W(d_2, d_3)$, on $\mathcal{P}(1,d_2)\times\mathcal{P}(1,d_3)/k$. On the dense open set $U(d_2, d_3)$, of $\mathcal{P}(1,d_2)\times\mathcal{P}(1,d_3)/k$, defined by the condition that $(g_2)^3 - 27(g_3)^2$ has $\mathrm{Max}(3d_2, 2d_3)$ distinct zeroes in $\overline{k}$, this perverse sheaf is a [shift and a Tate twist of a] single lisse sheaf, say $\mathcal{W}(d_2, d_3)$, which is mixed of weight ≤ 0. The weight zero quotient $\mathrm{Gr}^0\mathcal{W}(d_2, d_3)$ of $\mathcal{W}(d_2, d_3)$ is related to L-functions of elliptic curves over function fields as follows. For (g_2, g_3) in $U(d_2, d_3)(E)$, the Weierstrass equation

$$y^2 = 4x^3 - g_2(t)x - g_3(t)$$

defines an elliptic curve over the rational function field E(t), and its (unitarized) L-function is the local L-function of $\mathrm{Gr}^0\mathcal{W}(d_2, d_3)$ at the point (g_2, g_3) in $U(d_2, d_3)(E)$. It is the geometric monodromy of $\mathrm{Gr}^0\mathcal{W}(d_2, d_3)$ we wish to calculate.

This concludes our quick overview of the sorts of universal families we wish to treat. These families have in common some essential features.

The first feature is that, in each case, the parameter space is itself a large linear space of $\mathbb{A}^m$-valued functions (m = 1 in the first two sorts of families, $m = 2$ in the third sort) on some fixed variety V, i.e., in each case our parameter space is a large linear subspace $\mathcal{F}$ of the space $\mathrm{Hom}_{k\text{-scheme}}(V, \mathbb{A}^m)$. [It happens that V is itself an affine space in the examples we have given above ($\mathbb{A}^n$ in the first two sorts, $\mathbb{A}^1$ in the third sort), but this turns out to be a red herring.]

The second feature is that our family of sums has the following structure: for each E/k, we are given a function

$$K(E, \ ): \mathbb{A}^m(E) \to \mathbb{C},$$
$$x \mapsto K(E, x),$$

on the E-valued points of the target $\mathbb{A}^m$, and our family of sums is

$$f \text{ in } \mathcal{F}(E) \subset \mathrm{Hom}_{E\text{-scheme}}(V, \mathbb{A}^m) \mapsto \Sigma_{v \text{ in } V(E)} K(E, f(v)).$$

In the additive character case, we have $m=1$, and $x \mapsto K(E, x)$ is the function $x \mapsto \psi_E(x)$ on $\mathbb{A}^1(E) = E$. In the multiplicative character case, we have $m=1$, and $x \mapsto K(E, x)$ is the function $x \mapsto \chi_E(x)$ on $\mathbb{A}^1(E) = E$. In the case of L-functions of elliptic curves over function fields, we have $m=2$, and the function $(a, b) \mapsto K(E, a, b)$ on $\mathbb{A}^2(E) = E\times E$ is the function

$$(a, b) \mapsto \Sigma_{x \text{ in } E}\, \chi_{2,E}(4x^3 - ax - b).$$

[In these cases, the function $x \mapsto K(E, x)$ on $\mathbb{A}^m(E)$ also satisfies in addition the "integral zero" condition

$$\Sigma_{x \text{ in } \mathbb{A}^m(E)} K(E, x) = 0,$$

as the reader will easily check. This turns out to be an important condition, but one that can be somewhat relaxed.]

The third feature is that, in each case, the collection of functions

$$K(E, \ ): \mathbb{A}^m(E) \to \mathbb{C},$$
$$x \mapsto K(E, x),$$

is, up to sign, the trace function of a perverse sheaf K on $\mathbb{A}^m$.

Although not apparent from these examples, there is also interest in introducing, in addition to our perverse sheaf K on $\mathbb{A}^m$, a perverse sheaf L on the source variety V, with trace function

$$L(E, \ ): V(E) \to \mathbb{C},$$
$$v \mapsto L(E, v),$$

and considering the family of sums

$$f \text{ in } \mathcal{F}(E) \subset \mathrm{Hom}_{E\text{-scheme}}(V, \mathbb{A}^m) \mapsto \Sigma_{v \text{ in } V(E)} K(E, f(v))L(E, v).$$

Slightly more generally, one might fix a single function

$$h \text{ in } \mathrm{Hom}_{k\text{-scheme}}(V, \mathbb{A}^m),$$

and consider the family of sums "with an offset of h", namely

$$f \text{ in } \mathcal{F}(E) \subset \mathrm{Hom}_{E\text{-scheme}}(V, \mathbb{A}^m)$$
$$\mapsto \Sigma_{v \text{ in } V(E)} K(E, h(v) + f(v))L(E,v).$$

Let us now turn to a brief description of the "theoretical"

aspects of this book, which are mainly concentrated in the first two chapters.

In the first chapter, we show that, under very mild hypotheses on K and L, these sums are, for any fixed h, the trace function (up to sign) of a perverse sheaf Twist(L,K,$\mathcal{F}$,h) on the function space $\mathcal{F}$. This general construction is responsible for the perverse sheaves M(n, e, ψ) and M(n, e, χ) on the space $\mathcal{P}$(n, e)/k discussed in the additive and multiplicative character cases, and it is responsible for the perverse sheaf W(d_2, d_3), on $\mathcal{P}(1,d_2)\times\mathcal{P}(1,d_3)$/k discussed in the Weierstrass family case. We then formulate in diophantine terms a general orthogonality theorem for pure perverse sheaves, which is formally analogous to the orthogonality theorem for the characters of finite-dimensional representations of a compact Lie group. Proceeding along the same lines, we formulate in diophantine terms the theory of the Frobenius-Schur indicator for geometrically irreducible pure lisse sheaves. This theory is formally analogous to that of the Frobenius-Schur indicator for irreducible representations of a compact Lie group, which tells us whether a given irreducible representation is self dual or not, and tells us, in the autodual case, whether the autoduality is symplectic or orthogonal We then show that, given these diophantine invariants for suitable input perverse sheaves K on $\mathbb{A}^m$ and L on V, there is a simple rule for calculating them for (a suitable quotient of) the perverse sheaf Twist(L,K,$\mathcal{F}$,h) on the function space $\mathcal{F}$.

Up to this point in our theoretical analysis, we require relatively little of our space of functions $\mathcal{F}$, only that it contain the constant functions and that it separate points. We then formulate the notion of "higher moments" for pure perverse sheaves. [The orthogonality theorem is concerned with the "second moment".] To get results on the higher moments, we must require that the function space $\mathcal{F}$ be suitably large, more precisely, that it be "d-separating" for some d ≥ 4. Here d-separating means that given any field extension E/k, and any d distinct points v_1, ..., v_d in V(E), the E-linear map "simultaneous evaluation" at the points v_1, ..., v_d",

$$\mathcal{F}\otimes_k E \rightarrow (\mathbb{A}^m(E))^d,$$
$$f \mapsto (f(v_1), f(v_2), \ldots, f(v_d))$$

is surjective. [In the examples, the degrees ("e" in the first two cases, "d_2" and "d_3" in the Weierstrass case) are taken to be at least 3 in order to insure that our function spaces are at least 4-separating.]

We end the first chapter by proving a quite general "Higher Moment Theorem" . We suppose that the function space $\mathcal{F}$ is d-separating for some d ≥ 4. Then we get control of the even moments M_{2k}, for every positive even integer 2k ≤ d, of (a suitable quotient of) the perverse sheaf Twist(L,K,$\mathcal{F}$,h) on the function space $\mathcal{F}$. An immediate consequence of this control is the fact that the support of (this suitable quotient of) the perverse sheaf Twist(L,K,$\mathcal{F}$,h) is the entire space $\mathcal{F}$. Its restriction to an open dense set of $\mathcal{F}$ is a (shift of a single) lisse sheaf, whose geometric monodromy is what we wish to calculate, and whose higher moments we now control.

In chapter 2, we bring to bear some very important ideas of Michael Larsen, about the determination of classical groups through their higher moments. The idea which we exploit most extensively is "Larsen's Alternative", in which we are given an integer $N \geq 2$ and a reductive subgroup H of one of the classical groups $GL(N, \mathbb{C})$ or $O(N, \mathbb{C})$, or, when N is even and at least 4, $Sp(N, \mathbb{C})$, and we are told that H has the same fourth moment as the ambient group in the given N-dimensional representation (namely 2, 3, 3 in the three successive cases). Larsen's Alternative is the marvelous statement that either H is finite, or that, in the three successive cases, we have

H contains SL(N), in the GL(N) case,

H is either SO(N) or O(N), in the O(N) case,

H is Sp(N), in the Sp(N) case.

This very nearly reduces us to ruling out the possibility that H is finite. [We say very nearly, because we must still compute determinants, i.e., we must still distinguish between SO(N) and O(N), and we must still distinguish among the various groups between SL(N) and GL(N).] Fortunately, there is a great deal known about the possible finite groups which could arise in this context. For $N \geq 3$, any such finite group is, because of its low fourth moment, automatically a primitive subgroup of GL(N). We can then apply the plethora of known results on finite primitive irreducible subgroups of GL(N), due (in chronological order) to Blichfeldt, Mitchell, Huffman-Wales, Zalesskii, and Wales. We can apply all this theory to an H which is the geometric monodromy group attached to (a suitable quotient of the restriction to a dense open set of) the perverse sheaf $\mathrm{Twist}(L,K,\mathcal{F},h)$, thanks to the control over moments gained in the first chapter. For such an H, there are further tools we can bring to bear, both algebro-geometric (the theory of "sheaves of perverse origin") and diophantine in nature. All of this is explained in the second chapter.

A further idea of Michael Larsen is his unpublished "Eighth Moment Conjecture". Suppose $N \geq 8$, and suppose we are given a reductive subgroup H of one of $GL(N, \mathbb{C})$ or $O(N, \mathbb{C})$, or, when N is even, $Sp(N, \mathbb{C})$. Suppose H has the same eighth moment as the ambient group in the given N-dimensional representation. Then Larsen conjectured that, in the successive cases, we have

H contains SL(N), in the GL(N) case,

H is either SO(N) or O(N), in the O(N) case,

H is Sp(N), in the Sp(N) case.

In other words, if we have the correct eighth moment (which implies that the lower even moments are also "correct"), then the "H finite" case of Larsen's Alternative cannot arise. Larsen's Eighth Moment Conjecture has recently been proven by Guralnick and Tiep. Combining their result and the Higher Moment Theorem, we avoid the need to rule out the "H finite" case, provided only that our space of functions $\mathcal{F}$ is at least 8-separating. To see what this means in practice, consider the three examples of universal families we considered above. To have $\mathcal{F}$ 8-separating, we need to take the degree $e \geq 7$ in the cases of additive and multiplicative character sums, and we need to take the degrees d_2 and d_3 both ≥ 7 in the Weierstrass case. [But we still face the earlier mentioned problem of computing the determinant.]

With these tools at hand, we get down to concrete applications. Chapters 3 and 4 are devoted to additive character sums, first on $\mathbb{A}^n$ and then on more general varieties. In chapter 5, we study multiplicative character sums. The results we obtain in these chapters are nearly complete, except that in a number of cases we cannot distinguish whether we have $SO(N)$ or $O(N)$. In chapter 6, we apply the theory of middle additive convolution on the additive group $\mathbb{G}_a = \mathbb{A}^1$ to both additive and multiplicative character sums on $\mathbb{A}^n$. This theory allows us in many cases to compute determinants, and thus distinguish between the $O(N)$ and $SO(N)$ cases. It is in this use of middle additive convolution that we are falling back on the method of restricting to a suitable curve and then computing local monodromies, in order to show that our group contains pseudoreflections of specified determinant. In an appendix to chapter 6, we further develop some technical themes which appeared in the proof of a key technical result, Theorem 6.2.11, which was worked out jointly with Eric Rains.

In chapter 7, we work systematically with "pullback to a curve from $\mathbb{A}^1$" situations. A typical example of the situation we study is this. Take a finite field k of odd characteristic, and consider the rational function field in one variable $k(\lambda)$, over which we have the Legendre curve, defined by the equation

$$y^2 = x(x-1)(x - \lambda).$$

Fix an integer $e \geq 3$. For each finite extension E/k, and each each polynomial $f(\lambda)$ in $E[\lambda]$ of degree at most e (i.e., f lies in $\mathcal{P}(1, e)(E)$), we have the pullback equation

$$y^2 = x(x-1)(x - f(\lambda)).$$

The sums

$$\mathrm{Sum}(f, E) := \Sigma_{x,\lambda \text{ in } E}\ \chi_{2,E}(x(x-1)(x - f(\lambda)))$$

are, up to sign, the trace function of a perverse sheaf on $\mathcal{P}(1, e)$. For f in the dense open set U of $\mathcal{P}(1, e)$ consisting of those polynomials f such that $f(f-1)$ has $2e$ distinct roots in $\overline{k}$, this perverse sheaf is a (shift and a Tate twist of a) single lisse sheaf, whose rank N is

$2e - 2$, if e odd,

$2e - 3$, if e even,

and whose local L-function at f in $U(E)$ is precisely the unitarized L-function of the elliptic curve over $E(\lambda)$ defined by the pullback equation

$$y^2 = x(x-1)(x - f(\lambda)).$$

We prove that this lisse sheaf has geometric monodromy group the full orthogonal group O(N), provided that $N \geq 9$. At the very end of this chapter, we give some results on degeneration of Leray spectral sequences, which are certainly well known to the experts, but for which we know of no convenient reference.

In chapter 8, we indicate how the general theory of Twist(L,K,$\mathcal{F}$,h) developed here allows us to recover some of the results of [Ka-TLFM].

Chapters 9, 10, and 11 are devoted to a detailed study of families of L-functions of elliptic curves over function fields in one variable over finite constant fields. Chapter 9 is devoted to explaining how various classical families of elliptic curves provide appropriate input, namely a suitable perverse sheaf K on an $\mathbb{A}^m$, to the general theory. Chapter 10 works out what the general theory gives for various sorts of Weierstrass families, and Chapter 11 works it out for other, more neglected, universal families, which we call FJTwist families.

In chapter 12, we return to theoretical questions, developing some general if ad hoc methods which allow us to work "over $\mathbb{Z}$" instead of "just" over a finite field. These methods apply nicely to the case of multiplicative character sums, and to the various Weierstrass and FJTwist families. What they make possible is equidistribution statements where we are allowed to work over bigger and bigger finite fields, whose characteristics are allowed to vary, e.g., bigger and bigger prime fields, rather than the more restrictive setting of bigger and bigger finite fields of a fixed characteristic. Unfortunately, the methods do not apply at all to additive character sums. Nonetheless, we believe that the corresponding equidistribution statements, about additive character sums over bigger and bigger finite fields whose characteristics are allowed to vary, are in fact true statements. It is just that we are presently incapable of proving them.

In the final chapter 13, we make explicit the application of our results to the arithmetic of elliptic curves over function fields. We first give results on average analytic rank in our families. We then pass to the large-N limit, e.g., by taking Weierstrass families of type (d_2, d_3) as described earlier, and letting $\mathrm{Max}(3d_2, 2d_3)$ tend to infinity, and give results concerning low-lying zeroes as incarnated in the eigenvalue location measures of [Ka-Sar, RMFEM].

It is a pleasure to acknowledge the overwhelming influence on this book of the ideas and work of Deligne, Gabber, and Larsen. In the course of working on the book, I visited the Institute for Advanced Study, the University of Tokyo, the University of Minnesota, the University of Paris at Orsay, I.H.E.S., and the University of Paris VI. I thank all these institutions for their hospitality and support.

Chapter 1: Basic results on perversity and higher moments

(1.1) The notion of a d-separating space of functions

(1.1.1) Throughout this section, we work over a field k. For each integer $m \geq 1$, we denote by $\mathbb{A}^m{}_k$, or just $\mathbb{A}^m$ if no confusion is likely, the m-dimensional affine space over k.

(1.1.2) Let V be a separated k-scheme of finite type. The set

$$\mathrm{Hom}_{k\text{-schemes}}(V, \mathbb{A}^m)$$

of k-morphisms from V to $\mathbb{A}^m$ is naturally a k-vector space (addition and scalar multiplication on the target). Concretely, it is the k-vector space of m-tuples of regular functions on V. If V is nonempty, $\mathrm{Hom}_{k\text{-schemes}}(V, \mathbb{A}^m)$ contains $\mathbb{A}^m(k)$ as the subspace of constant maps.

(1.1.3) For technical reasons, we consider the following generalization of a finite-dimensional k-subspace of $\mathrm{Hom}_{k\text{-schemes}}(V, \mathbb{A}^m)$: a pair $(\mathcal{F}, \tau)$ consisting of a finite-dimensional k-vector space $\mathcal{F}$ and a k-linear map

$$\tau : \mathcal{F} \to \mathrm{Hom}_{k\text{-schemes}}(V, \mathbb{A}^m).$$

We will, by abuse of language, refer to such a pair $(\mathcal{F}, \tau)$ as a space of functions on V.

(1.1.4) Given such an $(\mathcal{F}, \tau)$, for any extension field L/k, and for any point v in V(L), we denote by eval(v) the L-linear "evaluation at v" map

$$\mathrm{eval}(v) : \mathcal{F}\otimes_k L \to \mathbb{A}^m(L),$$
$$f \mapsto (\tau(f))(v).$$

When the map τ is clear from the context, as it usually will be, we write simply f(v) for $(\tau(f))(v)$:

$$f(v) := (\tau(f))(v).$$

(1.1.5) Given an integer $d \geq 1$, we say that $(\mathcal{F}, \tau)$, or simply $\mathcal{F}$, if the map τ is clear from the context, is d-separating if the following condition holds: for any extension field L/k, and for any d distinct points $v_1, v_2,\ldots, v_d$ in V(L), the L-linear map "simultaneous evaluation"

$$\mathrm{eval}(v_1, v_2,\ldots, v_d) : \mathcal{F}\otimes_k L \to (\mathbb{A}^m(L))^d,$$
$$f \mapsto (f(v_1),\ldots, f(v_d)),$$

is surjective.

(1.1.6) A trivial but useful remark is this. Suppose we are given a space of functions $(\mathcal{F}, \tau)$ on V, and a k-vector subspace $\mathcal{F}_1 \subset \mathcal{F}$. Restrict τ to $\mathcal{F}_1$, and view $(\mathcal{F}_1, \tau|\mathcal{F}_1)$ as a space of functions on V. If

$(\mathcal{F}_1, \tau|\mathcal{F}_1)$ is d-separating, then $(\mathcal{F}, \tau)$ is d-separating.

(1.1.7) We say that $(\mathcal{F}, \tau)$ contains the constants if the subspace $\tau(\mathcal{F})$ of $\mathrm{Hom}_{k\text{-schemes}}(V, \mathbb{A}^m)$ contains the subspace $\mathbb{A}^m(k)$ of constant maps.

(1.1.8) For $m = 1$, the prototypical example of a d-separating $\mathcal{F}$ which contains the constants is this. Suppose that V is affine, or that, more generally, we are given an integer $n \geq 1$ and a radiciel (injective on field-valued points) k-morphism

$$\pi : V \to \mathbb{A}^n.$$

For each integer $e \geq 0$, denote by $\mathcal{P}_e$ the k-space of all polynomial functions P on $\mathbb{A}^n$ of degree at most e. Thus $\mathcal{P}_e$ contains the constants, and it is well known that $\mathcal{P}_e$ is d-separating on $\mathbb{A}^n$ for every $d \leq e+1$, cf. [Ka-LAMM, 2.2.10]. Then the pair $(\mathcal{P}_e, \pi^{-1})$:

$$\mathcal{P}_e \to \mathrm{Hom}_{k\text{-schemes}}(V, \mathbb{A}^1),$$
$$P \mapsto P\circ\pi,$$

is d-separating for every $d \leq e+1$, and it contains the constants.

(1.1.9) Another example to keep in mind, still with $m = 1$, is this. Suppose that C/k is a projective, smooth, geometrically connected curve over k of genus denoted g, $S \subset C$ is a finite nonempty reduced closed subscheme, and V is $C - S$. Take any effective divisor D on C whose support lies in S and such that $\deg(D) \geq 2g + d-1$. Then the Riemann-Roch space $L(D)$, viewed as a space of functions on $C - S$, contains the constants and is d-separating.

(1.1.10) For general $m \geq 1$, the prototypical example of a d-separating $\mathcal{F}$ which contains the constants is obtained by a product construction. For each i=1 to m, suppose we have pairs $(\mathcal{F}_i, \tau_i)$,

$$\tau_i : \mathcal{F}_i \to \mathrm{Hom}_{k\text{-schemes}}(V, \mathbb{A}^1),$$

which contain the constants and are d-separating. Then the product space $\mathcal{F} := \prod_i\mathcal{F}_i$ with the product map $\tau := \prod_i\tau_i$,

$$\prod_i\tau_i : \prod_i\mathcal{F}_i \to \prod_i \mathrm{Hom}_{k\text{-schemes}}(V, \mathbb{A}^1) = \mathrm{Hom}_{k\text{-schemes}}(V, \mathbb{A}^m)$$

is d-separating and contains the constants.

(1.1.11) For example, with $V = C - S$ as above, we might take the $\mathcal{F}_i$ each to be Riemann-Roch spaces $L(D_i)$, where the D_i are m possibly distinct effective divisors, each of degree at least $2g + d - 1$, and each with support contained in S.

(1.1.12) As another example, with V as in (1.1.8), we might be given various integers $n_i \geq 1$, i =1 to r, and radiciel k-morphisms

$$\pi_i : V \to \mathbb{A}^{n_i}.$$

We might then pick integers $e_i \geq d-1$, and take

$$(\mathcal{F}_i, \tau_i) := (\mathcal{P}_{e_i}, \pi_i^{-1}).$$

(1.1.13) Let us specialize this last example. If we take all n_i to have a common value n, and all π_i to be the same radiciel k-

morphism

$$\pi : V \to \mathbb{A}^n,$$

and all $e_i = 1$, then the product $\mathcal{F}$ is the space AffMaps($\mathbb{A}^n$, $\mathbb{A}^m$) of all affine maps from $\mathbb{A}^n$ to $\mathbb{A}^m$, and we are in the situation considered at length in [Ka-ACT, Chapter 1]. Of course the $\mathcal{F}$ we get this way has no reason to be more than 2-separating. If we keep all π_i equal to π, but now take all e_i to have some common value $e \geq d-1$, we obtain a situation which is d-separating, and which can be viewed as an instance of the previous AffMaps situation. Denote by $N(n,e)$ the number of monomials X^W in n variables whose degrees lie in [1, e], and replace π by the map $\pi_e : V \to \mathbb{A}^{N(n,e)}$ defined by all such monomials. The situation (all $\pi_i = \pi$, all $e_i = e$) is just the situation (all $\pi_i = \pi_e$, all $e_i = 1$).

(1.1.14) If $\dim(V) \geq 1$, and if $(\mathcal{F}, \tau)$ is d-separating, then $(\mathcal{F}, \tau)$ is r-separating for every integer r with $1 \leq r \leq d$. Indeed, given r distinct points in some V(L), we must show that the map

$$\mathrm{eval}(v_1,\ldots, v_r) : \mathcal{F}\otimes_k L \to (\mathbb{A}^m(L))^r$$

is surjective. This is a linear algebra question, which can be checked after an arbitrary field extension E/L. After some field extension E/L, we can find points $v_{r+1},\ldots, v_d$ in V(E) such that $v_1,\ldots, v_d$ are all distinct in V(E). Then $\mathrm{eval}(v_1,\ldots, v_d)$ maps $\mathcal{F}\otimes_k E$ onto $(\mathbb{A}^m(E))^d$. Composing this map with the projection "first r factors" of $(\mathbb{A}^m(E))^d$ onto $(\mathbb{A}^m(E))^r$, we find that $\mathrm{eval}(v_1,\ldots, v_r)$ maps $\mathcal{F}\otimes_k E$ onto $(\mathbb{A}^m(E))^r$, as required. [We need to assume $\dim(V) \geq 1$ because if $\dim(V) = 0$, then for $d > \mathrm{Card}(V(\bar{k}))$, there do not exist d distinct points in any V(L), so any $\mathcal{F}$, even {0}, is d-separating for such d.]

(1.1.15) There are two final notions we need to introduce, which will be used in the next section. We say that $(\mathcal{F}, \tau)$ is quasifinitely 2-separating if the following condition holds: for every extension field L/k, and for every point v in V(L), there are only finitely many points w in V(L) for which the simultaneous evaluation map

$$\mathrm{eval}(v, w) : \mathcal{F}\otimes_k L \to (\mathbb{A}^m(L))^2,$$
$$f \mapsto (f(v), f(w)),$$

fails to be surjective. This notion arises naturally as follows. Suppose we start with an $(\mathcal{F}, \tau)$ which is 2-separating for V. If $\pi : U \to V$ is a quasifinite k-morphism, then $(\mathcal{F}, \pi^{-1}\circ\tau)$ is quasifinitely 2-separating for U. Indeed, given two (not necessarily distinct points) u_1 and u_2 in U(L), we have

$$\mathrm{eval}(u_1, u_2) = \mathrm{eval}(\pi(u_1), \pi(u_2))$$

as maps from $\mathcal{F}\otimes_k L$ to $(\mathbb{A}^m(L))^2$. But $\mathrm{eval}(\pi(u_1), \pi(u_2))$ is surjective so long as $\pi(u_2) \neq \pi(u_1)$. For fixed u_1, there are only finitely many u_2 for which this fails, precisely because π is quasifinite.

(1.1.16) We say that $(\mathcal{F}, \tau)$ is quasifinitely difference-separating if the following condition holds: for every extension field L/k, and for every point v in V(L), there are only finitely many points w in V(L) for which the difference map

$$\mathrm{eval}(v) - \mathrm{eval}(w) : \mathcal{F}\otimes L \to \mathbb{A}^m(L)$$

fails to be surjective.

(1.1.17) Notice that if $(\mathcal{F}, \tau)$ is quasifinitely 2-separating, then it is quasifinitely difference-separating. Indeed, for given L and given v in V(L), there are only finitely many w in V(L) for which

$$\mathrm{eval}(v, w): \mathcal{F}\otimes L \to (\mathbb{A}^m(L))^2$$

fails to be surjective. Composing with the surjective subtraction map

$$(\mathbb{A}^m(L))^2 \to \mathbb{A}^m(L)$$
$$(x, y) \mapsto x - y,$$

we see that, except for finitely many exceptional w, the map

$$\mathrm{eval}(v) - \mathrm{eval}(w) : \mathcal{F}\otimes L \to \mathbb{A}^m(L)$$

is itself surjective.

(1.2) Review of semiperversity and perversity

(1.2.1) We work over a field k. We fix a prime number ℓ which is invertible in k. On variable separated k-schemes of finite type X, we work systematically with objects of $D^b{}_c(X, \overline{\mathbb{Q}}_\ell)$.

(1.2.2) Recall [BBD, 4.0.1] that an object K of $D^b{}_c(X, \overline{\mathbb{Q}}_\ell)$ is called semiperverse if its cohomology sheaves $\mathcal{H}^i(K)$ satisfy

$$\dim \mathrm{Supp}(\mathcal{H}^i(K)) \le -i, \text{ for every integer } i.$$

One way to think of semiperversity is this. Pick a stratification $\{Z_\alpha\}_\alpha$ of X, i.e., write X^{red} as the disjoint union of finitely many locally closed subschemes Z_α/k. Assume that the stratification is smooth, in the sense that each Z_α/k is connected and smooth. Assume further that the stratification is adapted to K, in the sense that for each i and each α, $\mathcal{H}^i(K)|Z_\alpha$ is lisse on Z_α. Then K is semiperverse if and only if we have, for all (i, α),

$$\mathcal{H}^i(K)|Z_\alpha = 0 \text{ if } \dim(Z_\alpha) > -i.$$

Lemma 1.2.3 Let K be an object of $D^b{}_c(X, \overline{\mathbb{Q}}_\ell)$, and $\{Z_\alpha\}_\alpha$ an arbitrary stratification of X. Then K is semiperverse on X if and only if, for each α, $K|Z_\alpha$ is semiperverse on Z_α.

proof Pick a smooth stratification $\{Z_{\alpha,\beta}\}_\beta$ of each Z_α which is adapted to $K|Z_\alpha$. Then $\{Z_{\alpha,\beta}\}_{\alpha,\beta}$ is a smooth stratification of X to which K is adapted. So both the semiperversity of K and the semiperversity of $K|Z_\alpha$ for each α are equivalent to the conditions

$$\mathcal{H}^i(K)|Z_{\alpha,\beta} = 0 \text{ if } \dim(Z_{\alpha,\beta}) > -i$$

for all (i,α,β). QED

(1.2.4) An object K is called perverse if both K and its Verdier dual $D_{X/k}K := RHom(K, f^!\overline{\mathbb{Q}}_\ell)$, $f: X \to \mathrm{Spec}(k)$ denoting the structural morphism, are semiperverse.

(1.3) A twisting construction: the object Twist(L,K,$\mathcal{F}$,h)

(1.3.1) The general setup is similar to that in [Ka-ACT, Chapter 1]. We give ourselves a separated k-scheme V of finite type, and a space of functions $(\mathcal{F}, \tau)$,

$$\tau : \mathcal{F} \to \mathrm{Hom}_{k\text{-scheme}}(V, \mathbb{A}^m),$$

on V. We also give ourselves a k-morphism

$$h : V \to \mathbb{A}^m.$$

We view $\mathcal{F}$ as an affine space over k (i.e., the k-scheme whose A-valued points, for any k-algebra A, are $\mathcal{F}\otimes_k A$). We have a k-morphism

$$h_{aff} : V\times\mathcal{F} \to \mathbb{A}^m,$$
$$(v, f) \mapsto h(v) + f(v).$$

We also have the two projections

$$pr_1 : V\times\mathcal{F} \to V,$$
$$pr_2 : V\times\mathcal{F} \to \mathcal{F}.$$

(1.3.2) Given objects

K in $D^b_c(\mathbb{A}^m, \overline{\mathbb{Q}}_\ell)$,

L in $D^b_c(V, \overline{\mathbb{Q}}_\ell)$,

we have the objects $h_{aff}{}^*K[\dim\mathcal{F} - m]$ and $pr_1{}^*L$ on $V\times\mathcal{F}$. We then form their tensor product $pr_1{}^*L\otimes h_{aff}{}^*K[\dim\mathcal{F} - m]$ on $V\times\mathcal{F}$. We then form the object

$$M := Rpr_{2!}(pr_1{}^*L\otimes h_{aff}{}^*K[\dim\mathcal{F} - m])$$

on $\mathcal{F}$. We view this M as a "twist" of L by K, via $\mathcal{F}$ and h. When we wish to emphasize its genesis, we will denote it Twist(L,K,$\mathcal{F}$,h):

(1.3.3) $M = \mathrm{Twist}(L,K,\mathcal{F},h) := Rpr_{2!}(pr_1{}^*L\otimes h_{aff}{}^*K[\dim\mathcal{F} - m])$.

(1.4) The basic theorem and its consequences

(1.4.1) We assume henceforth that our ground field k has positive characteristic $p > 0$. [We make this assumption because the proof of the basic theorem below uses the Fourier Transform on the ℓ-adic derived category, an operation which only makes sense in positive characteristic.]

Semiperversity Theorem 1.4.2 (compare [Ka-ACT, 1.5]) Hypotheses and notations as in (1.3) and (1.4.1) above, suppose in addition that the following four conditions hold:

1) $(\mathcal{F}, \tau)$ is quasifinitely difference-separating on V, and contains the constants,

2) K in $D^b_c(\mathbb{A}^m, \overline{\mathbb{Q}}_\ell)$ is semiperverse on $\mathbb{A}^m$,

3) L in $D^b_c(V, \overline{\mathbb{Q}}_\ell)$ is semiperverse on V,

4) the graded vector space

$$H^*{}_c(\mathbb{A}^m\otimes\bar{k}, K[m])\otimes H^*{}_c(V\otimes\bar{k}, L)$$

is concentrated in degree ≤ 0, i.e., we have

$$H^i{}_c(\mathbb{A}^m\otimes\bar{k}, K[m])\otimes H^j{}_c(V\otimes\bar{k}, L) = 0 \text{ if } i + j \geq 1,$$

or equivalently, we have

$$H^i{}_c((V\times\mathbb{A}^m)\otimes_k\bar{k}, pr_1{}^*L\otimes pr_2{}^*K) = 0 \text{ if } i \geq m+1.$$

Then the object

$$M = \mathrm{Twist}(L,K,\mathcal{F},h) := Rpr_{2!}(pr_1{}^*L\otimes h_{aff}{}^*K[\dim\mathcal{F} - m])$$

on $\mathcal{F}$ is semiperverse, and

$$H_c{}^*(\mathcal{F}\otimes_k\bar{k}, M[\dim\mathcal{F}]) = H^*{}_c(\mathbb{A}^m\otimes\bar{k}, K[m])\otimes H^*{}_c(V\otimes\bar{k}, L)(m -\dim\mathcal{F}).$$

proof The proof of the theorem is similar to, but simpler than, the proof of [Ka-ACT, Theorem 1.5].

Because $(\mathcal{F}, \tau)$ contains the constants, we can pick a subspace $\mathbb{A}^m(k)$ in $\mathcal{F}$ which maps by τ to the $\mathbb{A}^m(k)$ of constant maps in $\mathrm{Hom}_{k\text{-schemes}}(V, \mathbb{A}^m)$. Having made this choice, we make the further choice of a k-subspace $\mathcal{F}_0 \subset \mathcal{F}$ which is a vector space complement to $\mathbb{A}^m(k)$. [For example, if V(k) is nonempty, we can pick a point v in V(k) and take for $\mathcal{F}_0$ those f in $\mathcal{F}$ such that f(v) = 0 in $\mathbb{A}^m(k)$.] Thus we have a direct sum decomposition

$$\mathcal{F} = \mathcal{F}_0 \oplus \mathbb{A}^m(k)$$

of k-spaces, or, what is the same, a product decomposition

$$\mathcal{F} = \mathcal{F}_0\times\mathbb{A}^m$$

of affine spaces over k. [This decomposition is the analogue of the decomposition

$$\mathrm{AffMaps}(\mathbb{A}^n, \mathbb{A}^m) = \mathrm{LinMaps}(\mathbb{A}^n, \mathbb{A}^m) \times \mathbb{A}^m$$

of [Ka-ACT, 1.3].]

Exactly as in [Ka-ACT, 1.7.2], it suffices to show that the Fourier Transform $FT_\psi(M)$, ψ a chosen nontrivial $\bar{\mathbb{Q}}_\ell{}^\times$-valued additive character of the prime subfield $\mathbb{F}_p$ of k, is semiperverse on the linear dual space

$$\mathcal{F}^\vee = \mathcal{F}_0{}^\vee \times \mathbb{A}^{m\vee}.$$

Let us recall how this semiperversity is established.

For L/k any extension field, and for any v in V(L), we denote by

$$\mathrm{eval}_0(v) : \mathcal{F}_0\otimes_k L \to \mathbb{A}^m(L)$$

the restriction to the subspace $\mathcal{F}_0\otimes_k L$ of the evaluation map

$$\mathrm{eval}(v) : \mathcal{F}\otimes_k L \to \mathbb{A}^m(L).$$

For any $a^\vee$ in $\mathbb{A}^{m\vee}(L)$, the composite

$$a^\vee\circ\mathrm{eval}_0(v)$$

is thus an L-valued point of $\mathcal{F}_0{}^\vee$. So we get a k-morphism

$$\rho : V \times \mathbb{A}^{m\vee} \to \mathcal{F}_0{}^\vee \times \mathbb{A}^{m\vee},$$

$$\rho(v, \beta^\vee) := (a^\vee \circ \mathrm{eval}_0(v), a^\vee).$$

On $V \times \mathbb{A}^{m\vee}$, we have the external tensor product

$$\mathrm{pr}_1{}^*L \otimes \mathrm{pr}_2{}^*FT_\psi(K),$$

which is semiperverse, each external factor being semiperverse. We then tensor this object with the Tate-twisted lisse, rank one Artin-Schreir sheaf

$$\mathcal{L}_{\psi(-\beta^\vee(h(v)))}(-\dim\mathcal{F}_0),$$

cf. [Ka-ESDE, 7.2.1], placed in degree zero. The resulting object

$$S := (\mathrm{pr}_1{}^*L \otimes \mathrm{pr}_2{}^*FT_\psi(K)) \otimes \mathcal{L}_{\psi(-\beta^\vee(h(v)))}(-\dim\mathcal{F}_0)$$

is still semiperverse on $V \times \mathbb{A}^{m\vee}$.

The key observation is that on $\mathcal{F}_0{}^\vee \times \mathbb{A}^{m\vee}$, we have

$$FT_\psi(M) = R\rho_! S.$$

To see this, we first calculate $FT_\psi(M)$ by looking at the diagram

$$\begin{array}{ccccc} & & \mathrm{pr}_{1,2} & & \\ & V \times \mathcal{F} \times \mathcal{F}^\vee & \to & V \times \mathcal{F} & \\ \mathrm{pr}_{2,3} & \downarrow & & \downarrow & \mathrm{pr}_2 \\ & & \mathrm{pr}_1 & & \\ & \mathcal{F} \times \mathcal{F}^\vee & \to & \mathcal{F} & \\ \mathrm{pr}_2 & \downarrow & & & \\ & \mathcal{F}^\vee & & & \end{array}$$

the upper square of which is Cartesian.

Recall that $M := R\mathrm{pr}_{2!}N$, for N the object on $V \times \mathcal{F}$ given by

$$N := \mathrm{pr}_1{}^*L \otimes h_{\mathrm{aff}}{}^*K[\dim\mathcal{F} - m].$$

By definition of Fourier Transform, we have

$$FT_\psi(M) = R\mathrm{pr}_{2!}((\mathrm{pr}_1{}^*M) \otimes \mathcal{L}_{\psi(f^\vee f)})[\dim\mathcal{F}].$$

By proper base change, we obtain

$$FT_\psi(M) = R\mathrm{pr}_{2!}R\mathrm{pr}_{2,3!}((\mathrm{pr}_{1,2}{}^*N) \otimes \mathrm{pr}_{2,3}{}^*(\mathcal{L}_{\psi(f^\vee f)})[\dim\mathcal{F}]$$

$$= R(\mathrm{pr}_3 : V \times \mathcal{F} \times \mathcal{F}^\vee \to \mathcal{F}^\vee)_!((\mathrm{pr}_{1,2}{}^*N) \otimes \mathrm{pr}_{2,3}{}^*(\mathcal{L}_{\psi(f^\vee f)}))[\dim\mathcal{F}]$$

$$= R(\mathrm{pr}_3 : V \times \mathcal{F} \times \mathcal{F}^\vee \to \mathcal{F}^\vee)_! P,$$

for P the object on $V \times \mathcal{F} \times \mathcal{F}^\vee$ given by

$$P := (\mathrm{pr}_{1,2}{}^*N) \otimes \mathrm{pr}_{2,3}{}^*(\mathcal{L}_{\psi(f^\vee f)})[\dim\mathcal{F}].$$

To calculate, expand out

$$V \times \mathcal{F} \times \mathcal{F}^\vee = V \times \mathcal{F}_0 \times \mathbb{A}^m \times \mathcal{F}_0{}^\vee \times \mathbb{A}^{m\vee},$$

and factor the projection of $V \times \mathcal{F} \times \mathcal{F}^\vee$ onto $\mathcal{F}^\vee$ as the composite

$$V \times \mathcal{F}_0 \times \mathbb{A}^m \times \mathcal{F}_0{}^\vee \times \mathbb{A}^{m\vee}$$
$$\downarrow pr_{1,2,4,5}$$
$$V \times \mathcal{F}_0 \times \mathcal{F}_0{}^\vee \times \mathbb{A}^{m\vee}$$
$$\downarrow pr_{1,3,4}$$
$$V \times \mathcal{F}_0{}^\vee \times \mathbb{A}^{m\vee}$$
$$\downarrow pr_{2,3}$$
$$\mathcal{F}_0{}^\vee \times \mathbb{A}^{m\vee}.$$

In "functional" notation, the object P on the space $V \times \mathcal{F}_0 \times \mathbb{A}^m \times \mathcal{F}_0{}^\vee \times \mathbb{A}^{m\vee}$, with "coordinates" $(v, f_0, a, f_0{}^\vee, a^\vee)$, is

$$L(v)\otimes K(h(v)+f_0(v)+a)\otimes \mathcal{L}_\psi(f_0{}^\vee f_0 + a^\vee a)[2\dim\mathcal{F} - m].$$

To apply $Rpr_{1,2,4,5!}$ to this object, i.e., to "integrate out" the variable "a", we may first apply the automorphism σ of $V \times \mathcal{F}_0 \times \mathbb{A}^m \times \mathcal{F}_0{}^\vee \times \mathbb{A}^{m\vee}$ as scheme over $V \times \mathcal{F}_0 \times \mathcal{F}_0{}^\vee \times \mathbb{A}^{m\vee}$ given by translation of the "a" variable:

$$\sigma(v, f_0, a, f_0{}^\vee, a^\vee) := (v, f_0, a - h(v) - f_0(v), f_0{}^\vee, a^\vee).$$

After this automorphism, the object P becomes

$$\sigma^*P = L(v)\otimes K(a)\otimes \mathcal{L}_\psi(f_0{}^\vee f_0 + a^\vee a - a^\vee f_0(v) -a^\vee h(v))[2\dim\mathcal{F} - m]$$
$$=K(a)\otimes \mathcal{L}_\psi(a^\vee a)[m]$$
$$\otimes L(v)\otimes \mathcal{L}_\psi(-a^\vee h(v))\otimes \mathcal{L}_\psi(f_0{}^\vee f_0 - a^\vee f_0(v))[2\dim\mathcal{F}_0].$$

When we apply $Rpr_{1,2,4,5!}$, we notice that the second factor is a pullback from the base, while the effect of integrating out the "a" in the first factor $K(a)\otimes \mathcal{L}_\psi(a^\vee a)[m]$ is just taking the Fourier Transform of K on $\mathbb{A}^m$. So by the projection formula we get

$$Rpr_{1,2,4,5!}P \cong Rpr_{1,2,4,5!}(\sigma^*P)$$
$$= FT_\psi(K)(a^\vee)\otimes L(v)\otimes \mathcal{L}_\psi(-a^\vee h(v))\otimes \mathcal{L}_\psi(f_0{}^\vee f_0 - a^\vee f_0(v))[2\dim\mathcal{F}_0]$$

on $V \times \mathcal{F}_0 \times \mathcal{F}_0{}^\vee \times \mathbb{A}^{m\vee}$.

The next step is to apply $Rpr_{1,3,4!}$, i.e., to integrate out the "f_0". Here it is only the factor $\mathcal{L}_\psi(f_0{}^\vee f_0 - a^\vee f_0(v))[2\dim\mathcal{F}_0]$ which involves the variable "f_0", the rest comes from the base. When we integrate out the "f_0" from $\mathcal{L}_\psi(f_0{}^\vee f_0 - a^\vee f_0(v))[2\dim\mathcal{F}_0]$, we are forming the Fourier Transform of the shifted constant sheaf $\overline{\mathbb{Q}}_{\ell,\mathcal{F}_0}[\dim\mathcal{F}_0]$ on $\mathcal{F}_0$, and evaluating at the point

$$f_0{}^\vee = a^\vee\circ \mathrm{eval}_0(v)$$

in $\mathcal{F}_0{}^\vee$. Since the Fourier Transform of the constant sheaf is the delta function, i.e., since

$$FT_\psi(\overline{\mathbb{Q}}_{\ell,\mathcal{F}_0}[\dim\mathcal{F}_0]) = \delta_0(-\dim\mathcal{F}_0),$$

we find that

$$Rpr_{1,3,4!}(\mathcal{L}_\psi(f_0{}^\vee f_0 - a^\vee f_0(v))[2\dim\mathcal{F}_0])$$

= (the constant sheaf $\overline{\mathbb{Q}}_\ell(-\dim\mathcal{F}_0)$ on the locus $f_0{}^\vee = a^\vee\circ\mathrm{eval}_0(v)$), extended by zero to all of $V\times\mathcal{F}_0\times\mathcal{F}_0{}^\vee\times\mathbb{A}^{m\vee}$.

Let us denote by Z the closed subscheme of $V\times\mathcal{F}_0{}^\vee\times\mathbb{A}^{m\vee}$ consisting of those points $(v, f_0{}^\vee, a^\vee)$ with $f_0{}^\vee = a^\vee\circ\mathrm{eval}_0(v)$. If we identify this closed subscheme Z with $V\times\mathbb{A}^{m\vee}$ by the map

$$i : (v, a^\vee) \to (v, f\ a^\vee\circ\mathrm{eval}_0(v), a^\vee),$$

we find that $Rpr_{1,3,4!}(Rpr_{1,2,4,5!}(\sigma^*P))$ is supported in Z, where it is given by

$$L(v)\otimes\mathcal{L}_\psi(-a^\vee h(v))\otimes FT_\psi(K)(a^\vee)(-\dim\mathcal{F}_0).$$

The composite

$$\begin{array}{l} Z \\ \downarrow \\ V\times\mathcal{F}_0{}^\vee\times\mathbb{A}^{m\vee} \\ \downarrow pr_{2,3} \\ \mathcal{F}_0{}^\vee\times\mathbb{A}^{m\vee} \end{array}$$

is precisely the morphism ρ. Thus we find that

$$FT_\psi(M) = R\rho_!S,$$

for S the object on $V\times\mathbb{A}^{m\vee}$ given by

$$S := L(v)\otimes\mathcal{L}_\psi(-a^\vee h(v))\otimes FT_\psi(K)(a^\vee)(-\dim\mathcal{F}_0),$$

as asserted.

If the morphism ρ were quasifinite, then we would be done: $R\rho_!$ preserves semiperversity if ρ is quasifinite.

Although ρ is not quasifinite, we claim that ρ is in fact quasifinite over the open set

$$\mathcal{F}_0{}^\vee \times \mathbb{A}^{m\vee} - \{0, 0\}.$$

To see this, we argue as follows. Suppose that over some extension field L/k, we have a point $(a^\vee\circ\mathrm{eval}_0(v), a^\vee) \neq (0, 0)$ in the image $\rho(V(L)\times\mathbb{A}^{m\vee}(L))$. Then $a^\vee$ must itself be nonzero. We must show that there are at most finitely many points w in V(L) such that

$$(a^\vee\circ\mathrm{eval}_0(v), a^\vee) = (a^\vee\circ\mathrm{eval}_0(w), a^\vee),$$

i.e., that there are at most finitely many points w in V(L) such that

$$a^\vee\circ(\mathrm{eval}_0(v) - \mathrm{eval}_0(w)) = 0 \text{ on } \mathcal{F}_0\otimes L.$$

This last condition is equivalent to the condition

$$a^\vee\circ(\mathrm{eval}(v) - \mathrm{eval}(w)) = 0 \text{ on } \mathcal{F}\otimes L,$$

simply because $\mathcal{F} = \mathcal{F}_0 \oplus \mathbb{A}^m(k)$, and $\mathrm{eval}(v) - \mathrm{eval}(w)$ tautologically kills constants. By the hypothesis that $(\mathcal{F}, \tau)$ is quasifinitely difference-separating, we know that except for finitely many w in V(L), the map

$$\mathrm{eval}(v) - \mathrm{eval}(w) : \mathcal{F}\otimes L \to \mathbb{A}^m(L)$$

is surjective. When this map is surjective, $a^\vee\circ(\mathrm{eval}(v) - \mathrm{eval}(w))$ is nonzero, because $a^\vee$ is nonzero. So for given L/k and given v in V(L), there are only finitely many w in V(L) for which

$$(a^\vee\circ\mathrm{eval}_0(v), a^\vee) = (a^\vee\circ\mathrm{eval}_0(w), a^\vee).$$

Thus ρ is quasifinite over the open set $\mathcal{F}_0{}^\vee \times \mathbb{A}^{m\vee} - \{0, 0\}$.

Therefore the restriction to $\mathcal{F}_0{}^\vee \times \mathbb{A}^{m\vee} - \{0, 0\}$ of $FT_\psi(M)$ is indeed semiperverse. Once we know this, then $FT_\psi(M)$ is semiperverse on all of $\mathcal{F}_0{}^\vee \times \mathbb{A}^{m\vee}$ if and only if its restriction to the single missing point (0, 0) is semiperverse on that point, i.e., is concentrated in degree ≤ 0. But this restriction is just

$$FT_\psi(M)_{(0,0)} = R\Gamma_c(\mathbb{A}^m\otimes\bar{k}, K[m])\otimes R\Gamma_c(V\otimes\bar{k}, L)(-\dim\mathcal{F}_0).$$

Thus M is indeed semiperverse on $\mathcal{F} = \mathcal{F}_0 \times \mathbb{A}^m$. For any M on $\mathcal{F}$, the value at the origin of $FT_\psi(M)$ on $\mathcal{F}^\vee$ is $H_c{}^*(\mathcal{F}\otimes_k\bar{k}, M[\dim\mathcal{F}])$, so we find

$$H_c{}^*(\mathcal{F}\otimes_k\bar{k}, M[\dim\mathcal{F}]) = H^*{}_c(\mathbb{A}^m\otimes\bar{k}, K[m])\otimes H^*{}_c(V\otimes\bar{k}, L)(-\dim\mathcal{F}_0),$$

as asserted. QED

(1.4.3) Remark on hypothesis 4) in Theorem 1.4.2 For any K on $\mathbb{A}^m$, $R\Gamma_c(\mathbb{A}^m\otimes\bar{k}, K[m])$ is the stalk at the origin of $FT_\psi(K)$. As FT_ψ preserves semiperversity, we see that if K is semiperverse on $\mathbb{A}^m$, then $R\Gamma_c(\mathbb{A}^m\otimes\bar{k}, K[m])$ is concentrated in degree ≤ 0. So hypothesis 4) of the theorem holds if either

$$R\Gamma_c(V\otimes\bar{k}, L) \text{ is concentrated in degree } \le 0,$$

or if

$$R\Gamma_c(\mathbb{A}^m\otimes\bar{k}, K[m]) = 0.$$

See the Exactness Corollary 1.4.5 below for a development of this remark.

Perversity Corollary 1.4.4 (compare [Ka-ACT, 1.6]) Hypotheses and notations as in Theorem 1.4.2 (and in its proof, for the notion of $\mathcal{F}_0$), we have the following results.

1) The object $A := pr_1{}^*L\otimes h_{aff}{}^*K[\dim\mathcal{F}_0]$ on $V\times\mathcal{F}$ is semiperverse, and its Verdier dual $DA := D_{V\times\mathcal{F}/k}(A)$ is given by

$$DA = pr_1{}^*(DL)\otimes h_{aff}{}^*(DK)[\dim\mathcal{F}_0](\dim\mathcal{F}_0),$$

where we have written $DL := D_{V/k}(L)$, $DK := D_{\mathbb{A}^m/k}(K)$.

2) The Verdier dual $DM := D_{\mathcal{F}/k}(M)$ of $M := Rpr_{2!}(A)$ is $Rpr_{2*}(DA)$.

3) If K is perverse on $\mathbb{A}^m$ and L is perverse on V, then A and DA are

perverse on $V \times \mathcal{F}$.

4) Suppose that V is affine, that K is perverse on $\mathbb{A}^m$, and that L is perverse on V. Then M and DM are perverse on $\mathcal{F}$.

proof 1) Exactly as in [Ka-ACT, 1.6], consider the automorphism σ of $V \times \mathcal{F} = V \times \mathcal{F}_0 \times \mathbb{A}^m$ given by

$$\sigma(v, f_0, a) := (v, f_0, h(v) + f_0(v) + a).$$

Then A is the σ^* pullback of the external tensor product of L on V, $\overline{\mathbb{Q}}_\ell[\dim \mathcal{F}_0]$ on $\mathcal{F}_0$, and K on $\mathbb{A}^m$. The external tensor product of semiperverses is semiperverse. Formation of the dual commutes with σ^* and with external tensor product. The dual of $\overline{\mathbb{Q}}_\ell[\dim \mathcal{F}_0]$ on $\mathcal{F}_0$ is $\overline{\mathbb{Q}}_\ell[\dim \mathcal{F}_0](\dim \mathcal{F}_0)$.

2) This is an instance of the fact that duality interchanges $Rpr_{2!}$ and Rpr_{2*}.

3) If K and L are perverse, then DK and DL are perverse, so are semiperverse, so by 1), applied not only to K and L but also to DL and DK, both A and DA are semiperverse. So both A and DA are perverse.

4) If V is affine, then pr_2 is an affine morphism, so preserves semiperversity. If both K and L are perverse, then, by 3), DA is semiperverse. So $Rpr_{2*}(DA)$ is semiperverse. As its dual $M := Rpr_{2!}(A)$ is semiperverse by the theorem, both M and DM are perverse. QED

Exactness Corollary 1.4.5 Fix an affine k-scheme V of finite type, an integer $m \geq 1$, a space of functions $(\mathcal{F}, \tau)$,

$$\tau : \mathcal{F} \to \mathrm{Hom}_{k\text{-scheme}}(V, \mathbb{A}^m),$$

on V which is quasifinitely difference-separating and contains the constants, and a k-morphism

$$h : V \to \mathbb{A}^m.$$

1) Suppose K is perverse on $\mathbb{A}^m$, and suppose

$$H_c^i(\mathbb{A}^m \otimes \overline{k}, K) = 0 \text{ for } i > m - \dim(V).$$

Then for any perverse sheaf L on V, the object Twist(L,K,$\mathcal{F}$,h) on $\mathcal{F}$ is perverse, and the functor $L \mapsto$ Twist(L,K,$\mathcal{F}$,h) from the category of perverse sheaves on V to the category of perverse sheaves on $\mathcal{F}$ is exact.

2) Suppose L is perverse on V, and suppose

$$H_c^i(V \otimes \overline{k}, L) = 0 \text{ for } i > 0.$$

Then for any perverse sheaf K on $\mathbb{A}^m$, the object Twist(L,K,$\mathcal{F}$,h) on $\mathcal{F}$ is perverse, and the functor $K \mapsto$ Twist(L,K,$\mathcal{F}$,h) from the category of perverse sheaves on $\mathbb{A}^m$ to the category of perverse sheaves on $\mathcal{F}$ is exact.

3) Suppose we are given a short exact sequence of perverse sheaves on $\mathbb{A}^m$,

$$0 \to K_1 \to K_2 \to K_3 \to 0,$$

and suppose that each K_j satisfies

$$H_c^i(\mathbb{A}^m \otimes \bar{k}, K_j) = 0 \text{ for } i > m - \dim(V).$$

Then for any perverse L on V, the objects $\mathrm{Twist}(L,K_j,\mathcal{F},h)$ on $\mathcal{F}$ are perverse, and we have a short exact sequence of perverse sheaves on $\mathcal{F}$,

$$0 \to \mathrm{Twist}(L,K_1,\mathcal{F},h) \to \mathrm{Twist}(L,K_2,\mathcal{F},h) \to \mathrm{Twist}(L,K_3,\mathcal{F},h) \to 0.$$

4) Suppose we are given a short exact sequence of perverse sheaves on V,

$$0 \to L_1 \to L_2 \to L_3 \to 0,$$

and suppose that each L_j satisfies

$$H_c^i(V \otimes \bar{k}, L_j) = 0 \text{ for } i > 0.$$

Then for any perverse K on $\mathbb{A}^m$, the objects $\mathrm{Twist}(L_j,K,\mathcal{F},h)$ on $\mathcal{F}$ are perverse, and we have a short exact sequence of perverse sheaves on $\mathcal{F}$,

$$0 \to \mathrm{Twist}(L_1,K,\mathcal{F},h) \to \mathrm{Twist}(L_2,K,\mathcal{F},h) \to \mathrm{Twist}(L_3,K,\mathcal{F},h) \to 0.$$

proof There are two key points. The first is that the formation of $\mathrm{Twist}(L,K,\mathcal{F},h)$ is a functor

$$D^b_c(V, \overline{\mathbb{Q}}) \times D^b_c(\mathbb{A}^m, \overline{\mathbb{Q}}_\ell) \to D^b_c(\mathcal{F}, \overline{\mathbb{Q}}_\ell),$$
$$(L, K) \mapsto \mathrm{Twist}(L,K,\mathcal{F},h),$$

which is triangulated (i.e., carries distinguished triangles to distinguished triangles) in each variable separately. This is clear from the fact that $R(pr_2)_!$ is triangulated, and the description in the previous result of $pr_1{}^*L \otimes h_{\mathrm{aff}}{}^*K[\dim \mathcal{F}_0]$ on $V \times \mathcal{F}$ as the pullback by an automorphism of the external tensor product of L on V, $\overline{\mathbb{Q}}_\ell[\dim \mathcal{F}_0]$ on $\mathcal{F}_0$, and K on $\mathbb{A}^m$, the formation of which is visibly bi-triangulated. The second key point is that on any separated scheme X/k, a short exact sequence of perverse sheaves on X,

$$0 \to N_1 \to N_2 \to N_3 \to 0,$$

is precisely a distinguished triangle in $D^b_c(X, \overline{\mathbb{Q}}_\ell)$ whose terms happen to be perverse, cf. [BBD 1.2.3 and 1.3.6].

With these points in mind, the assertions 1) through 4) are obvious. For instance, to prove 1), we note that for any perverse L on V, indeed for any semiperverse L on V, we have

$$H_c^i(V \otimes \bar{k}, L) = 0 \text{ for } i > \dim(V).$$

This vanishing follows from the dimension inequalities

$$\dim \mathrm{Supp}\, \mathcal{H}^i(L) \le -i \text{ for every integer } i$$

defining semiperversity, and the spectral sequence

$$E_2^{a,b} = H_c^{a}(V\otimes\bar{k}, \mathcal{H}^b(L)) \Rightarrow H_c^{a+b}(V\otimes\bar{k}, L),$$

cf. the proof of 1.10.5. So under the hypotheses of 1), we get the vanishing

$$H^i_c((V\times\mathbb{A}^m)\otimes_k\bar{k}, pr_1{}^*L\otimes pr_2{}^*K) = 0 \text{ if } i \geq m+1.$$

So for any perverse L on V, Twist(L,K,$\mathcal{F}$,h) is perverse, by the Perversity Corollary 1.4.4. A short exact sequence of perverse sheaves on V,

$$0 \to L_1 \to L_2 \to L_3 \to 0,$$

gives a distinguished triangle in $D^b_c(V, \overline{\mathbb{Q}}_\ell)$, so we get a distinguished triangle on $\mathcal{F}$,

$$\to \mathrm{Twist}(L_1,K,\mathcal{F},h) \to \mathrm{Twist}(L_2,K,\mathcal{F},h) \to \mathrm{Twist}(L_3,K,\mathcal{F},h) \to,$$

which, having perverse terms, is a short exact sequence of perverse sheaves on $\mathcal{F}$. The proofs of 2), 3), and 4) are similar, and left to the reader. QED

(1.5) Review of weights

(1.5.1) In this section, we work over a finite field k, on a separated scheme X/k of finite type. We fix a prime number $\ell \neq \mathrm{char}(k)$, and a field embedding

$$\iota : \overline{\mathbb{Q}}_\ell \subset \mathbb{C}.$$

We denote by $|z|$ the complex absolute value of a complex number z. For α in $\overline{\mathbb{Q}}_\ell$, we will write the complex absolute value of $\iota(\alpha)$ as

$$|\alpha|_\iota := |\iota(\alpha)|,$$

or simply as

$$|\alpha| := |\alpha|_\iota$$

when no confusion is likely.

(1.5.2) For an object N in $D^b_c(X, \overline{\mathbb{Q}}_\ell)$, its trace function is the $\overline{\mathbb{Q}}_\ell$-valued function on pairs (a finite extension E/k, a point x in X(E)) defined by

$$(E, x) \mapsto N(E, x) := \Sigma_i\ (-1)^i\mathrm{Trace}(\mathrm{Frob}_{E,x} \mid \mathcal{H}^i(N)).$$

[Here and throughout, $\mathrm{Frob}_{E,x}$ is the geometric Frobenius attached to the E-valued point x in X(E).] We view N(E, x), via the fixed ι, as a $\mathbb{C}$-valued function, and denote by

$$(E, x) \mapsto \overline{N}(E, x)$$

the complex conjugate $\mathbb{C}$-valued function.

(1.5.3) Recall [De-Weil II, 1.2.2] that for w a real number, a constructible $\overline{\mathbb{Q}}_\ell$-sheaf $\mathcal{G}$ is said to be punctually ι-pure of weight w if, for each finite extension E/k, and for each point x in X(E), all the eigenvalues α of $\mathrm{Frob}_{E,x} \mid \mathcal{G}$ have $|\alpha| = (\#E)^{w/2}$. A constructible $\overline{\mathbb{Q}}_\ell$-sheaf $\mathcal{G}$ is said to be ι-mixed of weight $\leq$ w if it is a successive extension of finite many constructible $\overline{\mathbb{Q}}_\ell$-sheaves, each of which is punctually ι-pure of some weight $\leq$ w.

(1.5.4) Let us say that a real number w "occurs via ι" if there exists an ℓ-adic unit α in $\overline{\mathbb{Q}}_\ell$ with $|\alpha| = (\mathrm{char}(k))^{w/2}$, or equivalently (take powers or roots, $\overline{\mathbb{Q}}_\ell$ being algebraically closed), if for some finite extension E/k, there exists an ℓ-adic unit β in $\overline{\mathbb{Q}}_\ell$ with $|\beta| = (\#E)^{w/2}$. Thus any rational number w occurs via ι. And if a nonzero constructible $\overline{\mathbb{Q}}_\ell$-sheaf $\mathcal{G}$ is punctually ι-pure of weight w, then w "occurs in ι".

Lemma 1.5.5 Let w be a real number, and $\mathcal{G}$ a nonzero constructible $\overline{\mathbb{Q}}_\ell$-sheaf on X/k which is ι-mixed of weight $\le$ w. The set of real numbers $w_0 \le w$ such that $\mathcal{G}$ is ι-mixed of weight $\le w_0$ has a least element, and that element occurs via ι.

proof Write $\mathcal{G}$ as a successive extension of finitely many nonzero constructible $\overline{\mathbb{Q}}_\ell$-sheaves $\mathcal{G}_i$, with $\mathcal{G}_i$ punctually ι-pure of some weight $w_i \le w$. As remarked above, each w_i "occurs in ι". The largest of the w_i is the least w_0. QED

(1.5.6) Recall [De-Weil II, 6.2.2] that an object N in $D^b{}_c(X, \overline{\mathbb{Q}}_\ell)$ is said to be ι-mixed of weight $\le$ w if, for each integer i, the cohomology sheaf $\mathcal{H}^i(N)$ is ι-mixed of weight $\le$ w+i. The object N is said to be ι-pure of weight w if N is ι mixed of weight $\le$ w, and if its Verdier dual $D_{X/k}N$ is ι-mixed of weight $\le$ -w.

Lemma 1.5.7 Let w be a real number, and N a nonzero object in $D^b{}_c(X, \overline{\mathbb{Q}}_\ell)$ which is ι-mixed of weight $\le$ w. The set of real numbers $w_0 \le w$ such that N is ι-mixed of weight $\le w_0$ has a least element, and that element occurs via ι.

proof For each i such that $\mathcal{H}^i(N)$ is nonzero, apply the previous lemma to $\mathcal{H}^i(N)$, which is mixed of weight $\le$ w + i. Denote by $w_{0,i}$ the least real number such that $\mathcal{H}^i(N)$ is ι-mixed of weight $\le w_{0,i} + i$. Then each $w_{0,i}$ occurs via ι, and the largest of the $w_{0,i}$ is the least w_0. QED

Lemma 1.5.8 Let w be a real number, N a nonzero object in $D^b{}_c(X, \overline{\mathbb{Q}}_\ell)$ which is ι-pure of weight w. Then w occurs via ι.

proof With no loss of generality, we may assume that the support of N is X (because for a closed immersion i : Z $\to$ X, we have $i_*\circ D_{Z/k} = D_{X/k}\circ i_*$). Then there exists a dense open set U of X which is smooth over k, with N|U nonzero, ι-pure of weight w, and with lisse cohomology sheaves. Then for any i with $\mathcal{H}^i(N|U)$ nonzero, $\mathcal{H}^i(N|U)$ is punctually ι-pure of weight w+i, and hence w occurs via ι. QED

(1.5.9) The main theorem of Deligne's Weil II is that for $f: X \to Y$ a k-morphism between separated schemes of finite type, for any N in $D^b{}_c(X, \overline{\mathbb{Q}}_\ell)$ which is ι-mixed of weight $\le w$, the object $Rf_!N$ on Y is ι-mixed of weight $\le w$.

(1.5.10) Here is a simple application, to the object M constructed in the Semiperversity Theorem.

Weight Corollary 1.5.11 (compare [Ka-ACT, 1.6]) Hypotheses and notations as in the Semiperversity Theorem 1.4.2 (and in its proof, for the notion of $\mathcal{F}_0$), fix an embedding ι of $\overline{\mathbb{Q}}_\ell$ into $\mathbb{C}$, and real numbers a and b. Suppose in addition that

1) k is a finite field,
2) L is ι-mixed of weight $\le a$,
3) K is ι-mixed of weight $\le b$.

Then we have the following results.

1) The object $A := pr_1{}^*L \otimes h_{aff}{}^*K[\dim \mathcal{F}_0]$ on $V \times \mathcal{F}$ is ι-mixed of weight $\le a + b + \dim(\mathcal{F}_0)$.

2) The object $M := Rpr_{2!}(A)$ on $\mathcal{F}$ is ι-mixed of weight $\le a + b + \dim(\mathcal{F}_0)$.

proof 1) By means of the automorphism σ of $V \times \mathcal{F} = V \times \mathcal{F}_0 \times \mathbb{A}^m$ given by

$$\sigma(v, f_0, a) := (v, f_0, h(v) + f_0(v) + a),$$

A is the σ^*-pullback of the external tensor product of L on V, $\overline{\mathbb{Q}}_\ell[\dim \mathcal{F}_0]$ on $\mathcal{F}_0$, and K on $\mathbb{A}^m$. The object $\overline{\mathbb{Q}}_\ell[\dim \mathcal{F}_0]$ on $\mathcal{F}_0$ is ι-pure of weight $\dim \mathcal{F}_0$, and weights add for external tensor products.

2) This is a special case of Deligne's main theorem [De-Weil II, 3.3.1] in Weil II. QED

(1.5.12) We now resume our review of weights. A perverse sheaf N on X is called ι-mixed (resp. ι-pure) if it ι-mixed (resp. ι-pure) as an object of $D^b{}_c(X, \overline{\mathbb{Q}}_\ell)$. One knows [BBD, 5.3.1 and 5.3.4] that if a perverse sheaf N is ι-mixed, then every simple constituent of N as perverse sheaf is ι-pure of some weight. More precisely, if a perverse sheaf N is ι-mixed, then for any finite set of real numbers $w_1 < w_2 \ldots < w_r$ which includes the weights of all the simple constituents of N, N admits a unique increasing filtration as a perverse sheaf

$$0 \subset N_{\le w_1} \subset N_{\le w_2} \ldots \subset N_{\le w_r} = N$$

such that for each i, $N_{\le w_i}$ is mixed of weight $\le w_i$ and the associated graded object $N_{\le w_i}/N_{\le w_{i-1}}$ ($:= N_{\le w_1}$ for $i=0$) is either zero or is ι-pure of weight w_i.

Lemma 1.5.13 Suppose M (resp. N) in $D^b{}_c(X, \overline{\mathbb{Q}}_\ell)$ is semiperverse and ι-mixed of weight $\le a$ (resp. $\le b$). Then for variable finite extensions E of k, we have

$$\Sigma_{x \text{ in } X(E)} |M(E, x)N(E, x)| = O((\#E)^{(a+b)/2}).$$

proof By Lemma 1.5.7 above, we reduce immediately to the case where the weights a and b both occur via ι. Replacing M and N by suitable constant field twists $M\otimes\alpha^{\deg}$ and $N\otimes\beta^{\deg}$, we reduce to the case where a = b = 0. Pick a smooth stratification $\{Z_\alpha\}$ of X to which both M and N are adapted. On each strat Z_α, $M|Z_\alpha$ and $N|Z_\alpha$ remain semiperverse, and ι-mixed of weight ≤ 0. Break the sum over X(E) into sums over the individual Z_α. So it suffices to treat the case where X is a Z_α. Thus X/k is connected and smooth over k of some dimension d := dim(X) ≥ 0, and both M and N have all their cohomology sheaves lisse on X. By semiperversity, $\mathcal{H}^{-i}(M) = 0$ for i < d. So the (at most finitely many) nonvanishing $\mathcal{H}^{-i}(M)$ all have i ≥ d, and hence are ι-mixed of weight ≤ -d. Thus for any finite extension E/k, we have the estimate

$$|M(E, x)| \le (\Sigma_i \operatorname{rank}(\mathcal{H}^i(M)))(\#E)^{-d/2}.$$

Similarly for N. Thus we get

$$|M(E, x)N(E, x)| = O((\#E)^{-d}).$$

The number of terms in the sum is #X(E), which is trivially $O((\#E)^d)$. QED

(1.6) Remarks on the various notions of mixedness

(1.6.1) Suppose instead of fixing a single field embedding ι of $\overline{\mathbb{Q}}_\ell$ into $\mathbb{C}$, we fix some nonempty collection $\mathcal{I}$ of such field embeddings. Given a real number w, we say that a constructible $\overline{\mathbb{Q}}_\ell$-sheaf $\mathcal{G}$ is punctually $\mathcal{I}$-pure of weight w if it is punctually ι-pure of weight w for every ι in $\mathcal{I}$. We say that a constructible $\overline{\mathbb{Q}}_\ell$-sheaf $\mathcal{G}$ is $\mathcal{I}$-mixed of weight ≤ w if it is a successive extension of finitely many constructible $\overline{\mathbb{Q}}_\ell$-sheaves, each of which is punctually $\mathcal{I}$-pure of some weight ≤ w. We say that a constructible $\overline{\mathbb{Q}}_\ell$-sheaf $\mathcal{G}$ is $\mathcal{I}$-mixed of **integer** weight ≤ w if it is a successive extension of finitely many constructible $\overline{\mathbb{Q}}_\ell$-sheaves, each of which is punctually $\mathcal{I}$-pure of some integer weight ≤ w. An object N in $D^b{}_c(X, \overline{\mathbb{Q}}_\ell)$ is said to be $\mathcal{I}$-mixed of weight ≤ w [resp. $\mathcal{I}$-mixed of integer weight ≤ w] if, for each integer i, the cohomology sheaf $\mathcal{H}^i(N)$ is $\mathcal{I}$-mixed of weight ≤ w+i [resp. $\mathcal{I}$-mixed of integer weight ≤ w + i].

(1.6.2) The main theorem [De-Weil II, 3.3.1] of Deligne's Weil II asserts that for f: X → Y a k-morphism between separated schemes of finite type, for N in $D^b{}_c(X, \overline{\mathbb{Q}}_\ell)$ which is $\mathcal{I}$-mixed of weight ≤ w [resp. $\mathcal{I}$-mixed of integer weight ≤ w], the object $Rf_!N$ on Y is $\mathcal{I}$-mixed of weight ≤ w [resp. $\mathcal{I}$-mixed of integer weight ≤ w]. Strictly speaking, the theorem is stated there "only" for the hardest case when $\mathcal{I}$ consists of all embeddings, and when N is $\mathcal{I}$-mixed of integer weight (i.e., the case when N is "mixed" in the sense of Weil II), but the proof is done "ι by ι" and establishes, for any $\mathcal{I}$, the $Rf_!$-stability of both $\mathcal{I}$-mixedness of weight $\le w$ and of $\mathcal{I}$-mixedness of integer weight $\le w$.

(1.6.3) The object N is said to be $\mathcal{I}$-pure of weight w if N is $\mathcal{I}$-mixed of weight $\le w$, and if its Verdier dual $D_{X/k}N$ is $\mathcal{I}$-mixed of weight $\le -w$. [It is a non-obvious fact that if w is an integer, and if N is $\mathcal{I}$-pure of weight w, then N is $\mathcal{I}$-mixed of integer weight $\le w$, and its Verdier dual is $\mathcal{I}$-mixed of integer weight $\le -w$.]

(1.6.4) If a perverse sheaf N is $\mathcal{I}$-mixed, then every simple constituent of N is $\mathcal{I}$-pure of some weight, and N admits a unique increasing weight filtration as above. For any finite set of real numbers $w_1 < w_2 ... < w_r$ which includes the weights of all the simple constituents of N, N admits a unique increasing filtration as a perverse sheaf

$$0 \subset N_{\le w_1} \subset N_{\le w_2} ... \subset N_{\le w_r} = N$$

such that for each i, $N_{\le w_i}$ is mixed of weight $\le w_i$ and the associated graded object $N_{\le w_i}/N_{\le w_{i-1}}$ ($:= N_{\le w_1}$ for $i=0$) is either zero or is $\mathcal{I}$-pure of weight w_i. If N is $\mathcal{I}$-mixed of integer weight, then all the weights w_i are integers.

(1.6.5) Although all the objects one encounters "in Nature" are mixed in the sense of Weil II, i.e., $\mathcal{I}$-mixed of integer weight for $\mathcal{I}$ the collection of all embeddings, we will nonetheless work systematically with ι-mixed objects for our single chosen ι. We will not even make the (altogether reasonable) assumption that our ι-mixed objects are ι-mixed of integer weight, although doing so would allow us to replace, in the following pages, every single $\varepsilon > 0$ (each of which has its genesis as a positive difference of distinct weights) by $\varepsilon = 1$. Caveat emptor.

(1.7) The Orthogonality Theorem

(1.7.1) We now prove orthogonality relations for the trace functions of perverse sheaves.

Orthogonality Theorem 1.7.2 Let M and N on X/k be perverse, and ι-pure of weight zero. Write the pullbacks M^{geom} and N^{geom} of M and N to $X\otimes_k \bar{k}$ as sums of perverse irreducibles with multiplicities, say

$$M^{geom} = \Sigma_i m_i V_i, \quad N^{geom} = \Sigma_i n_i V_i,$$

with $\{V_i\}_i$ a finite set of pairwise non-isomorphic perverse irreducibles on $X\otimes_k \bar{k}$, and with non-negative integers m_i and n_i. [This is possible by [BBD, 5.3.8].]

1) For any integer $n \ge 1$, denoting by k_n/k the extension field of degree n, we have

$$\Sigma_i m_i n_i = \mathrm{limsup}_{E/k_n} |\Sigma_{x \text{ in } X(E)} M(E, x)\overline{N}(E, x)|,$$

the limsup taken over all finite extensions E/k_n.

2) If $\Sigma_i m_i n_i = 0$, i.e., if M^{geom} and N^{geom} have no common constituent, then for variable finite extensions E/k, we have

$$|\Sigma_{x \text{ in } X(E)} M(E, x)\overline{N}(E, x)| = O((\#E)^{-1/2}).$$

3)The following conditions a) and b) are equivalent.

3a) For variable finite extensions E/k, we have

$$\Sigma_{x \text{ in } X(E)} |M(E, x)|^2 = 1 + O((\#E)^{-1/2}).$$

3b) M is geometrically irreducible, i.e., its pullback M^{geom} to $X\otimes_k\overline{k}$ is an irreducible perverse sheaf on $X\otimes_k\overline{k}$.

proof The assertions concern the trace functions of M and N, and the objects M^{geom} and N^{geom}. The latter are semisimple on $X\otimes_k\overline{k}$, and the former depend only on the semisimplifications of M and N on X. So we may replace M and N by their semisimplifications on X, and reduce to the case where both M and N are sums of perverse irreducibles on X, each ι-pure of weight zero. Let $\{Z_\alpha\}$ be a smooth stratification of X to which $N\oplus M$ is adapted. Then any direct factor of $N\oplus M$, in particular every simple constituent of $N\oplus M$, is also adapted to this stratification.

Step 1. Fix a simple constituent W of $N\oplus M$, and a strat Z_α. Our first task is to show that either

case a) $W|Z_\alpha$ is perverse irreducible and ι-pure of weight zero on Z_α, i.e., of the form $\mathcal{G}[\dim(Z_\alpha)]$ for an irreducible lisse sheaf $\mathcal{G}$ on Z_α which is ι-pure of weight $-\dim(Z_\alpha)$, and W is the middle extension of $\mathcal{G}[\dim(Z_\alpha)]$, or

case b) the trace function of $W|Z_\alpha$ satisfies

$$|W(E, x)| = O((\#E)^{(-\dim(Z_\alpha) - 1)/2})$$

for variable finite extensions E/k, and variable points x in $Z_\alpha(E)$.

To see this we argue as follows. Because W is perverse, with lisse cohomology sheaves on Z_α, we have

$$\mathcal{H}^{-i}(W)|Z_\alpha = 0 \text{ for } i < \dim(Z_\alpha).$$

Because W is ι-pure of weight zero,

$$\mathcal{H}^{-i}(W)|Z_\alpha \text{ is mixed of weight} \le -i.$$

So the possibly nonvanishing among the $\mathcal{H}^{-i}(W)|Z_\alpha$ have $i \ge \dim(Z_\alpha)$, and those with $i \ge 1+\dim(Z_\alpha)$ are all mixed of weight $\le -1 - \dim(Z_\alpha)$. If $\mathcal{H}^{-\dim(Z_\alpha)}(W)|Z_\alpha$ vanishes, we are in case b). Thus it suffices to show that either $\mathcal{H}^{-\dim(Z_\alpha)}(W)|Z_\alpha$ vanishes, or that we have case a).

Because W is perverse, its only possibly nonzero cohomology

sheaves are $\mathcal{H}^{-i}(W)$ for $0 \le i \le \dim(\mathrm{Supp}(W))$. Because W is perverse irreducible, we have the enhanced inequality

$$\dim(\mathrm{Supp}(\mathcal{H}^{-i}(W))) \le i-1 \text{ for } i < \dim(\mathrm{Supp}(W)).$$

Consider first the case in which $\dim(\mathrm{Supp}(W)) > \dim(Z_\alpha)$. Then the enhanced inequality shows that $\mathcal{H}^{-\dim(Z_\alpha)}(W)$ has support of dimension at most $\dim(Z_\alpha) - 1$. Therefore $\mathcal{H}^{-\dim(Z_\alpha)}(W)|Z_\alpha$ vanishes, and we have case b).

Consider next the case in which $\dim(\mathrm{Supp}(W)) < \dim(Z_\alpha)$. As $W|Z_\alpha$ has lisse cohomology sheaves, all its cohomology sheaves vanish, and we have case b).

Consider now the case in which $\dim(\mathrm{Supp}(W)) = \dim(Z_\alpha)$. Then $\mathrm{Supp}(W)$ is an irreducible closed subscheme of X. If $\mathcal{H}^{-\dim(Z_\alpha)}(W)|Z_\alpha$ vanishes, we are in case b). If not, then Z_α lies in $\mathrm{Supp}(W)$. In this case, as $\dim(\mathrm{Supp}(W)) = \dim(Z_\alpha)$, and $Z_\alpha = Z_\alpha \cap \mathrm{Supp}(W)$ is locally closed in $\mathrm{Supp}(W)$, we see that Z_α is a dense open set of $\mathrm{Supp}(W)$. But W has lisse cohomology sheaves on Z_α. Since W is perverse irreducible and ι-pure of weight zero, there is an irreducible lisse sheaf $\mathcal{G}$ on Z_α which is ι-pure of weight $-\dim(Z_\alpha)$, such that W is the middle extension of $\mathcal{G}[\dim(Z_\alpha)]$, and $W|Z_\alpha = \mathcal{G}[\dim(Z_\alpha)]$ is perverse irreducible and ι-pure of weight zero. This ends Step 1.

Step 2. For a given simple constituent W of $N \oplus M$, there is a unique index $\alpha = \alpha(W)$ such that the pair (W, Z_α) is in case a) of Step 1, namely the unique index α such that Z_α contains the generic point of $\mathrm{Supp}(W)$. Indeed, for this α, $\mathcal{H}^{-\dim(\mathrm{Supp}(W))}(W)|Z_\alpha$ is nonzero (and lisse). Therefore, Z_α lies in $\mathrm{Supp}(W)$, so Z_α is a dense open set of $\mathrm{Supp}(W)$. Hence $\dim(\mathrm{Supp}(W))) = \dim(Z_\alpha)$, so $\mathcal{H}^{-\dim(Z_\alpha)}(W)|Z_\alpha$ is nonzero, and we are in case a). And if $\mathcal{H}^{-\dim(Z_\beta)}(W)|Z_\beta$ is nonzero for some β, the argument of Step 1 shows that Z_β is a dense open set of $\mathrm{Supp}(W)$, so contains the generic point of W, and hence $\beta = \alpha$.

Step 3 Let W_1 and W_2 be irreducible constituents of $N \oplus M$. Fix a strat Z_α. If either (W_1, Z_α) or (W_2, Z_α) is case b) of Step 1, we have

$$|\Sigma_{x \text{ in } Z_\alpha(E)} W_1(E, x)\overline{W}_2(E, x)| = O((\#E)^{-1/2}).$$

If both (W_1, Z_α) and (W_2, Z_α) are case a) of Step 1, i.e., if

$$\alpha(W_1) = \alpha(W_2) = \alpha,$$

then

$$W_i|Z_\alpha = \mathcal{W}_i[\dim(Z_\alpha)] \text{ with } \mathcal{W}_i \text{ lisse and } \iota\text{-pure of weight } -\dim(Z_\alpha),$$

for $i = 1, 2$. Denote by $\overline{\mathcal{W}}_2$ the lisse, ι-pure of weight $-\dim(Z_\alpha)$ sheaf on Z_α given by

$\overline{\mathcal{W}}_2 := (\mathcal{W}_2)^\vee(\dim(Z_\alpha))$.

Then by the Lefschetz Trace Formula we have

$$\Sigma_{x \text{ in } Z_\alpha(E)} W_1(E, x)\overline{W}_2(E, x)$$

$$= \Sigma_{i=0 \text{ to } 2\dim(Z_\alpha)} \text{Trace}(\text{Frob}_E \mid H_c{}^i(Z_\alpha \otimes_k \bar{k}, \mathcal{W}_1 \otimes \overline{\mathcal{W}}_2)).$$

The sheaf $\mathcal{W}_1 \otimes \overline{\mathcal{W}}_2$ is lisse, and ι-pure of weight $-2\dim(Z_\alpha)$. By Deligne's fundamental result [De-Weil II, 3.3.1], all but the $H_c{}^{2\dim(Z_\alpha)}$ terms are $O((\#E)^{-1/2})$, and the $H_c{}^{2\dim(Z_\alpha)}$ is ι-pure of weight zero, so we get

$$\Sigma_{x \text{ in } Z_\alpha(E)} W_1(E, x)\overline{W}_2(E, x)$$

$$= \text{Trace}(\text{Frob}_E \mid H_c{}^{2\dim(Z_\alpha)}(Z_\alpha \otimes_k \bar{k}, \mathcal{W}_1 \otimes \overline{\mathcal{W}}_2)) + O((\#E)^{-1/2}),$$

with $H_c{}^{2\dim(Z_\alpha)}(Z_\alpha \otimes_k \bar{k}, \mathcal{W}_1 \otimes \overline{\mathcal{W}}_2)$ ι-pure of weight zero.

Step 4 Write M and N as sums of perverse irreducibles W_i with multiplicities, say

$$M = \Sigma_i a_i W_i,\ N = \Sigma_i b_i W_i,$$

with non-negative integers a_i and b_i.Then

$$\Sigma_{x \text{ in } X(E)} M(E, x)\overline{N}(E, x)$$

$$= \Sigma_{i,\,j}\, a_i b_j \Sigma_\alpha\, \Sigma_{x \text{ in } Z_\alpha(E)} W_i(E, x)\overline{W}_j(E, x).$$

The innermost sum is $O((\#E)^{-1/2})$ unless $\alpha(W_1) = \alpha(W_2) = \alpha$.

For each α, denote by $\mathcal{P}(\alpha)$ the set of those indices i such that $\alpha(W_i) = \alpha$. Then we get

$$\Sigma_{x \text{ in } X(E)} M(E, x)\overline{N}(E, x)$$

$$= \Sigma_\alpha\, \Sigma_{i,\,j \text{ in } \mathcal{P}(\alpha)}\, a_i b_j \text{Trace}(\text{Frob}_E \mid H_c{}^{2\dim(Z_\alpha)}(Z_\alpha \otimes_k \bar{k}, \mathcal{W}_i \otimes \overline{\mathcal{W}}_j))$$
$$+ O((\#E)^{-1/2}).$$

$$= \text{Trace}(\text{Frob}_E \mid \oplus_\alpha \oplus_{i,j \text{ in } \mathcal{P}(\alpha)} H_c{}^{2\dim(Z_\alpha)}(Z_\alpha \otimes_k \bar{k}, \mathcal{W}_i \otimes \overline{\mathcal{W}}_j)^{a_i b_j})$$
$$+ O((\#E)^{-1/2}).$$

The direct sum

$$T := \oplus_\alpha \oplus_{i,j \text{ in } \mathcal{P}(\alpha)} H_c{}^{2\dim(Z_\alpha)}(Z_\alpha \otimes_k \bar{k}, \mathcal{W}_i \otimes \overline{\mathcal{W}}_j)^{a_i b_j}$$

is ι-pure of weight zero. So replacing $\text{Frob}_k \mid T$ by its semisimplification, we get, via ι, a unitary operator

$$A := (\text{Frob}_k \mid T)^{\text{s.s.}}$$

on a finite-dimensional $\mathbb{C}$-space $\iota T := T \otimes \mathbb{C}$ such that for any finite extension E/k,

$$\iota\text{Trace}(\text{Frob}_E \mid T) = \text{Trace}(A^{\deg(E/k)} \mid \iota T).$$

Thus we find

$$\Sigma_{x \text{ in } X(E)} M(E, x)\overline{N}(E, x) = \text{Trace}(A^{\deg(E/k)} \mid \iota T) + O((\#E)^{-1/2}).$$

Because A is unitary, we get the estimate

$$|\text{Trace}(A^{\deg(E/k)} \mid \iota T)| \le \dim(T)$$

for any finite extension E/k.

For any $n \geq 1$, the subgroup of $\mathrm{Aut}_{\mathbb{C}}(\iota T)$ generated by powers of the unitary operator A^n has compact closure, so the sequence of operators $\{A^{nd}\}_{d \geq 1}$ has the identity operator as a cluster point. So for every $n \geq 1$, we get

$$\limsup_{E/k_n} |\Sigma_{x \text{ in } X(E)} M(E, x)\overline{N}(E, x)| = \dim(T).$$

Step 5 We now prove assertions 1) and 2) of the orthogonality theorem. Recall that we have written

$$M^{geom} = \Sigma_i m_i V_i,\ N^{geom} = \Sigma_i n_i V_i,$$

with $\{V_i\}_i$ a finite set of pairwise non-isomorphic perverse irreducibles on $X\otimes_k \overline{k}$, and with non-negative integers m_i and n_i. In view of the results of the previous section, it suffices to prove that

$$\dim(T) = \Sigma_i n_i m_i.$$

To indicate the dependence on (N, M), we will write this as

$$\dim(T(N, M)) = \Sigma_i n_i(N) m_i(M).$$

To see this, we argue as follows. Both sides are bilinear in (N, M) for direct sum decompositions of the arguments, so it suffices to treat the case when N is a single perverse irreducible W_i on X, and when M is a single perverse irreducible W_j on X. Put

$$\alpha_i := \alpha(W_i).$$

Any irreducible constituent V_1 of $W_i{}^{geom}$ on $X\otimes_k \overline{k}$ is the middle extension of an object $\mathcal{V}_1[\dim(Z_{\alpha_i})]$, for an irreducible lisse sheaf $\mathcal{V}_1$ on a connected component of $Z_{\alpha_i}\otimes_k \overline{k}$. Similarly, any irreducible constituent V_2 of $W_j{}^{geom}$ on $X\otimes_k \overline{k}$ is the middle extension of an object $\mathcal{V}_2[\dim(Z_{\alpha_j})]$, for an irreducible lisse sheaf $\mathcal{V}_2$ on a connected component of $Z_{\alpha_j}\otimes_k \overline{k}$.

Suppose first $\alpha_i \neq \alpha_j$. Then $T(W_i, W_j) = 0$. So if $\alpha_i \neq \alpha_j$, V_1 and V_2 are certainly non-isomorphic, because open dense sets of their supports are disjoint. So we find the desired equality in this case: both sides vanish.

Suppose now $\alpha_i = \alpha_j$, say with common value α. Then there exist lisse sheaves $\mathcal{W}_i$ and $\mathcal{W}_j$ on Z_α, ι-pure of weight zero, such that W_i is the middle extension of $\mathcal{G}_i[-\dim(Z_\alpha)]$, and W_j is the middle extension of $\mathcal{G}_j[-\dim(Z_\alpha)]$. Denote by

$$Z_{\alpha,1}, \ldots, Z_{\alpha,r}$$

the connected components of $Z_\alpha\otimes_k \overline{k}$. On each connected component, the pullbacks of $\mathcal{G}_i$ and $\mathcal{G}_j$ are semisimple (by ι-purity), so decompose

$$\mathcal{W}_i|Z_{\alpha,\nu} = \oplus_\mu\, n_{i,\nu,\mu}\, \mathcal{I}_{\nu,\mu},$$
$$\mathcal{W}_j|Z_{\alpha,\nu} = \oplus_\mu\, m_{j,\nu,\mu}\, \mathcal{I}_{\nu,\mu},$$

where the $\mathcal{I}_{\nu,\mu}$ are pairwise non-isomorphic irreducible lisse sheaves on $Z_{\alpha,\nu}$, and the integers $n_{i,\nu,\mu}$ and $m_{j,\nu,\mu}$ are non-negative. The irreducible decomposition of $W_i{}^{geom}$ is then

$$W_i{}^{geom} = \oplus_\nu \oplus n_{i,\nu,\mu} \text{ (middle extension of } \mathcal{F}_{\nu,\mu}[-\dim(Z_\alpha)]),$$

and similarly for $W_j{}^{geom}$. Thus for the data (W_i, W_j), we have

$$\Sigma_a n_a(W_i)m_a(W_j) = \Sigma_\nu \Sigma_\mu n_{i,\nu,\mu} m_{j,\nu,\mu}.$$

On the other hand, the space $T(W_i, W_j)$ is given by

$$T(W_i, W_j) := H_c{}^{2\dim(Z_\alpha)}(Z_\alpha \otimes_k \bar{k}, \mathcal{W}_i \otimes \bar{\mathcal{W}}_j)$$
$$= H_c{}^{2\dim(Z_\alpha)}(Z_\alpha \otimes_k \bar{k}, \underline{\mathrm{Hom}}(\mathcal{W}_j, \mathcal{W}_i)(\dim(Z_\alpha))$$
$$= \oplus_\nu H_c{}^{2\dim(Z_\alpha)}(Z_{\alpha,\nu}, \underline{\mathrm{Hom}}(\mathcal{W}_j, \mathcal{W}_i)(\dim(Z_\alpha))$$
$$= \oplus_\nu \oplus_{\mu,\tau} H_c{}^{2\dim(Z_\alpha)}(Z_{\alpha,\nu}, \underline{\mathrm{Hom}}(\mathcal{F}_{\nu,\tau}, \mathcal{F}_{\nu,\mu})(\dim(Z_\alpha))^{n_{i,\nu,\mu} m_{j,\nu,\tau}}.$$

Because both $\mathcal{F}_{\nu,\tau}$ and $\mathcal{F}_{\nu,\mu}$ are $\pi_1(Z_{\alpha,\nu})$-irreducible, we have

$$H_c{}^{2\dim(Z_\alpha)}(Z_{\alpha,\nu}, \underline{\mathrm{Hom}}(\mathcal{F}_{\nu,\tau}, \mathcal{F}_{\nu,\mu})(\dim(Z_\alpha)) = \mathrm{Hom}(\mathcal{F}_{\nu,\tau}, \mathcal{F}_{\nu,\mu}),$$

a space of dimension $\delta_{\mu,\tau}$. Thus we get

$$\dim T(W_i, W_j) = \Sigma_\nu \Sigma_\mu n_{i,\nu,\mu} m_{j,\nu,\mu},$$

as required. This completes the proof of assertions 1) and 2) of the orthogonality theorem.

Step 6 We now prove assertion 3). Thus M on X is perverse, and ι-pure of weight zero. Recall that

$$M^{geom} = \Sigma_i m_i V_i,$$

with $\{V_i\}_i$ a finite set of pairwise non-isomorphic perverse irreducibles on $X \otimes_k \bar{k}$, and with non-negative integers m_i. We have proven that for any integer $n \geq 1$, we have

$$\Sigma_i (m_i)^2 = \mathrm{limsup}_{E/k_n} \Sigma_{x \text{ in } X(E)} |M(E, x)|^2.$$

So if condition 3a) holds, i.e., if for variable finite extensions E/k, we have

$$\Sigma_{x \text{ in } X(E)} |M(E, x)|^2 = 1 + O((\#E)^{-1/2}),$$

then $\Sigma_i (m_i)^2 = 1$, i.e., M is geometrically irreducible. Conversely, suppose M is geometrically irreducible. Then for $\alpha = \alpha(M)$, Z_α is geometrically irreducible, and $M|Z_\alpha$ is $\mathcal{M}[\dim(Z_\alpha)]$ with $\mathcal{M}$ a lisse, geometrically irreducible lisse sheaf on Z_α which is ι-pure of weight $-\dim(Z_\alpha)$. The space $T(M, M)$ is given by

$$T(M, M) = H_c{}^{2\dim(Z_\alpha)}(Z_\alpha \otimes_k \bar{k}, \underline{\mathrm{End}}(\mathcal{M})(\dim(Z_\alpha))$$
$$= \mathrm{End}_{\pi_1(Z_\alpha \otimes_k \bar{k})}(\mathcal{M}).$$

It is one-dimensional (by the geometric irreducibility of $\mathcal{M}$), spanned by the ($\pi_1(Z_\alpha)$-equivariant) identity endomorphism, so Frob_k acts on it with eigenvalue 1. Thus we find, for every finite extension E/k,

$$\mathrm{Trace}(\mathrm{Frob}_E \mid T(M,M)) = 1.$$

But we have seen above that for any finite extension E/k, we have

$$\Sigma_{x \text{ in } X(E)} |M(E, x)|^2 - \mathrm{Trace}(\mathrm{Frob}_E \mid T(M,M)) = + O((\# E)^{-1/2}).$$

QED

(1.8) First Applications of the Orthogonality Theorem

Duality Lemma 1.8.1 Suppose M and N on X are both perverse, geometrically irreducible (i.e., perverse irreducible on $X\otimes_k\bar{k}$), and ι-pure of weight zero. Denote by $D_{X/k}M$ the Verdier dual of M. Then we have the following results.

1) M and its Verdier dual $D_{X/k}M$ have, via ι, complex conjugate trace functions.

2) If N and M are geometrically isomorphic (i.e., pulled back to $X\otimes_k\bar{k}$, they are isomorphic), then there is a unique α in $\bar{\mathbb{Q}}_\ell^\times$ for which there exists an isomorphism $N \cong M\otimes\alpha^{\deg}$ on X. This unique α has $|\alpha| = 1$.

3) If M is geometrically isomorphic to its Verdier dual $D_{V/k}M$, there exists a β in $\bar{\mathbb{Q}}_\ell^\times$, such that $M\otimes\beta^{\deg}$ on X is isomorphic to its own Verdier dual. This β, unique up to sign, has $|\beta| = 1$.

4) If M has, via ι, a real-valued trace function, then M is isomorphic to $D_{X/k}M$.

proof 1) Because M is perverse and geometrically irreducible on X, it is a middle extension. More precisely, its support is a geometrically irreducible closed subscheme Z of X, inclusion denoted i: $Z \to X$. There exists a dense affine open set j: $U \to Z$, such that U/k is smooth and geometrically connected, of some dimension $d \geq 0$, and a lisse $\bar{\mathbb{Q}}_\ell$-sheaf $\mathcal{M}$ on U, which is geometrically irreducible and ι-pure of weight -dimU, such that M is $i_\star j_{!\star}\mathcal{M}[\dim U]$. Its Verdier dual $D_{X/k}M$ is $i_\star j_{!\star}\mathcal{M}^\vee(\dim U)[\dim U]$, for $\mathcal{M}^\vee$ the linear dual (contragredient representation of $\pi_1(U)$). Because $\mathcal{M}$ is ι-pure of weight -dimU, $\mathcal{M}^\vee(\dim U)[\dim U]$ and $\mathcal{M}[\dim U]$ have complex conjugate trace functions on U. By a fundamental theorem of Ofer Gabber [Fuj-Indep, Theorem 3], their middle extensions $i_\star j_{!\star}\mathcal{M}[\dim U]$ and $i_\star j_{!\star}\mathcal{M}^\vee(\dim U)[\dim U]$ have complex conjugate trace functions on X. [Gabber's result is this. Suppose we have two pairs (ℓ_1, ι_1) and (ℓ_2, ι_2), in which ℓ_i is a prime invertible in k, and in which ι_i is a field embedding of $\bar{\mathbb{Q}}_{\ell_i}$ into $\mathbb{C}$. Suppose we have objects N_i in $D^b{}_c(U, \bar{\mathbb{Q}}_{\ell_i})$ which are both perverse, geometrically irreducible, and ι_i-pure of weight zero. Suppose that, via ι_1 and ι_2, their trace functions agree, i.e., we have an equality in $\mathbb{C}$,

$$\iota_1 N_1(E, u) = \iota_2 N_2(E, u),$$

for every finite extension E/k and every u in U(E). Then via ι_1 and

ι_2, the trace functions of $i_*j_{!*}N_1$ and $i_*j_{!*}N_2$ agree: we have an equality in $\mathbb{C}$

$$\iota_1 i_*j_{!*}N_1(E, x) = \iota_2\, i_*j_{!*}N_2(E, x)$$

for every finite extension E/k and every x in $X(E)$. We apply this with $\ell_1 = \ell_2 = \ell$, with $\iota_1 = \iota$, and with $\iota_2 = C\circ\iota$, for C the complex conjugation automorphism of $\mathbb{C}$, with $N_1 = \mathcal{M}[\dim U]$ and with $N_2 = \mathcal{M}^\vee(\dim U)[\dim U]$.]

2) If M and N are geometrically isomorphic, they have the same support Z, and we can choose a common affine open U in Z which is lisse and geometrically connected over k, on which both M and N are lisse. Then there exist lisse $\overline{\mathbb{Q}}_\ell$-sheaves $\mathcal{M}$ and $\mathcal{N}$ on U, both geometrically irreducible and ι-pure of weight $-\dim U$, such that $\mathcal{M}$ is $i_*j_{!*}\mathcal{M}[\dim U]$, and N is $i_*j_{!*}\mathcal{N}[\dim U]$. The sheaves $\mathcal{M}$ and $\mathcal{N}$ are both geometrically irreducible on U, and they are geometrically isomorphic, i.e., they are isomorphic irreducible $\pi_1(U\otimes_k\overline{k})$-representations. Since $\pi_1(U\otimes_k\overline{k})$ is a normal subgroup of $\pi_1(U)$, these representations are obtained one from the other by tensoring with a linear character of $\pi_1(U)$ which is trivial on $\pi_1(U\otimes_k\overline{k})$, and such characters are precisely those of the form $\alpha^{\deg}$. So we find

$$\mathcal{N} \cong \mathcal{M}\otimes\alpha^{\deg} \text{ on } U.$$

Since $\mathcal{M}$ and $\mathcal{N}$ are both ι-pure of the same weight, we must have $|\alpha| = 1$. Taking middle extensions commutes with tensoring with a lisse sheaf on the ambient space, so we get

$$N = i_*j_{!*}\mathcal{N}[\dim U] \cong i_*j_{!*}(\mathcal{M}\otimes\alpha^{\deg}) = M\otimes\alpha^{\deg},$$

as required. To show the uniqueness of an α for which there exists an isomorphism $N \cong M\otimes\alpha^{\deg}$, suppose we also have an isomorphism $N \cong M\otimes\gamma^{\deg}$. Then we have an isomorphism $M \cong M\otimes(\alpha/\gamma)^{\deg}$, and hence, restricting to U, an isomorphism $\mathcal{M} \cong \mathcal{M}\otimes(\alpha/\gamma)^{\deg}$ of geometrically irreducible lisse sheaves on U. Tensoring with $\mathcal{M}^\vee$, we get

$$\underline{\mathrm{End}}(\mathcal{M}) \cong \underline{\mathrm{End}}(\mathcal{M})\otimes(\alpha/\gamma)^{\deg}.$$

Taking invariants under $\pi_1(U\otimes_k\overline{k})$, we find an isomorphism

$$\overline{\mathbb{Q}}_\ell \cong (\alpha/\gamma)^{\deg}$$

of one-dimensional representations of $\mathrm{Gal}(\overline{k}/k)$, whence $\alpha = \gamma$.

3) Applying 2), we get α in $\overline{\mathbb{Q}}_\ell^\times$, which via ι has complex absolute value one, and an isomorphism $M\otimes\alpha^{\deg} \cong D_{X/k}M$. Take for β a square root of α. Then

$$D_{X/k}(M\otimes\beta^{\deg}) \cong D_{X/k}(M)\otimes\beta^{-\deg} \cong M\otimes\alpha^{\deg}\otimes\beta^{-\deg}$$
$$= M\otimes(\alpha/\beta)^{\deg} = M\otimes\beta^{\deg}.$$

To show uniqueness of β up to sign, suppose that for some γ in $\overline{\mathbb{Q}}_\ell^\times$, we have

$$D_{X/k}(M\otimes\gamma^{\deg}) \cong M\otimes\gamma^{\deg}.$$

As $D_{X/k}(M\otimes\gamma^{\deg}) \cong D_{X/k}(M)\otimes\gamma^{-\deg}$, we get

$$D_{X/k}(M) \cong M\otimes(\gamma^2)^{\deg}.$$

By the unicity in part 2), applied with $N = D_{X/k}(M)$, we get $\gamma^2 = \beta^2$.
4) In view of 1), M and $D_{X/k}(M)$ have the same trace function. Restricting to U, we find that $\mathcal{M}[\dim U]$ and $\mathcal{M}^\vee(\dim U)[\dim U]$ have the same trace function. As both are irreducible, by Chebotarev they are isomorphic. Hence their middle extensions M and $D_{X/k}(M)$ are isomorphic. QED

First Corollary 1.8.2 Let X/k be a separated k-scheme of finite type with k a finite field, ℓ a prime invertible in k, and $\iota : \overline{\mathbb{Q}}_\ell \subset \mathbb{C}$ a field embedding. Suppose M and N on X are both perverse, geometrically irreducible (i.e., perverse irreducible on $X\otimes_k\overline{k}$), and ι-pure of weight zero.
1) For variable finite extensions E/k, we have

$$\Sigma_{x \text{ in } X(E)} |M(E, x)|^2 = 1 + O((\#E)^{-1/2}).$$

2) If M and N are not geometrically isomorphic, then for variable finite extensions E of k, we have

$$|\Sigma_{x \text{ in } X(E)} M(E, x)\overline{N}(E, x)| = O((\#E)^{-1/2}).$$

3) If M and N are geometrically isomorphic, then for variable finite extensions E/k, we have

$$|\Sigma_{x \text{ in } X(E)} M(E, x)\overline{N}(E, x)| = 1 + O((\#E)^{-1/2}).$$

4) If M is isomorphic to $D_{X/k}M$, then its trace function takes, via ι, real values, and for variable finite extensions E/k, we have

$$\Sigma_{x \text{ in } X(E)} M(E, x)^2 = 1 + O((\#E)^{-1/2}).$$

5) Suppose that there exists real ε with $1 \geq \varepsilon > 0$ such that for variable finite extensions E/k we have

$$\Sigma_{x \text{ in } X(E)} M(E, x)^2 = 1 + O((\#E)^{-\varepsilon/2}).$$

Then M is isomorphic to $D_{X/k}M$.
6) If M is geometrically isomorphic to $D_{X/k}M$, then for variable finite extensions E/k, we have

$$|\Sigma_{x \text{ in } X(E)} M(E, x)^2| = 1 + O((\#E)^{-1/2}).$$

7) If M is not geometrically isomorphic to $D_{X/k}M$, then for variable finite extensions E/k, we have

$$|\Sigma_{x \text{ in } X(E)} M(E, x)^2| = O((\#E)^{-1/2}).$$

proof Assertions 1) and 2), special cases of the orthogonality theorem, are "mise pour mémoire".
3) use 1), and part 2) of the Duality Lemma 1.8.1.
4) use 1), and part 1) of the Duality Lemma 1.8.1.
5) By part 4) of the Duality Lemma 1.8.1, there exists β in $\overline{\mathbb{Q}}_\ell^\times$, such

that $M\otimes\beta^{\deg}$ on X is isomorphic to its own Verdier dual. This β, unique up to sign, has $|\beta| = 1$. So by part 4), applied to $M\otimes\beta^{\deg}$, we have

$$\beta^{2\deg(E/k)}\Sigma_{x \text{ in } X(E)} M(E, x)^2 = 1 + O((\#E)^{-1/2}).$$

By hypothesis, we have

$$\Sigma_{x \text{ in } X(E)} M(E, x)^2 = 1 + O((\#E)^{-\varepsilon/2}).$$

So we find

$$\beta^{2\deg(E/k)} = 1 + O((\#E)^{-\varepsilon/2}).$$

Consider the complex power series in one variable T defined by

$$\Sigma_{n\geq 0}\, \beta^{2n}T^n = 1/(1 - \beta^2T).$$

It satisfies

$$1/(1 - \beta^2T) - 1/(1 - T) = \text{a series convergent in } |T| < (\#k)^{\varepsilon/2}.$$

Therefore the left hand side has no poles on the unit circle, i.e., we have $\beta^2 = 1$. Thus β is ±1. Since β is unique up to sign, we may choose $\beta = 1$, and find that M is already self dual.
6) use 4), and part 4) of the Duality Lemma 1.8.1.
7) use 2), and part 1) of the Duality Lemma 1.8.1. QED

Second Corollary 1.8.3 Let X/k be a separated k-scheme of finite type with k a finite field, ℓ a prime invertible in k, and $\iota : \overline{\mathbb{Q}}_\ell \subset \mathbb{C}$ a field embedding. Suppose M and N on X are perverse, and ι-mixed of weight ≤ 0. So for all sufficiently small real $w > 0$, we have short exact sequences of perverse sheaves

$$0 \to M_{\leq -w} \to M \to \mathrm{Gr}^0(M) \to 0,$$

$$0 \to N_{\leq -w} \to N \to \mathrm{Gr}^0(N) \to 0.$$

Here $\mathrm{Gr}^0(M)$ and $\mathrm{Gr}^0(N)$ are both ι-pure of weight 0, and $M_{\leq -w}$ and $N_{\leq -w}$ are both ι-mixed of weight $\leq -w$. Fix one such w, with $1 \geq w > 0$.

Write the pullbacks $\mathrm{Gr}^0(M)^{\mathrm{geom}}$ and $\mathrm{Gr}^0(N)^{\mathrm{geom}}$ of $\mathrm{Gr}^0(M)$ and $\mathrm{Gr}^0(N)$ to $X\otimes_k\overline{k}$ as sums of perverse irreducibles with multiplicities, say

$$\mathrm{Gr}^0(M)^{\mathrm{geom}} = \Sigma_i\, m_iV_i,\ \mathrm{Gr}^0(N)^{\mathrm{geom}} = \Sigma_i\, n_iV_i,$$

with $\{V_i\}_i$ a finite set of pairwise non-isomorphic perverse irreducibles on $X\otimes_k\overline{k}$, and with non-negative integers m_i and n_i. [This is possible by [BBD, 5.3.8].]
Then we have the following results.
1) For any integer $n \geq 1$, denoting by k_n/k the extension field of degree n, we have

$$\Sigma_i\, m_in_i = \limsup_{E/k_n} |\Sigma_{x \text{ in } X(E)} M(E, x)\overline{N}(E, x)|,$$

$$\Sigma_i (m_i)^2 = \mathrm{limsup}_{E/k_n} \Sigma_{x \text{ in } X(E)} |M(E, x)|^2,$$

the limsup taken over all finite extensions E/k_n.

2) If $\Sigma_i (m_i)^2 = 0$, i.e., if $Gr^0(M) = 0$, then there exists real $\varepsilon > 0$ such that for variable finite extensions E/k, we have

$$\Sigma_{x \text{ in } X(E)} |M(E, x)|^2 = O((\#E)^{-\varepsilon/2}).$$

3) The following conditions a) and b) are equivalent.

3a) There exists real $\varepsilon > 0$ such that for variable finite extensions E/k, we have

$$\Sigma_{x \text{ in } X(E)} |M(E, x)|^2 = 1 + O((\#E)^{-\varepsilon/2}).$$

3b) $Gr^0(M)$ is geometrically irreducible.

proof From the identities

$$Gr^0(M)(E, x) = M(E, x) - M_{\le -w}(E, x),$$

$$Gr^0(N)(E, x) = N(E, x) - N_{\le -w}(E, x),$$

we get

$$\begin{aligned} &Gr^0(M)(E, x)\overline{Gr}^0(N)(E, x) - M(E, x)\overline{N}(E, x) \\ &= -M_{\le -w}(E, x)\overline{N}(E, x) - M(E, x)\overline{N}_{\le -w}(E, x) \\ &\qquad + M_{\le -w}(E, x)\overline{N}_{\le -w}(E, x). \end{aligned}$$

By Lemma 1.5.13, we have

$$\Sigma_{x \text{ in } X(E)} |M_{\le -w}(E, x)\overline{N}_{\le -w}(E, x)| = O((\#E)^{-w}),$$

$$\Sigma_{x \text{ in } X(E)} |M_{\le -w}(E, x)\overline{N}(E, x)| = O((\#E)^{-w/2}),$$

$$\Sigma_{x \text{ in } X(E)} |M(E, x)\overline{N}_{\le -w}(E, x)| = O((\#E)^{-w/2}).$$

So we find

$$\begin{aligned} &\Sigma_{x \text{ in } X(E)} M(E, x)\overline{N}(E, x) \\ &= \Sigma_{x \text{ in } X(E)} Gr^0(M)(E, x)\overline{Gr}^0(N)(E, x) + O((\#E)^{-w/2}). \end{aligned}$$

So the corollary is simply the Orthogonality Theorem 1.7.2, applied to $Gr^0(M)$ and to $Gr^0(N)$. QED

Third Corollary 1.8.4 Hypotheses and notations as in the Second Corollary 1.8.3, suppose in addition that $Gr^0(M)$ is geometrically irreducible. Then we have the following results.

1) If $Gr^0(M)$ is isomorphic to $D_{X/k}Gr^0(M)$, there exists real $\varepsilon > 0$ such that for variable finite extensions E/k, we have

$$\Sigma_{x \text{ in } X(E)} M(E, x)^2 = 1 + O((\#E)^{-\varepsilon/2}).$$

2) Suppose there exists real $\varepsilon > 0$ such that for variable finite extensions E/k, we have

$$\Sigma_{x \text{ in } X(E)} M(E, x)^2 = 1 + O((\#E)^{-\varepsilon/2}).$$

Then $Gr^0(M)$ is isomorphic to $D_{X/k}Gr^0(M)$.

3) Suppose that M has, via ι, a real-valued trace function. Then $Gr^0(M)$ is isomorphic to $D_{X/k}Gr^0(M)$.

4) If $Gr^0(M)$ is geometrically isomorphic to $D_{X/k}Gr^0(M)$, then there

exists real $\varepsilon > 0$ such that for variable finite extensions E/k, we have

$$|\Sigma_{x \text{ in } X(E)} M(E, x)^2| = 1 + O((\#E)^{-\varepsilon/2}).$$

5) If $Gr^0(M)$ is not geometrically isomorphic to $D_{X/k}Gr^0(M)$, then there exists real $\varepsilon > 0$ such that for variable finite extensions E/k, we have

$$|\Sigma_{x \text{ in } X(E)} M(E, x)^2| = O((\#E)^{-\varepsilon/2}).$$

proof Just as in the proof of the Second Corollary 1.8.3, we have

$$\Sigma_{x \text{ in } X(E)} M(E, x)^2 = \Sigma_{x \text{ in } X(E)} Gr^0(M)(E, x)^2 + O((\#E)^{-w/2}).$$

So parts 1), 2), 4), and 5) result from the First Corollary 1.8.2, applied to $Gr^0(M)$. For part 3), we notice that if M has a real valued trace function, we have

$$\Sigma_{x \text{ in } X(E)} M(E, x)^2 = \Sigma_{x \text{ in } X(E)} |M(E, x)|^2.$$

The right hand side is, for some real $\varepsilon > 0$, $1 + O((\#E)^{-\varepsilon/2}$, thanks to part 3) of the Second Corollary 1.8.3. Now apply part 2). QED

(1.9) Questions of autoduality: the Frobenius-Schur indicator theorem

(1.9.1) Let K be an algebraically closed field of characteristic zero. When a group G operates irreducibly on a finite-dimensional K-vector space V, we have the following trichotomy: either the representation V of G is not self dual, or it is orthogonally self dual, or it is symplectically self dual. The Frobenius-Schur indicator of the G-representation V, denoted FSI(G, V), is defined as

$$\begin{aligned} FSI(G, V) &:= 0, \text{ if } V \text{ is not self dual,} \\ &= 1, \text{ if } V \text{ is orthogonally self dual,} \\ &= -1, \text{ if } V \text{ is symplectically self dual.} \end{aligned}$$

[When K is $\mathbb{C}$ and G is compact, Frobenius and Schur discovered in 1906 their integral formula for the Frobenius-Schur indicator:

$$FSI(G, V) = \int_G \mathrm{Trace}(g^2 \mid V)dg,$$

for dg the total mass one Haar measure on G.]

(1.9.2) Now let k be a field, ℓ a prime number invertible in k, and U/k a separated k-scheme of finite type, which is smooth and connected, of dimension $d = \dim U \geq 0$. A lisse, irreducible $\overline{\mathbb{Q}}_\ell$-sheaf $\mathcal{G}$ on U "is" an irreducible $\overline{\mathbb{Q}}_\ell$-representation of $\pi_1(U)$, and so we may speak of its Frobenius-Schur indicator $FSI(\pi_1(U), \mathcal{G})$. We will sometimes write

$$FSI(U, \mathcal{G}) := FSI(\pi_1(U), \mathcal{G}).$$

If U/k is geometrically connected, and if $\mathcal{G}$ is geometrically irreducible, i.e., irreducible as a representation of $\pi_1(U\otimes_k\overline{k})$, we may also speak of its Frobenius-Schur indicator as a representation of $\pi_1(U\otimes_k\overline{k})$, which we call the geometric Frobenius-Schur indicator of $\mathcal{G}$ on U:

$$FSI^{geom}(U, \mathcal{G}) := FSI(U\otimes_k\overline{k}, \mathcal{G}) := FSI(\pi_1(U\otimes_k\overline{k}), \mathcal{G}).$$

These indicators are birational invariants, in the sense that for any dense open set $U_1 \subset U$, we have

$$\mathrm{FSI}(U, \mathcal{G}) = \mathrm{FSI}(U_1, \mathcal{G}|U_1),$$

$$\mathrm{FSI}^{\mathrm{geom}}(U, \mathcal{G}) = \mathrm{FSI}^{\mathrm{geom}}(U_1, \mathcal{G}|U_1).$$

These equalities hold simply because $\pi_1(U_1)$ maps onto $\pi_1(U)$, and $\pi_1(U_1 \otimes_k \overline{k})$ maps onto $\pi_1(U \otimes_k \overline{k})$.

(1.9.3) Frobenius-Schur indicator for perverse sheaves Let X/k be a separated k-scheme of finite type, and let M on X/k be perverse and geometrically irreducible. Its support is a geometrically irreducible closed subscheme Z of X, inclusion denoted $i: Z \to X$. There exists a dense affine open set $j: U \to Z$, such that U/k is smooth and geometrically connected, of some dimension $d \geq 0$, and a lisse $\overline{\mathbb{Q}}_\ell$-sheaf $\mathcal{M}$ on U, which is geometrically irreducible, such that M is $i_\star j_{!\star}\mathcal{M}[\dim U]$. We define

$$\mathrm{FSI}(X, M) := (-1)^d \mathrm{FSI}(U, \mathcal{M}),$$

$$\mathrm{FSI}^{\mathrm{geom}}(X, M) := (-1)^d \mathrm{FSI}^{\mathrm{geom}}(U, \mathcal{G}).$$

(1.9.4) A diophantine version of the Frobenius-Schur integral (compare [Ka-GKM 4.2, pp. 54-55]) We now return to working over a finite field k. As before, X/k is separated of finite type, and ℓ is invertible in k. Let N be an object in $D^b_c(X, \overline{\mathbb{Q}}_\ell)$. For each finite extension E/k, denote by E_2/E the quadratic extension of E. For x in $X(E)$, we may view x as lying in $X(E_2)$, and speak of $N(E_2, x)$:

$$N(E_2, x) = \mathrm{Trace}(\mathrm{Frob}_{E_2, x} \mid N) = \mathrm{Trace}((\mathrm{Frob}_{E, x})^2 \mid N)$$

$$= \Sigma_i (-1)^i \mathrm{Trace}((\mathrm{Frob}_{E, x})^2 \mid \mathcal{H}^i(N)).$$

For each finite extension E/k, we define the Frobenius-Schur sum attached to N to be

$$\mathrm{FS}(X, N, E) := \Sigma_{x \text{ in } X(E)} N(E_2, x).$$

Lemma 1.9.5 If N is semiperverse and ι-mixed of weight $\leq a$, then for variable finite extensions E/k, we have

$$\mathrm{FS}(X, M, E) = O((\#E)^a).$$

proof By Lemma 1.5.7 above, we reduce to the case where the weight a occurs via ι. Replacing M by a suitable constant field twist $M \otimes \alpha^{\deg}$, we reduce to the case where $a = 0$. Pick a smooth stratification $\{Z_\alpha\}$ of X to which M is adapted. On each strat Z_α, $M|Z_\alpha$ remains semiperverse, and ι-mixed of weight ≤ 0. Break the sum over $X(E)$ into sums over the individual Z_α. So it suffices to treat the case where X is a Z_α. Thus X/k is connected and smooth over k of some dimension $d := \dim(X) \geq 0$, and M has all its cohomology sheaves lisse on X. By semiperversity, $\mathcal{H}^{-i}(M) = 0$ for $i < d$. So the (at most finitely many) nonvanishing $\mathcal{H}^{-i}(M)$ all have

$i \geq d$, and hence are ι-mixed of weight $\leq -d$. Thus for any finite extension E/k, and any x in X(E), we have the estimate

$$|M(E, x)| \leq (\Sigma_i \operatorname{rank}(\mathcal{H}^i(M)))(\#E)^{-d/2}.$$

Keeping the same x but replacing E by E_2, we have, in particular,

$$|M(E_2, x)| \leq (\Sigma_i \operatorname{rank}(\mathcal{H}^i(M)))(\#E)^{-d}.$$

The number of terms in the sum for FS(X, M, E) is $\#X(E)$, which is trivially $O((\#E)^d)$. QED

Frobenius-Schur Indicator Theorem 1.9.6 Let M on X/k be perverse, geometrically irreducible, and ι-pure of weight zero.

1) For variable finite extensions E/k, we have

$$|FS(X, M, E)| = |FSI^{geom}(X, M)| + O((\#E)^{-1/2}).$$

2) If there exists an isomorphism $M \cong D_{X/k}(M)$ on X, then for variable finite extensions E/k, we have

$$FS(X, M, E) = FSI^{geom}(X, M) + O((\#E)^{-1/2}).$$

3) Let ς be a square root of 1 in $\mathbb{C}$, i.e., ς is ± 1. Suppose that there exists a real ε with $1 \geq \varepsilon > 0$ such that for variable finite extensions E/k, we have

$$FS(X, M, E) = \varsigma + O((\#E)^{-\varepsilon/2}).$$

Then there exists an isomorphism $M \cong D_{X/k}(M)$ on X, and ς is $FSI^{geom}(X, M)$.

proof We reduce immediately to the case that X is Supp(M). We then take a smooth stratification $\{Z_\alpha\}$ of X to which M is adapted. Because X = Supp(M) is geometrically irreducible, the unique strat which contains the generic point of X is a dense open set, say U, which is smooth and geometrically connected over k, of dimension d = dimX. We first reduce to the case when X = U. Breaking up the sum by strats, we have

$$FS(X, M, E) = FS(U, M, E) + \Sigma_{\text{strats } Z_\alpha \neq U} FS(Z_\alpha, M, E)$$

On any strat $Z_\alpha \neq U$, M on Z_α is "case b)" of Step 1 of the proof of the Orthogonality Theorem 1.7.2, i.e., we have

$$|M(E, x)| = O((\#E)^{(-\dim(Z_\alpha) - 1)/2})$$

for variable finite extensions E/k, and variable points x in $Z_\alpha(E)$. So a fortiori we have

$$|M(E_2, x)| = O((\#E)^{-\dim(Z_\alpha) - 1})$$

for variable finite extensions E/k, and variable points x in $Z_\alpha(E)$. So on any such strat Z_α, we have, for variable finite extensions E/k,

$$|FS(Z_\alpha, M, E)| = O((\#E)^{-1/2}).$$

Thus we find

$$FS(X, M, E) = FS(U, M, E) + O((\#E)^{-1/2}).$$

From the definition of the Frobenius-Schur indicator, we have

$$FSI^{geom}(X, M) = FSI^{geom}(U, M).$$

So it suffices to treat the case when $X = U$ is itself smooth and geometrically connected, of dimension $d \geq 0$. Thus M is $\mathcal{M}[d]$ for $\mathcal{M}$ a lisse, geometrically irreducible $\overline{\mathbb{Q}}_\ell$-sheaf on U, ι-pure of weight $-d$. By definition, we have

$$\mathrm{FS}(U, M, E) = (-1)^d \mathrm{FS}(U, \mathcal{M}, E),$$

and

$$\mathrm{FSI}^{\mathrm{geom}}(U, M) = (-1)^d \mathrm{FSI}^{\mathrm{geom}}(U, \mathcal{M}).$$

Recall the linear algebra identity: for g in $GL(V)$,

$$\mathrm{Trace}(g^2|V) = \mathrm{Trace}(g|\mathrm{Sym}^2(V)) - \mathrm{Trace}(g|\Lambda^2(V)).$$

Applying this to g a Frobenius, we find that for E/k a finite extension, and x in $U(E)$, we have the linear algebra identity

$$\mathcal{M}(E_2, x) = (\mathrm{Sym}^2\mathcal{M})(E, x) - (\Lambda^2\mathcal{M})(E, x).$$

Thus we have

$$\begin{aligned}
&\mathrm{FS}(U, \mathcal{M}, E)\\
&= \Sigma_{x \text{ in } U(E)} (\mathrm{Sym}^2\mathcal{M})(E, x) - \Sigma_{x \text{ in } U(E)} (\Lambda^2\mathcal{M})\mathcal{M}(E, x)\\
&= \Sigma_{i=0 \text{ to } 2d} (-1)^i \mathrm{Trace}(\mathrm{Frob}_E \mid H_c^i(U\otimes_k\bar{k}, \mathrm{Sym}^2\mathcal{M}))\\
&\qquad - \Sigma_{i=0 \text{ to } 2d} (-1)^i \mathrm{Trace}(\mathrm{Frob}_E \mid H_c^i(U\otimes_k\bar{k}, \Lambda^2\mathcal{M})),
\end{aligned}$$

by the Lefschetz trace formula.

Notice that both $\mathrm{Sym}^2\mathcal{M}$ and $\Lambda^2\mathcal{M}$ are ι-pure of weight $-2d$. By Deligne's fundamental estimate, the summands with $i \leq 2d - 1$ are $O((\#E)^{-1/2})$, so we find

$$\begin{aligned}
\mathrm{FS}(U, \mathcal{M}, E) &= \mathrm{Trace}(\mathrm{Frob}_E \mid H_c^{2d}(U\otimes_k\bar{k}, \mathrm{Sym}^2\mathcal{M}))\\
&\quad - \mathrm{Trace}(\mathrm{Frob}_E \mid H_c^{2d}(U\otimes_k\bar{k}, \Lambda^2\mathcal{M}))\\
&\quad + O((\#E)^{-1/2}).
\end{aligned}$$

The groups $H_c^{2d}(U\otimes_k\bar{k}, \mathrm{Sym}^2\mathcal{M})$ and $H_c^{2d}(U\otimes_k\bar{k}, \Lambda^2\mathcal{M})$ are both ι-pure of weight zero.

The cohomology group $H_c^{2d}(U\otimes_k\bar{k}, \mathrm{Sym}^2\mathcal{M})$ is the space of $\pi_1(U\otimes_k\bar{k})$-equivariant symmetric bilinear forms on $\mathcal{M}$ with values in $\overline{\mathbb{Q}}_\ell(d)$. Because $\mathcal{M}$ is $\pi_1(U\otimes_k\bar{k})$-irreducible, this space is one-dimensional if $\mathcal{M}$ is orthogonally self dual as $\pi_1(U\otimes_k\bar{k})$-representation, and vanishes otherwise.

Similarly, the cohomology group $H_c^{2d}(U\otimes_k\bar{k}, \Lambda^2\mathcal{M})$ is the space of $\pi_1(U\otimes_k\bar{k})$-equivariant alternating bilinear forms on $\mathcal{M}$ with values in $\overline{\mathbb{Q}}_\ell(d)$. It is one-dimensional if $\mathcal{M}$ is symplectically self dual as $\pi_1(U\otimes_k\bar{k})$-representation, and vanishes otherwise.

We now prove part 1) of the theorem.

Suppose first that $\mathcal{M}$ is not geometrically self dual, i.e., that $\mathrm{FSI}^{\mathrm{geom}}(U, \mathcal{M}) = 0$. Then both cohomology groups vanish, and we have

$$\mathrm{FS}(U, \mathcal{M}, E) = O((\#E)^{-1/2}).$$

Suppose next that $\mathcal{M}$ is orthogonally self dual as $\pi_1(U\otimes_k\bar{k})$-representation, i.e., that $\mathrm{FSI}^{\mathrm{geom}}(U, \mathcal{M}) = 1$. Then $H_c^{2d}(U\otimes_k\bar{k}, \mathrm{Sym}^2\mathcal{M})$ is one-dimensional, pure of weight zero, so of the form $(\alpha_+)^{\deg}$ for some α_+ in $\bar{\mathbb{Q}}_\ell^\times$ with $|\alpha_+| = 1$, and $H_c^{2d}(U\otimes_k\bar{k}, \Lambda^2\mathcal{M})$ vanishes. So in this case we get

$$\mathrm{FS}(U, \mathcal{M}, E) = (\alpha_+)^{\deg(E/k)} + O((\#E)^{-1/2}).$$

Suppose finally that $\mathcal{M}$ is symplectically self dual as $\pi_1(U\otimes_k\bar{k})$-representation. Then $H_c^{2d}(U\otimes_k\bar{k}, \Lambda^2\mathcal{M})$ is one-dimensional, pure of weight zero, so of the form $(\alpha_-)^{\deg}$ for some α_- in $\bar{\mathbb{Q}}_\ell^\times$ with $|\alpha_-| = 1$, and $H_c^{2d}(U\otimes_k\bar{k}, \mathrm{Sym}^2\mathcal{M})$ vanishes. So in this case we get

$$\mathrm{FS}(U, \mathcal{M}, E) = -(\alpha_-)^{\deg(E/k)} + O((\#E)^{-1/2}).$$

So in all cases, we find that for variable finite extensions E/k, we have

$$|\mathrm{FS}(U, M, E)| = |\mathrm{FSI}^{\mathrm{geom}}(U, M)| + O((\#E)^{-1/2}).$$

We now prove part 2). Suppose there exists an isomorphism $M \cong D_{X/k}(M)$ on $X = U$, i.e., an isomorphism $\mathcal{M}[d] \cong \mathcal{M}^\vee(d)[d]$ on U. This means exactly that $\mathcal{M}$ and $\mathcal{M}^\vee(d)$ are isomorphic lisse sheaves on U. Thus $\mathcal{M}$ is geometrically self dual. The canonical $\pi_1(U)$-equivariant pairing

$$\mathcal{M} \times \mathcal{M}^\vee(d) \to \bar{\mathbb{Q}}_\ell(d)$$

is visibly nonzero (indeed, it is a perfect pairing). Composing with the given isomorphism from $\mathcal{M}$ to $\mathcal{M}^\vee(d)$, we get a $\pi_1(U)$-equivariant perfect pairing

$$\mathcal{M} \times \mathcal{M} \to \bar{\mathbb{Q}}_\ell(d).$$

This pairing, viewed as an element of $H_c^{2d}(U\otimes_k\bar{k}, \mathcal{M}\otimes\mathcal{M})$, is a basis, fixed by Frob_k, of whichever of the one-dimensional spaces $H_c^{2d}(U\otimes_k\bar{k}, \mathrm{Sym}^2\mathcal{M})$ or $H_c^{2d}(U\otimes_k\bar{k}, \Lambda^2\mathcal{M})$ is nonzero. In other words, if $\mathrm{FSI}^{\mathrm{geom}}(U, \mathcal{M}) = 1$, then $\alpha_+ = 1$; if $\mathrm{FSI}^{\mathrm{geom}}(U, \mathcal{M}) = -1$, then $\alpha_- = 1$. So we find that for variable finite extensions E/k, we have

$$\mathrm{FS}(U, M, E) = \mathrm{FSI}^{\mathrm{geom}}(U, M) + O((\#E)^{-1/2}).$$

To prove 3), we argue as follows. We have $\varsigma = \pm 1$, and we are told that for variable finite extensions E/k, we have

$$\mathrm{FS}(U, M, E) = \varsigma + O((\#E)^{-1/2}).$$

From part 1), we see that $\mathrm{FSI}^{\mathrm{geom}}(U, \mathcal{M})$ is ± 1. In other words, M is geometrically self dual on U. As proven in part 3) of the Duality lemma 1.8.1, there exists an α with $|\alpha| = 1$, unique up to sign, such that $M\otimes\alpha^{\deg}$ is self dual on X. So by part 2) applied to $M\otimes\alpha^{\deg}$, we get

$$FS(U, M\otimes\alpha^{\deg}, E) = FSI^{geom}(U, M\otimes\alpha^{\deg}) + O((\#E)^{-1/2}).$$

But M and $M\otimes\alpha^{\deg}$ are geometrically isomorphic, so we have

$$FS(U, M\otimes\alpha^{\deg}, E) = FSI^{geom}(U, M) + O((\#E)^{-1/2}).$$

But we have the identity

$$FS(U, M\otimes\alpha^{\deg}, E) = \alpha^{2\deg(E/k)}FS(U, M, E).$$

Combine this with the hypothesis, namely that for some real ε with $1 \ge \varepsilon > 0$ we have

$$FS(U, M, E) = \varsigma + O((\#E)^{-\varepsilon/2}).$$

We get

$$\alpha^{2\deg(E/k)}\varsigma = FSI^{geom}(U, M) + O((\#E)^{-\varepsilon/2}).$$

Since ς is ± 1, we may rewrite this as

$$\alpha^{2\deg(E/k)} = \varsigma FSI^{geom}(U, M) + O((\#E)^{-\varepsilon/2}).$$

Consider the complex power series in one variable T defined by

$$\Sigma_{n\ge 0}\ \alpha^{2n}T^n = 1/(1 - \alpha^2 T).$$

It satisfies

$$1/(1 - \alpha^2 T) - \varsigma FSI^{geom}(U, M)/(1 - T)$$

$$= \text{a series convergent in } |T| < (\#k)^{\varepsilon/2}.$$

Therefore the left hand side has no poles on the unit circle. This in turn implies that

$$\alpha^2 = 1,\ \varsigma FSI^{geom}(U, M) = 1.$$

Thus α is ± 1, and $\varsigma = FSI^{geom}(U, M)$. Since α is unique up to sign, we may choose $\alpha = 1$, and find that M is already self dual. QED

Corollary 1.9.7 Let X/k be a separated k-scheme of finite type with k a finite field, ℓ a prime invertible in k, and $\iota : \overline{\mathbb{Q}}_\ell \subset \mathbb{C}$ a field embedding. Suppose M on X is perverse, and ι-mixed of weight ≤ 0. So for all sufficiently small real $w > 0$, we have a short exact sequence of perverse sheaves

$$0 \to M_{\le -w} \to M \to Gr^0(M) \to 0,$$

with $Gr^0(M)$ ι-pure of weight 0, and $M_{\le -w}$ ι-mixed of weight $\le -w$. Fix one such w, with $1 \ge w > 0$. Suppose that $Gr^0(M)$ is geometrically irreducible. Then we have the following results.

1) For variable finite extensions E/k, we have

$$|FS(X, M, E)| = |FSI^{geom}(X, Gr^0(M))| + O((\#E)^{-w/2}).$$

2) If there exists an isomorphism $Gr^0(M) \cong D_{X/k}(Gr^0(M))$ on X (e.g., if M has a real valued trace function, cf. Corollary 1.8.4, part 3)), then for variable finite extensions E/k, we have

$$FS(X, M, E) = FSI^{geom}(X, Gr^0(M)) + O((\#E)^{-w/2}).$$

3) Let ς be a square root of 1 in $\mathbb{C}$. Suppose that there exists a real $\varepsilon > 0$ such that for variable finite extensions E/k, we have

$$FS(X, M, E) = \varsigma + O((\#E)^{-\varepsilon/2}).$$

Then there exists an isomorphism $Gr^0(M) \cong D_{X/k}(Gr^0(M))$ on X, and

ς is $\mathrm{FSI}^{\mathrm{geom}}(X, \mathrm{Gr}^0(M))$.

proof We have

$$FS(X, M, E) = FS(X, \mathrm{Gr}^0(M), E) + FS(X, M_{\leq -w}, E).$$

By Lemma 1.9.5, we have

$$FS(X, M_{\leq -w}, E) = O((\#E)^{-w}).$$

Now apply the previous theorem 1.9.6 to $\mathrm{Gr}^0(M)$. QED

(1.10) Dividing out the "constant part" of an ι-pure perverse sheaf

(1.10.1) In this section, we work on an X/k which is smooth and geometrically connected, of some dimension $d \geq 0$. The object $\overline{\mathbb{Q}}_\ell[d](d/2)$ is geometrically irreducible and ι-pure of weight zero on X/k. We will refer to it as the constant perverse sheaf on X/k. We will refer to its pullback to $X\otimes_k\bar{k}$ as the constant perverse sheaf on $X\otimes_k\bar{k}$

(1.10.2) A perverse sheaf N on X/k is called geometrically constant if N^{geom} is isomorphic to the direct sum of finitely many copies of the constant perverse sheaf on $X\otimes_k\bar{k}$. Equivalently, N is geometrically constant if and only if it is of the form $\mathcal{G}[d]$ for $\mathcal{G}$ a lisse $\overline{\mathbb{Q}}_\ell$-sheaf on X which is geometrically constant. A perverse sheaf N on X/k is called geometrically totally nonconstant if N^{geom} is semisimple and if none of its simple constitutents is isomorphic to the constant perverse sheaf on $X\otimes_k\bar{k}$.

Lemma 1.10.3 Let N be perverse and ι-pure of weight zero on X/k. Then N has a unique direct sum decomposition

$$N = N_{\mathrm{cst}} \oplus N_{\mathrm{ncst}}$$

with N_{cst} geometrically constant (i.e., of the form $\mathcal{G}[d]$ for $\mathcal{G}$ a lisse $\overline{\mathbb{Q}}_\ell$-sheaf on X which is ι-pure of weight -d and geometrically constant) and with N_{ncst} geometrically totally nonconstant.

proof Because N is ι-pure, N^{geom} is semisimple [BBD 5.3.8]. In its isotypical decomposition, separate out the isotypical component $(N^{\mathrm{geom}})_{\mathrm{cst}}$ of the constant perverse sheaf. We get

$$N^{\mathrm{geom}} = (N^{\mathrm{geom}})_{\mathrm{cst}} \oplus (N^{\mathrm{geom}})_{\mathrm{ncst}},$$

with $(N^{\mathrm{geom}})_{\mathrm{ncst}}$ a sum of nonconstant irreducibles. Each summand is stable by Frobenius pullback, so the projections of N^{geom} onto the two factors are, by [BBD 5.1.2], endomorphisms of N which are a pair of orthogonal idempotents of N. This gives the existence. Uniqueness is clear, again by [BBD 5.1.2], since the pullback to $X\otimes_k\bar{k}$ of any such decomposition must be the decomposition of N^{geom} we started with. QED

Lemma 1.10.4 Let N be perverse and ι-pure of weight zero on X/k.

Then $H_c^d(X\otimes_k\bar{k}, N_{ncst}) = 0$.

proof On $X\otimes_k\bar{k}$, $(N_{ncst})^{geom}$ is the direct sum of nonconstant perverse irreducibles. So the lemma results from

Sublemma 1.10.5 Let M be perverse on $X\otimes_k\bar{k}$.

1) $H_c^i(X\otimes_k\bar{k}, M)$ vanishes for i outside the closed interval [-d, d].

2) If in addition M is irreducible and nonconstant on $X\otimes_k\bar{k}$, then $H_c^d(X\otimes_k\bar{k}, M) = 0$.

proof of sublemma Immediate reduction to the case when M is perverse irreducible. If d = dim(X) is zero, the first assertion is obvious and the second is vacuous: $X\otimes_k\bar{k}$ is a point and every perverse sheaf on $X\otimes_k\bar{k}$ is constant. So suppose $d \geq 1$, and let M have dimension of support $d(M) \geq 0$. Look at the spectral sequence

$$E_2^{p,q} = H_c^p(X\otimes_k\bar{k}, \mathcal{H}^q(M)) \Rightarrow H_c^{p+q}(X\otimes_k\bar{k}, M).$$

The only possibly nonvanishing $\mathcal{H}^q(M)$ have $0 \geq q \geq -d(M)$, and we have

$$\dim\mathrm{Supp}\mathcal{H}^{-d(M)}(M) = d(M),$$
$$\dim\mathrm{Supp}\mathcal{H}^{-i}(M) \leq i, \text{ for } 0 \leq i \leq d(M) - 1.$$

So we see that $H_c^i(X\otimes_k\bar{k}, M)$ vanishes for i outside the closed interval [-d(M), d(M)]. This proves the first assertion. Suppose now that M is perverse irreducible and nonconstant. If $d(M) < d$, we are done. If $d(M) = d$, the spectral sequence shows that

$$H_c^d(X\otimes_k\bar{k}, M) = E_\infty^{2d, -d}$$

is a quotient of

$$E_2^{2d, -d} = H_c^{2d}(X\otimes_k\bar{k}, \mathcal{H}^{-d}(M)).$$

This last group is a birational invariant. But on some dense open set U of $X\otimes_k\bar{k}$, M|U is $\mathcal{G}[d]$ for a lisse sheaf $\mathcal{G}$ on U which is irreducible and nonconstant, and so has $H_c^{2d}(U, \mathcal{G}) = 0$. By the birational invariance, $H_c^{2d}(X\otimes_k\bar{k}, \mathcal{H}^{-d}(M)) = 0$. QED

Lemma 1.10.6 Let N be perverse and ι-pure of weight zero on X/k. Then $H_c^d(X\otimes_k\bar{k}, N)$ is ι-pure of weight d. View $H_c^d(X\otimes_k\bar{k}, N)(d)$ as a geometrically constant lisse sheaf $\mathcal{G}$ on X which is ι-pure of weight -d. Then N_{cst}, the constant part of N, is given by $N_{cst} \cong \mathcal{G}[d]$.

proof Write N as $N_{cst} \oplus N_{ncst}$. By the previous lemma,

$$H_c^d(X\otimes_k\bar{k}, N_{ncst}) = 0.$$

Now write $N_{cst} \cong \mathcal{H}[d]$ for some lisse, geometrically constant sheaf $\mathcal{H}$ on X which is ι-pure of weight -d. Then

$$H_c^d(X\otimes_k\bar{k}, N) = H_c^d(X\otimes_k\bar{k}, N_{cst} \oplus N_{ncst})$$
$$= H_c^d(X\otimes_k\bar{k}, N_{cst}) = H_c^d(X\otimes_k\bar{k}, \mathcal{H}[d])$$

$$= H_c^{2d}(X\otimes_k\bar{k}, \mathcal{H}) = \mathcal{H}(-d),$$

the last equality because $\mathcal{H}$ is geometrically constant. Thus $H_c^d(X\otimes_k\bar{k}, N) = \mathcal{H}(-d)$ is ι-pure of weight d, and the sheaf $\mathcal{H}$ is $H_c^d(X\otimes_k\bar{k}, N)(d)$, as asserted. QED

Corollary 1.10.7 $N_{cst} = 0$ if and only if $H_c^d(X\otimes_k\bar{k}, N) = 0$.

Corollary 1.10.8 Let N be perverse and ι-pure of weight zero on X/k. Then the trace function of N_{ncst} is related to that of N as follows. For any finite extension E/k, and for any point x in X(E),

$$N_{ncst}(E, x) = N(E, x) - (-1)^d(\#E)^{-d}\mathrm{Trace}(\mathrm{Frob}_E | H_c^d(X\otimes_k\bar{k}, N)).$$

The trace function of N_{cst} is given by

$$N_{cst}(E, x) = (-1)^d(\#E)^{-d}\mathrm{Trace}(\mathrm{Frob}_E | H_c^d(X\otimes_k\bar{k}, N)).$$

(1.11) The subsheaf N_{ncst0} in the mixed case

(1.11.1) We continue to work on an X/k which is smooth and geometrically connected, of some dimension d ≥ 0. Now suppose we have a perverse sheaf N on X/k which is ι-mixed of weight ≤ 0. So for some w > 0 we have a short exact sequence

$$0 \to N_{\le -w} \to N \to \mathrm{Gr}^0(N) \to 0,$$

with $N_{\le -w}$ perverse and ι-mixed of weight ≤ -w, and with $\mathrm{Gr}^0(N)$ perverse and ι-pure of weight 0. Our ultimate object of interest is $\mathrm{Gr}^0(N)_{ncst}$, the nonconstant part of $\mathrm{Gr}^0(N)$, which is a natural quotient of N. To deal with it, we now define a perverse subsheaf

$$N_{ncst0} \subset N,$$

with the property that the inclusion induces an equality

$$\mathrm{Gr}^0(N_{ncst0}) = \mathrm{Gr}^0(N)_{ncst}.$$

(1.11.2) To define N_{ncst0}, first apply Lemma 1.10.3 to $\mathrm{Gr}^0(N)$, to get a decomposition

$$\mathrm{Gr}^0(N) = \mathrm{Gr}^0(N)_{cst} \oplus \mathrm{Gr}^0(N)_{ncst}$$

Then view $\mathrm{Gr}^0(N)_{cst}$ as a quotient of $\mathrm{Gr}^0(N)$, and define

$$N_{ncst0} := \mathrm{Ker}(\text{the composite map } N \to \mathrm{Gr}^0(N) \to \mathrm{Gr}^0(N)_{cst}).$$

(1.11.3) So we have short exact sequences

$$0 \to N_{ncst0} \to N \to \mathrm{Gr}^0(N)_{cst} \to 0$$

and

$$0 \to N_{\le -w} \to N_{ncst0} \to \mathrm{Gr}^0(N)_{ncst} \to 0.$$

In particular, we have the desired property

(1.11.4) $$\mathrm{Gr}^0(N_{ncst0}) \cong \mathrm{Gr}^0(N)_{ncst}.$$

Lemma 1.11.5 Denote by $H_c^d(X\otimes_k\bar{k}, N)_{wt=d}$ the weight d quotient of $H_c^d(X\otimes_k\bar{k}, N)$ (which is a priori ι-mixed of weight ≤ d). The natural map $N \to Gr^0(N)$ induces an isomorphism

$$H_c^d(X\otimes_k\bar{k}, N)_{wt=d} \cong H_c^d(X\otimes_k\bar{k}, Gr^0(N)).$$

proof As $H_c^{d+1}(X\otimes_k\bar{k}, \text{perverse}) = 0$, the end of the long cohomology sequence for $H_c^*(X\otimes_k\bar{k}, -)$ and the short exact sequence

$$0 \to N_{\le -w} \to N \to Gr^0(N) \to 0$$

is a right exact sequence

$$H_c^d(X\otimes_k\bar{k}, N_{\le -w}) \to H_c^d(X\otimes_k\bar{k}, N) \to H_c^d(X\otimes_k\bar{k}, Gr^0(N)) \to 0,$$

in which the first term is ι-mixed of weight $\le d - w$, and in which the last term is ι-pure of weight $-d$. QED

Corollary 1.11.6 $N = N_{ncst0}$ if and only if $H_c^d(X\otimes_k\bar{k}, N)$ is ι-mixed of weight $\le d - \varepsilon$ for some real $\varepsilon > 0$.

proof Indeed, $N = N_{ncst0}$ if and only if $Gr^0(N)_{cst} = 0$, if and only if (by Corollary 1.10.7) $H_c^d(X\otimes_k\bar{k}, Gr^0(N)) = 0$, if and only if the weight d quotient of $H_c^d(X\otimes_k\bar{k}, N)$ vanishes. QED

Lemma 1.11.7 Denote by $H_c^d(X\otimes_k\bar{k}, N)_{wt=d}$ the weight d quotient of $H_c^d(X\otimes_k\bar{k}, N)$.

1) The constant object $Gr^0(N)_{cst}$ is $\mathcal{G}[d](d)$, for $\mathcal{G}$ the constant sheaf

$$H_c^d(X\otimes_k\bar{k}, N)_{wt=d}.$$

2) The trace function of $Gr^0(N)_{cst}$ is given in terms of N by

$Gr^0(N)_{cst}(E, x) = (-1)^d(\#E)^{-d}\mathrm{Trace}(Frob_E | H_c^d(X\otimes_k\bar{k}, N)_{wt=d}).$

3) The trace function of N_{ncst0} is related to that of N by

$N_{ncst0}(E, x) = N(E, x) - (-1)^d(\#E)^{-d}\mathrm{Trace}(Frob_E | H_c^d(X\otimes_k\bar{k}, N)_{wt=d}).$

proof Assertions 1) and 2) result from Lemma 1.10.6 and Corollary 1.10.8, applied to $Gr^0(N)$, and Lemma 1.11.5. Assertion 3) is then immediate from the short exact sequence

$$0 \to N_{ncst0} \to N \to Gr^0(N)_{cst} \to 0. \qquad \text{QED}$$

(1.12) Interlude: abstract trace functions; approximate trace functions

(1.12.1) In this section, we work on an X/k which is separated of finite type, of dimension $d = \dim(X) \ge 0$. By an **abstract trace function** F on X/k, we mean a rule which attaches to each pair (E, x), consisting of a finite extension E/k and a point x in X(E), a number F(E, x) in $\mathbb{C}$. In practice, the abstract trace functions we will encounter usually start life with values in $\overline{\mathbb{Q}}_\ell$, which is then viewed as embedded in $\mathbb{C}$ via ι.

(1.12.2) The abstract trace functions form a $\mathbb{C}$-algebra, with pointwise operations:

$$(F+G)(E, x) := F(E, x) + G(E, x),$$
$$(FG)(E, x) := F(E, x)G(E, x),$$
$$(\lambda F)(E, x) = \lambda F(E, x).$$

(1.12.3) Given an abstract trace function F, we denote by $\overline{F}$ its complex conjugate, and by $|F|$ its absolute value:

$$\overline{F}(E, x) := \text{the complex conjugate of } F(E, x),$$
$$|F|(E, x) := |F(E, x)|.$$

(1.12.4) Given an abstract trace function F, we define the Frobenius-Schur sums

$$FS(X, F, E) := \Sigma_{x \text{ in } X(E)} F(E_2, x).$$

(1.12.5) We adapt the notation

(1.12.6) $$\Sigma_E F := \Sigma_{x \text{ in } X(E)} F(E, x).$$

Thus for a, b non-negative integers, this notation gives

$$\Sigma_E F^a\overline{F}^b := \Sigma_{x \text{ in } X(E)} F(E, x)^a\overline{F}(E, x)^b,$$
$$\Sigma_E |F|^a := \Sigma_{x \text{ in } X(E)} |F(E, x)|^a,$$

and, given a second abstract trace function G,

$$\Sigma_E FG := \Sigma_{x \text{ in } X(E)} F(E, x)G(E, x).$$

(1.12.7) Now suppose we are given a perverse sheaf N on X/k, which is ι-mixed of weight ≤ 0, and an abstract trace function N'. We say that N' is an **approximate trace function** for N if there exists a real number $\varepsilon > 0$ such that for variable finite extensions E/k, and for variable points x in X(E), we have

$$N(E, x) - N'(E, x) = O((\#E)^{-\varepsilon/2 - d/2}).$$

Lemma 1.12.8 Suppose N and M are perverse on X/k, both ι-mixed of weight ≤ 0. Suppose N' and M' are approximate trace functions for N and M respectively. Then there exists $\varepsilon > 0$ such that for variable finite extensions E/k, we have

1) $\Sigma_E N'\overline{M}' - \Sigma_E N\overline{M} = O((\#E)^{-\varepsilon/2})$,

2) $\Sigma_E N'M' - \Sigma_E NM = O((\#E)^{-\varepsilon/2})$,

3) $FS(X, N', E) - FS(X, N, E) = O((\#E)^{-\varepsilon})$.

proof Take a stratification $\{Z_\alpha\}$ of X by connected smooth locally closed subschemes to which both N and M are adapted. On a strat Z_α of dimension $d_\alpha \le d = \dim(X)$,

$$\#Z_\alpha(E) = O((\#E)^{d_\alpha}),$$

and we have the estimates

$$N(E, x \text{ in } Z_\alpha(E)) = O((\#E)^{-d_\alpha/2}),$$
$$M(E, x \text{ in } Z_\alpha(E)) = O((\#E)^{-d_\alpha/2}),$$

simply because N and M are semiperverse, and ι-mixed of weight ≤ 0. So we have

$$\Sigma_E |N| = O(\Sigma_\alpha \Sigma_{x \text{ in } Z_\alpha(E)} (\#E)^{-d_\alpha/2})$$
$$= O(\Sigma_\alpha (\#E)^{d_\alpha/2}) = O((\#E)^{d/2}).$$

Expanding out the sum we are to estimate, we get

$$\Sigma_E N'\bar{M}' - \Sigma_E N\bar{M}$$
$$= \Sigma_E (N' - N)M + \Sigma_E N(\bar{M}' - \bar{M}) + \Sigma_E (N' - N)(\bar{M}' - \bar{M})$$
$$= O((\#E)^{-\varepsilon/2 - d/2})\Sigma_E |M| + O((\#E)^{-\varepsilon/2 - d/2})\Sigma_E |N|$$
$$+ O((\#E)^{-\varepsilon - d})\#X(E)$$
$$= O((\#E)^{-\varepsilon/2}).$$

This proves 1). The proof of 2) is entirely similar. For 3), we use the tautological estimate

$$N'(E_2, x) - N(E_2, x) = O((\#E_2)^{-\varepsilon/2-d/2}) = O((\#E)^{-\varepsilon-d}),$$

and the trivial bound $\#X(E) = O((\#E)^d)$. QED

(1.13) Two uniqueness theorems

Theorem 1.13.1 Let X/k be a separated scheme of finite type, of dimension $d \geq 0$, F an abstract trace function on X/k, and M and N two perverse sheaves on X. Suppose that both M and N are ι-mixed of weight ≤ 0, and that F is an approximate trace function for both M and N. Then $Gr^0(M)$ and $Gr^0(N)$ are geometrically isomorphic, i.e., we have $Gr^0(M) \cong Gr^0(N)$ as perverse sheaves on $X\otimes_k\bar{k}$.

proof Since F is an approximate trace function for both M and for N, we have, thanks to the previous lemma, the estimates

$$\Sigma_E |F|^2 - \Sigma_E |N|^2 = O((\#E)^{-\varepsilon/2}),$$
$$\Sigma_E |F|^2 - \Sigma_E |M|^2 = O((\#E)^{-\varepsilon/2}),$$
$$\Sigma_E |F|^2 - \Sigma_E N\bar{M} = O((\#E)^{-\varepsilon/2}),$$

for some $\varepsilon > 0$. Both $Gr^0(M)$ and $Gr^0(N)$ are geometrically semisimple (because they are ι-pure of weight 0). Write their pullbacks $Gr^0(M)^{geom}$ and $Gr^0(N)^{geom}$ to $X\otimes_k\bar{k}$ as sums of perverse irreducibles with multiplicities, say

$$Gr^0(M)^{geom} = \Sigma_i m_iV_i,\ Gr^0(N)^{geom} = \Sigma_i n_iV_i,$$

with $\{V_i\}_i$ a finite set of pairwise non-isomorphic perverse irreducibles on $X\otimes_k\bar{k}$, and with non-negative integers m_i and n_i. Then by Second Corollary 1.8.3, we have

$$\limsup_E \Sigma_E |N|^2 = \Sigma_i (n_i)^2,$$
$$\limsup_E \Sigma_E |M|^2 = \Sigma_i (m_i)^2,$$
$$\limsup_E |\Sigma_E N\bar{M}| = \Sigma_i n_im_i.$$

In view of the above estimates, these three limsup's are all equal to $\limsup_E \Sigma_E |F|^2$. So we find

$$\Sigma_i (n_i)^2 = \Sigma_i (m_i)^2 = \Sigma_i n_im_i.$$

Therefore we get

$$\Sigma_i (n_i - m_i)^2 = 0,$$

so $n_i = m_i$ for each i, as required. QED

Here is an arithmetic sharpening of this uniqueness result.

Theorem 1.13.2 Let X/k be a separated scheme of finite type, of dimension $d \geq 0$, F an abstract trace function on X/k, and M and N two perverse sheaves on X. Suppose that both M and N are ι-mixed of weight ≤ 0, that both are semisimple objects in the category of perverse sheaves on X, and that F is an approximate trace function for both M and N. Then $Gr^0(M) \cong Gr^0(N)$ as perverse sheaves on X.

proof Let $\{Z_\alpha\}$ be a smooth stratification of X to which $N \oplus M$ is adapted. Thus each Z_α is a smooth and connected k-scheme, of dimension denoted d_α. Any direct factor of $N \oplus M$, in particular any simple constituent of $N \oplus M$, is also adapted to this stratification. Given a simple constituent W of $N \oplus M$, there is a unique strat Z_α, $\alpha = \alpha(W)$, which contains the generic point of its support. As $N \oplus M$ is ι-mixed of weight ≤ 0, any simple constituent W is ι-pure, of some weight $w(W) \leq 0$. As we have seen in the proof of the Orthogonality Theorem 1.7.2, for $\alpha = \alpha(W)$, we have $W|Z_\alpha = \mathcal{W}_\alpha[d_\alpha]$, for $\mathcal{W}_\alpha$ a semisimple, lisse, $\overline{\mathbb{Q}}_\ell$-sheaf on Z_α which is ι-pure of weight $-w(W) - d_\alpha$; moreover, W is the middle extension from Z_α of $W|Z_\alpha$. For $\beta \neq \alpha(W)$, $W|Z_\beta$ has a trace function which satisfies

$$(W|Z_\beta)(E, x) = O((\#E)^{(-d_\alpha - w(W) - 1)/2}).$$

Let us denote by $N(\alpha)$ the direct factor of N consisting of the sum (with multiplicity) of those of its simple constituents whose supports have generic point in Z_α. Let us denote by $N(\alpha, 0)$ the direct factor of $N(\alpha)$ which is the sum (with multiplicity) of those of its simple constituents which are ι-pure of weight 0. In other words, $N(\alpha, 0)$ is just $Gr^0(N(\alpha))$, viewed as a direct factor of $N(\alpha)$, or equivalently, $N(\alpha, 0)$ is, in this notation, $(Gr^0(N))(\alpha)$. Let us denote by $N(\alpha, < 0)$ the direct factor of $N(\alpha)$ which is the sum (with multiplicity) of those of its simple constituents which are ι-pure of weight < 0.

So we have direct sum decompositions

$$N = \oplus_\alpha N(\alpha),\ Gr^0(N) = \oplus_\alpha N(\alpha, 0),\ N(\alpha) = N(\alpha, 0) \oplus N(\alpha, < 0).$$

We have

$$N(\alpha)|Z_\alpha = \mathcal{N}_\alpha[d_\alpha],\ \mathcal{N}_\alpha \text{ lisse on } Z_\alpha,\ \iota\text{-mixed of weight} \leq -d_\alpha,$$

$$N(\alpha, 0)|Z_\alpha = \mathcal{N}_{\alpha,0}[d_\alpha],\ \mathcal{N}_{\alpha,0} = Gr^{-d_\alpha}(\mathcal{N}_\alpha),$$

$$N(\alpha, < 0)|Z_\alpha = \mathcal{N}_{\alpha, < 0}[d_\alpha],\ \mathcal{N}_{\alpha,0} \text{ lisse, } \iota\text{-mixed of weight} < -d_\alpha.$$

For $\beta \neq \alpha$, $N(\beta)|Z_\alpha$ has a trace function which satisfies

$$(N(\beta)|Z_\alpha)(E, x) = O((\#E)^{(-d_\alpha - 1)/2}).$$

Moreover, $N(\alpha, 0)$ is the middle extension from Z_α of $N(\alpha, 0)|Z_\alpha$.

The key point is that if F is an approximate trace function for N on X, then $F|Z_\alpha$ is an approximate trace function for $N(\alpha, 0)|Z_\alpha$ on Z_α. [Remember that the dimension of the ambient space figures in the definition of "approximate trace function".] Indeed, $F|Z_\alpha$ approximates to within $O((\#E)^{(-\dim(X) - \varepsilon)/2})$, for some $\varepsilon > 0$, the trace function of $N|Z_\alpha$. The trace funciton of $N|Z_\alpha$ is itself approximated to within $O((\#E)^{(-d_\alpha - \varepsilon)/2})$ by the trace function of $N(\alpha, 0)|Z_\alpha$, because the trace functions of both

$$N(\alpha, < 0)|Z_\alpha$$

and of each

$$N(\beta)|Z_\alpha, \text{ for } \beta \neq \alpha,$$

are $O((\#E)^{(-d_\alpha - \varepsilon)/2})$, for some $\varepsilon > 0$. Since $d_\alpha \le \dim(X)$, we indeed find that $F|Z_\alpha$ is an approximate trace function for $N(\alpha, 0)|Z_\alpha$ on Z_α.

Of course, we could repeat all of this with N replaced by M, and we would conclude that $F|Z_\alpha$ is an approximate trace function for $M(\alpha, 0)|Z_\alpha$ on Z_α. It now suffices to show that

$$N(\alpha, 0)|Z_\alpha \cong M(\alpha, 0)|Z_\alpha.$$

For taking middle extensions, we get $N(\alpha, 0) \cong M(\alpha, 0)$. Summing over α, we get $Gr^0(N) \cong Gr^0(M)$.

So our situation is this. We are given a smooth connected k-scheme Z_α of dimension $d_\alpha \ge 0$, and two semisimple, lisse $\overline{\mathbb{Q}}_\ell$-sheaves $\mathcal{G}$ and $\mathcal{H}$ on Z_α, each of which is ι-pure of weight $-d_\alpha$, and we are told that the perverse sheaves $\mathcal{G}[d_\alpha]$ and $\mathcal{H}[d_\alpha]$ on Z_α admit a common approximate trace function F on Z_α. We must show that $\mathcal{G}[d_\alpha] \cong \mathcal{H}[d_\alpha]$ as perverse sheaves on Z_α, or equivalently that $\mathcal{G} \cong \mathcal{H}$ as lisse sheaves on Z_α. Since we are given that $\mathcal{G}$ and $\mathcal{H}$ are both semisimple, it suffices to show that $\mathcal{G}$ and $\mathcal{H}$ have the same trace function.

To do this, we will show that $\mathcal{G}$ and $\mathcal{H}$ have the same local L-function everywhere. Thus let E/k be a finite extension, and x in $Z_\alpha(E)$ an E-valued point. For each integer $n \ge 1$, denote by E_n/E the extension of degree n inside $\overline{k}$. Consider the complex (via ι) power series

$$L(E, x, \mathcal{G})(T) := \exp(\Sigma_{n\ge1}\ \mathcal{G}(E_n, x)T^n/n),$$

$$L(E, x, \mathcal{H})(T) := \exp(\Sigma_{n\ge1}\ \mathcal{H}(E_n, x)T^n/n).$$

On the one hand, we have the identities

$$L(E, x, \mathcal{G})(T) = 1/\det(1 - T\mathrm{Frob}_{E,x}|\mathcal{G}),$$

$$L(E, x, \mathcal{H})(T) = 1/\det(1 - T\mathrm{Frob}_{E,x}|\mathcal{H}).$$

On the other hand, the fact that $\mathcal{G}[d_\alpha]$ and $\mathcal{H}[d_\alpha]$ on Z_α admit a common approximate trace function F on Z_α gives us the estimate

$$\mathcal{G}(E_n, x) - \mathcal{H}(E_n, x) = O((\#E)^{-n(d_\alpha + \varepsilon)/2}).$$

Therefore the series

$$\Sigma_{n\geq 1} (\mathcal{G}(E_n, x) - \mathcal{H}(E_n, x))T^n/n$$

converges absolutely in $|T| < (\#E)^{(d_\alpha + \varepsilon)/2}$, and hence its exponential is invertible in this region, i.e., the ratio

$$\det(1 - T\mathrm{Frob}_{E,x}|\mathcal{H})/\det(1 - T\mathrm{Frob}_{E,x}|\mathcal{G})$$

is invertible in the region $|T| < (\#E)^{(d_\alpha + \varepsilon)/2}$. But both $\mathcal{G}$ and $\mathcal{H}$ are ι-pure of weight $-d_\alpha$, so both numerator and denominator in this last ratio are finite products of factors of the form

$$1 - T\gamma, |\gamma| = (\#E)^{-d_\alpha/2},$$

say

$$\det(1 - T\mathrm{Frob}_{E,x}|\mathcal{H}) = \Pi_\gamma (1 - T\gamma)^{h(\gamma)},$$

$$\det(1 - T\mathrm{Frob}_{E,x}|\mathcal{G}) = \Pi_\gamma (1 - T\gamma)^{g(\gamma)},$$

the product over all γ with $|\gamma| = (\#E)^{-d_\alpha/2}$, with integer exponents $h(\gamma)$ and $g(\gamma)$ which are each zero for all but finitely many γ. The invertibility in the region $|T| < (\#E)^{(d_\alpha + \varepsilon)/2}$ of the ratio implies that we have $h(\gamma) = g(\gamma)$ for all γ with $|\gamma| = (\#E)^{-d_\alpha/2}$, hence that we have

$$\det(1 - T\mathrm{Frob}_{E,x}|\mathcal{H}) = \det(1 - T\mathrm{Frob}_{E,x}|\mathcal{G}),$$

hence that we have

$$\mathcal{H}(E_n, x) = \mathcal{G}(E_n, x)$$

for all n, in particular that $\mathcal{H}(E, x) = \mathcal{G}(E, x)$. Thus $\mathcal{G}$ and $\mathcal{H}$ have the same trace function, as required. QED

(1.14) The central normalization F_0 of a trace function F

(1.14.1) We first explain the terminology. In probability theory, one starts with a probability space (X, μ). Given a bounded random variable f on X, with expectation $E(f) := \int_X f d\mu$, one forms the random variable

$$f_0 := f - E(f).$$

Thus f_0 is the unique random variable which differs from f by a constant and which has $E(f_0) = 0$. The variance of f is defined to be the expectation of $(f_0)^2$. More generally, the "central moments" of f are defined as the expectations of the powers of f_0, and the "absolute central moments" are defined as the expectations of the powers of $|f_0|$. However, there does not seem to be a standard name for the function f_0, nor for the rule which attaches it to f. We propose to call f_0 the central normalization of f.

(1.14.2) Having given the motivation, we now pass to the context of abstract trace functions. Given an abstract trace function F on

X/k, we define its central normalization F_0 to be the abstract trace function given by the following rule. For E/k a finite extension, and for x in $X(E)$, we define

$$F_0(E, x) := F(E, x) - (1/\#X(E))\Sigma_{x \text{ in } X(E)} F(E, x)$$
$$= F(E, x) - (1/\#X(E))\Sigma_E F.$$

Thus for any finite extension E/k, we have

$$\Sigma_E F_0 = 0.$$

[For those E with $X(E)$ nonempty, this vanishing is obvious from the definition of F_0. If $X(E)$ is empty, then $\Sigma_E F_0$ is the empty sum.]

(1.14.3) A basic compatibility

Lemma 1.14.4 Let X/k be smooth and geometrically connected, of dimension $d \geq 0$. Let N be a perverse sheaf on X/k which is ι-mixed of weight ≤ 0. Then we have the following results.

1) The central normalization N_0 of the trace function of N is an approximate trace function for N_{ncst0}.

2) N_0 is an approximate trace function for N if and only if $H_c^d(X\otimes_k\overline{k}, N)$ is ι-mixed of weight $\leq d - \varepsilon$ for some $\varepsilon > 0$.

proof 1) The trace function of N_{ncst0} is given by Lemma 1.11.7:

$N_{ncst0}(E, x) = N(E, x) - (-1)^d(\#E)^{-d}\mathrm{Trace}(\mathrm{Frob}_E \mid H_c^d(X\otimes_k\overline{k}, N)_{wt=d})$.

The central normalization N_0 is given by

$$N_0(E, x) = N(E, x) - (1/\#X(E))\Sigma_E N.$$

So we must show that there exists $\varepsilon > 0$ such that for any finite extension E/k with $X(E)$ nonempty, we have

$$(1/\#X(E))\Sigma_E N - (-1)^d(\#E)^{-d}\mathrm{Trace}(\mathrm{Frob}_E \mid H_c^d(X\otimes_k\overline{k}, N)_{wt=d})$$
$$= O((\#E)^{-\varepsilon/2}).$$

Because X/k is smooth and geometrically connected of dimension d, it is geometrically irreducible, so by Lang-Weil (cf. [LW], [Ka-Sar-RMFEM, 9.0.15.1]) we have

$$\#X(E) = (\#E)^d(1 + O((\#E)^{-1/2}).$$

So for $\#E$ large, we have

$$1/\#X(E) = (\#E)^{-d}(1 + O((\#E)^{-1/2}).$$

By the Lefschetz trace formula,

$$\Sigma_E N = \Sigma_i (-1)^i\mathrm{Trace}(\mathrm{Frob}_E \mid H_c^i(X\otimes_k\overline{k}, N)).$$

As noted in the proof of 1.10.5 above, $H_c^i(X\otimes_k\overline{k}, N)$ vanishes for i outside the closed interval $[-d, d]$. And by Deligne [De-Weil II, 3.3.1], $H_c^i(X\otimes_k\overline{k}, N)$ is ι-mixed of weight $\leq i$. So for some $\varepsilon > 0$ we get

$$\Sigma_E N = (-1)^d\mathrm{Trace}(\mathrm{Frob}_E \mid H_c^d(X\otimes_k\overline{k}, N)_{wt=d}) + O((\#E)^{d - \varepsilon/2}).$$

Replacing if necessary ε by $\mathrm{Min}(\varepsilon, 1)$, we get

$$(1/\#X(E))\Sigma_E N$$
$$= (-1)^d(\#E)^{-d}\mathrm{Trace}(\mathrm{Frob}_E \mid H_c^d(X\otimes_k\overline{k}, N)_{wt=d}) + O((\#E)^{-\varepsilon/2}),$$

as required. This proves 1). For 2), the condition that $H_c^d(X\otimes_k\overline{k}, N)$

is ι-mixed of weight $\leq d - \varepsilon$ for some $\varepsilon > 0$ is precisely the condition that $N = N_{ncst0}$, cf. Corollary 1.11.6. QED

(1.15) First applications to the objects Twist(L, K, $\mathcal{F}$, h): the notion of standard input

(1.15.1) We continue to work over a finite field k. As earlier, we fix a prime number $\ell \neq \mathrm{char}(k)$, and a field embedding

$$\iota : \overline{\mathbb{Q}}_\ell \subset \mathbb{C}.$$

We also choose a square root of char(k) in $\overline{\mathbb{Q}}_\ell$, so that we can form Tate twists by half-integers.

(1.15.2) We will have repeated occasion to consider the following general situation. We fix

an integer $m \geq 1$,

a perverse sheaf K on $\mathbb{A}^m/k$,

an affine k-scheme V/k of finite type,

a k-morphism $h : V \to \mathbb{A}^m$,

a perverse sheaf L on V/k,

an integer $d \geq 2$,

a space of functions $(\mathcal{F}, \tau)$ on V, i.e., a finite-dimensional k-vector space $\mathcal{F}$ and a k-linear map

$$\tau : \mathcal{F} \to \mathrm{Hom}_{k\text{-schemes}}(V, \mathbb{A}^m).$$

(1.15.3) We make the following hypotheses on these data.

1) K is ι-mixed of weight ≤ 0, and $\mathrm{Gr}^0(K)$, the weight 0 quotient of K, is geometrically irreducible on $\mathbb{A}^m/k$.

2) L is ι-mixed of weight ≤ 0, and $\mathrm{Gr}^0(L)$, the weight 0 quotient of L, is geometrically irreducible on V/k.

3) $(\mathcal{F}, \tau)$ is d-separating, and contains the constants.

4) the $\mathbb{Z}$-graded vector space

$$H^*_c((V \times \mathbb{A}^m)\otimes \overline{k}, pr_1{}^*L \otimes pr_2{}^*K)$$

is concentrated in degree $\leq m$.

(1.15.4) We will say that data $(m \geq 1, K, V, h, L, d \geq 2, (\mathcal{F}, \tau))$ as above, which satisfies hypotheses 1) through 4), is "**standard input**".

Lemma 1.15.5 Given standard input

$$(m \geq 1, K, V, h, L, d \geq 2, (\mathcal{F}, \tau)),$$

consider the object M = Twist(L, K, $\mathcal{F}$, h) on the space $\mathcal{F}$. We have the following results.

1) The Tate-twisted object $M(\dim \mathcal{F}_0/2)$ is perverse, and ι-mixed of weight ≤ 0.

2) The constant object $\mathrm{Gr}^0(M(\dim \mathcal{F}_0/2))_{cst}$ is $\mathcal{G}[\dim \mathcal{F}]$, for $\mathcal{G}$ the constant sheaf

$$H^m_c((V \times \mathbb{A}^m)\otimes \overline{k}, pr_1{}^*L \otimes pr_2{}^*K)_{wt=m}(-\dim \mathcal{F}_0/2).$$

3) $\mathrm{Gr}^0(M(\dim \mathcal{F}_0/2)) = \mathrm{Gr}^0(M(\dim \mathcal{F}_0/2))_{ncst}$ if and only if

$$H_c^m((V\times\mathbb{A}^m)\otimes\bar{k},\ pr_1{}^*L\otimes pr_2{}^*K)$$

is ι-mixed of weight $\le m-\varepsilon$, for some $\varepsilon>0$.

4) $Gr^0(K) = Gr^0(K)_{ncst}$ if and only if

$$H_c^m(\mathbb{A}^m\otimes_k\bar{k},\ K)$$

is ι-mixed of weight $\le m-\varepsilon$, for some $\varepsilon>0$.

5) If $Gr^0(K)$ is geometrically constant, i.e., if $Gr^0(K)_{ncst} = 0$, then $Gr^0(M(\dim\mathcal{F}_0/2))$ is geometrically constant, i.e., $Gr^0(M(\dim\mathcal{F}_0/2))_{ncst} = 0$.

proof 1) We have already shown, in the Perversity and Weight Corollaries 1.4.4 and 1.5.11, that M is perverse, and ι-mixed of weight $\le \dim\mathcal{F}_0$. So its Tate twist $M(\dim\mathcal{F}_0/2)$ is perverse, and ι-mixed of weight ≤ 0.

2) We have also shown in the Semiperversity Theorem 1.4.2 that

$$H_c{}^*(\mathcal{F}\otimes_k\bar{k},\ M[\dim\mathcal{F}]) = H^*{}_c(\mathbb{A}^m\otimes\bar{k},\ K[m])\otimes H^*{}_c(V\otimes\bar{k},\ L)(m-\dim\mathcal{F})$$
$$= H_c{}^*((V\times\mathbb{A}^m)\otimes\bar{k},\ pr_1{}^*L\otimes pr_2{}^*K[m])(m-\dim\mathcal{F}).$$

Thus we have

$$H_c{}^{\dim\mathcal{F}}(\mathcal{F}\otimes_k\bar{k},\ M)$$
$$= H^m{}_c((V\times\mathbb{A}^m)\otimes\bar{k},\ pr_1{}^*L\otimes pr_2{}^*K)(m-\dim\mathcal{F}),$$

i.e., we have

$$H_c{}^{\dim\mathcal{F}}(\mathcal{F}\otimes_k\bar{k},\ M(\dim\mathcal{F}_0/2))$$
$$= H^m{}_c((V\times\mathbb{A}^m)\otimes\bar{k},\ pr_1{}^*L\otimes pr_2{}^*K)(-\dim\mathcal{F}_0/2).$$

By Lemma 1.11.7, $Gr^0(M(\dim\mathcal{F}_0/2))_{cst}$ is $\mathcal{G}[\dim\mathcal{F}]$ for $\mathcal{G}$ the constant sheaf

$$H_c{}^{\dim\mathcal{F}}(\mathcal{F}\otimes_k\bar{k},\ M(\dim\mathcal{F}_0/2))_{wt\,=\dim\mathcal{F}}$$
$$= H^m{}_c((V\times\mathbb{A}^m)\otimes\bar{k},\ pr_1{}^*L\otimes pr_2{}^*K)_{wt=m}(-\dim\mathcal{F}_0/2).$$

3) From 2), we see that $Gr^0(M(\dim\mathcal{F}_0/2)) = Gr^0(M(\dim\mathcal{F}_0/2))_{ncst}$ if and only if

$$H^m{}_c((V\times\mathbb{A}^m)\otimes\bar{k},\ pr_1{}^*L\otimes pr_2{}^*K)$$

is ι-mixed of weight $\le m-\varepsilon$.

4) This is just 1.11.6, applied to K.

5) Suppose $Gr^0(K)$ is geometrically constant. So by an $\alpha^{\deg}$ twist, we may assume that $Gr^0(K)$ is $\overline{\mathbb{Q}}_\ell[m](m/2)$. Then by Lemma 1.11.7, the group $H^m{}_c(\mathbb{A}^m\otimes\bar{k},\ K)$ is certainly nonzero, because already its weight m quotient is nonzero. By condition 4) of the notion of standard input,

$$H^*{}_c((V\times\mathbb{A}^m)\otimes\bar{k},\ pr_1{}^*L\otimes pr_2{}^*K)$$

is concentrated in degree $\le m$. From the Kunneth formula, we infer that $H^*_c(V\otimes\bar{k}, L)$ is concentrated in degree ≤ 0. Therefore for any perverse sheaf K' on $\mathbb{A}^m$, we see that

$$H^*_c((V\times\mathbb{A}^m)\otimes\bar{k}, pr_1^*L\otimes pr_2^*K')$$

is concentrated in degree $\le m$ (simply because $H^*_c(\mathbb{A}^m\otimes\bar{k}, K')$ is concentrated in degree $\le m$, for any perverse K'). So from the perversity theorem, we see that for any perverse K' on $\mathbb{A}^m$, the derived category object Twist(L, K', $\mathcal{F}$, h) is a perverse sheaf on $\mathcal{F}$. Moreover, for our fixed L, the functor

$$K' \mapsto \mathrm{Twist}(L, K', \mathcal{F}, h)$$

from perverse sheaves on $\mathbb{A}^m$ to perverse sheaves on $\mathcal{F}$ is exact. For if

$$0 \to A \to B \to C \to 0$$

is a short exact sequence of perverse sheaves on $\mathbb{A}^m$, it is also a distinguished triangle in the derived category on $\mathbb{A}^m$, which is transformed into a distinguished triangle

$$\mathrm{Twist}(L, A, \mathcal{F}, h) \to \mathrm{Twist}(L, B, \mathcal{F}, h) \to \mathrm{Twist}(L, C, \mathcal{F}, h) \to$$

in the derived category on $\mathcal{F}$ whose terms are perverse, and this is precisely a short exact sequence of perverse sheaves on $\mathcal{F}$.

Now apply this discussion to the short exact sequence

$$0 \to K_{\le -w} \to K \to Gr^0(K) = \overline{\mathbb{Q}}_\ell[m](m/2) \to 0,$$

for some $w > 0$. We get a short exact sequence

$$0 \to \mathrm{Twist}(L, K_{\le -w}, \mathcal{F}, h) \to \mathrm{Twist}(L, K, \mathcal{F}, h)$$
$$\to \mathrm{Twist}(L, \overline{\mathbb{Q}}_\ell[m](m/2), \mathcal{F}, h) \to 0.$$

By 1.5.11, Twist(L, $K_{\le -w}$, $\mathcal{F}$, h)($\dim\mathcal{F}_0/2$) is ι-mixed of weight $\le -w$, so we have

$$Gr^0(M(\dim\mathcal{F}_0/2)) \cong Gr^0(\mathrm{Twist}(L, \overline{\mathbb{Q}}_\ell[m](m/2), \mathcal{F}, h)(\dim\mathcal{F}_0/2)).$$

Thus we are reduced to treating the case when K is itself equal to $\overline{\mathbb{Q}}_\ell[m](m/2)$. But in this case, the pullback h_{aff}^*K is itself the constant sheaf, Tate-twisted and shifted, and $M(\dim\mathcal{F}_0/2)$ is the geometrically constant sheaf on $\mathcal{F}$ given by $\mathcal{G}[\dim\mathcal{F}](\dim\mathcal{F}/2)$, for $\mathcal{G}$ the geometrically constant sheaf $H_c^0(V\otimes_k\bar{k}, L)$. QED

Standard Input Theorem 1.15.6 Given standard input

$$(m \ge 1, K, V, h, L, d \ge 2, (\mathcal{F}, \tau)),$$

suppose in addition that $Gr^0(K)$ is not geometrically constant, i.e., that

$$H_c^m(\mathbb{A}^m\otimes_k\bar{k}, K)$$

is ι-mixed of weight $\le m - \varepsilon$, for some $\varepsilon > 0$.

Consider the object M = Twist(L, K, $\mathcal{F}$, h) on the space $\mathcal{F}$. We have the following results.

1) The Tate-twisted object $M(\dim\mathcal{F}_0/2)$ is perverse, and ι-mixed of weight ≤ 0.

2) The nonconstant part $Gr^0(M(\dim\mathcal{F}_0/2))_{ncst}$ of the weight zero

quotient $\mathrm{Gr}^0(M(\dim\mathcal{F}_0/2))$ is geometrically irreducible.

3) The Frobenius-Schur indicator of $\mathrm{Gr}^0(M(\dim\mathcal{F}_0/2))_{ncst}$ is given by the product formula

$$\mathrm{FSI}^{\mathrm{geom}}(\mathcal{F}, \mathrm{Gr}^0(M(\dim\mathcal{F}_0/2))_{ncst})$$
$$= ((-1)^{\dim\mathcal{F}_0})\times\mathrm{FSI}^{\mathrm{geom}}(\mathbb{A}^m, \mathrm{Gr}^0(K))\times\mathrm{FSI}^{\mathrm{geom}}(V, \mathrm{Gr}^0(L)).$$

4) If both $\mathrm{Gr}^0(L)$ on V/k and $\mathrm{Gr}^0(K)$ on $\mathbb{A}^m/k$ are self dual, then $\mathrm{Gr}^0(M(\dim\mathcal{F}_0/2))_{ncst}$ on $\mathcal{F}/k$ is self dual.

proof Assertion 1), already proven in Lemma 1.15.5, is simply "mis pour mémoire".

The proof of the remaining assertions makes use of approximate trace functions. Let us denote by K_0 and M_0 the central normalizations of the trace functions of K and of M on $\mathbb{A}^m$ and on $\mathcal{F}$ respectively. Then $M_0(\dim\mathcal{F}_0/2)$ is the central normalization of the trace function of $M(\dim\mathcal{F}_0/2)$. Notice that as $\mathrm{Gr}^0(K)$ is both geometrically irreducible and geometrically nonconstant, we have $K = K_{ncst}$. The key point is that, by Lemma 1.14.4, K_0 is an approximate trace function for K, and $M_0(\dim\mathcal{F}_0/2)$ is an approximate trace function for $M(\dim\mathcal{F}_0/2)_{ncst0}$.

So in order to prove 2), the geometric irreducibility of

$$\mathrm{Gr}^0(M(\dim\mathcal{F}_0/2))_{ncst} = \mathrm{Gr}^0(M(\dim\mathcal{F}_0/2)_{ncst0}),$$

it suffices, by part 3) of Second Corollary 1.8.3, to show there exists real $\varepsilon > 0$ such that for variable finite extensions E/k, we have

$$\Sigma_E |M(\dim\mathcal{F}_0/2)_{ncst0}|^2 = 1 + O((\#E)^{-\varepsilon/2}).$$

By Lemma 1.12.8, part 1), we know that, for some $\varepsilon > 0$, we have

$$\Sigma_E |M(\dim\mathcal{F}_0/2)_{ncst0}|^2 = \Sigma_E |M_0(\dim\mathcal{F}_0/2)|^2 + O((\#E)^{-\varepsilon/2}).$$

So it suffices to show that

$$\Sigma_E |M_0(\dim\mathcal{F}_0/2)|^2 = 1 + O((\#E)^{-\varepsilon/2}),$$

i.e.,

$$((\#E)^{-\dim\mathcal{F}_0})\Sigma_E |M_0|^2 = 1 + O((\#E)^{-\varepsilon/2}).$$

To show this, choose a real $\varepsilon > 0$ such that

$$\Sigma_E|K|^2 = 1 + O((\#E)^{-\varepsilon/2}),$$
$$\Sigma_E|L|^2 = 1 + O((\#E)^{-\varepsilon/2}).$$

This is possible by hypotheses 1) and 2), and criterion 3) of the Second Corollary 1.8.3. Again invoking Lemma 1.12.8, part 1), for a possibly smaller $\varepsilon > 0$ we will have

$$\Sigma_E|K_0|^2 = 1 + O((\#E)^{-\varepsilon/2}).$$

From the definition of M,

$$M := Rpr_{2!}(pr_1{}^*L\otimes h_{aff}{}^*K[\dim\mathcal{F} - m]),$$

the Lefschetz trace formula, and proper base change, we have

$$M(E, f) = ((-1)^{\dim\mathcal{F}_0})\Sigma_{v \text{ in } V(E)} L(E, v)K(E, h(v) + f(v)).$$

Key Lemma 1.15.7 The abstract trace function M_0 is given in terms of the trace function of L and the abstract trace function K_0 by the formula

$$M_0(E, f) = ((-1)^{\dim \mathcal{F}_0})\Sigma_{v \text{ in } V(E)} L(E, v)K_0(E, h(v) + f(v)).$$

proof of Key Lemma 1.15.7 By definition, we have

$$M_0(E, f) = M(E, f) - (1/\#\mathcal{F}(E))\Sigma_E M.$$

Our first task is to compute $\Sigma_E M$. We have

$$\Sigma_E M = \Sigma_{f \text{ in } \mathcal{F}(E)}((-1)^{\dim \mathcal{F}_0})\Sigma_{v \text{ in } V(E)} L(E, v)K(E, h(v) + f(v))$$
$$= ((-1)^{\dim \mathcal{F}_0})\Sigma_{v \text{ in } V(E)} L(E, v)\Sigma_{f \text{ in } \mathcal{F}(E)}K(E, h(v) + f(v)).$$

Because $(\mathcal{F}, \tau)$ is 2-separating, it is 1-separating, and so for fixed v in V(E), the E-linear map

$$\mathrm{eval}(v) : \mathcal{F}(E) \to \mathbb{A}^m(E),$$
$$f \mapsto f(v),$$

is surjective. So the inner sum is given by

$$\Sigma_{f \text{ in } \mathcal{F}(E)}K(E, h(v) + f(v))$$
$$= \#\mathrm{Ker}(\mathrm{eval}(v))\Sigma_{a \text{ in } \mathbb{A}^m(E)} K(E, h(v) + a)$$
$$= \#\mathrm{Ker}(\mathrm{eval}(v))\Sigma_{a \text{ in } \mathbb{A}^m(E)} K(E, a)$$
$$= (\#\mathcal{F}(E))((\#E)^{-m})\Sigma_E K.$$

So we find

$$\Sigma_E M = ((-1)^{\dim \mathcal{F}_0})(\Sigma_E L)(\#\mathcal{F}(E))((\#E)^{-m})\Sigma_E K.$$

Thus we have

$$(1/\#\mathcal{F}(E))\Sigma_E M = ((-1)^{\dim \mathcal{F}_0})(\Sigma_E L)((\#E)^{-m})\Sigma_E K.$$

Therefore we find

$$M_0(E, f) = M(E, f) - ((-1)^{\dim \mathcal{F}_0})(\Sigma_E L)((\#E)^{-m})\Sigma_E K$$
$$= ((-1)^{\dim \mathcal{F}_0})\Sigma_{v \text{ in } V(E)} L(E, v)K(E, h(v) + f(v))$$
$$- ((-1)^{\dim \mathcal{F}_0})\Sigma_{v \text{ in } V(E)} L(E, v)((\#E)^{-m})\Sigma_E K$$
$$= ((-1)^{\dim \mathcal{F}_0})\Sigma_{v \text{ in } V(E)} L(E, v)K_0(E, h(v) + f(v)). \text{ QED}$$

We now calculate in closed form the sum $\Sigma_E |M_0|^2$.

Lemma 1.15.8 We have the identity

$$\Sigma_E |M_0|^2 = (\Sigma_E|L|^2)\times((\#E)^{\dim \mathcal{F}_0})\times\Sigma_E|K_0|^2.$$

proof We have

$$|M_0(E, f)|^2$$
$$= (\Sigma_{v \text{ in } V(E)} L(E, v)K_0(E, h(v) + f(v)))$$
$$\times(\Sigma_{w \text{ in } V(E)} \overline{L}(E, w)\overline{K}_0(E, h(w) + f(w))).$$

Summing over f in $\mathcal{F}(E)$, we get

$$\Sigma_E |M_0|^2$$
$$= \Sigma_{v,w \text{ in } V(E)} L(E, v)\overline{L}(E, w)$$
$$\times\Sigma_{f \text{ in } \mathcal{F}(E)}K_0(E, h(v) + f(v))\overline{K}_0(E, h(w) + f(w)).$$

We now break up the sum according to whether v=w, or not. We get

$$\Sigma_E |M_0|^2$$
$$= \Sigma_{v \text{ in } V(E)} L(E, v)\overline{L}(E, v)$$
$$\times\Sigma_{f \text{ in } \mathcal{F}(E)}K_0(E, h(v) + f(v))\overline{K}_0(E, h(v) + f(v))$$
$$+ \Sigma_{v\neq w \text{ in } V(E)} L(E, v)\overline{L}(E, w)$$
$$\times\Sigma_{f \text{ in } \mathcal{F}(E)}K_0(E, h(v) + f(v))\overline{K}_0(E, h(w) + f(w)).$$

We will show that the first sum gives the main term, and the second sum vanishes.

Consider the first sum. Because $(\mathcal{F}, \tau)$ is 2-separating, it is certainly 1-separating, so for fixed v in V(E), the E-linear map

$$\text{eval}(v) : \mathcal{F}(E) \rightarrow \mathbb{A}^m(E),$$
$$f \mapsto f(v),$$

is surjective. So the innermost sum in the first sum simplifies to

$$\Sigma_{f \text{ in } \mathcal{F}(E)}K_0(E, h(v) + f(v))\overline{K}_0(E, h(v) + f(v))$$
$$= \#(\text{Ker}(\text{eval}(v)))\Sigma_{a \text{ in } \mathbb{A}^m(E)} K_0(E, h(v) + a)\overline{K}_0(E, h(v) + a).$$

By an additive change of variable $a \mapsto a + h(v)$, we rewrite this as

$$= \#(\text{Ker}(\text{eval}(v)))\Sigma_{a \text{ in } \mathbb{A}^m(E)} K_0(E, a)\overline{K}_0(E, a)$$
$$= \#(\text{Ker}(\text{eval}(v)))\Sigma_E|K_0|^2$$
$$= ((\#E)^{\dim\mathcal{F} - m})\times\Sigma_E|K_0|^2$$
$$= ((\#E)^{\dim\mathcal{F}_0})\times\Sigma_E|K_0|^2.$$

So the first term is the product

$$(\Sigma_E|L|^2)\times((\#E)^{\dim\mathcal{F}_0})\times\Sigma_E|K_0|^2.$$

We will now show that the second sum vanishes.

Because $(\mathcal{F}, \tau)$ is 2-separating, for a given pair $v\neq w$ of distinct points in V(E), the E-linear map

$$(\text{eval}(v), \text{eval}(w)) : \mathcal{F}(E) \rightarrow \mathbb{A}^m(E)\times \mathbb{A}^m(E),$$
$$f \mapsto (f(v), f(w)),$$

is surjective. So the innermost sum in the second sum simplifies to

$$\Sigma_{f \text{ in } \mathcal{F}(E)}K_0(E, h(v) + f(v))\overline{K}_0(E, h(w) + f(w))$$
$$= \#(\text{Ker}((\text{eval}(v), \text{eval}(w))))$$
$$\times\Sigma_{a,b \text{ in} \mathbb{A}^m(E)} K_0(E, h(v) + a)\overline{K}_0(E, h(w) + b)$$
$$= \#(\text{Ker}((\text{eval}(v), \text{eval}(w))))\Sigma_{a,b \text{ in} \mathbb{A}^m(E)} K_0(E, a)\overline{K}_0(E, b)$$
$$= \#(\text{Ker}((\text{eval}(v), \text{eval}(w))))|\Sigma_E K_0|^2$$
$$= 0, \text{ cf. } 1.14.2.$$

QED for Lemma 1.15.8

From the above Lemma 1.15.8, we get

$$((\#E)^{-\dim\mathcal{F}_0})\Sigma_E |M_0|^2 = (\Sigma_E|L|^2)(\Sigma_E|K_0|^2)$$
$$= (1 + O((\#E)^{-\varepsilon/2}))(1 + O((\#E)^{-\varepsilon/2}))$$
$$= 1 + O((\#E)^{-\varepsilon/2}),$$

which proves part 2) of the Standard Input Theorem.

To prove parts 3) and 4), we calculate the Frobenius-Schur

sums for $M_0(\dim\mathcal{F}_0/2)$.

Lemma 1.15.9 We have the identity

$FS(\mathcal{F}, M_0(\dim\mathcal{F}_0/2), E) = ((-1)^{\dim\mathcal{F}_0})\times FS(V, L, E)\times FS(\mathbb{A}^m, K_0, E)$.

proof By definition, we have

$$FS(\mathcal{F}, M_0(\dim\mathcal{F}_0/2), E) := ((\# E)^{-\dim\mathcal{F}_0})\times FS(\mathcal{F}, M_0, E)$$

$$= ((\# E)^{-\dim\mathcal{F}_0})\times\Sigma_{f \text{ in } \mathcal{F}(E)} M_0(E_2, f)$$

$$= ((-\# E)^{-\dim\mathcal{F}_0})\times\Sigma_{f \text{ in } \mathcal{F}(E)} \Sigma_{v \text{ in } V(E_2)} L(E_2, v)K_0(E_2, h(v) + f(v))$$

$$= ((-\# E)^{-\dim\mathcal{F}_0})\times\Sigma_{v \text{ in } V(E_2)} L(E_2, v)\Sigma_{f \text{ in } \mathcal{F}(E)} K_0(E_2, h(v) + f(v)).$$

We break the sum into two sums, according to whether v in $V(E_2)$ lies in V(E), or not. We will show that the first sum, over v in V(E), is the main term, and the second sum vanishes.

The first sum is

$$((-\# E)^{-\dim\mathcal{F}_0})\times\Sigma_{v \text{ in } V(E)} L(E_2, v)\Sigma_{f \text{ in } \mathcal{F}(E)} K_0(E_2, h(v) + f(v)).$$

Because $(\mathcal{F}, \tau)$ is 1-separating, the inner sum is simply

$$\Sigma_{f \text{ in } \mathcal{F}(E)} K_0(E_2, h(v) + f(v))$$

$$= \#(\mathrm{Ker}(\mathrm{eval}(v)))\times\Sigma_{a \text{ in } \mathbb{A}^m(E)} K_0(E_2, h(v) + a)$$

$$= ((\# E)^{\dim\mathcal{F}_0})\times\Sigma_{a \text{ in } \mathbb{A}^m(E)} K_0(E_2, a)$$

$$= ((\# E)^{\dim\mathcal{F}_0})\times FS(\mathbb{A}^m, K_0, E).$$

So the first sum is the product

$$(((-\# E)^{-\dim\mathcal{F}_0})\times\Sigma_{v \text{ in } V(E)} L(E_2, v))$$

$$\times((\# E)^{\dim\mathcal{F}_0})\times FS(\mathbb{A}^m, K_0, E)$$

$$= ((-1)^{\dim\mathcal{F}_0})\times FS(V, L, E)\times FS(\mathbb{A}^m, K_0, E).$$

It remains to show that the second sum vanishes. For v in $V(E_2)$, denote by $\bar{v}$ its image under the nontrivial automorphism of E_2/E. Thus we are summing over points v in $V(E_2)$ with $v \neq \bar{v}$. So the second sum is

$$= ((-\# E)^{-\dim\mathcal{F}_0})\Sigma_{v\neq\bar{v} \text{ in } V(E_2)} L(E_2, v)\Sigma_{f \text{ in } \mathcal{F}(E)} K_0(E_2, h(v) + f(v)).$$

We will show that already its innermost sum vanishes. For this, we need the following

Sublemma 1.15.10 For a point $v \neq \bar{v}$ in $V(E_2)$, the E-linear evaluation map eval(v), viewed as a map from $\mathcal{F}(E)$ to $\mathbb{A}^m(E_2)$, $f \mapsto f(v)$, is surjective.

proof of Sublemma 1.15.10 Because $(\mathcal{F}, \tau)$ is 2-separating, and $v \neq \bar{v}$, the map

$$(\mathrm{eval}(v), \mathrm{eval}(\bar{v})) : \mathcal{F}(E_2) \to \mathbb{A}^m(E_2)\times\mathbb{A}^m(E_2),$$

$$\varphi \mapsto (\varphi(v), \varphi(\bar{v}))$$

is surjective. Take a point of the form $(a, \bar{a})$ in $\mathbb{A}^m(E_2)\times\mathbb{A}^m(E_2)$. Then there exists φ in $\mathcal{F}(E_2)$ such that

$$\varphi(v) = a, \varphi(\bar{v}) = \bar{a}.$$

Pick γ in E_2 with $\gamma \neq 0$ and $\overline{\gamma} \neq \gamma$. Then $\{1, \gamma\}$ is an E-basis of E_2, and we can write φ in $\mathcal{F}(E_2)$ uniquely as $f + \gamma g$, with f, g in $\mathcal{F}(E)$. Thus we get the identities

$$f(v) + \gamma g(v) = a,$$
$$f(\overline{v}) + \gamma g(\overline{v}) = \overline{a}.$$

Because f, g both lie in $\mathcal{F}(E)$, they are $\mathrm{Gal}(E_2/E)$-equivariant maps from $V(E_2)$ to $\mathbb{A}^m(E_2)$. So "conjugating" the second identity, we get

$$f(v) + \overline{\gamma} g(v) = a.$$

Subtracting from the first, we find $(\gamma - \overline{\gamma})g(v) = 0$, so $g(v) = 0$, and $f(v) = a$. QED for Sublemma 1.15.10.

Thanks to Sublemma 1.15.10, we can evaluate the innermost sum

$$\Sigma_{f \text{ in } \mathcal{F}(E)} K_0(E_2, h(v) + f(v))$$
$$= (\#\mathrm{Ker}(\mathrm{eval}(v): \mathcal{F}(E) \to \mathbb{A}^m(E_2))) \times \Sigma_{a \text{ in } \mathbb{A}^m(E_2)} K_0(E_2, h(v) + a)$$
$$= ((\#E)^{\dim\mathcal{F} - 2m}) \times \Sigma_{a \text{ in } \mathbb{A}^m(E_2)} K_0(E_2, a)$$
$$= ((\#E)^{\dim\mathcal{F}_0 - m}) \times \Sigma_{E_2} K_0$$
$$= 0, \text{ cf. } 1.14.2.$$

We now make use of Lemma 1.15.9, according to which we have the identity

$$FS(\mathcal{F}, M_0(\dim\mathcal{F}_0/2), E) = ((-1)^{\dim\mathcal{F}_0}) \times FS(V, L, E) \times FS(\mathbb{A}^m, K_0, E).$$

Suppose first that at least one of $\mathrm{Gr}^0(K)$ or $\mathrm{Gr}^0(L)$ is not geometrically self dual. Applying part 1) of Corollary 1.9.7 to K and L shows that for some real $\varepsilon > 0$, we have

$$FS(V, L, E) \times FS(\mathbb{A}^m, K, E) = O((\#E)^{-\varepsilon/2}).$$

Then applying Lemma 1.12.8, part 3), to K and K_0, we get

$$FS(V, L, E) \times FS(\mathbb{A}^m, K_0, E) = O((\#E)^{-\varepsilon/2}).$$

Therefore we have

$$FS(V, M_0(\dim\mathcal{F}_0/2), E) = O((\#E)^{-\varepsilon/2}).$$

Again by Lemma 1.12.8, part 3), now applied to $M(\dim\mathcal{F}_0/2)_{\mathrm{ncst}0}$ and to $M_0(\dim\mathcal{F}_0/2)$, this now gives

$$FS(V, M(\dim\mathcal{F}_0/2)_{\mathrm{ncst}0}, E) = O((\#E)^{-\varepsilon/2}).$$

Applying part 1) of Corollary 1.9.7 to $M(\dim\mathcal{F}_0/2)_{\mathrm{ncst}0}$ shows now that $\mathrm{Gr}^0(M(\dim\mathcal{F}_0/2))_{\mathrm{ncst}}$ is not geometrically self dual.

If both of $\mathrm{Gr}^0(K)$ and $\mathrm{Gr}^0(L)$ are geometrically self dual, then replacing each of K and L by a suitable unitary $\alpha^{\deg}$ twist of itself, we reduce to the case when both $\mathrm{Gr}^0(K)$ and $\mathrm{Gr}^0(L)$ are self dual.

Then part 2) of Corollary 1.9.7, applied to both K and L, together with Lemma 1.12.8, part 3), show that for some real $\varepsilon > 0$ we have

$$\begin{aligned} &\mathrm{FS}(\mathcal{F}, \mathrm{M}_0(\dim\mathcal{F}_0/2), \mathrm{E}) \\ &= (((-1)^{\dim\mathcal{F}_0})\times\mathrm{FSI}^{\mathrm{geom}}(\mathrm{V}, \mathrm{Gr}^0(\mathrm{L}))\times\mathrm{FSI}^{\mathrm{geom}}(\mathbb{A}^m, \mathrm{Gr}^0(\mathrm{K})) \\ &\qquad + O((\#\mathrm{E})^{-\varepsilon/2}). \end{aligned}$$

Applying Lemma 1.12.8, part 3) and part 3) of Corollary 1.9.7 to $\mathrm{M}(\dim\mathcal{F}_0/2)_{\mathrm{ncst0}}$, we infer that $\mathrm{Gr}^0(\mathrm{M}(\dim\mathcal{F}_0/2))_{\mathrm{ncst}}$ is itself self dual on $\mathcal{F}/k$, and that we have

$$\begin{aligned} &\mathrm{FSI}^{\mathrm{geom}}(\mathcal{F}, \mathrm{Gr}^0(\mathrm{M}(\dim\mathcal{F}_0/2))_{\mathrm{ncst}}) \\ &= (((-1)^{\dim\mathcal{F}_0})\times\mathrm{FSI}^{\mathrm{geom}}(\mathrm{V}, \mathrm{Gr}^0(\mathrm{L}))\times\mathrm{FSI}^{\mathrm{geom}}(\mathbb{A}^m, \mathrm{Gr}^0(\mathrm{K})). \text{ QED} \end{aligned}$$

(1.16) Review of higher moments

(1.16.1) Let K be an algebraically closed field of characteristic zero. Suppose a group G operates completely reducibly (e.g., irreducibly) on a finite-dimensional K-vector space V. For each pair (a, b) of non-negative integers, we denote by $\mathrm{M}_{a,b}(\mathrm{G}, \mathrm{V})$ the dimension of the space of G-invariant vectors in $\mathrm{V}^{\otimes a}\otimes(\mathrm{V}^\vee)^{\otimes b}$:

$$\mathrm{M}_{a,b}(\mathrm{G}, \mathrm{V}) := \dim_K (\mathrm{V}^{\otimes a}\otimes(\mathrm{V}^\vee)^{\otimes b})^{\mathrm{G}}.$$

We call $\mathrm{M}_{a,b}(\mathrm{G}, \mathrm{V})$ the (a, b)'th moment of (G, V). For each even integer $2n \geq 2$, we denote by $\mathrm{M}_{2n}(\mathrm{G}, \mathrm{V})$ the 2n'th absolute moment, defined by

$$\mathrm{M}_{2n}(\mathrm{G}, \mathrm{V}) := \mathrm{M}_{n,n}(\mathrm{G}, \mathrm{V}).$$

(1.16.2) The terminology "moments" comes about as follows. When K is $\mathbb{C}$ and G is compact, there are integral formulas for $\mathrm{M}_{a,b}(\mathrm{G}, \mathrm{V})$ and for $\mathrm{M}_{2n}(\mathrm{G}, \mathrm{V})$. Denote by

$$\chi : \mathrm{G} \to \mathbb{C}$$
$$\chi(g) := \mathrm{Trace}(g|\mathrm{V}),$$

the character of the representation. Then we have

$$\mathrm{M}_{a,b}(\mathrm{G}, \mathrm{V}) = \int_{\mathrm{G}} \chi(g)^a\overline{\chi}(g)^b dg,$$

$$\mathrm{M}_{2n}(\mathrm{G}, \mathrm{V}) = \int_{\mathrm{G}} |\chi(g)|^{2n} dg,$$

for dg the total mass one Haar measure on G. [Thus the terminology "moments" and "absolute moments".]

(1.16.3) There is one elementary inequality we will need later.

Lemma 1.16.4 Let K be an algebraically closed field of characteristic zero. Suppose a group G operates completely reducibly on a finite-dimensional K-vector space V. If $\mathrm{V} \neq 0$, then $\mathrm{M}_{2n}(\mathrm{G}, \mathrm{V}) \geq 1$ for all $n \geq 1$.

proof $\mathrm{M}_{2n}(\mathrm{G}, \mathrm{V})$ is the dimension of the G-invariants in $\mathrm{V}^{\otimes n}\otimes(\mathrm{V}^\vee)^{\otimes n} = \underline{\mathrm{End}}(\mathrm{V}^{\otimes n})$, i.e., $\mathrm{M}_{2n}(\mathrm{G}, \mathrm{V})$ is the dimension of $\mathrm{End}_{\mathrm{G}}(\mathrm{V}^{\otimes n})$, which always contains the scalars. QED

(1.17) Higher moments for geometrically irreducible lisse sheaves

(1.17.1) We continue to work over a finite field k. As earlier, we fix a prime number $\ell \neq \mathrm{char}(k)$, a field embedding

$$\iota : \overline{\mathbb{Q}}_\ell \subset \mathbb{C},$$

and a square root of char(k) in $\overline{\mathbb{Q}}_\ell$.

(1.17.2) Let U/k be a separated k-scheme of finite type, which is smooth and geometrically connected, of dimension $d = \dim U \geq 0$. Suppose $\mathcal{G}$ on U is a lisse, geometrically irreducible $\overline{\mathbb{Q}}_\ell$-sheaf on U. Then $\mathcal{G}$ "is" an irreducible representation of $\pi_1(U\otimes_k\overline{k})$, and we may speak of its higher moments $M_{a,b}((\pi_1(U\otimes_k\overline{k}), \mathcal{G})$ and $M_{2n}((\pi_1(U\otimes_k\overline{k}), \mathcal{G})$. We call these the geometric higher moments of $\mathcal{G}$ on U, and write

$$M_{a,b}{}^{\mathrm{geom}}(U, \mathcal{G}) := M_{a,b}(\pi_1(U\otimes_k\overline{k}), \mathcal{G}),$$

$$M_{2n}{}^{\mathrm{geom}}(U, \mathcal{G}) := M_{2n}(\pi_1(U\otimes_k\overline{k}), \mathcal{G}).$$

These moments are birational invariants, in the sense that for any dense open set $U_1 \subset U$, we have

$$M_{a,b}{}^{\mathrm{geom}}(U, \mathcal{G}) = M_{a,b}{}^{\mathrm{geom}}(U_1, \mathcal{G}|U_1),$$

$$M_{2n}{}^{\mathrm{geom}}(U, \mathcal{G}) = M_{2n}{}^{\mathrm{geom}}(U_1, \mathcal{G}|U_1).$$

These equalities hold simply because $\pi_1(U_1\otimes_k\overline{k})$ maps onto $\pi_1(U\otimes_k\overline{k})$.

(1.17.3) When $\mathcal{G}$ is ι-pure of some weight w, there is a diophantine analogue of the classical integral formulas for higher moments.

Theorem 1.17.4 Let U/k be a separated k-scheme of finite type, which is smooth and geometrically connected, of dimension $d = \dim U \geq 0$. Suppose $\mathcal{G}$ on U is a lisse, geometrically irreducible $\overline{\mathbb{Q}}_\ell$-sheaf on U, which is ι-pure of some weight w. Fix a pair (a, b) of non-negative integers. For each finite extension E/k, consider the sum $\Sigma_E\, \mathcal{G}^a\overline{\mathcal{G}}^b$ Then we have the following results.

1) We have the estimate

$$|\Sigma_E\, \mathcal{G}^a\overline{\mathcal{G}}^b| = O((\#E)^{\dim U + (a+b)w/2}).$$

2) We have the limit formula

$$M_{a,b}{}^{\mathrm{geom}}(U, \mathcal{G}) = \mathrm{limsup}_E\, |\Sigma_E\, \mathcal{G}^a\overline{\mathcal{G}}^b|/(\#E)^{\dim U + (a+b)w/2}.$$

3) We have the limit formula

$$\dim U + (a+b)w/2 = \mathrm{limsup}_E\, \log(|\Sigma_E\, \mathcal{G}^a\overline{\mathcal{G}}^b|)/\log(\#E).$$

proof 1) Each summand is trivially bounded in absolute value by $(\mathrm{rank}\mathcal{G})^{a+b}(\#E)^{(a+b)w/2}$, and there are $\#U(E) = O((\#E)^{\dim U})$ terms.
2) and 3) By an $\alpha^{\deg}$ twist, we reduce immediately to the case when $w = 0$. In this case, the contragredient $\mathcal{G}^\vee$ of $\mathcal{G}$ has complex-conjugate trace function to that of $\mathcal{G}$. So by the Lefschetz trace

formula, we have

$$\Sigma_E \ \mathcal{G}^a\bar{\mathcal{G}}^b$$
$$= \Sigma_{i=0 \text{ to } 2\dim U} (-1)^i \mathrm{Trace}(\mathrm{Frob}_E | H_c{}^i(U\otimes_k \bar{k}, \mathcal{G}^{\otimes a}\otimes(\mathcal{G}^\vee)^{\otimes b})).$$

Now $\mathcal{G}^{\otimes a}\otimes(\mathcal{G}^\vee)^{\otimes b}$ is ι-pure of weight zero, so by Deligne's result [De-Weil II, 3.3.1], we have

$$\Sigma_E \ \mathcal{G}^a\bar{\mathcal{G}}^b$$
$$= \mathrm{Trace}(\mathrm{Frob}_E | H_c{}^{2d}(U\otimes_k \bar{k}, \mathcal{G}^{\otimes a}\otimes(\mathcal{G}^\vee)^{\otimes b})) + O((\#E)^{\dim U - 1/2}).$$

Moreover, the group $H_c{}^{2d}(U\otimes_k \bar{k}, \mathcal{G}^{\otimes a}\otimes(\mathcal{G}^\vee)^{\otimes b})$ is ι-pure of weight 2dimU, and its dimension is precisely $M_{a,b}{}^{geom}(U, \mathcal{G})$. So if we view this $H_c{}^{2d}$ as an $M_{a,b}{}^{geom}(U, \mathcal{G})$-dimensional $\mathbb{C}$-vector space T via ι, then the semisimplification of $\mathrm{Frob}_k/(\#E)^{\dim U}$ is a unitary operator A on T, and

$$|\Sigma_E \ \mathcal{G}^a\bar{\mathcal{G}}^b| = ((\#E)^{\dim U})\times(|\mathrm{Trace}(A^{\deg(E/k)})| + O((\#E)^{-1/2})).$$

Assertions 2) and 3) now follow, by the same compactness argument already used in the proof of the Orthogonality Theorem 1.7.2. QED

(1.18) Higher moments for geometrically irreducible perverse sheaves

(1.18.1) Let X/k be a separated k-scheme of finite type, and let N on X/k be perverse and geometrically irreducible. Its support is a geometrically irreducible closed subscheme Z of X, inclusion denoted i: Z → X. There exists a dense affine open set j: U → Z, such that U/k is smooth and geometrically connected, of some dimension $d \ge 0$, and a lisse $\bar{\mathbb{Q}}_\ell$-sheaf $\mathcal{N}$ on U, which is geometrically irreducible, such that N is $i_\star j_{!\star}\mathcal{N}[\dim U]$. We define

$$M_{a,b}{}^{geom}(X, N) := M_{a,b}{}^{geom}(U, \mathcal{N}),$$
$$M_{2n}{}^{geom}(X, N) := M_{2n}{}^{geom}(U, \mathcal{N}).$$

This definition is independent of the auxiliary choice of smooth, geometrically connected dense open set U of Supp(M) on which M is lisse, thanks to the birational invariance of the quantites $M_{a,b}{}^{geom}(U, \mathcal{N})$ and $M_{2n}{}^{geom}(U, \mathcal{N})$.

(1.19) A fundamental inequality

Theorem 1.19.1 Let X/k be smooth and geometrically connected, of non-negative dimension, and let M on X/k be perverse, and ι-mixed of weight ≤ 0. Write its weight filtration

$$0 \to M_{\le -\varepsilon} \to M \to \mathrm{Gr}^0(M) \to 0,$$

for some $\varepsilon > 0$. Let $\tilde{M}$ be an abstract trace function which is an approximate trace function for M. Suppose that $\mathrm{Gr}^0(M)$ is geometrically irreducible. Suppose that for some real $\varepsilon > 0$, some integer $n \geq 2$ and some real numbers λ and A_{2n}, we have an inequality

$$\Sigma_E |\tilde{M}|^{2n} \leq A_{2n}(\#E)^{(1-n)\lambda} + O((\#E)^{(1-n)\lambda - \varepsilon/2}),$$

for E/k a variable finite extension. Then we have the following results.

1) $\dim(\mathrm{Supp}(\mathrm{Gr}^0(M))) \geq \lambda$.

2) If $\dim(\mathrm{Supp}(\mathrm{Gr}^0(M))) = \lambda$, then $M_{2n}{}^{\mathrm{geom}}(X, \mathrm{Gr}^0(M)) \leq A_{2n}$.

proof Let us put

$$d := \dim(\mathrm{Supp}(\mathrm{Gr}^0(M))).$$

Take a smooth stratification $\{Z_\alpha\}$ of X which is adapted to all three objects $M_{\leq -\varepsilon}$, M, and $\mathrm{Gr}^0(M)$. There is precisely one strat Z_α which contains the generic point of $\mathrm{Supp}(\mathrm{Gr}^0(M))$. Exactly as in Step 2 of the proof of the Orthogonality Theorem 1.7.2 this strat Z_α has dimension d, it is a dense open set of $\mathrm{Supp}(\mathrm{Gr}^0(M))$, it is geometrically connected, and $\mathrm{Gr}^0(M)|Z_\alpha$ is of the form $\mathcal{G}[d]$ for $\mathcal{G}$ a lisse, geometrically irreducible lisse $\overline{\mathbb{Q}}_\ell$-sheaf on Z_α which is ι-pure of weight -d.

The restriction to Z_α of $M_{\leq -\varepsilon}$ has lisse cohomology sheaves, so by semiperversity of $M_{\leq -\varepsilon}$ we have $\mathcal{H}^{-i}(M_{\leq -\varepsilon})|Z_\alpha = 0$ if $i \leq d-1$. So the possibly nonzero $\mathcal{H}^{-i}(M_{\leq -\varepsilon})|Z_\alpha$ all have $i \geq d$, and so all are ι-mixed of weight $\leq -d - \varepsilon$. Thus we have the estimate

$$|M_{\leq -\varepsilon}(E, x)| = O((\#E)^{(-d-\varepsilon)/2}).$$

From the identities

$$M(E, x) = \mathrm{Gr}^0(M)(E, x) + M_{\leq -\varepsilon}(E, x),$$

$$\tilde{M}(E, x) = M(E, x) + O((\#E)^{-\varepsilon - \dim(X)/2})$$

we see that for x in $Z_\alpha(E)$, we have

$$\tilde{M}(E, x) = \mathrm{Gr}^0(M)(E, x) + O((\#E)^{(-d-\varepsilon)/2}).$$

On Z_α, we have

$$\mathrm{Gr}^0(M)(E, x) = O((\#E)^{-d/2}).$$

Thus we have

$$\Sigma_E |\tilde{M}|Z_\alpha|^{2n} = \Sigma_E |\mathrm{Gr}^0(M)|Z_\alpha|^{2n} + O(\#Z_\alpha(E)(\#E)^{(-2nd-\varepsilon)/2})$$
$$= \Sigma_E |\mathcal{G}|^{2n} + O((\#E)^{(1-n)d - \varepsilon/2}).$$

We now prove 1). By Lemma 1.16.4, we have the inequality

$$M_{2n}{}^{\mathrm{geom}}(Z_\alpha, \mathcal{G}) \geq 1.$$

So for an infinity of finite extensions E/k, we get

$$\Sigma_E |\mathcal{G}|^{2n} \geq (1/2)\times(\#E)^{(1-n)d}.$$

So for these E, we get

$$\Sigma_E |\tilde{M}|Z_\alpha|^{2n} \geq (1/2)\times(\#E)^{(1-n)d} + O((\#E)^{(1-n)d - \varepsilon/2}),$$

and, a fortiori,

$$\Sigma_E |\tilde{M}|^{2n} \geq \Sigma_E |\tilde{M}|Z_\alpha|^{2n}$$
$$\geq (1/2)\times(\# E)^{(1-n)d} + O((\# E)^{(1-n)d - \varepsilon/2}).$$

From the inequality

$$\Sigma_E |\tilde{M}|^{2n} \leq A_{2n}(\# E)^{(1-n)\lambda} + O((\# E)^{(1-n)\lambda - \varepsilon/2}),$$

we retain only

$$\Sigma_E |\tilde{M}|^{2n} = O((\# E)^{(1-n)\lambda}).$$

Then we have

$$(\# E)^{(1-n)d} = O((\# E)^{(1-n)\lambda})$$

for an infinity of E, whence (as $n \geq 2$) we have $d \geq \lambda$.

2) If $d = \lambda$, we have

$$A_{2n}(\# E)^{(1-n)d} + O((\# E)^{(1-n)d - \varepsilon/2})$$
$$\geq \Sigma_E |\tilde{M}|^{2n}$$
$$\geq \Sigma_E |\tilde{M}|Z_\alpha|^{2n} = \Sigma_E |\mathcal{G}|^{2n} + O((\# E)^{(1-n)d -\varepsilon/2}).$$

Divide through by $(\# E)^{(1-n)d}$ and take the limsup over E. QED

Corollary 1.19.2 Let X/k be smooth and geometrically connected of dimension $\dim X \geq 0$. Let M on X/k be perverse, and ι-mixed of weight ≤ 0. Write its weight filtration

$$0 \to M_{\leq -\varepsilon} \to M \to \mathrm{Gr}^0(M) \to 0,$$

for some $\varepsilon > 0$. Let $\tilde{M}$ be an abstract trace function which is an approximate trace function for M. Suppose that $\mathrm{Gr}^0(M)$ is geometrically irreducible. Suppose that for some real $\varepsilon > 0$, some integer $n \geq 2$ and some integer A_{2n}, we have an inequality

$$\Sigma_E |\tilde{M}|^{2n} \leq A_{2n}(\# E)^{(1-n)\dim X} + O((\# E)^{(1-n)\dim X - \varepsilon/2}),$$

for E/k a variable finite extension. Then the support of $\mathrm{Gr}^0(M)$ is X itself, and $M_{2n}{}^{\mathrm{geom}}(X, \mathrm{Gr}^0(M)) \leq A_{2n}$.

proof By part 1) of Theorem 1.19.1, $\dim(\mathrm{Supp}(\mathrm{Gr}^0(M))) \geq \dim X$. As X is geometrically irreducible, we have equality. Then by part 2) of Theorem 1.19.1, we have the asserted inequality

$$M_{2n}{}^{\mathrm{geom}}(X, \mathrm{Gr}^0(M)) \leq A_{2n}. \quad \text{QED}$$

(1.20) Higher moment estimates for Twist(L,K,$\mathcal{F}$,h)

(1.20.1) Recall that for an even integer $2n \geq 2$, $2n!!$ is the product

$$2n!! := (2n-1)(2n-3)\ldots(1)$$

of the odd integers in the interval $[0, 2n]$.

Higher Moment Theorem 1.20.2 Suppose we are given standard input

$$(m \geq 1, K, V, h, L, d \geq 2, (\mathcal{F}, \tau)).$$

Suppose in addition that $\mathrm{Gr}^0(K)$ is not geometrically constant, and

that the following three additional hypotheses hold:
1) The perverse sheaf L is $\mathcal{L}[\dim V]$ for some constructible $\overline{\mathbb{Q}}_\ell$-sheaf $\mathcal{L}$ on V.
2) There exists a closed subscheme $W \subset \mathbb{A}^m$, inclusion $i: W \to \mathbb{A}^m$, such that the perverse sheaf K on $\mathbb{A}^m$ is $i_*\mathcal{K}[\dim W]$ for some some constructible $\overline{\mathbb{Q}}_\ell$-sheaf $\mathcal{K}$ on W.
3) We have the inequality $\dim V + \dim W \geq m+1$.

Denote by M the perverse sheaf $M := \mathrm{Twist}(L,K,\mathcal{F},h)$ on $\mathcal{F}$. Denote by M_0 the central normalization of its trace function. For each integer $n \geq 1$ with $2n \leq d$, there exists a real $\varepsilon > 0$ such that we have the following results.
1) If $\mathrm{Gr}^0(M(\dim\mathcal{F}_0/2))_{ncst}$ is not geometrically self dual, then

$$\Sigma_E |M_0(\dim\mathcal{F}_0/2)|^{2n} = (n!)((\#E)^{(1-n)\dim\mathcal{F}})(1 + O((\#E)^{-\varepsilon/2})).$$

2) If $\mathrm{Gr}^0(M(\dim\mathcal{F}_0/2))_{ncst}$ is geometrically self dual, then

$$\Sigma_E |M_0(\dim\mathcal{F}_0/2)|^{2n} = (2n!!)((\#E)^{(1-n)\dim\mathcal{F}})(1 + O((\#E)^{-\varepsilon/2})).$$

Before giving the proof of the theorem, let us give its main consequence.

Corollary 1.20.3 In the situation of Theorem 1.20.2, suppose in addition that $d \geq 4$. Then we have the following results concerning the perverse sheaf $M := \mathrm{Twist}(L,K,\mathcal{F},h)$ on $\mathcal{F}$.
1) The support of $\mathrm{Gr}^0(M(\dim\mathcal{F}_0/2))_{ncst}$ is all of $\mathcal{F}$.
2) For any dense open set $U \subset \mathcal{F}$ on which M is lisse, $M(\dim\mathcal{F}_0/2)|U$ is of the form $\mathcal{M}(\dim\mathcal{F}/2)[\dim\mathcal{F}]$, for a lisse $\overline{\mathbb{Q}}_\ell$-sheaf $\mathcal{M}$ on U which is ι-mixed of weight ≤ 0. The nonconstant part $\mathrm{Gr}^0(\mathcal{M})_{ncst}$ of the highest weight quotient $\mathrm{Gr}^0(\mathcal{M})$ of $\mathcal{M}$ as lisse sheaf on U is geometrically irreducible, and $\mathrm{Gr}^0(\mathcal{M})_{ncst}[\dim\mathcal{F}]$ on U is the Tate-twisted restriction $(\mathrm{Gr}^0(M(\dim\mathcal{F}_0/2))_{ncst}|U)(-\dim\mathcal{F}/2)$.
3) The necessary and sufficient condition (cf. Lemma 1.15.5, part 3)) for the equality

$$\mathrm{Gr}^0(M(\dim\mathcal{F}_0/2)) = \mathrm{Gr}^0(M(\dim\mathcal{F}_0/2))_{ncst}$$

of perverse sheaves on $\mathcal{F}$, namely that

$$H_c^m((V\times\mathbb{A}^m)\otimes\bar{k}, pr_1^*L\otimes pr_2^*K)$$

is ι-mixed of weight $\leq m - \varepsilon$, for some $\varepsilon > 0$ is also a necessary and sufficient condition for the equality

$$\mathrm{Gr}^0(\mathcal{M}) = \mathrm{Gr}^0(\mathcal{M})_{ncst}$$

of lisse sheaves on U.
4) Fix $n \geq 1$ with $2n \leq d$. We have the moment estimates

$M_{2n}{}^{geom}(U, Gr^0(\mathcal{M})_{ncst}) \le n!$, if $Gr^0(\mathcal{M})_{ncst}$ is not geometrically self dual on U,

$M_{2n}{}^{geom}(U, Gr^0(\mathcal{M})_{ncst}) \le 2n!!$ if $Gr^0(\mathcal{M})_{ncst}$ is geometrically self dual on U.

5) Fix $n \ge 1$ with $2n \le d$. Suppose that $\mathrm{rank}(Gr^0(\mathcal{M}))_{ncst} \ge n$, and that $Gr^0(\mathcal{M})$ is not geometrically self dual on U. Then we have

$$M_{2n}{}^{geom}(U, Gr^0(\mathcal{M})_{ncst}) = n!.$$

6) Fix $n \ge 1$ with $2n \le d$. Suppose that $\mathrm{rank}(Gr^0(\mathcal{M})_{ncst}) \ge 2n$, and that $Gr^0(\mathcal{M})_{ncst}$ is geometrically self dual on U. Then we have

$$M_{2n}{}^{geom}(U, Gr^0(\mathcal{M})_{ncst}) = 2n!!.$$

7) The geometric Frobenius-Schur indicator of $Gr^0(\mathcal{M})_{ncst}$ on U is given by the product formula

$$FSI^{geom}(U, Gr^0(\mathcal{M})_{ncst})$$
$$= (-1)^m FSI^{geom}(V, Gr^0(L)) FSI^{geom}(\mathbb{A}^m, Gr^0(K)).$$

proof of Corollary 1.20.3 1) We have $d \ge 4$, so we may take $n = 2$ in Theorem 1.20.1. Now apply Corollary 1.19.3.

2) On any dense open set $U \subset \mathcal{F}$ on which any perverse sheaf N is lisse, $N|U$ is of the form $\mathcal{N}(\dim\mathcal{F}/2)[\dim\mathcal{F}]$ for a lisse $\overline{\mathbb{Q}}_\ell$-sheaf $\mathcal{N}$ on U. If N is ι-mixed of weight ≤ 0, then $\mathcal{N}$ is ι-mixed of weight ≤ 0, and the highest weight quotient $Gr^0(\mathcal{N})$ of $\mathcal{N}$ as lisse sheaf on U is related to the highest weight quotient $Gr^0(N)$ of N as perverse sheaf by $Gr^0(N)|U = (Gr^0(\mathcal{N})[\dim\mathcal{F}])(\dim\mathcal{F}/2)$. Similarly, the nonconstant part $Gr^0(\mathcal{N})_{ncst}$ of $Gr^0(\mathcal{N})$ as lisse sheaf on U is related to the nonconstant part $Gr^0(N)_{ncst}$ of $Gr^0(N)$ as perverse sheaf by

$$Gr^0(N)_{ncst}|U = (Gr^0(\mathcal{N})_{ncst}[\dim\mathcal{F}])(\dim\mathcal{F}/2).$$

If $Gr^0(N)_{ncst}$ is geometrically irreducible and its support is all of $\mathcal{F}$, then $Gr^0(N)_{ncst}|U$ is still geometrically irreducible, and hence $Gr^0(\mathcal{N})_{ncst}$ is geometrically irreducible as lisse sheaf on U. Take for N the perverse sheaf $M(\dim\mathcal{F}_0/2)$.

3) Indeed, $Gr^0(M(\dim\mathcal{F}_0/2))$, being pure of weight zero, is geometrically (i.e., on $\mathcal{F}\otimes_k\overline{k}$) semisimple [BBD 5.3.8], so the direct sum of $Gr^0(M(\dim\mathcal{F}_0/2))_{ncst}$ and of $Gr^0(M(\dim\mathcal{F}_0/2))_{cst}$, the latter geometrically isomorphic to (a constant sheaf $\mathcal{G}$)[$\dim\mathcal{F}$]. Thus on $\mathcal{F}\otimes_k\overline{k}$ we have

$$Gr^0(M(\dim\mathcal{F}_0/2)) \cong Gr^0(M(\dim\mathcal{F}_0/2))_{ncst} \oplus \mathcal{G}[\dim\mathcal{F}].$$

So on any dense open set U on which M is lisse, we have a direct sum decomposition of lisse sheaves on $U\otimes_k\overline{k}$

$$\mathrm{Gr}^0(\mathcal{M}) = \mathrm{Gr}^0(\mathcal{M})_{\mathrm{ncst}} \oplus \mathcal{G}.$$

Thus we have $\mathrm{Gr}^0(\mathcal{M}) = \mathrm{Gr}^0(\mathcal{M})_{\mathrm{ncst}}$ if and only if $\mathcal{G} = 0$, if and only if we have

$$\mathrm{Gr}^0(M(\dim\mathcal{F}_0/2)) = \mathrm{Gr}^0(M(\dim\mathcal{F}_0/2))_{\mathrm{ncst}}.$$

4) Simply apply Corollary 1.19.3.

5) We will reverse the inequality. For any finite-dimensional representation V of any group G, and any non-negative integers a and b, we have the a priori inequality

$$M_{a,b}(G, V) \geq M_{a,b}(GL(V), V).$$

In characteristic zero, we have

$$M_{a,a}(GL(V), V) = a! \text{ if } \dim V \geq a.$$

So if $\mathrm{rank}(\mathrm{Gr}^0(\mathcal{M})_{\mathrm{ncst}}) \geq n$, we have

$$M_{2n}{}^{\mathrm{geom}}(U, \mathrm{Gr}^0(\mathcal{M})_{\mathrm{ncst}}) \geq n!.$$

6) We reverse the inequality. For any finite-dimensional symplectic (resp. orthogonal) representation V of any group G, and any non-negative integers a and b, we have the a priori inequality

$$M_{a,b}(G, V) \geq M_{a,b}(Sp(V), V),$$

$$(\text{resp. } M_{a,b}(G, V) \geq M_{a,b}(O(V), V)).$$

In characteristic zero, we have

$$M_{a,a}(Sp(V), V) = M_{a,a}(O(V), V) = 2a!! \text{ if } \dim V \geq 2a.$$

So if $\mathrm{rank}(\mathrm{Gr}^0(\mathcal{M})_{\mathrm{ncst}}) \geq 2n$, and $\mathrm{Gr}^0(\mathcal{M})$ is geometrically self dual (and hence either orthogonally or symplectically self dual), we have

$$M_{2n}{}^{\mathrm{geom}}(U, \mathrm{Gr}^0(\mathcal{M})_{\mathrm{ncst}}) \geq 2n!!.$$

7) This is just a rewriting of the already established (cf. Theorem 1.15.6, part 3)) multiplicative formula for the Frobenius-Schur indicator, namely

$$\mathrm{FSI}^{\mathrm{geom}}(\mathcal{F}, \mathrm{Gr}^0(M(\dim\mathcal{F}_0/2))_{\mathrm{ncst}})$$
$$= ((-1)^{\dim\mathcal{F}_0}) \times \mathrm{FSI}^{\mathrm{geom}}(\mathbb{A}^m, \mathrm{Gr}^0(K)) \times \mathrm{FSI}^{\mathrm{geom}}(V, \mathrm{Gr}^0(L)).$$

Since $\mathrm{Gr}^0(M(\dim\mathcal{F}_0/2))_{\mathrm{ncst}}$ has support all of $\mathcal{F}$, and $\mathrm{FSI}^{\mathrm{geom}}$ does not see Tate twists, we have

$$\mathrm{FSI}^{\mathrm{geom}}(\mathcal{F}, \mathrm{Gr}^0(M(\dim\mathcal{F}_0/2))_{\mathrm{ncst}})$$
$$:= (-1)^{\dim\mathcal{F}}\mathrm{FSI}^{\mathrm{geom}}(U, \mathrm{Gr}^0(\mathcal{M})_{\mathrm{ncst}}(\dim\mathcal{F}/2))$$
$$= (-1)^{\dim\mathcal{F}}\mathrm{FSI}^{\mathrm{geom}}(U, \mathrm{Gr}^0(\mathcal{M})_{\mathrm{ncst}}).$$

QED

(1.21) Proof of the Higher Moment Theorem 1.20.2: combinatorial preliminaries

(1.21.1) We begin by recalling [Stan-ECI, 3.7.1] a version of the Moebius inversion formula. Suppose we are given a finite, partially ordered set P. Then there exists a unique assignment of integers $\mu(p, q)$, one for each pair (p, q) of elements of P with $p \geq q$, with the following property:

for any abelian group A, and any map $g : P \to A$, if we define a map $f : P \to A$ by the rule

$$f(p) = \Sigma_{q\le p}\, g(q),$$

then we have the inversion formula

$$g(p) = \Sigma_{q\le p}\, \mu(p, q)f(q).$$

We will not need the precise values of the integers $\mu(p, q)$, but we will make constant use of the fact [Stan-ECI, 3.7] that

$$\mu(p, p) = 1, \text{ for all p in P}.$$

(1.21.2) In our application, we fix an integer $n \ge 1$, and we consider the set $\{1, 2,..., 2n\}$. The partially ordered set P will be the set of all partitions $\mathcal{P}$ of the set $\{1, 2,..., 2n\}$, or, what is the same, the set of all equivalence relations on the set $\{1, 2,..., 2n\}$. Given two partitions $\mathcal{P}$ and $\mathcal{Q}$ of the set $\{1, 2,..., 2n\}$, we say that $\mathcal{P} \ge \mathcal{Q}$ if $\mathcal{Q}$ is a coarsening of $\mathcal{P}$ in the sense that each set $\mathcal{Q}_\alpha$ in the partition $\mathcal{Q}$ is a union of sets $\mathcal{P}_\beta$ in the partition $\mathcal{P}$. In terms of equivalence relations, $\mathcal{P} \ge \mathcal{Q}$ means that $\mathcal{P}$-equivalence implies $\mathcal{Q}$-equivalence.

(1.21.3) Given a partition (i.e., an equivalence relation) $\mathcal{P}$, we denote by $\lambda = \lambda(\mathcal{P})$ the number of subsets (i.e., the number of equivalence classes) into which $\{1, 2,..., 2n\}$ is divided by $\mathcal{P}$.

$$\lambda = \lambda(\mathcal{P}) := \#\mathcal{P}$$

We label these subsets $\mathcal{P}_1,..., \mathcal{P}_\lambda$ by the following convention: $\mathcal{P}_1$ is the subset containing 1, and, if $\lambda > 1$, $\mathcal{P}_2$ is the subset containing the least integer not in $\mathcal{P}_1$, et cetera (i.e., for $1 \le i \le \lambda-1$, $\mathcal{P}_{i+1}$ is the subset containing the least integer not in the union of those $\mathcal{P}_j$ with $j \le i$). We denote by $\nu = \nu(\mathcal{P})$ the number of singletons in $\mathcal{P}$, i.e., ν is the number of one-element equivalence classes, or the number of indices i for which $\#\mathcal{P}_i = 1$.

$$\nu := \#\{\text{singletons among the } \mathcal{P}_i\}.$$

Each subset $\mathcal{P}_i$ has a "type" (a_i, b_i) in $\mathbb{Z}_{\ge 0}\times\mathbb{Z}_{\ge 0}$, defined by

$$a_i := \#(\mathcal{P}_i\cap\{1, 2,..., n\}),$$
$$b_i := \#(\mathcal{P}_i\cap\{n+1, n+2,..., 2n\}).$$

Thus we have

$$\nu := \#\{i \text{ such that } a_i + b_i = 1\}.$$

Lemma 1.21.4 Let $\mathcal{P} \ge \mathcal{P}'$ be partitions of the set $\{1, 2,..., 2n\}$. Put

$$\lambda = \lambda(\mathcal{P}),\ \lambda' := \lambda(\mathcal{P}'),$$
$$\nu = \nu(\mathcal{P}),\ \nu' = \nu(\mathcal{P}').$$

Then we have the following inequalities.

1) $2n \ge 2\lambda - \nu$, i.e., $\lambda - n - \nu/2 \le 0$, with equality if and only if $\#\mathcal{P}_i \le 2$ for all i.

2) $\lambda \ge \lambda'$, with equality if and only if $\mathcal{P} = \mathcal{P}'$.

3) $\nu \ge \nu'$.

4) $\lambda - \nu/2 \ge \lambda' - \nu'/2$, i.e., $\lambda' - \lambda - (\nu' - \nu)/2 \le 0$.

proof 1) Since $\mathcal{P}$ is a partition of $\{1, 2,..., 2n\}$, we have

$$2n = \Sigma_{i=1 \text{ to } \lambda} \#\mathcal{P}_i = \nu + \Sigma_{i \text{ with } \#\mathcal{P}_i \geq 2} \#\mathcal{P}_i.$$

There are $\lambda - \nu$ indices i for which $\#\mathcal{P}_i \geq 2$, so we have

$$\nu + \Sigma_{i \text{ with } \#\mathcal{P}_i \geq 2} \#\mathcal{P}_i \geq \nu + 2(\lambda - \nu) = 2\lambda - \nu.$$

Thus $2n \geq 2\lambda - \nu$, with equality if and only if every i with $\#\mathcal{P}_i \geq 2$ has $\#\mathcal{P}_i = 2$.

2), 3), and 4) Since $\mathcal{P} \geq \mathcal{P}'$, $\mathcal{P}'$ is obtained from $\mathcal{P}$ by collapsing together various of the $\mathcal{P}_i$. So either $\mathcal{P} = \mathcal{P}'$, in which case there is nothing to prove, or we can pass from $\mathcal{P}$ to $\mathcal{P}'$ through a sequence of intermediate coarsenings where at each step we collapse precisely two sets into one.

Thus we may reduce to the case where $\mathcal{P}'$ is obtained from $\mathcal{P}$ by collapsing precisely two sets, say $\mathcal{P}_i$ and $\mathcal{P}_j$. Thus λ' is $\lambda - 1$.

If neither $\mathcal{P}_i$ nor $\mathcal{P}_j$ is a singleton, then $\nu = \nu'$. In this case,

$$\lambda' - \nu'/2 = \lambda - 1 - \nu/2 < \lambda - \nu/2.$$

If exactly one of $\mathcal{P}_i$ or $\mathcal{P}_j$ is a singleton, then $\nu' = \nu - 1$. In this case,

$$\lambda' - \nu'/2 = \lambda -1 - (\nu-1)/2 = \lambda - 1/2 - \nu/2 < \lambda - \nu/2.$$

If both $\mathcal{P}_i$ and $\mathcal{P}_j$ are singletons, then $\nu' = \nu - 2$. In this case,

$$\lambda' - \nu'/2 = \lambda -1 - (\nu-2)/2 = \lambda - \nu/2. \qquad \text{QED}$$

(1.21.5) What do these combinatorics have to do with the Higher Moment Theorem? To see the relation, we first restate that theorem in terms of the sums $\Sigma_E |M_0|^{2n}$, rather than $\Sigma_E |M_0(\dim\mathcal{F}_0/2)|^{2n}$.

Higher Moment Theorem bis 1.21.6 Hypotheses and notations as in the Higher Moment Theorem 1.20.2, denote by M the perverse sheaf M := Twist(L,K,$\mathcal{F}$,h) on $\mathcal{F}$. Denote by M_0 its centrally normalized trace function. For each integer $n \geq 1$ with $2n \leq d$, there exists a real $\varepsilon > 0$ such that we have

1) $\Sigma_E |M_0|^{2n} = (n!)((\#E)^{\dim\mathcal{F} - nm})(1 + O((\#E)^{-\varepsilon/2}))$,

if $Gr^0(M(\dim\mathcal{F}_0/2))_{ncst}$ is not geometrically self dual,

2) $\Sigma_E |M_0|^{2n} = (2n!!)((\#E)^{\dim\mathcal{F} - nm})(1 + O((\#E)^{-\varepsilon/2}))$,

if $Gr^0(M(\dim\mathcal{F}_0/2))_{ncst}$ is geometrically self dual.

(1.21.7) How do we calculate $\Sigma_E |M_0|^{2n}$? By definition, we have

$$\Sigma_E |M_0|^{2n} := \Sigma_{f \text{ in } \mathcal{F}(E)} |M_0(E, f)|^{2n}$$

$$= \Sigma_{f \text{ in } \mathcal{F}(E)} |\Sigma_{v \text{ in } V(E)} L(E, v)K_0(h(v) + f(v))|^{2n}$$

$$= \Sigma_{f \text{ in } \mathcal{F}(E)} \Sigma_{v_1, v_2,\ldots, v_{2n} \text{ in } V(E)}$$

$$(\Pi_{i=1 \text{ to } n} L(E, v_i))\times(\Pi_{i=n+1 \text{ to } 2n} \overline{L}(E, v_i))$$

$$\times(\Pi_{i=1 \text{ to } n} K_0(h(v_i) + f(v_i)))\times(\Pi_{i=n+1 \text{ to } 2n} \overline{K}_0(h(v_i) + f(v_i)))$$

$= \Sigma_{v_1, v_2,..., v_{2n} \text{ in } V(E)} (\Pi_{i=1 \text{ to } n} L(E, v_i)) \times (\Pi_{i=n+1 \text{ to } 2n} \overline{L}(E, v_i))$
$\times \Sigma_{f \text{ in } \mathcal{F}(E)} (\Pi_{i=1 \text{ to } n} K_0(h(v_i) + f(v_i)))$
$\times (\Pi_{i=n+1 \text{ to } 2n} \overline{K}_0(h(v_i) + f(v_i)))$.

(1.21.8) Now fix an outer summand, a 2n-tuple $(v_1, v_2,..., v_{2n})$ of points in V(E). Such an n-tuple determines an equivalence relation $\mathcal{P}$ on the index set {1, 2,..., 2n}, by decreeing that two indices i and j are $\mathcal{P}$-equivalent if and only if $v_i = v_j$. Thus among $v_1, v_2,..., v_{2n}$, precisely $\lambda = \lambda(\mathcal{P})$ are distinct. For each of the $\lambda = \lambda(\mathcal{P})$ sets $\mathcal{P}_i$ into which {1, 2,..., 2n} is partitioned, let us denote by $v_{\mathcal{P}_i}$ the common value of v_j for those indices j in $\mathcal{P}_i$. Denote by (a_i, b_i) the type of $\mathcal{P}_i$. The inner summand may be rewritten as

$$\Sigma_{f \text{ in } \mathcal{F}(E)} (\Pi_{i=1 \text{ to } n} K_0(h(v_i) + f(v_i))) \times (\Pi_{i=n+1 \text{ to } 2n} \overline{K}_0(h(v_i) + f(v_i)))$$
$$= \Sigma_{f \text{ in } \mathcal{F}(E)} \Pi_{i=1 \text{ to } \lambda} (\Pi_{\text{those } j = 1 \text{ to } n \text{ in } \mathcal{P}_i} K_0(h(v_j) + f(v_j)))$$
$$\times (\Pi_{\text{those } j = n+1 \text{ to } 2n \text{ in } \mathcal{P}_i} \overline{K}_0(h(v_j) + f(v_j)))$$

$$= \Sigma_{f \text{ in } \mathcal{F}(E)} \Pi_{i=1 \text{ to } \lambda} (K_0(h(v_{\mathcal{P}_i}) + f(v_{\mathcal{P}_i}))^{a_i} \times \overline{K}_0(h(v_j) + f(v_j))^{b_i}).$$

(1.21.9) Since $\lambda \le 2n \le d$, and $(\mathcal{F}, \tau)$ is d-separating, the E-linear map

$$(\text{eval}(v_{\mathcal{P}_1}),..., \text{eval}(v_{\mathcal{P}_\lambda})) : \mathcal{F}(E) \to \mathbb{A}^m(E)^\lambda,$$
$$f \mapsto (f(v_{\mathcal{P}_1}),..., f(v_{\mathcal{P}_\lambda})),$$

is surjective. So we may continue this rewriting
$= ((\#E)^{\dim \mathcal{F} - \lambda m}) \Pi_{i=1 \text{ to } \lambda} (\Sigma_{a \text{ in } \mathbb{A}^m(E)} K_0(E, a)^{a_i} \times \overline{K}_0(E, a)^{b_i})$
$= ((\#E)^{\dim \mathcal{F} - \lambda m}) \Pi_{i=1 \text{ to } \lambda} (\Sigma_E K_0^{a_i} \times \overline{K}_0^{b_i})$.

(1.21.10) If we put this back into the entire sum for $\Sigma_E |M_0|^{2n}$, and we separate outer summands by the partitions $\mathcal{P}$ to which they give rise, we find

$$\Sigma_E |M_0|^{2n} = \Sigma_{\mathcal{P}} ((\#E)^{\dim \mathcal{F} - \lambda m}) \Pi_{i=1 \text{ to } \lambda} (\Sigma_E K_0^{a_i} \times \overline{K}_0^{b_i})$$
$$\times \Sigma_{v_1, v_2,..., v_{2n} \text{ in } V(E) \text{ of type } \mathcal{P}}$$
$$(\Pi_{i=1 \text{ to } n} L(E, v_i)) \times (\Pi_{i=n+1 \text{ to } 2n} \overline{L}(E, v_i)).$$

The inner sum we can rewrite as

$$\Sigma_{v_1, v_2,..., v_{2n} \text{ in } V(E) \text{ of type } \mathcal{P}}$$
$$(\Pi_{i=1 \text{ to } n} L(E, v_i)) \times (\Pi_{i=n+1 \text{ to } 2n} \overline{L}(E, v_i))$$

$$= \Sigma_{v_{\mathcal{P}_1}, v_{\mathcal{P}_2}, \ldots, v_{\mathcal{P}_\lambda} \text{ all distinct in } V(E)} \prod_{i=1 \text{ to } \lambda} (L(E, v_{\mathcal{P}_i})^{a_i} \times \overline{L}(E, v_{\mathcal{P}_i})^{b_i}).$$

(1.21.11) At this point, the combinatorics enter. For fixed E/k, consider the $\mathbb{C}$-valued function of partitions $\mathcal{P} \mapsto g(\mathcal{P})$ defined by

$$g(\mathcal{P}) := \Sigma_{v_1, v_2, \ldots, v_{2n} \text{ in } V(E) \text{ of type } \mathcal{P}} (\prod_{i=1 \text{ to } n} L(E, v_i)) \times (\prod_{i=n+1 \text{ to } 2n} \overline{L}(E, v_i))$$

$$= \Sigma_{v_{\mathcal{P}_1}, v_{\mathcal{P}_2}, \ldots, v_{\mathcal{P}_\lambda} \text{ all distinct in } V(E)} \prod_{i=1 \text{ to } \lambda} (L(E, v_{\mathcal{P}_i})^{a_i} \times \overline{L}(E, v_{\mathcal{P}_i})^{b_i}).$$

Its Moebius partner function,

$$f(\mathcal{P}) := \Sigma_{\mathcal{P}' \leq \mathcal{P}} g(\mathcal{P}'),$$

is then given by

$$f(\mathcal{P}) = \Sigma_{v_{\mathcal{P}_1}, v_{\mathcal{P}_2}, \ldots, v_{\mathcal{P}_\lambda} \text{ in } V(E)} \prod_{i=1 \text{ to } \lambda} (L(E, v_{\mathcal{P}_i})^{a_i} \times \overline{L}(E, v_{\mathcal{P}_i})^{b_i})$$

$$= \prod_{i=1 \text{ to } \lambda} (\Sigma_E L^{a_i} \times \overline{L}^{b_i}).$$

(1.21.12) To go further, it will be convenient to introduce the following notations. For a partition $\mathcal{P}$, with sets $\mathcal{P}_i$ of type (a_i, b_i) for $i = 1$ to $\lambda = \lambda(\mathcal{P})$, we define

$$S_E(\mathcal{P}, L) := \Sigma_{v_1, v_2, \ldots, v_{2n} \text{ in } V(E) \text{ of type } \mathcal{P}} (\prod_{i=1 \text{ to } n} L(E, v_i)) \times (\prod_{i=n+1 \text{ to } 2n} \overline{L}(E, v_i))$$

$$= \Sigma_{v_{\mathcal{P}_1}, v_{\mathcal{P}_2}, \ldots, v_{\mathcal{P}_\lambda} \text{ all distinct in } V(E)} \prod_{i=1 \text{ to } \lambda} (L(E, v_{\mathcal{P}_i})^{a_i} \times \overline{L}(E, v_{\mathcal{P}_i})^{b_i}).$$

We further define

$$\Sigma_E(\mathcal{P}, L) = \prod_{i=1 \text{ to } \lambda} (\Sigma_E L^{a_i} \times \overline{L}^{b_i}),$$

$$\Sigma_E(\mathcal{P}, K_0) = \prod_{i=1 \text{ to } \lambda} (\Sigma_E K_0^{a_i} \times \overline{K}_0^{b_i}).$$

We have the tautologous relation

$$\Sigma_E(\mathcal{P}, L) = \Sigma_{\mathcal{P}' \leq \mathcal{P}} S_E(\mathcal{P}', L).$$

Moebius inversion gives

$$S_E(\mathcal{P}, L) = \Sigma_{\mathcal{P}' \leq \mathcal{P}} \mu(\mathcal{P}, \mathcal{P}')\Sigma_E(\mathcal{P}', L)$$

$$= \Sigma_E(\mathcal{P}, L) + \Sigma_{\mathcal{P}' < \mathcal{P}} \mu(\mathcal{P}, \mathcal{P}')\Sigma_E(\mathcal{P}', L).$$

The sum $\Sigma_E |M_0|^{2n}$ is given by

$$\Sigma_E |M_0|^{2n} = \Sigma_{\mathcal{P}} ((\#E)^{\dim \mathcal{F} - \lambda(\mathcal{P})m})\Sigma_E(\mathcal{P}, K_0)S_E(\mathcal{P}, L)$$

$$= \Sigma_{\mathcal{P}} ((\#E)^{\dim \mathcal{F} - \lambda(\mathcal{P})m})\Sigma_E(\mathcal{P}, K_0)\Sigma_E(\mathcal{P}, L)$$

$$+ \Sigma_{(\mathcal{P}', \mathcal{P}) \text{ with } \mathcal{P}' < \mathcal{P}} \mu(\mathcal{P}, \mathcal{P}')((\#E)^{\dim \mathcal{F} - \lambda(\mathcal{P})m})\Sigma_E(\mathcal{P}, K_0)\Sigma_E(\mathcal{P}', L).$$

(1.21.13) The Higher Moment Theorem bis 1.21.6 asserts that there exists a real $\varepsilon > 0$ such that we have the estimate

$$\Sigma_E |M_0|^{2n} = A_{2n}(\#E)^{\dim \mathcal{F} - nm})(1 + O((\#E)^{-\varepsilon/2})),$$

with

$A_{2n} := n!$, if $Gr^0(M(\dim \mathcal{F}_0/2))_{ncst}$ is not geometrically self dual,

$A_{2n} := 2n!!$, if $\mathrm{Gr}^0(M(\dim\mathcal{F}_0/2))_{ncst}$ is geometrically self dual.

In order to prove this, we will examine in turn each of the summands. A few will be equal to

$$(\#E)^{\dim\mathcal{F} - nm)}(1 + O((\#E)^{-\varepsilon/2})),$$

many will vanish, and all the rest will be

$$O((\#E)^{\dim\mathcal{F} - nm - \varepsilon/2}).$$

Lemma 1.21.14 If $\mathcal{P}$ contains a singleton, then $\Sigma_E(\mathcal{P}, K_0) = 0$.

proof By definition, we have

$$\Sigma_E(\mathcal{P}, K_0) = \Pi_{i=1 \text{ to } \lambda} (\Sigma_E K_0{}^{a_i} \times \bar{K}_0{}^{b_i}).$$

If $\mathcal{P}$ contains a singleton, one of the factors is either $\Sigma_E K_0 = 0$ or $\Sigma_E \bar{K}_0 = 0$. QED

Lemma 1.21.15 For any pair of non-negative integers (a, b) with $a+b \geq 2$, we have the estimates

$$|\Sigma_E L^a \bar{L}^b| \leq \Sigma_E |L|^{a+b} = O((\#E)^{\dim(V)(1 - (a+b)/2)}),$$

$$|\Sigma_E K_0{}^a \bar{K}_0{}^b| \leq \Sigma_E |K_0|^{a+b} = O((\#E)^{\dim(W)(1 - (a+b)/2)}).$$

proof We know that L is ι-mixed of weight ≤ 0. By hypothesis, its only nonvanishing cohomology group is $\mathcal{H}^{-\dim(V)}(L)$, which is ι-mixed of weight $\leq -\dim(V)$. So we have the estimate

$$|L(E, v)| = O((\#E)^{-\dim(V)/2}).$$

Hence we have

$$|L(E, v)^a \bar{L}(E, v)^b| = |L(E, v)|^{a+b} = O((\#E)^{-(a+b)\times\dim(V)/2}),$$

and there are $\#V(E) = O((\#E)^{\dim(V)})$ terms in the sum $|\Sigma_E L^a \bar{L}^b|$. Similarly, the trace function of K is supported in W, and we have

$$|K(E, x)| = O((\#E)^{-\dim(W)/2}), \text{ for } x \text{ in } W(E),$$
$$= 0, \text{ for } x \text{ in } \mathbb{A}^m(E) - W(E).$$

Since $\dim(W) \leq \dim(\mathbb{A}^m)$, and K_0 is an approximate trace function for K, we have

$$|K_0(E, x)| = O((\#E)^{-\dim(W)/2}), \text{ for } x \text{ in } W(E),$$

$$|K_0(E, x)| = O((\#E)^{-m/2}), \text{ for } x \text{ in } \mathbb{A}^m(E) - W(E).$$

Summing separately over $W(E)$ and its complement, we find

$$|\Sigma_E K_0{}^a \bar{K}_0{}^b|$$
$$= O((\#E)^{\dim(W)(1 - (a+b)/2)}) + O((\#E)^{m(1 - (a+b)/2)})$$
$$= O((\#E)^{\dim(W)(1 - (a+b)/2)}). \text{ QED}$$

Lemma 1.21.16 For any partition $\mathcal{P}$ with no singletons, with $\lambda = \lambda(\mathcal{P})$ the number of equivalence classes, we have the estimates

$$|\Sigma_E(\mathcal{P}, L)| = O((\#E)^{\dim(V)(\lambda - n)}),$$

$$|\Sigma_E(\mathcal{P}, K_0)| = O((\#E)^{\dim(W)(\lambda - n)}).$$

proof By definition, we have

$$|\Sigma_E(\mathcal{P}, L)| = \prod_{i=1 \text{ to } \lambda} |\Sigma_E L^{a_i} \times \bar{L}^{b_i}|.$$

Each of the λ $\mathcal{P}_i$'s has type (a_i, b_i) with $a_i + b_i \geq 2$, so gives a factor

$$|\Sigma_E L^{a_i} \times \bar{L}^{b_i}| = O((\#E)^{\dim(V)(1 - (a_i+b_i)/2)}),$$

by the previous lemma. So we get

$$|\Sigma_E(\mathcal{P}, L)| = O((\#E)^{\dim(V)(\Sigma_{\mathcal{P}_i} (1 - (a_i+b_i)/2)})$$

$$= O((\#E)^{\dim(V)((\lambda - n)}).$$

Similarly for $\Sigma_E(\mathcal{P}, K_0)$. QED

(1.21.17) With these lemmas established, we can analyze the individual summands in

$$\Sigma_E |M_0|^{2n} = \Sigma_{\mathcal{P}} ((\#E)^{\dim\mathcal{F} - \lambda(\mathcal{P})m})\Sigma_E(\mathcal{P}, K_0)S_E(\mathcal{P}, L)$$

$$= \Sigma_{\mathcal{P}} ((\#E)^{\dim\mathcal{F} - \lambda(\mathcal{P})m})\Sigma_E(\mathcal{P}, K_0)\Sigma_E(\mathcal{P}, L)$$

$$+ \Sigma_{(\mathcal{P}',\mathcal{P}) \text{ with } \mathcal{P}' < \mathcal{P}} \mu(\mathcal{P}, \mathcal{P}')((\#E)^{\dim\mathcal{F} - \lambda(\mathcal{P})m})\Sigma_E(\mathcal{P}, K_0)\Sigma_E(\mathcal{P}', L).$$

In view of the vanishing (cf. Lemma 1.21.14) of $\Sigma_E(\mathcal{P}, K_0)$ whenever $\mathcal{P}$ contains a singleton, we have

$$\Sigma_E |M_0|^{2n}$$

$$= \Sigma_{\mathcal{P} \text{ with no singleton}} ((\#E)^{\dim\mathcal{F} - \lambda(\mathcal{P})m})\Sigma_E(\mathcal{P}, K_0)S_E(\mathcal{P}, L)$$

$$= \Sigma_{\mathcal{P} \text{ with no singleton}} ((\#E)^{\dim\mathcal{F} - \lambda(\mathcal{P})m})\Sigma_E(\mathcal{P}, K_0)\Sigma_E(\mathcal{P}, L)$$

$$+ \Sigma_{\mathcal{P} \text{ with no singleton}} \Sigma_{\mathcal{P}' < \mathcal{P}}$$

$$\mu(\mathcal{P}, \mathcal{P}')((\#E)^{\dim\mathcal{F} - \lambda(\mathcal{P})m})\Sigma_E(\mathcal{P}, K_0)\Sigma_E(\mathcal{P}', L).$$

(1.21.18) We first show that all the summands in the second line, and many in the first, are negligible with respect to the target magnitude $(\#E)^{\dim\mathcal{F} - nm}$.

Lemma 1.21.19 Suppose $\mathcal{P}$ contains no singleton and $\mathcal{P}' \leq \mathcal{P}$. If either $\lambda(\mathcal{P}) < n$ or if $\mathcal{P}' < \mathcal{P}$, then we have the estimate

$$|((\#E)^{\dim\mathcal{F} - \lambda(\mathcal{P})m})\Sigma_E(\mathcal{P}, K_0)\Sigma_E(\mathcal{P}', L)|$$

$$= O((\#E)^{\dim\mathcal{F} - nm - 1}).$$

proof Since $\mathcal{P}' \leq \mathcal{P}$, we have $\lambda' := \lambda(\mathcal{P}') \leq \lambda := \lambda(\mathcal{P})$. By the above lemma, we have

$$|((\#E)^{\dim\mathcal{F} - \lambda(\mathcal{P})m})\Sigma_E(\mathcal{P}, K_0)\Sigma_E(\mathcal{P}', L)|$$

$$= O((\#E)^{\dim\mathcal{F} - \lambda m + \dim(V)(\lambda' - n) + \dim(W)(\lambda - n)}).$$

The exponent of $\#E$ is

$$\dim\mathcal{F} - \lambda m + \dim(V)(\lambda' - n) + \dim(W)(\lambda - n)$$

$$= \dim\mathcal{F} - \lambda m + \dim(V)(\lambda - n) + \dim(W)(\lambda - n) - \dim(V)(\lambda - \lambda')$$

$$= \dim\mathcal{F} - nm + (n-\lambda)m + \dim(V)(\lambda - n) + \dim(W)(\lambda - n) - \dim(V)(\lambda - \lambda')$$

$$= \dim\mathcal{F} - nm - (n-\lambda)(\dim(V) + \dim(W) - m) - \dim(V)(\lambda - \lambda').$$

Since $\mathcal{P}$ has no singletons, we have $\lambda \leq n$. By hypothesis, we have $\dim(V) + \dim(W) - m \geq 1$. Therefore we have

$$(n-\lambda)(\dim(V) + \dim(W) - m) \geq n - \lambda.$$

Because $\dim(V) + \dim(W) \geq m+1$, we have $\dim(V) \geq 1$, hence we have

$$\dim(V)(\lambda - \lambda') \geq \lambda - \lambda'.$$

So we get the asserted estimate if either $\lambda > \lambda'$ or if $n > \lambda$. Since we are given that $\mathcal{P} \leq \mathcal{P}'$, we have $\mathcal{P}' < \mathcal{P}$ if and only if $\lambda' < \lambda$. QED

(1.21.20) At this point, we have

$$\Sigma_E |M_0|^{2n}$$

$$= \Sigma_{\mathcal{P} \text{ with no singleton and } \lambda = n} ((\#E)^{\dim \mathcal{F} - \lambda(\mathcal{P})m}) \Sigma_E(\mathcal{P}, K_0) \Sigma_E(\mathcal{P}, L)$$

$$+ O((\#E)^{\dim \mathcal{F} - nm - 1}).$$

(1.21.21) The $\mathcal{P}$ with no singletons and with $\lambda = n$ are precisely the partitions of $\{1, 2, \ldots, 2n\}$ into n sets of pairs. For such a partition $\mathcal{P}$ of $\{1, 2, \ldots, 2n\}$ into n sets of pairs, say $\mathcal{P}_1, \mathcal{P}_2, \ldots, \mathcal{P}_n$, each $\mathcal{P}_i$ has a "type", which is either (2, 0), (1, 1), or (0, 2). We denote by $h^{2,0}$, $h^{1,1}$, $h^{0,2}$ respectively the number of each type:

$$h^{2,0} = h^{2,0}(\mathcal{P}) := \#\{i \text{ such that } \mathcal{P}_i \text{ is of type } (2, 0)\},$$

$$h^{1,1} = h^{1,1}(\mathcal{P}) := \#\{i \text{ such that } \mathcal{P}_i \text{ is of type } (1, 1)\},$$

$$h^{0,2} = h^{0,2}(\mathcal{P}) := \#\{i \text{ such that } \mathcal{P}_i \text{ is of type } (0, 2)\}.$$

We call these the "hodge numbers" of the partition $\mathcal{P}$ of $\{1, 2, \ldots, 2n\}$ into n sets of pairs.

Hodge Symmetry Lemma 1.21.22 For a partition $\mathcal{P}$ of $\{1, 2, \ldots, 2n\}$ into n sets of pairs $\mathcal{P}_1, \mathcal{P}_2, \ldots, \mathcal{P}_n$, its hodge numbers satisfy $h^{2,0} = h^{0,2}$.

proof We have

$$n = \#\{1, 2, \ldots, n\} = \Sigma_i \#(\mathcal{P}_i \cap \{1, 2, \ldots, n\}) = 2h^{2,0} + 1h^{1,1} + 0h^{0,2},$$

and

$$n = \#\{n+1, n+2, \ldots, 2n\} = \Sigma_i \#(\mathcal{P}_i \cap \{n+1, n+2, \ldots, 2n\})$$

$$= 0h^{2,0} + 1h^{1,1} + 2h^{0,2}.$$

Thus $2h^{2,0} = n - h^{1,1} = 2h^{0,2}$. QED

Lemma 1.21.23 1) Suppose $\mathrm{Gr}^0(M(\dim \mathcal{F}_0/2))_{ncst}$ is geometrically self dual. Then there exists a real $\varepsilon > 0$ such that for every partition $\mathcal{P}$ of $\{1, 2, \ldots, 2n\}$ into n sets of pairs, we have

$$((\#E)^{\dim \mathcal{F} - \lambda m}) \Sigma_E(\mathcal{P}, K_0) \Sigma_E(\mathcal{P}, L)$$

$$= ((\#E)^{\dim \mathcal{F} - nm})(1 + O((\#E)^{-\varepsilon/2})).$$

2) Suppose $\mathrm{Gr}^0(M(\dim \mathcal{F}_0/2))_{ncst}$ is not geometrically self dual. Then there exists a real $\varepsilon > 0$ such that for a partition $\mathcal{P}$ of $\{1, 2, \ldots, 2n\}$ into n sets of pairs, we have the following results.

2a) If $\mathcal{P}$ is entirely of type (1, 1), i.e., if $h^{2,0} = h^{0,2} = 0$, then

$$((\#E)^{\dim \mathcal{F} - \lambda m}) \Sigma_E(\mathcal{P}, K_0) \Sigma_E(\mathcal{P}, L)$$

$$= ((\#E)^{\dim \mathcal{F} - nm})(1 + O((\#E)^{-\varepsilon/2})).$$

2b) If $\mathcal{P}$ is not entirely of type (1, 1), then

$$|((\#E)^{\dim\mathcal{F} - \lambda m})\Sigma_E(\mathcal{P}, K_0)\Sigma_E(\mathcal{P}, L)|$$
$$= O((\#E)^{\dim\mathcal{F} - \lambda m - \varepsilon/2})$$

proof By the Standard Input Theorem 1.15.6, $\mathrm{Gr}^0(M(\dim\mathcal{F}_0/2))_{ncst}$ is geometrically self dual if and only if both $\mathrm{Gr}^0(L)$ and $\mathrm{Gr}^0(K)$ are geometrically self dual. By Corollary 1.8.4 and Lemma 1.12.8, there exists a real $\varepsilon > 0$ such that

$$\Sigma_E |L|^2 = 1 + O((\#E)^{-\varepsilon/2}),$$
$$\Sigma_E |K_0|^2 = 1 + O((\#E)^{-\varepsilon/2}),$$
$$|\Sigma_E L^2| = 1 + O((\#E)^{-\varepsilon/2}), \text{ if } \mathrm{Gr}^0(L) \text{ is geometrically self dual,}$$
$$|\Sigma_E L^2| = O((\#E)^{-\varepsilon/2}), \text{ if } \mathrm{Gr}^0(L) \text{ is not geometrically self dual,}$$
$$|\Sigma_E K_0{}^2| = 1 + O((\#E)^{-\varepsilon/2}), \text{ if } \mathrm{Gr}^0(K) \text{ is geometrically self dual,}$$
$$|\Sigma_E K_0{}^2| = O((\#E)^{-\varepsilon/2}), \text{ if } \mathrm{Gr}^0(K) \text{ is not geometrically self dual.}$$

Now by the definitions of $\Sigma_E(\mathcal{P}, K_0)$ and of $\Sigma_E(\mathcal{P}, L)$, we have

$$\Sigma_E(\mathcal{P}, K_0) = (\Sigma_E K_0{}^2)^{h^{2,0}} \times (\Sigma_E |K_0|^2)^{h^{1,1}} \times (\Sigma_E \overline{K}_0{}^2)^{h^{0,2}},$$
$$\Sigma_E(\mathcal{P}, L) = (\Sigma_E L^2)^{h^{2,0}} \times (\Sigma_E |L|^2)^{h^{1,1}} \times (\Sigma_E \overline{L}^2)^{h^{0,2}}.$$

By the Hodge Symmetry Lemma 1.21.22, we have

$$\Sigma_E(\mathcal{P}, K_0) = |\Sigma_E K_0{}^2|^{2h^{2,0}} \times (\Sigma_E |K_0|^2)^{h^{1,1}},$$
$$\Sigma_E(\mathcal{P}, L) = |\Sigma_E L^2|^{2h^{2,0}} \times (\Sigma_E |L|^2)^{h^{1,1}}.$$

If $\mathrm{Gr}^0(M(\dim\mathcal{F}_0/2))_{ncst}$ is geometrically self dual, then every factor in the above products is $1 + O((\#E)^{-\varepsilon/2})$, so we have

$$|\Sigma_E(\mathcal{P}, K_0)||\Sigma_E(\mathcal{P}, L)| = 1 + O((\#E)^{-\varepsilon/2}),$$

and 1) is proven.

If $\mathrm{Gr}^0(M(\dim\mathcal{F}_0/2))_{ncst}$ is not geometrically self dual, the factors $\Sigma_E |K_0|^2$ and $\Sigma_E |L|^2$ are both $1 + O((\#E)^{-\varepsilon/2})$, but at least one of the factors $|\Sigma_E K_0{}^2|$ or $|\Sigma_E L^2|$ is $O((\#E)^{-\varepsilon/2})$ and the other is $O(1)$. So the product $|\Sigma_E K_0{}^2| \times |\Sigma_E L^2|$ is $O((\#E)^{-\varepsilon/2})$. If $\mathcal{P}$ is purely of type (1, 1), these terms do not occur, and we have

$$\Sigma_E(\mathcal{P}, K_0)\Sigma_E(\mathcal{P}, L) = (\Sigma_E |K_0|^2)^n(\Sigma_E |L|^2)^n = 1 + O((\#E)^{-\varepsilon/2}).$$

But if $h^{2,0} + h^{0,2} \geq 1$, we get

$$\Sigma_E(\mathcal{P}, K_0)\Sigma_E(\mathcal{P}, L)$$
$$= (|\Sigma_E K_0{}^2||\Sigma_E L^2|)^{2h^{2,0}} \times ((\Sigma_E |K_0|^2)(\Sigma_E |L|^2))^{h^{1,1}}$$
$$= O((\#E)^{-\varepsilon(2h^{2,0})/2}) = O((\#E)^{-\varepsilon}). \qquad \text{QED}$$

(1.21.24) At this point, we can conclude the proof of the Higher

Moment Theorem bis 1.21.6, and so also the proof of the Higher Moment Theorem 1.20.2. If $Gr^0(M(\dim\mathcal{F}_0/2))_{ncst}$ is not geometrically self dual, precisely those partitions $\mathcal{P}$ of $\{1, 2,..., 2n\}$ into n sets of pairs which are entirely of type (1, 1) contribute, each 1, to the leading term in the moment estimate. Such $\mathcal{P}$ are the same as the bijections π from the set $\{1, 2,..., n\}$ with the set $\{n+1, n+2,..., 2n\}$, the $\mathcal{P}$ corresponding to π having $\mathcal{P}_i = \{i, \pi(i)\}$. There are n! such bijections, so n! such $\mathcal{P}$. If $Gr^0(M(\dim\mathcal{F}_0/2))_{ncst}$ is geometrically self dual, then every partition of $\{1, 2,..., 2n\}$ into n sets of pairs contributes, each 1, to the leading term in the moment estimate. There are 2n!! such partitions, cf. [Weyl, Section 5 of Chapter V, pp. 147-149], [Ka-LAMM, 2.3.3]. QED for the Higher Moment Theorem.

(1.22) Variations on the Higher Moment Theorem

(1.22.1) In this section, we give some variant formulations of the theorem. In the first, we drop the hypothesis that L be a single sheaf, placed in suitable degree, but instead require L to be geometrically irreducible.

Higher Moment Theorem-first variant 1.22.2 Suppose we are given standard input

$$(m \geq 1, K, V, h, L, d \geq 2, (\mathcal{F}, \tau)).$$

Suppose in addition that $Gr^0(K)$ is not geometrically constant, and make the following three additional hypotheses:

1) The perverse sheaf L on V is geometrically irreducible, and its support is all of V.

2) There exists a closed subscheme $W \subset \mathbb{A}^m$, inclusion $i: W \to \mathbb{A}^m$, such that the perverse sheaf K on $\mathbb{A}^m$ is $i_*\mathcal{K}[\dim W]$ for some constructible $\overline{\mathbb{Q}}_\ell$-sheaf $\mathcal{K}$ on W.

3) We have the inequalities

$$\dim W \geq m-1,$$
$$\dim V + \dim W \geq m + 1.$$

Denote by M the perverse sheaf $M := \mathrm{Twist}(L,K,\mathcal{F},h)$ on $\mathcal{F}$. Denote by M_0 the central normalization of its trace function. For each integer $n \geq 1$ with $2n \leq d$, there exists a real $\varepsilon > 0$ such that we have

$$\Sigma_E |M_0(\dim\mathcal{F}_0/2)|^{2n} = (n!)((\#E)^{(1-n)\dim\mathcal{F}})(1 + O((\#E)^{-\varepsilon/2}))$$

if $Gr^0(M(\dim\mathcal{F}_0/2))_{ncst}$ is not geometrically self dual, and we have

$$\Sigma_E |M_0(\dim\mathcal{F}_0/2)|^{2n} = (2n!!)((\#E)^{(1-n)\dim\mathcal{F}})(1 + O((\#E)^{-\varepsilon/2})),$$

if $Gr^0(M(\dim\mathcal{F}_0/2))_{ncst}$ is geometrically self dual.

proof Exactly as in the proof of the Higher Moment Theorem 1.20.2, we have

$$\Sigma_E |M_0|^{2n}$$
$$= \Sigma_{\mathcal{P} \text{ with no singleton}} ((\#E)^{\dim\mathcal{F} - \lambda(\mathcal{P})m})\Sigma_E(\mathcal{P}, K_0)S_E(\mathcal{P}, L)$$
$$= \Sigma_{\mathcal{P} \text{ with no singleton}} ((\#E)^{\dim\mathcal{F} - \lambda(\mathcal{P})m})\Sigma_E(\mathcal{P}, K_0)\Sigma_E(\mathcal{P}, L)$$
$$+ \Sigma_{\mathcal{P} \text{ with no singleton}} \Sigma_{\mathcal{P}' < \mathcal{P}}$$
$$\mu(\mathcal{P}, \mathcal{P}')((\#E)^{\dim\mathcal{F} - \lambda(\mathcal{P})m})\Sigma_E(\mathcal{P}, K_0)\Sigma_E(\mathcal{P}', L).$$

Those terms
$$((\#E)^{\dim\mathcal{F} - \lambda(\mathcal{P})m})\Sigma_E(\mathcal{P}, K_0)\Sigma_E(\mathcal{P}, L),$$
with $\mathcal{P}$ a partition of $\{1, 2,..., 2n\}$ into n pairs, are analyzed as in the proof of the Higher Moment Theorem 1.20.2.

We will show that every other term is $O((\#E)^{\dim\mathcal{F} - nm - \varepsilon/2})$ for some $\varepsilon > 0$. For this, we need some lemmas.

Lemma 1.22.3 For L perverse, geometrically irreducible, and ι-pure of weight 0 on V, with support all of V, we have the following estimates. For any integer $r \geq 2$, and any approximate trace function $\tilde{L}$ for L, we have
$$\Sigma_E |\tilde{L}|^r = O((\#E)^{\dim(V)(1 - r/2)} + (\#E)^{-r/2}).$$

proof of Lemma 1.22.3 Any irreducible perverse sheaf is a middle extension. If its support is all of V, its only possibly nonvanishing cohomology sheaves $\mathcal{H}^{-i}(L)$ have i in the interval $[1, \dim(V)]$, and for these we have
$$\dim(\mathrm{Supp}(\mathcal{H}^{-\dim(V)}(L))) = \dim(V),$$
$$\dim(\mathrm{Supp}(\mathcal{H}^{-i}(L))) \leq i-1, \text{ for for } i < \dim(V).$$

So we have
$$|L(E, v)| = O((\#E)^{-\dim(V)/2}) \text{ on a set of dim} = \dim(V),$$
$$|L(E, v)| = O((\#E)^{(1-\dim(V))/2}) \text{ on a set of dim} = \dim(V)-2,$$
$$|L(E, v)| = O((\#E)^{(2-\dim(V))/2}) \text{ on a set of dim} = \dim(V)-3,$$
...
$$|L(E, v)| = O((\#E)^{-1/2}) \text{ on a set of dim} = 0.$$

Since $\tilde{L}$ is an approximate trace function for L, we have
$$\tilde{L}(E, v) = L(E, v) + O((\#E)^{-\varepsilon/2-\dim(V)/2}).$$

So we have the same estimates for $\tilde{L}$:
$$|\tilde{L}(E, v)| = O((\#E)^{-\dim(V)/2}) \text{ on a set of dim} = \dim(V),$$
$$|\tilde{L}(E, v)| = O((\#E)^{(1-\dim(V))/2}) \text{ on a set of dim} = \dim(V)-2,$$
$$|\tilde{L}(E, v)| = O((\#E)^{(2-\dim(V))/2}) \text{ on a set of dim} = \dim(V)-3,$$
...
$$|\tilde{L}(E, v)| = O((\#E)^{-1/2}) \text{ on a set of dim} = 0.$$

So we have
$$|\tilde{L}(E, v)|^r = O((\#E)^{-\dim(V)r/2}) \text{ on a set of dim} = \dim(V),$$
$$|\tilde{L}(E, v)|^r = O((\#E)^{(1-\dim(V))r/2}) \text{ on a set of dim} = \dim(V)-2,$$

$|\tilde{L}(E, v)|^r = O((\#E)^{(2-\dim(V))r/2})$ on a set of dim = dim(V)-3,

...

$|\tilde{L}(E, v)|^r = O((\#E)^{-r/2})$ on a set of dim = 0.

So adding up, we find that $\Sigma_E |L|^r$ is "big O" of the sum

$$(\#E)^{\dim(V)}(\#E)^{-\dim(V)r/2}$$
$$+ (\#E)^{\dim(V)-2}(\#E)^{(1-\dim(V))r/2}$$
$$+ (\#E)^{\dim(V)-3}(\#E)^{(2-\dim(V))r/2}$$
$$+...+ (\#E)^{-r/2}.$$

In this sum, after the first term, the terms are increasing: the exponents increase by r/2 - 1 at each step. So the sum is "big O" of the sum of its first and last terms. QED

Refined Lemma 1.22.4 Let L be perverse, geometrically irreducible, and ι-pure of weight 0 on V, with support all of V. For a partition $\mathcal{P}$ without singletons, define non-negative integers c_i as follows. Each $\mathcal{P}_i$ has a type (a_i, b_i) with $a_i + b_i \geq 2$: write

$$a_i + b_i = 2 + c_i.$$

Then define, for such a partition, subsets T and R of $\{1,..., \lambda\}$ by

$$T := \{i \text{ such that } c_i = 1\},$$
$$R := \{i \text{ such that } c_i \geq 2\}.$$

Then for $\tilde{L}$ any approximate trace function for L, we have the following estimates.

1) If dimV = 1, then

$$|\Sigma_E(\mathcal{P}, \tilde{L})| = O((\#E)^{\lambda-n}).$$

2) If dimV = 2, then we have

$$|\Sigma_E(\mathcal{P}, \tilde{L})| = O((\#E)^{\lambda-n - (1/2)\#T - \#R}).$$

3) If dimV $\geq$ 3, then we have

$$|\Sigma_E(\mathcal{P}, \tilde{L})| = O((\#E)^{\lambda-n - \#T - \#R}).$$

proof If dimV = 1, then L, being perverse irreducible with support all of V, must be of the form $\mathcal{L}[1]$ for a constructible $\overline{\mathbb{Q}}_\ell$-sheaf $\mathcal{L}$ on V, and we simply apply Lemma 1.21.16.

If dim V $\geq$ 2, then as before we write out the definition,

$$\Sigma_E(\mathcal{P}, \tilde{L}) = \prod_{i=1 \text{ to } \lambda} (\Sigma_E \tilde{L}^{a_i}\overline{\tilde{L}}^{b_i}).$$

The factors with $a_i + b_i = 2$ are O(1), so we can ignore them. We then apply Lemma 1.22.3 above to each factor with $c_i \geq 1$. We get

$$|\Sigma_E(\mathcal{P}, \tilde{L})|$$
$$= O(\prod_{i \text{ in } T\cup R} ((\#E)^{-c_i\dim(V)/2} + (\#E)^{-1-c_i/2})).$$

If dim(V) $\geq$ 3, then for each $c_i \geq 1$, we have

$$(\#E)^{-c_i\dim(V)/2} \leq (\#E)^{-1-c_i/2},$$

so we get

$$\Sigma_E(\mathcal{P}, \tilde{L}) = O(\prod_{i \text{ in } T\cup R} (\#E)^{-1-c_i/2}).$$

The exponent is

$$\Sigma_{i \text{ in } R\cup T} (-1 - c_i/2) = \Sigma_{i \text{ with } \#\mathcal{P}_i > 2} (-1 - c_i/2)$$
$$= \Sigma_i (-1 - c_i/2) + \#\{i \text{ with } \#\mathcal{P}_i = 2\}$$
$$= (-1/2)\Sigma_i (2 + c_i) + \lambda - \#T - \#R$$
$$= (-1/2)\Sigma_i (a_i + b_i) + \lambda - \#T - \#R$$
$$= -n + \lambda - \#T - \#R.$$

If $\dim(V) = 2$, there is a slight modification. For $c_i \geq 2$, we still have

$$(\#E)^{-c_i\dim(V)/2} \leq (\#E)^{-1-c_i/2},$$

but for $c_i = 1$, we have rather

$$(\#E)^{-1-c_i/2} = (\#E)^{-3/2} < (\#E)^{-1} = (\#E)^{-c_i\dim(V)/2}$$

So we get

$$\Sigma_E(\mathcal{P}, \tilde{L})$$
$$= O((\Pi_{i \text{ in } T} (\#E)^{-c_i\dim(V)/2}) \times (\Pi_{i \text{ in } R} (\#E)^{-1-c_i/2})).$$

Recall now that $\dim(V) = 2$, and that $c_i = 1$ for i in T. Using the calcuation just above, namely

$$\Sigma_{i \text{ in } R\cup T} (-1 - c_i/2) = -n + \lambda - \#T - \#R,$$

we see that the exponent is

$$- \Sigma_{i \text{ in } T} c_i\dim(V)/2 - \Sigma_{i \text{ in } R} (1 + c_i/2)$$
$$= - \Sigma_{i \text{ in } T} c_i - \Sigma_{i \text{ in } R} (1 + c_i/2)$$
$$= - \Sigma_{i \text{ in } T} 1 - \Sigma_{i \text{ in } R} (1 + c_i/2)$$
$$= \Sigma_{i \text{ in } T} 1/2 - \Sigma_{i \text{ in } T} (1 + c_i/2) - \Sigma_{i \text{ in } R} (1 + c_i/2)$$
$$= (1/2)\#T - \Sigma_{i \text{ in } R\cup T} (1 + c_i/2)$$
$$= (1/2)\#T - n + \lambda - \#T - \#R$$
$$= -n + \lambda - (1/2)\#T - \#R.$$

QED

Recall from Lemma 1.21.16 that for K we have

Lemma 1.22.5 For any partition $\mathcal{P}$ without singletons, with $\lambda = \lambda(\mathcal{P})$ the number of equivalence classes, we have the estimate

$$|\Sigma_E(\mathcal{P}, K_0)| = O((\#E)^{\dim W(\lambda - n)}).$$

(1.22.6) We now turn to the final stage of the proof of the theorem. If $\dim(V) = 1$, then as noted in the Refined Lemma 1.22.4 above, L is $\mathcal{L}[\dim V]$, and our theorem is a special case of the Higher Moment Theorem 1.20.2. So we may assume that $\dim(V) \geq 2$. In this case, for any $\mathcal{P}$ with all $\#\mathcal{P}_i \geq 2$, we have the uniform estimate

$$|\Sigma_E(\mathcal{P}, L)| = O((\#E)^{\lambda - n - (1/2)\#T - \#R}),$$

thanks to the Refined Lemma 1.22.4.

Lemma 1.22.7 Suppose $\dim(V) \geq 2$. For $\mathcal{P} > \mathcal{P}'$, and $\mathcal{P}$ containing no singleton, we have the estimate

$$|((\#E)^{\dim \mathcal{F} - \lambda(\mathcal{P})m}) \Sigma_E(\mathcal{P}, K_0)\Sigma_E(\mathcal{P}', L)|$$
$$= O((\#E)^{\dim \mathcal{F} - nm - 1/2}).$$

proof Multiply through by the inverse of $(\#E)^{\dim \mathcal{F} - \lambda(\mathcal{P})m}$. Put

$$\lambda := \lambda(\mathcal{P}),\ \lambda' := \lambda(\mathcal{P}').$$

We must show that

$$|\Sigma_E(\mathcal{P}, K_0)||\Sigma_E(\mathcal{P}', L)| = O((\#E)^{(\lambda - n)m - 1/2}).$$

In view of the bound

$$|\Sigma_E(\mathcal{P}, K_0)| = O((\#E)^{\dim W(\lambda - n)}),$$

it suffices to show that

$$|\Sigma_E(\mathcal{P}', L)| = O((\#E)^{(\lambda - n)(m - \dim W) - 1/2}).$$

Since $\mathcal{P}$ contains no singleton, each of its $\mathcal{P}_i$ has type (a_i, b_i) with $a_i + b_i \geq 2$. Now $\mathcal{P}'$ is obtained from $\mathcal{P}$ by a sequence of collapsing together various of the $\mathcal{P}_i$. So $\mathcal{P}'$ contains no singletons, and at least one of the sets $\mathcal{P}'_i$ will have type (a'_i, b'_i) with $a'_i + b'_i \geq 4$. Thus $\mathcal{P}'$ has $T' \cup R'$ nonempty. By the Refined Lemma 1.22.4, we have

$$|\Sigma_E(\mathcal{P}', L)| = O((\#E)^{\lambda - n - (1/2)\#T' - \#R'}).$$

So it suffices to show that

$$(\#E)^{\lambda - n - (1/2)\#T' - \#R'} \leq (\#E)^{(\lambda - n)(m - \dim W) - 1/2},$$

i.e., to show that

$$\lambda - n - (1/2)\#T' - \#R' \leq (\lambda - n)(m - \dim W) - 1/2,$$

i.e., to show that

$$(\lambda - n)(1 + \dim W - m) - (1/2)\#T' - \#R' \leq -1/2.$$

This is clear, since

$\lambda - n \leq 0$ (as $\mathcal{P}$ has all $\#\mathcal{P}_i \geq 2$),

$1 + \dim W - m \geq 0$, by hypothesis,

$T' \cup R'$ is nonempty. QED

Lemma 1.22.8 Suppose $\dim(V) \geq 2$. For $\mathcal{P}$ containing no singleton, and containing some $\mathcal{P}_i$ with $\#\mathcal{P}_i \geq 3$, we have

$$|((\#E)^{\dim \mathcal{F} - \lambda(\mathcal{P})m}) \Sigma_E(\mathcal{P}, K_0)\Sigma_E(\mathcal{P}, L)|$$
$$= O((\#E)^{\dim \mathcal{F} - nm - 1/2}).$$

proof Repeat the above argument, but with $\mathcal{P} = \mathcal{P}'$. We must show

$$|\Sigma_E(\mathcal{P}, L)| = O((\#E)^{(\lambda - n)(m - \dim W) - 1/2}).$$

By the Refined Lemma 1.22.4, we have.

$$|\Sigma_E(\mathcal{P}, L)| = O((\#E)^{\lambda - n - (1/2)\#T - \#R}).$$

So we reduce to showing that

$$\lambda - n - (1/2)\#T - \#R \leq (\lambda - n)(m - \dim W) - 1/2,$$

i.e.,

$$(\lambda - n)(1 + \dim W - m) - (1/2)\#T - \#R \leq -1/2.$$

which now holds because $T \cup R$ is nonempty. QED

These last two lemmas take care of all terms except those of the form $((\#E)^{\dim \mathcal{F}} - \lambda(\mathcal{P})m)\Sigma_E(\mathcal{P}, K_0)\Sigma_E(\mathcal{P}, L)$, for $\mathcal{P}$ a partition of $\{1, 2,..., 2n\}$ into n sets of pairs. As already remarked above, their analysis is exactly the same as that given at the end of the proof of the Higher Moment Theorem 1.20.2. QED

Corollary 1.22.9 Hypotheses as in the Higher Moment Theorem-first variant 1.22.2, suppose $d \geq 4$. Then we have all the conclusions of Corollary 1.20.3
proof Simply repeat the proof of Corollary 1.20.3. QED

(1.22.10) In the next variant, we drop the hypothesis that K be a single sheaf, placed in suitable degree, but instead require K to be geometrically irreducible and geometrically nonconstant.
Higher Moment Theorem-second variant 1.22.11 Suppose we are given standard input

$$(m \geq 1, K, V, h, L, d \geq 2, (\mathcal{F}, \tau)).$$

Make the following three additional hypotheses.

1) The perverse sheaf L is $\mathcal{L}[\dim V]$ for some constructible $\overline{\mathbb{Q}}_\ell$-sheaf $\mathcal{L}$ on V.

2) The perverse sheaf K on $\mathbb{A}^m$ is geometrically irreducible and geometrically nonconstant, with support denoted W.
3) We have the inequalities

$$\dim V \geq m-1,$$
$$\dim V + \dim W \geq m + 1.$$

Denote by M the perverse sheaf $M := \mathrm{Twist}(L,K,\mathcal{F},h)$ on $\mathcal{F}$. Denote by M_0 the central normalization of its trace function. For each integer $n \geq 1$ with $2n \leq d$, there exists a real $\varepsilon > 0$ such that we have

$$\Sigma_E |M_0(\dim \mathcal{F}_0/2)|^{2n} = (n!)(\#E)^{(1-n)\dim \mathcal{F}})(1 + O((\#E)^{-\varepsilon/2}))$$

if $\mathrm{Gr}^0(M(\dim \mathcal{F}_0/2))_{ncst}$ is not geometrically self dual,
and we have

$$\Sigma_E |M_0(\dim \mathcal{F}_0/2)|^{2n} = (2n!!)(\#E)^{(1-n)\dim \mathcal{F}})(1 + O((\#E)^{-\varepsilon/2})),$$

if $\mathrm{Gr}^0(M(\dim \mathcal{F}_0/2))_{ncst}$ is geometrically self dual.
proof The proof is very similar to that of the first variant 1.22.2, except that the relevant estimates for K_0 and L are interchanged.

If $\dim W \leq 1$, with inclusion $i : W \to \mathbb{A}^m$, then K is $i_*\mathcal{K}[\dim W]$ for a constructible $\overline{\mathbb{Q}}_\ell$-sheaf $\mathcal{K}$ on W which is ι-mixed of weight $\leq -\dim W$. So in this case, as we have assumed

$$\dim V + \dim W \geq m + 1,$$

we reduce to the original Higher Moment Theorem 1.20.2

Suppose now $\dim W \geq 2$. Then by the Refined Lemma 1.22.4, applied to K on W, we have

Lemma 1.22.12 For K perverse, geometrically irreducible, and

ι-pure of weight 0 on $\mathbb{A}^m$, with support W, we have the following estimates. For any integer $r \geq 2$, and any approximate trace function $\tilde{K}$ for K, we have

$$\Sigma_E |\tilde{K}|^r = O((\#E)^{\dim(W)(1 - r/2)} + (\#E)^{-r/2}).$$

proof Since $\tilde{K}$ is an approximate trace funtion for K on $\mathbb{A}^m$, $\tilde{K} \mid W$ is certainly an approximate trace funtion for K on W. So by Lemma 1.22.3, applied to $K \mid W$, we have

$$\Sigma_E |\tilde{K} \mid W|^r = O((\#E)^{\dim(W)(1 - r/2)} + (\#E)^{-r/2}).$$

On $\mathbb{A}^m(E) - W(E)$, $\tilde{K}(E, x)$ is $O((\#E)^{-\varepsilon/2 - m/2})$, so we get

$$\Sigma_E |\tilde{K} \mid \mathbb{A}^m - W|^r = O((\#E)^{m-r\varepsilon/2 - rm/2}) = O((\#E)^{m(1-r/2)}).$$

QED

Exactly as in the proof of the Refined Lemma 1.22.4, this implies

Refined Lemma bis 1.22.13 Suppose K is perverse, geometrically irreducible, and ι-pure of weight 0 on $\mathbb{A}^m$, with support W, with $\dim W \geq 2$. For a partition $\mathcal{P}$ without singletons, define non-negative integers c_i as follows. Each $\mathcal{P}_i$ has a type (a_i, b_i) with $a_i + b_i \geq 2$: write

$$a_i + b_i = 2 + c_i.$$

Then define, for such a partition, subsets T and R of $\{1,..., \lambda\}$ by

$$T := \{i \text{ such that } c_i = 1\}$$

$$R := \{i \text{ such that } c_i \geq 2\}.$$

Then for $\tilde{K}$ any approximate trace function for K, we have the following estimates.

1) If $\dim W = 2$, then we have

$$|\Sigma_E(\mathcal{P}, \tilde{K})| = O((\#E)^{\lambda-n - (1/2)\#T - \#R}).$$

2) If $\dim W \geq 3$, then we have

$$|\Sigma_E(\mathcal{P}, \tilde{K})| = O((\#E)^{\lambda-n - \#T - \#R}).$$

(1.22.14) We now turn to the proof of the second variant 1.22.11, in the remaining case $\dim W \geq 2$. Exactly as in the proof of the Higher Moment Theorem 1.20.2, we have

$$\Sigma_E |M_0|^{2n}$$
$$= \Sigma_{\mathcal{P} \text{ with no singleton}} ((\#E)^{\dim \mathcal{F} - \lambda(\mathcal{P})m})\Sigma_E(\mathcal{P}, K_0)S_E(\mathcal{P}, L)$$
$$= \Sigma_{\mathcal{P} \text{ with no singleton}} ((\#E)^{\dim \mathcal{F} - \lambda(\mathcal{P})m})\Sigma_E(\mathcal{P}, K_0)\Sigma_E(\mathcal{P}, L)$$
$$+ \Sigma_{\mathcal{P} \text{ with no singleton}} \Sigma_{\mathcal{P}' < \mathcal{P}}$$
$$\mu(\mathcal{P}, \mathcal{P}')((\#E)^{\dim \mathcal{F} - \lambda(\mathcal{P})m})\Sigma_E(\mathcal{P}, K_0)\Sigma_E(\mathcal{P}', L).$$

Those terms

$$((\#E)^{\dim \mathcal{F} - \lambda(\mathcal{P})m})\Sigma_E(\mathcal{P}, K_0)\Sigma_E(\mathcal{P}, L),$$

with $\mathcal{P}$ a partition of $\{1, 2,..., 2n\}$ into n pairs, are analyzed as in the proof of the Higher Moment Theorem 1.20.2. We will show that every other term is $O((\#E)^{\dim\mathcal{F} - nm - 1/2})$.

(1.22.15) Suppose $\mathcal{P}' \le \mathcal{P}$, and write $\lambda = \lambda(\mathcal{P})$, $\lambda' = \lambda'(\mathcal{P})$. Is

$$((\#E)^{\dim\mathcal{F} - \lambda m})\Sigma_E(\mathcal{P}, K_0)\Sigma_E(\mathcal{P}', L)$$
$$= O((\#E)^{\dim\mathcal{F} - nm - 1/2}),$$

i.e., is

$$\Sigma_E(\mathcal{P}, K_0)\Sigma_E(\mathcal{P}', L) = O((\#E)^{(\lambda - n)m - 1/2})?$$

Plug in the estimates

$$\Sigma_E(\mathcal{P}, K_0) = O((\#E)^{\lambda - n - (1/2)\#T - \#R}),$$
$$\Sigma_E(\mathcal{P}', L) = O((\#E)^{\dim V(\lambda' - n)}).$$

The answer is certainly yes if

$$((\#E)^{\lambda - n - (1/2)\#T - \#R})(\#E)^{\dim V(\lambda' - n)}$$
$$= O((\#E)^{(\lambda - n)m - 1/2}),$$

i.e., if (since every quantity is either an integer or a half integer)

$$\lambda - n - (1/2)\#T - \#R + \dim V(\lambda' - n) < (\lambda - n)m,$$

i.e., if

$$(\lambda - n)(1 + \dim V - m) + (\lambda' - \lambda)\dim V < (1/2)\#T + \#R.$$

The key point is that both $\lambda - n$ and $\lambda' - \lambda$ are ≤ 0, while by hypothesis we have

$$1 + \dim V - m \ge 0,$$
$$\dim V \ge m+1 - \dim(W) \ge 1.$$

So in all cases the left hand side is ≤ 0 and the right hand side is ≥ 0. If $\mathcal{P}' < \mathcal{P}$, then $\lambda - \lambda' \ge 1$, so the left hand side is $\le -\dim V \le -1$. If $\mathcal{P}' = \mathcal{P}$ but if $\mathcal{P}$, which has no singletons, is not a partition into n pairs, then $T \cup R$ is nonempty, so the right hand side is strictly positive. So in both of these cases, we have the desired strict inequality.

(1.22.16) This takes care of all terms except those of the form

$$((\#E)^{\dim\mathcal{F} - \lambda(\mathcal{P})m})\Sigma_E(\mathcal{P}, K_0)\Sigma_E(\mathcal{P}, L),$$

for $\mathcal{P}$ a partition of {1, 2,..., 2n} into n sets of pairs. Their analysis is exactly the same as that given at the end of the proof of the Higher Moment Theorem 1.20.2. QED

Corollary 1.22.17 Hypotheses as in the Higher Moment second variant 1.22.11, suppose $d \ge 4$. Then we have all the conclusions of Corollary 1.20.3

proof Simply repeat the proof of Corollary 1.20.3. QED

(1.22.18) In the next variant, we drop the hypothesis that L or K be a single sheaf, placed in suitable degree, but instead require L to be geometrically irreducible, and require K to be geometrically irreducible and geometrically nonconstant.

Higher Moment Theorem-third variant 1.22.19 Suppose we are given standard input

$$(m \ge 1, K, V, h, L, d \ge 2, (\mathcal{F}, \tau)).$$

Make the following three additional hypotheses.

1) The perverse sheaf L on V is geometrically irreducible, and its support is all of V.

2) The perverse sheaf K on $\mathbb{A}^m$ is geometrically irreducible and geometrically nonconstant, with support denoted W.
3) We have the inequalities

$$m \le 2,$$
$$\dim W \ge m-1,$$
$$\dim V + \dim W \ge m + 1.$$

Denote by M the perverse sheaf M := Twist(L,K,$\mathcal{F}$,h) on $\mathcal{F}$. Denote by M_0 the central normalization of its trace function. For each integer $n \ge 1$ with $2n \le d$, there exists a real $\varepsilon > 0$ such that we have

$$\Sigma_E |M_0(\dim\mathcal{F}_0/2)|^{2n} = (n!)((\#E)^{(1-n)\dim\mathcal{F}})(1 + O((\#E)^{-\varepsilon/2}))$$

if $\mathrm{Gr}^0(M(\dim\mathcal{F}_0/2))_{\mathrm{ncst}}$ is not geometrically self dual,
and we have

$$\Sigma_E |M_0(\dim\mathcal{F}_0/2)|^{2n} = (2n!!)((\#E)^{(1-n)\dim\mathcal{F}})(1 + O((\#E)^{-\varepsilon/2})),$$

if $\mathrm{Gr}^0(M(\dim\mathcal{F}_0/2))_{\mathrm{ncst}}$ is geometrically self dual.

proof If m = 1, then dimW ≤ 1, so K is $i_*\mathcal{K}[\dim W]$ for a constructible $\overline{\mathbb{Q}}_\ell$-sheaf $\mathcal{K}$ on W which is ι-mixed of weight ≤ -dimW. So this case is a special case of the first variant 1.22.2.

Suppose now that m=2. If dimW ≤ 1, then K is $i_*\mathcal{K}[\dim W]$ for a constructible $\overline{\mathbb{Q}}_\ell$-sheaf $\mathcal{K}$ on W which is ι-mixed of weight ≤ -dimW. Once again we have a special case of the first variant 1.22.2 We cannot have dimW = 0, because by hypothesis dimW ≥ m-1.

It remains to treat the case where m=2 and W = $\mathbb{A}^2$. From the inequality dimV + dimW ≥ m + 1, we find dimV ≥ 1. If dimV = 1, then L is $\mathcal{L}[\dim V]$ for a constructible $\overline{\mathbb{Q}}_\ell$-sheaf $\mathcal{L}$ on V which is ι-mixed of weight ≤ -dimV. The inequality dimV ≥ m-1 trivially holds, so we have a special case of the second variant 1.22.11.

It remains now to treat the case m = 2 = dimW, and dimV ≥ 2. In this case, for any $\mathcal{P}$ with no singletons, we have the inequalities

$$|\Sigma_E(\mathcal{P}, K_0)| = O((\#E)^{\lambda-n - (1/2)\#T - \#R}),$$
$$|\Sigma_E(\mathcal{P}, L)| = O((\#E)^{\lambda-n - (1/2)\#T - \#R}),$$

thanks to the Refined Lemma bis 1.22.13, applied to both K and L.

Exactly as in the proof of the Higher Moment Theorem 1.20.2, we have

$$\Sigma_E |M_0|^{2n}$$
$$= \Sigma_{\mathcal{P}\text{ with no singleton}} ((\#E)^{\dim\mathcal{F}} - \lambda(\mathcal{P})m)\Sigma_E(\mathcal{P}, K_0)S_E(\mathcal{P}, L)$$
$$= \Sigma_{\mathcal{P}\text{ with no singleton}} ((\#E)^{\dim\mathcal{F}} - \lambda(\mathcal{P})m)\Sigma_E(\mathcal{P}, K_0)\Sigma_E(\mathcal{P}, L)$$
$$+ \Sigma_{\mathcal{P}\text{ with no singleton}} \Sigma_{\mathcal{P}' < \mathcal{P}}$$
$$\mu(\mathcal{P}, \mathcal{P}')((\#E)^{\dim\mathcal{F}} - \lambda(\mathcal{P})m)\Sigma_E(\mathcal{P}, K_0)\Sigma_E(\mathcal{P}', L).$$

Those terms

$$((\#E)^{\dim \mathcal{F} - \lambda(\mathcal{P})m})\Sigma_E(\mathcal{P}, K_0)\Sigma_E(\mathcal{P}, L),$$

with $\mathcal{P}$ a partition of {1, 2,..., 2n} into n pairs, are analyzed as in the proof of the Higher Moment Theorem 1.20.2. We will show that every other term is $O((\#E)^{\dim \mathcal{F} - nm - 1/2})$.

Suppose $\mathcal{P}' \le \mathcal{P}$, and write $\lambda = \lambda(\mathcal{P})$, $\lambda' = \lambda'(\mathcal{P})$. If $\mathcal{P}' < \mathcal{P}$, then $\lambda' < \lambda$. If $\mathcal{P} = \mathcal{P}'$ but $\mathcal{P}$ is not a partition into n pairs, then $T \cup R$ is nonempty. We will show that in both of these cases, we have

$$((\#E)^{\dim \mathcal{F} - \lambda m})\Sigma_E(\mathcal{P}, K_0)\Sigma_E(\mathcal{P}', L)$$
$$= O((\#E)^{\dim \mathcal{F} - nm - 1/2}),$$

i.e., we have

$$\Sigma_E(\mathcal{P}, K_0)\Sigma_E(\mathcal{P}', L) = O((\#E)^{(\lambda - n)m - 1/2}).$$

Plug in the estimates

$$\Sigma_E(\mathcal{P}, K_0) = O((\#E)^{\lambda - n - (1/2)\#T - \#R}),$$
$$\Sigma_E(\mathcal{P}', L) = O((\#E)^{\lambda' - n - (1/2)\#T' - \#R'}).$$

It suffices to show that

$$\lambda - n - (1/2)\#T - \#R + \lambda' - n - (1/2)\#T' - \#R' < (\lambda - n)m,$$

i.e., that

$$(\lambda - n)2 + \lambda' - \lambda - (1/2)\#T - \#R - (1/2)\#T' - \#R' < (\lambda - n)2,$$

i.e., that

$$-(\lambda - \lambda') - (1/2)\#T - \#R - (1/2)\#T' - \#R' < 0,$$

i.e., that

$$(\lambda - \lambda') + (1/2)\#T + \#R + (1/2)\#T' + \#R' > 0.$$

This visibly holds if either $\lambda' < \lambda$ or if $T \cup R$ is nonempty.

This takes care of all terms except those of the form

$$((\#E)^{\dim \mathcal{F} - \lambda(\mathcal{P})m})\Sigma_E(\mathcal{P}, K_0)\Sigma_E(\mathcal{P}, L),$$

for $\mathcal{P}$ a partition of {1, 2,..., 2n} into n sets of pairs. Their analysis is, as noted above, exactly the same as that given at the end of the proof of the Higher Moment Theorem 1.20.2. QED

Corollary 1.22.20 Hypotheses as in the Higher Moment Theorem-third variant 1.22.19, suppose $d \ge 4$. Then we have all the conclusions of Corollary 1.20.3.

proof Simply repeat the proof of Corollary 1.20.3. QED

(1.22.21) Here is another version, where we allow m=3 in the previous variant, but where we only get the second and fourth moments.

Higher Moment Theorem-fourth variant 1.22.22 Suppose we are given standard input

$$(m \ge 1, K, V, h, L, d \ge 2, (\mathcal{F}, \tau)).$$

which satisfies the following three additional hypotheses.

1) The perverse sheaf L on V is geometrically irreducible, and its support is all of V.

2) The perverse sheaf K on $\mathbb{A}^m$ is geometrically irreducible and

geometrically nonconstant, with support denoted W.
3) We have the inequalities

$$d \geq 4,$$
$$m \leq 3,$$
$$\dim V + \dim W \geq m + 1,$$
$$\dim(W) \geq m - 1,$$
$$\dim(V) \geq m - 1.$$

Denote by M the perverse sheaf $M := \mathrm{Twist}(L, K, \mathcal{F}, h)$ on $\mathcal{F}$. Denote by M_0 the central normalization of its trace function. For each integer $n \geq 1$ with $2n \leq 4$, i.e., for $n = 1$ or 2, there exists a real $\varepsilon > 0$ such that we have

$$\Sigma_E |M_0(\dim\mathcal{F}_0/2)|^{2n} = (n!)((\#E)^{(1-n)\dim\mathcal{F}})(1 + O((\#E)^{-\varepsilon/2}))$$

if $\mathrm{Gr}^0(M(\dim\mathcal{F}_0/2))_{ncst}$ is not geometrically self dual,
and we have

$$\Sigma_E |M_0(\dim\mathcal{F}_0/2)|^{2n} = (2n!!)((\#E)^{(1-n)\dim\mathcal{F}})(1 + O((\#E)^{-\varepsilon/2})),$$

if $\mathrm{Gr}^0(M(\dim\mathcal{F}_0/2))_{ncst}$ is geometrically self dual.

proof If $m \leq 2$, this is a special case of the third variant 1.22.19. Suppose now that $m = 3$. Then $\dim V \geq 2$ and $\dim W \geq 2$. In this case, for any $\mathcal{P}$ with no singletons, we have the inequalities

$$|\Sigma_E(\mathcal{P}, K_0)| = O((\#E)^{\lambda - n - (1/2)\#T - \#R}),$$
$$|\Sigma_E(\mathcal{P}, L)| = O((\#E)^{\lambda - n - (1/2)\#T - \#R}),$$

thanks to the Refined Lemma bis 1.22.13, applied to both K and L.

Exactly as in the proof of the Higher Moment Theorem 1.20.2 we have

$$\Sigma_E |M_0|^{2n}$$
$$= \Sigma_{\mathcal{P}\text{ with no singleton}} ((\#E)^{\dim\mathcal{F} - \lambda(\mathcal{P})m})\Sigma_E(\mathcal{P}, K_0)S_E(\mathcal{P}, L)$$
$$= \Sigma_{\mathcal{P}\text{ with no singleton}} ((\#E)^{\dim\mathcal{F} - \lambda(\mathcal{P})m})\Sigma_E(\mathcal{P}, K_0)\Sigma_E(\mathcal{P}, L)$$
$$+ \Sigma_{\mathcal{P}\text{ with no singleton}} \Sigma_{\mathcal{P}' < \mathcal{P}}$$
$$\mu(\mathcal{P}, \mathcal{P}')((\#E)^{\dim\mathcal{F} - \lambda(\mathcal{P})m})\Sigma_E(\mathcal{P}, K_0)\Sigma_E(\mathcal{P}', L).$$

Those terms

$$((\#E)^{\dim\mathcal{F} - \lambda(\mathcal{P})m})\Sigma_E(\mathcal{P}, K_0)\Sigma_E(\mathcal{P}, L),$$

with $\mathcal{P}$ a partition of $\{1, 2, \ldots, 2n\}$ into n pairs, are analyzed as in the proof of the Higher Moment Theorem 1.20.2. We will show that every other term is $O((\#E)^{\dim\mathcal{F} - nm - 1/2})$. If $n=1$, there are no other terms: there is only one partition $\mathcal{P}$ of $\{1, 2\}$ with no singletons. For $n=2$, we argue as follows.

Suppose $\mathcal{P}' \leq \mathcal{P}$, and write $\lambda = \lambda(\mathcal{P})$, $\lambda' = \lambda'(\mathcal{P})$. If $\mathcal{P}' < \mathcal{P}$, then $\lambda' < \lambda$. If $\mathcal{P} = \mathcal{P}'$ but $\mathcal{P}$ is not a partition into n pairs, then $T \cup R$ is nonempty. We will show that in both of these cases, we have

$$((\#E)^{\dim\mathcal{F} - \lambda m})\Sigma_E(\mathcal{P}, K_0)\Sigma_E(\mathcal{P}', L)$$
$$= O((\#E)^{\dim\mathcal{F} - nm - 1/2}).$$

Exactly as in the proof of the third variant 1.22.19, it suffices to show that

$$\lambda\text{-}n - (1/2)\#T - \#R + \lambda'\text{-}n - (1/2)\#T' - \#R' < (\lambda - n)m,$$

i.e., that

$$(\lambda\text{-}n)2 + \lambda'\text{-}\lambda - (1/2)\#T - \#R - (1/2)\#T' - \#R' < (\lambda - n)3,$$

i.e., that

$$n\text{-}\lambda < (\lambda\text{-}\lambda') + (1/2)\#T + \#R + (1/2)\#T' + \#R'.$$

As $n=2$, then either $\lambda = 2$ and the inequality holds if either $\lambda > \lambda'$ or if $T \cup R$ is nonempty, or $\lambda = 1$. In this $\lambda = 1$ case, $\mathcal{P}$ is the one set partition of $\{1, 2, 3,4\}$, which has type $(a, b) = (2, 2)$ and c $(:= a+b-2) = 2$. Thus T is empty, and R has one element. Furthermore, $\mathcal{P}' = \mathcal{P}$ in this case, so T' is empty and R' has one element. So the inequality we need reduces to

$$1 < 0 + 0 + 1 + 0 + 1,$$

which indeed holds. QED

Corollary 1.22.23 Hypotheses as in the Higher Moment Theorem-fourth variant 1.22.22, we have all the conclusions of the $d=4$ case of Corollary 1.20.3.
proof Simply repeat the proof of Corollary 1.20.3. QED

(1.23) Counterexamples

(1.23.1) In this section, we give examples to show that the dimension hypothesis,

$$\dim(V) + \dim(W) \geq m+1,$$

in the higher moment theorem 1.20.2 and in its variants 1.22.2, 1.22.11, 1.22.19, and 1.22.22, is essential. We fix a nontrivial $\overline{\mathbb{Q}}_\ell$-valued additive character ψ of k, and form the Artin-Schreier sheaf $\mathcal{L}_\psi$ on $\mathbb{A}^1$. The object $\mathcal{L}_\psi[1](1/2)$ on $\mathbb{A}^1$ is perverse, geometrically irreducible, ι-pure of weight zero, and $H_c{}^*(\mathbb{A}^1 \otimes_k \overline{k}, \mathcal{L}_\psi[1](1/2)) = 0$. For any finite extension E/k, we denote by $\psi_E : E \to \overline{\mathbb{Q}}_\ell{}^\times$ the nontrivial additive character $\psi_E := \psi \circ \mathrm{Trace}_{E/k}$. On any $\mathbb{A}^m$, the object

$$\mathcal{L}_{\psi(x_1 + x_2 + \ldots + x_m)}[m](m/2)$$

is perverse, geometrically irreducible, ι-pure of weight zero, and $H_c{}^*(\mathbb{A}^m \otimes_k \overline{k}, \mathcal{L}_{\psi(x_1 + x_2 + \ldots + x_m)}[m](m/2)) = 0$. We denote by δ_0 the perverse sheaf on $\mathbb{A}^m$ which is the constant sheaf at the origin, extended by zero.

First example 1.23.2 For the first example we take as data

$m \geq 1$ arbitrary,

$K = \delta_0$ on $\mathbb{A}^m$,

$V = \mathbb{A}^m$,

$h : V = \mathbb{A}^m \to \mathbb{A}^m$ the constant map zero.

$L = \overline{\mathbb{Q}}_\ell[m](m/2)$ on V,

$d \geq 4$,

$(\mathcal{F}, \tau) :=$ (all maps $f = (f_1,\ldots, f_m)$ with each f_i an arbitrary polynomial in m variables of degree $\leq d-1$, the inclusion).

In this example, K and L are both perverse, geometrically irreducible, and ι-pure of weight zero. What about cohomology of K and L? The group $H_c{}^*(\mathbb{A}^m\otimes_k\bar{k}, K) = H_c{}^*(\mathbb{A}^m\otimes_k\bar{k}, \delta_0)$ is $\bar{\mathbb{Q}}_\ell$, placed in degree 0, and

$$H^i{}_c((V\times\mathbb{A}^m)\otimes\bar{k}, pr_1{}^*L\otimes pr_2{}^*K) = H^i{}_c(V\otimes_k\bar{k}, L)$$

$$= H^i{}_c(\mathbb{A}^m\otimes_k\bar{k}, \bar{\mathbb{Q}}_\ell[m](m/2)) = \bar{\mathbb{Q}}_\ell(-m/2), \text{ for } i = m,$$

$$= 0, \text{ for } i \neq m.$$

So we have standard input, but we violate the dimension inequality

$$\dim(V) + \dim(W) \geq m+1.$$

Let us see what the object $M := \mathrm{Twist}(L,K,\mathcal{F},h)$ looks like. Consider the dense open set U of $\mathcal{F}$ consisting of those $f = (f_1,\ldots, f_m)$ such that the equations $f_1 = f_2 = \ldots = f_m = 0$ define a closed subscheme Z_f of $\mathbb{A}^m$ which is finite etale over the base of degree $(d-1)^m$. In $\mathbb{A}^m\times U$, define Z to be the closed subscheme of those points (v, f) where f(v) = 0, inclusion denoted $i : Z \to \mathbb{A}^m\times U$. Thus we have a diagram

$$\begin{array}{ccccc} & i & & pr_1 & \\ Z & \to & \mathbb{A}^m\times U & \to & \mathbb{A}^m \\ & \pi\searrow & \downarrow pr_2 & & \\ & & U. & & \end{array}$$

Then $\pi_*\bar{\mathbb{Q}}_\ell$ is lisse on U of rank $(d-1)^m$, and M|U is

$$(\pi_*\bar{\mathbb{Q}}_\ell)[\dim\mathcal{F}](m/2).$$

Thus $M(\dim\mathcal{F}_0/2)|U$ is ι-pure of weight zero. The nonconstant part $(\pi_*\bar{\mathbb{Q}}_\ell)_{ncst}$ can be described as either the quotient of $\pi_*\bar{\mathbb{Q}}_\ell$ by the constant sheaf it contains by adjunction, or as the kernel of the trace map for the finite etale morphism π,

$$\mathrm{Trace}_\pi : \pi_*\bar{\mathbb{Q}}_\ell \to \bar{\mathbb{Q}}_\ell.$$

By the Standard Input Theorem 1.15.6, $(\pi_*\bar{\mathbb{Q}}_\ell)_{ncst}$ is geometrically irreducible, and orthogonally self dual, of rank $(d-1)^m - 1$. Because $d \geq 4$, the family $(\mathcal{F}, \tau)$ is 4-separating. So if the conclusions of the Higher Moment Theorem 1.20.2 and its Corollary 1.20.3 held here, we would find that, for $(d-1)^m - 1 \geq 4$, the fourth moment of $(\pi_*\bar{\mathbb{Q}}_\ell)_{ncst}$ is 3. But this is false, because $\pi_1(U\otimes_k\bar{k})$ acts on $(\pi_*\bar{\mathbb{Q}}_\ell)_{ncst}$ through what is a priori a subgroup (though in fact it is well known to be the entire group) of the symmetric group S_N, $N = (d-1)^m - 1$, acting in the (N-1)-dimensional irreducible subrepresentation Aug_{N-1} of its tautological N-dimensional representation. One knows [Ka-LAMM, 2.4.3] that

$$M_4(S_N, \mathrm{Aug}_{N-1}) = 4, \text{ for all } N \geq 4.$$

Since a smaller group has larger moments, the fourth moment of $(\pi_* \overline{\mathbb{Q}}_\ell)_{\mathrm{ncst}}$ is at least 4. Thus the dimension hypothesis cannot be dropped in the Higher Moment Theorem 1.20.2.

Second example 1.23.3 For the second example, we take as input the data

$$(m \geq 1, K, V, h, L, d \geq 2, (\mathcal{F}, \tau))$$

as follows:

$m \geq 1$ arbitrary,

$K = \delta_0$ on $\mathbb{A}^m$,

$V = \mathbb{A}^m$,

$h : V = \mathbb{A}^m \to \mathbb{A}^m$ the constant map zero,

$L = \mathcal{L}_{\psi(x_1 + x_2 + \ldots + x_m)}[m](m/2)$ on $V = \mathbb{A}^m$,

$d \geq 5$,

$(\mathcal{F}, \tau) :=$ (all maps $f = (f_1, \ldots, f_m)$ with each f_i an arbitrary polynomial in m variables of degree $\leq d-1$, the inclusion).

This is standard input, because $H_c^*(\mathbb{A}^m \otimes_k \overline{k}, \delta_0)$ is $\overline{\mathbb{Q}}_\ell$, placed in degree 0, and $H_c^*(\mathbb{A}^m \otimes_k \overline{k}, L) = 0$. From this vanishing, we get

$$H_c^*((V \times \mathbb{A}^m) \otimes \overline{k}, \mathrm{pr}_1^* L \otimes \mathrm{pr}_2^* K) = 0,$$

by Kunneth. So we have $M(\dim \mathcal{F}_0/2) = M(\dim \mathcal{F}_0/2)_{\mathrm{ncst}0}$, and hence $\mathrm{Gr}^0(M(\dim \mathcal{F}_0/2)) = \mathrm{Gr}^0(M(\dim \mathcal{F}_0/2))_{\mathrm{ncst}}$.

Once again, all the hypotheses of the Higher Moment Theorem 1.20.2 are satisfied **except** for the dimension inequality

$$\dim(V) + \dim(W) \geq m+1.$$

For $M := \mathrm{Twist}(L,K,\mathcal{F},h)$, its trace function is

$$M(E, f) = ((-1)^{\dim \mathcal{F}_0}) \Sigma_{v \text{ in } V(E)} L(E, v) K(E, h(v) + f(v))$$
$$= ((-1)^{\dim \mathcal{F}_0}) \Sigma_{v \text{ in } V(E) = \mathbb{A}^m(E)} (-1)^m \psi_E(\Sigma_i v_i)(\#E)^{-m/2} \delta_0(f(v))$$
$$= (-1)^{\dim \mathcal{F}} (\#E)^{-m/2} \Sigma_{v \text{ in } \mathbb{A}^m(E) \text{ with } f(v) = 0} \psi_E(\Sigma_i v_i).$$

The perverse sheaf M is lisse on the dense open set $U \subset \mathcal{F}$ given in the previous example. On U, $M(\dim \mathcal{F}_0/2)|U$ is $\mathcal{M}(\dim \mathcal{F}/2)[\dim \mathcal{F}]$, for $\mathcal{M}$ the lisse sheaf on U, of rank $(d-1)^m$ and ι-pure of weight zero, whose trace function is

$$\mathcal{M}(E, f) = \Sigma_{v \text{ in } \mathbb{A}^m(E) \text{ with } f(v) = 0} \psi_E(\Sigma_i v_i).$$

So $\mathcal{M} = \mathcal{M}_{\mathrm{ncst}}$ is geometrically irreducible, by the Standard Input Theorem 1.15.6.

The key point is that the lisse sheaf $\mathcal{M}$ is monomial, i.e., induced as a representation of $\pi_1(U \otimes_k \overline{k})$ from a lisse sheaf of rank one on a finite etale covering of U of degree $(d-1)^m$. Indeed, in $\mathbb{A}^m \times U$, define Z to be the closed subscheme of those points (v, f) where $f(v) = 0$, inclusion denoted $i : Z \to \mathbb{A}^m \times U$. Thus we have a diagram

$$\begin{array}{ccccc} & i & & pr_1 & \\ Z & \to & \mathbb{A}^m \times U & \to & \mathbb{A}^m \\ & \pi\searrow & \downarrow pr_2 & & \\ & & U. & & \end{array}$$

On Z, we have the lisse rank one sheaf

$$\mathcal{N} := i^*pr_1{}^*\mathcal{L}_{\psi(x_1 + x_2 + \ldots + x_m)}.$$

The map π makes Z a finite etale covering of degree $(d-1)^m$, and $\mathcal{M}$ is just $\pi_*\mathcal{N}$. Now $\mathcal{N}$, being of finite order char(k) on Z, certainly has finite geometric monodromy. Therefore $\mathcal{M}$ on U has finite geometric monodromy, and its monodromy is induced, i.e., not primitive. This is a contradiction, as follows.

Suppose first that char(k) is odd. Then K is geometrically self dual, but L is not. So $\mathcal{M}$ is not geometrically self dual. If the Higher Moment Theorem 1.20.2 held, then $\mathcal{M}$, which has rank

$$(d-1)^m \geq 4^m \geq 2,$$

would have $M_4{}^{geom}(U, \mathcal{M}) = 2$. But for a finite-dimensional representation

$$\rho : G \to GL(V)$$

in characteristic zero with finite image, $M_4(G, V) = 2$ forces the representation to be primitive, cf. [Ka-LAMM, 1.3.2].

Suppose next that char(k) is 2. Then both K and L are geometrically self dual, so $\mathcal{M}$ is geometrically self dual, in fact orthogonally self dual. If the Higher Moment Theorem 1.20.2 held, then $\mathcal{M}$, whose rank is $(d-1)^m \geq 4^m \geq 4$, would have $M_4{}^{geom}(U, \mathcal{M}) = 3$. But for a finite-dimensional representation

$$\rho : G \to O(V)$$

in characteristic zero with finite image and $\dim(V) \geq 3$, $M_4(G, V) = 3$ forces the representation to be primitive, cf. [Ka-LAMM, 1.3.2].

Third example 1.23.4 In this example, we take as input

$m \geq 2$ arbitrary,

$K = \mathcal{L}_{\psi(x_1)}[1](1/2)\otimes\delta_0(x_2)\otimes\delta_0(x_3)\otimes\ldots\otimes\delta_0(x_m)$ on $\mathbb{A}^m$,

$V = \mathbb{A}^{m-1}$,

$h : V = \mathbb{A}^{m-1} \to \mathbb{A}^m$ the constant map zero,

$L = \overline{\mathbb{Q}}_\ell[m-1]((m-1)/2)$ on $V = \mathbb{A}^{m-1}$,

$d \geq 5$,

$(\mathcal{F}, \tau)$:= (all maps $f = (f_1,\ldots, f_m)$ with each f_i an arbitrary polynomial in m-1 variables of degree $\leq d-1$, the inclusion).

In this example, we have $H_c{}^*(\mathbb{A}^m\otimes_k\overline{k}, K) = 0$, so we have standard input, and we have

$$H_c{}^*((V\times\mathbb{A}^m)\otimes\bar{k}, pr_1{}^*L\otimes pr_2{}^*K) = 0.$$

So $Gr^0(M(\dim\mathcal{F}_0/2)) = Gr^0(M(\dim\mathcal{F}_0/2))_{ncst}$. Once again, all the hypotheses of the Higher Moment Theorem 1.20.2 are satisfied, **except** for the dimension inequality

$$\dim(V) + \dim(W) \geq m+1.$$

And if we take $m=2$, then all the hypotheses of all the variants of the Higher Moment Theorem are satisfied, except for the same dimension inequality.

Consider the dense open set $U \subset \mathcal{F}$ consisting of those $f = (f_1,..., f_m)$ such that the equations $f_2 = ... = f_m = 0$ define a closed subscheme of $\mathbb{A}^{m-1}$ which is finite etale of degree $(d-1)^{m-1}$. Then $M(\dim\mathcal{F}_0/2)|U$ is $\mathcal{M}(\dim\mathcal{F}/2)[\dim\mathcal{F}]$, for $\mathcal{M}$ the lisse sheaf on U, of rank $(d-1)^{m-1}$ and ι-pure of weight zero, whose trace function is

$$\mathcal{M}(E, f) = \Sigma_{v \text{ in } \mathbb{A}^{m-1}(E) \text{ with } f_2(v) = f_3(v) = ... = f_m(v) = 0}\ \psi_E(f_1(v)).$$

Just as in the previous example, the lisse sheaf $\mathcal{M}$ is monomial, and we arrive at the same contradiction concerning the fourth moment.

Chapter 2: How to apply the results of Chapter 1

(2.1) How to apply the Higher Moment Theorem

(2.1.1) When we apply the Higher Moment Theorem 1.20.2, or one of its variants, in a situation with $d \geq 4$, its output is a perverse sheaf M on the affine space $\mathcal{F}$ about which, by Corollary 1.20.3 or one of its variants, we have the following information:
(2.1.1.1) $M(\dim\mathcal{F}_0/2)$ is ι-mixed of weight ≤ 0,
(2.1.1.2) The support of $\mathrm{Gr}^0(M(\dim\mathcal{F}_0/2))_{ncst}$ is all of $\mathcal{F}$,
(2.1.1.3) For any dense open set $U \subset \mathcal{F}$ on which M is lisse, $M(\dim\mathcal{F}_0/2)|U$ is of the form $\mathcal{M}(\dim\mathcal{F}/2)[\dim\mathcal{F}]$, for a lisse $\overline{\mathbb{Q}}_\ell$-sheaf $\mathcal{M}$ on U which is ι-mixed of weight ≤ 0.
(2.1.1.4) The nonconstant part $\mathrm{Gr}^0(\mathcal{M})_{ncst}$ of the highest weight quotient $\mathrm{Gr}^0(\mathcal{M})$ of $\mathcal{M}$ as lisse sheaf on U is geometrically irreducible.
(2.1.1.5) We know a necessary and sufficient condition for the equality

$$\mathrm{Gr}^0(\mathcal{M}) = \mathrm{Gr}^0(\mathcal{M})_{ncst}$$

of lisse sheaves on U, namely that

$$H_c^m((V\times\mathbb{A}^m)\otimes\bar{k}, \mathrm{pr}_1{}^*L\otimes\mathrm{pr}_2{}^*K)$$

is ι-mixed of weight $\leq m - \varepsilon$, for some $\varepsilon > 0$. More precisely, we know that $\mathrm{Gr}^0(\mathcal{M})_{cst}$ is the constant sheaf

$$H_c^m((V\times\mathbb{A}^m)\otimes\bar{k}, \mathrm{pr}_1{}^*L\otimes\mathrm{pr}_2{}^*K)_{wt=m}(m/2).$$

(2.1.1.6) We know the Frobenius-Schur indicator $\mathrm{FSI}^{geom}(U, \mathrm{Gr}^0(\mathcal{M})_{ncst})$.
(2.1.1.7) We know that if both $\mathrm{Gr}^0(K)$ and $\mathrm{Gr}^0(L)$ are arithmetically self dual as perverse sheaves on $\mathbb{A}^m$ and V respectively, then $\mathrm{Gr}^0(\mathcal{M})_{ncst})$ is arithmetically self dual as a lisse sheaf on U.
(2.1.1.8) We know that if $\mathrm{Gr}^0(\mathcal{M})_{ncst}$ is not geometrically self dual, then for any $n \geq 1$ with $2n \leq d$, we have the inequality

$$M_{2n}{}^{geom}(U, \mathrm{Gr}^0(\mathcal{M})_{ncst}) \leq n!,$$

with equality if $\mathrm{rank}(\mathrm{Gr}^0(\mathcal{M})_{ncst}) \geq n$. In particular, if $\mathrm{Gr}^0(\mathcal{M})_{ncst}$ is not geometrically self dual, and if $\mathrm{rank}(\mathrm{Gr}^0(\mathcal{M})_{ncst}) \geq 2$, then

$$M_4{}^{geom}(U, \mathrm{Gr}^0(\mathcal{M})_{ncst}) = 2.$$

(2.1.1.9) We know that if $\mathrm{Gr}^0(\mathcal{M})_{ncst}$ is geometrically self dual, then for any $n \geq 1$ with $2n \leq d$, we have the inequality

$$M_{2n}{}^{geom}(U, Gr^0(\mathcal{M})_{ncst}) \le 2n!!,$$

with equality if $\mathrm{rank}(Gr^0(\mathcal{M})_{ncst}) \ge 2n$. In particular, if $Gr^0(\mathcal{M})_{ncst}$ is geometrically self dual, and if $\mathrm{rank}(Gr^0(\mathcal{M})_{ncst}) \ge 4$, then

$$M_4{}^{geom}(U, Gr^0(\mathcal{M})_{ncst}) = 3.$$

[This equality $M_4{}^{geom}(U, Gr^0(\mathcal{M})_{ncst}) = 3$ also holds if $Gr^0(\mathcal{M})_{ncst}$ is geometrically self dual with an orthogonal autoduality, and $\mathrm{rank}(Gr^0(\mathcal{M})_{ncst}) = 2$ or 3: one checks by hand that

$$M_4(O(2), std_2) = M_4(O(3), std_3) = 3.]$$

Remark 2.1.1.9.1 Indeed, according to [Rains, 3.4], we have $M_{2n}(O(r), std_r) = 2n!!$ if $r \ge n$. This result of Rains clarifies a question raised by Weyl, who notes that this equality holds for $r \ge 2n$ but fails for $r < n$, cf. [Weyl, page 149, lines 8-13].

(2.1.2) What conclusions can we draw about the geometric monodromy of $Gr^0(\mathcal{M})_{ncst}$? Let us denote by G_{geom} the Zariski closure of the image of $\pi_1(U\otimes_k \bar{k})$ in the $\overline{\mathbb{Q}}_\ell$-representation V given by $Gr^0(\mathcal{M})_{ncst}$. Because $Gr^0(\mathcal{M})_{ncst}$ is ι-pure of weight zero, we know [De-Weil II, 1.3.9 and 3.4.1 (iii)] that G_{geom} is a semisimple (by which we mean that its identity component $G_{geom}{}^0$ is semisimple), not necessarily connected, subgroup of GL(V).

For $Gr^0(\mathcal{M})_{ncst}$ of very low rank, there are very few possibilities. For $Gr^0(\mathcal{M})_{ncst}$ of rank one, G_{geom} must be finite, as GL(1) has no nontrivial connected semisimple subgroups. For $Gr^0(\mathcal{M})_{ncst}$ of rank two, G_{geom} must either be finite or contain SL(2), because SL(2) is the unique nontrivial connected semisimple subgroup of GL(2). For $Gr^0(\mathcal{M})_{ncst}$ of rank three and, geometrically, orthogonally self dual, G_{geom} must be either finite or SO(3) or O(3), because these are the only semisimple subgroups of O(3).

(2.2) Larsen's Alternative

(2.2.1) To go further, we use the following remarkable result of Larsen, which we apply, after embedding $\overline{\mathbb{Q}}_\ell$ into $\mathbb{C}$ via ι, to the group $G = G_{geom}$ acting on V.

Theorem 2.2.2 (Larsen's Alternative, cf. [Lar-Char], [Lar-Normal], [Ka-LFM, page 113], [Ka-LAMM, 1.1.6]) Let V be a $\mathbb{C}$-vector space of dimension $N \ge 2$, $G \subset GL(V)$ a (not necessarily connected) Zariski closed reductive subgroup of GL(V).

0) If $M_4(G, V) \le 5$, then V is G-irreducible.

1) If $M_4(G, V) = 2$, then either $G \supset SL(V)$, or $G/(G\cap\text{scalars})$ is finite. If

in addition G∩scalars is finite (e.g., if G is semisimple), then either $G^0 = SL(V)$, or G is finite.
2) Suppose ⟨,⟩ is a nondegenerate symmetric bilinear form on V, and suppose G lies in the orthogonal group O(V) := Aut(V, ⟨,⟩). If $M_4(G, V) = 3$, then either G = O(V), or G = SO(V), or G is finite. If dim(V) is 2 or 4, then G is not contained in SO(V).
3) Suppose ⟨,⟩ is a nondegenerate alternating bilinear form on V (such a form exists only if dim(V) is even), suppose G lies in the symplectic group Sp(V) := Aut(V, ⟨,⟩), and suppose dim(V) > 2. If $M_4(G, V) = 3$, then either G = Sp(V), or G is finite.

(2.2.3) So when we apply the Higher Moment Theorem 1.20.2, or one of its variants, in a situation with $d \geq 4$, we have the following possibilities for the group G_{geom} for the $\overline{\mathbb{Q}}_\ell$-representation V of $\pi_1(U\otimes_k\bar{k})$ given by $Gr^0(\mathcal{M})_{ncst}$.

(2.2.3.1) If $Gr^0(\mathcal{M})_{ncst}$ is not geometrically self dual, and of rank at least 2, then either G_{geom} is finite, or G_{geom} contains SL(V). [For dim(V) = 1, G_{geom} is finite, being a semisimple subgroup of GL(1).]

(2.2.3.2) If $Gr^0(\mathcal{M})_{ncst}$ is, geometrically, symplectically self dual of rank at least 4, then either G_{geom} is finite, or G_{geom} = Sp(V). [If dim(V) = 2, it is also true that G_{geom} is either finite or Sp(V), just by the paucity of choice of semisimple subgroups of Sp(V) = SL(2).]

(2.2.3.3) If $Gr^0(\mathcal{M})_{ncst}$ is, geometrically, orthogonally self dual of rank at least 3, then either G_{geom} is finite, or G_{geom} = SO(V), or G_{geom} = O(V). [If $\dim(V) \leq 2$, then G_{geom} is finite, because O(1) is {±1}, and because in O(2) the only semisimple subgroups are finite.]

(2.2.4.) The upshot of this is that whenever in 2.2.3 above we can also show that G_{geom} is not finite, then we have very nearly determined G_{geom}:

(2.2.4.1) If $Gr^0(\mathcal{M})_{ncst}$ is not geometrically self dual, and if G_{geom} is not finite, then G_{geom} contain SL(V). We have "only" to compute the determinant of our representation to know G_{geom} exactly.

(2.2.4.2) If $Gr^0(\mathcal{M})_{ncst}$ is, geometrically, symplectically self dual, and if G_{geom} is not finite (for instance because it contains a unipotent element $A \neq 1$), then G_{geom} is Sp(V), cf. [Ka-LAMM, 1.4.7].

(2.2.4.3) If $Gr^0(\mathcal{M})_{ncst}$ is, geometrically, orthogonally self dual, and if G_{geom} is not finite, then G_{geom} is SO(V) or O(V), and we have "only" to compute the determinant of our representation to know G_{geom} exactly.

(2.3) Larsen's Eighth Moment Conjecture

(2.3.1) How can we analyze the question of whether G_{geom} is finite? There is a remarkable (unpublished!) conjecture of Larsen, according to which we can detect a finite G_{geom} just by knowing its second, fourth, sixth, and eighth absolute moments.

Eighth Moment Conjecture 2.3.2 (Larsen) Let V be a $\mathbb{C}$-vector space of dimension $N \geq 8$, G one of the groups GL(V), O(V), or, if N is even, Sp(V). Let Γ be a finite subgroup of G. Then we have a strict inequality

$$\Sigma_{k=1 \text{ to } 4} M_{2k}(\Gamma, V) > \Sigma_{k=1 \text{ to } 4} M_{2k}(G, V).$$

(2.3.2.1) Since we have the a priori inequality

$$M_{2k}(\Gamma, V) \geq M_{2k}(G, V)$$

for every k, an equivalent formulation is this.

Eighth Moment Conjecture 2.3.2.2 (= 2.3.2 bis) (Larsen) Let V be a $\mathbb{C}$-vector space of dimension $N \geq 8$. If Γ is a finite subgroup of G, for G one of the groups GL(V), O(V), or, if N is even, Sp(V), then for some k in {1, 2, 3, 4}, we have $M_{2k}(\Gamma, V) > M_{2k}(G, V)$.

(2.4) Remarks on Larsen's Eighth Moment Conjecture

(2.4.1) We could "abstract" the substance of Larsen's Conjecture by saying that a reductive subgroup G of GL(V), V any finite-dimensional $\mathbb{C}$-vector space, "has the Larsen Eighth Moment property", if it is true that for any finite subgroup Γ of G, we have a strict inequality

$$\Sigma_{k=1 \text{ to } 4} M_{2k}(\Gamma, V) > \Sigma_{k=1 \text{ to } 4} M_{2k}(G, V).$$

The conjecture then asserts that if $\dim(V) \geq 8$, the groups GL(V), O(V), and, if dim(V) is even, Sp(V), all have the Larsen Eighth Moment property.

(2.4.2) If the conjecture holds, does it imply that SO(V) also has the Larsen Eighth Moment property? The answer is yes provided that $\dim(V) \geq 9$, simply because for $\dim(V) \geq 9$, we have

$$M_{2k}(SO(V), V) = M_{2k}(O(V), V) = 2k!!$$

for any k in {1, 2, 3, 4}. Indeed, one knows more generally that

$$M_{2k}(SO(V), V) = M_{2k}(O(V), V) = 2k!!$$

so long as $2k < \dim(V)$.

(2.4.3) As noted in 2.1.1.9.1 above, the equality

$$M_{2k}(O(V), V) = 2k!!$$

remains true so long as $\dim(V) \geq k$. And for dim(V) odd, we have

$$M_{2k}(SO(V), V) = M_{2k}(O(V), V)$$

for every integer k (simply because $O(V) = \pm SO(V)$ for dim(V) odd).

(2.4.4) But for dim(V) even, and any k with $\dim(V) \leq 2k$, we have

$$M_{2k}(SO(V), V) > M_{2k}(O(V), V).$$

Indeed, this inequality is obvious for $2k = \dim(V)$, because the determinant $\Lambda^{\dim(V)}(V)$ is a constituent of $V^{\otimes 2k}$ ($\cong \mathrm{End}(V^{\otimes k})$ as $O(V)$-representation) which is invariant under $SO(V)$ but not under $O(V)$. It follows in the general case from repeated applications of the following lemma.

Lemma 2.4.5 ([GT, Lemma 2.1]) Let $k \geq 1$ be an integer, V a finite-dimensional $\mathbb{C}$-vector space, H and G subgroups of $GL(V)$ with $H \subset G$. If $M_{2k}(H, V) > M_{2k}(G, V)$, then $M_{2k+2}(H, V) > M_{2k+2}(G, V)$.

proof We have

$$M_{2k+2}(G, V) = \dim\mathrm{Hom}_G(\mathbb{1}, \mathrm{End}(V)^{\otimes k+1}).$$

Now form the $GL(V)$-equivariant decomposition

$$\mathrm{End}(V) = \mathbb{1} \oplus \mathrm{End}^0(V)$$

of $\mathrm{End}(V)$ as the direct sum of scalar endomorphisms and endomorphisms of trace zero. We get

$$\begin{aligned} &M_{2k+2}(G, V) \\ &= \dim\mathrm{Hom}_G(\mathbb{1}, \mathrm{End}(V)^{\otimes k}) + \dim\mathrm{Hom}_G(\mathbb{1}, \mathrm{End}^0(V)\otimes\mathrm{End}(V)^{\otimes k}) \\ &= M_{2k}(G, V) + \dim\mathrm{Hom}_G(\mathbb{1}, \mathrm{End}^0(V)\otimes\mathrm{End}(V)^{\otimes k}). \end{aligned}$$

The same argument applied to H gives

$$M_{2k+2}(H, V) = M_{2k}(H, V) + \dim\mathrm{Hom}_H(\mathbb{1}, \mathrm{End}^0(V)\otimes\mathrm{End}(V)^{\otimes k}).$$

Since $H \subset G$, we have an a priori inequality

$$\begin{aligned} &\dim\mathrm{Hom}_H(\mathbb{1}, \mathrm{End}^0(V)\otimes\mathrm{End}(V)^{\otimes k}) \\ &\qquad \geq \dim\mathrm{Hom}_G(\mathbb{1}, \mathrm{End}^0(V)\otimes\mathrm{End}(V)^{\otimes k}). \end{aligned}$$

So the assertion is obvious. QED

(2.4.6) Using Lemma 2.4.5, we can restate Larsen's Eighth Moment Conjecture as

Eighth Moment Conjecture 2.4.7 (=2.3.2 ter) (Larsen) Let V be a $\mathbb{C}$-vector space of dimension $N \geq 8$. If Γ is a finite subgroup of G, for G one of the groups $GL(V)$, $O(V)$, or, if N is even, $Sp(V)$, then $M_8(\Gamma, V) > M_8(G, V)$.

(2.4.8) For ease of later reference, we also formulate a twelfth moment conjecture, which, in view of Lemma 2.4.5, trivially implies Larsen's Eighth Moment Conjecture.

Twelfth Moment Conjecture 2.4.9 Let V be a $\mathbb{C}$-vector space of dimension $N \geq 8$. If Γ is a finite subgroup of G, for G one of the groups $GL(V)$, $O(V)$, or, if N is even, $Sp(V)$, then we have

$$M_{12}(\Gamma, V) > M_{12}(G, V).$$

(2.5) How to apply Larsen's Eighth Moment Conjecture; its current status

(2.5.1) Let us first make explicit the relevance of Larsen's conjecture to our situation.

Theorem 2.5.2 Suppose that Larsen's Eighth Moment Conjecture 2.3.2 holds. Take any instance of the Higher Moment Theorem 1.20.2, or of one of its first three variants, in a situation with $d \geq 8$, and with

$$N := \mathrm{rank}(\mathrm{Gr}^0(\mathcal{M})_{ncst}) \geq 8.$$

Then we have the following results.

1) $G_{geom} \supset SL(N)$, if $\mathrm{Gr}^0(\mathcal{M})_{ncst}$ is not geometrically self dual.

2) $G_{geom} = Sp(N)$, if $\mathrm{Gr}^0(\mathcal{M})_{ncst}$ is, geometrically, symplectically self dual.

3) $G_{geom} = SO(N)$ or $O(N)$, if $\mathrm{Gr}^0(\mathcal{M})_{ncst}$ is, geometrically, orthogonally self dual.

proof In any $d \geq 8$ instance of the Higher Moment Theorem, or of one of its first three variants, the semisimple group G_{geom} for $\mathrm{Gr}^0(\mathcal{M})_{ncst}$ has the same absolute moments M_2, M_4, M_6, and M_8 as the relevant ambient group $GL(N)$, or $O(N)$, or $Sp(N)$. So if Larsen's Eighth Moment Conjecture is true, G_{geom} is not finite. Once G_{geom} is not finite, Larsen's Alternative 2.2.2 gives the asserted conclusion. QED

(2.5.3) Guralnick and Tiep [GT, Theorem 1.4] have recently proven both the Twelfth Moment Conjecture 2.4.9 and Larsen's Eighth Moment Conjecture 2.3.2, in the following very strong form.

Theorem 2.5.4 (Guralnick-Tiep) Let V be a finite-dimensional $\mathbb{C}$-vector space with $\dim(V) \geq 5$, G one of the groups $GL(V)$, $O(V)$, or, if $\dim(V)$ is even, $Sp(V)$.

1) If Γ is any finite subgroup of G, we have

$$M_{12}(\Gamma, V) > M_{12}(G, V).$$

2) In the case $\dim(V) = 6$, $G = Sp(V)$, the subgroup $\Gamma = 2J_2$ has

$$M_{2k}(\Gamma, V) = M_{2k}(Sp(V), V) \text{ for } k=1,2,3,4,5.$$

3) Except for the exceptional case given in 2) above, if Γ is any finite subgroup of G, we have

$$M_8(\Gamma, V) > M_8(G, V).$$

(2.6) Other tools to rule out finiteness of G_{geom}

(2.6.1) When we began writing, in January of 2002, Larsen's Eighth Moment Conjecture was still only a conjecture, so we needed other tools to rule out the possibility that G_{geom} be finite. Even now that Larsen's Eighth Moment Conjecture is no longer a conjecture, such tools are still needed to treat instances of the Higher Moment Theorem, or of one of its variants, where the family of functions $\mathcal{F}$ is not 8-separating, but is only d-separating for d with $4 \leq d \leq 7$. Indeed, we rely almost entirely on these tools in the rest of this book. Those results that depend on the truth of Larsen's Eighth

Moment Conjecture are so labeled.

(2.6.2) We begin by recalling some known group-theoretic tools.

Primitivity Theorem 2.6.3 ([Ka-LAMM, 1.3.2]) Let V be a $\mathbb{C}$-vector space of dimension $N \geq 2$, $G \subset GL(V)$ a finite subgroup of GL(V). Suppose that one of the following conditions 1), 2), or 3) holds.

1) $M_4(G, V) = 2$.

2) G lies in O(V), $\dim(V) \geq 3$, and $M_4(G, V) = 3$.

3) G lies in Sp(V), $\dim(V) \geq 4$, and $M_4(G, V) = 3$.

Then G is a finite irreducible primitive subgroup of GL(V), i.e., there exists no proper subgroup H of G such that V is induced from a representation of H.

Tensor-Indecomposability Lemma 2.6.4 ([Ka-LAMM, 1.3.6]) Let V be a $\mathbb{C}$-vector space of dimension $N \geq 2$, $G \subset GL(V)$ a finite subgroup of GL(V). Suppose that $M_4(G, V) \leq 3$. Then V is tensor-indecomposable in the following (strong) sense. There exists no expression of the $\mathbb{C}$-vector space V as a tensor product

$$V = V_1 \otimes V_2$$

of $\mathbb{C}$-vector spaces X and Y in such a way that all three of the following conditions are satisfied:

$\dim(V_1) \geq 2$,

$\dim(V_2) \geq 2$,

every element g in G, viewed as lying in $GL(V) = GL(V_1 \otimes V_2)$, can be written in the form $A \otimes B$ with A in $GL(V_1)$ and with B in $GL(V_2)$.

Normal Subgroup Corollary 2.6.5 ([Ka-LAMM, 1.3.7], cf. also [Lar-Char, 1.6]) Let V be a $\mathbb{C}$-vector space of dimension $N \geq 2$, $G \subset GL(V)$ a finite subgroup of GL(V). Let H be a proper normal subgroup of G. Suppose that one of the following conditions 1), 2), or 3) holds.

1) $M_4(G, V) = 2$.

2) G lies in O(V), $\dim(V) \geq 3$, and $M_4(G, V) = 3$.

3) G lies in Sp(V), $\dim(V) \geq 4$, and $M_4(G, V) = 3$.

Then either H acts on V as scalars and lies in the center Z(G), or V is H-irreducible.

(2.6.6) We next combine these results with some classical results of Blichfeld [Blich] (cf. [Dorn, 29.8] for a modern exposition of Blichfeld's 60° theorem), and of Mitchell [Mit], and with recent results of Wales [Wales] and Zalesskii [Zal], to give criteria which force G to be big. Recall that an element A in GL(V) is called a pseudoreflection if $\mathrm{Ker}(A - 1)$ has codimension 1 in V. A pseudoreflection of order 2 is called a reflection. Recall [Ka-TLFM,

1.0.2] that for a given integer $r \geq 0$, an element A in GL(V) is said to have drop r if Ker(A - 1) has codimension r in V. Thus

$$\text{drop of } A = \dim(V/\mathrm{Ker}(A-1)).$$

Given an integer r with $1 \leq r < \dim(V)$, an element A of GL(V) is called quadratic of drop r if it has drop r and if it satisfies the following two conditions: its minimal polynomial is $(T-1)(T-\lambda)$ for some nonzero λ, and it acts on the r-dimensional space V/Ker(A - 1) as the scalar λ. Thus a quadratic element of drop 1 is precisely a pseudoreflection.

Theorem 2.6.7 ([Ka-LAMM, 1.4.2]) Let V be a $\mathbb{C}$-vector space of dimension $N \geq 2$, G in GL(V) a (not necessarily connected) Zariski closed reductive subgroup of GL(V) with $M_4(G, V) = 2$. Fix an integer r with $1 \leq r < \dim(V)$. If any of the following conditions is satisfied, then $G \supset SL(V)$.

0) G contains a unipotent element $A \neq 1$.

1) G contains a quadratic element A of drop r which has finite order $n \geq 6$.

2) G contains a quadratic element A of drop r which has finite order 4 or 5, and $\dim(V) > 2r$.

3) G contains a quadratic element A of drop r which has finite order 3, and $\dim(V) > 4r$.

4) G contains a reflection A, and $\dim(V) > 8$.

Theorem 2.6.8 (Mitchell) ([Ka-LAMM, 1.4.4]) Let V be a $\mathbb{C}$-vector space of dimension $N > 8$, $G \subset GL(V)$ a finite irreducible primitive subgroup of $GL(V) \cong GL(N, \mathbb{C})$ which contains a reflection A. Let $\Gamma \subset G$ denote the normal subgroup of G generated by all the reflections in G. Then we have the following results.

1) Γ is (conjugate in GL(V) to) the group S_{N+1}, viewed as a subgroup of $GL(N, \mathbb{C})$ by the deleted permutation representation of S_{N+1} (:= its "permutation of coordinates" action on the hyperplane Aug_N in $\mathbb{C}^{N+1}$ consisting of those vectors whose coordinates sum to zero).

2) G is the product of Γ with the group $G \cap (\text{scalars})$.

3) $M_4(G, V) > 3$.

Theorem 2.6.9 ([Ka-LAMM, 1.4.6]) Let V be a $\mathbb{C}$-vector space of dimension $N > 8$ equipped with a nondegenerate quadratic form. Let $G \subset O(V)$ be a (not necessarily connected) Zariski closed reductive subgroup of O(V) with $M_4(G, V) = 3$. If G contains a reflection, then $G = O(V)$.

(2.6.10) What becomes of the above theorem when N is 7 or 8? We have the following result.

Theorem 2.6.11 Let V be a $\mathbb{C}$-vector space of dimension N = 7 or 8 equipped with a nondegenerate quadratic form. Let $G \subset O(V)$ be a (not necessarily connected) Zariski closed reductive subgroup of O(V).

with $M_4(G, V) = 3$. If G contains a reflection, then either $G = O(V)$ or G is the Weyl group $W(E_N)$ of the root system E_N.

proof By Larsen's Alternative 2.2.2, G is either O(V) or SO(V) or is a finite subgroup of O(V). The SO(V) case cannot occur, because G contains a reflection. If G is finite, it is primitive as a subgroup of GL(V), has $M_4(G, V) = 3$, and contains a reflection.

When N is 7 or 8, then by results of Blichfeld [Blich] and Mitchell [Mit], a finite primitive subgroup of GL(N) which contains a reflection has image in PGL(N) either the image of $W(E_N)$ in its reflection representation, or the image of the symmetric group S_{N+1} in its deleted permutation representation. Any subgroup of GL(N) with image S_{N+1} in PGL(N) has fourth moment 4, cf. [Ka-LAMM, 1.4.5 and 2.4.3]. As our group has fourth moment 3, we conclude that G_{geom} is a subgroup of O(N) which contains a reflection and which has the same image in PGL(N), and hence the same image in $O(N)/\pm1$, as does $W(E_N)$. Since $W(E_N)$ contains ±1, $W(E_N)$ is the complete inverse image in O(N) of its image in $O(N)/\pm1$. Thus G_{geom} is a subgroup Γ of $W(E_N)$, which contains a reflection and which maps onto $W(E_N)/\pm1$. We wish to conclude that Γ is $W(E_N)$. This is given by the following lemma.

Lemma 2.6.11.1 For N = 7 or 8, let Γ be a subgroup of $W(E_N)$, which contains a reflection and which maps onto $W(E_N)/\pm1$. Then Γ is $W(E_N)$.

proof Unfortunately, we must resort to ad hoc arguments, treating the two cases N = 7 and N = 8 separately. In both cases we define

$$W^+(E_N) := W(E_N)\cap SO(N).$$

For N = 7, we argue as follows. Since $W(E_7)$ contains -1 and does not lie in SO(7), we have a product decomposition

$$W(E_7) \cong (\pm1)\times W^+(E_7),\ g \mapsto (\det(g), g/\det(g)).$$

It is known that $W^+(E_7)$ is a simple group (it is the group $Sp(6, \mathbb{F}_2)$, or $S_6(2)$ in ATLAS notation, cf. [Bbki-Lie VI, §4, ex. 3, page 229]). Thus Γ is a subgroup of $(\pm1)\times W^+(E_7)$, the product of two non-isomorphic simple groups, and Γ maps onto each factor. [It maps onto $W(E_7)/(\pm1)$ by hypothesis, and it maps onto ±1 because it contains a reflection.] So by Goursat's Lemma, Γ is the entire group.

For N = 8, the argument is a bit different. Here ±1 lies in SO(8), and the key fact is that the quotient $W^+(E_8)/(\pm1)$ is a simple group (it isthe group $O_8(2)^+$ in ATLAS notation, cf. [Bbki-Lie VI, §4, ex. 1, page 228]). Now consider the group Γ. It is a subgroup of $W(E_8)$ which contains a reflection and which maps onto $W(E_8)/\pm1$. So by cardinality we see that either $\Gamma = W(E_8)$, and we are done, or Γ has index 2 in $W(E_8)$. If Γ has index 2 in $W(E_8)$, we reach a contradiction as follows. Since Γcontains a reflection, Γ cannot be the group $W^+(E_8)$. Therefore the group $\Gamma\cap W^+(E_8)$ is normal in $W(E_8)$ (being the intersection of two normal subgroups) of index 4, and it is normal in $W^+(E_8)$ of index two. So its image in $W^+(E_8)/(\pm 1)$ is a normal subgroup of index at most two. As $W^+(E_8)/(\pm 1)$ is a simple group of large order, we conclude that $\Gamma\cap W^+(E_8)$ maps onto $W^+(E_8)/(\pm 1)$, and hence, by cardinality, that this map is an isomorphism $\Gamma\cap W^+(E_8) \cong W^+(E_8)/(\pm 1)$. Thus we find that the simple group $O_8(2)^+$ sits in $W(E_8)$ as a normal subgroup of index 4. Since the eight-dimensional reflection representation ρ of $W(E_8)$ has fourth moment 3, it is primitive, so its restriction to the normal subgroup $O_8(2)^+$ must be isotypical. This restriction cannot be trivial, otherwise ρ itself factors through the abelian quotient group, so cannot be irreducible. But the lowest-dimensional nontrivial irreducible representation of $O_8(2)^+$ has dimension 28. QED for Lemma 2.6.11.1 and, with it, Theorem 2.6.11.

(2.6.12) This concludes our review of the relevant group-theoretic tools of which we are aware.

(2.7) Some conjectures on drops

Drop Ratio Conjecture 2.7.1 Let V be a $\mathbb{C}$-vector space of dimension $N \geq 2$, $G \subset GL(V)$ a finite subgroup of $GL(V)$. Suppose that one of the following conditions 1), 2), or 3) holds.

1) $M_4(G, V) = 2$.

2) G lies in $O(V)$, $\dim(V) \geq 3$, and $M_4(G, V) = 3$.

3) G lies in $Sp(V)$, $\dim(V) \geq 4$, and $M_4(G, V) = 3$.

Then for any $A\neq 1$ in G, we have the inequality

$$\mathrm{drop}(A)/\dim(V) \geq 1/8.$$

Equivalently, for any $A \neq 1$ in G we have the inequality

$$\dim(\mathrm{Ker}(A-1))/\dim(V) \leq 7/8.$$

(2.7.2) What is known in the direction of the Drop Ratio Conjecture? One key point is that, as noted above, the conjecture concerns a finite irreducible G which is primitive and strongly tensor-indecomposable, cf. 2.6.4 above. For $\dim(V) = n \geq 3$, the group G cannot be contained in $\mathbb{C}^\times S_{n+1}$ in its standard n-dimensional deleted permutation representation, because such a group has fourth moment at least 4, cf. [Ka-LAMM, 1.4.5, 1.4.5.1, 2.4.5]. The Drop Ratio Conjecture holds for groups containing an element of drop 1, i.e., for groups containing a pseudoreflection, by classical results of Mitchell [Mit]. Indeed, if a primitive finite group G with

$M_4(G, V) \le 3$ contains a reflection, then $\dim(V) \le 8$, cf. 2.6.8 above. If a primitive G contains a pseudoreflection of order 3, then $\dim(V) \le 4$, and if G contains a pseudoreflection of order 4 or more, then $\dim(V) \le 2$, cf. [Mit]. By Wales [Wales], it holds for primitive groups containing a quadratic element of any drop r, so long as that element has order at least 3. By Wales [Wales-Inv] and Huffman-Wales [Huff-Wales-Inv], it holds for primitive, tensor-indecomposable groups containing an involution of drop 2. By Huffman-Wales [Huff-Wales-Equal] and Huffman [Huff-Eig], it holds for primitive groups containing an element of drop 2 and order ≥ 3.

(2.7.3) We now state two more optimistic versions of this conjecture. For A in GL(V), define its projective drop $\mathrm{drop}_{pr}(A)$ to be

$$\mathrm{drop}_{pr}(A) := \mathrm{Min}_{\lambda \text{ in } \mathbb{C}^\times}\ \mathrm{drop}(\lambda A)$$
$$= \mathrm{Min}_{\lambda \text{ in } \mathbb{C}^\times}\ \dim(V/\mathrm{Ker}(A-\lambda)).$$

For A in GL(V) with image in PGL(V) of finite order, define its projective order $\mathrm{order}_{pr}(A)$ to be

$$\mathrm{order}_{pr}(A) := \text{order of } A \text{ in } PGL(V).$$

More Optimistic Drop Ratio Conjecture 2.7.4 Hypotheses and notations as in the Drop Ratio Conjecture, for any non-scalar A in G, we have the inequality

$$\mathrm{drop}_{pr}(A)/\dim(V) \ge 1/8.$$

Equivalently, for any non-scalar A in G, and any λ in $\mathbb{C}^\times$, we have

$$\dim(\mathrm{Ker}(A - \lambda))/\dim(V) \le 7/8.$$

(2.7.5) In the case of a finite group $G \subset GL(V)$ with $M_4(G, V) = 2$, the More Optimistic Drop Ratio Conjecture for G is equivalent to the Drop Ratio Conjecture for the slightly larger group

$$G_1 := (\text{roots of unity of order } \#G)G,$$

which also has $M_4(G_1, V) = 2$. Indeed, any group G_{int} between G and $\mathbb{C}^\times G$ has the same M_4 (indeed, the same M_{2k} for every k) as G:

$$M_4(G_{int}, V) = M_4(G, V).$$

This is obvious from the description of $M_4(G, V)$ as the dimension of the G-invariants in $\mathrm{End}(V)^{\otimes 2}$, a space on which the scalars in GL(V) operate trivially.

(2.7.6) On the other hand, the More Optimistic Drop Ratio Conjecture for finite subgroups G of either O(V) or Sp(V) with fourth moment 3 is apparently stronger than the Drop Ratio Conjecture: the trick used above of replacing G by G_1 cannot be used here, because G_1 will no longer lie in O(V) or Sp(V) respectively, though it will have $M_4(G_1, V) = 3$.

Remark 2.7.6.1 In late 2004, proofs of the More Optimistic Drop

Ratio Conjecture 2.7.4 were announced independently by Gluck and Maagard, working together, and by Tiep.

Most Optimistic Drop Ratio Conjecture 2.7.7 Hypotheses and notations as in the Drop Ratio Conjecture, for any A in G we have the inequality

$$\mathrm{drop}_{\mathrm{pr}}(A)/\dim(V) \geq (\mathrm{order}_{\mathrm{pr}}(A) - 1)/(\mathrm{order}_{\mathrm{pr}}(A) + 6).$$

Equivalently, for any A in G, and any λ in $\mathbb{C}^\times$, we have

$$\dim(\mathrm{Ker}(A - \lambda))/\dim(V) \leq 7/(\mathrm{order}_{\mathrm{pr}}(A) + 6).$$

(2.7.8) We have essentially no evidence whatever for this most optimistic conjecture! The right hand side has been cooked up to be compatible with what is known for elements of drop 1 or 2, and with what is known for quadratic elements of any drop which have order at least 3.

(2.8) More tools to rule out finiteness of G_{geom}: sheaves of perverse origin and their monodromy

(2.8.1) Let us now leave the realm of conjecture. A second approach to proving G_{geom} is not finite is based on the theory of sheaves of perverse origin, cf. [Ka-SMD]. To motivate this, remember the genesis of G_{geom} as the geometric monodromy group attached to (the restriction to a dense open set $U \subset \mathcal{F}$ of) the perverse sheaf $\mathrm{Gr}^0(M(\dim\mathcal{F}_0/2))_{\mathrm{ncst}}$ on $\mathcal{F}$. Concretely, we pass to a dense open set U of $\mathcal{F}$ on which M is lisse. There $M(\dim\mathcal{F}_0/2)|U$ is of the form $\mathcal{M}(\dim\mathcal{F}/2)[\dim\mathcal{F}]$, for a lisse $\overline{\mathbb{Q}}_\ell$-sheaf $\mathcal{M}$ on U which is ι-mixed of weight ≤ 0. We are concerned with the group G_{geom} attached to $\mathrm{Gr}^0(\mathcal{M})_{\mathrm{ncst}}$, the nonconstant part of the highest weight quotient $\mathrm{Gr}^0(\mathcal{M})$ of $\mathcal{M}$.

(2.8.2) So as not to obscure the underlying ideas, let us suppose, for simplicity of exposition, that the lisse sheaf $\mathcal{M}$ on U is itself ι-pure of weight zero, and geometrically irreducible. Then $\mathrm{Gr}^0(\mathcal{M})_{\mathrm{ncst}}$ is just $\mathcal{M}$ itself. [Recall that $\mathrm{Gr}^0(\mathcal{M})_{\mathrm{ncst}}$ is geometrically irreducible.]

(2.8.3) So in this case, our situation is that we have a perverse sheaf N (here $M(-m/2)$) on $\mathcal{F}$, we are told that on an open dense set U of $\mathcal{F}$, $N|U = \mathcal{N}[\dim\mathcal{F}]$ for a lisse sheaf $\mathcal{N}$ on U which is geometrically irreducible on U. A key (though apparently trivial) observation is that, given N, the lisse sheaf $\mathcal{N}$ on U has a canonical prolongation to a constructible $\overline{\mathbb{Q}}_\ell$-sheaf $\mathcal{N}$ on all of $\mathcal{F}$, namely

$$\mathcal{N} := \mathcal{H}^{-\dim\mathcal{F}}(N).$$

The constructible $\overline{\mathbb{Q}}_\ell$-sheaf $\mathcal{N}$ on $\mathcal{F}$ is what we call a sheaf of perverse origin.

(2.8.4) Given a field k in which ℓ is invertible, and a smooth connected k-scheme T/k of dimension $\dim T \geq 0$, we say that a constructible $\overline{\mathbb{Q}}_\ell$-sheaf $\mathcal{N}$ on T is of perverse origin if there exists a perverse sheaf N on T and an isomorphism

$$\mathcal{N} \cong \mathcal{H}^{-\dim T}(N).$$

It is equivalent to require only the existence of an object N in ${}^pD^{\geq 0}$ on T for which we have $\mathcal{N} \cong \mathcal{H}^{-\dim T}(N)$, cf. [Ka-SMD, Prop 4].

(2.8.5) A propos of this notion, there are two standard facts about perversity that one should keep in mind.

(2.8.5.1) The first is that if an object N on T lies in ${}^pD^{\geq 0}$, then for any connected smooth k-scheme S of dimension $\dim S \geq 0$, and for any k-morphism $f : S \to T$, the shifted pullback $f^*N[\dim S - \dim T]$ lies in ${}^pD^{\geq 0}$ on S, cf. [Ka-SMD, proof of Proposition 7]. This has as consequence that for $\mathcal{N}$ of perverse origin on T, $f^*\mathcal{N}$ is of perverse origin on S, cf. [Ka-SMD, Proposition 7].

(2.8.5.2) The second fact to keep in mind is that for any separated k-scheme of finite type X/k, and for any **affine** morphism

$$f: X \to T,$$

if an object K on X lies in ${}^pD^{\geq 0}$ on X, then the object $Rf_!K$ on T lies in ${}^pD^{\geq 0}$ on Y. Consequently, if K on X lies in ${}^pD^{\geq 0}$ on X, e.g., if K is perverse on X, then

$$\mathcal{H}^{-\dim T}(Rf_!K) = R^{-\dim T}f_!(K)$$

is a sheaf of perverse origin on T.

(2.8.6) What is the relevance of sheaves of perverse origin to monodromy? Given a constructible $\overline{\mathbb{Q}}_\ell$-sheaf $\mathcal{N}$ on our smooth connected k-scheme T, let Z be a reduced irreducible closed subscheme of T, and assume that a dense open set $V_1 \subset Z$ is smooth over k (a condition which is automatic if k is perfect). The restriction of $\mathcal{N}$ to V_1 is constructible, so there exists a dense open set $V \subset V_1$ such that $\mathcal{N}|V$ is lisse on V. On the other hand, there is a dense open set $U \subset T - Z$ on which $\mathcal{N}$ is lisse. Thus we have a lisse sheaf $\mathcal{N}|U$ on U, and a lisse sheaf $\mathcal{N}|V$ on V.

(2.8.7) For $\mathcal{N}$ of perverse origin on T, the monodromy of $\mathcal{N}|V$ is "smaller" than the monodromy of $\mathcal{N}|U$. To make this precise, let us pick geometric points u of U and v of V. We have the monodromy homomorphisms

$$\rho_U: \pi_1(U, u) \to \mathrm{Aut}_{\overline{\mathbb{Q}}_\ell}(\mathcal{N}_u)$$

and

$$\rho_V: \pi_1(V, v) \to \mathrm{Aut}_{\overline{\mathbb{Q}}_\ell}(\mathcal{N}_v)$$

attached to $\mathcal{N}|U$ and to $\mathcal{N}|V$ respectively. We define compact subgroups

$$\Gamma_U := \rho_U(\pi_1(U, u)) \subset \mathrm{Aut}_{\overline{\mathbb{Q}}_\ell}(\mathcal{N}_u),$$

$$\Gamma_V := \rho_V(\pi_1(V, v)) \subset \mathrm{Aut}_{\overline{\mathbb{Q}}_\ell}(\mathcal{N}_v).$$

Semicontinuity Theorem 2.8.8 ([Ka-SMD, Theorem 1]) For T/k smooth and connected of dimension $\dim T \geq 1$, and for $\mathcal{N}$ a constructible $\overline{\mathbb{Q}}_\ell$-sheaf on T of perverse origin, the group Γ_V is isomorphic to a subquotient of the group Γ_U. More precisely, there exists a compact group D, a continuous group homomorphism

$$D \to \Gamma_U,$$

a closed normal subgroup $I \subset D$, and a $\overline{\mathbb{Q}}_\ell$-linear embedding

$$\mathcal{N}_v \subset (\mathcal{N}_u)^I,$$

with the following property: if we view $\mathcal{N}_u$ as a representation of D via the given homomorphism $D \to \Gamma_U$, and if we then view $(\mathcal{N}_u)^I$ as a representation of D/I, then the subspace $\mathcal{N}_v$ is D/I-stable, and under the induced action of D/I on $\mathcal{N}_v$, the image of D/I in the group $\mathrm{Aut}_{\overline{\mathbb{Q}}_\ell}(\mathcal{N}_v)$ is the group Γ_V.

Semicontinuity Corollary 2.8.9 ([Ka-SMD, Corollary 10]) Hypotheses and notations as in Theorem 2.8.8, denote by N_U (resp. N_V) the rank of the lisse $\overline{\mathbb{Q}}_\ell$-sheaf $\mathcal{N}|U$ (resp. of the lisse $\overline{\mathbb{Q}}_\ell$-sheaf $\mathcal{N}|V$). Denote by G_U (resp. G_V) the Zariski closure of Γ_U (resp. of Γ_V) in $\mathrm{Aut}_{\overline{\mathbb{Q}}_\ell}(\mathcal{N}_u) \cong GL(N_U, \overline{\mathbb{Q}}_\ell)$ (resp. in $\mathrm{Aut}_{\overline{\mathbb{Q}}_\ell}(\mathcal{N}_v) \cong GL(N_V, \overline{\mathbb{Q}}_\ell)$).

1) We have the inequality of ranks

$$N_V \leq N_U.$$

2) The algebraic group G_V is a subquotient of G_U.

In particular, we have

2a) if G_U is finite (or equivalently if Γ_U is finite) then G_V is finite (or equivalently Γ_V is finite),

2b) $\dim(G_V) \leq \dim(G_U)$,

2c) $\mathrm{rank}(G_V) \leq \mathrm{rank}(G_U)$.

Proposition 2.8.10 ([Ka-SMD, Proposition 12]) Hypotheses and notations as in the Semicontinuity Theorem 2.8.8, let $\mathcal{N}$ be of perverse origin on T. The integer-valued function on T given by

$$t \mapsto \mathrm{rank}(\mathcal{N}_t)$$

is lower semicontinuous, i.e., for every integer $r \geq 0$, there exists a reduced closed subscheme $T_{\leq r} \subset T$ such that a geometric point t of T lies in $T_{\leq r}$ if and only if the stalk $\mathcal{N}_t$ has rank $\leq r$. If we denote by N the generic rank of $\mathcal{N}$, then $T = T_{\leq N}$, and $T - T_{\leq N-1}$ is the largest open set on which $\mathcal{N}$ is lisse.

(2.8.11) In order to apply the semicontinuity theorem, we next recall the following Twisting Lemma, cf. [Ka-ESDE, 8.14.3 and 8.14.3.1]. In this lemma, we work over a finite field k in which ℓ is invertible. We are given a smooth, geometrically connected k-scheme U/k, and a lisse $\overline{\mathbb{Q}}_\ell$-sheaf $\mathcal{N}$ on U. We fix a geometric point u in U. We have $\pi_1(U, u)$, which we denote $\pi_1^{arith}(U, u)$, and its closed normal subgroup $\pi_1^{geom}(U, u) := \pi_1(U\otimes_k\overline{k}, u)$. We view $\mathcal{N}$ as a $\overline{\mathbb{Q}}_\ell$-representation

$$\rho : \pi_1^{arith}(U, u) \rightarrow GL(\mathcal{N}_u),$$

and define

$$G_{arith,\mathcal{N}} := \text{the Zariski closure of } \rho(\pi_1^{arith}(U, u)) \text{ in } GL(\mathcal{N}_u),$$

$$G_{geom.\mathcal{N}} := \text{the Zariski closure of } \rho(\pi_1^{geom}(U, u)) \text{ in } GL(\mathcal{N}_u).$$

Twisting Lemma 2.8.12 In the situation 2.8.11 above, consider the following two conditions.

1) $G_{geom,\mathcal{N}}$ is finite.

2) There exists a unit α in $\overline{\mathbb{Q}}_\ell^\times$ such that for the twisted sheaf $\mathcal{N}\otimes\alpha^{deg}$, $G_{arith,\mathcal{N}\otimes\alpha^{deg}}$ is finite.

We always have the implication

$$2) \Rightarrow 1).$$

If in addition $\mathcal{N}$ is geometrically irreducible, i.e., if $G_{geom,\mathcal{N}}$ is an irreducible subgroup of $GL(\mathcal{N}_u)$, then 1) and 2) are equivalent.

proof The implication 2) $\Rightarrow$ 1) is trivial, since G_{geom} does not see twisting, so $G_{geom,\mathcal{N}} = G_{geom,\mathcal{N}\otimes\alpha^{deg}}$, which is a subgroup of $G_{arith,\mathcal{N}\otimes\alpha^{deg}}$. To show that 1) $\Rightarrow$ 2) when $\mathcal{N}$ is geometrically irreducible, we argue as follows. For any lisse $\overline{\mathbb{Q}}_\ell$-sheaf $\mathcal{N}$ on U, $\det(\mathcal{N})$ is geometrically of finite order, cf. [De-Weil II, 1.3.4]. So some power $\det(\mathcal{N})^{\otimes N}$ of $\det(\mathcal{N})$ is geometrically trivial, so of the form β^{deg} for some unit β in $\overline{\mathbb{Q}}_\ell^\times$. So if we choose an N'th root α of $1/\beta$, then $\mathcal{N}\otimes\alpha^{deg}$ has its determinant arithmetically of finite order. So the Twisting Lemma 2.8.12 results from the following variant, applied to $\mathcal{N}\otimes\alpha^{deg}$.

Twisting Lemma 2.8.12.1 (= 2.8.12 bis) In the situation of the Twisting Lemma 2.8.12, suppose that $\mathcal{N}$ is geometrically irreducible and that $\det(\mathcal{N})$ is arithmetically of finite order. Then the following conditions are equivalent.

1) $G_{geom,\mathcal{N}}$ is finite.

2) $G_{arith,\mathcal{N}}$ is finite.

proof Denote by n the rank of the lisse sheaf $\mathcal{N}$. It is trivial that 2) implies 1), since $G_{geom,\mathcal{N}}$ is a subgroup of $G_{arith,\mathcal{N}}$. To see that 1) implies 2), we argue as follows. Because $\pi_1^{geom}(U, u)$ is a normal subgroup of $\pi_1^{\mathrm{arith}}(U, u)$, G_{geom} is a normal subgroup of G_{arith}. If G_{geom} is finite, then a fixed power of every element g in G_{arith} is a scalar [g normalizes G_{geom}, but as G_{geom} is finite, $\mathrm{Aut}(G_{geom})$ is finite, say of order m_0, so g^{m_0} centralizes G_{geom}, and hence is scalar, because G_{geom} is irreducible]. On the other hand, $\det(\mathcal{N})$ is of finite order, say of order m_1, so any element g in G_{arith} has $\det(g)^{m_1} = 1$. Thus if g in G_{arith} is a scalar, then $g^{nm_1} = 1$. So G_{arith} is a $\overline{\mathbb{Q}}_\ell$ -algebraic group in which every element satisfies $g^{m_0 m_1 n} = 1$. So $\mathrm{Lie}(G_{arith})$ is a $\overline{\mathbb{Q}}_\ell$-vector space killed by $m_0 m_1 n$, and so vanishes. Therefore G_{arith} is finite, as required. QED

Scalarity Corollary 2.8.13 Let k be a finite field in which ℓ is invertible, T/k a smooth, geometrically connected k-scheme, and $\mathcal{N}$ a constructible $\overline{\mathbb{Q}}_\ell$-sheaf on T of perverse origin. Suppose that for some dense open set $U \subset T$, $\mathcal{N}|U$ is lisse and geometrically irreducible on U. Suppose further that $G_{geom,\mathcal{N}|U}$ is finite. Then we have the following results.
1) For any irreducible closed subscheme $Z \subset T$, and for any dense open set $V \subset Z$ such that $\mathcal{N}|V$ is lisse, of rank denoted N_V, the image of $G_{arith,\mathcal{N}|V}$ in $\mathrm{PGL}(N_V, \overline{\mathbb{Q}}_\ell)$ is finite.
2) For any finite extension field E/k, and for any E-valued point t in T(E), a power of $\mathrm{Frob}_{E,t} \mid \mathcal{N}_t$ is a scalar.

proof To prove 1), we argue as follows. Denote by N_U the rank of $\mathcal{N}|U$. The largest open set of T on which $\mathcal{N}$ is lisse is, by 2.6.10 above, the set

$$U_1 := T - T_{\leq N_U - 1}$$

of points where the stalk has rank N_U. The group G_{geom} for $\mathcal{N}|U_1$ is equal to the group G_{geom} for $\mathcal{N}|U$, simply because $\pi_1^{\mathrm{geom}}(U_1)$ is a quotient of $\pi_1^{\mathrm{geom}}(U)$, for any dense open set U of U_1. So with no loss of generality, we may assume that $U = U_1$.

By the Twisting Lemma 2.8.12, there is a unit α such that $G_{arith,\mathcal{N}\otimes\alpha^{\deg}|U_1}$ is finite.

Consider first the case when $N_V = N_U$. Then $V \subset U_1$. So if we choose a geometric point u of U_1 which lies in V, then the monodromy representation of $\mathcal{N}\otimes\alpha^{\deg}|V$ is obtained from that of $\mathcal{N}\otimes\alpha^{\deg}|U_1$ by composition with the map of arithmetic fundamental groups

$$\pi_1^{\mathrm{arith}}(V, u) \to \pi_1^{\mathrm{arith}}(U_1, u)$$

induced by the inclusion of V into U. So a fortiori, if $\mathcal{N}\otimes\alpha^{\deg}|U_1$ has finite arithmetic monodromy, so does $\mathcal{N}\otimes\alpha^{\deg}|V$.

Consider now the case when $N_V < N_U$. Then $V \subset T_{\le N_V} \subset T_{<N_U}$, and hence $Z \subset T_{<N_U}$. Thus U_1 is a dense open set of $T - Z$. By part (2a) of Corollary 2.8.9, applied to the sheaf $\mathcal{N}\otimes\alpha^{\deg}$ of perverse origin on T, the finiteness of $G_{\mathrm{arith},\mathcal{N}\otimes\alpha^{\deg}|U_1}$ implies the finiteness of $G_{\mathrm{arith},\mathcal{N}\otimes\alpha^{\deg}|V}$. So the image of $G_{\mathrm{arith},\mathcal{N}|V}$ in $PGL(N_V, \overline{\mathbb{Q}}_\ell)$ is finite.

To prove 2), simply apply part 1) to the case Z = {the closed point underlying the E-valued point t}. QED

Punctual Purity Corollary 2.8.14 Let k be a finite field in which ℓ is invertible, T/k a smooth, geometrically connected k-scheme, $\mathcal{N}$ a constructible $\overline{\mathbb{Q}}_\ell$-sheaf on T of perverse origin, and $\iota : \overline{\mathbb{Q}}_\ell \subset \mathbb{C}$ a field embedding. Suppose that for some dense open set $U \subset T$, $\mathcal{N}|U$ is lisse and geometrically irreducible on U. Suppose further that $G_{\mathrm{geom},\mathcal{N}|U}$ is finite. Then we have the following results.
1) There exists a real number w such that $\mathcal{N}|U$ is ι-pure of some weight w.
2) For any irreducible closed subscheme $Z \subset T$, and for any dense open set $V \subset Z$ such that $\mathcal{N}|V$ is lisse, $\mathcal{N}|V$ is ι-pure of weight w.
3) For any finite extension field E/k, and for any E-valued point t in T(E), all the eigenvalues of $\mathrm{Frob}_{E,t} \mid \mathcal{N}_t$ are ι-pure of weight w.

proof The assertion is invariant under replacing $\mathcal{N}$ by an $\alpha^{\deg}$ twist of itself. By the Twisting Lemma 2.8.12, there exists a unit α in $\overline{\mathbb{Q}}_\ell^\times$ such that $(\mathcal{N}\otimes\alpha^{\deg})|U$ has finite G_{arith} on U. So replacing $\mathcal{N}$ by $\mathcal{N}\otimes\alpha^{\deg}$, we reduce to the case when $\mathcal{N}|U$ has finite G_{arith}. In this case, $\mathcal{N}|U$ is ι-pure of weight zero (since a power of every Frobenius acts as the identity on $\mathcal{N}|U$, and so has roots of unity as eigenvalues). This proves 1). To prove 2), apply part (2a) of Corollary 2.8.9, to the sheaf $\mathcal{N}$ of perverse origin on T: the finiteness of $G_{\mathrm{arith},\mathcal{N}|U}$ implies the finiteness of $G_{\mathrm{arith},\mathcal{N}|V}$, which in turn implies that $\mathcal{N}|V$ is ι-pure of weight zero. To prove 3), simply apply 2) to the case when Z is the closed point underlying the E-valued point t. QED

Chapter 3: Additive character sums on $\mathbb{A}^n$

(3.1) The $\mathcal{L}_\psi$ theorem

(3.1.1) In this section, we will consider in detail the following general class of "standard inputs", cf. 1.15.4. We work over a finite field k of characteristic p, in which the prime ℓ is invertible. We take

$m=1$,

a nontrivial $\overline{\mathbb{Q}}_\ell^\times$-valued additive character ψ of k,

$K = \mathcal{L}_\psi(1/2)[1]$ on $\mathbb{A}^1$,

an integer $n \geq 1$,

$V = \mathbb{A}^n$,

$h : V \to \mathbb{A}^1$ the function $h = 0$,

L on V a perverse, geometrically irreducible sheaf which is ι-pure of weight zero, which in a Zariski open neighborhood U_0 of the origin in $\mathbb{A}^n$ is of the form $\mathcal{L}[n]$ for $\mathcal{L}$ a nonzero lisse $\overline{\mathbb{Q}}_\ell$-sheaf on U_0,

an integer $e \geq 3$,

$(\mathcal{F}, \tau) = (\mathcal{P}_e$, evaluation), for $\mathcal{P}_e$ the space of all k-polynomial functions on $\mathbb{A}^n$ of degree $\leq e$.

$\mathcal{L}_\psi$ Theorem 3.1.2 Take standard input as in 3.1.1 above. Then we have the following results concerning $M = \mathrm{Twist}(L, K, \mathcal{F}, h)$.

1) The object $M(\dim\mathcal{F}_0/2)$ on $\mathcal{F} = \mathcal{P}_e$ is perverse, geometrically irreducible and geometrically nonconstant, and ι-pure of weight zero.

2) The Frobenius-Schur indicator of $M(\dim\mathcal{F}_0/2)$ is given by

$$\begin{aligned} &\mathrm{FSI}^{\mathrm{geom}}(\mathcal{P}_e, M(\dim\mathcal{F}_0/2)) \\ &\quad = 0, \text{ if p is odd}, \\ &\quad = ((-1)^{1+\dim\mathcal{F}_0})\times\mathrm{FSI}^{\mathrm{geom}}(\mathbb{A}^n, L), \text{ if } p = 2. \end{aligned}$$

3) The restriction of $M(\dim\mathcal{F}_0/2)$ to some dense open set U of $\mathcal{P}_e$ is of the form $\mathcal{M}(\dim\mathcal{F}/2)[\dim\mathcal{F}]$ for $\mathcal{M}$ a lisse sheaf on U of rank $N := \mathrm{rank}(\mathcal{M}|U)$, with

$N \geq (e-1)^n\mathrm{rank}(\mathcal{L}|U_0)$, if e is prime to p,

$N \geq \mathrm{Max}((e-2)^n, (1/e)((e-1)^n + (-1)^n(e-1)))\mathrm{rank}(\mathcal{L}|U_0)$, if $p|e$.

4) The Frobenius-Schur indicator of $\mathcal{M}|U$ is given by

$$\begin{aligned} &\mathrm{FSI}^{\mathrm{geom}}(U, \mathcal{M}|U) \\ &\quad = 0, \text{ if } p \text{ is odd}, \\ &\quad = \mathrm{FSI}^{\mathrm{geom}}(\mathbb{A}^n, L) = (-1)^n \mathrm{FSI}^{\mathrm{geom}}(U_0, \mathcal{L}), \text{ if } p = 2. \end{aligned}$$

5) Suppose in addition that one of the following five conditions holds:
a) $p \geq 7$,
b) $n \geq 3$,
c) $p = 5$ and $e \geq 4$,
d) $p = 3$ and $e \geq 7$,
e) $p = 2$ and $e \geq 7$.
Then we have the following results concerning the group $G_{\mathrm{geom},\mathcal{M}|U}$.

5A) If $\mathrm{FSI}^{\mathrm{geom}}(U, \mathcal{M}|U) = 0$, the group $G_{\mathrm{geom},\mathcal{M}|U}$ contains $SL(N)$.

5B) If $\mathrm{FSI}^{\mathrm{geom}}(U, \mathcal{M}|U) = -1$, the group $G_{\mathrm{geom},\mathcal{M}|U}$ is $Sp(N)$.

5C) If $\mathrm{FSI}^{\mathrm{geom}}(U, \mathcal{M}|U) = 1$, the group $G_{\mathrm{geom},\mathcal{M}|U}$ is either $SO(N)$ or $O(N)$.

(3.2) Proof of the $\mathcal{L}_\psi$ Theorem 3.1.2

(3.2.1) Recall from 1.1.8 that $\mathcal{P}_e$ is d-separating for $d=e+1 \geq 4$, and that

$$H_c{}^*(\mathbb{A}^1 \otimes_k \bar{k}, K) = 0.$$

Because the lisse, rank one sheaf $\mathcal{L}_\psi$ is of order $p = \mathrm{char}(k)$ as a character of $\pi_1(\mathbb{A}^1 \otimes_k \bar{k})$, it is geometrically self dual if and only if $p = 2$, in which case it is orthogonally self dual. So the Frobenius-Schur indicator of $K = \mathcal{L}_\psi(1/2)[1]$ on $\mathbb{A}^1$ is given by

$$\begin{aligned} \mathrm{FSI}^{\mathrm{geom}}(\mathbb{A}^1, K) &= 0, \text{ if } p = \mathrm{char}(k) \text{ is odd}, \\ &= -1, \text{ if } p = \mathrm{char}(k) = 2. \end{aligned}$$

(3.2.2) We first prove 1), that $M(\dim\mathcal{F}_0/2)$ on $\mathcal{F} = \mathcal{P}_e$ is perverse, geometrically irreducible, and ι-pure of weight zero. This depends on the following compatibility.

Compatibility Lemma 3.2.3 For any $e \geq 1$, the k-morphism

$$\begin{aligned} \mathrm{eval} : \mathbb{A}^n &\to (\mathcal{P}_e)^\vee, \\ v &\mapsto \mathrm{eval}(v), \end{aligned}$$

is a closed immersion. In terms of the Fourier Transform on the target space $(\mathcal{P}_e)^\vee$,

$$FT_\psi : D^b{}_c((\mathcal{P}_e)^\vee, \bar{\mathbb{Q}}_\ell) \to D^b{}_c(\mathcal{P}_e, \bar{\mathbb{Q}}_\ell),$$

we have

$$\mathrm{Twist}(L, K, \mathcal{F}, h) = FT_\psi(\mathrm{eval}_*(L))(1/2),$$

i.e.,

$$\mathrm{Twist}(L, K, \mathcal{F}, h)(\dim\mathcal{F}_0/2) = FT_\psi(\mathrm{eval}_*(L))(\dim\mathcal{F}/2).$$

proof Pick coordinate functions $x_1,\ldots, x_n$ on $\mathbb{A}^n$. Then the map eval sends a point v in $\mathbb{A}^n$ with coordinates $(v_1,\ldots, v_n)$ to the vector of all monomials in the v_i of degree at most e. Just looking at the monomials of degree one, the v_i themselves, shows that we have a closed immersion. As $K = \mathcal{L}_\psi(1/2)[1]$ on $\mathbb{A}^1$, the statement concerning Fourier Transform is a tautology. QED

(3.2.4) We exploit this compatibility as follows. The object L is perverse and geometrically irreducible and ι-pure of weight zero on $\mathbb{A}^n$, with support all of $\mathbb{A}^n$. Since $n \geq 1$, L does not have punctual support. Since eval is a closed immersion, the object eval_*L is perverse and geometrically irreducible and ι-pure of weight zero on $(\mathcal{P}_e)^\vee$, and does not have punctual support. Therefore, by the miraculous properties of Fourier Transform FT_ψ (cf. [Ka-Lau 2.2.1, 2.3.1] and [Lau-TF, 1.2.2.1, 1.2.3.1, 1.3.2.4]) $FT_\psi(\mathrm{eval}_*(L))$ is perverse and geometrically irreducible and ι-pure of weight $\dim\mathcal{P}_e = \dim\mathcal{F}$ on $\mathcal{P}_e$, and it is not geometrically constant. This proves 1).

(3.2.5) Statements 2), 3), and 4), **except** for the rank estimate in 3), result immediately from applying the Higher Moment Theorem-first variant 1.22.2, with $d = e+1 \geq 4$, to the standard input 3.1.1, cf. 2.2.3. In view of Larsen's Alternative 2.2.2, the conclusions 5A), 5B), and 5C) hold whenever $G_{\mathrm{geom},\mathcal{M}|U}$ is not finite. So in order to prove 5), it suffices to show that $G_{\mathrm{geom},\mathcal{M}|U}$ is not finite if any of the conditions 5a) through 5e) holds.

We now explain how to prove the rank estimate in part 3), and how to show that $G_{\mathrm{geom},\mathcal{M}|U}$ is not finite if any one of the conditions 5a) through 5e) holds. The proof is based upon exploiting a homothety contraction argument and the Semicontinuity Theorem 2.8.8 to reduce to the special case when the object L on $\mathbb{A}^n$ is geometrically constant. We then treat that case separately.

(3.3) Interlude: the homothety contraction method

(3.3.1) With an eye to later applications, we will give the homothety contraction method in slightly greater generality than is required for the $\mathcal{L}_\psi$ theorem.

(3.3.2) We work over an arbitrary field k in which ℓ is invertible. We fix integers $m \geq 1$ and $n \geq 1$, and a perverse sheaf K on $\mathbb{A}^m$. We assume that

$$H_c^i(\mathbb{A}^m\otimes\bar{k}, K[m]) = 0 \text{ for } i > -n.$$

We take

$$V := \mathbb{A}^n,$$

$$h : V \to \mathbb{A}^m \text{ the function } h=0.$$

(3.3.3) We take for $(\mathcal{F}, \tau)$ a space of $\mathbb{A}^m$-valued functions on $\mathbb{A}^n$ which contains the constants and is quasifinitely difference-separating on $\mathbb{A}^n$. We assume further that $(\mathcal{F}, \tau)$ is "stable by homothety" in the sense that we are given a $\mathbb{Z}$-grading of the k-space $\mathcal{F}$, say $\mathcal{F} = \oplus_i \mathcal{F}^i$, which satisfies the following condition. Using the grading, for every k-algebra R we define an action of the group $R^\times$ on $\mathcal{F}\otimes_k R$ as follows: for t in $R^\times$, and for $f = \Sigma f^{[i]}$ in $\mathcal{F}\otimes_k R = \oplus_i \mathcal{F}^i\otimes_k R$, we define

$$f_t := \Sigma_i t^i f^{[i]} \text{ in } \mathcal{F}\otimes_k R.$$

We require that under the natual homothety action of $R^\times$ on $\mathbb{A}^n(R)$, for any point v in $\mathbb{A}^n(R)$, for any point t in $R^\times$, and for any f in $\mathcal{F}\otimes_k R$, we have

$$f(tv) = f_t(v).$$

(3.3.4) For example, if m=1, then for any $e \geq 1$, we can take for $\mathcal{F}$ the space $\mathcal{P}_e$ of all polynomial functions on $\mathbb{A}^n$ of degree at most e, with the usual grading by degree. And for general m, if we pick m integers $e_i \geq 1$, we can take for $\mathcal{F}$ the space $\Pi_i \mathcal{P}_{e_i}$ of m-tuples of polynomial functions on $\mathbb{A}^n$, the i'th being of degree at most e_i. In this case, an m-tuple $(f_1,..., f_m)$ of functions is said to be homogeneous of some given degree d if each f_i is homogeneous of that degree. [Remember that we perform addition and scalar multiplication of m-tuples of functions componentwise, so that

$$(\Sigma_i f_1^{[i]},..., \Sigma_i f_m^{[i]}) = \Sigma_i (f_1^{[i]},..., f_m^{[i]}).]$$

(3.3.5) We fix a perverse sheaf L on $V := \mathbb{A}^n$. In view of the hypothesis on K, namely

$$H_c^i(\mathbb{A}^m\otimes\bar{k}, K[m]) = 0 \text{ for } i > -n,$$

and the fact that $H_c^i(\mathbb{A}^n\otimes\bar{k}, L[n]) = 0$ for $i > 0$, we see that whatever the perverse sheaf L, the object $M = \mathrm{Twist}(L,K,\mathcal{F},h=0)$ is perverse on $\mathcal{F}$, cf. 1.4.2 and 1.4.4, part 4).

(3.3.6) Now consider the homothety action of $\mathbb{A}^1$ on $\mathbb{A}^n$,

$$\mathrm{hmt}: \mathbb{A}^1\times\mathbb{A}^n \to \mathbb{A}^n,$$
$$(t, v) \mapsto tv.$$

Lemma 3.3.7 For L a perverse sheaf on $V := \mathbb{A}^n$, the object

$$L(tv)[1] := \mathrm{hmt}^*(L)[1]$$

on $\mathbb{A}^1\times\mathbb{A}^n$ lies in ${}^pD^{\geq 0}$.

proof This is an instance of 2.8.5.1. QED

Lemma 3.3.8 For L perverse on $V := \mathbb{A}^n$, and for K perverse on $\mathbb{A}^m$, the object

$$L(tv)\otimes K(f(v))[1+\dim\mathcal{F}_0]$$

on $\mathbb{A}^1\times\mathbb{A}^n\times\mathcal{F}$ lies in ${}^pD^{\geq 0}$.

proof Write $\mathcal{F}$ as $\mathcal{F}_0 \times \mathbb{A}^m$, with coordinates (f_0, a). In coordinates (t, v, f_0, a) on

$$\mathbb{A}^1 \times \mathbb{A}^n \times \mathcal{F} = \mathbb{A}^1 \times \mathbb{A}^n \times \mathcal{F}_0 \times \mathbb{A}^m,$$

$L(tv)\otimes K(f(v))[1+\dim \mathcal{F}_0]$ is

$$L(tv)\otimes K(f_0(v) + a)[1+\dim \mathcal{F}_0].$$

Under the automorphism σ of $\mathbb{A}^1 \times \mathbb{A}^n \times \mathcal{F}_0 \times \mathbb{A}^m$ given by

$$(t, v, f_0, a) \mapsto (t, v, f_0, a - f_0(v)),$$

the pullback by σ of $L(tv)\otimes K(f(v))[1+\dim \mathcal{F}_0]$ is the object

$$L(tv)\otimes K(a)[1+\dim \mathcal{F}_0],$$

which is the external tensor product of

$L(tv)[1]$ on $\mathbb{A}^1 \times \mathbb{A}^n$,

$\overline{\mathbb{Q}}_\ell[\dim \mathcal{F}_0]$ on $\mathcal{F}_0$,

K on $\mathbb{A}^m$.

Each external tensoree is in ${}^pD^{\geq 0}$, hence their external tensor product is as well. Since the property of lying in ${}^pD^{\geq 0}$ is invariant under pullback by automorphisms, we find that $L(tv)\otimes K(f(v))[1+\dim \mathcal{F}_0]$ lies in ${}^pD^{\geq 0}$. QED

(3.3.9) Consider the morphism

$$pr_{1,3} : \mathbb{A}^1 \times \mathbb{A}^n \times \mathcal{F} \to \mathbb{A}^1 \times \mathcal{F},$$
$$(t, v, f) \mapsto (t, f),$$

and the object

$$L(tv)\otimes K(f(v))[1+\dim \mathcal{F}_0]$$

in ${}^pD^{\geq 0}$ on $\mathbb{A}^1 \times \mathbb{A}^n \times \mathcal{F}$. Define the object $M_{def} = M_{def}(t, f)$ on $\mathbb{A}^1 \times \mathcal{F}$ by

$$M_{def} := R(pr_{1,3})_!(L(tv)\otimes K(f(v))[1+\dim \mathcal{F}_0]).$$

[We view M_{def} as a deformation of the object M on $\mathcal{F}$, whence the name.]

Corollary 3.3.10 The object M_{def} on $\mathbb{A}^1 \times \mathcal{F}$ lies in ${}^pD^{\geq 0}$.

proof This is just the fact that for π an affine morphism, $R\pi_!$ preserves ${}^pD^{\geq 0}$. QED

(3.3.11) What is the relation of the object M_{def} on $\mathbb{A}^1 \times \mathcal{F}$ to the perverse object

$$M = \mathrm{Twist}(L,K,\mathcal{F},h=0)$$

on $\mathcal{F}$? If we pull back M_{def} to $\{1\} \times \mathcal{F}$, we recover M, up to a shift:

$$M = M_{def}(1, f)[-1].$$

This is just proper base change. Similarly, if we pull back M_{def} to $\{0\} \times \mathcal{F}$, we recover the Twist object attached to $L(0)$:

$$\mathrm{Twist}(L(0),K,\mathcal{F},h=0) = M_{def}(0, f)[-1].$$

Lemma 3.3.12 Consider the automorphism λ of $\mathbb{G}_m \times \mathcal{F}$ given by

$$\lambda : \mathbb{G}_m \times \mathcal{F} \to \mathbb{G}_m \times \mathcal{F}, \quad (t, f) \mapsto (t, f_t),$$

and the projection

$$pr_2 : \mathbb{G}_m \times \mathcal{F} \to \mathcal{F}.$$

Then we have

$$\lambda^*(M_{def}|\mathbb{G}_m \times \mathcal{F}) \cong pr_2{}^* M[1].$$

proof By definition, $\lambda^*(M_{def}|\mathbb{G}_m \times \mathcal{F})$ is the restriction to $\mathbb{G}_m \times \mathcal{F}$ of

$$R(pr_{1,3})_!(L(tv) \otimes K(f(tv))[1+\dim \mathcal{F}_0]).$$

By means of the $\mathbb{G}_m \times \mathcal{F}$-linear scale automorphism of $\mathbb{G}_m \times \mathbb{A}^n \times \mathcal{F}$ given by

$$\text{scale} : \mathbb{G}_m \times \mathbb{A}^n \times \mathcal{F} \to \mathbb{G}_m \times \mathbb{A}^n \times \mathcal{F}, \quad (t, v, f) \mapsto (t, tv, f),$$

this last object is just

$$R(pr_{1,3})_!(\text{scale}^*(L(v) \otimes K(f(v))[1+\dim \mathcal{F}_0]))$$
$$\cong R(pr_{1,3})_!(L(v) \otimes K(f(v))[1+\dim \mathcal{F}_0])$$
$$= (pr_2 : \mathbb{G}_m \times \mathcal{F} \to \mathcal{F})^* M[1],$$

the last equality by proper base change. QED

(3.3.12.1) Now let us pass from the objects M_{def}, M, and $\mathrm{Twist}(L(0),K,\mathcal{F},h=0)$, all of which lie in ${}^pD^{\geq 0}$, to the sheaves of perverse origin

$$\mathcal{M}_{def} := \mathcal{H}^{-1-\dim \mathcal{F}}(M_{def})(-m/2), \text{ on } \mathbb{A}^1 \times \mathcal{F},$$
$$\mathcal{M} := \mathcal{H}^{-\dim \mathcal{F}}(M)(-m/2), \text{ on } \mathcal{F},$$
$$\mathcal{M}_0 = \mathcal{H}^{-\dim \mathcal{F}}(\mathrm{Twist}(L(0),K,\mathcal{F},h=0))(-m/2), \text{ on } \mathcal{F}.$$

The Tate twists $(-m/2)$ are put in the definition so as to be compatible with the $\mathcal{M}$ which occurs throughout the discussion 2.1.1.3 of the Higher Moment Theorem and its Corollary, cf. 2.2.3.

Homothety Contraction Theorem 3.3.13 Suppose we are in the setting of the homothety contraction construction. Thus

K on $\mathbb{A}^m$ is perverse, and satisfies

$$H_c{}^i(\mathbb{A}^m \otimes \bar{k}, K[m]) = 0 \text{ for } i > -n,$$

L on $\mathbb{A}^n$ is perverse,

$(\mathcal{F}, \tau)$ is quasifinitely difference-separating and stable by homothety.

Let U be a dense open set of $\mathcal{F}$ on which both $\mathcal{M}$ and $\mathcal{M}_0$ are lisse. Pick a geometric point u of U, and denote by ρ and ρ_0 the corresponding $\overline{\mathbb{Q}}_\ell$-representations of $\pi_1(U, u)$. Denote by

$$N := \mathrm{rank}(\mathcal{M}|U),$$
$$N_0 := \mathrm{rank}(\mathcal{M}_0|U),$$

$\Gamma := \rho(\pi_1(U, u)) \subset GL(\mathcal{M}_u) \cong GL(N, \overline{\mathbb{Q}}_\ell)$,
$\Gamma_0 := \rho_0(\pi_1(U, u)) \subset GL(\mathcal{M}_{0u}) \cong GL(N_0, \overline{\mathbb{Q}}_\ell)$,
$G :=$ the Zariski closure of Γ in $GL(N, \overline{\mathbb{Q}}_\ell)$,
$G_0 :=$ the Zariski closure of Γ_0 in $GL(N_0, \overline{\mathbb{Q}}_\ell)$.

Then we have the following results.
1) We have the inequality of ranks

$$N_0 \le N.$$

2) The group Γ_0 is isomorphic to a subquotient of Γ.
3) The algebraic group G_0 is a subquotient of G.
In particular, we have
3a) if G is finite (or equivalently if Γ is finite) then G_0 is finite (or equivalently Γ_0 is finite),
3b) $\dim(G_0) \le \dim(G)$,
3c) $\mathrm{rank}(G_0) \le \mathrm{rank}(G)$.

proof Consider the sheaf $\mathcal{M}_{def}$ of perverse origin on $\mathbb{A}^1 \times \mathcal{F}$. It is related to the sheaves $\mathcal{M}$ and $\mathcal{M}_0$ of perverse origin on $\mathcal{F}$ as follows. We have

$$\mathcal{M} = \mathcal{M}_{def}(1, f),$$
$$\mathcal{M}_0 = \mathcal{M}_{def}(0, f),$$
$$\lambda^*(\mathcal{M}_{def}|\mathbb{G}_m \times \mathcal{F}) \cong pr_2^*\mathcal{M}.$$

Since a sheaf of perverse origin is lisse precisely where it has maximum rank, we see that $\lambda^*(\mathcal{M}_{def}|\mathbb{G}_m \times \mathcal{F}) \cong pr_2^*\mathcal{M}|\mathbb{G}_m \times \mathcal{F}$ is lisse precisely at those points $(t\neq 0, f)$ in $\mathbb{G}_m \times \mathcal{F}$ such that $\mathcal{M}$ has maximal rank N at f. Therefore $\mathcal{M}_{def}|\mathbb{G}_m \times \mathcal{F}$ is lisse precisely at those points $(t\neq 0, f)$ such that $\mathcal{M}$ has maximum rank N at f_t. In particular, both $\lambda^*(\mathcal{M}_{def}|\mathbb{G}_m \times \mathcal{F}) \cong pr_2^*\mathcal{M}|\mathbb{G}_m \times \mathcal{F}$ and $\mathcal{M}_{def}|\mathbb{G}_m \times \mathcal{F}$ are lisse at every point of $\{1\}\times U$. So both are lisse at the geometric point $(1,u)$. Pick a dense open set U_{def} of $\mathbb{G}_m \times \mathcal{F}$ which contains $(1, u)$, on which both $\mathcal{M}_{def}|\mathbb{G}_m \times \mathcal{F}$ and $\lambda^*(\mathcal{M}_{def}|\mathbb{G}_m \times \mathcal{F}) \cong pr_2^*\mathcal{M}|\mathbb{G}_m \times \mathcal{F}$ are lisse.

Denote by ρ_{def} the $\overline{\mathbb{Q}}_\ell$-representations of $\pi_1(U_{def}, (1, u))$ corresponding to $\mathcal{M}_{def}$, and define

$\Gamma_{def} := \rho_{def}(\pi_1(U_{def}, (1, u))) \subset GL(\mathcal{M}_{def, (1,u)}) \cong GL(N, \overline{\mathbb{Q}}_\ell)$,
$G_{def} :=$ the Zariski closure of Γ_{def} in $GL(N, \overline{\mathbb{Q}}_\ell)$.

Similarly denote by $\rho_{def,\lambda}$ the $\overline{\mathbb{Q}}_\ell$-representations of $\pi_1(U_{def}, (1, u))$ corresponding to $\lambda^*(\mathcal{M}_{def}|\mathbb{G}_m \times \mathcal{F}) \cong pr_2^*\mathcal{M}|\mathbb{G}_m \times \mathcal{F}$. Notice that $\lambda^*(\mathcal{M}_{def}|\mathbb{G}_m \times \mathcal{F})$ and $\mathcal{M}_{def}$ have the same stalk at $(1, u)$. Define

$\Gamma_{def,\lambda} := \rho_{def,\lambda}(\pi_1(U_{def}, (1, u))) \subset GL(\mathcal{M}_{def, (1,u)}) \cong GL(N, \overline{\mathbb{Q}}_\ell)$,
$G_{def,\lambda} :=$ the Zariski closure of $\Gamma_{def,\lambda}$ in $GL(N, \overline{\mathbb{Q}}_\ell)$.

Now Γ_{def} (resp. $\Gamma_{def,\lambda}$) is independent of the particular choice of dense open set U_{def} on which $\mathcal{M}_{def}|\mathbb{G}_m\times\mathcal{F}$ (resp. $\lambda^*(\mathcal{M}_{def}|\mathbb{G}_m\times\mathcal{F})$) is lisse. If we use U_{def} for $\mathcal{M}_{def}|\mathbb{G}_m\times\mathcal{F}$ and $\lambda^{-1}U_{def}$ for $\lambda^*(\mathcal{M}_{def}|\mathbb{G}_m\times\mathcal{F})$, we see that we have an equality of image groups

$$\Gamma_{def} = \Gamma_{def,\lambda}.$$

On the other hand, the map

$$pr_2 : \mathbb{G}_m\times U \to U$$

induces a surjection on fundamental groups, so we have

$$\Gamma = \Gamma_{def,\lambda}.$$

So we have corresponding equalities of Zariski closures:

$$G_{def} = G_{def,\lambda} = G.$$

We apply the Semicontinuity Theorem 2.8.8 to the sheaf $\mathcal{M}_{def}$ of perverse origin on $\mathbb{A}^1\times\mathcal{F}$. We take for $Z \subset \mathbb{A}^1\times\mathcal{F}$ the irreducible closed set $\{0\}\times\mathcal{F}$. The sheaf $\mathcal{M}_{def}$ is lisse of rank N on the dense open set U_{def} in $\mathbb{A}^1\times\mathcal{F} - Z$, and $\mathcal{M}_{def}|Z = \mathcal{M}_0$ is lisse of rank N_0 on U. The Semicontinuity Theorem tells us that $N_0 \le N$, and that G_0 is a subquotient of $G_{def} = G_{def,\lambda} = G$. QED

Remark 3.3.14 The sheaf $\mathcal{M}_0$ vanishes unless $\mathcal{H}^{-n}(L(0)) = \mathcal{H}^{-n}(L)_0$, the stalk at 0 of $\mathcal{H}^{-n}(L)$, is nonzero. Indeed, the object L(0) is a successive extension of the objects

(the constant sheaf on $\mathbb{A}^n$ with value $\mathcal{H}^{-i}(L)_0)[i]$,

for i running from n down to 0. If $\mathcal{H}^{-n}(L(0)) = 0$, then L(0) lies in ${}^pD^{\ge 1}$. This implies that the entire object $L(0)\otimes K(f(v))[\dim\mathcal{F}_0]$ lies in ${}^pD^{\ge 1}$ on $\mathbb{A}^n\times\mathcal{F}$, which in turn implies that

$$\mathrm{Twist}(L(0), K, \mathcal{F}, h=0) := Rpr_{2!}(L(0)\otimes K(f(v))[\dim\mathcal{F}_0])$$

lies in ${}^pD^{\ge 1}$ on $\mathcal{F}$. This in turn implies that

$$\mathcal{M}_0 := \mathcal{H}^{-\dim\mathcal{F}}(\mathrm{Twist}(L(0), K, \mathcal{F}, h=0))(-m/2)$$
$$= \mathcal{H}^{-\dim\mathcal{F}}(\text{an object in } {}^pD^{\ge 1} \text{ on } \mathcal{F})$$
$$= 0.$$

Remark 3.3.15 Suppose on the contrary that in an open neighborhood U_0 of the origin in $\mathbb{A}^n$, $L|U_0$ is $\mathcal{L}[n]$ for some nonzero lisse sheaf $\mathcal{L}$ on U_0. Then L(0) is perverse. Indeed, L(0) is the geometrically constant perverse sheaf

(the constant sheaf of value $\mathcal{L}_0)[n]$.

First Homothety Contraction Corollary 3.3.16 Hypotheses and notations as in the Homothety Contraction Theorem 3.3.13, suppose **in addition** that, in an open neighborhood U_0 of the origin, $L|U_0$ is $\mathcal{L}[n]$ for some nonzero lisse sheaf $\mathcal{L}$ on U_0. Consider the following perverse sheaves on $\mathcal{F}$:

$$M := \mathrm{Twist}(L, K, \mathcal{F}, h=0),$$
$$M_0 := \mathrm{Twist}(L(0), K, \mathcal{F}, h=0),$$
$$M_{00} := \mathrm{Twist}(\overline{\mathbb{Q}}_\ell[n](n/2), K, \mathcal{F}, h=0),$$

and the sheaves of perverse origin on $\mathcal{F}$ to which they give rise:

$$\mathcal{M} := \mathcal{H}^{-\dim\mathcal{F}}(M)(-m/2),$$
$$\mathcal{M}_0 := \mathcal{H}^{-\dim\mathcal{F}}(M_0)(-m/2),$$
$$\mathcal{M}_{00} := \mathcal{H}^{-\dim\mathcal{F}}(M_{00})(-m/2).$$

Then we have the following results.

1) The perverse sheaves M_0 and M_{00} on $\mathcal{F}$ are related by

$$M_0 = M_{00}\otimes(\text{the geometrically constant sheaf } \mathcal{L}_0(-n/2)).$$

2) The sheaves of perverse origin $\mathcal{M}_0$ and $\mathcal{M}_{00}$ on $\mathcal{F}$ are related by

$$\mathcal{M}_0 = \mathcal{M}_{00}\otimes(\text{the geometrically constant sheaf } \mathcal{L}_0(-n/2)).$$

3) Let U be a dense open set of $\mathcal{F}$ on which $\mathcal{M}$, $\mathcal{M}_0$ and $\mathcal{M}_{00}$ are all lisse. Denote by

$$N := \mathrm{rank}(\mathcal{M}|U),$$
$$N_0 := \mathrm{rank}(\mathcal{M}_0|U),$$
$$N_{00} := \mathrm{rank}(\mathcal{M}_{00}|U).$$

Then we have equality

$$N_0 = N_{00}\times\dim(\mathcal{L}_0),$$

and the inequality

$$N \geq N_0.$$

proof Assertions 1) and 2) are trivial, since L(0) is the geometrically constant perverse sheaf $\mathcal{L}_0[n]$ on $\mathbb{A}^n$, so is isomorphic to the tensor product of the constant perverse sheaf $\overline{\mathbb{Q}}_\ell[n](n/2)$ with the geometrically constant lisse sheaf $\mathcal{L}_0(-n/2)$. Thanks to the hypotheses made on K, the functor

$$L \mapsto \mathrm{Twist}(L, K, \mathcal{F}, h=0)$$

is an exact functor from perverse sheaves on $\mathbb{A}^n$ to perverse sheaves on $\mathcal{F}$ (cf. Exactness Corollary 1.4.5), which visibly commutes with tensoring with geometrically constant (i.e., pulled back from Spec(k)) lisse sheaves. This proves 1), and 2) follows by applying the functor $\mathcal{H}^{-\dim\mathcal{F}}(?)(-m/2)$, which also commutes with tensoring with geometrically constant lisse sheaves.

The equality asserted in 3) is immediate from 2). The inequality $N \geq N_0$, "mise pour memoire", was already proven in part 1) of the Homothety Contraction Theorem 3.3.13. QED

Second Homothety Contraction Corollary 3.3.17 Hypotheses and notations as in the First Homothety Contraction Corollary 3.3.16, suppose **in addition** that

k is a finite field

K on $\mathbb{A}^m$ is perverse, geometrically irreducible, geometrically nonconstant, ι-pure of weight zero, and satisfies

$$n + \dim\mathrm{Supp}(K) \geq m+1.$$

L on $\mathbb{A}^n$ is perverse, geometrically irreducible, ι-pure of weight zero, and in an open neighborhood U_0 of the origin, $L|U_0$ is $\mathcal{L}[n]$ for some nonzero lisse sheaf $\mathcal{L}$ on U_0,

$(\mathcal{F}, \tau)$ is d-separating for some $d \geq 4$.

Consider the following perverse sheaves on $\mathcal{F}$:

$$M := \mathrm{Twist}(L, K, \mathcal{F}, h=0),$$
$$M_0 := \mathrm{Twist}(L(0), K, \mathcal{F}, h=0),$$
$$M_{00} := \mathrm{Twist}(\overline{\mathbb{Q}}_\ell[n](n/2), K, \mathcal{F}, h=0),$$

and the sheaves of perverse origin on $\mathcal{F}$ to which they give rise:

$$\mathcal{M} := \mathcal{H}^{-\dim\mathcal{F}}(M)(-m/2),$$
$$\mathcal{M}_0 := \mathcal{H}^{-\dim\mathcal{F}}(M_0)(-m/2),$$
$$\mathcal{M}_{00} := \mathcal{H}^{-\dim\mathcal{F}}(M_{00})(-m/2).$$

Then we have the following results.

1) The perverse sheaves M_0 and M_{00} on $\mathcal{F}$ are related by

$$M_0 = M_{00}\otimes(\text{the geometrically constant sheaf } \mathcal{L}_0(-n/2)).$$

2) The sheaves of perverse origin $\mathcal{M}_0$ and $\mathcal{M}_{00}$ on $\mathcal{F}$ are related by

$$\mathcal{M}_0 = \mathcal{M}_{00}\otimes(\text{the geometrically constant sheaf } \mathcal{L}_0(-n/2)).$$

3) Let U be a dense open set of $\mathcal{F}$ on which $\mathcal{M}$, $\mathcal{M}_0$, and $\mathcal{M}_{00}$ are all lisse. Denote by

$$N := \mathrm{rank}(\mathcal{M}|U),$$
$$N_{pure} := \mathrm{rank}(\mathrm{Gr}^0(\mathcal{M}|U)),$$
$$N_{pure,ncst} := \mathrm{rank}(\mathrm{Gr}^0(\mathcal{M}|U)_{ncst}),$$
$$N_0 := \mathrm{rank}(\mathcal{M}_0|U),$$
$$N_{0,pure,ncst} := \mathrm{rank}(\mathrm{Gr}^0(\mathcal{M}_0|U)_{ncst}),$$
$$N_{00} := \mathrm{rank}(\mathcal{M}_{00}|U),$$
$$N_{00,pure,ncst} := \mathrm{rank}(\mathrm{Gr}^0(\mathcal{M}_{00}|U)_{ncst}).$$

Then we have the equalities

$$N_0 = N_{00}\times\dim(\mathcal{L}_0),$$
$$N_{0,pure,ncst} = N_{00,pure,ncst}\times\dim(\mathcal{L}_0),$$

and the inequality

$$N \geq N_0.$$

4) Suppose that G_{geom} for $\mathrm{Gr}^0(\mathcal{M}_{00}|U)_{ncst}$ is not finite, and that we have the inequality

$$(N_{00,pure,ncst} -1)/2 \geq N - N_{pure}.$$

Then G_{geom} for $\mathrm{Gr}^0(\mathcal{M}|U)_{ncst}$ is not finite.

proof Assertions 1), 2), and 3) are proven exactly as in the proof of the First Homothety Contraction Corollary 3.3.16.

For 4), we argue as follows. By 2), $\mathcal{M}_0|U$ is geometrically isomorphic to the direct sum of $\dim(\mathcal{L}_0)$ copies of $\mathcal{M}_{00}|U$. So the group G_{geom} for $Gr^0(\mathcal{M}_0|U)_{ncst}$ is isomorphic (diagonal embedding) to the group G_{geom} for the geometrically irreducible (by the Higher Moment Theorem-second variant 1.22.11, applied to $\mathrm{Twist}(\overline{\mathbb{Q}}_\ell[n](n/2), K, \mathcal{F}, h=0)])$ lisse sheaf $Gr^0(\mathcal{M}_{00}|U)_{ncst}$. If this last group is not finite, we claim that its rank (dimension of a maximal torus) satisfies the inequality

$$\mathrm{rank}(G_{geom} \text{ for } Gr^0(\mathcal{M}_{00}|U)_{ncst}) \geq (N_{00,pure,ncst} - 1)/2.$$

To see this, we argue as follows. If this G_{geom} is not finite, then, being semisimple, it has rank at least one. So the asserted inequality holds trivially if $N_{00,pure,ncst} \leq 3$. When $N_{00,pure,ncst} \geq 4$, then by the Higher Moment Theorem-second variant 1.22.11 and Larsen's Alternative 2.2.2, it either contains $SL(N_{00,pure,ncst})$, or is one of $SO(N_{00,pure,ncst})$, $O(N_{00,pure,ncst})$, or $Sp(N_{00,pure,ncst})$. So looking case by case, we see that in each case the rank (dimension of a maximal torus) of this group satisfies the asserted inequality

$$\mathrm{rank}(G_{geom} \text{ for } Gr^0(\mathcal{M}_{00}|U)_{ncst}) \geq (N_{00,pure,ncst} - 1)/2.$$

On the other hand, $Gr^0(\mathcal{M}_{00}|U)_{ncst}$ is a quotient of $\mathcal{M}_{00}|U$, so its G_{geom} is a quotient of the group G_{geom} for $\mathcal{M}_{00}|U$. Thus

$$\mathrm{rank}(G_{geom} \text{ for } \mathcal{M}_{00}|U) \geq \mathrm{rank}(G_{geom} \text{ for } Gr^0(\mathcal{M}_{00}|U)_{ncst}).$$

So we have the inequality

$$\mathrm{rank}(G_{geom} \text{ for } \mathcal{M}_{00}|U) \geq (N_{00,pure,ncst} - 1)/2.$$

On the other hand, we have already proven in part 3c) of the Homothety Contraction Theorem 3.3.13 the inequality

$$\mathrm{rank}(G_{geom} \text{ for } \mathcal{M}|U) \geq \mathrm{rank}(G_{geom} \text{ for } \mathcal{M}_{00}|U).$$

So we have the inequality

$$\mathrm{rank}(G_{geom} \text{ for } \mathcal{M}|U) \geq (N_{00,pure,ncst} - 1)/2.$$

But we have assumed that we have

$$(N_{00,pure,ncst} - 1)/2 \geq N - N_{pure}.$$

So we have the inequality

$$\mathrm{rank}(G_{geom} \text{ for } \mathcal{M}|U) \geq N - N_{pure}.$$

Suppose that G_{geom} for $Gr^0(\mathcal{M}|U)_{ncst}$ is finite. We derive a contradiction to this last inequality as follows. The sheaf $Gr^0(\mathcal{M}|U)_{ncst}$ is a quotient of $\mathcal{M}|U$. We have a short exact sequence

$$0 \to (\mathcal{M}|U)_{wt < 0} \to \mathcal{M}|U \to Gr^0(\mathcal{M}|U) \to 0,$$

and a direct sum decomposition

$$Gr^0(\mathcal{M}|U) = Gr^0(\mathcal{M}|U)_{cst} \oplus Gr^0(\mathcal{M}|U)_{ncst},$$

cf. Lemma 1.10.3. So the action of G_{geom} for $\mathcal{M}|U$, in a basis adapted to this filtration, has the block upper triangular shape

$$\begin{matrix} A & * & * \\ 0 & Id & * \\ 0 & 0 & C, \end{matrix}$$

where the A block is the action of G_{geom} on $(\mathcal{M}|U)_{wt<0}$, the Id block is the trivial action of G_{geom} on $Gr^0(\mathcal{M}|U)_{cst}$, and the C block is the action of G_{geom} on $Gr^0(\mathcal{M}|U)_{ncst}$. So if G_{geom} for $Gr^0(\mathcal{M}|U)_{ncst}$ is finite, then the identity component $(G_{geom})^0$ has block shape

$$\begin{matrix} A & * & * \\ 0 & Id & * \\ 0 & 0 & Id, \end{matrix}$$

in which case we have the obvious equality

$$\mathrm{rank}(G_{geom} \text{ for } \mathcal{M}|U) = \mathrm{rank}(G_{geom} \text{ for } (\mathcal{M}|U)_{wt<0}).$$

The group G_{geom} for $(\mathcal{M}|U)_{wt<0}$ has determinant of finite order, because $(\mathcal{M}|U)_{wt<0}$ is a lisse sheaf on the lisse k-scheme U, so its determinant is geometrically of finite order, cf. [De-Weil II, 1.3.4]. Therefore we have the inclusion

$$(G_{geom} \text{ for } (\mathcal{M}|U)_{wt<0})^0$$
$$\subset SL(\mathrm{rank}((\mathcal{M}|U)_{wt<0})) = SL(N - N_{pure}).$$

So we have the inequality

$$\mathrm{rank}(G_{geom} \text{ for } (\mathcal{M}|U)_{wt<0}) < N - N_{pure},$$

which is the desired contradiction. QED

(3.4) Return to the proof of the $\mathcal{L}_\psi$ theorem

(3.4.1) We now apply the Second Homothety Contraction Corollary 3.3.17 with $m=1$, $K = \mathcal{L}_\psi[1](1/2)$ on $\mathbb{A}^1$, and with the given L on $\mathbb{A}^n$. Consider the perverse sheaf M on $\mathcal{F}$ given by

$$M = \mathrm{Twist}(L, \mathcal{L}_\psi[1](1/2), \mathcal{P}_e, h=0).$$

As before, we define the sheaf $\mathcal{M}$ of perverse origin on $\mathcal{F}$

$$\mathcal{M} := \mathcal{H}^{-\dim \mathcal{F}}(M)(-m/2).$$

Now consider the perverse sheaf on $\mathcal{F}$ given by

$$M_{00} := \mathrm{Twist}(\overline{\mathbb{Q}}_\ell[n](n/2), \mathcal{L}_\psi[1](1/2), \mathcal{P}_e, h=0),$$

and the sheaf of perverse origin on $\mathcal{F}$ defined by

$$\mathcal{M}_{00} := \mathcal{H}^{-\dim \mathcal{F}}(M_{00})(-m/2).$$

(3.4.2) We know, by the Higher Moment Theorem-second variant 1.22.11, that on any dense open set U of $\mathcal{F}$ on which both $\mathcal{M}$ and $\mathcal{M}_{00}$ are lisse, both are geometrically irreducible (and hence nonzero), geometrically nonconstant, and ι-pure of weight zero. So in the notations

$$N := \mathrm{rank}(\mathcal{M}|U),$$

$$N_{pure} := \mathrm{rank}(Gr^0(\mathcal{M}|U)),$$

$$N_{pure,ncst} := \mathrm{rank}(Gr^0(\mathcal{M}|U)_{ncst}),$$

$$N_{00} := \mathrm{rank}(\mathcal{M}_{00}|U),$$

$N_{00,\text{pure,ncst}} := \text{rank}(\text{Gr}^0(\mathcal{M}_{00}|U)_{\text{ncst}})$,

we have

$$N = N_{\text{pure}} = N_{\text{pure,ncst}} \geq 1,$$
$$N_{00} = N_{00,\text{pure,ncst}} \geq 1.$$

In particular, we have the inequality

$$(N_{00,\text{pure,ncst}} - 1)/2 \geq N - N_{\text{pure}}.$$

So if we knew that G_{geom} for $\text{Gr}^0(\mathcal{M}_{00}|U)_{\text{ncst}} = \mathcal{M}_{00}|U$ were not finite, we would conclude, from part 4) of the Second Homothety Contraction Corollary 3.3.17 that G_{geom} for $\text{Gr}^0(\mathcal{M}|U)_{\text{ncst}} = \mathcal{M}|U$ is not finite.

(3.4.3) We also know that

$$N \geq N_0 = \text{rank}(\mathcal{L}_0)N_{00}.$$

So it remains to prove two things.

(3.4.4) First, we must prove that G_{geom} for $\mathcal{M}_{00}|U$ is not finite, provided that any one of the following five conditions holds:

a) $p \geq 7$,
b) $n \geq 3$,
c) $p = 5$ and $e \geq 4$,
d) $p = 3$ and $e \geq 7$,
e) $p = 2$ and $e \geq 7$.

(3.4.5) Second, we must prove that we have the rank estimates

$$\text{rank}(\mathcal{M}_{00}|U) = (e-1)^n, \text{ if } e \text{ is prime to } p,$$

$$\text{rank}(\mathcal{M}_{00}|U) \geq \text{Max}((e-2)^n, (1/e)((e-1)^n + (-1)^n(e-1))), \text{ if } p|e.$$

(3.4.6) We will prove both of these statements in Theorem 3.8.2.

(3.5) Monodromy of exponential sums of Deligne type on $\mathbb{A}^n$

(3.5.1) In the previous section, we reduced the proof of the $\mathcal{L}_\psi$ theorem to the proof of certain results on the perverse sheaf M_{00} on $\mathcal{F} = \mathcal{P}_e$ given by

$$M_{00} := \text{Twist}(\overline{\mathbb{Q}}_\ell[n](n/2), \mathcal{L}_\psi[1](1/2), \mathcal{P}_e, h=0),$$

and the sheaf of perverse origin on $\mathcal{P}_e$ defined by

$$\mathcal{M}_{00} := \mathcal{H}^{-\dim \mathcal{F}}(M_{00})(-1/2).$$

(3.5.2) In the discussion which is to follow, we will need to pay attention to the parameters n, e, ψ, which had previously been fixed and which did not figure explicitly in the notation. So what we were calling M_{00} and $\mathcal{M}_{00}$ we now rename

$$M(n,e,\psi) := \text{Twist}(\overline{\mathbb{Q}}_\ell[n](n/2), \mathcal{L}_\psi[1](1/2), \mathcal{P}_e, h=0),$$

$$\mathcal{M}(n,e,\psi) := \mathcal{H}^{-\dim \mathcal{F}}(M(n,e,\psi))(-1/2).$$

And what we previously denoted $\mathcal{P}_e$ we now denote

$\mathcal{P}(n,e)$:= the space of polynomials in n variables of degree $\leq$ e, to make explicit "n".

Dependence on ψ Lemma 3.5.3 Given α in $k^{\times}$, denote by ψ_{α} the additive character $x \mapsto \psi(\alpha x)$, and denote by $f \mapsto \alpha f$ the homothety "α" on $\mathcal{P}(n,e)$. We have

$$M(n,e,\psi_{\alpha}) = [f \mapsto \alpha f]^* M(n,e,\psi),$$

$$\mathcal{M}(n,e,\psi_{\alpha}) = [f \mapsto \alpha f]^* \mathcal{M}(n,e,\psi).$$

proof This is a trivial instance of proper base change, reflecting the identity $\psi_{\alpha}(f(v)) = \psi(\alpha f(v))$. QED

(3.5.4) Before proceeding, let us relate the objects to the exponential sums they were built to incarnate. Given a finite extension E/k, and an additive character ψ of k, we denote by ψ_E the additive character of E defined by

$$\psi_E(x) := \psi(\mathrm{Trace}_{E/k}(x)).$$

Recall that we have also fixed a square root of p := char(k), allowing us to form Tate twists by half integers, and allowing us to give unambiguous meaning to half-integral powers of $\# E$.

(3.5.5) In down to earth terms, on the space $\mathbb{A}^n \times \mathcal{P}(n,e)$, with coordinates (v, f), we have the lisse sheaf $\mathcal{L}_{\psi(f(v))}$. Under the second projection $pr_2 : \mathbb{A}^n \times \mathcal{P}(n,e) \to \mathcal{P}(n,e)$, we form $Rpr_{2!}\mathcal{L}_{\psi(f(v))}$. For E/k a finite extension, and for f in $\mathcal{P}_e(E)$, the stalk of $Rpr_{2!}\mathcal{L}_{\psi(f(v))}$ at f is the object $R\Gamma_c(\mathbb{A}^n \otimes_k \overline{E}, \mathcal{L}_{\psi(f)})$, whose cohomology objects are the groups

$$\mathcal{H}^i(R\Gamma_c(\mathbb{A}^n \otimes_k \overline{E}, \mathcal{L}_{\psi(f)})) = H_c{}^i(\mathbb{A}^n \otimes_k \overline{E}, \mathcal{L}_{\psi(f)}).$$

The perverse sheaf M_{00} is just a Tate twist and a shift of $Rpr_{2!}\mathcal{L}_{\psi(f(v))}$; we have

$$Rpr_{2!}\mathcal{L}_{\psi(f(v))} = M(n,e,\psi)[-n - \dim \mathcal{P}(n,e)]((-n-1)/2).$$

In particular, we have

$$\mathcal{M}(n,e,\psi) = R^n pr_{2!}\mathcal{L}_{\psi(f(v))}(n/2).$$

Lemma 3.5.6 For E/k a finite extension, and for f in $\mathcal{F}(E) = \mathcal{P}(n,e)(E)$, i.e., for f an E-polynomial in n variables of degree at most e, we have the identity

$$\mathrm{Trace}(\mathrm{Frob}_{E,f} \mid M(n,e,\psi))$$
$$= (-1)^{n+\dim \mathcal{F}}(\# E)^{-(n+1)/2}\Sigma_{v \text{ in } \mathbb{A}^n(E)}\, \psi_E(f(v)).$$

proof Immediate from the Lefschetz Trace Formula, proper base change, and the definition of M_{00}. QED

Lemma 3.5.7 For $U \subset \mathcal{P}(n,e)$ an open dense set on which $M(n,e,\psi)$ has lisse cohomology sheaves, $M(n,e,\psi)|U$ is the lisse sheaf $\mathcal{M}(n,e,\psi)(1/2)|U$, placed in degree $-\dim \mathcal{F}$:

$$M(n,e,\psi)|U = (\mathcal{M}(n,e,\psi)(1/2)|U)[\dim \mathcal{F}].$$

For E/k a finite extension, and for f in U(n,e)(E), we have

$$\mathrm{Trace}(\mathrm{Frob}_{E,f} \mid \mathcal{M}(n,e,\psi)) = (-1)^n(\#E)^{-n/2}\Sigma_{v \text{ in } \mathbb{A}^n(E)} \psi_E(f(v)).$$

proof A perverse sheaf L on a smooth connected k-scheme U with lisse cohomology sheaves is of the form $\mathcal{L}[\dim U]$ for $\mathcal{L}$ a lisse sheaf on U, which can be recovered from L as $\mathcal{H}^{-\dim U}(L)$. Applied to $M_{00}|U$, we find the first assertion. The second assertion is then immediate from the previous lemma. QED

(3.5.8) At this point, we must recall a key result of Deligne [De-Weil I 8.4, Ka-SE 5.1.1] concerning exponential sums on $\mathbb{A}^n$. Let us say that an n-variable polynomial f in $\mathcal{P}(n,e)(\bar{k})$ is a Deligne polynomial if it satisfies the following two conditions:
D1) When we write $f = \Sigma_{i \le e} F_i$ as a sum of homogeneous forms, F_e is nonzero, and, in the case $n \ge 2$, the closed subscheme of $\mathbb{P}^{n-1}$ defined by the vanishing of F_e is smooth of codimension one.
D2) The integer e is prime to p.

(3.5.9) For a fixed integer e which is prime to p, the Deligne polynomials, those where the discriminant $\Delta(F_e)$ is invertible, form a dense open set $\mathcal{D}(n,e)$ of $\mathcal{P}(n,e)$. [And for any e, the polynomials in $\mathcal{P}(n,e)$ satisfying D1, i.e., those where $\Delta(F_e)$ is invertible, form a dense open set, but we will not have occasion to use that set.]

Theorem 3.5.10 ([De-Weil I, 8.4], [Ka-SE 5.1.1]) Fix an integer $e \ge 1$ prime to p. For any finite extension E/k, and any Deligne polynomial f in $\mathcal{D}(n,e)(E)$, we have the following results.
1) The "forget supports" maps

$$H_c^i(\mathbb{A}^n \otimes_k \bar{E}, \mathcal{L}_{\psi(f)}) \to H^i(\mathbb{A}^n \otimes_k \bar{E}, \mathcal{L}_{\psi(f)})$$

are all isomorphisms.
2) The groups $H_c^i(\mathbb{A}^n \otimes_k \bar{E}, \mathcal{L}_{\psi(f)})$ vanish for $i \neq n$.
3) The group $H_c^n(\mathbb{A}^n \otimes_k \bar{E}, \mathcal{L}_{\psi(f)})$ is pure of weight n, and has dimension $(e-1)^n$.

Corollary 3.5.11 Suppose $e \ge 1$ is prime to p. Then $M(n,e,\psi)$ has lisse cohomology sheaves on the dense open set $\mathcal{D}(n,e) \subset \mathcal{P}(n,e)$ consisting of Deligne polynomials. We have

$$M(n,e,\psi)|\mathcal{D}(n,e) = \mathcal{M}(n,e,\psi)(1/2)[\dim \mathcal{P}(n,e)]|\mathcal{D}(n,e),$$

$$\mathcal{M}(n,e,\psi) = R^n pr_{2!}\mathcal{L}_{\psi(f(v))}(n/2).$$

In particular, $\mathcal{M}(n,e,\psi) = R^n pr_{2!}\mathcal{L}_{\psi(f(v))}(n/2)$ is lisse on $\mathcal{D}(n,e)$ of rank $(e-1)^n$, and the other $R^i pr_{2!}\mathcal{L}_{\psi(f(v))}$ vanish on $\mathcal{D}(n,e)$.
proof Looking fibre by fibre, we see that $R^i pr_{2!}\mathcal{L}_{\psi(f(v))}|\mathcal{D}(n,e)$ vanishes for $i \neq n$. Therefore $\mathcal{H}^{-i}(M(n,e,\psi))|\mathcal{D}(n,e)$ vanishes for

$i \neq \dim\mathcal{F}$. The remaining cohomology sheaf

$$\mathcal{H}^{-\dim\mathcal{F}}(M(n,e,\psi)) = R^n pr_{2!}\mathcal{L}_{\psi(f(v))}((n+1)/2)$$

is of perverse origin on $\mathcal{P}(n,e)$. As it has constant rank $(e-1)^n$ on $\mathcal{D}(n,e)$, it is lisse on $\mathcal{D}(n,e)$. QED

Corollary 3.5.12 For $e \geq 2$ prime to p, E/k any finite extension, and f in $\mathcal{D}(n,e)(E)$, we have

$$\mathrm{Trace}(\mathrm{Frob}_{E,f} \mid \mathcal{M}(n,e,\psi))$$
$$= (-1)^n(\# E)^{-n/2}\Sigma_{v \text{ in } \mathbb{A}^n(E)} \psi_E(f(v)).$$

Determinant Lemma 3.5.13 Suppose $e \geq 2$ is prime to p. Then we have the following results.

1) For any finite extension E/k, and for any f in $\mathcal{D}(n,e)(E)$, we have

$$\det(-\mathrm{Frob}_E \mid H_c^n(\mathbb{A}^n \otimes_k \overline{E}, \mathcal{L}_{\psi(f)}))$$
$$= (\text{a root of unity}) \times (\# E)^{(n/2)(e-1)^n}.$$

2) The determinant of the lisse sheaf $\mathcal{M}(n,e,\psi)|\mathcal{D}(n,e)$ is arithmetically of finite order.

proof Any lisse $\overline{\mathbb{Q}}_\ell$-sheaf on a geometrically connected smooth k-scheme has its determinant geometrically of finite order, cf. [De-Weil II, 1.3.4]. So to show that the determinant of $\mathcal{M}(n,e,\psi)|\mathcal{D}(n,e)$ is arithmetically of finite order, it suffices to show, for a single finite extension E/k and for a single f in $\mathcal{D}(n,e)(E)$, that $\det(\mathrm{Frob}_{E,f}|\mathcal{M}(n,e,\psi))$ is a root of unity. For any such f, we have

$$\mathcal{M}(n,e,\psi)_f = H_c^n(\mathbb{A}^n \otimes_k \overline{E}, \mathcal{L}_{\psi(f)})(n/2),$$

and $H_c^n(\mathbb{A}^n \otimes_k \overline{E}, \mathcal{L}_{\psi(f)})$ has dimension $(e-1)^n$. So we have

$$\det(\mathrm{Frob}_{E,f}|\mathcal{M}(n,e,\psi))$$
$$= \det(\mathrm{Frob}_E \mid H_c^n(\mathbb{A}^n \otimes_k \overline{E}, \mathcal{L}_{\psi(f)}))/(\# E)^{(n/2)(e-1)^n}.$$

So it suffices to prove 1). For this, we argue as follows. Since we are claiming an equality up to roots of unity, we may replace E by its quadratic extension, which has the effect of "squaring both sides". This reduces us to the case where $\# E$ is a square, and so where

$$(\# E)^{(n/2)(e-1)^n}$$

lies in $\mathbb{Z}$. Think of ψ as having values in the cyclotomic field $\mathbb{Q}(\zeta_p)$. Then for every ℓ not p, and for every embedding of $\mathbb{Q}(\zeta_p)$ into $\overline{\mathbb{Q}}_\ell$, the L-function of $\mathbb{A}^n$ with coefficients in $\mathcal{L}_{\psi(f)}$, a rational function in $\mathbb{Q}(\zeta_p)(T)$ which is either a polynomial or a reciprocal polynomial, is given, in $\overline{\mathbb{Q}}_\ell(T)$, by

$$(L(\mathbb{A}^n/E, \mathcal{L}_{\psi(f)})(T))^{(-1)^{n+1}} = \det(1-T\mathrm{Frob}_E \mid H_c^n(\mathbb{A}^n \otimes_k \overline{E}, \mathcal{L}_{\psi(f)})),$$

where, via the given embedding, we view ψ as having values in $\overline{\mathbb{Q}}_\ell$. So we see that

$$A := \det(-\mathrm{Frob}_E \mid H_c^n(\mathbb{A}^n \otimes_k \overline{E}, \mathcal{L}_{\psi(f)}))$$

lies in $\mathbb{Q}(\zeta_p)$, because it is the highest degree coefficient in the $\mathbb{Q}(\zeta_p)$-polynomial

$$(L(\mathbb{A}^n/E, \mathcal{L}_{\psi(f)})(T))^{(-1)^{n+1}}.$$

On the one hand, the ℓ-adic interpretation of A shows that for every finite place λ of $\mathbb{Q}(\zeta_p)$ of residue characteristic not p, A is a λ-adic unit. And Deligne's theorem 3.5.10 that $H_c^n(\mathbb{A}^n\otimes_k\bar{E}, \mathcal{L}_{\psi(f)})$ is pure of weight n and of dimension $(e-1)^n$ shows that A has absolute value $(\#E)^{(n/2)(e-1)^n}$ at every archimedean place. So the ratio

$$A/(\#E)^{(n/2)(e-1)^n}$$

is a nonzero element in $\mathbb{Q}(\zeta_p)$ which is a unit at every place, finite or infinite, except possibly at places lying over p. But in $\mathbb{Q}(\zeta_p)$, there is a unique place lying over p. By the product formula, the ratio must be also be a unit at the unique place over p. Being a unit everywhere, the ratio is (Kronecker's theorem!) a root of unity. QED

Corollary 3.5.14 Suppose $e \geq 2$ is prime to p. Given a finite extension E/k, and f in $\mathcal{D}(n,e)(E)$, suppose some power of

$$\mathrm{Frob}_E|\ H_c^n(\mathbb{A}^n\otimes_k\bar{E}, \mathcal{L}_{\psi(f)})$$

has all equal eigenvalues. Then we have the following results.

1) Every eigenvalue of

$$\mathrm{Frob}_E|\ H_c^n(\mathbb{A}^n\otimes_k\bar{E}, \mathcal{L}_{\psi(f)})$$

is of the form

$$(\text{a root of unity})\times(\#E)^{n/2}.$$

2) The sum $\Sigma_{v \text{ in } \mathbb{A}^n(E)}\ \psi_E(f(v))$ is of the form

$$(\#E)^{n/2}\times(\text{an algebraic integer}).$$

proof 1) Indeed, if some power $(\mathrm{Frob}_E)^r$ has equal eigenvalues, say all equal to w^r, then

$$w^{r(e-1)^n} = \det((\mathrm{Frob}_E)^r|\ H_c^n(\mathbb{A}^n\otimes_k\bar{E}, \mathcal{L}_{\psi(f)}))$$
$$= (\#E)^{(rn/2)(e-1)^n}.$$

But all the eigenvalues of Frob_E are of the form (a root of unity)$\times$w. Since $\pm\Sigma_{v \text{ in } \mathbb{A}^n(E)}\ \psi_E(f(v))$ is the sum of the eigenvalues of

$$\mathrm{Frob}_E|\ H_c^n(\mathbb{A}^n\otimes_k\bar{E}, \mathcal{L}_{\psi(f)}),$$

assertion 2) is obvious from 1). QED

Given an integer $e \geq 1$, whether or not prime to p, let us denote by

$$U(n,e,\psi) \subset \mathcal{P}(n,e)$$

a dense open set on which $M(n,e,\psi)$ has lisse cohomology sheaves. For e prime to p, we may take $U(n,e,\psi) = \mathcal{D}(n,e)$. For e divisible by p, we know much less. If n=1, then the open set

$$U(1,e) := \{\Sigma_{i\le e} a_i x^i \mid a_{e-1} \text{ is invertible}\} \subset \mathcal{P}(1,e)$$

is such a $U(1,e,\psi)$ for every ψ, and

$$\mathrm{rank}(\mathcal{M}(1,e,\psi)|U(1,e,\psi)) = e-2.$$

But for e divisible by p and $n \ge 2$, we do not know an explicit description of any $U(n,e,\psi)$. Nor do we know if we can choose a single $U(n,e)$ which, after arbitrary extension of scalars from k to a finite extension E/k, serves as a $U(n,e,\psi)$ for every nontrivial ψ on E. Nor do we know the rank of $\mathcal{M}(n,e\ \psi)|U(n,e,\psi)$. See [Ka-Lau, Remark 5.5.2] for similar conundra.

(3.5.15) We now consider the effect of varying e, keeping n fixed.

Lemma 3.5.16 Let $e \ge 2$, and consider the natural inclusion

$$\mathcal{P}(n,e-1) \subset \mathcal{P}(n,e).$$

We have

$$M(n,e,\psi)[\dim\mathcal{P}(n,e-1) - \dim\mathcal{P}(n,e)]|\mathcal{P}(n,e-1) = M(n,e-1,\psi),$$
$$\mathcal{M}(n,e,\psi)|\mathcal{P}(n,e-1) = \mathcal{M}(n,e-1,\psi).$$

proof. An instance of proper base change. QED

Corollary 3.5.17 For $e \ge 2$ divisible by p,

$$\mathrm{rank}(\mathcal{M}(n,e,\psi)|U(n,e,\psi)) \ge (e-2)^n.$$

proof Since $\mathcal{M}(n,e,\psi)$ on $\mathcal{P}(n,e)$ is of perverse origin, its generic rank is at least the generic rank of any pullback. Its pullback to $\mathcal{P}(n,e-1)$ has generic rank $(e-2)^n$, by Corollary 3.5.11, since e-1 is prime to p. QED

(3.5.18) For $n \ge 2$, and $e \ge 1$, denote by

$$\mathrm{Homog}(n,e) \subset \mathcal{P}(n,e)$$

the linear space of homogeneous forms of degree e in n variables, and by

$$\mathrm{NSHomog}(n,e) \subset \mathrm{Homog}(n,e)$$

the dense open set consisting of nonsingular (NS) forms, i.e., those forms F of degree e such that the closed subscheme of $\mathbb{P}^{n-1}$ defined by the vanishing of F is smooth of codimension one.

Lemma 3.5.19 For $n \ge 2$ and $e \ge 1$, the pullback

$$M(n,e,\psi)[\dim\mathrm{Homog}(n,e) - \dim\mathcal{P}(n,e)]|\mathrm{Homog}(n,e)$$

is perverse on $\mathrm{Homog}(n,e)$.

proof Indeed, evaluation of homogeneous forms of any given degree defines a finite map

$$\mathrm{eval} : \mathbb{A}^n \to \mathrm{Homog}(n,e)^\vee.$$

So for any perverse L on $\mathbb{A}^n$, $\mathrm{eval}_\star L$ is perverse on $\mathrm{Homog}(n,e)^\vee$. The object

$$M(n,e,\psi)[\dim\mathrm{Homog}(n,e) - \dim\mathcal{P}(n,e)]|\mathrm{Homog}(n,e)$$

is none other than $\mathrm{FT}_\psi(\mathrm{eval}_\star \overline{\mathbb{Q}}_\ell[n]((n+1)/2))$, cf. the proof of Compatibility Lemma 3.2.3, and hence is itself perverse. QED

(3.6) Interlude: an exponential sum calculation

(3.6.1) The following calculation grew out of a discussion with Steve Sperber. We fix $e \geq 2$, $n \geq 2$. We denote by e_0 the "prime to p part" of e, i.e., write e as

$$e = e_0 p^a,$$

with e_0 prime to p, and $a \geq 0$. For a nontrivial $\overline{\mathbb{Q}}_\ell^\times$-valued multiplicative character χ of $k^\times$, we extend χ by zero to k. For any finite extension E/k, we denote by χ_E the multiplicative character of $E^\times$ defined by

$$\chi_E(x) := \chi(N_{E/k}(x)).$$

We recall that the gauss sum $g(\psi, \chi)$ is defined by

$$g(\psi, \chi) = \Sigma_{x \text{ in } k^\times} \psi(x)\chi(x),$$

and satisfies the Hasse-Davenport relation: for a finite extension E/k,

$$-g(\psi_E, \chi_E) = (-g(\psi, \chi))^{\deg(E/k)}.$$

(3.6.2) For a homogeneous form F of degree e, a nontrivial multiplicative character χ of order dividing e (i.e., of order dividing e_0), and a point $v \neq 0$ in $\mathbb{A}^n(k)$, the value $\chi(F(v))$ depends only on the image of v in $\mathbb{P}^{n-1}(k)$ (because if we replace v by λv, λ in $k^\times$, $F(\lambda v) = \lambda^e F(v)$, and $\chi^e = \mathbb{1}$). So we may view $\chi(F(v))$ as a function on $\mathbb{P}^{n-1}(k)$. More precisely, denote by

$$X_F \subset \mathbb{P}^{n-1}$$

the hypersurface defined by the vanishing of F, and by $\mathbb{P}^{n-1}[1/F]$ its complement. As explained in [Ka-ENSMCS, section 8, page 11], we can speak of the lisse, rank one Kummer sheaf $\mathcal{L}_{\chi(F)}$ on $\mathbb{P}^{n-1}[1/F]$.

(3.6.3) Suppose now that F in NSHomog(n,e)(k) is a nonsingular form. Denote by L in $H^2(X_F \otimes_k \overline{k}, \overline{\mathbb{Q}}_\ell)(1)$ the cohomology class of a hyperplane section, and define

$$\mathrm{Prim}^{n-2}(X_F \otimes_k \overline{k}, \overline{\mathbb{Q}}_\ell)$$
$$:= H^{n-2}(X_F \otimes_k \overline{k}, \overline{\mathbb{Q}}_\ell), \text{ if n is odd,}$$
$$:= \text{orthogonal complement of } L^a \text{ in } H^{n-2}(X_F \otimes_k \overline{k}, \overline{\mathbb{Q}}_\ell), \text{ if}$$

$n-2 = 2a$.

One knows that

$$\dim\mathrm{Prim}^{n-2}(X_F \otimes_k \overline{k}, \overline{\mathbb{Q}}_\ell) = ((e-1)^n + (-1)^n(e-1))/e.$$

It is well known [Ka-Sar-RMFEM, 11.4.2] that the cohomology of X_F, as a graded Frob_k-module, is given by

$$H^*(X_F \otimes_k \overline{k}, \overline{\mathbb{Q}}_\ell) \cong \mathrm{Prim}^{n-2}(X_F \otimes_k \overline{k}, \overline{\mathbb{Q}}_\ell) \oplus H^*(\mathbb{P}^{n-2} \otimes_k \overline{k}, \overline{\mathbb{Q}}_\ell).$$

We also recall [Ka-ENSMCS, section 8, pages 11-12] that for any nontrivial multiplicative χ with $\chi^e = \mathbb{1}$, and for any F in NSHomog(n,e)(k), we have

$H_c^i(\mathbb{P}^{n-1}[1/F]\otimes_k\overline{k}, \mathcal{L}_{\chi(F)}) = 0$ for $i \neq n-1$,

$H_c^{n-1}(\mathbb{P}^{n-1}[1/F]\otimes_k\overline{k}, \mathcal{L}_{\chi(F)})$ is pure of weight $n-1$,

$\dim H_c^{n-1}(\mathbb{P}^{n-1}[1/F]\otimes_k\overline{k}, \mathcal{L}_{\chi(F)}) = ((e-1)^n + (-1)^{n+1})/e$.

Theorem 3.6.4 Let $e \geq 2$, $n \geq 2$. Suppose that E/k is a finite extension which contains the e_0'th roots of unity. Let F in NSHomog(n,e)(E) be a nonsingular form of degree e in n variables. For any finite extension L/E, we have the identity

$$(-1)^n\Sigma_{v \text{ in } \mathbb{A}^n(L)} \psi_L(F(v))$$
$$= \mathrm{Trace}(\mathrm{Frob}_L, \mathrm{Prim}^{n-2}(X_F\otimes_E\overline{E}, \overline{\mathbb{Q}}_\ell)(-1))$$
$$+\Sigma_{\chi \neq \mathbb{1},\ \chi^{e_0} = \mathbb{1}} (-g(\psi_E, \overline{\chi}_E))\mathrm{Trace}(\mathrm{Frob}_L, H_c^{n-1}(\mathbb{P}^{n-1}[1/F]\otimes_E\overline{E}, \mathcal{L}_{\chi(F)})).$$

proof Extending scalars, it suffices to treat universally the case where $L = E = k$, and k contains the e_0'th roots of unity. In view of the Frob_k-module isomorphism

$$H^*(X_F\otimes_k\overline{k}, \overline{\mathbb{Q}}_\ell) \cong \mathrm{Prim}^{n-2}(X_F\otimes_k\overline{k}, \overline{\mathbb{Q}}_\ell) \oplus H^*(\mathbb{P}^{n-2}\otimes_k\overline{k}, \overline{\mathbb{Q}}_\ell)$$

and the Lefschetz Trace formula, applied to both X_F and $\mathbb{P}^{n-2}$, we have

$$\#X_F(k) = (-1)^{n-2}\mathrm{Trace}(\mathrm{Frob}_k, \mathrm{Prim}^{n-2}(X_F\otimes_k\overline{k}, \overline{\mathbb{Q}}_\ell)) + \#\mathbb{P}^{n-2}(k).$$

We rewrite this as

$$\mathrm{Trace}(\mathrm{Frob}_k, \mathrm{Prim}^{n-2}(X_F\otimes_k\overline{k}, \overline{\mathbb{Q}}_\ell)(-1))$$
$$= (-1)^{n-2}(\#k)(\#X_F(k) - \#\mathbb{P}^{n-2}(k)).$$

Similarly, the vanishing of $H_c^i(\mathbb{P}^{n-1}[1/F]\otimes_k\overline{k}, \mathcal{L}_{\chi(F)})$ for $i\neq n-1$ gives

$$\mathrm{Trace}(\mathrm{Frob}_k, H_c^{n-1}(\mathbb{P}^{n-1}[1/F]\otimes_k\overline{k}, \mathcal{L}_{\chi(F)})$$
$$=(-1)^{n-1}\Sigma_{x \text{ in } \mathbb{P}^{n-1}(k)} \chi(F(x)).$$

With these preparations , we calculate directly. For each point x in $\mathbb{P}^{n-1}(k)$, choose a lift $\tilde{x}$ in $\mathbb{A}^n(k)$. Then each nonzero point v in $\mathbb{A}^n(k)$ can be written uniquely as $\lambda\tilde{x}$, with λ in $k^\times$ and x in $\mathbb{P}^{n-1}(k)$. Now $F(0) = 0$ because F is homogeneous, so

$$\Sigma_{v \text{ in } \mathbb{A}^n(k)} \psi(F(v))$$
$$= 1 + \Sigma_{v \text{ in } \mathbb{A}^n(k)} \psi(F(v))$$
$$= 1 + \Sigma_{\lambda\neq 0 \text{ in } k} \Sigma_{x \text{ in } \mathbb{P}^{n-1}(k)} \psi(F(\lambda\tilde{x}))$$
$$= 1 + \Sigma_{\lambda\neq 0 \text{ in } k} \Sigma_{x \text{ in } \mathbb{P}^{n-1}(k)} \psi(\lambda^e F(\tilde{x}))$$
$$= 1 + \Sigma_{x \text{ in } \mathbb{P}^{n-1}(k)} \Sigma_{\lambda\neq 0 \text{ in } k} \psi(\lambda^e F(\tilde{x}))$$
$$= 1 + \Sigma_{x \text{ in } \mathbb{P}^{n-1}(k)} \Sigma_{\rho\neq 0 \text{ in } k} \psi(\rho F(\tilde{x}))(\#\{e\text{'th roots of } \rho \text{ in } k^\times\})$$
$$= 1 + \Sigma_{x \text{ in } \mathbb{P}^{n-1}(k)} \Sigma_{\rho\neq 0 \text{ in } k} \psi(\rho F(\tilde{x}))(\#\{e_0\text{'th roots of } \rho \text{ in } k^\times\})$$
$$= 1 + \Sigma_{x \text{ in } \mathbb{P}^{n-1}(k)} \Sigma_{\rho\neq 0 \text{ in } k} \psi(\rho F(\tilde{x}))(1 + \Sigma_{\chi \neq \mathbb{1},\ \chi^{e_0} = \mathbb{1}} \chi(\rho))$$

$= 1 + \Sigma_{x \text{ in } \mathbb{P}^{n-1}(k)} \Sigma_{\rho\neq 0 \text{ in } k} \psi(\rho F(\tilde{x}))$

$+ \Sigma_{\chi \neq \mathbb{1},\ \chi^{e_0} = \mathbb{1}} \Sigma_{x \text{ in } \mathbb{P}^{n-1}(k)} \Sigma_{\rho\neq 0 \text{ in } k} \psi(\rho F(\tilde{x}))\chi(\rho)$

$= 1 - (\Sigma_{x \text{ in } \mathbb{P}^{n-1}(k)} 1) + \Sigma_{x \text{ in } \mathbb{P}^{n-1}(k)} \Sigma_{\rho \text{ in } k} \psi(\rho F(\tilde{x}))$

$+ \Sigma_{\chi \neq \mathbb{1},\ \chi^{e_0} = \mathbb{1}} \Sigma_{x \text{ in } \mathbb{P}^{n-1}(k)} \Sigma_{\rho\neq 0 \text{ in } k} \psi(\rho F(\tilde{x}))\chi(\rho).$

The innermost sums are standard:

$\Sigma_{\rho \text{ in } k} \psi(\rho F(\tilde{x})) = \#k$, if $F(\tilde{x}) = 0$, i.e., if x lies in X_F,

$= 0$, if not.

$$\Sigma_{\rho\neq 0 \text{ in } k} \psi(\rho F(\tilde{x}))\chi(\rho) = \overline{\chi}(F(\tilde{x}))g(\psi, \chi).$$

So we find

$\Sigma_{v \text{ in } \mathbb{A}^n(k)} \psi(F(v))$

$= 1 - \#\mathbb{P}^{n-1}(k) + (\#k)\#X_F(k)$

$+ \Sigma_{\chi \neq \mathbb{1},\ \chi^{e_0} = \mathbb{1}} g(\psi, \overline{\chi})\Sigma_{x \text{ in } \mathbb{P}^{n-1}(k)} \chi(F(\tilde{x})).$

Notice that

$$1 - \#\mathbb{P}^{n-1}(k) = -(\#k)\#\mathbb{P}^{n-2}(k).$$

So we find

$\Sigma_{v \text{ in } \mathbb{A}^n(k)} \psi(F(v))$

$= (\#k)(\#X_F(k) - \#\mathbb{P}^{n-2}(k))$

$+ \Sigma_{\chi \neq \mathbb{1},\ \chi^{e_0} = \mathbb{1}} g(\psi, \overline{\chi})\Sigma_{x \text{ in } \mathbb{P}^{n-1}(k)} \chi(F(\tilde{x})).$

Now using our preparatory identities, valid when F is nonsingular, we find

$(-1)^n\Sigma_{v \text{ in } \mathbb{A}^n(k)} \psi(F(v))$

$= \mathrm{Trace}(\mathrm{Frob}_k, \mathrm{Prim}^{n-2}(X_F\otimes_k\overline{k}, \overline{\mathbb{Q}}_\ell)(-1))$

$+\Sigma_{\chi \neq \mathbb{1},\ \chi^{e_0} = \mathbb{1}} (-g(\psi_E, \overline{\chi}_E))\mathrm{Trace}(\mathrm{Frob}_k, H_c^{n-1}(\mathbb{P}^{n-1}[1/F]\otimes_k\overline{k}, \mathcal{L}_{\chi(F)})),$

as required. QED

Theorem 3.6.5 Let $e \geq 2$, $n \geq 2$. Let

$$UH(n,e,\psi) \subset \mathrm{Homog}(n,e)$$

be a dense open set on which

$$M(n,e,\psi)[\dim\mathrm{Homog}(n,e) - \dim\mathcal{P}(n,e)]|\mathrm{Homog}(n,e)$$

has lisse cohomology sheaves. Then we have the following results.

1) $\mathcal{M}(n,e,\psi)|UH(n,e,\psi)$ is lisse of rank

$$(1/e)((e-1)^n + (-1)^n(e-1)) + ((e_0 - 1)/e)((e-1)^n + (-1)^{n+1}),$$

and pure of weight zero.

2) For any finite extension E/k containing the e_0'th roots of unity, and for any homogeneous form F in $UH(n,e,\psi)(E)$ which also lies in NSHomog(n,e), we have the following explicit formula:

$$\det(1 - (\#E)^{n/2}T\mathrm{Frob}_{E,F} \mid \mathcal{M}(n,e,\psi))$$
$$= \det(1 - (\#E)T\mathrm{Frob}_E \mid \mathrm{Prim}^{n-2}(X_F\otimes_E\overline{E}, \overline{\mathbb{Q}}_\ell))$$
$$\times\prod_{\chi\neq\mathbb{1},\ \chi^{e_0}=\mathbb{1}}$$
$$\det(1 - (-g(\psi_E, \chi_E))T\mathrm{Frob}_E \mid H_c^{n-1}(\mathbb{P}^{n-1}[1/F]\otimes_k\overline{k}, \mathcal{L}_{\chi(F)})).$$

proof We know that $\mathcal{M}(n,e,\psi)|UH(n,e,\psi)$ is lisse and mixed, and that it is (a Tate twist of) the only nonzero cohomology sheaf of

$$M(n,e,\psi)[\dim\mathrm{Homog}(n,e) - \dim\mathcal{P}(n,e)]|UH(n,e).$$

For any finite extension E/k, and any F in $UH(n,e,\psi)(E)$, we have

$$\mathrm{Trace}(\mathrm{Frob}_{E,f} \mid \mathcal{M}(n,e,\psi)) = (-1)^n(\#E)^{-n/2}\Sigma_{v \text{ in } \mathbb{A}^n(E)}\ \psi_E(f(v)),$$

cf. 3.5.7. So for F in $UH(n,e,\psi)(E)$ which also lies in NSHomog(n,e), we get the explicit formula asserted in 2). From this formula, we compute that the stalk $\mathcal{M}(n,e,\psi)_F$ has rank equal to

$$\dim \mathrm{Prim}^{n-2}(X_F\otimes_E\overline{E}, \overline{\mathbb{Q}}_\ell)$$
$$+ \Sigma_{\chi\neq\mathbb{1},\ \chi^{e_0}=\mathbb{1}} \dim H_c^{n-1}(\mathbb{P}^{n-1}[1/F]\otimes_k\overline{k}, \mathcal{L}_{\chi(F)}),$$

which gives the asserted value for the rank.

The explicit formula 2) also shows that on the smaller open set $UH(n,e,\psi)\cap\mathrm{NSHomog}(n,e)$, $\mathcal{M}(n,e,\psi)$ is indeed pure of weight zero. As $\mathcal{M}(n,e,\psi)$ is lisse on $UH(n,e,\psi)$, it is pure of weight zero on all of $UH(n,e,\psi)$. [On $UH(n,e,\psi)$, $\mathcal{M}(n,e,\psi)$ is both lisse and mixed, so has a weight filtration by lisse subsheaves which are each pure of some weight, and that weight can be read on the smaller open set.] QED

Corollary 3.6.6 For $e \geq 2$ divisible by p, and $n \geq 2$, we have the estimate

$$\mathrm{rank}(\mathcal{M}(n,e,\psi)|U(n,e,\psi))$$
$$\geq (1/e)((e-1)^n + (-1)^n(e-1)) + ((e_0 - 1)/e)((e-1)^n + (-1)^{n+1})$$
$$\geq (1/e)((e-1)^n + (-1)^n(e-1)).$$

proof Since $\mathcal{M}(n,e,\psi)$ on $\mathcal{P}(n,e)$ is of perverse origin, its generic rank exceeds the generic rank of any pullback. Its pullback to Homog(n,e) has generic rank

$$(1/e)((e-1)^n + (-1)^n(e-1)) + ((e_0 - 1)/e)((e-1)^n + (-1)^{n+1})$$

by the above Corollary. QED

Theorem 3.6.7 In Theorem 3.6.5, we may take for

$$UH(n,e,\psi) \subset \mathrm{Homog}(n,e)$$

the set NSHomog(n,e).

proof For brevity, we will write NS for NSHomog(n,e). Consider the product space $\mathbb{A}^n\times\mathrm{NS}$ with coordinates (x, F), and the lisse sheaf $\mathcal{L}_{\psi(F(x))}$. We must show that all the sheaves $R^i(pr_2)_!\mathcal{L}_{\psi(F(x))}$ on NS are lisse. To see this, we first perform three preliminary reductions.

First reduction: The restriction of $\mathcal{L}_{\psi(F(x))}$ to $\{0\}\times\mathrm{NS}$ is the constant sheaf $\overline{\mathbb{Q}}_\ell$, simply because F is homogeneous of strictly positive degree. Consider the excision sequence for

$$(\mathbb{A}^n - \{0\})\times NS \subset \mathbb{A}^n\times NS \supset \{0\}\times NS$$

and the sheaf $\mathcal{L}_{\psi(F(x))}$ on $\mathbb{A}^n\times NS$. Denote by

$$\varphi : (\mathbb{A}^n - \{0\})\times NS \to NS$$

and

$$id : \{0\}\times NS \to NS$$

the projections onto their second factors. We have

$$R^0(id)_!\overline{\mathbb{Q}}_\ell = \overline{\mathbb{Q}}_\ell, \text{ all other } R^i(id)_!\overline{\mathbb{Q}}_\ell = 0.$$

In particular, all the sheaves $R^i(id)_!\overline{\mathbb{Q}}_\ell$ on NS are lisse. By the long exact excision sequence

$$R^{i-1}(id)_!\overline{\mathbb{Q}}_\ell \to R^i\varphi_!\mathcal{L}_{\psi(F(x))} \to R^i(pr_2)_!\mathcal{L}_{\psi(F(x))} \to R^i(id)_!\overline{\mathbb{Q}}_\ell \to ,$$

to show that all the sheaves $R^i(pr_2)_!\mathcal{L}_{\psi(F(x))}$ on NS are lisse, it suffices to show that all the $R^i\varphi_!\mathcal{L}_{\psi(F(x))}$ are lisse on NS.

Second reduction: Denote by Blow the blow up of the origin in $\mathbb{A}^n$. Explicitly, Blow is the closed subscheme of $\mathbb{A}^n\times\mathbb{P}^{n-1}$ defined by the equations $x_iZ_j = x_jZ_i$ for all $i < j$, where the Z_i are the homogeneous coordinates in $\mathbb{P}^{n-1}$. The two projections give maps

$$bl : Blow \to \mathbb{A}^n$$

and

$$\pi : Blow \to \mathbb{P}^{n-1}.$$

The first map has

$$bl^{-1}(\{0\}) = \mathbb{P}^{n-1},$$

and induces

$$Blow - \mathbb{P}^{n-1} \cong \mathbb{A}^n - \{0\}.$$

The second map is the tautological line bundle $\mathcal{O}(1)$. Take the product of this "constant" siutation with NS. We get maps

$$bl\times id : Blow\times NS \to \mathbb{A}^n\times NS,$$

and

$$\pi\times id: Blow\times NS \to \mathbb{P}^{n-1}\times NS.$$

The pullback $(bl\times id)^*\mathcal{L}_{\psi(F(x))}$ on Blow×NS is a lisse sheaf of rank one, whose restriction to

$$(bl\times id)^{-1}(\{0\}\times NS) = \mathbb{P}^{n-1}\times NS$$

is the constant sheaf $\overline{\mathbb{Q}}_\ell$ on that space. So we have an excision situation with the schemes

$$(Blow - \mathbb{P}^{n-1})\times NS \subset Blow\times NS \supset \mathbb{P}^{n-1}\times NS,$$

and the sheaf $(bl\times id)^*\mathcal{L}_{\psi(F(x))}$ on Blow×NS.

On the open set

$$(Blow - \mathbb{P}^{n-1})\times NS \cong (\mathbb{A}^n - \{0\})\times NS,$$

the sheaf $(bl\times id)^*\mathcal{L}_{\psi(F(x))}$ is just the restriction of $\mathcal{L}_{\psi(F(x))}$. Denote by

ρ : Blow×NS → NS,

p_2 : $\mathbb{P}^{n-1}$×NS → NS,

the second projections. Then we have a long exact excision sequence

$$\ldots \to R^i\varphi_!\mathcal{L}_{\psi(F(x))} \to R^i(\pi\times\mathrm{id})_!((\mathrm{bl}\times\mathrm{id})^*\mathcal{L}_{\psi(F(x))}) \to R^i(p_2)_!\overline{\mathbb{Q}}_\ell \to \ldots.$$

Now p_2 is a proper smooth map, so the sheaves $R^i(p_2)_!\overline{\mathbb{Q}}_\ell$ on NS are all lisse. So in order to show that all the sheaves $R^i\varphi_!\mathcal{L}_{\psi(F(x))}$ on NS are lisse, it suffices to show that all the sheaves $R^i\rho_!((\mathrm{bl}\times\mathrm{id})^*\mathcal{L}_{\psi(F(x))})$ on NS are lisse.

Third reduction: Recall the canonical morphism

$$\pi : \mathrm{Blow} \to \mathbb{P}^{n-1},$$

and its product with NS:

$$\pi\times\mathrm{id} : \mathrm{Blow}\times\mathrm{NS} \to \mathbb{P}^{n-1}\times\mathrm{NS}.$$

We have a Leray spectral sequence of sheaves on NS

$$E_2^{a,b} = R^a(p_2)_!R^b(\pi\times\mathrm{id})_!((\mathrm{bl}\times\mathrm{id})^*\mathcal{L}_{\psi(F(x))})$$
$$\Rightarrow R^{a+b}\rho_!((\mathrm{bl}\times\mathrm{id})^*\mathcal{L}_{\psi(F(x))}).$$

So it suffices to show that every term $E_2^{a,b}$ is lisse on NS. For this, we may and will extend scalars to a finite extension $E/\mathbb{F}_p$ which contains the e_0'th roots of unity.

We need some basic understanding of the sheaves $R^b(\pi\times\mathrm{id})_!((\mathrm{bl}\times\mathrm{id})^*\mathcal{L}_{\psi(F(x))})$ on $\mathbb{P}^{n-1}$×NS. Cover $\mathbb{P}^{n-1}$ with the Zariski open sets $U_i := \mathbb{P}^{n-1}[1/Z_i]$ where Z_i is invertible. To fix ideas, take $i=1$. Write points as $(1, z)$, with z in $\mathbb{A}^{n-1}$. Over this open set U_1, Blow is the product $\mathbb{A}^1\times U_1$, π is the projection, and the restriction to $\mathbb{A}^1\times U_1\times\mathrm{NS}$, with coordinates (λ, z, F), of the lisse sheaf $(\mathrm{bl}\times\mathrm{id})^*\mathcal{L}_{\psi(F(x))}$ is the lisse sheaf $\mathcal{L}_{\psi(F(\lambda(1, z)))} = \mathcal{L}_{\psi(\lambda^eF(1, z))}$.

Over every point in the open set $\mathcal{U}$ of $\mathbb{P}^{n-1}$×NS where $F(Z)$ is invertible, the lisse sheaf $\mathcal{L}_{\psi(\lambda^eF(1, z))}$ on the $\mathbb{A}^1$ fibre has constant $\mathrm{Swan}_\infty = e_0$ (:= the prime to p part of e). By Deligne's semicontinuity theorem [Lau-SC, 2.1.2], it follows that all the sheaves $R^b(\pi\times\mathrm{id})_!((\mathrm{bl}\times\mathrm{id})^*\mathcal{L}_{\psi(F(x))})$ are lisse on $\mathcal{U}$. Looking fibre by fibre, we see that, on $\mathcal{U}$, only $R^1(\pi\times\mathrm{id})_!((\mathrm{bl}\times\mathrm{id})^*\mathcal{L}_{\psi(F(x))})$ is possibly nonzero. Invoking the Euler Poincaré formula fibre by fibre, we see that, on $\mathcal{U}$, the sheaf $R^1(\pi\times\mathrm{id})_!((\mathrm{bl}\times\mathrm{id})^*\mathcal{L}_{\psi(F(x))})$ is lisse of rank $e_0 - 1$. Decomposing it under the action of μ_{e_0} on the $\mathbb{A}^1$ variable λ, we see further a direct sum decomposition into μ_{e_0}-eigenspaces. Looking fibre by fibre, we see that the trivial eigenspace vanishes, and that each of the others has rank one. Moreover, by Chebotarev, we see that the eigenspace attached to a nontrivial character $\overline{\chi}$ of μ_{e_0} is the lisse sheaf

$$(-g(\psi_E, \overline{\chi}_E))^{\deg} \otimes \mathcal{L}_{\chi(F(Z))}$$

on $\mathcal{U}$. Thus we obtain

$$R^1(\pi \times id)_!((bl \times id)^* \mathcal{L}_{\psi(F(x))}) \mid \mathcal{U}$$

$$\cong \oplus_{\chi \neq \mathbb{1},\, \chi^{e_0} = \mathbb{1}} (-g(\psi_E, \overline{\chi}_E))^{\deg} \otimes \mathcal{L}_{\chi(F(Z))} \text{ on } \mathcal{U}.$$

Over the closed set $\mathcal{X}$ in $\mathbb{P}^{n-1} \times NS$ where $F(Z) = 0$, the lisse sheaf $\mathcal{L}_{\psi(\lambda^e F(1,\, z))}$ on the $\mathbb{A}^1$ fibre is the constant sheaf $\overline{\mathbb{Q}}_\ell$. So again by (a trivial instance of) Deligne's semicontinuity theorem [Lau-SC, 2.1.2], it follows that all the sheaves $R^b(\pi \times id)_!((bl \times id)^* \mathcal{L}_{\psi(F(x))})$ are lisse on $\mathcal{X}$. Looking fibre by fibre, we see that, on $\mathcal{X}$, only $R^2(\pi \times id)_!((bl \times id)^* \mathcal{L}_{\psi(F(x))})$ is nonzero, and that it is lisse of rank one. By Chebotarev, we see that

$$R^2(\pi \times id)_!((bl \times id)^* \mathcal{L}_{\psi(F(x))}) \mid \mathcal{X} \cong \overline{\mathbb{Q}}_\ell(-1) \text{ on } \mathcal{X}.$$

So the upshot of this analysis is that we have

$$R^1(\pi \times id)_!((bl \times id)^* \mathcal{L}_{\psi(F(x))})$$

$$\cong \oplus_{\chi \neq \mathbb{1},\, \chi^{e_0} = \mathbb{1}} (-g(\psi_E, \overline{\chi}_E))^{\deg} \otimes \mathcal{L}_{\chi(F(Z))} \text{ on } \mathcal{U}, \text{ extended by } 0$$

and

$$R^2(\pi \times id)_!((bl \times id)^* \mathcal{L}_{\psi(F(x))}) \cong \overline{\mathbb{Q}}_\ell(-1) \text{ on } \mathcal{X}, \text{ extended by } 0;$$

all other R^i vanish.

The projection of $\mathcal{X}$ onto NS, say

$$\tau : \mathcal{X} \to NS$$

is proper and smooth (indeed this map τ is the universal family of smooth hypersurfaces in $\mathbb{P}^{n-1}$ of degree e). So all the terms $E_2{}^{a,2}$ in our spectral sequence are lisse on NS, because they are given concretely by

$$E_2{}^{a,2} \cong R^a \tau_! \overline{\mathbb{Q}}_\ell(-1).$$

Denote by

$$\sigma : \mathcal{U} \to NS$$

the projection. Over a point $\mathcal{U}$ in NS, the fibre is $\mathbb{P}^{n-1}[1/F]$. Looking fibre by fibre, we see from [Ka-ENSMCS, section 8, pages 11-12] that

$$E_2{}^{a,1} = 0 \text{ for } a \neq n-1,$$

and that

$$E_2{}^{n-1,1} \cong \oplus_{\chi \neq \mathbb{1},\, \chi^{e_0} = \mathbb{1}} (-g(\psi_E, \overline{\chi}_E))^{\deg} \otimes R^{n-1}\sigma_! \mathcal{L}_{\chi(F)}.$$

It remains only to explain why the sheaves $R^{n-1}\sigma_!\mathcal{L}_{\chi(F)}$.on NS are lisse. We know from [Ka-ENSMCS, section 8, pages 11-12] that each of these sheaves has constant fibre dimension. But the morphism σ is affine and lisse of relative dimension n-1, and $\mathcal{L}_{\chi(F)}$ is lisse on $\mathcal{U}$. By {Ka-SMD, Cor. 6], $R^{n-1}\sigma_!\mathcal{L}_{\chi(F)}$ is a sheaf of perverse origin on

NS. As it has constant fibre dimension, it is lisse, by [Ka-SMD, Prop. 11]. QED

(3.7) Interlude: separation of variables

(3.7.1) We first give the motivation. Suppose we are looking at an exponential sum

$$\Sigma_{v \text{ in } \mathbb{A}^n(E)} \psi_E(f(v))$$

in $n \geq 2$ variables. Suppose that f in $\mathcal{P}(n,e)(E)$ is the sum of two polynomials f_1 and f_2 in disjoint sets of variables, i.e., we have

$$n = a + b, \text{ with } a, b \text{ both } \geq 1,$$

and polynomials

$$f_1 \text{ in } \mathcal{P}(a,e)(E),\ f_2 \text{ in } \mathcal{P}(b,e)(E)$$

such that in coordinates

$$(y_1, y_2,..., y_a, z_1, z_2,..., z_b)$$

for $\mathbb{A}^n$, we have

$$f(y, z) = f_1(y) + f_2(z).$$

Then the exponential sum factors:

$$\Sigma_{v \text{ in } \mathbb{A}^n(E)} \psi_E(f(v))$$
$$= (\Sigma_{y \text{ in } \mathbb{A}^a(E)} \psi_E(f_1(y)))\times(\Sigma_{z \text{ in } \mathbb{A}^b(E)} \psi_E(f_2(z))).$$

(3.7.2) The cohomological explanation is that on

$$\mathbb{A}^n = \mathbb{A}^a \times \mathbb{A}^b,$$

the sheaf $\mathcal{L}_{\psi(f)}$ is the external tensor product

$$\mathcal{L}_{\psi(f)} = pr_1^*\mathcal{L}_{\psi(f_1)} \otimes pr_2^*\mathcal{L}_{\psi(f_2)}.$$

So the Kunneth formula [SGA 4, Expose XVII, 5.4.3] gives

$$R\Gamma_c(\mathbb{A}^n \otimes_k \overline{E}, \mathcal{L}_{\psi(f)})$$
$$= R\Gamma_c(\mathbb{A}^a \otimes_k \overline{E}, \mathcal{L}_{\psi(f_1)}) \otimes R\Gamma_c(\mathbb{A}^b \otimes_k \overline{E}, \mathcal{L}_{\psi(f_2)}),$$

so in particular

$$H_c^*(\mathbb{A}^n \otimes_k \overline{E}, \mathcal{L}_{\psi(f)})$$
$$= H_c^*(\mathbb{A}^a \otimes_k \overline{E}, \mathcal{L}_{\psi(f_1)}) \otimes H_c^*(\mathbb{A}^b \otimes_k \overline{E}, \mathcal{L}_{\psi(f_2)}).$$

So we get the following lemma.

Lemma 3.7.3 Suppose $f(y, z) = f_1(y) + f_2(z)$ as above. Then we have

$$R\Gamma_c(\mathbb{A}^n \otimes_k \overline{E}, \mathcal{L}_{\psi(f)})$$
$$= R\Gamma_c(\mathbb{A}^a \otimes_k \overline{E}, \mathcal{L}_{\psi(f_1)}) \otimes R\Gamma_c(\mathbb{A}^b \otimes_k \overline{E}, \mathcal{L}_{\psi(f_2)}).$$

Suppose in addition that

$$H_c^i(\mathbb{A}^a \otimes_k \overline{E}, \mathcal{L}_{\psi(f_1)}) = 0 \text{ for } i \neq a,$$

$$H_c^i(\mathbb{A}^b \otimes_k \overline{E}, \mathcal{L}_{\psi(f_2)}) = 0 \text{ for } i \neq b.$$

Then

$$H_c^i(\mathbb{A}^n \otimes_k \overline{E}, \mathcal{L}_{\psi(f)}) = 0 \text{ for } i \neq n,$$

and

$$H_c^n(\mathbb{A}^n\otimes_k\bar{E}, \mathcal{L}_{\psi(f)})$$
$$= H_c^a(\mathbb{A}^a\otimes_k\bar{E}, \mathcal{L}_{\psi(f_1)})\otimes H_c^b(\mathbb{A}^b\otimes_k\bar{E}, \mathcal{L}_{\psi(f_2)}).$$

(3.7.4) We can carry out the same analysis in the universal situation.

Lemma 3.7.5 For $a \geq 1$, $b \geq 1$, and $e \geq 1$, we have a closed immersion

$$\mathcal{P}(a,e) \times \mathcal{P}(b,e) \to \mathcal{P}(a+b,e),$$
$$(f_1, f_2) \mapsto f_1(y) + f_2(z).$$

Put

$$\delta := \dim\mathcal{P}(a,e) + \dim\mathcal{P}(b,e) - \dim\mathcal{P}(a+b,e).$$

The (shifted) pullback of $M(a+b,e,\psi)(-1/2)$ to $\mathcal{P}(a,e) \times \mathcal{P}(b,e)$ is the external tensor product

$$M(a+b,e,\psi)(-1/2)|(\mathcal{P}(a,e) \times \mathcal{P}(b,e))[\delta]$$
$$= pr_1{}^*M(a,e,\psi)(-1/2)\otimes pr_2{}^*M(b,e,\psi)(-1/2).$$

proof This is just the Kunneth formula [SGA 4, Expose XVII, 5.4.3]. QED

Lemma 3.7.6 For $a \geq 1$, $b \geq 1$, and $e \geq 1$, we have a direct sum decomposition of constructible $\bar{\mathbb{Q}}_\ell$- sheaves on $\mathcal{P}(a,e) \times \mathcal{P}(b,e)$,

$$\mathcal{M}(a+b,e,\psi)|(\mathcal{P}(a,e) \times \mathcal{P}(b,e))$$
$$\cong pr_1{}^*\mathcal{M}(a,e,\psi)\otimes pr_2{}^*\mathcal{M}(b,e,\psi) \oplus \text{(other terms)}.$$

The induced map

$$pr_1{}^*\mathcal{M}(a,e,\psi)\otimes pr_2{}^*\mathcal{M}(b,e,\psi) \to \mathcal{M}(a+b,e,\psi)|(\mathcal{P}(a,e) \times \mathcal{P}(b,e))$$

is the inclusion of a direct factor. This inclusion is an isomorphism at any point (f_1, f_2) in $\mathcal{P}(a,e)(E) \times \mathcal{P}(b,e)(E)$ at which both

$$H_c^i(\mathbb{A}^a\otimes_k\bar{E}, \mathcal{L}_{\psi(f_1)}) = 0 \text{ for } i \neq a,$$

and

$$H_c^i(\mathbb{A}^b\otimes_k\bar{E}, \mathcal{L}_{\psi(f_2)}) = 0 \text{ for } i \neq b.$$

proof The first statement is just spelling out what the Kunneth formula gives on cohomology sheaves. The second is, by proper base change, just a restatement of Lemma 3.7.3. QED

Corollary 3.7.7 Let $a \geq 1$, $b \geq 1$. If $e \geq 2$ is prime to p, then on the space

$$\mathcal{D}(a,e)\times\mathcal{D}(b,e)$$

of pairs (f_1, f_2) of Deligne polynomials, we have an isomorphism of lisse sheaves,

$$\mathcal{M}(a+b,e,\psi)|(\mathcal{D}(a,e) \times \mathcal{D}(b,e))$$
$$\cong pr_1{}^*(\mathcal{M}(a,e,\psi)|\mathcal{D}(a,e))\otimes pr_2{}^*(\mathcal{M}(b,e,\psi)|\mathcal{D}(b,e)).$$

Corollary 3.7.8 Let $a \geq 2$, $b \geq 1$. Suppose $e \geq 3$ is divisible by p. Let

$$UH(a,e,\psi) \subset NSHomog(a,e) \subset Homog(a,e)$$

be a dense open set of nonsingular homogeneous forms on which

$$M(a,e,\psi)[\dim Homog(a,e) - \dim \mathcal{P}(a,e)]|Homog(a,e)$$

has lisse cohomology sheaves. Then on the space

$$UH(a,e,\psi) \times \mathcal{D}(b,e-1),$$

we have an isomorphism of lisse sheaves,

$$\mathcal{M}(a+b,e,\psi)|(UH(a,e,\psi) \times \mathcal{D}(b,e-1))$$
$$\cong pr_1{}^*(\mathcal{M}(a,e,\psi)|UH(a,e,\psi)) \otimes pr_2{}^*(\mathcal{M}(b,e,\psi)|\mathcal{D}(b,e-1)).$$

(3.8) Return to the monodromy of exponential sums of Deligne type on $\mathbb{A}^n$

(3.8.1) With all these preliminaries out of the way, we can now prove the target theorem of this section.

Theorem 3.8.2 Let $n \geq 1$, $e \geq 3$. Denote by $U(n,e,\psi)$ a dense open set of $\mathcal{P}(n,e)$ on which $M(n,e,\psi)$ has lisse cohomology sheaves. If e is prime to p, take $U(n,e,\psi)$ to be $\mathcal{D}(n,e)$, the space of Deligne polynomials.

1) If e is prime to p, then $\mathcal{M}(n,e,\psi)|U(n,e,\psi)$ has rank $(e-1)^n$.

2) Is e is divisible by p, then $\mathcal{M}(n,e,\psi)|U(n,e,\psi)$ has rank at least

$$\mathrm{rank}(\mathcal{M}(n,e,\psi)|U(n,e,\psi))$$
$$\geq \mathrm{Max}((e-2)^n, (1/e)((e-1)^n + (-1)^n(e-1))).$$

3) Suppose that any of the following five conditions holds:

a) $p \geq 7$,

b) $n \geq 3$,

c) $p = 5$ and $e \geq 4$,

d) $p = 3$ and $e \geq 7$,

e) $p = 2$ and $e \geq 7$.

Then G_{geom} for $\mathcal{M}(n,e,\psi)|U(n,e,\psi)$ is not finite.

proof Assertion 1) was proven in 3.5.11. Assertion 2) was proven in 3.5.17 and 3.6.6. [The attentive reader will notice that while 3.6.6 only covers the case $n \geq 2$, that is the only case where there is something to prove, as the quantity $(1/e)((e-1)^n + (-1)^n(e-1))$ vanishes for $n=1$.]

To prove 5), we proceed case by case.

Case a) Here $p \geq 7$. Because $\mathcal{M}(n,e,\psi)$ is a sheaf of perverse origin on $\mathcal{P}(n,e)$, and $e \geq 3$, if the group G_{geom} for $\mathcal{M}(n,e,\psi)|U(n,e,\psi)$ is finite, then the group G_{geom} for $\mathcal{M}(n,e,\psi)|U(n,3,\psi) = \mathcal{M}(n,3,\psi)|U(n,3,\psi)$ is finite. So it suffices to show that G_{geom} for $\mathcal{M}(n,3,\psi)|U(n,3,\psi)$ is not finite. Since $p \geq 7 > 3$, $U(n,3,\psi) = \mathcal{D}(n,3)$ is the space of cubic Deligne polynomials. Now restrict to the subspace where the variables separate completely:

$$\mathcal{M}(n,3,\psi)|\mathcal{D}(1,3)^n \cong \otimes_{i=1 \text{ to } n}\, pr_i{}^*(\mathcal{M}(1,3,\psi)|\mathcal{D}(1,3)).$$

So we have the n-fold external product with itself of the lisse, rank two sheaf $\mathcal{M}(1,3,\psi)|\mathcal{D}(1,3)$ on $\mathcal{D}(1,3)$. Restrict further to the diagonal $\mathcal{D}(1,3)$ inside $\mathcal{D}(1,3)^n$. There we find

$$\mathcal{M}(n,3,\psi)|(\text{diagonal } \mathcal{D}(1,3) \text{ in } \mathcal{D}(1,3)^n) = \mathcal{M}(1,3,\psi)^{\otimes n}|\mathcal{D}(1,3).$$

So it suffices to show that $\mathcal{M}(1,3,\psi)^{\otimes n}|\mathcal{D}(1,3)$ does not have have finite G_{geom}. For this, it suffices to show that $\mathcal{M}(1,3,\psi)$ does not have finite G_{geom}. Restrict $\mathcal{M}(1,3,\psi)$ further, to the $\mathbb{A}^1$ in $\mathcal{D}(1,3)$ given by the one pararameter family

$$t \mapsto x^3 + tx.$$

There $\mathcal{M}(1,3,\psi)|\mathbb{A}^1$ is [a shift and Tate twist of] $FT_\psi(\mathcal{L}_{\psi(x^3)})$, so is lisse of rank two and geometrically irreducible on $\mathbb{A}^1$, cf. [Ka-MG, Theorem 17]. It suffices to prove that $\mathcal{M}(1,3,\psi)|\mathbb{A}^1$ does not have finite G_{geom}. As proven in [Ka-MG, Prop. 5], any lisse, geometrically irreducible sheaf of rank r on $\mathbb{A}^1$ over an algebraically closed field of characteristic $p > 2r+1$ is Lie-irreducible, i.e., the identity component $(G_{geom})^0$ acts irreducibly. In our case, namely $\mathcal{M}(1,3,\psi)|\mathbb{A}^1$, we have $r=2$ and $p \geq 7 > 2\times 2 + 1$, so we can only have $(G_{geom})^0 = SL(2)$.

Case b) Here $n \geq 3$. Just as in case a), it suffices to show that G_{geom} for $\mathcal{M}(n,3,\psi)|U(n,3,\psi)$ is not finite. For this, it suffices to find a single finite extension E/k, and a single f in $\mathcal{P}(n,3)$, such that $Frob_{E,f}$ acting on $\mathcal{M}(n,3,\psi)_f$ has no power a scalar, cf. Scalarity Corollary 2.8.13.

If $n=3$, pick a dense open set

$$UH(3,3,\psi) \subset NSHomog(3,3)$$

of nonsingular cubics over which

$$M(3,3,\psi)|UH(3,3,\psi)[\dim UH(3,3,\psi) - \dim \mathcal{P}(3,3)]$$

has lisse cohomology sheaves. Recall that for a nonsingular cubic form $F(x_1,x_2,x_3)$ in three variables over a field of characteristic p, the locus $X_F \subset \mathbb{P}^2$ defined by its vanishing is a curve of genus one. That curve is ordinary precisely when its Hasse invariant, which is itself a form of degree p-1 in the coefficients of F, namely

$$\text{Hasse}(X_F) = \text{coefficient of } (x_1x_2x_3)^{p-1} \text{ in } F(x_1,x_2,x_3)^{p-1},$$

is nonzero, a condition which visibly holds on a dense open set of cubic forms. Now take an F which is an E-valued point of the dense open set

$$UH(3,3,\psi)[1/\text{Hasse}],$$

for some finite extension E/k. Then $Frob_E$ acting on

$$H^1(X_F\otimes_E\overline{E}, \overline{\mathbb{Q}}_\ell) = Prim^1(X_F\otimes_E\overline{E}, \overline{\mathbb{Q}}_\ell)$$

has no power a scalar, indeed its two eigenvalues have different p-adic valuations (for any extension to $\overline{\mathbb{Q}}_\ell$ of the p-adic valuation on $\mathbb{Q}$). But we have seen, in Theorem 3.6.4 with $e=n=3$, that

$\det(1 - (\#E)^{3/2}T\mathrm{Frob}_{E,F} \mid \mathcal{M}(3,3,\psi))$

$= \det(1 - (\#E)T\mathrm{Frob}_E \mid \mathrm{Prim}^1(X_F\otimes_E\overline{E}, \overline{\mathbb{Q}}_\ell))$

$\times\prod_{\chi\neq\mathbb{1},\ \chi^{e_0} = \mathbb{1}}$

$\det(1 - (-g(\psi_E, \chi_E))T\mathrm{Frob}_E \mid H_c^2(\mathbb{P}^2[1/F]\otimes_k\overline{k}, \mathcal{L}_{\chi(F)})).$

Therefore the two eigenvalues of $\mathrm{Frob}_E \mid \mathrm{Prim}^1(X_F\otimes_E\overline{E}, \overline{\mathbb{Q}}_\ell)$ occur, after division by $(\#E)^{1/2}$, as eigenvalues of $\mathrm{Frob}_{E,F} \mid \mathcal{M}(3,3,\psi)$. So $\mathrm{Frob}_{E,F} \mid \mathcal{M}(3,3,\psi)$ has two eigenvalues with different p-adic valuations, and hence no power of $\mathrm{Frob}_{E,F} \mid \mathcal{M}(3,3,\psi)$ is scalar.

If $n \geq 4$, say $n = 3 + b$, keep the $F(x_1,x_2,x_3)$ used above, and take for f the polynomial in n variables $(x_1,x_2,x_3, y_1,\ldots, y_b)$ given by

$$f := F(x_1,x_2,x_3) + g(y),$$

with

$$g(y) = \Sigma_{i=1 \text{ to } b} (y_i)^3, \text{ if } p\neq 3,$$
$$= \Sigma_{i=1 \text{ to } b} (y_i)^2, \text{ if } p=3.$$

We get

$$\mathcal{M}(n,3,\psi)_f = \mathcal{M}(3,3,\psi)_F\otimes\mathcal{M}(b,3,\psi)_g, \text{ if } p \neq 3,$$
$$\mathcal{M}(n,3,\psi)_f = \mathcal{M}(3,3,\psi)_F\otimes\mathcal{M}(b,2,\psi)_g, \text{ if } p = 3,$$

with $\mathrm{Frob}_{E,f}$ acting as the tensor product $\mathrm{Frob}_{E,f} = \mathrm{Frob}_{E,F}\otimes\mathrm{Frob}_{E,g}$. In both cases, $\mathrm{Frob}_{E,F}$ on $\mathcal{M}(3,3,\psi)_F$ has two eigenvalues with different p-adic valuations, and the second factor $\mathrm{Frob}_{E,g}$ has a nonzero eigenvalue (because $\mathrm{Frob}_{E,g}$ acts invertibly, and the space is nonzero: $\dim\mathcal{M}(b,3,\psi)_g = 2^b$ if $p \neq 3$, $\dim\mathcal{M}(b,2,\psi)_g = 1$, if $p = 3$). Therefore $\mathrm{Frob}_{E,f}|\mathcal{M}(n,3)_f$ has two eigenvalues with different p-adic valuations, and hence no power of it is scalar.

Case c) Here $p=5$, and $e \geq 4$. We first use the Dependence on ψ Lemma 3.5.3 to reduce to the case where ψ is (the composition with $\mathrm{Trace}_{k/\mathbb{F}_p}$ of) a nontrivial additive character of $\mathbb{F}_p$, and so reduce to the case when k is $\mathbb{F}_p = \mathbb{F}_5$. Exactly as in case a), we reduce to showing that $\mathcal{M}(1,4,\psi)|\mathcal{D}(1,4)$ does not have finite G_{geom}. For this, it suffices to exhibit a single finite extension $E/\mathbb{F}_5$, and a single f in $\mathcal{D}(1,4,\psi)(E)$ such that no power of $\mathrm{Frob}_{E,f} \mid \mathcal{M}(1,4,\psi)$ is scalar. Recall from 3.5.11 that for any f in $\mathcal{D}(1,4)(E)$, we have

$$\mathcal{M}(1,4,\psi)(-1/2)_f = H_c^1(\mathbb{A}^1\otimes_E\overline{E}, \mathcal{L}_{\psi(f)}).$$

So we must exhibit an f such that no power of

$$\mathrm{Frob}_E|H_c^1(\mathbb{A}^1\otimes_E\overline{E}, \mathcal{L}_{\psi(f)})$$

is scalar. In view of Corollary 3.5.14, if some power is scalar, then we have

$$\Sigma_{v \text{ in } E}\ \psi_E(f(v)) = (\#E)^{1/2}(\text{an algebraic integer}).$$

Take $E = \mathbb{F}_p = \mathbb{F}_5$, and $f = x^4$. Since ψ is a nontrivial additive character of $\mathbb{F}_5$, $\psi(1)$ is a primitive fifth root of unity ζ_5. For v nonzero in $\mathbb{F}_5$, $v^4 = 1$. Meanwhile $\psi(0) = 1$. Thus

$$\Sigma_{v \text{ in } \mathbb{F}_5} \psi(v^4) = 4\psi(1) + 1 = 4\zeta_5 + 1 = 5\zeta_5 + (1-\zeta_5).$$

But $1-\zeta_5$ is a uniformizer for the unique 5-adic place of $\mathbb{Q}(\zeta_5)$, and $\mathrm{ord}_5(1-\zeta_5) = 1/4$. Therefore we have

$$\mathrm{ord}_5(\Sigma_{v \text{ in } \mathbb{F}_5} \psi(v^4)) = 1/4,$$

and hence $\Sigma_{v \text{ in } \mathbb{F}_5} \psi(v^4)$ is not divisible by Sqrt(5) as an algebraic integer.

Case d) Here p=3, and $e \geq 7$. Exactly as in the case above, we reduce to the case when $k = \mathbb{F}_3$, ψ is a nontrivial additive character of $\mathbb{F}_3$, and to showing that $\mathcal{M}(1,7,\psi)|\mathcal{D}(1,7)$ does not have finite G_{geom}. For this, it suffices to find a finite extension $E/\mathbb{F}_3$ and an f in $\mathcal{D}(1,7)(E)$ such that $\Sigma_{v \text{ in } E} \psi_E(f(v))$ is not divisible by $(\#E)^{1/2}$ as an algebraic integer. Here we take $E = \mathbb{F}_{27}$, viewed as $\mathbb{F}_3(\alpha)$, for α a root of

$$X^3 - X - 1,$$

and we take $f = \alpha^2x^7 + x^5$. We calculate

$$\Sigma_{v \text{ in } E} \psi_E(f(v))$$

as follows. Quite generally, for any prime, any finite field $E/\mathbb{F}_p$, and any polynomial f(x) in E[x], consider the affine Artin-Schreier curve

$$C_f : y^p - y = f(x).$$

Over a point v in $\mathbb{A}^1(E) = E$, there are p points in $C_f(E)$ if $\mathrm{Trace}_{E/\mathbb{F}_p}(f(v)) = 0$, and none otherwise. By definition, we have

$$\psi_E(f(v)) = \psi(\mathrm{Trace}_{E/\mathbb{F}_p}(f(v))).$$

By the orthogonality relations for characters, we thus find

$$\Sigma_{\lambda \text{ in } \mathbb{F}_p} \psi_E(\lambda f(v)) = p, \text{ if } \mathrm{Trace}_{E/\mathbb{F}_p}(f(v)) = 0,$$
$$= 0, \text{ if not.}$$

So we have

$$\#C_f(E) = \Sigma_{v \text{ in } E} \Sigma_{\lambda \text{ in } \mathbb{F}_p} \psi_E(\lambda f(v))$$
$$= \#E + \Sigma_{\lambda \neq 0 \text{ in } \mathbb{F}_p} \Sigma_{v \text{ in } E} \psi_E(\lambda f(v)).$$

Now return to characteristic p=3. The only $\lambda \neq 0$ in $\mathbb{F}_3$ are ± 1, so we get

$$\#C_f(E) = \#E + \Sigma_{v \text{ in } E} \psi_E(f(v)) + \Sigma_{v \text{ in } E} \psi_E(-f(v)).$$

As our $f = \alpha^2x^7 + x^5$ is an odd function, we have

$$\Sigma_{v \text{ in } E} \psi_E(f(v)) = \Sigma_{v \text{ in } E} \psi_E(-f(v)),$$

so we find

$$\#C_f(E) = \#E + 2\Sigma_{v \text{ in } E} \psi_E(f(v)).$$

In our case, direct calculation shows that

$$\#C_f(\mathbb{F}_{27}) = 21.$$

Thus we conclude that for $f = \alpha^2x^7 + x^5$ as above, we have

$$\Sigma_{v \text{ in } \mathbb{F}_{27}} \psi_{\mathbb{F}_{27}}(f(v)) = -3,$$

which is not divisible by Sqrt(27) as an algebraic integer.

Case e) Here $p=2$, and $e \geq 7$. As above, we reduce to the case when $k= \mathbb{F}_2$, ψ is the nontrivial character of $\mathbb{F}_2$, and to showing that $\mathcal{M}(1,7,\psi)|\mathcal{D}(1,7)$ does not have finite G_{geom}. For this, it suffices to find a finite extension $E/\mathbb{F}_2$ and an f in $\mathcal{D}(1,7)(E)$ such that

$$\Sigma_{v \text{ in } E} \psi_E(f(v))$$

is not divisible by $(\#E)^{1/2}$ as an algebraic integer. We take $E = \mathbb{F}_8$, and $f = x^7$. For $v \neq 0$ in $\mathbb{F}_8$, $v^7 = 1$, and

$$\psi_{\mathbb{F}_8}(v^7) = \psi(\text{Trace}_{\mathbb{F}_8/\mathbb{F}_2}(1)) = \psi(3) = \psi(1) = -1.$$

Thus

$$\Sigma_{v \text{ in } \mathbb{F}_8} \psi_{\mathbb{F}_8}(v^7) = 7(-1)+ 1 = -6 = 2\times 3,$$

which is not divisible by Sqrt(8) as an algebraic integer. QED

Remark 3.8.3 How sharp are the exclusions we have imposed for low $n = 1$ or 2 in low characteristic $p = 2, 3$, or 5? We do not know how sharp our exclusions are for $n = 2$. For $n=1$, we can do no better.

Supersingularity Lemma 3.8.4
1) In characteristic $p = 2$, any Artin-Schreier curve

$$y^p - y = f(x), \text{ f a polynomial in x of degree 3 or 5,}$$

is supersingular.
2) In characteristic $p = 3$, any Artin-Schreier curve

$$y^p - y = f(x), \text{ f a polynomial in x of degree 4 or 5,}$$

is supersingular.
3) In characteristic $p = 5$, any Artin-Schreier curve

$$y^p - y = f(x), \text{ f a polynomial in x of degree 3,}$$

is supersingular.

proof This is a consequence of the following result of van der Geer and van der Vlugt.

Theorem 3.8.5 ([vdG-vdV-RM, 5.4 and 13.4]) Fix a prime p, an integer $h \geq 0$, and an additive polynomial

$$A(x) = \Sigma_{i=0 \text{ to } h}\, a_i x^{p^i},$$

with coefficients in $\overline{\mathbb{F}}_p$ and $a_h \neq 0$. Then the Artin-Schreier curve of genus $g = (1/2)(p-1)p^h$ given by the equation

$$y^p - y = xA(x)$$

is supersingular.

Corollary 3.8.6 Fix a prime p, an integer $h \geq 1$, and an additive polynomial

$$A(x) = \Sigma_{i=0 \text{ to } h}\, a_i x^{p^i},$$

with coefficients in $\overline{\mathbb{F}}_p$ and $a_h \neq 0$. For any a_{-1} in $\overline{\mathbb{F}}_p$, the Artin-Schreier curve of equation

$$y^p - y = a_{-1}x + xA(x)$$

is supersingular.

proof The isomorphism class of an Artin-Schreier curve $y^p - y = f$ depends only on f up to Artin-Schreier equivalence (i.e., additive equivalence modulo $g^p - g$'s). We will show that the $\overline{\mathbb{F}}_p$-isomorphism class of the curve in question is independent of a_{-1}. To see this, begin with the curve

$$y^p - y = xA(x).$$

For any b in $\overline{\mathbb{F}}_p$, this curve is isomorphic to the curve

$$y^p - y = (x+b)A(x+b).$$

Because A is an additive polynomial, we have

$$(x+b)A(x+b) = (x+b)(A(x) + A(b))$$
$$= xA(x) + bA(x) + A(b)x + \beta A(b).$$

Now write

$$b = \beta^{p^h},$$

and

$$a_i = \alpha_i^{p^i} \text{ for } i = 0 \text{ to } h.$$

The term $bA(x) = \beta^{p^h}A(x)$ is easily reduced modulo Artin-Schreier equivalence:

$$\beta^{p^h}A(x) = \beta^{p^h}\Sigma_{i=0 \text{ to } h}\, a_i x^{p^i}$$
$$= \beta^{p^h}\Sigma_{i=0 \text{ to } h}\, (\alpha_i x)^{p^i}$$
$$= \Sigma_{i=0 \text{ to } h}\, (\beta^{p^{h-i}}\alpha_i x)^{p^i}$$
$$\approx \Sigma_{i=0 \text{ to } h}\, (\beta^{p^{h-i}}\alpha_i)x.$$

Here the coefficient of x is a polynomial in β of degree at most p^h. The coefficient of x in the term $A(b)x$ is equal to

$$A(b) = A(\beta^{p^h}) = \Sigma_{i=0 \text{ to } h}\, (\alpha_i\beta^{p^h})^{p^i}$$

and is thus a polynomial in β of degree p^{2h}. Thus $(x+b)A(x+b)$ is Artin-Schreier equivalent to

$$xA(x) + (\text{polynomial in } \beta \text{ of degree } p^{2h})x.$$

So we can choose β to achieve any desired a_{-1} as the coefficient of x. QED

proof of Supersingularity Lemma 3.8.4 If $p = 2$, we are looking at curves which, after Artin-Schreier reduction, are of the form

$$y^2 - y = ax + bx^3 + cx^5,$$

where either c or b is nonzero. These are supersingular, by the above Corollary 3.8.6.

If $p = 3$, we are looking at curves which, after Artin-Schreier reduction, are of the form

$$y^3 - y = ax + bx^2 + cx^4 + dx^5,$$

with either d or c nonzero. If d is nonzero, then by an additive translation of x we eliminate the x^4 term, after which an Artin-Schreier reduction gives us a curve of the form

$$y^3 - y = ax + bx^2 + dx^5.$$

This curve is supersingular, because it is covered by the curve

$$y^3 - y = ax^2 + bx^4 + dx^{10},$$

which is supersingular by the van der Geer van der Vlugt Theorem 3.8.5. If $d = 0$, then $c \neq 0$, and we have

$$y^3 - y = ax + bx^2 + cx^4,$$

which is itself supersingular by the van der Geer van der Vlugt Theorem 3.8.5.

If $p = 5$, then we are looking at curves which, after Artin-Schreier reduction, are of the form

$$y^5 - y = ax + bx^2 + cx^3,$$

with $c \neq 0$. By an additive translation of x, followed by an Artin-Schreier reduction, we get a curve of the form

$$y^5 - y = ax + cx^3.$$

This curve is supersingular, because it is covered by the curve

$$y^5 - y = ax^2 + cx^6,$$

which is itself supersingular by the van der Geer van der Vlugt Theorem 3.8.5. QED for the Supersingularity Lemma 3.8.4.

(3.9) Application to Deligne polynomials

(3.9.1) With this last theorem 3.8.2, we have now completed the proof of the $\mathcal{L}_\psi$ Theorem 3.1.2. We first state explicitly its application to universal families of Deligne polynomials.

Theorem 3.9.2 Let k be a finite field, $p := \mathrm{char}(k)$, $\ell \neq p$, and ψ a nontrivial additive $\overline{\mathbb{Q}}_\ell^\times$-valued of k. Fix $n \geq 1$, $e \geq 3$, with e prime to p. Denote by $\mathcal{D}(n,e)$ the space of Deligne polynomials, and denote by $\mathcal{M}(n,e,\psi)|\mathcal{D}(n,e)$ the lisse, geometrically irreducible, and pure of weight zero $\overline{\mathbb{Q}}_\ell$-sheaf of rank $(e-1)^n$ on $\mathcal{D}(n,e)$ whose trace function is given by

$$\mathrm{Trace}(\mathrm{Frob}_{E,f} \mid \mathcal{M}(n,e,\psi)) = (-1)^n(\#E)^{-n/2}\Sigma_{v \text{ in } \mathbb{A}^n(E)} \psi_E(f(v)).$$

Suppose that any of the following five conditions holds:

a) $p \geq 7$,

b) $n \geq 3$,

c) $p = 5$ and $e \geq 4$,

d) $p = 3$ and $e \geq 7$,

e) $p = 2$ and $e \geq 7$.

Then we have the following results concerning the group G_{geom} for the lisse sheaf $\mathcal{M}(n,e,\psi)|\mathcal{D}(n,e)$.

1) If $p \neq 2$, G_{geom} contains $SL((e-1)^n)$.

2) If $p = 2$ and n is odd, $G_{geom} = Sp((e-1)^n)$.

3) If $p = 2$ and n is even, G_{geom} is either $SO((e-1)^n)$ or $O((e-1)^n)$.

Remark 3.9.3 In Theorem 6.8.34, we will show that if in addition $e(e-1)$ is prime to p (which forces $p \neq 2$) and ne is even, then G_{geom} is the group

$$GL_{2p}((d-1)^n) := \{A \text{ in } GL((d-1)^n) \text{ with } (\det A)^{2p} = 1\}.$$

(3.9.4) We now explain how to apply the $\mathcal{L}_\psi$ theorem to the construction of "small" families of Deligne polynomials with big monodromy. The idea is quite simple. In any number $n \geq 1$ of variables, if we take an integer d prime to p, and a Deligne polynomial

$$F \text{ in } \mathcal{D}(n,d)(k)$$

of degree d, then for any integer $e < d$, and for any polynomial

$$f \text{ in } \mathcal{P}(n,e)(k)$$

of degree at most e, the sum $F + f$ is again a Deligne polynomial of degree d (simply because being a Deligne polynomial depends only on the highest degree term). So we have a closed immersion

$$\mathcal{P}(n,e) \to \mathcal{D}(n,d),$$
$$f \mapsto F+f.$$

When we restrict the lisse sheaf $\mathcal{M}(n,d,\psi)|\mathcal{D}(n,d)$ to $\mathcal{P}(n,e)$ by this closed immersion $f \mapsto F+f$, we get a lisse sheaf

$$\mathcal{M}(n,e,\psi,F) \text{ on } \mathcal{P}(n,e)$$

of rank $(d-1)^n$, whose trace function is given by

$$\mathrm{Trace}(\mathrm{Frob}_{E,f} \mid \mathcal{M}(n,e,\psi,F))$$
$$= (-1)^n(\#E)^{-n/2}\Sigma_{v \text{ in } \mathbb{A}^n(E)} \psi_E(F(v) + f(v)).$$

The key observation is that $\mathcal{M}(n,e,\psi,F)$ is an instance of the kind of object addressed in the $\mathcal{L}_\psi$ Theorem 3.1.2, where we take for L on $\mathbb{A}^n$ the object $\mathcal{L}_{\psi(F)}[n](n/2)$. The precise result is

Lemma 3.9.5 The perverse sheaf

$$M := \mathrm{Twist}(\mathcal{L}_{\psi(F)}[n](n/2), \mathcal{L}_\psi[1](1/2), \mathcal{P}_e, h=0)$$

on $\mathcal{P}_e = \mathcal{P}(n,e)$ is related to the lisse sheaf $\mathcal{M}(n,e,\psi,F)$ on $\mathcal{P}(n,e)$ by

$$M(-1/2) = \mathcal{M}(n,e,\psi,F)[\dim \mathcal{P}(n,e)].$$

proof This is just proper base change for the closed immersion of $\mathcal{P}(n,e)$ into $\mathcal{D}(n,d)$ given by $f \mapsto F+f$, together with the identity

$$\mathcal{L}_{\psi(F)} \otimes \mathcal{L}_{\psi(f)} \cong \mathcal{L}_{\psi(F+f)}.$$

QED

Theorem 3.9.6 Let k be a finite field, p := char(k), $\ell \neq p$, and ψ a nontrivial additive $\overline{\mathbb{Q}}_\ell^\times$-valued of k. Fix integers

$$n \geq 1,\ d > e \geq 3$$

with d prime to p. Fix a k-rational Deligne polynomial F in n variables of degree d,

$$F \text{ in } \mathcal{D}(n,d)(k).$$

Suppose that any one of the following five conditions holds:
a) $p \geq 7$,
b) $n \geq 3$,
c) $p = 5$ and $e \geq 4$,
d) $p = 3$ and $e \geq 7$,
e) $p = 2$ and $e \geq 7$.
Then we have the following results concerning the group G_{geom} for the lisse sheaf $\mathcal{M}(n,e,\psi,F)$ on $\mathcal{P}(n,e)$.
1) If $p \neq 2$, G_{geom} contains $SL((d-1)^n)$.
2) If $p = 2$ and n is odd, $G_{geom} = Sp((d-1)^n)$.
3) If $p = 2$ and n is even, G_{geom} is either $SO((d-1)^n)$ or $O((d-1)^n)$.
proof This is just a special case of the $\mathcal{L}_\psi$ theorem. QED

Remark 3.9.7 We will show in Theorem 6.8.15 that if in addition d(d-1) is prime to p, then G_{geom} is the group

$$GL_p((d-1)^n) := \{A \text{ in } GL((d-1)^n) \text{ with } (\det A)^p = 1\}.$$

(3.10) Self dual families of Deligne polynomials

(3.10.1) In the above discussion of the monodromy of families of Deligne polynomials, we only encountered self dual geometric monodromy in characteristic two, for the simple reason that $\mathcal{L}_\psi$, being of order p, is geometrically self dual precisely for p=2.

(3.10.2) However, in any characteristic, if we take an integer e prime to p and f in $\mathcal{D}(n,e)(k)$ a Deligne polynomial which is odd, i.e., an f which satisfies

$$f(-x) = -f(x),$$

then the cohomology group

$$H_c^n(\mathbb{A}^n\otimes_k\overline{k},\ \mathcal{L}_{\psi(f)})(n/2)$$

carries an autoduality

$$H_c^n(\mathbb{A}^n\otimes_k\overline{k},\ \mathcal{L}_{\psi(f)})(n/2)\times H_c^n(\mathbb{A}^n\otimes_k\overline{k},\ \mathcal{L}_{\psi(f)})(n/2) \to \overline{\mathbb{Q}}_\ell$$

which is symplectic for n odd, and orthogonal for n even. The pairing is defined as follows. By Poincaré duality, for any polynomial f, the cup product is a perfect pairing

$$H_c^n(\mathbb{A}^n\otimes_k\overline{k},\ \mathcal{L}_{\psi(f)})(n/2)\times H^n(\mathbb{A}^n\otimes_k\overline{k},\ \mathcal{L}_{\overline{\psi}(f)})(n/2)\to \overline{\mathbb{Q}}_\ell.$$

By Theorem 3.5.10, for any Deligne polynomial f the natural "forget supports" map is an isomorphism

$$H_c^n(\mathbb{A}^n\otimes_k\overline{k},\ \mathcal{L}_{\overline{\psi}(f)})(n/2) \cong H^n(\mathbb{A}^n\otimes_k\overline{k},\ \mathcal{L}_{\overline{\psi}(f)})(n/2).$$

So for f a Deligne polynomial, the cup product is a perfect pairing of compact cohomology groups

$$H_c^n(\mathbb{A}^n\otimes_k\bar{k}, \mathcal{L}_{\psi(f)})(n/2)\times H_c^n(\mathbb{A}^n\otimes_k\bar{k}, \mathcal{L}_{\bar{\psi}(f)})(n/2) \to \bar{\mathbb{Q}}_\ell,$$
$$(\alpha, \beta) \mapsto \alpha \cup \beta.$$

For f an odd polynomial, under the involution [-1] of $\mathbb{A}^n$,
$$[-1] : x \mapsto -x,$$
we have
$$[-1]^*\mathcal{L}_{\psi(f)} = \mathcal{L}_{\bar{\psi}(f)},$$
i.e.,
$$[-1]^*\mathcal{L}_{\psi(f(x))} := \mathcal{L}_{\psi(f(-x))} = \mathcal{L}_{\psi(-f(x))} = \mathcal{L}_{\bar{\psi}(f(x))}.$$
So we get an isomorphism
$$H_c^n(\mathbb{A}^n\otimes_k\bar{k}, \mathcal{L}_{\psi(f)})(n/2) \cong H_c^n(\mathbb{A}^n\otimes_k\bar{k}, \mathcal{L}_{\bar{\psi}(f)})(n/2),$$
$$\alpha \mapsto [-1]^*\alpha.$$
The desired autoduality on $H_c^n(\mathbb{A}^n\otimes_k\bar{k}, \mathcal{L}_{\psi(f)})(n/2)$ is given in terms of the cup product by the pairing
$$(\alpha, \beta) := \alpha\cup[-1]^*\beta.$$

(3.10.3) To see that this pairing has the asserted symmetry property, we resort to a global argument. Consider the Artin-Schreier covering
$$AS(f) := \{(x, y) \text{ in } \mathbb{A}^n\times\mathbb{A}^1 \text{with } y - y^{\#k} = f(x)\}$$
$$\mathrm{pr}_1\downarrow$$
$$\mathbb{A}^n,$$
with structural group the additive group of k, with λ in k acting as
$$(x, y) \mapsto (x, y + \lambda).$$
If we push out this (k, +) torsor by the character ψ, we get $\mathcal{L}_{\psi(f)}$.

When we decompose $H_c^*(AS(f)\otimes_k\bar{k}, \bar{\mathbb{Q}}_\ell)$ under the action of (k, +) into ψ-isotypical components, we get
$$H_c^*(AS(f)\otimes_k\bar{k}, \bar{\mathbb{Q}}_\ell)^\psi = H_c^*(\mathbb{A}^n\otimes_k\bar{k}, \mathcal{L}_{\psi(f)}).$$
Because f is odd, we can form the automorphism [-1] of AS(f) given b
$$[-1](x, y) := (-x, -y).$$
This automorphism carries the ψ-eigenspace to the $\bar{\psi}$-eigenspace. We can view the pairing
$$H_c^n(\mathbb{A}^n\otimes_k\bar{k}, \mathcal{L}_{\psi(f)})(n/2) \times H_c^n(\mathbb{A}^n\otimes_k\bar{k}, \mathcal{L}_{\psi(f)})(n/2) \to \bar{\mathbb{Q}}_\ell$$
as being the restriction to the ψ-eigenspace of the pairing
$$H_c^n(AS(f)\otimes_k\bar{k}, \bar{\mathbb{Q}}_\ell)(n/2) \times H_c^n(AS(f)\otimes_k\bar{k}, \bar{\mathbb{Q}}_\ell)(n/2) \to \bar{\mathbb{Q}}_\ell$$
given in terms of the cup product pairing on AS(f) by
$$(\alpha, \beta) := \alpha\cup[-1]^*\beta.$$
The advantage is that the cup product pairing on
$$H_c^n(AS(f)\otimes_k\bar{k}, \bar{\mathbb{Q}}_\ell)(n/2)$$
has known symmetry; it is symplectic if n is odd, and orthogonal if n is even. Remembering that [-1] is an involution, we readily calculate

$$(\beta, \alpha) := \beta \cup [-1]^* \alpha = [-1]^* \beta \cup [-1]^*[-1]^* \alpha$$
$$= [-1]^* \beta \cup \alpha = (-1)^n \alpha \cup [-1]^* \beta$$
$$= (-1)^n (\alpha, \beta).$$

(3.10.4) In characteristic two, every polynomial, in particular every Deligne polynomial, is odd. It is sometimes convenient to impose the stronger condition that f be "strongly odd", by which we mean that when we write $f = \Sigma_{i \le e} F_i$ as a sum of homogeneous forms, then only those F_i with i odd are possibly nonzero. In any odd characteristic, a polynomial is odd if and only if it is strongly odd. But in characteristic two, being strongly odd is more restrictive.

(3.10.5) If we construct the above pairing in the universal family, the same arguments give the following result.

Theorem 3.10.6 Let k be a finite field of characteristic p, ℓ a prime $\ell \neq p$, and ψ a nontrivial additive $\overline{\mathbb{Q}}_\ell^\times$-valued of k. Fix integers

$$n \ge 1,\ e \ge 3,$$

with e prime to p and odd. Denote by

$$\mathcal{D}(n,e,\mathrm{odd}) \subset \mathcal{D}(n,e)$$

the linear subspace of those Deligne polynomials which are strongly odd. Then the restriction $\mathcal{M}(n,e,\psi)|\mathcal{D}(n,e,\mathrm{odd})$ of the lisse sheaf $\mathcal{M}(n,e,\psi)|\mathcal{D}(n,e)$ carries an autoduality

$$\mathcal{M}(n,e,\psi)|\mathcal{D}(n,e,\mathrm{odd}) \times \mathcal{M}(n,e,\psi)|\mathcal{D}(n,e,\mathrm{odd}) \to \overline{\mathbb{Q}}_\ell$$

which is symplectic if n is odd, and orthogonal if n is even.

We now wish to establish the following two theorems, which are the analogues for the self dual case of Theorems 3.9.2 and 3.9.6 above.

Theorem 3.10.7 Let k be a finite field of characteristic p, ℓ a prime with $\ell \neq p$, and ψ a nontrivial additive $\overline{\mathbb{Q}}_\ell^\times$-valued of k. Fix integers

$$n \ge 1,\ e \ge 3,$$

with e prime to p and odd. Suppose that one of the following five conditions holds:

a) $p \ge 7$,
b) $p \neq 3$ and $n \ge 3$,
c) $p = 5$ and $e \ge 7$,
d) $p = 3$ and $e \ge 7$,
e) $p = 2$ and $e \ge 7$.

Then we have the following results concerning the group G_{geom} for the lisse sheaf $\mathcal{M}(n,e,\psi)|\mathcal{D}(n,e,\mathrm{odd})$.

1) If n is odd, $G_{geom} = Sp((e-1)^n)$.

2) If n is even, G_{geom} is either $SO((e-1)^n)$ or $O((e-1)^n)$.

Remark 3.10.8 We will show later, in Theorem 6.8.35, that when n is even, we in fact have $G_{geom} = O((e-1)^n)$.

Theorem 3.10.9 Let k be a finite field of characteristic p, ℓ a prime with $\ell \neq p$, and ψ a nontrivial additive $\overline{\mathbb{Q}}_\ell^\times$-valued of k. Fix integers

$$n \geq 1,\ d > e \geq 3$$

with both d and e prime to p and odd. Fix a k-rational Deligne polynomial F in n variables of degree d which is strongly odd,

$$F \text{ in } \mathcal{D}(n,d,\text{odd})(k).$$

Denote by

$$\mathcal{P}(n,e,\text{odd}) \subset \mathcal{P}(n,e)$$

the linear subspace of strongly odd polynomials of degree at most e. Suppose that one of the following five conditions holds:

a) $p \geq 7$,
b) $p \neq 3$ and $n \geq 3$,
c) $p = 5$ and $e \geq 7$,
d) $p = 3$ and $e \geq 7$,
e) $p = 2$ and $e \geq 7$.

Then we have the following results concerning the group G_{geom} for the lisse sheaf $\mathcal{M}(n,e,\psi,F)|\mathcal{P}(n,e,\text{odd})$.

1) If n is odd, $G_{geom} = Sp((e-1)^n)$.

2) If n is even and p is odd, $G_{geom} = SO((e-1)^n)$.

3) If n is even and $p = 2$, G_{geom} is either $SO((e-1)^n)$ or $O((e-1)^n)$.

(3.11) Proofs of the theorems on self dual families

(3.11.1) For any odd $e \geq 1$, evaluation on strongly odd polynomials defines a closed immersion

$$\text{eval} : \mathbb{A}^n \to (\mathcal{P}(n,e,\text{odd}))^\vee.$$

[Indeed, already evaluation on linear forms does so.] So for any perverse L on $\mathbb{A}^n$, eval_*L is perverse on $(\mathcal{P}(n,e,\text{odd}))^\vee$, and $FT_\psi(\text{eval}_*L)$ is perverse on $\mathcal{P}(n,e,\text{odd})$. If in addition L is geometrically irreducible and pure of weight zero on $\mathbb{A}^n$, then eval_*L is geometrically irreducible and pure of weight zero on $(\mathcal{P}(n,e,\text{odd}))^\vee$, and $FT_\psi(\text{eval}_*L)$ is geometrically irreducible and pure of weight $\dim\mathcal{P}(n,e,\text{odd})$ on $\mathcal{P}(n,e,\text{odd})$. The object

$$M(n,e,\psi,\text{odd})$$

on $\mathcal{P}(n,e,\text{odd})$ defined by

$$M(n,e,\psi,\text{odd}) := M(n,e,\psi)[\dim\mathcal{P}(n,e,\text{odd}) - \dim\mathcal{P}(n,e)]|\mathcal{P}(n,e,\text{odd})$$

is none other than $FT_\psi(\text{eval}_*\overline{\mathbb{Q}}_\ell[n]((n+1)/2))$, cf. 3.2.3. Therefore $M(n,e,\psi,\text{odd})(-1/2)$ is perverse, geometrically irreducible, and pure of weight $\dim\mathcal{P}(n,e,\text{odd})$. On the dense open set

$$\mathcal{D}(n,e,\text{odd}) \subset \mathcal{P}(n,e,\text{odd}),$$

we have

$$M(n,e,\psi,\text{odd})(-1/2)|\mathcal{D}(n,e,\text{odd}) = \mathcal{M}(n,e,\psi)|\mathcal{D}(n,e,\text{odd}).$$

And for any odd e_0 with $1 \leq e_0 \leq e$, we have

$$M(n,e_0,\psi,\text{odd})$$
$$= M(n,e,\psi,\text{odd})[\dim\mathcal{P}(n,e_0,\text{odd}) - \dim\mathcal{P}(n,e,\text{odd})]|\mathcal{P}(n,e_0,\text{odd}),$$

and

$$\mathcal{M}(n,e_0,\psi,\mathrm{odd}) = \mathcal{M}(n,e,\psi,\mathrm{odd})|\mathcal{P}(n,e_0,\mathrm{odd}).$$

Fourth Moment Theorem 3.11.2 Let $e \ge 3$ be an odd integer which is prime to p. If $p = 3$, assume also that $e \ge 7$.

1) We have

$$M_4{}^{\mathrm{geom}}(\mathcal{D}(n,e,\mathrm{odd}), \mathcal{M}(n,e,\psi)|\mathcal{D}(n,e,\mathrm{odd})) \le 3.$$

2) For any $d > e$ with d prime to p, and any F in $\mathcal{D}(n,d,\mathrm{odd})(k)$, we have

$$M_4{}^{\mathrm{geom}}(\mathcal{P}(n,e), \mathcal{M}(n,e,\psi,F)) \le 3.$$

proof Fix a finite extension E/k. For f in $\mathcal{D}(n,e,\mathrm{odd})(E)$, we have

$$\mathrm{Trace}(\mathrm{Frob}_{E,f} \mid \mathcal{M}(n,e,\psi)) = (-1)^n(\#E)^{-n/2}\Sigma_{v \text{ in } \mathbb{A}^n(E)} \psi_E(f(v)).$$

And for f in $\mathcal{D}(n,e,\mathrm{odd})(E)$, we have

$$\mathrm{Trace}(\mathrm{Frob}_{E,f} \mid \mathcal{M}(n,e,\psi,F))$$
$$= (-1)^n(\#E)^{-n/2}\Sigma_{v \text{ in } \mathbb{A}^n(E)} \psi_E(F(v) + f(v)).$$

Thanks to Theorem 1.17.4, it suffices to establish that for variable E/k, we have the two inequalities

$$\Sigma_{f \text{ in } \mathcal{D}(n,e,\mathrm{odd})(E)} |\mathrm{Trace}(\mathrm{Frob}_{E,f} \mid \mathcal{M}(n,e,\psi))|^4$$
$$\le 3(\#E)^{\dim \mathcal{P}(n,e,\mathrm{odd})}(1 + O(\#E)^{-1/2}),$$
$$\Sigma_{f \text{ in } \mathcal{P}(n,e,\mathrm{odd})(E)} |\mathrm{Trace}(\mathrm{Frob}_{E,f} \mid \mathcal{M}(n,e,\psi,F))|^4$$
$$\le 3(\#E)^{\dim \mathcal{P}(n,e,\mathrm{odd})}(1 + O(\#E)^{-1/2}).$$

We get both of these if we show that for **any** F in $\mathcal{P}(n,d,\mathrm{odd})(k)$, we have

$$\Sigma_{f \text{ in } \mathcal{P}(n,e,\mathrm{odd})(E)} |(\#E)^{-n/2}\Sigma_{v \text{ in } \mathbb{A}^n(E)} \psi_E(F(v) + f(v))|^4$$
$$\le 3(\#E)^{\dim \mathcal{P}(n,e,\mathrm{odd})}(1 + O(\#E)^{-1/2}),$$

simply take $F=0$ to get a strengthened form of the first inequality.

With E fixed and the domains of summation understood, we rewrite what we must show as

$$\Sigma_f |\Sigma_v \psi_E(F(v) + f(v))|^4$$
$$\le 3(\#E)^{2n+ \dim \mathcal{P}(n,e,\mathrm{odd})}(1 + O(\#E)^{-1/2}).$$

We expand out the left hand side, letting x, y, z, and w each run over $\mathbb{A}^n(E)$:

$$\Sigma_f |\Sigma_v \psi_E(F(v) + f(v))|^4$$
$$= \Sigma_f \Sigma_{x,y,z,w} \psi_E(F(x)+F(y)-F(z)-F(w)+f(x)+f(y)-f(z)-f(w))$$
$$= \Sigma_{x,y,z,w} \psi_E(F(x)+F(y)-F(z)-F(w)) \Sigma_f \psi_E(f(x)+f(y)-f(z)-f(w)).$$

For fixed x,y,z,w, $f \mapsto \psi_E(f(x)+f(y)-f(z)-f(w))$ is a $\overline{\mathbb{Q}}_\ell{}^\times$-valued character of the additive group $\mathcal{P}(n,e,\mathrm{odd})(E)$, which is trivial if and only if

$$f(x) + f(y) = f(z) + f(w) \text{ for all f in } \mathcal{P}(n,e,\mathrm{odd})(E).$$

So by orthogonality for finite abelian groups, the innermost sum is given by

$$\Sigma_f \psi_E(f(x)+f(y)-f(z)-f(w))$$
$$= (\#E)^{\dim \mathcal{P}(n,e,\mathrm{odd})},$$
$$\text{if } f(x) + f(y) = f(z) + f(w) \text{ for all } f \text{ in } \mathcal{P}(n,e,\mathrm{odd})(E),$$
$$= 0, \text{ if not.}$$

Lemma 3.11.3 (p ≠ 3 case) Let E be a field in which 3 is invertible, $e \geq 3$ an integer. Four points x, y, z, w in $\mathbb{A}^n(E)$ satisfy the condition (*E) below,

(*E) $\quad f(x) + f(y) = f(z) + f(w)$ for all f in $\mathcal{P}(n,e,\mathrm{odd})(E)$,

if and only if one of the following conditions holds:

$$x = z \text{ and } y = w,$$

or

$$x = w \text{ and } y = z,$$

or

$$x = -y \text{ and } z = -w.$$

proof The "if" assertion is trivial. To prove the "only if", we first consider the key case n=1.

Sublemma 3.11.4 (p ≠ 3 case)(compare [Ka-LFM, page 118]) Let E be a field in which 3 is invertible, a,b,c,d elements of E which satisfy the two equations

$$a + b = c + d,$$
$$a^3 + b^3 = c^3 + d^3.$$

Then either

$$a = c \text{ and } b = d,$$

or

$$a = d \text{ and } b = c,$$

or

$$a = -b \text{ and } c = -d.$$

proof If the common value of a+b=c+d is 0, we are in the third case. If not, then both a+b and c+d are nonzero, so we can divide, and get

$$(a^3 + b^3)/(a+b) = (c^3 + d^3)/(c+d),$$

i.e.,

$$a^2 - ab + b^2 = c^2 - cd + d^2.$$

But squaring the relation a + b=c + d gives

$$a^2 + 2ab + b^2 = c^2 + 2cd + d^2.$$

Subtracting, we find

$$3ab = 3cd.$$

Since 3 is invertible, we get ab=cd, which, together with a + b=c + d, tells us that

$$(X-a)(X-b) = (X-c)(X-d),$$

and hence we are in one of the first two cases. QED

We now turn to proving the "only if" in the general case in 3.11.3. If four points x, y, z, w in $\mathbb{A}^n(E)$ satisfy (*E), then for any extension field F/E, these points satisfy (*F), simply because $\mathcal{P}(n,e,\mathrm{odd})(F)$ is the F-span of $\mathcal{P}(n,e,\mathrm{odd})(E)$. So we may reduce to the case when E is infinite.

Pick a linear form L in $\mathcal{P}(n,1,\mathrm{odd})(E)$. Then both L and L^3 are in

$\mathcal{P}(n,e,odd)(E)$, so we have

$$L(x) + L(y) = L(z) + L(w),$$
$$L^3(x) + L^3(y) = L^3(z) + L^3(w).$$

By the sublemma, we have either

$$L(x) = L(z) \text{ and } L(y) = L(w),$$

or

$$L(x) = L(w) \text{ and } L(y) = L(z),$$

or

$$L(x) = -L(y) \text{ and } L(z) = -L(w).$$

So we have

$$L(x-z)L(x-w)L(x+y) = 0$$

for all L.

We must show that at least one of x-z, x-w, and x+y is zero. If not, each of

$$L \mapsto L(x-z),$$
$$L \mapsto L(x-w),$$
$$L \mapsto L(x+y)$$

is a nonzero linear polynomial function on $\mathcal{P}(n,1,odd)$, whose product, a cubic polymonial function, vanishes on all the E-valued points of $\mathcal{P}(n,1,odd)$. Since E is infinite and $\mathcal{P}(n,1odd)$ is an affine space, $\mathcal{P}(n,1,odd)(E)$ is Zariski dense in $\mathcal{P}(n,1,odd)$. So this product is the zero function. But the coordinate ring of $\mathcal{P}(n,1,odd)$ is an integral domain, contradiction. QED for 3.11.3.

Lemma 3.11.5 (p = 3 case) Let E be a field of characteristic 3, $e \geq 7$ an integer. Four points x, y, z, w in $\mathbb{A}^n(E)$ satisfy the condition $(\ast E)$ below,

$(\ast E)$ $\quad f(x) + f(y) = f(z) + f(w)$ for all f in $\mathcal{P}(n,e,odd)(E)$,

if and only if one of the following conditions holds:

$$x = z \text{ and } y = w,$$

or

$$x = w \text{ and } y = z,$$

or

$$x = -y \text{ and } z = -w$$

or

$$x = y = -z \text{ and } w = 0,$$

or

$$x = y = -w \text{ and } z = 0,$$

or

$$x = 0 \text{ and } -y = z = w,$$

or

$$y = 0 \text{ and } -x = z = w.$$

proof The "if" assertion is trivial, the last four cases working because of the characteristic three identity $1 + 1 = -1 + 0$. To prove the "only if", we first consider the key case n=1.

Sublemma 3.11.6A (p = 3 case) Let E be a field of characteristic 3, a,b,c,d elements of E which satisfy the three equations

$$a + b = c + d,$$

$$a^5 + b^5 = c^5 + d^5,$$
$$a^7 + b^7 = c^7 + d^7.$$

Then either

$$a=c \text{ and } b=d,$$

or

$$a=d \text{ and } b=c,$$

or

$$a = -b \text{ and } c=-d,$$

or

$$a = b = -c \text{ and } d = 0,$$

or

$$a = b= -d \text{ and } c = 0,$$

or

$$a = 0 \text{ and } -b = c = d,$$

or

$$b = 0 \text{ and } -a = c = d.$$

proof Denote by 2λ the common value of $a+b = c+d$. In the new variables A, C, λ defined by

$$a = \lambda + A,\ c = \lambda + C,$$

we readily solve for $b = \lambda - A$ and $d = \lambda - C$. The equations become

$$(\lambda+A)^5 + (\lambda-A)^5 = (\lambda+C)^5 + (\lambda-C)^5,$$
$$(\lambda+A)^7 + (\lambda-A)^7 = (\lambda+C)^7 + (\lambda-C)^7,$$

i.e.,

$$20\lambda^3A^2 + 10\lambda A^4 = 20\lambda^3C^2 + 10\lambda C^4,$$
$$42\lambda^5A^2 + 70\lambda^3A^4 + 14\lambda A^6 = 42\lambda^5C^2 + 70\lambda^3C^4 + 14\lambda C^6.$$

Because we are in characteristic 3, we may rewrite these as

$$2\lambda^3A^2 + \lambda A^4 = 2\lambda^3C^2 + \lambda C^4,$$
$$\lambda^3A^4 + 2\lambda A^6 = \lambda^3C^4 + 2\lambda C^6,$$

i.e.,

$$2\lambda^3(A^2 - C^2) + \lambda(A^4 - C^4) = 0,$$
$$\lambda^3(A^4 - C^4) + 2\lambda(A^6 - C^6) = 0,$$

i.e.,

$$2\lambda(A^2 - C^2)(2\lambda^2 + A^2 + C^2) = 0,$$
$$\lambda(A^2 - C^2)(\lambda^2(A^2 + C^2) + 2A^4 + 2A^2C^2 + 2C^4) = 0.$$

If $\lambda = 0$, we have the case $a= -b$, $c = -d$. If $A^2 = C^2$, we have either $A = C$ (which gives the case $a = c$ and $b = d$) or $A = - C$ (which gives the case $a = d$ and $b = c$). Suppose, then, that $\lambda(A^2 - C^2) \neq 0$. Then A, C, and λ satisfy the equations

$$2\lambda^2 + A^2 + C^2 = 0,$$
$$\lambda^2(A^2 + C^2) + 2A^4 + 2A^2C^2 + 2C^4 = 0,$$

i.e. (remember we are in characteristic 3),

$$A^2 + C^2 = \lambda^2,$$

$$\lambda^2(A^2 + C^2) + 2(A^2 + C^2)^2 - 2A^2C^2 = 0,$$

i.e.,

$$A^2 + C^2 = \lambda^2,$$
$$A^2C^2 = 0.$$

Thus A^2 and C^2 are the roots of the quadratic equation

$$X^2 - \lambda^2 X = 0.$$

So either $A = 0$ and $C = \pm\lambda$ (these are the cases $a=b=-c$, $d=0$, and $a=b=-d$, $c=0$), or $A = \pm\lambda$ and $C = 0$ (these are the cases $b=0$, $-a = c=d$, and $a=0$, $-b = c = d$). QED

Sublemma 3.11.6B (p = 3 case) Let E be a field of characteristic 3, a,b,c elements of E which satisfy the three equations

$$a + b = c,$$
$$a^5 + b^5 = c^5,$$
$$a^7 + b^7 = c^7.$$

Then either

$$a = 0,$$

or

$$b = 0,$$

or

$$c = 0,$$

or

$$a = b.$$

proof Simply take $d=0$ in the previous sublemma. QED

We now turn to proving the "only if" in the general case of 3.11.5. As in the $p \neq 3$ case, we reduce to the case when E is infinite.

Pick a linear form L in $\mathcal{P}(n,1,\text{odd})(E)$. Then L, L^5, and L^7 are in $\mathcal{P}(n,e,\text{odd})(E)$, so we have

$$L(x) + L(y) = L(z) + L(w),$$
$$L^5(x) + L^5(y) = L^5(z) + L^5(w),$$
$$L^7(x) + L^7(y) = L^7(z) + L^7(w).$$

By Sublemma 3.11.6A, we have either

$$L(x) = L(z) \text{ and } L(y) = L(w),$$

or

$$L(x) = L(w) \text{ and } L(y) = L(z),$$

or

$$L(x) = -L(y) \text{ and } L(z) = -L(w),$$

or

$$L(x) = L(y) = -L(z) \text{ and } L(w) = 0,$$

or

$$L(x) = L(y) = -L(w) \text{ and } L(z) = 0,$$

or

$$L(x) = 0 \text{ and } -L(y) = L(z) = L(w),$$

or

$$L(y) = 0 \text{ and } -L(x) = L(z) = L(w).$$

So we have

$$L(x-z)L(x-w)L(x+y)L(x)L(y)L(z)L(w) = 0$$

for all L. Exactly as in the proof of Lemma 3.11.3, we first infer that one of x-z, x-w, x+y, x, y, z, w is zero.

If any of x-z, x-w, x+y is zero, we are in one of the first three asserted cases.

If, say, w = 0, then we apply the second sublemma 3.11.6B to conclude now that

$$L(x-y)L(x)L(y)L(z) = 0$$

for all L, and hence that one of x-y, x, y, or z vanishes. If x=y, then from the condition (*E) we infer that z = -x, an asserted case. If x vanishes, then x=w, an asserted case. If y vanishes, then y=w, an asserted case. If z vanishes, then (*E) gives x + y = 0, an asserted case. QED for 3.11.5.

(3.11.7) We now return to the calculation of

$$\Sigma_f |\Sigma_v \psi_E(F(v) + f(v))|^4$$
$$= \Sigma_{x,y,z,w} \psi_E(F(x)+F(y)-F(z)-F(w)) \Sigma_f \psi_E(f(x)+f(y)-f(z)-f(w)).$$

Suppose first $p \neq 3$. By the lemma, the inner sum vanishes unless

x = z and y = w,

or

x = w and y = z,

or

x = -y and z = -w.

In each of these cases, the inner sum is $\#\mathcal{P}(n,e,odd)(E)$, and its coefficient in our sum is

$$\psi_E(F(x)+F(y)-F(z)-F(w)) = \psi_E(0) = 1.$$

So our sum is simply the product of $\#\mathcal{P}(n,e,odd)(E)$ with the number of four-tuples x, y, z, w of points in $\mathbb{A}^n(E)$ which satisfy

x = z and y = w,

or

x = w and y = z,

or

x = -y and z = -w.

The number of such four-tuples is readily computed by inclusion-exclusion to be

$$3(\#E)^{2n} - 3(\#E)^n + 1.$$

So we end up with the closed formula

$$\Sigma_f |\Sigma_v \psi_E(F(v) + f(v))|^4$$
$$= (\#E)^{\dim\mathcal{P}(n,e,odd)}(3(\#E)^{2n} - 3(\#E)^n + 1)$$
$$= 3(\#E)^{2n+ \dim\mathcal{P}(n,e,odd)}(1 + O(\#E)^{-1/2}).$$

Suppose now p=3. Then there are four additional cases in which the inner sum does not vanish, namely

x = y = -z, w = 0,

x = y = -w, z = 0,

x = 0, -y = z = w,

$y = 0, -x = z = w.$

In each of these cases as well, the inner sum is $\#\mathcal{P}(n,e,odd)(E)$, and its coefficient in our sum is 1. So the sum is the product of $\#\mathcal{P}(n,e,odd)(E)$ and

$$3(\#E)^{2n} - 3(\#E)^n + 1 + (\text{the number of cases special to } p=3)$$
$$= 3(\#E)^{2n} - 3(\#E)^n + 1 + 4(\#E - 1)^n.$$

So we end up in characteristic p=3 with the closed formula

$$\Sigma_f |\Sigma_v \psi_E(F(v) + f(v))|^4$$
$$= (\#E)^{\dim \mathcal{P}(n,e,odd)}(3(\#E)^{2n} - 3(\#E)^n + 1 + 4(\#E - 1)^n)$$
$$= 3(\#E)^{2n+ \dim \mathcal{P}(n,e,odd)}(1 + O(\#E)^{-n}).$$

This concludes the proof of the Fourth Moment Theorem 3.11.2. QED

Fourth Moment Corollary 3.11.8 ($p \neq 3$ case) Suppose $p \neq 3$.

1) For any odd $e \geq 3$ prime to p, we have

$$M_4{}^{geom}(\mathcal{D}(n,e,odd), \mathcal{M}(n,e,\psi)|\mathcal{D}(n,e,odd)) = 3,$$

unless n=1 and e=3.

2) For any $d > e \geq 3$ with both d and e odd and prime to p, and any F in $\mathcal{D}(n,d,odd)(k)$,we have

$$M_4{}^{geom}(\mathcal{P}(n,e), \mathcal{M}(n,e,\psi,F)) = 3.$$

proof Any completely reducible orthogonal or symplectic $\overline{\mathbb{Q}}_\ell$-representation of dimension at least 3 has fourth moment at least 3. The rank of $\mathcal{M}(n,e,\psi)|\mathcal{D}(n,e,odd)$ is $(e-1)^n$, and the rank of $\mathcal{M}(n,e,\psi,F)$ is $(d-1)^n$, both of which are at least 3 unless n=1 and e=3. QED

Fourth Moment Corollary 3.11.9 ($p = 3$ case) Suppose $p = 3$.

1) For any odd $e \geq 7$ prime to p, we have

$$M_4{}^{geom}(\mathcal{D}(n,e,odd), \mathcal{M}(n,e,\psi)|\mathcal{D}(n,e,odd)) = 3.$$

2) For any $d > e \geq 7$ with both d and e odd and prime to p, and any F in $\mathcal{D}(n,d,odd)(k)$,we have

$$M_4{}^{geom}(\mathcal{P}(n,e), \mathcal{M}(n,e,\psi,F)) = 3.$$

(3.12) Proof of Theorem 3.10.7

Consider first the case n=1 and e = 3. This case is only permitted in characteristic $p \geq 7$. In any such characteristic, already the one-parameter family of sums $t \mapsto \Sigma\psi(x^3 + tx)$ is a rank two lisse sheaf on $\mathbb{A}^1$ which is Lie-irreducible and symplectically self dual, so must have $G_{geom} = SL(2)$.

In the remaining cases, our self dual (by 3.10.6) lisse sheaves have fourth moment 3, thanks to the Corollaries 3.11.8 and 3.11.9 above. So by Larsen's Alternative 2.2.2, we have only to show that G_{geom} for $\mathcal{M}(n,e,\psi)|\mathcal{D}(n,e,odd)$ is not finite.

We do this as follows. Since $\mathcal{M}(n,e,\psi)$ on $\mathcal{P}(n,e)$ is of perverse origin, its restriction

$$\mathcal{M}(n,e,\psi,\text{odd}) := \mathcal{M}(n,e,\psi)|\mathcal{P}(n,e,\text{odd})$$

is of perverse origin on $\mathcal{P}(n,e,\text{odd})$. And by further restriction we obtain

$$\mathcal{M}(n,e,\psi,\text{odd})|\mathcal{P}(n,e_0,\text{odd}) = \mathcal{M}(n,e_0,\psi,\text{odd}).$$

So it suffices to show that G_{geom} for

$$\mathcal{M}(n,e_0,\psi,\text{odd})|\mathcal{D}(n,e_0,\text{odd})$$

is not finite, for e_0 the lowest value allowed in the theorem.

If $p \geq 7$, we need to show that G_{geom} for

$$\mathcal{M}(n,3,\psi,\text{odd})|\mathcal{D}(n,3,\text{odd})$$

is not finite. But we already know this. The n-parameter family

$$\Sigma_i (x_i)^3 + \Sigma_i t_i x_i$$

of cubic Deligne polynomials we used to show that G_{geom} for

$$\mathcal{M}(n,3,\psi)|\mathcal{D}(n,3)$$

is infinite is a family of strongly odd cubic Deligne polynomials.

If $n \geq 3$ and $p \neq 3$, we also need to show that G_{geom} for

$$\mathcal{M}(n,3,\psi,\text{odd})|\mathcal{D}(n,3,\text{odd})$$

is not finite. But we already know this as well, because we proved that G_{geom} for

$$\mathcal{M}(n,3,\psi)|\mathcal{D}(n,3)$$

with $n \geq 3$ was infinite by exhibiting a cleverly chosen homogeneous cubic Deligne polynomial (remember $p\neq 3$), namely

(nonsingular ternary cubic which is ordinary)
+ (sum of cubes of remaining variables, if any),

whose Frobenius has no power a scalar.

We now treat the last three cases ($p = 5, 3, 2$) by noting that the lowest value of e allowed is 7. We must show that G_{geom} for

$$\mathcal{M}(n,7,\psi,\text{odd})|\mathcal{D}(n,7,\text{odd})$$

is not finite. We reduce to the case when the additive character ψ is a nontrivial character of $\mathbb{F}_p$, and then to the case $n=1$. For each of $p = 2, 3, 5$, it suffices to exhibit a strongly odd polynomial of degree 7 over $E := \mathbb{F}_{p^3}$, such that

$$\Sigma_{x \text{ in } E}\ \psi_E(f(x))$$

is not divisible by $\text{Sqrt}(\#E) = \text{Sqrt}(p^3)$ as an algebraic integer. Then no power of its Frobenius is a scalar, cf. 3.5.14.

For $p = 2$, we have already seen, in the proof of case e) of Theorem 3.8.2, that x^7 is such an f. For $p = 3$, we have already seen, in the proof of case d) of Theorem 3.8.2, that $\alpha^2x^7 + x^5$ is such an f, for α in $\mathbb{F}_{27}$ satisfying $\alpha^3 - \alpha = 1$. For $p = 5$, we claim that $x^7 + x^3$ is such an f. To see this, consider, for any prime p, any finite extension $E/\mathbb{F}_p$, and any f in E[x], the affine Artin-Schreier curve

$$C_f : y^p - y = f(x).$$

As explained inside case d) in the proof of Theorem 3.8.2, we have

$$\#C_f(E) = \Sigma_{v \text{ in } E} \Sigma_{\lambda \text{ in } \mathbb{F}_p} \psi_E(\lambda f(v))$$
$$= \#E + \Sigma_{\lambda \neq 0 \text{ in } \mathbb{F}_p} \Sigma_{v \text{ in } E} \psi_E(\lambda f(v))$$
$$= \#E + \mathrm{Trace}_{\mathbb{Z}[\zeta_p]/\mathbb{Z}}(\Sigma_{v \text{ in } E} \psi_E(f(v))).$$

If $\Sigma_{v \text{ in } E} \psi_E(f(v))$ were divisible by Sqrt($\#E$) as an algebraic integer, then its trace to $\mathbb{Z}$ would be divisible by Sqrt($\#E$) as an algebraic integer, and so $\#C_f(E)$ would be divisible by Sqrt($\#E$) as an algebraic integer. But for $E = \mathbb{F}_{125}$, and $f(x) = x^7 + x^3$, direct calculation shows that

$$\#C_f(\mathbb{F}_{125}) = 305 = 5\times 61,$$

so $\#C_f(\mathbb{F}_{125})$ is not divisible by Sqrt(125). QED

(3.13) Proof of Theorem 3.10.9

This follows from Theorem 3.10.7 by a homothety contraction argument. By the fourth moment corollaries 3.11.8 and 3.11.9, our self dual pure sheaf $\mathcal{M}(n,e,\psi,F)$ has rank at least 3, and it has fourth moment three. So by Larsen's Alternative 2.2.2, it suffices to prove that G_{geom} is not finite. For then G_{geom} is the full symplectic group Sp for n odd, and it is either SO or O for n even. If $p \neq 2$, the O case cannot occur, because the determinant would be a character of order two of $\pi_1{}^{geom}$ of an affine space $\mathcal{P}(n,e,odd)$ in odd characteristic p. But

$$\mathrm{Hom}(\pi_1{}^{geom}(\mathcal{P}(n,e,odd)), \mathbb{F}_2)$$
$$= H^1(\mathcal{P}(n,e,odd)\otimes_k \bar{k}, \mathbb{F}_2) = 0, \text{ for } p\neq 2.$$

It remains to show that G_{geom} is not finite. We introduce a parameter t. On the product space

$$\mathbb{A}^n \times \mathbb{A}^1 \times \mathcal{P}(n,e,odd)$$

we have the lisse, rank one $\overline{\mathbb{Q}}_\ell$-sheaf

$$(x, t, f) \mapsto \mathcal{L}_\psi(F(tx) + f(x)).$$

The projection

$$pr_{2,3} : \mathbb{A}^n \times \mathbb{A}^1 \times \mathcal{P}(n,e,odd) \to \mathbb{A}^1 \times \mathcal{P}(n,e,odd)$$

is affine and smooth, of relative dimension n, so the $\overline{\mathbb{Q}}_\ell$-sheaf

$$\mathcal{M}_{def} := R^n(pr_{2,3})_! \mathcal{L}_\psi(F(tx) + f(x))$$

on $\mathbb{A}^1 \times \mathcal{P}(n,e,odd)$ is of perverse origin, cf. [Ka-SMD, Corollary 6]. Exactly as in the proof of the Homothety Contraction Theorem 3.3.13, we see that the restriction of $\mathcal{M}_{def}$ to $\mathbb{G}_m \times \mathcal{P}(n,e,odd)$ has the same G_{geom} as $\mathcal{M}(n,e,\psi,F)$, while its restriction to $\{0\}\times\mathcal{P}(n,e,odd)$ is $\mathcal{M}(e,n,\psi)|\mathcal{P}(n,e,odd)$. Since this last sheaf, restricted to the dense open set $\mathcal{D}(n,e,odd)$, has its G_{geom} infinite by the first theorem, we infer from the Semicontinuity Corollary 2.8.9 that G_{geom} for

$\mathcal{M}_{def}|(\mathbb{G}_m \times \mathcal{P}(n,e,odd))$ is infinite, and this last G_{geom} is the same as the G_{geom} for $\mathcal{M}(n,e,\psi,F)$. QED

Chapter 4: Additive character sums on more general X

(4.1) The general setting

(4.1.1) In the previous chapter, we considered in some detail the monodromy groups attached to exponential sums of Deligne type on affine space $\mathbb{A}^n$. In this chapter, we consider the analogous question for sums of Deligne type on more general varieties.

(4.1.2) Fix a finite field k of characteristic p. Fix also a projective, smooth, geometrically connected k-scheme X/k, of dimension $n \geq 1$, given with a projective embedding

$$X \subset \mathbb{P}^N := \mathbb{P}.$$

We fix an integer $r \geq 1$, an r-tuple $(d_1,\ldots, d_r)$ of strictly positive integers, and an r-tuple $(Z_1,\ldots, Z_r)$ of nonzero global sections

$$Z_i \text{ in } H^0(\mathbb{P}, \mathcal{O}_{\mathbb{P}}(d_i)).$$

When no ambiguity is likely, the hypersurface in $\mathbb{P}$ defined by the vanishing of Z_i will itself be denoted Z_i.

(4.1.3) We make the following transversality hypothesis (TrZ, X) concerning the Z_i with respect to X:

(TrZ, X): for every nonvoid subset $\mathcal{I}$ of $\{1,\ldots, r\}$, the closed subscheme

$$X\cap(\cap_{i \text{ in } \mathcal{I}} Z_i)$$

of X is smooth of codimension $\#\mathcal{I}$ in X. [For $\#\mathcal{I} \geq n+1$, this means only that $X \cap(\cap_{i \text{ in } \mathcal{I}} Z_i)$ is empty.]

(4.1.4) We denote by

$$V := X - X\cap(\cup_i Z_i) = X[1/(\Pi_i Z_i)]$$

the affine open set of X where all the Z_i are invertible.

(4.1.5) For an r-tuple $(e_1, e_2,\ldots, e_r)$ of integers $e_i \geq 0$, and an element

$$H \text{ in } H^0(\mathbb{P}, \mathcal{O}_{\mathbb{P}}(\Sigma_i e_i d_i)),$$

the ratio

$$h := H/\Pi_i Z_i^{e_i}$$

makes sense as a function on $\mathbb{P}[1/(\Pi_i Z_i)]$, and then, by restriction, as a function on V, i.e., as a morphism

$$h : V \to \mathbb{A}^1.$$

We denote

$$\mathcal{P}(e_1,\ldots, e_r) := H^0(\mathbb{P}, \mathcal{O}_{\mathbb{P}}(\Sigma_i e_i d_i)),$$

and we denote by

$$\tau : \mathcal{P}_{(e_1,\ldots,\, e_r)} \to \mathrm{Hom}_{k\text{-schemes}}(V, \mathbb{A}^1)$$

the linear map

$$H \mapsto h := H/\prod_i Z_i^{e_i} \mid V.$$

Thus

$$(\mathcal{P}_{(e_1,\ldots,\, e_r)}, \tau)$$

is a linear space of $\mathbb{A}^1$-valued functions on V.

Lemma 4.1.6 Notations as in 4.1.5 above, the space $(\mathcal{P}_{(e_1,\ldots,\, e_r)}, \tau)$ is $(1 + \Sigma_i e_i d_i)$-separating as a linear space of functions on V.

proof We must show that for any extension field L/k, and any

$$d := 1 + \Sigma_i e_i d_i$$

distinct points P_i in V(E), the simultaneous evaluation map

$$\mathrm{eval}(P_1,\ldots P_d) : \mathcal{P}_{(e_1,\ldots,\, e_r)}\otimes_k E \to \mathbb{A}^d(E)$$

is surjective, or equivalently, that in $H^0(\mathbb{P}, \mathcal{O}_{\mathbb{P}}(\Sigma_i e_i d_i))\otimes_k L$, the subspace of those sections H which vanish at all the P_i has codimension d. Extending the field L if necessary, we may find a linear form Lin in $H^0(\mathbb{P}, \mathcal{O}_{\mathbb{P}}(1))\otimes_k L$ which is nonzero at all the P_i. Then it is equivalent to show that among the functions H/Lin^{d-1} on $\mathbb{P}[1/\mathrm{Lin}]$, those which vanish at our d distinct points form a subspace of codimension d. This is in turn equivalent to showing the surjectivity of the simultaneous evaluation map

$$\mathrm{eval}(P_1,\ldots P_d) : \{H/\mathrm{Lin}^{d-1}\}_{H \text{ in } H^0(\mathbb{P},\, \mathcal{O}_{\mathbb{P}}(\Sigma_i e_i d_i))\otimes_k L} \to \mathbb{A}^d(L).$$

But this last surjectivity is just the fact (cf. 1.1.8) that on the affine space $\mathbb{P}[1/\mathrm{Lin}] \cong \mathbb{A}^N$, the space of polynomial functions of degree $\le d-1$ is d-separating. QED

Lemma 4.1.7 Notations as above, suppose that $e_i \ge 1$ for all i. Then we have the following results.

1) The tautological morphism

$$V \to \mathbb{P}(H^0(\mathbb{P}, \mathcal{O}_{\mathbb{P}}(\Sigma_i e_i d_i)))[1/(\prod_i Z_i)]$$
$$= \mathbb{P}(H^0(\mathbb{P}, \mathcal{O}_{\mathbb{P}}(\Sigma_i e_i d_i)))[1/(\prod_i Z_i^{e_i})]$$

is a closed immersion.

2) The space $\mathbb{P}(H^0(\mathbb{P}, \mathcal{O}_{\mathbb{P}}(\Sigma_i e_i d_i)))[1/(\prod_i Z_i^{e_i})]$ is an affine space $\mathbb{A}^M$, on which the polynomial functions of degree ≤ 1 are precisely the functions $H/\prod_i Z_i^{e_i}$, for H in $H^0(\mathbb{P}, \mathcal{O}_{\mathbb{P}}(\Sigma_i e_i d_i))$.

3) The natural map

$$\mathrm{eval} : V \to \mathcal{P}_{(e_1,\ldots,\, e_r)}{}^{\vee},$$
$$v \mapsto (H \mapsto (\tau H)(v))$$

is a closed immersion.

proof 1) Let us denote by

$$\alpha : V \to \mathbb{P}(H^0(\mathbb{P}, \mathcal{O}_{\mathbb{P}}(\Sigma_i e_i d_i)))[1/(\Pi_i Z_i)]$$

the morphism in question. Starting with the given projective embedding $X \subset \mathbb{P}$, we compose with the Segre embedding by hypersurface sections of degree $\Sigma_i e_i d_i$ to get a closed immersion

$$i_{\Sigma e_i d_i} : X \to \mathbb{P}(H^0(\mathbb{P}, \mathcal{O}_{\mathbb{P}}(\Sigma_i e_i d_i))).$$

In the target space, we have the affine open set

$$\mathbb{P}(H^0(\mathbb{P}, \mathcal{O}_{\mathbb{P}}(\Sigma_i e_i d_i)))[1/(\Pi_i Z_i^{e_i})] \subset \mathbb{P}(H^0(\mathbb{P}, \mathcal{O}_{\mathbb{P}}(\Sigma_i e_i d_i))),$$

whose inverse image in X is the affine open set V. Thus we have a cartesian diagram

$$\begin{array}{ccc} V & \subset & \mathbb{P}(H^0(\mathbb{P}, \mathcal{O}_{\mathbb{P}}(\Sigma_i e_i d_i)))[1/(\Pi_i Z_i^{e_i})] \\ \cap & & \cap \\ X & \subset & \mathbb{P}(H^0(\mathbb{P}, \mathcal{O}_{\mathbb{P}}(\Sigma_i e_i d_i))) \end{array}$$

where the bottom horizontal inclusion is the closed immersion $i_{\Sigma e_i d_i}$ and the top horizontal inclusion is the map α, which is therefore itself a closed immersion.

2) The description of

$$\mathbb{P}(H^0(\mathbb{P}, \mathcal{O}_{\mathbb{P}}(\Sigma_i e_i d_i)))[1/(\Pi_i Z_i^{e_i})]$$

as an affine space $\mathbb{A}^M$, on which the polynomial functions of degree ≤ 1 are precisely the functions $H/\Pi_i Z_i^{e_i}$, is a tautology.

3) This follows from 1), since we have a canonical isomorphism

$$\mathbb{P}(H^0(\mathbb{P}, \mathcal{O}_{\mathbb{P}}(\Sigma_i e_i d_i)))[1/(\Pi_i Z_i^{e_i})] \cong$$

the closed subscheme of $\mathcal{P}(e_1,\ldots, e_r)^\vee$ consisting of linear forms which take the value 1 on $\Pi_i Z_i^{e_i}$. QED

(4.1.8) Let us say that an element H in $H^0(\mathbb{P}, \mathcal{O}_{\mathbb{P}}(\Sigma_i e_i d_i))\otimes_k \bar{k}$ is of strong Deligne type, with respect to X and $(Z_1,\ldots, Z_r)$, and that the function

$$h := H/\Pi_i Z_i^{e_i} \text{ on } V$$

is a strong Deligne polynomial on V, if the following conditions D0), D1), and D2) are satisfied.

D0) $X\cap H$ is smooth of codimension one in X.

D1) For every nonvoid subset $\mathcal{I}$ of $\{1,\ldots, r\}$, the closed subscheme

$$X\cap H\cap(\cap_{i \text{ in } \mathcal{I}} Z_i)$$

of X is smooth of codimension $1 + \#\mathcal{I}$ in X. [For $\#\mathcal{I} \geq n$, this means only that $X\cap H\cap(\cap_{i \text{ in } \mathcal{I}} Z_i)$ is empty.]

D2) Every e_i is strictly positive and prime to p.

(4.1.9) We say say that H is of weak Deligne type, and that h is a weak Deligne polynomial, if the conditions D1) and D2) (but not necessarily D0)) are satisfied.

(4.1.10) For a fixed r-tuple $(e_1,\ldots, e_r)$ of strictly positive and prime-to-p integers, the forms of strong (resp. weak) Deligne type with respect to X and $(Z_1,\ldots, Z_r)$ form a dense open set of $H^0(\mathbb{P}, \mathcal{O}_{\mathbb{P}}(\Sigma_i e_i d_i))$. We denote by

$$\mathcal{D}(X, Z_1,\ldots, Z_r, e_1,\ldots, e_r) \subset \mathcal{P}(e_1,\ldots, e_r) := H^0(\mathbb{P}, \mathcal{O}_{\mathbb{P}}(\Sigma_i e_i d_i))$$

the dense open set consisting of forms of weak Deligne type with respect to X and $(Z_1,\ldots, Z_r)$.

(4.1.11) Recall the following theorem [Ka-SE, 5.4.1, and assertion (0) on page 169, lines 5-7].

Theorem 4.1.12 Hypotheses and notations as above, suppose all the e_i are strictly positive and prime to p. Let E/k be a finite extension. Suppose H in $H^0(\mathbb{P}, \mathcal{O}_{\mathbb{P}}(\Sigma_i e_i d_i))\otimes_k E$ is of weak Deligne type with respect to X and $(Z_1,\ldots, Z_r)$. Put $h := H/\Pi_i Z_i^{e_i}$. For any prime ℓ invertible in k, and any nontrivial $\overline{\mathbb{Q}}_\ell^\times$-valued additive character ψ of k, form the lisse, rank one $\overline{\mathbb{Q}}_\ell$-sheaf $\mathcal{L}_{\psi(h)}$ on $V\otimes_k E$. Then we have the following results for the weak Deligne polynomial h on V.

1) The "forget supports" maps

$$H^i{}_c(V\otimes_k \overline{E}, \mathcal{L}_{\psi(h)}) \to H^i(V\otimes_k \overline{E}, \mathcal{L}_{\psi(h)})$$

are all isomorphisms.

2) The groups $H^i{}_c(V\otimes_k \overline{E}, \mathcal{L}_{\psi(h)})$ vanish for $i \neq n$.

3) The group $H^n{}_c(V\otimes_k \overline{E}, \mathcal{L}_{\psi(h)})$ is pure of weight n.

4) If $H - \lambda\Pi_i Z_i^{e_i}$ is of strong Deligne type for some λ in $\overline{k}$ (a condition which is always satisfied, see Adolphson's result just below), then the dimension of $H^n{}_c(V\otimes_k \overline{E}, \mathcal{L}_{\psi(h)})$ is equal to the Chern class expression

$$(-1)^n \int_X c(X)/((1+(\Sigma e_i d_i)L)\Pi_{i=1 \text{ to } r}(1+d_i L)).$$

(4.1.13) I am indebted to Alan Adolphson for the statement and proof of the following result, which shows that the dimension formula in part 4) of the above theorem holds for every H of weak Deligne type, cf. [Ka-SE, Remark on page 172], where this question was raised but left open.

Lemma 4.1.14 (Adolphson) Hypotheses and notations as in the theorem above, if H is of weak Deligne type, then for all but finitely many λ in $\overline{k}$, $H - \lambda\Pi_i Z_i^{e_i}$ is of strong Deligne type.

proof The question is geometric, so we may extend scalars to $\overline{k}$. Let us write H_λ for $H - \lambda\Pi_i Z_i^{e_i}$. We must show that for almost all λ,

$X \cap H_\lambda$ is smooth of codimension one in X. As H is of weak Deligne type, we know that, for each i, $X \cap H \cap Z_i$ is smooth of codimension one in $X \cap Z_i$. For every λ, H and H_λ have the same intersection with Z_i. So $X \cap H_\lambda \cap Z_i$ is smooth of codimension one in $X \cap Z_i$ (and hence $X \cap H_\lambda$ is everywhere of codimension one in X, because otherwise $X \cap H_\lambda = X$, impossible if $X \cap H_\lambda \cap Z_i$ has codimension one in $X \cap Z_i$). Now $X \cap H_\lambda \cap Z_i$ is defined in $X \cap H_\lambda$ by one equation. So at every point of $X \cap H_\lambda \cap Z_i$, $X \cap H_\lambda$ is itself regular, and hence smooth. This shows two things:

a) for every λ, $\mathrm{Sing}(X \cap H_\lambda)$ has dimension ≤ 0 [because otherwise it would have a nonempty intersection with the projective hypersurface Z_i];

b) for every λ, every point of $\mathrm{Sing}(X \cap H_\lambda)$ lies in $V := X - X \cap (\cup_i Z_i)$.

So it suffices to show that for almost all λ,

$$V \cap H_\lambda = V \cap (h = \lambda)$$

is smooth. For this, it suffices to show that the closed subscheme $\mathrm{CritPt}(h, V)$ of V defined by the vanishing of dh as a section of $\Omega^1{}_{V/k}$ has dimension ≤ 0. For this, it suffices to show that its closure $\overline{\mathrm{C}}\mathrm{ritPt}(h, V)$ in X is disjoint from $X \cap Z_1$.

We argue by contradiction. Suppose x is a $\overline{k}$-valued point of $X \cap Z_1$ which lies in $\overline{\mathrm{C}}\mathrm{ritPt}(h, V)$. Renumbering the remaining Z_i, we may assume that this point x lies on Z_i for i=1 to s, and does not lie on Z_j for $j > s$. So we can find local coordinates $z_1, z_2, \ldots, z_n$ on X at x such that, near x, the function $h = H/\Pi_i Z_i{}^{e_i}$ is of the form

$$h = f/(\Pi_{i=1 \text{ to } s}\, z_i{}^{e_i}),$$

with f in the local ring $\mathcal{O}_{X,x}$, and such that, for i = 1 to s, the divisor Z_i is defined near x by the vanishing of z_i, and such that, for $i > s$, $z_i(s) \neq 0$. [Concretely, if we pick a linear form L which is invertible at x, we may take $z_i := Z_i/L^{d_i}$ for i = 1 to s, and then f is $(H/L^{\Sigma e_i d_i})/(\Pi_{i>s}\, (Z_i/L^{d_i})^{e_i})$.] We rewrite this as

$$f = h(\Pi_{i=1 \text{ to } s}\, z_i{}^{e_i}).$$

We readily calculate (in $\Omega^1{}_{\overline{k}(X)/\overline{k}}$)

$$df = dh(\Pi_{i=1 \text{ to } s}\, z_i{}^{e_i}) + h(\Pi_{i=1 \text{ to } s}\, z_i{}^{e_i})\Sigma_{i=1 \text{ to } s}\, e_i dz_i/z_i$$
$$= dh(\Pi_{i=1 \text{ to } s}\, z_i{}^{e_i}) + f\Sigma_{i=1 \text{ to } s}\, e_i dz_i/z_i.$$

Postmultiplying by $z_1(\Pi_{i=2 \text{ to } s}\, dz_i)$, we get

$$z_1 df(\Pi_{i=2 \text{ to } s}\, dz_i)$$
$$= z_1(\Pi_{i=1 \text{ to } s}\, z_i{}^{e_i})dh(\Pi_{i=2 \text{ to } s}\, dz_i) + e_1 f(\Pi_{i=1 \text{ to } s}\, dz_i).$$

We rewrite this relation as

$$z_1(\prod_{i=1 \text{ to } s} z_i^{e_i}) dh(\prod_{i=2 \text{ to } s} dz_i) = z_1 df(\prod_{i=2 \text{ to } s} dz_i) - e_1 f(\prod_{i=1 \text{ to } s} dz_i).$$

This shows that the product of dh by $z_1(\prod_{i=1 \text{ to } s} z_i^{e_i})(\prod_{i=2 \text{ to } s} dz_i)$ is a holomorphic s-form on an open neighborhood U of x. This product vanishes as a section of $\Omega^s_{U/k}$ on $\overline{\mathrm{Crit}}(h, V)\cap U$, because, being divisible by dh on $U\cap V$, it vanishes on $\mathrm{Crit}(h, V)\cap U$. But x lies in $\overline{\mathrm{Crit}}(h, V)$. So we conclude that the holomorphic s-form

$$z_1 df(\prod_{i=2 \text{ to } s} dz_i) - e_1 f(\prod_{i=1 \text{ to } s} dz_i)$$

vanishes at x. But z_1 vanishes at x, so we find that the holomorphic s-form $e_1 f(\prod_{i=1 \text{ to } s} dz_i)$ vanishes at x. But e_1 is prime to p, and $z_1,\ldots, z_s$ are part of a system of coordinates at x, so we infer that f vanishes at x. This in turn means that $H(x) = 0$. By the transversality of H to $X\cap Z_1\cap\ldots\cap Z_s$, $\{f, z_1,\ldots, z_s\}$ are part of a system of coordinates at x. Now premultiply the relation

$$z_1(\prod_{i=1 \text{ to } s} z_i^{e_i}) dh(\prod_{i=2 \text{ to } s} dz_i) = z_1 df(\prod_{i=2 \text{ to } s} dz_i) - e_1 f(\prod_{i=1 \text{ to } s} dz_i)$$

by dz_1/z_1. We find

$$-(\prod_{i=1 \text{ to } s} z_i^{e_i}) dh(\prod_{i=1 \text{ to } s} dz_i) = -df(\prod_{i=1 \text{ to } s} dz_i).$$

Again the left hand side vanishes on $\overline{\mathrm{Crit}}(h, V)\cap U$, so it vanishes at x. But the right hand side is nonzero at x, because $\{f, z_1,\ldots, z_s\}$ are part of a system of coordinates at x. Contradiction. QED

(4.2) The perverse sheaf $M(X, r, Z_i\text{'s}, e_i\text{'s}, \psi)$ on $\mathcal{P}(e_1,\ldots, e_r)$

(4.2.1) We continue with the general setup of 4.1. Thus k is a finite field k of characteristic p, in which the prime ℓ is invertible, and X/k is a projective, smooth, geometrically connected k-scheme, of dimension $n \geq 1$, given with a projective embedding

$$X \subset \mathbb{P}^N := \mathbb{P}.$$

We are given an r-tuple $(d_1,\ldots, d_r)$ of strictly positive integers, and an r-tuple $(Z_1,\ldots, Z_r)$ of nonzero global sections

$$Z_i \text{ in } H^0(\mathbb{P}, \mathcal{O}_{\mathbb{P}}(d_i))$$

which satisfy the transversality hypothesis (TrZ, X) of 4.1.3. We denote

$$V := X - X\cap(\cup_i Z_i) = X[1/(\prod_i Z_i)].$$

For an r-tuple $(e_1, e_2,\ldots, e_r)$ of integers $e_i \geq 0$, we get a linear space $(\mathcal{P}(e_1,\ldots, e_r), \tau)$ of $\mathbb{A}^1$-valued functions on V by taking

$$\mathcal{P}(e_1,\ldots, e_r) := H^0(\mathbb{P}, \mathcal{O}_{\mathbb{P}}(\Sigma_i e_i d_i)),$$

$$\tau : \mathcal{P}(e_1,\ldots, e_r) \to \mathrm{Hom}_{k\text{-schemes}}(V, \mathbb{A}^1)$$

the linear map

$$H \mapsto h := H/\Pi_i Z_i^{e_i} \mid V.$$

Recall from Lemma 4.1.6 that $(\mathcal{P}_{(e_1,\ldots, e_r)}, \tau)$ is $(1 + \Sigma_i e_i d_i)$-separating as a space of functions on V.

(4.2.2) We now consider the following "standard input", cf. 1.15.4:

$m=1$,

a nontrivial $\overline{\mathbb{Q}}_\ell^\times$-valued additive character ψ of k,

$K = \mathcal{L}_\psi(1/2)[1]$ on $\mathbb{A}^1$,

$V := X[1/(\Pi_i Z_i)]$ as above, of dimension $n \geq 1$,

$h: V \to \mathbb{A}^1$ the function $h = 0$,

$L = \overline{\mathbb{Q}}_\ell[n](n/2)$ on V,

$(\mathcal{F}, \tau) = (\mathcal{P}_{(e_1,\ldots, e_r)}, \tau)$, with $\Sigma e_i d_i \geq 1$.

(4.2.3) Attached to this input data is the perverse sheaf

$$\mathrm{Twist}(L, K, \mathcal{F}, h)$$

on $\mathcal{P}_{(e_1,\ldots, e_r)}$, which we call $M(X, r, Z_i\text{'s}, e_i\text{'s}, \psi)$:

$$M(X, r, Z_i\text{'s}, e_i\text{'s}, \psi) := \mathrm{Twist}(L, K, \mathcal{F}, h).$$

(4.2.4) We define the sheaf $\mathcal{M}(X, r, Z_i\text{'s}, e_i\text{'s}, \psi)$ of perverse origin on $\mathcal{P}_{(e_1,\ldots, e_r)}$ by

$$\mathcal{M}(X, r, Z_i\text{'s}, e_i\text{'s}, \psi) := \mathcal{H}^{-\dim \mathcal{F}}(M(X, r, Z_i\text{'s}, e_i\text{'s}, \psi))(-1/2).$$

(4.2.5) Exactly as in 3.5.5, the down to earth description of these objects is this. On the space $V \times \mathcal{P}_{(e_1,\ldots, e_r)}$, with coordinates v, H, we have the function $(\tau H)(v) := (H/\Pi Z_i^{e_i})(v)$, and we have the lisse sheaf $\mathcal{L}_{\psi((\tau H)(v))}$. Under the second projection

$$pr_2 : V \times \mathcal{P}_{(e_1,\ldots, e_r)} \to \mathcal{P}_{(e_1,\ldots, e_r)},$$

we form $Rpr_{2!}\mathcal{L}_{\psi((\tau H)(v))}$. For E/k a finite extension, and for H in $\mathcal{P}_{(e_1,\ldots, e_r)}(E)$, the stalk of $Rpr_{2!}\mathcal{L}_{\psi((\tau H)(v))}$ at H is the object $R\Gamma_c(\mathbb{A}^n \otimes_k \overline{E}, \mathcal{L}_{\psi((\tau H)(v))})$, whose cohomology objects are the groups

$$\mathcal{H}^i(R\Gamma_c(\mathbb{A}^n \otimes_k \overline{E}, \mathcal{L}_{\psi((\tau H)(v))})) = H_c^i(\mathbb{A}^n \otimes_k \overline{E}, \mathcal{L}_{\psi((\tau H)(v))}).$$

The perverse sheaf $M(X, r, Z_i\text{'s}, e_i\text{'s}, \psi)$ is just a Tate twist and a shift of $Rpr_{2!}\mathcal{L}_{\psi((\tau H)(v))}$. We have

$$Rpr_{2!}\mathcal{L}_{\psi((\tau H)(v))}$$
$$= M(X, r, Z_i\text{'s}, e_i\text{'s}, \psi)[-n - \dim \mathcal{P}_{(e_1,\ldots, e_r)}]((-n-1)/2),$$

and

$$\mathcal{M}(X, r, Z_i\text{'s}, e_i\text{'s}, \psi) = R^n pr_{2!}\mathcal{L}_{\psi((\tau H)(v))}(n/2).$$

(4.2.6) We have the following two lemmas, analogues of Lemmas 3.5.6 and 3.5.7.

Lemma 4.2.7 For E/k a finite extension, and H in $\mathcal{F}(E) = \mathcal{P}_{(e_1,\ldots, e_r)}(E)$, we have the identity

$$\mathrm{Trace}(\mathrm{Frob}_{E,H} \mid M(X, r, Z_i\text{'s}, e_i\text{'s}, \psi)))$$
$$= (-1)^{n+\dim \mathcal{F}}(\#E)^{-(n+1)/2}\Sigma_{v \text{ in } V(E)} \psi_E((\tau H)(v)).$$

Lemma 4.2.8 For $U \subset \mathcal{P}(e_1,\ldots, e_r)$ an open dense set on which $M(X, r, Z_i\text{'s}, e_i\text{'s}, \psi)$ has lisse cohomology sheaves, $M(X, r, Z_i\text{'s}, e_i\text{'s}, \psi)|U$ is the lisse sheaf $\mathcal{M}(X, r, Z_i\text{'s}, e_i\text{'s}, \psi)(1/2)|U$, placed in degree $-\dim\mathcal{F}$. For E/k a finite extension, and for H in U(E), we have

$$\mathrm{Trace}(\mathrm{Frob}_{E,H} \mid \mathcal{M}(X, r, Z_i\text{'s}, e_i\text{'s}, \psi))$$
$$= (-1)^{n}(\#E)^{-n/2}\Sigma_{v \text{ in } V(E)} \psi_E((\tau H)(v)).$$

Theorem 4.2.9 Take standard input as in 4.2.2 above. Suppose in addition that $e_i \geq 1$ for all i = 1 to r. Then we have the following results concerning the perverse sheaf

$$M := M(X, r, Z_i\text{'s}, e_i\text{'s}, \psi)$$

and the sheaf of perverse origin

$$\mathcal{M} := \mathcal{H}^{-\dim \mathcal{F}}(M)(-1/2)$$

on $\mathcal{F} = \mathcal{P}(e_1,\ldots, e_r)$.

1) The object $M(\dim\mathcal{F}_0/2)$ is perverse, geometrically irreducible and geometrically nonconstant, and ι-pure of weight zero.

2) The Frobenius-Schur indicator of $M(\dim\mathcal{F}_0/2)$ is given by

$$\mathrm{FSI}^{\mathrm{geom}}(\mathcal{P}(e_1,\ldots, e_r), M(\dim\mathcal{F}_0/2))$$
$$= 0, \text{ if p is odd,}$$
$$= ((-1)^{1+\dim\mathcal{F}_0})\times(-1)^n = (-1)^{n+\dim\mathcal{F}}, \text{ if p= 2.}$$

3) On any dense open set U of $\mathcal{P}(e_1,\ldots, e_r)$ on which M is lisse, we have $M(-1/2)|U = (\mathcal{M}|U)[\dim\mathcal{F}]$, and $\mathcal{M}|U$ is a lisse sheaf on U. If $\mathcal{M}|U$ is nonzero, then it is geometrically irreducible, geometrically nonconstant, and ι-pure of weight zero.

4) If all the e_i are prime to p, then M is lisse on the dense open set

$$\mathcal{D} := \mathcal{D}(X, Z_1,\ldots, Z_r, e_1,\ldots, e_r) \subset \mathcal{P}(e_1,\ldots, e_r) := H^0(\mathbb{P}, \mathcal{O}_{\mathbb{P}}(\Sigma_i e_i d_i))$$

consisting of all forms of weak Deligne type with respect to X and $(Z_1,\ldots, Z_r)$, cf. 4.1.10, and $\mathcal{M}|\mathcal{D}$ has rank given by

$$\mathrm{rank}(\mathcal{M}|U) = (-1)^n \textstyle\int_X c(X)/((1+(\Sigma e_i d_i)L)\Pi_{i=1 \text{ to } r}(1+d_i L)).$$

5) If $\Sigma e_i d_i \geq 3$, then $\mathcal{M}|U$ is nonzero, and its Frobenius-Schur indicator is given by

$$\mathrm{FSI}^{\mathrm{geom}}(U, \mathcal{M}|U)$$
$$= 0, \text{ if p is odd,}$$
$$= (-1)^n, \text{ if p = 2.}$$

6) If $\Sigma e_i d_i \geq 3$, then we have the following bound for the fourth

moment:

$$M_4{}^{geom}(U, \mathcal{M}) \le 2, \text{ if } p \ne 2,$$

$$M_4{}^{geom}(U, \mathcal{M}) \le 3, \text{ if } p = 2.$$

And for any $n \ge 3$ with $1 + \Sigma e_i d_i \ge 2n$, we have the following bound for the 2n'th moment:

$$M_{2n}{}^{geom}(U, \mathcal{M}) \le n!, \text{ if } p \ne 2,$$

$$M_{2n}{}^{geom}(U, \mathcal{M}) \le (2n)!! := (2n-1)(2n-3)...5.3.1, \text{ if } p = 2.$$

7) If $\mathcal{M}|U$ is nonzero, then $\det(\mathcal{M}|U)$ is arithmetically of finite order.

proof 1) We have the closed immersion of Lemma 4.1.7,

$$\text{eval} : V \to \mathcal{P}(e_1,..., e_r)^\vee,$$

$$v \mapsto (H \mapsto (\tau H)(v)).$$

In terms of the Fourier Transform on the target space $\mathcal{P}(e_1,..., e_r)^\vee$, we have, for any L on V,

$$\text{Twist}(L, K, \mathcal{F}, h) = FT_\psi(\text{eval}_*(L))(1/2),$$

cf. the proof of the Compatibility Lemma 3.2.3. The rest of the proof of 1) is the same as the proof of part 1) of the $\mathcal{L}_\psi$ Theorem 3.1.2. Statement 2) is just a spelling out of part 3) of the Standard Input Theorem 1.15.6. Statement 3) is a tautology, given Statement 1). Statement 4) results from Theorem 4.1.12, according to which on

$$\mathcal{D} := \mathcal{D}(X, Z_1,..., Z_r, e_1,..., e_r),$$

the only nonvanishing cohomology sheaf of $M|\mathcal{D}$ is $\mathcal{H}^{-\dim \mathcal{F}}(M)$, which has constant rank equal to the asserted rank on $\mathcal{D}$. As $\mathcal{H}^{-\dim \mathcal{F}}(M)$ is a sheaf of perverse origin, it is lisse on any open set where its rank is constant, cf. 2.8.10. In Statement 5), the nonvanishing of $\mathcal{M}|U$ results from part 1) of the Corollary 1.20.3 to the Higher Moment Theorem 1.20.2, applied with $d = \Sigma e_i d_i + 1 \ge 4$. The statement about the Frobenius-Schur indicator is then equivalent to that given in Statement 2). Statement 6) is just part 4) of Corollary 1.20.3. The proof of Statement 7) is entirely analogous to that of the Determinant Lemma 3.5.13. QED

Theorem 4.2.10 Hypotheses and notations as in Theorem 4.2.9 above, suppose further that we are in one of the following three situations:

a) $r \ge 4$,

b) $r = 3$ and $\Sigma e_i d_i \ge 4$,

c) $r = 2$ and $\Sigma e_i d_i \ge 5$.

Denote by N the rank of $\mathcal{M}|U$. Then $N \ge 2$, and we have the following results for the group $G_{geom} := G_{geom, \mathcal{M}|U}$.

1) If $p \ne 2$, then G_{geom} contains $SL(N)$.

2) If $p = 2$ and $n = \dim X$ is even, then G_{geom} is $Sp(N)$.

3) If $p = 2$ and $n = \dim X$ is odd, then $N \geq 3$, and G_{geom} is either $SO(N)$ or $O(N)$.

proof In all the cases considered, we certainly have $\Sigma e_i d_i \geq 3$. By Statement 5) of the previous theorem we know that $\mathcal{M}|U$ is nonzero. So by Statement 3) of the previous theorem, $\mathcal{M}|U$ is geometrically irreducible, geometrically nonconstant, and ι-pure of weight zero. Moreover, by Statement 6) of the previous theorem, $\mathcal{M}|U$ has very low fourth moment. So by Larsen's Alternative 2.2.2, to complete the proof of the theorem, it suffices to show that G_{geom} is not finite (which already forces $N \geq 2$, and forces $N \geq 3$ in case 3)), cf. the discussion in 2.1.2.

To show that G_{geom} is not finite, we use a degeneration argument, based on the Punctual Purity Corollary 2.8.14, combined with information on archimedean absolute values of eigenvalues of Frobenius.

We have, on the space $\mathcal{F} = \mathcal{P}_{(e_1,\ldots,\, e_r)}$, the sheaf of perverse origin $\mathcal{M}$, given explicitly by

$$\mathcal{M} = R^n pr_{2!}\mathcal{L}_{\psi((\tau H)(v))}(n/2),$$

cf. 4.2.5. We are trying to prove that $\mathcal{M}|U$ does not have a finite G_{geom}. We argue by contradiction. Thus suppose $\mathcal{M}|U$ has a finite G_{geom}. Since $\mathcal{M}|U$ is geometrically irreducible,we know by the Scalarity Corollary 2.8.13 that for any finite extension field E/k, and for any E-valued point t in $\mathcal{F}(E)$, some power of $Frob_{E,t} \mid \mathcal{M}_t$ is a scalar. Therefore the eigenvalues of $Frob_{E,t} \mid \mathcal{M}_t$ differ multiplicatively from each other by roots of unity, and hence all the eigenvalues of $Frob_{E,t} \mid \mathcal{M}_t$ must have the same (via ι) archimedean absolute values as each other. So to prove that G_{geom} is not finite, it suffices to exhibit a single finite extension field E/k, and a single E-valued point t in $\mathcal{F}(E)$, such that all the eigenvalues of $Frob_{E,t} \mid \mathcal{M}_t$ do not have the same (via ι) archimedean absolute values as each other.

To do this, we proceed as follows. After possibly renumbering the Z_i, we may assume that

$$e_1d_1 \geq e_2d_2 \geq \ldots \geq e_rd_r.$$

Then under any of the hypotheses a), b), c), we have the inequality

$$\Sigma_{i=1 \text{ to } r-1}\, e_id_i \geq 3.$$

Indeed, if $r \geq 4$, this is trivial, since each $e_i \geq 1$ and each $d_i \geq 1$. If $r=3$, we must have $e_1d_1 \geq 2$ (otherwise we have all $e_id_i = 1$, which is not allowed), and hence $e_1d_1 + e_2d_2 \geq 3$. Similarly, if $r = 2$, we must have $e_1d_1 \geq 3$.

The idea is to take as our special point t in $\mathcal{F}(E)$ a point H in $\mathcal{F}(E) = \mathcal{P}_{(e_1,\ldots,\, e_r)}(E)$ such that τH has no pole along Z_r, but which is otherwise as general as possible. Recall that

$$\tau H := H/\prod_{i=1 \text{ to } r} Z_i^{e_i}.$$

So we want to take H of the form

$$H = H_1 \times Z_r^{e_r},$$

for some

$$H_1 \text{ in } \mathcal{F}_1 := \mathcal{P}(e_1,\ldots, e_{r-1}, 0) = H^0(\mathbb{P}, \mathcal{O}_{\mathbb{P}}(\Sigma_{i=1 \text{ to } r-1} e_i d_i)).$$

There are three ways we can think of the space

$$\mathcal{F}_1 := \mathcal{P}(e_1,\ldots, e_{r-1}, 0) = H^0(\mathbb{P}, \mathcal{O}_{\mathbb{P}}(\Sigma_{i=1 \text{ to } r-1} e_i d_i)).$$

We can think of it as a space of functions on

$$V_1 := V = X[1/\prod_{i=1 \text{ to } r} Z_i],$$

by means of the map

$$\tau_1 : \mathcal{P}(e_1,\ldots, e_{r-1}, 0) \to \mathrm{Hom}_{k\text{-schemes}}(V_1, \mathbb{A}^1),$$

$$\tau_1 H_1 := H_1/\prod_{i=1 \text{ to } r-1} Z_i^{e_i} \mid V_1.$$

Or we can think of it as a space of functions on

$$V_2 := X[1/\prod_{i=1 \text{ to } r-1} Z_i],$$

by means of the map

$$\tau_2 : \mathcal{P}(e_1,\ldots, e_{r-1}, 0) \to \mathrm{Hom}_{k\text{-schemes}}(V_2, \mathbb{A}^1),$$

$$\tau_2 H_1 := H_1/\prod_{i=1 \text{ to } r-1} Z_i^{e_i} \mid V_2.$$

Or we can think of it as a space of functions on

$$V_3 := (X \cap Z_r)[1/\prod_{i=1 \text{ to } r-1} Z_i],$$

by means of the map

$$\tau_3 : \mathcal{P}(e_1,\ldots, e_{r-1}, 0) \to \mathrm{Hom}_{k\text{-schemes}}(V_3, \mathbb{A}^1),$$

$$\tau_3 H_1 := H_1/\prod_{i=1 \text{ to } r-1} Z_i^{e_i} \mid V_3.$$

The space $\mathcal{P}(e_1,\ldots, e_{r-1}, 0)$ thus carries three perverse sheaves, one from each of its above incarnations:

$$M_1 := M(X, r, Z_1,\ldots, Z_r, e_1,\ldots, e_{r-1}, 0, \psi),$$
$$M_2 := M(X, r-1, Z_1,\ldots, Z_{r-1}, e_1,\ldots, e_{r-1}, \psi),$$
$$M_3 := M(X \cap Z_r, r-1, Z_1,\ldots, Z_{r-1}, e_1,\ldots, e_{r-1}, \psi).$$

It also carries the three corresponding sheaves of perverse origin

$$\mathcal{M}_i := \mathcal{H}^{-\dim \mathcal{F}_1}(M_i)(-1/2).$$

Pick a dense open set U_{123} in $\mathcal{P}(e_1,\ldots, e_{r-1}, 0)$ on which all three of the M_i are lisse. Let E/k be a finite extension, large enough that $U_{123}(E)$ is nonempty. Pick a point H_1 in $U_{123}(E)$. So on U_{123}, we have

$$M_i(-1/2) \mid U_{123} = (\mathcal{M}_i \mid U_{123})[\dim \mathcal{F}_1],$$

for $i = 1, 2, 3$. By the down to earth description of these objects, cf. 4.2.6, and proper base change at H_1 in $U_{123}(E)$, we have

$$H_c^i(V_1 \otimes_k \bar{k}, \mathcal{L}_{\psi(\tau_1 H_1)}) = 0 \text{ for } i \neq n,$$

$$H_c^n(V_1\otimes_k\bar{k}, \mathcal{L}_{\psi(\tau_1 H_1)})(n/2) = \mathcal{M}_{1,H_1},$$

$$H_c^i(V_2\otimes_k\bar{k}, \mathcal{L}_{\psi(\tau_2 H_1)}) = 0 \text{ for } i \neq n,$$

$$H_c^n(V_2\otimes_k\bar{k}, \mathcal{L}_{\psi(\tau_2 H_1)})(n/2) = \mathcal{M}_{2,H_1},$$

$$H_c^i(V_3\otimes_k\bar{k}, \mathcal{L}_{\psi(\tau_3 H_1)}) = 0 \text{ for } i \neq n-1,$$

$$H_c^{n-1}(V_3\otimes_k\bar{k}, \mathcal{L}_{\psi(\tau_3 H_1)})((n-1)/2) = \mathcal{M}_{3,H_1}.$$

Now

$$V_1 := V = X[1/\Pi_{i=1 \text{ to } r} Z_i]$$

is open in

$$V_2 := X[1/\Pi_{i=1 \text{ to } r-1} Z_i],$$

with closed complement

$$V_3 := (X\cap Z_r)[1/\Pi_{i=1 \text{ to } r-1} Z_i].$$

And the lisse sheaf $\mathcal{L}_{\psi(\tau_2 H_1)}$ on V_2 restricts to the lisse sheaf $\mathcal{L}_{\psi(\tau_1 H_1)}$ on V_1, and to the lisse sheaf $\mathcal{L}_{\psi(\tau_3 H_1)}$ on V_3. So the excision long exact sequence of cohomology gives a short exact sequence

$$0 \to H_c^{n-1}(V_3\otimes_k\bar{k}, \mathcal{L}_{\psi(\tau_3 H_1)}) \to H_c^n(V_1\otimes_k\bar{k}, \mathcal{L}_{\psi(\tau_1 H_1)})$$
$$\to H_c^n(V_2\otimes_k\bar{k}, \mathcal{L}_{\psi(\tau_2 H_1)}) \to 0.$$

Tate-twisting by $\overline{\mathbb{Q}}_\ell(n/2)$, we get a short exact sequence

$$0 \to \mathcal{M}_{3,H_1}(1/2) \to \mathcal{M}_{1,H_1} \to \mathcal{M}_{2,H_1} \to 0.$$

[Indeed, this sequence is the stalk at H_1 of a short exact sequence of lisse sheaves on U_{123}.] By the previous theorem, applied to both V_2 and V_3, we know that both $\mathcal{M}_2|U_{123}$ and $\mathcal{M}_3|U_{123}$ are nonzero lisse sheaves, both ι-pure of weight zero. Therefore $\mathrm{Frob}_{E,H_1} \mid \mathcal{M}_{1,H_1}$ has eigenvalues that are ι-pure of weight zero, and it has eigenvalues that are ι-pure of weight -1.

On the other hand, for the form

$$H := H_1 \times Z_r^{e_r} \text{ in } \mathcal{P}(e_1,\ldots, e_r),$$

it is tautologous that the lisse sheaf $\mathcal{L}_{\psi(\tau H)}$ on $V = V_1$ is just the sheaf $\mathcal{L}_{\psi(\tau_1 H_1)}$ on V_1. Therefore we have

$$H_c^i(V\otimes_k\bar{k}, \mathcal{L}_{\psi(\tau H)}) = 0 \text{ for } i \neq n,$$

$$H_c^n(V\otimes_k\bar{k}, \mathcal{L}_{\psi(\tau H)})(n/2) = \mathcal{M}_H = \mathcal{M}_{1,H_1}.$$

Thus $\mathrm{Frob}_{E,H} \mid \mathcal{M}_H$ has eigenvalues that are ι-pure of weight zero, and it has eigenvalues that are ι-pure of weight -1. This is the desired contradiction. QED

(4.2.11) We now turn to the more difficult case when $r = 1$. We first consider the subcase when $n \geq 3$.

Theorem 4.2.12 Let X/k be projective, smooth, and geometrically connected, of dimension $n \geq 3$. Fix a projective embedding $X \subset \mathbb{P}$, an integer $d \geq 1$, and a section Z of $H^0(\mathbb{P}, \mathcal{O}_{\mathbb{P}}(d))$ such that $X \cap Z$ is smooth of codimension one in X. Fix an integer $e \geq 3$. If char(k) = 3, suppose in addition $e \geq 4$. Form the perverse sheaf

$$M = M(X, r=1, Z, e)$$

on $\mathcal{F} = \mathcal{P}_{(e)} := H^0(\mathbb{P}, \mathcal{O}_{\mathbb{P}}(ed))$, and form the sheaf of perverse origin $\mathcal{M} := \mathcal{H}^{-\dim \mathcal{F}}(M)(-1/2)$. Let $U \subset \mathcal{F}$ be a dense open set on which M is lisse. If e is prime to p, take for U the dense open set $\mathcal{D} := \mathcal{D}(X, Z, e) \subset \mathcal{F}$ consisting of all forms H of weak Deligne type with respect to X and Z (i.e., such that $X \cap Z \cap H$ is smooth of codimension one in $X \cap Z$). Denote by $N \geq 1$ the rank of $\mathcal{M}|U$. Then we have the following results concerning the group G_{geom} for $\mathcal{M}|U$.

1) If $p \neq 2$, then G_{geom} contains SL(N).

2) If $p = 2$ and $n = \dim X$ is odd, then G_{geom} is Sp(N).

3) If $p = 2$ and $n = \dim X$ is even, then G_{geom} is either SO(N) or O(N).

proof Exactly as in the proof of Theorem 4.2.9, it suffices to show that G_{geom} is not finite. By the Semicontinuity Corollary 2.8.9, it suffices to prove that G_{geom} is not finite for the lowest allowed value e_0 of e, namely $e_0 = 3$ if char(k) $\neq 3$, and $e_0 = 4$ if char(k) = 3. Thus e_0 satisfies both

$$e_0 \geq 3,$$

$$e_0 \text{ is prime to } p := \text{char}(k).$$

So it certainly suffices to prove that G_{geom} is not finite for any $e \geq 3$ which is prime to p, and this is what we now proceed to do.

Because e is prime to p, we may choose the open set $U \subset \mathcal{F}$ to be $\mathcal{D}(X, Z, e)$, consisting of all forms H of degree de of weak Deligne type with respect to X and Z (i.e., such that $X \cap Z \cap H$ is smooth of codimension one in $X \cap Z$). A key property of the open set $\mathcal{D}(X, Z, e) = U$ is that it is stable by homothety in the ambient $\mathcal{F}$. [It is (only!) to be sure of the stability of U by homothety that we require that e be prime to p.]

Recall from 4.2.8 that for E/k a finite extension, and for H in U(E), $\text{Trace}(\text{Frob}_{E,H} \mid \mathcal{M})$ is the exponential sum over the affine variety $V := X - X \cap Z$ given by

$$\text{Trace}(\text{Frob}_{E,H} \mid \mathcal{M})$$
$$= (-1)^n (\# E)^{-n/2} \Sigma_{v \text{ in } V(E)} \psi_E((H/Z^e)(v)).$$

We know that $\det(\mathcal{M}|U)$ is arithmetically of finite order, cf. Theorem 4.2.9, part 7).

Suppose now that G_{geom} is finite. Then by Twisting Lemma 2.8.12.1 (= 2.8.12 bis), G_{arith} is finite. So for E/k any finite extension, and for any H in U(E), a power of $\text{Frob}_{E,H}$ acts trivially on $\mathcal{M}$. Hence the eigenvalues of $\mathrm{Frob}_{E,H} \mid \mathcal{M}$ are roots of unity, and so $\mathrm{Trace}(\mathrm{Frob}_{E,H} \mid \mathcal{M})$ is an algebraic integer. Therefore we find the following divisibility result: for E/k any finite extension, and for any H in U(E),

$$\Sigma_{v \text{ in } V(E)} \psi_E((H/Z^e)(v)) = (-1)^n(\#E)^{n/2}\mathrm{Trace}(\mathrm{Frob}_{E,H} \mid \mathcal{M})$$
$$= (\#E)^{n/2}\times(\text{an algebraic integer}).$$

For any λ in $E^\times$, λH lies in U(E) as well, so we get the same divisibility for λH. Summing over λ in $E^\times$, we get the following divisbility: for E/k any finite extension, and for any H in U(E),

$$\Sigma_{\lambda \text{ in } E^\times} \Sigma_{v \text{ in } V(E)} \psi_E(\lambda(H/Z^e)(v))$$
$$= (\#E)^{n/2}\times(\text{an algebraic integer}).$$

This divisibility will be particularly powerful when H is a strong Deligne form with respect to X and Z, i.e., one such that $X\cap H$ is smooth of condimension one in X, and such that $X\cap Z\cap H$ is smooth of codimension one in $X\cap Z$. [Recall that $X\cap Z$ is smooth of codimension one in X by hypothesis on (X, Z).] The strong Deligne forms are a dense open set in the space of weak Deligne forms:

$$U_{str} := \mathcal{D}_{strong}(X, Z, e) \subset U := \mathcal{D}(X, Z, e).$$

The open set U_{str} is (visibly) stable by homothety in the ambient $\mathcal{F}$.

In order to proceed, we need an exponential sum identity.

(4.3) Interlude: An exponential sum identity

Lemma 4.3.1 Let k be a finite field, $\#k := q$, in which ℓ is invertible, and let ψ be a nontrivial $\overline{\mathbb{Q}}_\ell^\times$-valued additive character of k. Let X/k be projective, smooth, and geometrically connected, of dimension $n \geq 1$. Fix a projective embedding $X \subset \mathbb{P}$, an integer $d \geq 1$, and a section Z of $H^0(\mathbb{P}, \mathcal{O}_\mathbb{P}(d))$ such that $X\cap Z$ is smooth of codimension one in X. Denote by V the smooth affine open set of X given by

$$V := X[1/Z] := X - X\cap Z.$$

Fix an integer $e \geq 1$. Let E/k be a finite extension inside $\overline{k}$. Let H be a section of $H^0(\mathbb{P}, \mathcal{O}_\mathbb{P}(de))\otimes_k E$ such that $X\cap H$ is smooth of codimension one in X, and such that $X\cap Z\cap H$ is smooth of codimension one in $X\cap Z$. Denote by L in $H^n(X\otimes_k\overline{k}, \overline{\mathbb{Q}}_\ell)(1)$ the cohomology class of a hyperplane section. Consider the cohomology groups

$$\mathrm{Prim}^i(X) := H^i(X\otimes_k\overline{k}, \overline{\mathbb{Q}}_\ell)/LH^{i-2}(X\otimes_k\overline{k}, \overline{\mathbb{Q}}_\ell)(-1), \quad \text{any } i \leq n,$$

$$\mathrm{Prim}^i(X\cap Z) := H^i((X\cap Z)\otimes_k\overline{k}, \overline{\mathbb{Q}}_\ell)/LH^{i-2}((X\cap Z)\otimes_k\overline{k}, \overline{\mathbb{Q}}_\ell)(-1), \quad \text{any } i \leq n-1,$$

$$\mathrm{Ev}^{n-1}(X\cap Z) := H^{n-1}((X\cap Z)\otimes_E\overline{k}, \overline{\mathbb{Q}}_\ell)/H^{n-1}(X\otimes_k\overline{k}, \overline{\mathbb{Q}}_\ell),$$

$$\mathrm{Ev}^{n-1}(X\cap H) := H^{n-1}((X\cap H)\otimes_E\overline{k}, \overline{\mathbb{Q}}_\ell)/H^{n-1}(X\otimes_k\overline{k}, \overline{\mathbb{Q}}_\ell),$$

$$\mathrm{Ev}^{n-2}(X\cap Z\cap H) := H^{n-2}((X\cap Z\cap H)\otimes_E \bar{k},\ \bar{\mathbb{Q}}_\ell)/H^{n-2}(X\otimes_k \bar{k},\ \bar{\mathbb{Q}}_\ell).$$

Then for any finite extension F/E, we have the identity

$$(-1)^{n-1}\Sigma_{\lambda \text{ in } F^\times} \Sigma_{v \text{ in } V(F)}\ \psi_F(\lambda(H/Z^e)(v))$$
$$= \mathrm{Trace}(\mathrm{Frob}_F \mid \mathrm{Prim}^n(X))$$
$$+ \mathrm{Trace}(\mathrm{Frob}_F \mid \mathrm{Ev}^{n-1}(X\cap Z))$$
$$+ \mathrm{Trace}(\mathrm{Frob}_F \mid \mathrm{Ev}^{n-1}(X\cap H)(-1))$$
$$+ \mathrm{Trace}(\mathrm{Frob}_F \mid \mathrm{Ev}^{n-2}(X\cap Z\cap H)(-1)).$$

proof Extending scalars, it suffices to treat universally the case when F = E = k. We first "complete" the sum by adding on the $\lambda=0$ term, and using orthogonality of characters. We get

$$\Sigma_{\lambda \text{ in } k^\times} \Sigma_{v \text{ in } V(k)}\ \psi(\lambda(H/Z^e)(v))$$
$$= \Sigma_{\lambda \text{ in } k} \Sigma_{v \text{ in } V(k)}\ \psi(\lambda(H/Z^e)(v)) - \#V(k)$$
$$= q\#\{ v \text{ in } V(F) \text{ with } (H/Z^e)(v) = 0\} - \#V(k)$$
$$= q(\#(V\cap H)(k)) - \#V(k)$$
$$= q(\#(X\cap H)(k) - \#(X\cap Z\cap H)(k)) - (\#X(k) - \#(X\cap Z)(k))$$
$$= [q\#(X\cap H)(k) - \#X(k)] - [q\#(X\cap Z\cap H)(k) - \#(X\cap Z)(k)].$$

We now rewrite the first grouped term. Use the Lefschetz Trace Formula on $X\cap H$ to write $q\#(X\cap H)(k)$ as a sum of three terms:

$$q\#(X\cap H)(k) = \Sigma_{i \le n-2} (-1)^i\mathrm{Trace}(\mathrm{Frob}_k \mid H^i((X\cap H)\otimes_k \bar{k},\ \bar{\mathbb{Q}}_\ell)(-1))$$
$$+ (-1)^{n-1}\mathrm{Trace}(\mathrm{Frob}_k \mid H^{n-1}((X\cap H)\otimes_k \bar{k},\ \bar{\mathbb{Q}}_\ell)(-1))$$
$$+ \Sigma_{i \ge n} (-1)^i\mathrm{Trace}(\mathrm{Frob}_k \mid H^i((X\cap H)\otimes_k \bar{k},\ \bar{\mathbb{Q}}_\ell)(-1)).$$

Use the same formula to write $\#X(k)$ as the sum of three terms:

$$\#X(k) = \Sigma_{i \le n} (-1)^i\mathrm{Trace}(\mathrm{Frob}_k \mid H^i(X\otimes_k \bar{k},\ \bar{\mathbb{Q}}_\ell))$$
$$+ (-1)^{n+1}\mathrm{Trace}(\mathrm{Frob}_k \mid H^{n+1}(X\otimes_k \bar{k},\ \bar{\mathbb{Q}}_\ell))$$
$$+ \Sigma_{i \ge n+2} (-1)^i\mathrm{Trace}(\mathrm{Frob}_k \mid H^i(X\otimes_k k,\ \bar{\mathbb{Q}}_\ell)).$$

By the Poincaré dual of the weak Lefschetz theorem, the third terms in the two expressions are equal. To compute the difference of the second terms, use the strong Lefschetz theorem to write

$$H^{n-1}(X\otimes_k \bar{k},\ \bar{\mathbb{Q}}_\ell)(-1) \cong H^{n+1}(X\otimes_k \bar{k},\ \bar{\mathbb{Q}}_\ell).$$

Then the difference of the second terms is precisely

$$(-1)^{n-1}\mathrm{Trace}(\mathrm{Frob}_k \mid \mathrm{Ev}^{n-1}(X\cap H)(-1)).$$

The difference of the first terms is, by weak Lefschetz,

$$\Sigma_{i \le n-2} (-1)^i\mathrm{Trace}(\mathrm{Frob}_k \mid H^i(X\otimes_k \bar{k},\ \bar{\mathbb{Q}}_\ell)(-1))$$
$$- \Sigma_{i \le n} (-1)^i\mathrm{Trace}(\mathrm{Frob}_k \mid H^i(X\otimes_k k,\ \bar{\mathbb{Q}}_\ell)).$$
$$= - \Sigma_{i \le n} (-1)^i\mathrm{Trace}(\mathrm{Frob}_k \mid \mathrm{Prim}^i(X)).$$

For the last equality, use strong Lefschetz to get the injectivity of

$$L : H^i(X\otimes_k \bar{k},\ \bar{\mathbb{Q}}_\ell)(-1) \subset H^{i+2}(X\otimes_k \bar{k},\ \bar{\mathbb{Q}}_\ell)$$

for $i \le n-2$.

Thus we obtain

$$\begin{aligned}
&q\#(X\cap H)(k) - \#X(k)\\
&= (-1)^{n-1}\mathrm{Trace}(\mathrm{Frob}_k \mid \mathrm{Ev}^{n-1}(X\cap H)(-1))\\
&- \Sigma_{i \le n} (-1)^i \mathrm{Trace}(\mathrm{Frob}_k \mid \mathrm{Prim}^i(X)).
\end{aligned}$$

The same argument applied to the second grouped term gives

$$\begin{aligned}
&q\#(X\cap Z\cap H)(k) - \#(X\cap Z)(k)\\
&= (-1)^{n-2}\mathrm{Trace}(\mathrm{Frob}_k \mid \mathrm{Ev}^{n-2}(X\cap Z\cap H)(-1))\\
&- \Sigma_{i \le n-1} (-1)^i \mathrm{Trace}(\mathrm{Frob}_k \mid \mathrm{Prim}^i(X\cap Z)).
\end{aligned}$$

By weak Lefschetz, and the snake lemma applied for $i \le n-1$ to

$$\begin{array}{ccccccc}
0 \to & H^{i-2}(X\otimes_k\bar{k}, \bar{\mathbb{Q}}_\ell)(-1) & \xrightarrow{L} & H^i(X\otimes_k\bar{k}, \bar{\mathbb{Q}}_\ell) & \to & \mathrm{Prim}^i(X) \to 0\\
& \downarrow & & \downarrow L & & \downarrow\\
0 \to & H^{i-2}((X\cap Z)\otimes_k\bar{k}, \bar{\mathbb{Q}}_\ell)(-1) & \to & H^i((X\cap Z)\otimes_k\bar{k}, \bar{\mathbb{Q}}_\ell) & \to & \mathrm{Prim}^i(X\cap Z) \to 0,
\end{array}$$

we see that

$$\mathrm{Prim}^i(X) \cong \mathrm{Prim}^i(X\cap Z) \text{ for } i \le n-2,$$
$$\mathrm{Prim}^{n-1}(X) \subset \mathrm{Prim}^{n-1}(X\cap Z),$$

and

$$\mathrm{Ev}^{n-1}(X\cap Z) \cong \mathrm{Prim}^{n-1}(X\cap Z)/\mathrm{Prim}^{n-1}(X).$$

Thus we find

$$\begin{aligned}
&= [q\#(X\cap H)(k) - \#X(k)] - [q\#(X\cap Z\cap H)(k) - \#(X\cap Z)(k)].\\
&= (-1)^{n-1}\mathrm{Trace}(\mathrm{Frob}_k \mid \mathrm{Ev}^{n-1}(X\cap H)(-1))\\
&\quad + (-1)^{n-1}\mathrm{Trace}(\mathrm{Frob}_k \mid \mathrm{Ev}^{n-2}(X\cap Z\cap H)(-1))\\
&\quad + (-1)^{n-1}\mathrm{Trace}(\mathrm{Frob}_k \mid \mathrm{Prim}^n(X))\\
&\quad + (-1)^{n-1}\mathrm{Trace}(\mathrm{Frob}_k \mid \mathrm{Ev}^{n-1}(X\cap Z)). \qquad \text{QED}
\end{aligned}$$

Corollary 4.3.2 Hypotheses and notations as in Lemma 4.3.1 above, suppose that for every finite extension F/E, and for every λ in $F^\times$, the sum

$$\Sigma_{v \text{ in } V(F)} \psi_F(\lambda(H/Z^e)(v))$$

is divisible by $(\#F)^{n/2}$ as an algebraic integer. Then we have the following results.

1) $\mathrm{Ev}^{n-1}(X\cap Z) = 0$.

2) Every eigenvalue of Frob_E on $\mathrm{Ev}^{n-1}(X\cap H)$ is divisible, as an algebraic integer, by $(\#E)^{(n-2)/2}$.

3) $\mathrm{Prim}^n(X)$ is supersingular, in the sense that every eigenvalue of Frob_k on $\mathrm{Prim}^n(X)$ is of the form $(\#k)^{n/2}\times$(a root of unity).

4) $\mathrm{Ev}^{n-2}(X\cap Z\cap H)$ is supersingular, in the sense that every eigenvalue of Frob_E on $\mathrm{Ev}^{n-2}(X\cap Z\cap H)$ is of the form

$$(\#E)^{(n-2)/2}\times(\text{a root of unity}).$$

proof Consider the $\overline{\mathbb{Q}}_\ell$-finite-dimensional $\overline{\mathbb{Q}}_\ell[\mathrm{Gal}(\overline{k}/E)]$-module M defined as

$M := \mathrm{Prim}^n(X)\oplus\mathrm{Ev}^{n-1}(X\cap Z)\oplus\mathrm{Ev}^{n-1}(X\cap H)(-1)\oplus\mathrm{Ev}^{n-2}(X\cap Z\cap H)(-1).$

The previous lemma asserts that for F/E any finite extension, we have

$$\mathrm{Trace}(\mathrm{Frob}_F \mid M) = (-1)^{n-1}\Sigma_{\lambda \text{ in } F^\times} \Sigma_{v \text{ in } V(F)} \psi_F(\lambda(H/Z^e)(v)).$$

So the hypothesized divisibility of each inner sum by $(\#F)^{n/2}$ as an algebraic integer tells us that for F/E any finite extension, we have

$$\mathrm{Trace}(\mathrm{Frob}_F \mid M) = (\#F)^{n/2}\times(\text{an algebraic integer}).$$

In other words, the trace on M of every power of $(\#E)^{-/2}\mathrm{Frob}_E$ is an algebraic integer. It is standard (cf. [Ax, top of page 256]) to infer that every eigenvalue of Frob_E on M is of the form

$$(\#E)^{n/2}\times(\text{an algebraic integer}).$$

Since M is given as a direct sum of Frob_E-stable subspaces, we have the same information about each eigenvalue of Frob_E on each summand. Thanks to Deligne, we know that these eigenvalues are algebraic integers which are units at all finite places outside p, we know their archimedean absolute values, and we know that each lies in a CM field. We consider these summands one by one.

1) The summand $\mathrm{Ev}^{n-1}(X\cap Z)$ is pure of weight n-1. If $\mathrm{Ev}^{n-1}(X\cap Z)$ is nonzero, let α be an eigenvalue of Frob_E on it. By purity, we have $\alpha\overline{\alpha} = (\#E)^{n-1}$. But if we write α as $(\#E)^{n/2}\times\beta$ with β an algebraic integer, we get $\alpha\overline{\alpha} = (\#E)^n\beta\overline{\beta}$. Thus we find $\beta\overline{\beta} = 1/\#E$, which is impossible, because $\beta\overline{\beta}$ is an algebraic integer, while $1/\#E$ is not.

2) For α an eigenvalue of Frob_E on $\mathrm{Ev}^{n-1}(X\cap H)$, $(\#E)\alpha$ is an eigenvalue of Frob_E on the summand $\mathrm{Ev}^{n-1}(X\cap H)(-1)$, hence $(\#E)\alpha$ is divisible by $(\#E)^{n/2}$ as an algebraic integer.

3) Let α be an eigenvalue of Frob_E on $\mathrm{Prim}^n(X)$, and write α as $(\#E)^{n/2}\times\beta$ with β an algebraic integer. Because $\mathrm{Prim}^n(X)$ is pure of weight n, we get $\alpha\overline{\alpha} = (\#E)^n$, so we have $\beta\overline{\beta} = 1$. Thus β is a unit in the ring of all algebraic integers, and all its complex absolute values are 1. Hence (Kronecker's theorem) β is a root of unity. So every eigenvalue of Frob_E on $\mathrm{Prim}^n(X)$ is of the form $(\#E)^{n/2}\times$(a root of unity). Since $\mathrm{Frob}_E = (\mathrm{Frob}_k)^{\deg(E/k)}$ on $\mathrm{Prim}^n(X)$, every eigenvalue of Frob_k on $\mathrm{Prim}^n(X)$ is of the asserted form $(\#k)^{n/2}\times$(a root of

unity).

4) Let α be an eigenvalue of Frob_E on $\mathrm{Ev}^{n-2}(X\cap Z\cap H)$. Then $(\#E)\alpha$ is an eigenvalue of Frob_E on $\mathrm{Ev}^{n-2}(X\cap Z\cap H)(-1)$, and this space is pure of weight n. So the argument of 3) above shows that $(\#E)\alpha$ is of the form $(\#E)^{n/2}\times$(a root of unity), as required. QED

(4.4) Return to the proof of Theorem 4.2.12

(4.4.1) Recall that $e \geq 3$ is prime to p. Suppose that $\mathcal{M}|U$ has finite G_{geom}. Then by part 4) of the above corollary, for any finite extension E/k, and for H in $U_{str}(E)$ any strong Deligne form with respect to X and Z, $\mathrm{Ev}^{n-2}(X\cap Z\cap H)$ is supersingular. This leads to a contradiction as follows.

(4.4.2) Consider the universal family of smooth hypersurface sections of $X\cap Z$ of degree de. Its parameter space is the open set U of all weak Deligne forms H with respect to X and Z. Over U, we have the lisse sheaf $\mathcal{H}$ given by

$$H \mapsto \mathrm{Ev}^{n-2}(X\cap Z\cap H).$$

Its rank satisfies the inequality [Ka-Pan, Theorem 1]

$$\mathrm{rank}(\mathcal{H}) \geq \deg(X\cap Z)((de-1)/de)((de-1)^{n-1} - (-1)^{n-1}).$$

Because $d \geq 1$, $e \geq 3$, and $n \geq 3$, this inequality implies

$$\mathrm{rank}(\mathcal{H}) \geq 2.$$

Thus $\mathcal{H}$ is nonzero. One knows that $\mathcal{H}$ is pure of weight n-2, and geometrically irreducible. One knows further [Ka-LAMM, 2.2.4] that its geometric monodromy group $G_{geom,\mathcal{H}}$ is given by

$$G_{geom,\mathcal{H}} = Sp(\mathrm{rank}(\mathcal{H})), \text{ if n is odd,}$$
$$= O(\mathrm{rank}(\mathcal{H})), \text{ if n is even and } \mathrm{rank}(\mathcal{H}) > 8.$$

Since U_{str} is a dense open set in U, $\mathcal{H}|U_{str}$ remains geometrically irreducible, with the same geometric monodromy group.

(4.4.3) On the other hand, the supersingularity of $\mathrm{Ev}^{n-2}(X\cap Z\cap H)$ for every H in U_{str} implies [Ka-ESDE, 8.14.3] that $G_{geom,\mathcal{H}|U_{str}}$ is finite. This finiteness contradicts the above determination of $G_{geom,\mathcal{H}}$ unless $n \geq 3$ is even, and we have

$$\deg(X\cap Z)((de-1)/de)((de-1)^{n-1} - (-1)^{n-1}) \leq 8.$$

So except in this case, our sheaf $\mathcal{M}|U$ cannot have finite G_{geom}.

(4.4.4) It remains to treat the exceptional case. One easily checks that for $n \geq 4$ even, $d \geq 1$, and $e \geq 3$, the above inequality

$$\deg(X\cap Z)((de-1)/de)((de-1)^{n-1} - (-1)^{n-1}) \leq 8$$

can hold only if we have

$$\deg(X\cap Z) = 1,\ d = 1,\ e = 3,\ n = 4.$$

In this case, as $d = 1$, Z is a hyperplane section of X, and hence

$$\deg(X) = 1.$$

Thus X is $\mathbb{P}^4$, linearly embedded in $\mathbb{P}$, $V = X[1/L]$ is $\mathbb{A}^4$, $\mathrm{char}(k) \neq 3$, and H/Z^3 is a cubic Deligne polynomial on $\mathbb{A}^4$. In this case, the fact that G_{geom} for $\mathcal{M}|U$ is not finite is the special case n=3, e=3, of Theorem 3.8.2. QED

(4.5) The subcases n=1 and n=2

Theorem 4.5.1 Let X/k be projective, smooth, and geometrically connected, of dimension n = 1. Fix a projective embedding $X \subset \mathbb{P}$, an integer $d \geq 1$, and a section Z of $H^0(\mathbb{P}, \mathcal{O}_{\mathbb{P}}(d))$ such that $X \cap Z$ is smooth of codimension one in X. Fix an integer $e \geq 3$. If char(k) = 3, suppose in addition $e \geq 4$. Form the perverse sheaf

$$M = M(X, r=1, Z, e)$$

on $\mathcal{F} = \mathcal{P}_{(e)} := H^0(\mathbb{P}, \mathcal{O}_{\mathbb{P}}(ed))$, and form the sheaf of perverse origin $\mathcal{M} := \mathcal{H}^{-\dim \mathcal{F}}(M)(-1/2)$. Let $U \subset \mathcal{F}$ be a dense open set on which M is lisse. If e is prime to p, take for U the dense open set $\mathcal{D} := \mathcal{D}(X, Z, e) \subset \mathcal{F}$ consisting of all forms H of weak Deligne type with respect to X and Z (i.e., such that $X \cap Z \cap H$ is empty). Denote by $N \geq 1$ the rank of $\mathcal{M}|U$. Suppose that any of the following five conditions holds:

a) $d \times \deg(X) > 1$,
b) $p := \mathrm{char}(k) \geq 7$,
c) $p = 5$ and $e \geq 4$,
d) $p = 3$ and $e \geq 7$,
e) $p = 2$ and $e \geq 7$.

Then we have the following results concerning the group G_{geom} for $\mathcal{M}|U$.

1) If $p \neq 2$, then G_{geom} contains SL(N).
2) If $p = 2$, then G_{geom} is Sp(N).

proof Exactly as in the proof of the $n \geq 3$ case Theorem 4.2.12, it suffices to show that G_{geom} is not finite. If it is finite, then by Corollary 4.3.2, part 1), we infer that $Ev^0(X \cap Z) = 0$. But

$$Ev^0(X \cap Z) := H^0((X \cap Z) \otimes_k \bar{k}, \bar{\mathbb{Q}}_\ell)/H^0(X \otimes_k \bar{k}, \bar{\mathbb{Q}}_\ell)$$

has dimension $\#((X \cap Z)(\bar{k})) = d \times \deg(X) - 1$. So if $d \times \deg(X) > 1$, G_{geom} cannot be finite. If $d \times \deg(X) = 1$, then X is $\mathbb{P}^1$, embedded linearly, Z is a single point, $V = X[1/L]$ is $\mathbb{A}^1$, and H/Z^e is simply a polynomial of degree e in one variable. In this case, the fact that G_{geom} for $\mathcal{M}|U$ is not finite if any of b), c), d), or e) holds is the special case n=1 of Theorem 3.8.2. QED

(4.5.2) We next turn to the n=2 case of this theorem. The result is quite similar to that in the n=1 case, but the proof is quite different.

Theorem 4.5.3 Let X/k be projective, smooth, and geometrically connected, of dimension n = 2. Fix a projective embedding $X \subset \mathbb{P}$, an integer $d \geq 1$, and a section Z of $H^0(\mathbb{P}, \mathcal{O}_{\mathbb{P}}(d))$ such that $X \cap Z$ is smooth of codimension one in X. Fix an integer $e \geq 3$. Form the perverse sheaf

$$M = M(X, r=1, Z, e)$$

on $\mathcal{F} = \mathcal{P}_{(e)} := H^0(\mathbb{P}, \mathcal{O}_{\mathbb{P}}(ed))$, and form the sheaf of perverse origin $\mathcal{M} := \mathcal{H}^{-\dim \mathcal{F}}(M)(-1/2)$. Let $U \subset \mathcal{F}$ be a dense open set on which M is lisse. If e is prime to p, take for U the dense open set $\mathcal{D} := \mathcal{D}(X, Z, e) \subset \mathcal{F}$ consisting of all forms H of weak Deligne type with respect to X and Z (i.e., such that $X \cap Z \cap H$ is smooth of codimension one in $X \cap Z$). Denote by $N \geq 1$ the rank of $\mathcal{M}|U$. Suppose that any of the following five conditions holds:

a) $d \times \deg(X) > 1$,
b) $p := \mathrm{char}(k) \geq 7$,
c) $p = 5$ and $e \geq 4$,
d) $p = 3$ and $e \geq 7$,
e) $p = 2$ and $e \geq 7$.

Then we have the following results concerning the group G_{geom} for $\mathcal{M}|U$.

1) If $p \neq 2$, then G_{geom} contains $SL(N)$.
2) If $p = 2$, then G_{geom} is either $SO(N)$ or $O(N)$.

proof Exactly as in the proof of the $n \geq 3$ case, Theorem 4.2.12, it suffices to show that G_{geom} is not finite. Consider first the case in which a) does not hold, i.e., the case when $d \times \deg(X) = 1$. Then $d = 1$ and $\deg(X) = 1$. So X is $\mathbb{P}^2$, embedded linearly in $\mathbb{P}$, $V = X[1/Z]$ is $\mathbb{A}^2$, and $\{H/Z^e\}_H$ runs over polynomials of degree $\leq e$ on $\mathbb{A}^2$. That G_{geom} is not finite if any of b), c), d), or e) holds is the special case $n=2$ of Theorem 3.8.2.

We now treat the case when a) holds, i.e., the case in which $d \times \deg(X) > 1$. In this case, we have

$$d^2 \times \deg(X) > 1,$$

and it is this inequality which will be crucial below. We will use the Homothety Contraction Theorem 3.3.13 and a consideration of weights to show that G_{geom} is not finite. We argue by contradiction. Thus we suppose that G_{geom} is finite for $\mathcal{M}|U$.

Recall that $V := X[1/Z]$. Over the space

$$\mathcal{P}_{(e)} := H^0(\mathbb{P}, \mathcal{O}_{\mathbb{P}}(ed)),$$

we have the product space $V \times \mathcal{P}_{(e)}$, which carries the lisse sheaf

$$(v, H) \mapsto \mathcal{L}_{\psi((H/Z^e)(v))}.$$

The sheaf $\mathcal{M}$ of perverse origin on $\mathcal{P}_{(e)}$ is

$$\mathcal{M} := R^2 pr_{2!}\mathcal{L}_{\psi((H/Z^e)(v))}(1),$$

cf. 4.2.5. On any dense open set U of $\mathcal{P}_{(e)}$ over which $Rpr_{2!}\mathcal{L}_{\psi((H/Z^e)(v))}$ is lisse, $\mathcal{M}|U$ is nonzero, geometrically irreducible, and $\det(\mathcal{M}|U)$ is arithmetically of finite order, cf. Theorem 4.2.9, parts 3), 5), and 7). We have assumed that G_{geom} for $\mathcal{M}|U$ is finite. It then follows from Twisting Lemma 2.8.12.1 (= 2.8.12 bis) that

G_{arith} for $\mathcal{M}|U$ is finite.

After possibly replacing the finite field k by a finite extension, we may choose a degree d form

$$G \text{ in } H^0(\mathbb{P}, \mathcal{O}_{\mathbb{P}}(d))(k)$$

such that $X \cap G$ is lisse of codimension one in X, and such that $X \cap Z \cap G$ is lisse of codimension one in $X \cap Z$. Thus $X \cap Z \cap G$ is finite etale over k, of degree $d^2 \times \deg(X)$.

Having **fixed** such a choice of a form G of degree d, we restrict $\mathcal{M}$ to the linear subspace $\mathrm{Poly}_{\leq e}$ of $H^0(\mathbb{P}, \mathcal{O}_{\mathbb{P}}(ed))$ consisting of those forms H such that H/Z^e is a polynomial of degree at most e in G/Z. Concretely, $\mathrm{Poly}_{\leq e}$ is the image in $H^0(\mathbb{P}, \mathcal{O}_{\mathbb{P}}(ed))$ of the vector space $\mathrm{Poly}_{\leq e}$ of polynomials in one variable of degree $\leq$ e, under the injective linear map

$$f(x) := \Sigma_{i \leq e} a_i x^i \mapsto \Sigma_{i \leq e} a_i G^i Z^{e-i}.$$

Now we consider the restriction $\mathcal{M} \mid \mathrm{Poly}_{\leq e}$. For any dense open set $U_{poly} \subset \mathrm{Poly}_{\leq e}$ such that $\mathcal{M}|U_{poly}$ is lisse, G_{arith} for $\mathcal{M}|U_{poly}$ is finite, cf. Semicontinuity Corollary 2.8.9, which applies because G_{arith} for $\mathcal{M}|U$ is finite.

By proper base change, $\mathcal{M} \mid \mathrm{Poly}_{\leq e}$ can be described as follows. On $V \times \mathrm{Poly}_{\leq e}$, we have the lisse sheaf

$$(v, f) \mapsto \mathcal{L}_{\psi(f(G/Z)(v))},$$

and

$$\mathcal{M} \mid \mathrm{Poly}_{\leq e} = R^2 pr_{2!} \mathcal{L}_{\psi(f(G/Z)(v))}(1).$$

The next step is to exploit the fact that the function inside $\mathcal{L}_{\psi}$ is the composition of f with G/Z.

Denote by g the function on V defined by G/Z:

$$g := G/Z : V \to \mathbb{A}^1.$$

We factor the map

$$pr_2 : V \times \mathrm{Poly}_{\leq e} \to \mathrm{Poly}_{\leq e}$$

as the composite of

$$g \times id : V \times \mathrm{Poly}_{\leq e} \to \mathbb{A}^1 \times \mathrm{Poly}_{\leq e},$$
$$(v, f) \mapsto (g(v), f),$$

followed by the map

$$pr_{2,\mathbb{A}} : \mathbb{A}^1 \times \mathrm{Poly}_{\leq e} \to \mathrm{Poly}_{\leq e}.$$

The point now is that on the middle space $\mathbb{A}^1 \times \mathrm{Poly}_{\leq e}$, we have the lisse sheaf

$$(x, f) \mapsto \mathcal{L}_{\psi(f(x))},$$

whose pullback by $g \times id$ to $V \times \mathrm{Poly}_{\leq e}$ is the sheaf $\mathcal{L}_{\psi(f(G/Z)(v))}$. So we have a diagram of morphisms and sheaves

$$
\begin{array}{l}
V \times \mathrm{Poly}_{\leq e},\ \mathcal{L}_{\psi(f(G/Z)(v))} = (g \times \mathrm{id})^*(\mathcal{L}_{\psi(f(x))}) \\
\quad\downarrow \quad g \times \mathrm{id} \\
\mathbb{A}^1 \times \mathrm{Poly}_{\leq e},\ \mathcal{L}_{\psi(f(x))} \\
\quad\downarrow \quad \mathrm{pr}_{2,\mathbb{A}} \\
\quad \mathrm{Poly}_{\leq e}.
\end{array}
$$

We next examine the Leray spectral sequence attached to this diagram, using the projection formula to identify the E_2 terms:

$$
\begin{aligned}
E_2^{a,b} &:= R^a\mathrm{pr}_{2,\mathbb{A}!}R^b(g \times \mathrm{id})_!((g \times \mathrm{id})^*\mathcal{L}_{\psi(f(x))}) \\
&= R^a\mathrm{pr}_{2,\mathbb{A}!}(\mathcal{L}_{\psi(f(x))} \otimes R^b(g \times \mathrm{id})_!\overline{\mathbb{Q}}_\ell) \\
&= R^a\mathrm{pr}_{2,\mathbb{A}!}(\mathrm{pr}_1{}^*(R^b g_!\overline{\mathbb{Q}}_\ell \text{ on } \mathbb{A}^1) \otimes \mathcal{L}_{\psi(f(x))}) \\
&\qquad \Rightarrow R^{a+b}(\mathrm{pr}_2)_!\mathcal{L}_{\psi(f(G/Z)(v))}.
\end{aligned}
$$

Both the morphisms $g \times \mathrm{id}$ and $\mathrm{pr}_{2,\mathbb{A}}$ have all their fibres of dimension one, so certainly we have

$$E_2^{a,b} = 0 \text{ unless both } a, b \text{ lie in } [0, 2].$$

So we have

$$E_2^{1,1} = E_\infty^{1,1}.$$

Thus

$$\mathcal{N} := R^1\mathrm{pr}_{2,\mathbb{A}!}(\mathrm{pr}_1{}^*(R^1 g_!\overline{\mathbb{Q}}_\ell \text{ on } \mathbb{A}^1) \otimes \mathcal{L}_{\psi(f(x))})(1)$$

is a subquotient of

$$\mathcal{M} \mid \mathrm{Poly}_{\leq e} = R^2(\mathrm{pr}_2)_!\mathcal{L}_{\psi(f(G/Z)(v))}(1).$$

So on any dense open set $U_{\mathrm{poly}} \subset \mathrm{Poly}_{\leq e}$ on which both $\mathcal{M}$ and all the E_2 terms are lisse, we find that

$$\mathcal{N} \mid U_{\mathrm{poly}}$$

has finite G_{arith}.

We next bring to bear the Homothety Contraction Theorem 3.3.13.

Since V is affine and smooth of dimension 2, the sheaf $R^1 g_!\overline{\mathbb{Q}}_\ell$ on $\mathbb{A}^1$ is of perverse origin [Ka-SMD, Corollary 6]. From the known structure of perverse sheaves on a smooth curve [namely that a derived category object K on a smooth curve is perverse if and only if $\mathcal{H}^i(K)$ vanishes for i not in $[-1, 0]$, $\mathcal{H}^{-1}(K)$ has no nonzero punctual sections, and $\mathcal{H}^0(K)$ has at most punctual support], we see that a constructible $\overline{\mathbb{Q}}_\ell$-sheaf $\mathcal{G}$ is of perverse origin if and only if $\mathcal{G}[1]$ is a perverse sheaf.

Thus

$$L := R^1 g_!\overline{\mathbb{Q}}_\ell[1](1)$$

is a perverse sheaf on $\mathbb{A}^1$. And $(\mathcal{F}, \tau) := (\mathrm{Poly}_{\leq e}, \text{evaluation})$ is a space of $\mathbb{A}^1$-valued functions on $\mathbb{A}^1$ which is stable by homothety, contains the constants, and is quasifinitely difference-separating (indeed, is at least 4-separating, because $e \geq 3$). Take $K := \mathcal{L}_{\psi}[1](1/2)$, also a perverse sheaf on $\mathbb{A}^1$. Because $H_c{}^*(\mathbb{A}^1 \otimes \bar{k}, K[1]) = 0$, the Homothety Contraction Theorem 3.3.13 applies to this situation.

The perverse object

$$N = \mathrm{Twist}(L, K, \mathcal{F}, h=0)$$

on $\mathcal{F} := \mathrm{Poly}_{\leq e}$ gives rise to the sheaf of perverse origin

$$\mathcal{N} := \mathcal{H}^{-\dim \mathcal{F}}(N)(-1/2),$$

which is, by construction, none other than

$$\mathcal{N} := R^1 pr_{2,\mathbb{A}!}(pr_1{}^*(R^1 g_! \overline{\mathbb{Q}}_\ell \text{ on } \mathbb{A}^1) \otimes \mathcal{L}_{\psi(f(x))})(1).$$

Because L is a single sheaf, placed in degree -1, the object L(0) is itself perverse, equal to the object

$$(\text{const. sheaf on } \mathbb{A}^1 \text{ with value } H_c{}^1((V \cap (g=0)) \otimes_k \bar{k}, \overline{\mathbb{Q}}_\ell)[1](1).$$

The perverse object

$$N_0 = \mathrm{Twist}(L(0), K, \mathcal{F}, h=0)$$

on $\mathcal{F} := \mathrm{Poly}_{\leq e}$ gives rise to the sheaf of perverse origin

$$\mathcal{N}_0 := \mathcal{H}^{-\dim \mathcal{F}}(N_0)(-1/2).$$

The sheaf $\mathcal{N}_0$ is given explicitly by

$$\mathcal{N}_0 := R^1 pr_{2,\mathbb{A}!}(\mathcal{L}_{\psi(f(x))}) \otimes H_c{}^1((V \cap (g=0)) \otimes_k \bar{k}, \overline{\mathbb{Q}}_\ell)(1).$$

Since $\mathcal{N} \mid U_{poly}$ has finite G_{arith} for U_{poly} any dense open set of $\mathrm{Poly}_{\leq e}$ on which $\mathcal{N}$ is lisse, it follows from the Homothety Contraction Theorem 3.3.13 that $\mathcal{N}_0 \mid U_{poly}$ has finite G_{arith} for U_{poly} any dense open set on which $\mathcal{N}_0$ is lisse. In particular, $\mathcal{N}_0 \mid U_{poly}$ is pure of weight zero.

But on any open set on which it is lisse, $R^1 pr_{2,\mathbb{A}!}(\mathcal{L}_{\psi(f(x))})$ is nonzero (its rank is e-1 if e is prime to p, and e-2 otherwise) and it is pure of weight one, as is standard from the classical theory of exponential sums in one variable. Therefore the cohomology group

$$H_c{}^1((V \cap (g=0)) \otimes_k \bar{k}, \overline{\mathbb{Q}}_\ell)(1)$$

must itself be pure of weight -1. In other words, the cohomology group

$$H_c{}^1((V \cap (g=0)) \otimes_k \bar{k}, \overline{\mathbb{Q}}_\ell)$$

must be pure of weight one. But this is **false**. The variety $V \cap (g=0)$ is the variety $(X[1/Z]) \cap G = X \cap G - X \cap Z \cap G$, which is thus a projective smooth geometrically connected curve $X \cap G$ from which $d^2 \times \deg(X)$ points have been deleted. As

$$d^2 \times \deg(X) > 1,$$

$H_c{}^1((V \cap (g=0)) \otimes_k \bar{k}, \overline{\mathbb{Q}}_\ell)$ has a weight zero part of dimension

$$d^2 \times \deg(X) - 1 > 0.$$

This contradiction concludes the proof. QED

Chapter 5: Multiplicative character sums on $\mathbb{A}^n$

(5.1) The general setting

(5.1.1) In this chapter, we consider the monodromy groups attached to families of multiplicative character sums on $\mathbb{A}^n$ which are (the multiplicative character sum analogue of additive character sums) of Deligne type on $\mathbb{A}^n$.

(5.1.2) Let k be a finite field of characteristic p, ℓ a prime number invertible in k, and $\chi : k^\times \to \overline{\mathbb{Q}}_\ell^\times$ a nontrivial multiplicative character of $k^\times$, extended by zero to all of k. On $\mathbb{G}_m$, we have the corresponding Kummer sheaf $\mathcal{L}_\chi$. Using the inclusion $j : \mathbb{G}_m \to \mathbb{A}^1$, we form its middle extension

$$j_!\mathcal{L}_\chi \cong j_*\mathcal{L}_\chi \cong Rj_*\mathcal{L}_\chi$$

on $\mathbb{A}^1$.

(5.1.3) For any scheme X/k, and for any function f in $H^0(X, \mathcal{O}_X)$, we view f as a morphism $f: X \to \mathbb{A}^1$, and we form the $\overline{\mathbb{Q}}_\ell$-sheaf on X

$$\mathcal{L}_{\chi(f)} := f^*(j_!\mathcal{L}_\chi).$$

Thus $\mathcal{L}_{\chi(f)}$ is lisse of rank one on the open set X[1/f], and it vanishes outside this open set.

(5.1.4) One knows that $j_!\mathcal{L}_\chi[1](1/2)$ on $\mathbb{A}^1$ is perverse and geometrically irreducible, pure of weight zero, and

$$H_c{}^i(\mathbb{A}^1\otimes_k\overline{k}, j_!\mathcal{L}_\chi[1](1/2)) = 0 \text{ for all } i.$$

(5.1.5) Consider the following general class of "standard inputs", cf. 1.15.4. We take

- $m=1$,
- $\chi : k^\times \to \overline{\mathbb{Q}}_\ell^\times$ a nontrivial multiplicative character of $k^\times$,
- $K = j_!\mathcal{L}_\chi[1](1/2)$ on $\mathbb{A}^1$,
- an integer $n \geq 1$,
- $V = \mathbb{A}^n$,
- $h : V \to \mathbb{A}^1$ the function $h = 0$,
- $L := \overline{\mathbb{Q}}_\ell[n](n/2)$ on $V = \mathbb{A}^n$,
- an integer $e \geq 3$,
- $(\mathcal{F}, \tau) = (\mathcal{P}(n,e), \text{evaluation})$, for $\mathcal{P}(n,e)$ the space of all k-polynomial functions on $\mathbb{A}^n$ of degree $\leq e$.

(5.1.6) From this input data, we construct on $\mathcal{P}(n,e)$ the perverse sheaf

$$\mathrm{M}(n,e,\chi) := \mathrm{Twist}(\overline{\mathbb{Q}}_\ell[n](n/2),\ j_!\mathcal{L}_\chi[1](1/2), \mathcal{P}(n,e),\ h=0),$$

and the sheaf of perverse origin

$$\mathcal{M}(n,e,\chi) := \mathcal{H}^{-\dim \mathcal{F}}(\mathrm{M}(n,e,\chi))(-1/2).$$

(5.1.7) Before proceeding, let us relate the objects to the exponential sums they were built to incarnate. Given a finite extension E/k, and a nontrivial multiplicative character χ of $k^\times$, extended by zero to all of k, we denote by χ_E the nontrivial multiplicative character of E defined by

$$\chi_E(x) := \chi(\mathrm{Norm}_{E/k}(x)).$$

Recall that we have also fixed a square root of p := char(k), allowing us to form Tate twists by half-integers, and allowing us to give unambiguous meaning to half-integral powers of $\#E$.

(5.1.8) In down to earth terms, on the space $\mathbb{A}^n \times \mathcal{P}(n,e)$, with coordinates (v, f), we have the sheaf

$$\mathcal{L}_{\chi(f(v))} := f(v)^*(j_!\mathcal{L}_\chi),$$

which is lisse on the open set $(\mathbb{A}^n \times \mathcal{P}(n,e))[1/f(v)]$, and vanishes outside. Under the second projection $pr_2 : \mathbb{A}^n \times \mathcal{P}(n,e) \to \mathcal{P}(n,e)$, we form $Rpr_{2!}\mathcal{L}_{\chi(f(v))}$. For E/k a finite extension, and for f in $\mathcal{P}_e(E)$, the stalk of $Rpr_{2!}\mathcal{L}_{\chi(f(v))}$ at f is the object $R\Gamma_c(\mathbb{A}^n\otimes_k\overline{E},\ \mathcal{L}_{\chi(f)})$, whose cohomology objects are the groups

$$\mathcal{H}^i(R\Gamma_c(\mathbb{A}^n\otimes_k\overline{E},\ \mathcal{L}_{\chi(f)})) = H_c{}^i(\mathbb{A}^n\otimes_k\overline{E},\ \mathcal{L}_{\chi(f)}).$$

The perverse sheaf $\mathrm{M}(n,e,\chi)$ is just a Tate twist and a shift of $Rpr_{2!}\mathcal{L}_{\chi(f(v))}$; we have

$$Rpr_{2!}\mathcal{L}_{\chi(f(v))} = \mathrm{M}(n,e,\chi)[-n - \dim\mathcal{P}(n,e)]((-n-1)/2).$$

In particular, we have

$$\mathcal{M}(n,e,\chi) = R^npr_{2!}\mathcal{L}_{\chi(f(v))}(n/2).$$

Lemma 5.1.9 For E/k a finite extension, and for f in $\mathcal{F}(E) = \mathcal{P}(n,e)(E)$, i.e., for f an E-polynomial in n variables of degree at most e, we have the identity

$$\mathrm{Trace}(\mathrm{Frob}_{E,f} \mid \mathrm{M}(n,e,\chi))$$
$$= (-1)^{n+\dim\mathcal{F}}(\#E)^{-(n+1)/2}\Sigma_{v \text{ in } \mathbb{A}^n(E)}\ \chi_E(f(v)).$$

proof Immediate from the Lefschetz Trace Formula, proper base change, and the definition of $\mathrm{M}(n,e,\chi)$. QED

(5.1.10) At this point, we must recall a key result concerning exponential sums on $\mathbb{A}^n$. Let us say that an n-variable polynomial f in $\mathcal{P}(n,e)(\overline{k})$ is a strong Deligne polynomial if it satisfies the following three conditions D0), D1), and D2).

D0) The closed subscheme f=0 in $\mathbb{A}^n$ is smooth of codimension one.

D1) When we write $f = \Sigma_{i\le e} F_i$ as a sum of homogeneous forms, F_e is

nonzero, and, in the case $n \geq 2$, the closed subscheme of $\mathbb{P}^{n-1}$ defined by the vanishing of F_e is smooth of codimension one.
D2) The integer e is prime to p.

(5.1.11) For a fixed integer e which is prime to p, the strong Deligne polynomials form a dense open set $\mathcal{SD}(n,e)$ of $\mathcal{P}(n,e)$.

Theorem 5.1.12 ([Ka-ENSMCS, 5.1]) Fix an integer $e \geq 1$ prime to p, and a nontrivial multiplicative character χ of $k^\times$. Suppose that $\chi^e \neq \mathbb{1}$. For any finite extension E/k, and any strong Deligne polynomial f in $\mathcal{D}(n,e)(E)$, we have the following results.
1) The "forget supports" maps

$$H_c^i((\mathbb{A}^n \otimes_k \bar{E})[1/f], \mathcal{L}_{\chi(f)}) \to H^i((\mathbb{A}^n \otimes_k \bar{E})[1/f], \mathcal{L}_{\chi(f)})$$

are all isomorphisms.
2) The groups $H_c^i((\mathbb{A}^n \otimes_k \bar{E})[1/f], \mathcal{L}_{\chi(f)}) = H_c^i(\mathbb{A}^n \otimes_k \bar{E}, \mathcal{L}_{\chi(f)})$ vanish for $i \neq n$.
3) The group $H_c^n(\mathbb{A}^n \otimes_k \bar{E}, \mathcal{L}_{\chi(f)})$ is pure of weight n, and has dimension $(e-1)^n$.

Theorem 5.1.13 ([Ka-ENSMCS, 2.2, 6.2]) Fix an integer $e \geq 1$ prime to p, and a nontrivial multiplicative character χ of $k^\times$. Suppose that $\chi^e = \mathbb{1}$. For any finite extension E/k, and any strong Deligne polynomial f in $\mathcal{D}(n,e)(E)$, we have the following results.
1) The groups $H_c^i((\mathbb{A}^n \otimes_k \bar{E})[1/f], \mathcal{L}_{\chi(f)}) = H_c^i(\mathbb{A}^n \otimes_k \bar{E}, \mathcal{L}_{\chi(f)})$ vanish for $i \neq n$.
2) The group $H_c^n(\mathbb{A}^n \otimes_k \bar{E}, \mathcal{L}_{\chi(f)})$ is mixed of weights n and n-1, and has dimension $(e-1)^n$. It has

$$(1/e)((e-1)^{n+1} - (-1)^{n+1})$$

eigenvalues which are pure of weight n, and it has

$$(1/e)((e-1)^n - (-1)^n)$$

eigenvalues which are pure of weight n-1.

Corollary 5.1.14 Suppose $e \geq 1$ is prime to p. Then $M(n,e,\chi)$ has lisse cohomology sheaves on the dense open set $\mathcal{SD}(n,e) \subset \mathcal{P}(n,e)$ consisting of strong Deligne polynomials. We have

$$M(n,e,\chi)|\mathcal{SD}(n,e) = \mathcal{M}(n,e,\chi)(1/2)[\dim\mathcal{P}(n,e)]|\mathcal{SD}(n,e),$$

$$\mathcal{M}(n,e,\chi) = R^n pr_{2!}\mathcal{L}_{\chi(f(v))}(n/2).$$

The sheaf $\mathcal{M}(n,e,\chi) = R^n pr_{2!}\mathcal{L}_{\chi(f(v))}(n/2)$ is lisse on $\mathcal{SD}(n,e)$ of rank $(e-1)^n$, and the other $R^i pr_{2!}\mathcal{L}_{\chi(f(v))}$ vanish on $\mathcal{SD}(n,e)$. If in addition $\chi^e \neq \mathbb{1}$, then $\mathcal{M}(n,e,\chi)|\mathcal{SD}(n,e)$ is pure of weight zero. On the other hand, if $\chi^e = \mathbb{1}$, then $\mathcal{M}(n,e,\chi)|\mathcal{SD}(n,e)$ is mixed of weights 0 and -1, and sits in a short exact sequence of lisse sheaves

$$0 \to \mathrm{Gr}^{-1}(\mathcal{M}(n,e,\chi)|\mathcal{SD}(n,e)) \to \mathcal{M}(n,e,\chi)|\mathcal{SD}(n,e)$$
$$\to \mathrm{Gr}^{0}(\mathcal{M}(n,e,\chi)|\mathcal{SD}(n,e)) \to 0,$$

with Gr^{-1} pure of weight -1 and lisse of rank $(1/e)((e-1)^n - (-1)^n)$, and with Gr^0 pure of weight 0 and lisse of rank $(1/e)((e-1)^{n+1} - (-1)^{n+1})$.

proof Looking fibre by fibre, we see from the above two theorems that $R^i pr_{2!}\mathcal{L}_{\chi(f(v))}|\mathcal{SD}(n,e)$ vanishes for $i \neq n$.

Therefore $\mathcal{H}^{-i}(M(n,e,\chi))|\mathcal{SD}(n,e)$ vanishes for $i \neq \dim\mathcal{F}$. The remaining cohomology sheaf

$$\mathcal{H}^{-\dim\mathcal{F}}(M(n,e,\chi)) = R^n pr_{2!}\mathcal{L}_{\chi(f(v))}((n+1)/2)$$
$$= \mathcal{M}(n,e,\chi)(1/2)$$

is of perverse origin on $\mathcal{P}(n,e)$. As it has constant rank $(e-1)^n$ on $\mathcal{SD}(n,e)$, it is lisse on $\mathcal{SD}(n,e)$.

Looking fibre by fibre, we see that if $\chi^e \neq \mathbb{1}$, then $R^n pr_{2!}\mathcal{L}_{\chi(f(v))}|\mathcal{SD}(n,e)$ is punctually pure of weight n, while if $\chi^e = \mathbb{1}$, then $R^n pr_{2!}\mathcal{L}_{\chi(f(v))}|\mathcal{SD}(n,e)$ is mixed of weights n and n-1, with associated gradeds of the asserted ranks. QED

Corollary 5.1.15 For $e \geq 2$ prime to p, E/k any finite extension, and f in $\mathcal{SD}(n,e)(E)$, we have

$$\mathrm{Trace}(\mathrm{Frob}_{E,f} \mid \mathcal{M}(n,e,\chi))$$
$$= (-1)^n(\#E)^{-n/2}\Sigma_{v \text{ in } \mathbb{A}^n(E)}\, \chi_E(f(v)).$$

(5.2) First main theorem: the case when χ^e is nontrivial

(5.2.1) With all these preliminaries out of the way, we can now state and prove the first main theorem of this chapter.

Theorem 5.2.2 Suppose k is a finite field of characteristic p. Let $n \geq 1$ and $e \geq 3$ be integers, and let χ be a nontrivial multiplicative character of $k^\times$. Suppose further that e is prime to p, and that $\chi^e \neq \mathbb{1}$. Then we have the following results.

1) If χ does not have order two, then G_{geom} for $\mathcal{M}(n,e,\chi)|\mathcal{SD}(n,e)$ contains $SL((e-1)^n)$, except in the following cases.

$n = 1$, χ has order 4, $e = 3$, and $p \geq 5$,
$n = 1$, χ has order 6, $e = 3$, and $p \geq 5$,
$n = 1$, χ has order 6, $e = 4$, and $p \geq 5$,
$n = 1$, χ has order 6, $e = 5$, and $p \geq 7$,
$n = 1$, χ has order 10, $e = 3$, and $p \geq 7$.

In each of the exceptional cases, G_{geom} is finite.

2) If χ has order two, and n is odd, then G_{geom} for $\mathcal{M}(n,e,\chi)|\mathcal{S}\mathcal{D}(n,e)$ is $Sp((e-1)^n)$.

3) If χ has order two, and n is even, then G_{geom} for $\mathcal{M}(n,e,\chi)|\mathcal{S}\mathcal{D}(n,e)$ is either $SO((e-1)^n)$ or $O((e-1)^n)$, except in the case $(n = 2, e = 3, p \geq 5)$, in which case G_{geom} is finite.

Remark 5.2.3 We will show later, in Theorem 6.7.19, part 3), that if χ has order 2 and n is even, then G_{geom} always contains a reflection, and hence that G_{geom} is $O((e-1)^n)$, except in the case $(n = 2, e = 3, p \geq 5)$, in which case it is a finite primitive subgroup of $O(4)$ which contains a reflection. We will also give, in Theorem 6.7.21, quite precise results about G_{geom} in the case when χ does not have order 2.

proof of Theorem 5.2.2 On the space $\mathcal{P}(n,e)$ of polynomials of degree $\leq$ e in n variables, we have the perverse sheaf

$$M(n,e,\chi) := \mathrm{Twist}(\overline{\mathbb{Q}}_\ell[n](n/2),\ j_!\mathcal{L}_\chi[1](1/2), \mathcal{P}(n,e),\ h=0),$$

and the sheaf of perverse origin

$$\mathcal{M}(n,e,\chi) := \mathcal{H}^{-\dim \mathcal{F}}(M(n,e,\chi))(-1/2).$$

Combining 1.15.5, part 3), the vanishing

$$H_c{}^*(\mathbb{A}^1\otimes_k\bar{k},\ j_!\mathcal{L}_\chi) = 0,$$

and 1.20.3, part 3), we see that we have an equality of perverse sheaves on $\mathcal{P}(n,e)$,

$$Gr^0(M(n,e,\chi))_{ncst} = Gr^0(M(n,e,\chi)),$$

and an equality of pure of weight zero lisse sheaves on $\mathcal{S}\mathcal{D}(n,e)$,

$$Gr^0(\mathcal{M}(n,e,\chi)|\mathcal{S}\mathcal{D}(n,e))_{ncst} = Gr^0(\mathcal{M}(n,e,\chi)|\mathcal{S}\mathcal{D}(n,e)).$$

Because $e \geq 3$, $\mathcal{P}(n,e)$ is 4-separating; 1.20.3, part 2) then tells us that $Gr^0(\mathcal{M}(n,e,\chi)|\mathcal{S}\mathcal{D}(n,e))$ is geometrically irreducible. Because χ^e is nontrivial, we know by 5.1.14 that $\mathcal{M}(n,e,\chi)|\mathcal{S}\mathcal{D}(n,e)$ is pure of weight zero, so we have the further equality

$$\mathcal{M}(n,e,\chi)|\mathcal{S}\mathcal{D}(n,e) = Gr^0(\mathcal{M}(n,e,\chi)|\mathcal{S}\mathcal{D}(n,e)).$$

Applying 1.20.3, part 7), we see that the geometric Frobenius-Schur indicator of $\mathcal{M}(n,e,\chi)|\mathcal{S}\mathcal{D}(n,e)$ is given by

$$FSI^{geom}(\mathcal{S}\mathcal{D}(n,e),\ \mathcal{M}(n,e,\chi)) = 0, \text{ if order}(\chi) \neq 2,$$
$$= (-1)^n, \text{ if order}(\chi) = 2.$$

Since $\mathcal{M}(n,e,\chi)|\mathcal{S}\mathcal{D}(n,e)$ has rank $(e-1)^n$, we then see from 1.20.3, parts 4) and 5) that $\mathcal{M}(n,e,\chi)|\mathcal{S}\mathcal{D}(n,e)$ has fourth moment given by

$$M_4{}^{geom}(\mathcal{S}\mathcal{D}(n,e),\ \mathcal{M}(n,e,\chi)) = 2, \text{ if order}(\chi) \neq 2,$$
$$M_4{}^{geom}(\mathcal{S}\mathcal{D}(n,e),\ \mathcal{M}(n,e,\chi)) \leq 3, \text{ if order}(\chi) = 2,$$
$$M_4{}^{geom}(\mathcal{S}\mathcal{D}(n,e),\ \mathcal{M}(n,e,\chi)) = 3, \text{ if order}(\chi) = 2 \text{ and } (e-1)^n \geq 4.$$

We now bring to bear Larsen's Alternative 2.2.2, which gives us the following results.

If order(χ) $\neq$ 2, then either G_{geom} for $\mathcal{M}(n,e,\chi)|\mathcal{SD}(n,e)$ contains $SL((e-1)^n)$, or G_{geom} is a finite primitive irreducible subgroup of $GL((e-1)^n)$. If order(χ) = 2 and n is odd and $(e - 1)^n \geq 4$, then either G_{geom} is $Sp((e-1)^n)$, or G_{geom} is a finite primitive irreducible subgroup of $Sp((e-1)^n)$. If order(χ) = 2, n=1, and e=3, (the only case in our pantheon with order(χ) = 2, n odd, and with $(e-1)^n < 4$) then G_{geom} is a semisimple subgroup of Sp(2) = SL(2), so is either Sp(2) or is finite. If order(χ) = 2 and n is even, then G_{geom} is either $SO((e-1)^n)$, or $O((e-1)^n)$, or G_{geom} is a finite primitive irreducible subgroup of $O((e-1)^n)$.

There is one final general principle we can bring to bear, that of degeneration. For any integer e_0 with $2 \leq e_0 < e$ which is prime to p, the space $\mathcal{SD}(n, e_0)$ is a dense open set in $\mathcal{P}(n, e_0)$, and $\mathcal{P}(n, e_0)$ is an irreducible closed subscheme of $\mathcal{P}(n, e)$. Moreover, we have

$$\mathcal{M}(n,e,\chi)|\mathcal{P}(n, e_0) = \mathcal{M}(n, e_0, \chi),$$

an equality of sheaves of perverse origin on $\mathcal{P}(n, e_0)$. And we have seen above in 5.1.14 that $\mathcal{M}(n, e_0, \chi)|\mathcal{SD}(n, e_0)$ is lisse of rank $(e_0 - 1)^n$. It now follows from 2.8.9, part 2a), 2.8.13, and 2.8.14 that if G_{geom} for $\mathcal{M}(n,e,\chi)|\mathcal{SD}(n,e)$ is finite, then the following three statements hold:

1) G_{geom} for $\mathcal{M}(n, e_0, \chi)|\mathcal{SD}(n, e_0)$ is finite,

2) on $\mathcal{M}(n, e_0, \chi)|\mathcal{SD}(n, e_0)$, a power of every Frobenius is scalar, and

3) $\mathcal{M}(n, e_0, \chi)|\mathcal{SD}(n, e_0)$ is punctually pure of weight n.

Let us now apply these principles. Consider first the case

$$\text{order}(\chi) = 2, \text{ n odd}.$$

This case occurs only in odd characteristic p. We must show that for any $e \geq 3$ prime to p, G_{geom} for $\mathcal{M}(n,e,\chi)|\mathcal{SD}(n,e)$ is $Sp((e-1)^n)$. If not, we take e_0 = 2 (which is prime to p) in the degeneration discussion just above, and infer that

$$\mathcal{M}(n, 2, \chi)|\mathcal{SD}(n, 2)$$

is punctually pure of weight n. But according to 5.1.13, part 2), for n odd and order(χ) = 2, $\mathcal{M}(n, 2, \chi)|\mathcal{SD}(n, 2)$ is a rank one lisse sheaf, which is pure of weight n-1. This contradiction shows that G_{geom} for $\mathcal{M}(n,e,\chi)|\mathcal{SD}(n,e)$ is $Sp((e-1)^n)$, as asserted.

Let us also consider the case

$$\text{order}(\chi) = 3.$$

This case occurs only in characteristic p $\neq$ 3. We must show that for any $e \geq 4$ prime to p, G_{geom} for $\mathcal{M}(n,e,\chi)|\mathcal{SD}(n,e)$ is not finite. If

G_{geom} is finite, we take $e_0 = 3$ (which is prime to p) in the degeneration discussion above, and infer that

$$\mathcal{M}(n, 3, \chi)|\mathcal{SD}(n, 3)$$

is punctually pure of weight n. But according to 5.1.13, part 2), for order(χ) = 3, $\mathcal{M}(n, 3, \chi)|\mathcal{SD}(n, 3)$ is punctually mixed, with both weights n and n-1 occurring. This contradiction shows that G_{geom} for $\mathcal{M}(n,e,\chi)|\mathcal{SD}(n,e)$ is not finite, as asserted.

(5.3) Continuation of the proof of Theorem 5.2.2 for n=1

(5.3.1) We now turn to a detailed discussion of the hardest case

$$n = 1, \text{order}(\chi) \geq 3, e \geq 3 \text{ prime to p}, \chi^e \neq \mathbb{1}.$$

We know that either G_{geom} contains SL(e-1), or G_{geom} is a primitive irreducible finite subgroup of GL(e-1). The key is to observe that G_{geom} contains pseudoreflections of a quite specific type. Here is the precise result.

Pseudoreflection Theorem 5.3.2 (cf. [Ka-ACT, 5.7, 5)] and [Ka-TLFM, 4.2.2, proof of 5.6.1]) Suppose we are in the case

$$n = 1, \text{order}(\chi) \geq 3, e \geq 3 \text{ prime to p}, \chi^e \neq \mathbb{1}.$$

Then we have the following results.

1) If p ≠ 2, denote by χ_2 the quadratic character. Then G_{geom} for $\mathcal{M}(1, e, \chi)|\mathcal{SD}(1, e)$ contains pseudoreflections of order = order($\chi\chi_2$). More precisely, for any weakly supermorse polynomial f(X) (cf. [Ka-ACT, 5.5.2]) of degree e, the pullback of $\mathcal{M}(1, e, \chi)$ to the one-parameter family

$$t \mapsto t - f(X),$$

parameterized by t in $\mathbb{A}^1$ - {critical values of f}, is lisse, and its local monodromy at each critical value f(α) of f is a pseudoreflection of determinant $\chi\chi_2$, viewed as a tame character of the inertia group I(f(α)).

2) If p = 2, then G_{geom} for $\mathcal{M}(1, e, \chi)|\mathcal{SD}(1, e)$ contains pseudoreflections of order = 2×order(χ). More precisely, there is a dense open set U in $\mathcal{SD}(1, e)$ such that for any f in U, f has (e-1)/2 critical points, (e-1)/2 critical values, and over each critical value of f, the local monodromy of the sheaf $f_*\overline{\mathbb{Q}}_\ell/\overline{\mathbb{Q}}_\ell$ is a reflection of Swan conductor 1. Fix any such f. Over ∞, the local monodromy of $f_*\overline{\mathbb{Q}}_\ell/\overline{\mathbb{Q}}_\ell$ is tame; it is the direct sum of the e-1 nontrivial characters of I(∞) of order dividing e. The sheaf $f_*\overline{\mathbb{Q}}_\ell/\overline{\mathbb{Q}}_\ell$ is a geometrically irreducible middle extension on $\mathbb{A}^1$, with geometric monodromy group the full symmetric group S_e, in its deleted permutation representation. The pullback of $\mathcal{M}(1, e, \chi)$ to the one-parameter family

$$t \mapsto t - f(X),$$

parameterized by t in $\mathbb{A}^1$ - {critical values of f}, is lisse and geometrically irreducible, and is geometrically isomorphic to the restriction to $\mathbb{A}^1$ - {critical values of f} of the middle convolution

$$MC_\chi((f_*\overline{\mathbb{Q}}_\ell/\overline{\mathbb{Q}}_\ell)[1])$$

on $\mathbb{A}^1$. For any critical value $f(\alpha)$ of f, there exists a character χ_2 of order 2 and Swan conductor 1 of the inertia group $I(f(\alpha))$, such that the local monodromy of $MC_\chi((f_*\overline{\mathbb{Q}}_\ell/\overline{\mathbb{Q}}_\ell)[1])$ at the critical value $f(\alpha)$ of f is a pseudoreflection of determinant $\chi\chi_2$, viewed as a character of the inertia group $I(f(\alpha))$.

proof Assertion 1) is proven in [Ka-ACT, 5.7].

To prove assertion 2), we argue as follows. We first use [Ka-TLFM, 2.7.1 and 2.7.2] to get U and f with the asserted local monodromies at finite distance of $f_*\overline{\mathbb{Q}}_\ell/\overline{\mathbb{Q}}_\ell$. Fix one such f.

The local monodromy of $f_*\overline{\mathbb{Q}}_\ell/\overline{\mathbb{Q}}_\ell$ at ∞ is as asserted because f is a polynomial of degree e, and e is prime to p.

Notice that $f_*\overline{\mathbb{Q}}_\ell$ is a geometrically semisimple middle extension, and hence that $f_*\overline{\mathbb{Q}}_\ell/\overline{\mathbb{Q}}_\ell$ is a geometrically semisimple middle extension. We now show that $f_*\overline{\mathbb{Q}}_\ell$ has geometric monodromy group Γ_f equal to the full symmetric group S_e. Because f viewed as a finite etale map

$$f : \mathbb{A}^1 - \{\text{critical points of } f\} \to \mathbb{A}^1 - \{\text{crtical values of } f\}$$

has geometrically connected source space, Γ_f is a transitive subgroup of S_e. Because $\mathbb{A}^1\otimes_k\overline{k}$ is tamely simply connected, and $f_*\overline{\mathbb{Q}}_\ell$ is tamely ramified over ∞, Γ_f is generated by all conjugates of all local monodromies at finite distance. As these local monodromies at finite distance are all reflections, Γ_f must be all of S_e (because it is a transitive subgroup of S_e which is generated by reflections).

Thus $f_*\overline{\mathbb{Q}}_\ell/\overline{\mathbb{Q}}_\ell$ is an irreducible (because S_e acts irreducibly in its "deleted permutation representation") middle extension on $\mathbb{A}^1$, of generic rank e-1. Since $e-1 \geq 2$, $(f_*\overline{\mathbb{Q}}_\ell/\overline{\mathbb{Q}}_\ell)[1]$ as irreducible perverse sheaf on $\mathbb{A}^1$ has $\mathcal{P}$, and is of type 2d) in the sense of [Ka-RLS, 3.3.3]. Therefore the middle convolution $MC_\chi((f_*\overline{\mathbb{Q}}_\ell/\overline{\mathbb{Q}}_\ell)[1])$ is itself irreducible and of type 2d) by [Ka-RLS, 3.3.3]. Comparing trace functions, we see that the middle convolution $MC_\chi((f_*\overline{\mathbb{Q}}_\ell/\overline{\mathbb{Q}}_\ell)[1])$ is geometrically isomorphic to the pullback of $\mathcal{M}(1, e, \chi)$ to the one-parameter family $t \mapsto t - f(X)$. We now use [Ka-TLFM, 4.1.10, 1) and 4.2.2] to compute the local monodromies of the middle convolution $MC_\chi((f_*\overline{\mathbb{Q}}_\ell/\overline{\mathbb{Q}}_\ell)[1])$. QED

(5.3.3) In order to apply this result, we must recall some classical group-theoretic results of Mitchell.

Theorem 5.3.4 [Mit] Let Γ be a primitive irreducible finite subgroup of $GL(r, \mathbb{C})$, for some $r \geq 2$, and let γ in Γ be a pseudoreflection. Then we have

$\text{order}(\gamma) \le 5$, if $r = 2$,
$\text{order}(\gamma) \le 3$, if $r = 3$ or 4,
$\text{order}(\gamma) \le 2$, if $r \ge 5$.

proof Recall that for any primitive irreducible finite subgroup Γ of $GL(2, \mathbb{C})$, its image $\overline{\Gamma}$ in $PGL(2, \mathbb{C})$ is isomorphic to one of the three groups A_4, S_4, or A_5. In these groups, no element has order exceeding 5. On the other hand, the order of a pseudoreflection in $GL(2, \mathbb{C})$ is equal to the order of its image in $PGL(2, \mathbb{C})$. Thus for any primitive irreducible finite subgroup $\Gamma$of $GL(2, \mathbb{C})$, any pseudoreflection γ in Γ has order at most 5.

In more variables, the (much deeper) result is due to Mitchell [Mit]. Let Γ be a primitive irreducible finite subgroup of $GL(r, \mathbb{C})$, for some $r \ge 3$, and let γ in Γ be a pseudoreflection. Then we have

$\text{order}(\gamma) \le 3$, if $r = 3$ or 4,
$\text{order}(\gamma) \le 2$, if $r \ge 5$.

For a generalization of Michell's result, see [Wales]. QED

(5.3.5) We now return to the case

$$n = 1,\ \text{order}(\chi) \ge 3,\ e \ge 3 \text{ prime to } p,\ \chi^e \ne \mathbb{1}.$$

(5.3.6) If $p = 2$, then G_{geom} contains a pseudoreflection of order $2\times\text{order}(\chi) \ge 6$. By the previous theorem 5.3.4, G_{geom} cannot be a primitive irreducible finite subgroup of $GL(e-1)$ for any $e \ge 3$.

(5.3.7) Suppose now that p is odd. As χ has $\text{order}(\chi) \ge 3$, $\chi\chi_2$ is never trivial, and never of order 2. We have

$\text{order}(\chi\chi_2) = 3$ if and only if $\text{order}(\chi) = 6$,
$\text{order}(\chi\chi_2) = 4$ if and only if $\text{order}(\chi) = 4$,
$\text{order}(\chi\chi_2) = 5$ if and only if $\text{order}(\chi) = 10$.

By the previous theorem 5.3.4, G_{geom} for $\mathcal{M}(1, e, \chi)|\mathcal{SD}(1, e)$ cannot be a primitive irreducible finite subgroup of $GL(e-1)$ except possibly in the following cases:

$\text{order}(\chi) = 4$, $e = 3$, $p \ge 5$,
$\text{order}(\chi) = 6$, $e = 3$, $p \ge 5$,
$\text{order}(\chi) = 6$, $e = 4$, $p \ge 5$,
$\text{order}(\chi) = 6$, $e = 5$, $p \ge 7$,
$\text{order}(\chi) = 10$, $e = 3$, $p \ge 7$.

(5.3.8) To complete our discussion of the n=1 case, we must show that in each of the cases listed above, G_{geom} for $\mathcal{M}(1, e, \chi)|\mathcal{SD}(1, e)$ is in fact finite. In anticipation of later applications, we also state a result for the highest weight part $Gr^0(\mathcal{M}(1, e, \chi))$, in case χ^e is trivial.

Theorem 5.3.9 The group G_{geom} for $\mathcal{M}(1, e, \chi)|\mathcal{SD}(1, e)$ is finite in each of the following cases:

$\text{order}(\chi) = 4$, $e = 3$, $p \ge 5$,

order(χ) = 6, e = 3, p ≥ 5,
order(χ) = 6, e = 4, p ≥ 5,
order(χ) = 6, e = 5, p ≥ 7,
order(χ) = 10, e = 3, p ≥ 7.

The group G_{geom} for $Gr^0(\mathcal{M}(1, e, \chi))|\mathcal{SD}(1, e)$ is finite in each of the following cases:

order(χ) = 4, e = 4, p ≥ 3,
order(χ) = 6, e = 6, p ≥ 5.

proof We will show this finiteness by a transcendental argument. Let us first consider a more general situation. Fix an integer e ≥ 3, a prime ℓ, and an integer r ≥ 3 (which will be the order of χ).

Denote by $\Phi_r(X)$ in $\mathbb{Z}[X]$ the r'th cyclotomic polynomial, and put

$$\mathbb{Z}[\zeta_r] := \mathbb{Z}[X]/\Phi_r(X).$$

Thus $\mathbb{Z}[\zeta_r]$ is the ring of integers in the field of r'th roots of unity. Let us denote by R the ring

$$R := \mathbb{Z}[\zeta_r][1/\ell er],$$

and by S its spectrum:

$$S := \mathrm{Spec}(R).$$

Over S, we may form the affine space $\mathcal{P}(1, e)_S$ of polynomials of degree ≤ e in one variable. In $\mathcal{P}(1, e)_S$, we have the open set $\mathcal{SD}(1, e)_S$, consisting of those polynomials $\Sigma_{i \le e} a_i X^i$ whose leading coefficient a_e is invertible, and whose discriminant $\Delta(f)$ is invertible. This open set is smooth over S with geometrically connected fibres of dimension e+1. Its fibres over finite-field valued points of S are precisely the spaces $\mathcal{SD}(1, e)$ on which we have been working.

Over the space $\mathcal{SD}(1, e)_S$, we have the affine curve

$$\begin{array}{l} \mathcal{C} \\ \downarrow \pi \\ \mathcal{SD}(1, e)_S \\ \downarrow \\ S. \end{array}$$

whose fibre over a point "f" in $\mathcal{SD}(1, e)_S$ is the curve of equation

$$Y^r = f(X).$$

We compactify $\mathcal{C}/\mathcal{SD}(1, e)_S$ to a proper smooth curve

$$\begin{array}{l} \overline{\mathcal{C}} \\ \downarrow \overline{\pi} \\ \mathcal{SD}(1, e)_S \\ \downarrow \\ S, \end{array}$$

by adding to $\mathcal{C}/\mathcal{SD}(1, e)_S$ a divisor at ∞, D_∞, which is finite etale over $\mathcal{SD}(1, e)_S$ of degree gcd(e, r). The group $\mu_r(R)$ acts on the affine curve $\mathcal{C}/\mathcal{SD}(1, e)_S$, through its action on Y alone:

$$\zeta : (X, Y) \mapsto (X, \zeta Y).$$

This action extends to an action of $\mu_r(R)$ on $\overline{\mathcal{C}}/\mathcal{SD}(1, e)_S$. The group $\mu_r(R)$ acts transitively on each geometric fibre of $D_\infty/\mathcal{SD}(1, e)_S$.

On the space $\mathcal{SD}(1, e)_S$, we have the lisse sheaves

$$R^1\pi_!\overline{\mathbb{Q}}_\ell,\ R^1\overline{\pi}_*\mathbb{Q}_\ell = R^1\overline{\pi}_!\overline{\mathbb{Q}}_\ell,\ R^0(\overline{\pi}|D_\infty)_!\overline{\mathbb{Q}}_\ell,\ R^0\overline{\pi}_!\overline{\mathbb{Q}}_\ell = \overline{\mathbb{Q}}_\ell,$$

which fall into an excision short exact sequence

$$0 \to R^0\overline{\pi}_!\overline{\mathbb{Q}}_\ell = \overline{\mathbb{Q}}_\ell \to R^0(\overline{\pi}|D_\infty)_!\overline{\mathbb{Q}}_\ell \to R^1\pi_!\overline{\mathbb{Q}}_\ell \to R^1\overline{\pi}_*\overline{\mathbb{Q}}_\ell \to 0,$$

which is equivariant for the action of the group $\mu_r(R)$. Using the action of this group, we can decompose each of our sheaves into isotypical components, for the various $\overline{\mathbb{Q}}_\ell^\times$-valued characters χ of $\mu_r(R)$. We put, for any $\overline{\mathbb{Q}}_\ell$-sheaf $\mathcal{G}$ on which $\mu_r(R)$ operates,

$$\mathcal{G}(\chi) := (\mathcal{G} \otimes \chi)^{\mu_r(R)}.$$

Thus $\mathcal{G}(\chi)$ is the χ^{-1}-isotypical component of $\mathcal{G}$.

Fix a $\overline{\mathbb{Q}}_\ell^\times$-valued character χ of $\mu_r(R)$ which has $\mathrm{order}(\chi) = r$. Because $\mathrm{Gal}(\mathbb{Q}(\zeta_r)/\mathbb{Q})$ acts transitively on the primitive r'th roots of unity, there is a unique embedding $R \subset \overline{\mathbb{Q}}_\ell$ which carries any ζ in $\mu_r(R)$ to $\chi(\zeta)$. The group $\mu_r(R)$ operates transitively on each geometric fibre of $D_\infty/\mathcal{SD}(1, e)_S$, and each geometric fibre has cardinality $\gcd(e, r)$. So if r does not divide e, then we have

$$(R^0(\overline{\pi}|D_\infty)_!\overline{\mathbb{Q}}_\ell)(\chi) = 0,$$

and hence an isomorphism of lisse sheaves on $\mathcal{SD}(1, e)_S$,

$$(R^1\pi_!\overline{\mathbb{Q}}_\ell)(\chi) \cong (R^1\overline{\pi}_*\overline{\mathbb{Q}}_\ell)(\chi).$$

However, if r divides e, then

$$(R^0(\overline{\pi}|D_\infty)_!\overline{\mathbb{Q}}_\ell)(\chi) \text{ has rank one,}$$

and we have a short exact sequence of lisse sheaves on $\mathcal{SD}(1, e)_S$,

$$0 \to (R^0(\overline{\pi}|D_\infty)_!\overline{\mathbb{Q}}_\ell)(\chi) \to (R^1\pi_!\overline{\mathbb{Q}}_\ell)(\chi) \to (R^1\overline{\pi}_*\overline{\mathbb{Q}}_\ell)(\chi) \to 0.$$

For each finite field k and each ring homomorphism $\varphi : R \to k$, the induced map of r'th roots of unity is an isomorphism $\mu_r(R) \cong \mu_r(k)$. We have a canonical surjective homomorphism

$$k^\times \to \mu_r(k),$$
$$t \mapsto t^{(\#k - 1)/r}.$$

So for each finite field k and each ring homomorphism $\varphi : R \to k$, we obtain from χ a character χ_k of $k^\times$, defined as the composite

$$k^\times \to \mu_r(k) \cong \mu_r(R) \to \overline{\mathbb{Q}}_\ell^\times.$$

For each finite field k and each ring homomorphism $\varphi : R \to k$, the restriction of $(R^1\pi_!\overline{\mathbb{Q}}_\ell)(\chi)$ to the fibre $\mathcal{SD}(1, e)_k$ of $\mathcal{SD}(1, e)_S/S$ at φ is the sheaf $\mathcal{M}(1, e, \chi_k)(-1/2)|\mathcal{SD}(1, e)_k$:

$$(R^1\pi_*\overline{\mathbb{Q}}_\ell)(\chi)|\mathcal{SD}(1, e)_k \cong \mathcal{M}(1, e, \chi_k)(-1/2)|\mathcal{SD}(1, e)_k,$$

and we have

$$(R^1\overline{\pi}_*\overline{\mathbb{Q}}_\ell)(\chi)|\mathcal{SD}(1, e)_k \cong \mathrm{Gr}^0(\mathcal{M}(1, e, \chi_k))(-1/2)|\mathcal{SD}(1, e)_k.$$

Let us pick an embedding of the ring R into $\mathbb{C}$. We now apply to the lisse sheaf $(R^1\overline{\pi}_!\overline{\mathbb{Q}}_\ell)(\chi)$ on $\mathcal{SD}(1, e)_S$ the general specialization theorem [Ka-ESDE, 8.18.2], according to which the image of $\pi_1{}^{geom}(\mathcal{SD}(1, e)_k)$ in the ℓ-adic representation given by

$$(R^1\overline{\pi}_*\overline{\mathbb{Q}}_\ell)(\chi)|\mathcal{SD}(1, e)_k$$

is (conjugate in $GL(e-1, \overline{\mathbb{Q}}_\ell)$ to) a subgroup of the image of $\pi_1{}^{geom}(\mathcal{SD}(1, e)_{\mathbb{C}})$ in the ℓ-adic representation given by

$$(R^1\overline{\pi}_*\overline{\mathbb{Q}}_\ell)(\chi)|\mathcal{SD}(1, e)_{\mathbb{C}}$$

The upshot of all this is that, in order to prove that the lisse sheaf

$$(R^1\overline{\pi}_*\overline{\mathbb{Q}}_\ell)(\chi)|\mathcal{SD}(1, e)_k \cong \mathrm{Gr}^0(\mathcal{M}(1, e, \chi_k))(-1/2)|\mathcal{SD}(1, e)_k$$

has finite geometric monodromy on $\mathcal{SD}(1, e)_k$ for every finite field k, and for every ring homomorphism $\varphi : R \to k$, it suffices to show that

$$(R^1\overline{\pi}_*\overline{\mathbb{Q}}_\ell)(\chi)|\mathcal{SD}(1, e)_{\mathbb{C}}$$

has finite geometric monodromy on $\mathcal{SD}(1, e)_{\mathbb{C}}$.

By the fundamental comparison theorem relating ℓ-adic and topological cohomology for complex varieties [SGA 4, XVI, 4.1], we know that

$$(R^1\overline{\pi}_*\overline{\mathbb{Q}}_\ell)(\chi)|(\mathcal{SD}(1, e)_{\mathbb{C}})^{an} \cong ((R^1\overline{\pi}^{an}{}_*R)(\chi))\otimes_R\overline{\mathbb{Q}}_\ell,$$

where $\overline{\mathbb{Q}}_\ell$ is an R-algebra by the embedding $R \subset \overline{\mathbb{Q}}_\ell$ corresponding to the character χ of order r.

So to show that

$$(R^1\overline{\pi}_*\overline{\mathbb{Q}}_\ell)(\chi)|\mathcal{SD}(1, e)_{\mathbb{C}}$$

has finite geometric monodromy on $\mathcal{SD}(1, e)_{\mathbb{C}}$, it suffices to show that $(R^1\overline{\pi}^{an}{}_*R)(\chi)$ has finite monodromy on $(\mathcal{SD}(1, e)_{\mathbb{C}})^{an}$, or, equivalently, to show that $(R^1\overline{\pi}^{an}{}_*\mathbb{C})(\chi)$ has finite monodromy on $(\mathcal{SD}(1, e)_{\mathbb{C}})^{an}$.

Because $\overline{\pi}$ is a proper smooth curve, the local system $R^1\overline{\pi}^{an}{}_*\mathbb{Z}$ on $(\mathcal{SD}(1, e)_{\mathbb{C}})^{an}$ underlies a polarized family of Hodge structures. The finite cyclic group $\mu_r(R)$ acts on this polarized family. Notice that the $\mathbb{C}^\times$-valued characters of $\mu_r(R)$ of order r are all $\mathrm{Aut}(\mathbb{C})$-conjugate. According to [Ka-ASDE, 4.4.2], the following two conditions are equivalent.

1) For every $\mathbb{C}^\times$-valued character of $\mu_r(R)$ which has order r (all of

which are Aut($\mathbb{C}$)-conjugate), $(R^1\overline{\pi}^{an}{}_*\mathbb{C})(\chi)$ has finite monodromy on $(\mathcal{S}\mathcal{D}(1, e)_{\mathbb{C}})^{an}$.

2) For every $\mathbb{C}^\times$-valued character of $\mu_r(R)$ which has order r, the Hodge filtration F induced on $H_{DR}{}^1(\overline{\mathcal{C}}_{\mathbb{C}}/\mathcal{S}\mathcal{D}(1, e)_{\mathbb{C}})(\chi)$ is horizontal for the Gauss-Manin connection.

In the case at hand, a proper smooth family of curves, the Hodge filtration has $\mathrm{Fil}^2 = 0$, $\mathrm{Fil}^0 = $ all, so we can restate 2) as

2bis) For every $\mathbb{C}^\times$-valued character of $\mu_r(R)$ which has order r, the intersection

$$H_{DR}{}^1(\overline{\mathcal{C}}_{\mathbb{C}}/\mathcal{S}\mathcal{D}(1, e)_{\mathbb{C}})(\chi) \cap \mathrm{Fil}^1 H_{DR}{}^1(\overline{\mathcal{C}}_{\mathbb{C}}/\mathcal{S}\mathcal{D}(1, e)_{\mathbb{C}})$$

(taken inside $H_{DR}{}^1(\overline{\mathcal{C}}_{\mathbb{C}}/\mathcal{S}\mathcal{D}(1, e)_{\mathbb{C}})$) is stable under the Gauss-Manin connection on $H_{DR}{}^1(\overline{\mathcal{C}}_{\mathbb{C}}/\mathcal{S}\mathcal{D}(1, e)_{\mathbb{C}})$.

This stability can be checked over the generic point of the base, so we may reformulate 2bis) as

2ter) Denote by K the function field of $\mathcal{S}\mathcal{D}(1, e)_{\mathbb{C}}$, and by $\overline{\mathcal{C}}_K/K$ the generic fibre of $\overline{\mathcal{C}}_{\mathbb{C}}/\mathcal{S}\mathcal{D}(1, e)_{\mathbb{C}}$. For every $\mathbb{C}^\times$-valued character of $\mu_r(R)$ which has order r, the intersection

$$H_{DR}{}^1(\overline{\mathcal{C}}_K/K)(\chi) \cap H^0(\overline{\mathcal{C}}_K, \Omega^1{}_{\overline{\mathcal{C}}_K/K})$$

(taken inside $H_{DR}{}^1(\overline{\mathcal{C}}_K/K)$) is stable under the Gauss-Manin connection on $H_{DR}{}^1(\overline{\mathcal{C}}_K/K)$.

By functoriality, each subspace $H_{DR}{}^1(\overline{\mathcal{C}}_K/K)(\chi)$ is stable under the Gauss-Manin connection on $H_{DR}{}^1(\overline{\mathcal{C}}_K/K)$. So a sufficient condition for 2ter) to hold is the following rather draconian "all or nothing" condition:

(all or nothing condition) For every $\mathbb{C}^\times$-valued character of $\mu_r(R)$ which has order r, the intersection

$$H_{DR}{}^1(\overline{\mathcal{C}}_K/K)(\chi) \cap H^0(\overline{\mathcal{C}}_K, \Omega^1{}_{\overline{\mathcal{C}}_K/K})$$

(taken inside $H_{DR}{}^1(\overline{\mathcal{C}}_K/K)$) is either zero, or it is the whole space $H_{DR}{}^1(\overline{\mathcal{C}}_K/K)(\chi)$.

Since $\mathrm{Gr}_0(\mathcal{M}(1, e, \chi))$ has rank

$$e-1, \text{ if } \chi^e \neq \mathbb{1},$$
$$e-2 \text{ if } \chi^e = \mathbb{1},$$

the K-space $H_{DR}{}^1(\overline{C}_K/K)(\chi)$ has dimension

$e-1$, if $\chi^e \neq 1\!\!1$,

$e-2$ if $\chi^e = 1\!\!1$.

So the all or nothing condition can be rephrased in terms of dimensions of intersections:

(all or nothing condition bis) For every $\mathbb{C}^\times$-valued character of $\mu_r(R)$ which has order r, the dimension intersection

$$\dim_K(H_{DR}{}^1(\overline{C}_K/K)(\chi) \cap H^0(\overline{C}_K, \Omega^1{}_{\overline{C}_K/K}))$$

is

if $\chi^e \neq 1\!\!1$, either 0 or $e-1$.

if $\chi^e = 1\!\!1$, either 0 or $e-2$.

We will now check explicitly that this dimensional version of the all or nothing condition is satisfied in each of the listed cases:

$r = \mathrm{order}(\chi) = 4$, $e = 3$,
$r = \mathrm{order}(\chi) = 6$, $e = 3$,
$r = \mathrm{order}(\chi) = 6$, $e = 4$,
$r = \mathrm{order}(\chi) = 6$, $e = 5$,
$r = \mathrm{order}(\chi) = 10$, $e = 3$,
$r = \mathrm{order}(\chi) = 4$, $e = 4$,
$r = \mathrm{order}(\chi) = 6$, $e = 6$.

Case (r=4, e = 3) We are looking at the complete nonsingular model of the curve $y^4 = f_3(x)$, f_3 a complex cubic with no repeated roots. The genus is 3, and a basis of the space of differentials of the first kind is given by

$$dx/y^3,\ xdx/y^3,\ dx/y^2.$$

The first two transform by the same character of order 4, the last by a character of order 2. So for the two characters of order 4, one intersection has dimension 2, the other has dimension 0.

Case (r=6, e = 3) We are looking at the complete nonsingular model of the curve $y^6 = f_3(x)$, f_3 a complex cubic with no repeated roots. The genus is 4, and a basis of the space of differentials of the first kind is given by

$$dx/y^5,\ xdx/y^5,\ dx/y^4,\ dx/y^3.$$

The first two transform by the same character of order 6, the third by a character of order 3, the last by a character of order 2. So for the two characters of order 6, one intersection has dimension 2, the other has dimension 0.

Case (r=6, e = 4) We are looking at the complete nonsingular model of the curve $y^6 = f_4(x)$, f_4 a complex quartic with no repeated roots. The genus is 7, and a basis of the space of differentials of the first kind is given by

$$dx/y^5,\ xdx/y^5,\ x^2dx/y^5,\ xdx/y^4,\ dx/y^4,\ dx/y^3,\ dx/y^2.$$

The first three transform by the same character of order 6, the fourth and fifth by the same character of order 3, the sixth by a character of order 2, and the last by the other character of order 3. So for the two characters of order 6, one intersection has dimension 3, the other has dimension 0.

Case (r=6, e = 5) We are looking at the complete nonsingular model of the curve $y^6 = f_5(x)$, f_5 a complex quintic with no repeated roots. The genus is 10, and a basis of the space of differentials of the first kind is given by

$$dx/y^5,\ xdx/y^5,\ x^2dx/y^5,\ x^3dx/y^5,\ dx/y^4,\ xdx/y^4,\ x^2dx/y^4,$$
$$dx/y^3,\ xdx/y^3,\ dx/y^2.$$

The first four transform by the same character of order 6, the fifth, sixth, and seventh by the same character of order 3, the eighth and ninth by a character of order 2, and the last by the other character of order 3. So for the two characters of order 6, one intersection has dimension 4, the other has dimension 0.

Case (r=10, e = 3) We are looking at the complete nonsingular model of the curve $y^{10} = f_3(x)$, f_3 a complex cubic with no repeated roots. The genus is 9, and a basis of the space of differentials of the first kind is given by

$$dx/y^9,\ xdx/y^9,\ dx/y^8,\ xdx/y^8,\ dx/y^7,\ xdx/y^7,\ dx/y^6,$$
$$xdx/y^5,\ xdx/y^4.$$

The first two transform by the same character of order 10, the next two by a character of order 5, the next two by second character of order 10, the remaining three by characters of order 5, 2, and 5 respectively. So for the four characters of order 10, two intersections have dimension 2, and the other two have dimension 0.

Case (r=6, e = 5) We are looking at the complete nonsingular model of the curve $y^6 = f_5(x)$, f_5 a complex quintic with no repeated roots. The genus is 10, and a basis of the space of differentials of the first kind is given by

$$dx/y^5,\ xdx/y^5,\ x^2dx/y^5,\ x^3dx/y^5,\ dx/y^4,\ xdx/y^4,\ x^2dx/y^4,$$
$$dx/y^3,\ xdx/y^3,\ dx/y^2.$$

The first four transform by the same character of order 6, the fifth, sixth, and seventh by the same character of order 3, the eighth and ninth by the character of order 2, and the last by the other character of order 3. So for the two characters of order 6, one intersection has dimension 4, the other has dimension 0.

Case (r=4, e = 4) We are looking at the complete nonsingular model of the curve $y^4 = f_4(x)$, f_4 a complex quartic with no repeated roots. The genus is 3, and a basis of the space of differentials of the first kind is given by

$$dx/y^3,\ xdx/y^3,\ dx/y^2.$$

The first two transform by the same character of order 4, the last by a character of order 2. So for the two characters of order 4, one intersection has dimension 2, the other has dimension 0.

Case (r=6, e = 6) We are looking at the complete nonsingular model of the curve $y^6 = f_6(x)$, f_6 a complex sextic with no repeated roots. The genus is 10, and a basis of the space of differentials of the first kind is given by

$$dx/y^5,\ xdx/y^5,\ x^2dx/y^5,\ x^3dx/y^5,\ dx/y^4,\ xdx/y^4,\ x^2dx/y^4,$$
$$dx/y^3,\ xdx/y^3,\ dx/y^2.$$

The first four transform by the same character of order 6, the fifth, sixth, and seventh by the same character of order 3, the eighth and ninth by the character of order 2, and the last by the other character of order 3. So for the two characters of order 6, one intersection has dimension 4, the other has dimension 0. QED

(5.4) Continuation of the proof of Theorem 5.2.2 for general n

(5.4.1) We now pass from the n=1 case to the general case, by an induction on n. The induction is based on the observation that, for any integer $e \geq 1$, if k is a field in which e is invertible, and if an n-variable polynomial

$$f(x_1,..., x_n)$$

lies in $\mathcal{SD}(n, e)(k)$, then the polynomial in n+1 variables

$$f(x_1,..., x_n) - y^n$$

lies in $\mathcal{SD}(n+1, e)(k)$. Let us denote by

$$\text{AddedVar} : \mathcal{SD}(n, e) \to \mathcal{SD}(n+1, e)$$
$$f(x_1,..., x_n) \mapsto f(x_1,..., x_n) - y^e$$

this closed immersion.

Theorem 5.4.2 Let $n \geq 1$ and $e \geq 3$ be integers. Let k be a finite field of characteristic p, in which e is invertible, and which contains the e'th roots of unity. Let ℓ be a prime with $\ell \neq p$, and let

$$\chi : k^\times \to \overline{\mathbb{Q}}_\ell^\times$$

be a nontrivial multiplicative character. Suppose that $\chi^e \neq \mathbb{1}$. For each nontrivial multiplicative character ρ of $k^\times$ with $\rho^e = \mathbb{1}$, consider the Jacobi sum

$$J(\chi, \rho) := \Sigma_{u \text{ in } k^\times} \chi(1-u)\rho(u).$$

1) We have an isomorphism of lisse sheaves on $\mathcal{SD}(n, e)_k$

$$\text{AddedVar}^*\mathcal{M}(n+1, e, \chi)$$
$$\cong \oplus_{\rho \neq \mathbb{1},\ \rho^e = \mathbb{1}}\ \mathcal{M}(n, e, \chi\rho)(1/2) \otimes (-J(\chi,\rho))^{\deg}.$$

2) We have an isomorphism of lisse sheaves on $\mathcal{SD}(n, e)_{\overline{k}}$

$$\text{AddedVar}^*\mathcal{M}(n+1, e, \chi) \cong \oplus_{\rho \neq \mathbb{1},\ \rho^e = \mathbb{1}}\ \mathcal{M}(n, e, \chi\rho).$$

proof The second assertion is immediate from the first, since the constant field twists (1/2) and $(-J(\chi,\rho))^{\deg}$ are geometrically trivial. To prove the first assertion, the idea is to use the action of $\mu_e(k)$ on the pullback sheaf AddedVar*$\mathcal{M}(n+1, e, \chi)$ to break it into the named pieces.

The sheaf AddedVar*$\mathcal{M}(n+1, e, \chi)(-(n+1)/2)$ can be described as follows. On the product space

$$\mathbb{A}^{n+1}\times\mathcal{SD}(n, e) = \mathbb{A}^n\times\mathbb{A}^1\times\mathcal{SD}(n, e)$$

with coordinates $(x = (x_1,..., x_n), y, f)$, we have the sheaf

$$\mathcal{L}_{\chi(f(x) - y^e)},$$

which is lisse of rank one on the open set where $f(x) - y^e$ is invertible, and which vanishes outside this open set. From the definition of $\mathcal{M}(n+1, e, \chi)$ and proper base change, we see that

$$\text{AddedVar}^*\mathcal{M}(n+1, e, \chi)(-(n+1)/2) = R^{n+1}pr_{2!}\mathcal{L}_{\chi(f(v) - y^e)},$$

and

$$R^ipr_{2!}\mathcal{L}_{\chi(f(v) - y^e)} = 0 \text{ for } i \neq n+1.$$

The coefficient sheaf $\mathcal{L}_{\chi(f(x) - y^e)}$ is itself the pullback of the sheaf

$$\mathcal{L}_{\chi(f(x) - y)}$$

on the space $\mathbb{A}^n\times\mathbb{A}^1\times\mathcal{SD}(n, e)$ by the finite flat endomorphism

$$[e]: \mathbb{A}^n\times\mathbb{A}^1\times\mathcal{SD}(n, e) \to \mathbb{A}^n\times\mathbb{A}^1\times\mathcal{SD}(n, e)$$

given by

$$(x, y, f) \mapsto (x, y^e, f).$$

Thus we have

$$\begin{aligned}&\text{AddedVar}^*\mathcal{M}(n+1, e, \chi)(-(n+1)/2)\\ &= R^{n+1}pr_{2!}\mathcal{L}_{\chi(f(v) - y^e)}\\ &= R^{n+1}pr_{2!}[e]^*\mathcal{L}_{\chi(f(v) - y)}.\end{aligned}$$

Now use the fact that

$$pr_2 = pr_2\circ[e]$$

and the projection formula for [e] to rewrite

$$\begin{aligned}&R^{n+1}pr_{2!}[e]^*\mathcal{L}_{\chi(f(v) - y)}\\ &= R^{n+1}pr_{2!}[e]_*[e]^*\mathcal{L}_{\chi(f(v) - y)}\\ &= R^{n+1}pr_{2!}(\mathcal{L}_{\chi(f(v) - y)}\otimes[e]_*\overline{\mathbb{Q}}_\ell)\\ &= R^{n+1}pr_{2!}(\mathcal{L}_{\chi(f(v) - y)}\otimes\oplus_{\rho^e=\mathbb{1}}\mathcal{L}_{\rho(y)})\\ &= \oplus_{\rho^e=\mathbb{1}} R^{n+1}pr_{2!}(\mathcal{L}_{\chi(f(v) - y)}\otimes\mathcal{L}_{\rho(y)}).\end{aligned}$$

This direct sum decomposition shows that each term

$$R^{n+1}pr_{2!}(\mathcal{L}_{\chi(f(v) - y)}\otimes\mathcal{L}_{\rho(y)})$$

is lisse on $\mathcal{SD}(n, e)$, being a direct factor of a lisse sheaf.

The same calculation, applied with $i \neq n+1$, gives

$$\oplus_{\rho^e=\mathbb{1}} R^ipr_{2!}(\mathcal{L}_{\chi(f(v) - y)}\otimes\mathcal{L}_{\rho(y)}) = 0, \text{ for } i \neq n+1.$$

This vanishing allows us to compute the trace function of each individual sheaf

$$R^{n+1}pr_{2!}(\mathcal{L}_{\chi(f(v)-y)}\otimes\mathcal{L}_{\rho(y)})$$

as a character sum. It is this calculation that will be essential for the rest of the argument.

What we claim is that, for $\rho = \mathbb{1}$, we have

$$R^{n+1}pr_{2!}(\mathcal{L}_{\chi(f(v)-y)}) = 0,$$

while for $\rho \neq \mathbb{1}$, but $\rho^e = \mathbb{1}$, we have

$$R^{n+1}pr_{2!}(\mathcal{L}_{\chi(f(v)-y)}\otimes\mathcal{L}_{\rho(y)}) \cong \mathcal{M}(n, e, \chi\rho)(-n/2)\otimes(-J(\chi,\rho))^{deg}.$$

If this claim is correct, then we get

$$\begin{aligned}&\mathrm{AddedVar}^{*}\mathcal{M}(n+1, e, \chi)(-(n+1)/2)\\ &= \oplus_{\rho^e=\mathbb{1}} R^{n+1}pr_{2!}(\mathcal{L}_{\chi(f(v)-y)}\otimes\mathcal{L}_{\rho(y)})\\ &\cong \oplus_{\rho\neq\mathbb{1},\ \rho^e=\mathbb{1}} \mathcal{M}(n, e, \chi\rho)(-n/2)\otimes(-J(\chi,\rho))^{deg},\end{aligned}$$

which after untwisting gives the asserted isomorphism

$$\begin{aligned}&\mathrm{AddedVar}^{*}\mathcal{M}(n+1, e, \chi)\\ &\cong \oplus_{\rho\neq\mathbb{1},\ \rho^e=\mathbb{1}} \mathcal{M}(n, e, \chi\rho)(1/2)\otimes(-J(\chi,\rho))^{deg}.\end{aligned}$$

It remains to show that for $\rho = \mathbb{1}$, we have

$$R^{n+1}pr_{2!}(\mathcal{L}_{\chi(f(v)-y)}) = 0,$$

while for $\rho \neq \mathbb{1}$, but $\rho^e = \mathbb{1}$, we have

$$R^{n+1}pr_{2!}(\mathcal{L}_{\chi(f(v)-y)}\otimes\mathcal{L}_{\rho(y)}) \cong \mathcal{M}(n, e, \chi\rho)(-n/2)\otimes(-J(\chi,\rho))^{deg}.$$

To prove this, we argue as follows. For each ρ, we are given two lisse sheaves on a smooth, geometrically connected scheme over a finite field, the second of which is either 0 (when ρ is trivial) or is geometrically (and hence a fortiori arithmetically) irreducible (when ρ is nontrivial). It suffices, in such a situation, to show that the two sheaves in question have the same trace function. For then by Chebotarev, the two sheaves will have isomorphic semisimplifications (as representations of π_1^{arith}). But the second of them is either zero or arithmetically irreducible. By Chebotarev, the first has semisimplification either 0 or arithmetically irreducible, and hence the first is its own semisimplification, so Chebotarev provides the desired isomorphism.

It remains to show that the trace functions match. Take a finite extension E/k, and an E-valued point f in $\mathcal{SD}(n, e)(E)$. From the vanishing of the R^i with $i \neq n+1$ and the Lefschetz trace formula, we get

$$(-1)^{n+1}\mathrm{Trace}(\mathrm{Frob}_{E,f} \mid R^{n+1}pr_{2!}(\mathcal{L}_{\chi(f(v) - y)}\otimes\mathcal{L}_{\rho(y)}))$$
$$= \Sigma_{(x, y) \text{ in } \mathbb{A}^{n+1}(E)} \chi_E(f(x) - y)\rho_E(y).$$

Suppose first $\rho = \mathbb{1}$. Then we make the change of variable

$$(x, y) \mapsto (x, y + f(x))$$

to see that the sum vanishes:

$$\Sigma_{(x, y) \text{ in } \mathbb{A}^{n+1}(E)} \chi_E(f(x) - y) = \Sigma_{(x, y) \text{ in } \mathbb{A}^{n+1}(E)} \chi_E(-y)$$
$$= (\#E)^n\Sigma_{y \text{ in } E} \chi_E(-y) = 0.$$

Suppose next that $\rho \neq \mathbb{1}$. Because $\rho^e \neq \mathbb{1}$, while $\chi^e = \mathbb{1}$, we know that $\chi\rho$ is nontrivial. By the Hasse-Davenport relations [Dav-Has], we know that the Jacobi sums over k and over E are related by

$$-J(\chi_E, \rho_E) = (-J(\chi, \rho))^{\deg(E/k)}.$$

The trace identity we need is

$$\Sigma_{(x, y) \text{ in } \mathbb{A}^{n+1}(E)} \chi_E(f(x) - y)\rho_E(y)$$
$$= J(\chi_E, \rho_E)\Sigma_{x \text{ in } \mathbb{A}^n(E)} (\chi\rho)_E(f(x)).$$

In the first sum, it suffices to sum over those (x, y) with $f(x) \neq 0$. For if $f(x_0) = 0$, the sum over all (x_0, y) vanishes:

$$\Sigma_{y \text{ in } E} \chi_E(-y)\rho_E(y) = \chi_E(-1)\Sigma_{y \text{ in } E} (\chi\rho)_E(y) = 0,$$

simply because $\chi\rho$ is nontrivial. So the first sum is

$$\Sigma_{(x, y) \text{ in } \mathbb{A}^{n+1}(E) \text{ with } f(x) \neq 0} \chi_E(f(x) - y)\rho_E(y).$$

For each fixed x with $f(x) \neq 0$, make the change of variable

$$y \mapsto f(x)y.$$

Then the first sum becomes

$$\Sigma_{(x, y) \text{ in } \mathbb{A}^{n+1}(E) \text{ with } f(x) \neq 0} \chi_E(f(x) - f(x)y)\rho_E(f(x)y)$$
$$= \Sigma_{(x, y) \text{ in } \mathbb{A}^{n+1}(E) \text{ with } f(x) \neq 0} (\chi\rho)_E(f(x))\chi_E(1 - y)\rho_E(y)$$
$$= J(\chi_E, \rho_E)\Sigma_{x \text{ in } \mathbb{A}^n(E)} (\chi\rho)_E(f(x)).$$ QED

(5.4.3) With this result available, we can easily pass from the n=1 case to the general case

$$n \geq 2,\ e \geq 3 \text{ prime to } p,\ \chi^e \neq \mathbb{1}.$$

The idea is extremely simple. Suppose $\mathcal{M}(n+1, e, \chi)$ has finite G_{geom} on $\mathcal{SD}(n+1, e)$. Then certainly AddedVar$^*\mathcal{M}(n+1, e, \chi)$ has finite G_{geom} on $\mathcal{SD}(n, e)$. By the previous theorem, $\mathcal{M}(n, e, \chi\rho)$ has finite G_{geom} on $\mathcal{SD}(n, e)$, for **every** nontrivial ρ of order dividing e. If we already know that there is **some** nontrivial ρ of order dividing e such that $\mathcal{M}(n, e, \chi\rho)$ does not have finite G_{geom} on $\mathcal{SD}(n, e)$, we have a contradiction: we find that in fact $\mathcal{M}(n+1, e, \chi)$ does not have finite G_{geom} on $\mathcal{SD}(n+1, e)$, which, thanks to Larsen's Alternative, is all we need prove.

(5.4.4) Let us first treat the case n = 2, since here we will find one last exceptional case. Fix a value of $e \geq 3$ prime to p, and χ with

$\chi^e \neq \mathbb{1}$. We have already proven that $\mathcal{M}(1, e, \Lambda)$ has infinite G_{geom} on $\mathcal{SD}(1, e)$ except in five exceptional cases:

order(Λ) = 4, e = 3, p ≥ 5,
order(Λ) = 6, e = 3, p ≥ 5,
order(Λ) = 6, e = 4, p ≥ 5,
order(Λ) = 6, e = 5, p ≥ 7,
order(Λ) = 10, e = 3, p ≥ 7.

(5.4.5) If e ≥ 6, then $\mathcal{M}(1, e, \chi\rho)$ has infinite G_{geom}, because all the n=1 exceptional cases have e ≤ 5. So $\mathcal{M}(2, e, \chi)$ has infinite G_{geom} in the case e ≥ 6.

(5.4.6) If e = 5, then $\mathcal{M}(1, 5, \chi\rho)$ has infinite monodromy unless $\chi\rho$ has order 6. But remember ρ is nontrivial of order e = 5. So if $\chi\rho$ has order 6, then $\chi\rho^2$ has order 30, and so $\mathcal{M}(1, 5, \chi\rho^2)$ has infinite G_{geom}. So one of the two sheaves $\mathcal{M}(1, 5, \chi\rho)$ or $\mathcal{M}(1, 5, \chi\rho^2)$ has infinite G_{geom}, and hence $\mathcal{M}(2, 5, \chi)$ has infinite G_{geom}.

(5.4.7) If e = 4, then $\mathcal{M}(1, 4, \chi\rho)$ has infinite monodromy unless $\chi\rho$ has order 6. If $\chi\rho$ has order 6, and ρ has order 4, then $\chi\rho^3$ has order 3, and so $\mathcal{M}(1, 4, \chi\rho^3)$ has infinite G_{geom}. If $\chi\rho$ has order 6 and ρ has order 2, then χ has order 3, and $\chi\Lambda$ will have order 12 for either Λ of order 4, in which case $\mathcal{M}(1, 4, \chi\Lambda)$ will have infinite G_{geom}. So we conclude that $\mathcal{M}(2, 4, \chi)$ has infinite G_{geom}.

(5.4.8) If e = 3, then $\mathcal{M}(1, 3, \chi\rho)$ has infinite monodromy unless $\chi\rho$ has order 4, 6, or 10. Remember ρ has order 3. If $\chi\rho$ has order 4, then $\chi\rho^2$ has order 12, $\mathcal{M}(1, 3, \chi\rho^2)$ has infinite G_{geom}, and hence $\mathcal{M}(2, 3, \chi)$ has infinite G_{geom}. If $\chi\rho$ has order 10, then $\chi\rho^2$ has order 30, $\mathcal{M}(1, 3, \chi\rho^2)$ has infinite G_{geom}, and hence $\mathcal{M}(2, 3, \chi)$ has infinite G_{geom}. So the only possible exceptional case is when $\chi\rho$ has order 6. This happens precisely when χ has order 2, and gives us the remaining exceptional case.

Theorem 5.4.9 In characteristic p ≥ 5, with χ_2 denoting the quadratic character, $\mathcal{M}(2, 3, \chi_2)$ has finite G_{geom} on $\mathcal{SD}(2, 3)$.

proof Because we are dealing with the quadratic character, for any n ≥ 1, any odd e ≥ 3 prime to p, and for any f(x's) in $\mathcal{SD}(n, e)(k)$, the stalk of $\mathcal{M}(n, e, \chi_2)(-n/2)$ is

$$H_c^n(\mathbb{A}^n\otimes_k\bar{k}, \mathcal{L}_{\chi_2(f(x's))})$$

$$= H_c^n((\text{the affine hypersurface } X_f : z^2 = f(x's))\otimes_k\bar{k}, \overline{\mathbb{Q}}_\ell),$$

and this cohomology group is pure of weight n.

Since both 2 and e are prime to p, the affine hypersurface X_f is the quotient of the affine hypersurface

$$Y_f : (w^e)^2 = f((y_1)^2,..., (y_n)^2)$$

by the finite group $\Gamma := \mu_e \times \mu_2 \times ... \times \mu_2$, acting by

$$(\zeta, \varepsilon_1,..., \varepsilon_n) : (w, y_1,..., y_n) \mapsto (\zeta W, \varepsilon_1 y_1,..., \varepsilon_n y_n).$$

The affine hypersurface Y_f is the finite part of a projective smooth hypersurface Z_f in $\mathbb{P}^{n+1}$ of degree 2e (simply homogenize the equation of Y_f) whose hyperplane section at infinity, D_f, is itself smooth. Explicitly, let

$$F(X_0,..., X_n) := (X_0)^e f(X_1/X_0,..., X_n/X_0)$$

be the homogeneous form of degree e in n+1 homogeneous variables corresponding to f. Then Z_f has homogeneous equation

$$(W^e)^2 = F((Y_0)^2, (Y_1)^2,..., (Y_n)^2),$$

and D_f is the hyperplane section $Y_0 = 0$.

The group Γ acts on Z_f, extending its action on Y_f and respecting D_f. The excision sequence for

$$Y_f \subset Z_f \supset D_f,$$

$$...H^{n-1}(D_f \otimes_k \bar{k}, \bar{\mathbb{Q}}_\ell) \to H_c^n(Y_f \otimes_k \bar{k}, \bar{\mathbb{Q}}_\ell) \to H^n(Z_f \otimes_k \bar{k}, \bar{\mathbb{Q}}_\ell)$$
$$\to H^n(D_f \otimes_k \bar{k}, \bar{\mathbb{Q}}_\ell)...$$

shows that, for the weight filtration, we have

$$\mathrm{Gr}_W{}^n(H_c{}^n(Y_f \otimes_k \bar{k}, \bar{\mathbb{Q}}_\ell)) \cong \mathrm{Prim}^n(Z_f \otimes_k \bar{k}, \bar{\mathbb{Q}}_\ell).$$

Taking invariants under Γ, we find

$$\mathrm{Gr}_W{}^n(H_c{}^n(X_f \otimes_k \bar{k}, \bar{\mathbb{Q}}_\ell)) \cong (\mathrm{Prim}^n(Z_f \otimes_k \bar{k}, \bar{\mathbb{Q}}_\ell))^\Gamma.$$

But we know that $H_c{}^n(X_f \otimes_k \bar{k}, \bar{\mathbb{Q}}_\ell)$ is pure of weight n, so we find

$$H_c{}^n(X_f \otimes_k \bar{k}, \bar{\mathbb{Q}}_\ell) \cong (\mathrm{Prim}^n(Z_f \otimes_k \bar{k}, \bar{\mathbb{Q}}_\ell))^\Gamma.$$

Now work over the ground ring $R := \mathbb{Z}[\zeta_e][1/2e\ell]$, with the parameter space $\mathcal{S}\mathcal{D}(n, e)_R$. As f varies over $\mathcal{S}\mathcal{D}(n, e)_R$, we get a projective smooth hypersurface $Z_{\mathrm{univ}}/\mathcal{S}\mathcal{D}(n, e)_R$ in $\mathbb{P}^{n+1} \times \mathcal{S}\mathcal{D}(n, e)_R$, with an action of the finite group Γ, and an isomorphism of lisse sheaves on $\mathcal{S}\mathcal{D}(n, e)_R$

$$\mathcal{M}(n, e, \chi_2)(-n/2) \cong (\mathrm{Prim}^n(Z_{\mathrm{univ}}/\mathcal{S}\mathcal{D}(n, e)))^\Gamma.$$

Exactly as in the proof of Theorem 5.3.9 to show that $\mathcal{M}(n, e, \chi_2)$ has finite G_{geom} over some finite field k in which $2e\ell$ is invertible and which contains the e'th roots of unity, it suffices to choose an embedding $R \subset \mathbb{C}$, and to show $(\mathrm{Prim}^n(Z_{\mathrm{univ},\mathbb{C}}/\mathcal{S}\mathcal{D}(n, e)_\mathbb{C}))^\Gamma$ has finite G_{geom}.

Now $\mathrm{Prim}^n(Z_{\mathrm{univ},\mathbb{C}}/\mathcal{S}\mathcal{D}(n, e)_\mathbb{C})$ underlies a polarized variation of Hodge structure on $\mathcal{S}\mathcal{D}(n, e)_\mathbb{C}$, and so does its direct summand $(\mathrm{Prim}^n(Z_{\mathrm{univ},\mathbb{C}}/\mathcal{S}\mathcal{D}(n, e)_\mathbb{C}))^\Gamma$. So a sufficient condition for

$(\mathrm{Prim}^n(Z_{\mathrm{univ},\mathbb{C}}/\mathcal{SD}(n, e)_{\mathbb{C}}))^\Gamma$ to have finite G_{geom} is for n to be even and for $(\mathrm{Prim}^n(Z_{\mathrm{univ},\mathbb{C}}/\mathcal{SD}(n, e)_{\mathbb{C}}))^\Gamma$ to be entirely of Hodge type (n/2, n/2).

We will now check that for n = 2 and e = 3, $(\mathrm{Prim}^2(Z_{\mathrm{univ},\mathbb{C}}/\mathcal{SD}(2, 3)_{\mathbb{C}})))^\Gamma$ is entirely of type (1, 1). The Hodge numbers are constant in a variation of Hodge structure over a connected base, so it suffices to check at a single point. So we select the point

$$f(x_1, x_2) = (x_1)^3 + (x_2)^3 + 1.$$

Then our affine surface X_f is

$$z^2 = (x_1)^3 + (x_2)^3 + 1,$$

our affine covering surface Y is

$$w^6 = (y_1)^6 + (y_2)^6 + 1,$$

and our projective surface Z is the degree six Fermat surface

$$\mathrm{Fermat}_6 : W^6 = (Y_1)^6 + (Y_2)^6 + (Y_0)^6,$$

with Γ the group $\mu_3 \times \mu_2 \times \mu_2$, acting in the obvious way:

$$(\zeta_3, \varepsilon_1, \varepsilon_2) : (W, Y_1, Y_2, Y_0) \mapsto (\zeta_3 W, \varepsilon_1 Y_1, \varepsilon_2 Y_2, Y_0).$$

What must be checked is that the Γ-invariants in $\mathrm{Prim}^2(\mathrm{Fermat}_6)$ are entirely of type (1, 1). Since the Γ-invariants form a sub-Hodge structure, it suffices to check that

$$H^{2,0}(\mathrm{Fermat}_6) \cap (\mathrm{Prim}^2(\mathrm{Fermat}_6))^\Gamma = 0.$$

But one readily computes the holomorphic 2-forms on Fermat_6. They are, in affine coordinates (w, y_1, y_2),

$$\{\text{polynomials in } (w, y_1, y_2) \text{ of degree} \le 2\} dy_1 dy_2/w^5.$$

Under the action of the subgroup $\mu_3 \times \{1\} \times \{1\}$, the only invariants are the $\mathbb{C}$-multiples of the class

$$dy_1 dy_2/w^3,$$

and this class is not invariant under the subgroup $\{1\} \times \mu_2 \times \{1\}$. QED

(5.4.10) We now return to the proof of Theorem 5.2.2, in the remaining cases

$$n \ge 3,\ e \ge 3 \text{ prime to } p,\ \chi^e \ne \mathbb{1}.$$

We have just seen that for n=2, G_{geom} is never finite, except in the one case $\mathcal{M}(2, 3, \chi_2)$ in characteristic $p \ge 5$.

Let us next treat separately the case n=3, because it is the first case of Theorem 5.2.2 where G_{geom} is never finite. As explained in 5.4.3 above, in order to show that $\mathcal{M}(3, e, \chi)$ has infinite G_{geom}, it suffices to find some nontrivial character ρ of order dividing e such that $\mathcal{M}(2, e, \chi\rho)$ has infinite G_{geom}. This last condition is automatic if $e \ge 4$. If e = 3, then $\mathcal{M}(2, 3, \chi\rho)$ has infinite G_{geom} unless $p \ge 5$

and $\chi\rho$ is the quadratic character χ_2. But then $\chi\rho^2$ has order 6 (remember, ρ is nontrivial of order 3), in which case $\mathcal{M}(2, 3, \chi\rho^2)$ has infinite G_{geom}.

(5.4.11) Now that there are no exceptions for n=3, the general fact that $\mathcal{M}(n+1, e, \chi)$ has infinite G_{geom} if $\mathcal{M}(n, e, \chi\rho)$ has infinite G_{geom} for some nontrivial ρ of order dividing e shows, inductively, that there are no exceptions for any $n \geq 3$. This concludes the proof of Theorem 5.2.2. QED

(5.5) Analysis of $Gr^0(\mathcal{M}(n, e, \chi))$, for e prime to p but $\chi^e = \mathbb{1}$

(5.5.1) Suppose k is a finite field of characteristic p. Let $n \geq 1$ and $e \geq 3$ be integers, and let χ be a nontrivial multiplicative character of $k^\times$. Suppose further that e is prime to p, and that $\chi^e = \mathbb{1}$. We have seen that, under these conditions, $\mathcal{M}(n, e, \chi)|\mathcal{SD}(n,e)$ is mixed of weights -1 and 0, and its highest weight quotient $Gr^0(\mathcal{M}(n, e, \chi)|\mathcal{SD}(n,e))$ is lisse of rank $(1/e)((e-1)^{n+1} - (-1)^{n+1})$. For brevity in what follows, let us define

$$N(n, e) := (1/e)((e-1)^{n+1} - (-1)^{n+1}).$$

Theorem 5.5.2 Suppose k is a finite field of characteristic p. Let $n \geq 1$ and $e \geq 3$ be integers, and let χ be a nontrivial multiplicative character of $k^\times$. Suppose that e is prime to p, that $\chi^e = \mathbb{1}$. Suppose further that $N(n, e) \geq 2$ (or, what is the same, suppose that $e \geq 4$ if $n = 1$). Then we have the following results concerning the group G_{geom} for the highest weight quotient $Gr^0(\mathcal{M}(n, e, \chi)|\mathcal{SD}(n,e))$.

1) If χ does not have order 2, G_{geom} contains $SL(N(n, e))$, except in the following cases.

$n = 1$, χ has order 4, $e = 4$, and $p \geq 3$,
$n = 1$, χ has order 6, $e = 6$, and $p \geq 5$,
$n = 2$, χ has order 3, $e = 3$, and $p \geq 5$.

In each of the exceptional cases, G_{geom} is finite.

2) If χ has order 2, and n is odd, then G_{geom} is $Sp(N(n, e))$.

3) If χ has order 2 and n is even, then G_{geom} is either $SO(N(n, e))$ or $O(N(n, e))$, except in the case $(n = 2, e = 4, p \geq 3)$, in which case G_{geom} is finite.

Remark 5.5.3 We will show later, in Theorem 6.7.19, part 3), that if χ has order 2 and n is even, then G_{geom} always contains a reflection, and hence that G_{geom} is $O(N(n, e))$ except in the case $(n = 2, e = 4, p \geq 3)$. We will also show later, in Theorem 5.6.14 and again, by a different argument, in Theorem 6.7.17, part 2), that in the case $(n = 2, e = 4, p \geq 3)$, G_{geom} is the Weyl group of E_7 in its reflection representation. We will also give, in Theorem 6.7.21, quite precise results about G_{geom} in the case when χ does not have order 2.

proof Exactly as in the first paragraph of the proof of Theorem 5.2.2, we see that $Gr^0(\mathcal{M}(n, e, \chi)|\mathcal{S}\mathcal{D}(n,e))$ is geometrically irreducible, and that its Frobenius-Schur indicator is given by

$$\mathrm{FSI}^{geom}(\mathcal{S}\mathcal{D}(n,e), Gr^0(\mathcal{M}(n,e,\chi))) = 0, \text{ if order}(\chi) \neq 2,$$
$$= (-1)^n, \text{ if order}(\chi) = 2,$$

cf. Corollary 1.20.3, part 7).

Since $Gr^0(\mathcal{M}(n, e, \chi)|\mathcal{S}\mathcal{D}(n,e))$ has rank $N(n, e)$, we see from Corollary 1.20.3, parts 4) and 5), that $Gr^0(\mathcal{M}(n,e,\chi)|\mathcal{S}\mathcal{D}(n,e))$ has fourth moment given by

$$M_4{}^{geom}(\mathcal{S}\mathcal{D}(n,e), Gr^0(\mathcal{M}(n,e,\chi))) = 2, \text{ if order}(\chi) \neq 2,$$
$$\leq 3, \text{ if order}(\chi) = 2,$$
$$= 3, \text{ if order}(\chi) = 2 \text{ and } N(n, e) \geq 4.$$

We now bring to bear Larsen's Alternative 2.2.2, which gives us the following results concerning the group G_{geom} for the highest weight quotient $Gr^0(\mathcal{M}(n, e, \chi)|\mathcal{S}\mathcal{D}(n,e))$.

If order$(\chi) \neq 2$, then either G_{geom} contains $SL(N(n, e))$, or G_{geom} is a finite primitive irreducible subgroup of $GL(N(n, e))$. If order$(\chi) = 2$ and n is odd and $N(n, e) \geq 4$, then either G_{geom} is $Sp(N(n, e))$, or G_{geom} is a finite primitive irreducible subgroup of $Sp(N(n, e))$. If order$(\chi) = 2$ and n is even and $N(n, e) \geq 4$, then G_{geom} is either $SO(N(n, e))$, or $O(N(n, e))$, or a finite primitive irreducible subgroup of $O(N(n, e))$.

If order$(\chi) = 2$ but $N(n, e) \leq 3$, we claim that $n=1$ and $e=4$. Indeed, recall that by assumption $\chi^e = \mathbb{1}$, so e is even, and hence we have $e \geq 4$. Clearly for fixed n, $N(n, e)$ is monotone increasing in e. So if $N(n, e) \leq 3$, then $N(n, 4) \leq 3$, i.e.,

$$(1/4)(3^{n+1} - (-1)^{n+1}) \leq 3,$$

i.e.,

$$3^{n+1} - (-1)^{n+1} \leq 12,$$

i.e.,

$$3^{n+1} \leq 12 + (-1)^{n+1} \leq 13,$$

and this holds only for $n=1$. For $n=1$, $N(1, e) = e - 2$, so $N(1, e) \leq 3$ holds precisely when $e \leq 5$. As e is even and $e \geq 3$, we must have $e=4$.

Let us now treat the case $n = 1$. If order$(\chi) = 2$, then e is even and $e \geq 4$, and the characteristic p is not 2. It is proven in [Ka-ACT, 5.17], that, in this case, G_{geom} is $Sp(e-2)$. Let us now treat the case

$$n = 1, \text{ order}(\chi) \geq 3, e \geq 4 \text{ is prime to } p, \chi^e = \mathbb{1}.$$

In this case, we know that either G_{geom} contains SL(e-2), or G_{geom} is a primitive irreducible finite subgroup of GL(e-2). The key once again is to observe that G_{geom} contains pseudoreflections of a quite specific type. Here is the precise result.

Pseudoreflection Theorem 5.5.4 (cf. [Ka-ACT, 5.7, 5)], [Ka-TLFM, 4.2.2, proof of 5.6.1]) Suppose we are in the case

$$n = 1, \text{order}(\chi) \geq 3, e \geq 4 \text{ prime to } p, \chi^e = \mathbb{1}.$$

1) If $p \neq 2$, denote by χ_2 the quadratic character. Then G_{geom} for $Gr^0(\mathcal{M}(1, e, \chi)|\mathcal{SD}(1, e))$ contains pseudoreflections of order equal to order($\chi\chi_2$). More precisely, for any weakly supermorse polynomial f(X) of degree e, the pullback of $Gr^0(\mathcal{M}(1, e, \chi)|\mathcal{SD}(1, e))$ to the one-parameter family

$$t \mapsto t - f(X),$$

parameterized by t in $\mathbb{A}^1$ - {critical values of f}, is lisse, and its local monodromy at each critical value $f(\alpha)$ of f is a pseudoreflection of determinant $\chi\chi_2$, viewed as a tame character of the inertia group $I(f(\alpha))$.

2) If $p = 2$, then G_{geom} for $Gr^0(\mathcal{M}(1, e, \chi)|\mathcal{SD}(1, e))$ contains pseudoreflections of order = $2\times$order(χ). More precisely, there is a dense open set U in $\mathcal{SD}(1, e)$ such that for any f in U, f has (e-1)/2 critical points, (e-1)/2 critical values, and over each critical value of f, the local monodromy of the sheaf $f_*\overline{\mathbb{Q}}_\ell/\overline{\mathbb{Q}}_\ell$ is a reflection of Swan conductor 1. Fix any such f. Over ∞, the local monodromy of $f_*\overline{\mathbb{Q}}_\ell/\overline{\mathbb{Q}}_\ell$ is tame; it is the direct sum of the e-1 nontrivial characters of $I(\infty)$ of order dividing e. The sheaf $f_*\overline{\mathbb{Q}}_\ell/\overline{\mathbb{Q}}_\ell$ is a geometrically irreducible middle extension on $\mathbb{A}^1$, with geometric monodromy group the full symmetric group S_e, in its deleted permutation representation. The pullback of $Gr^0(\mathcal{M}(1, e, \chi)|\mathcal{SD}(1, e))$ to the one-parameter family

$$t \mapsto t - f(X),$$

parameterized by t in $\mathbb{A}^1$ - {critical values of f}, is lisse and geometrically irreducible, and is geometrically isomorphic to the restriction to $\mathbb{A}^1$ - {critical values of f} of the middle convolution

$$MC_\chi((f_*\overline{\mathbb{Q}}_\ell/\overline{\mathbb{Q}}_\ell)[1])$$

on $\mathbb{A}^1$. For any critical value $f(\alpha)$ of f, there exists a character χ_2 of order 2 and Swan conductor 1 of the inertia group $I(f(\alpha))$, such that the local monodromy of $MC_\chi((f_*\overline{\mathbb{Q}}_\ell/\overline{\mathbb{Q}}_\ell)[1])$ at the critical value $f(\alpha)$ of f is a pseudoreflection of determinant $\chi\chi_2$, viewed as a character of the inertia group $I(f(\alpha))$.

proof Assertion 1) is proven in [Ka-ACT, 5.18].

To prove assertion 2), most of the argument is the same as in the proof of Theorem 5.3.2. We first use [Ka-TLFM, 2.7.1 and 2.7.2] to get U and f with the asserted local monodromies at finite distance of

$f_*\overline{\mathbb{Q}}_\ell/\overline{\mathbb{Q}}_\ell$. Fix one such f. Exactly as in the proof of Theorem 5.3.2, we show that $f_*\overline{\mathbb{Q}}_\ell$ has geometric monodromy group Γ_f equal to the full symmetric group S_e, and hence that $f_*\overline{\mathbb{Q}}_\ell/\overline{\mathbb{Q}}_\ell$ is an irreducible middle extension on $\mathbb{A}^1$, with generic rank $e-1 \geq 3$. Thus $(f_*\overline{\mathbb{Q}}_\ell/\overline{\mathbb{Q}}_\ell)[1]$ as irreducible perverse sheaf on $\mathbb{A}^1$ has $\mathcal{P}$, and is of type 2d) in the sense of [Ka-RLS, 3.3.3]. Therefore the middle convolution $MC_\chi((f_*\overline{\mathbb{Q}}_\ell/\overline{\mathbb{Q}}_\ell)[1])$ is itself irreducible and of type 2d) by [Ka-RLS, 3.3.3]. Comparing trace functions, we see that the middle convolution $MC_\chi((f_*\overline{\mathbb{Q}}_\ell/\overline{\mathbb{Q}}_\ell)[1])$ is geometrically isomorphic to the pullback of $Gr^0(\mathcal{M}(1, e, \chi)|\mathcal{SD}(1, e))$ to the one-parameter family $t \mapsto t - f(X)$. We now use [Ka-TLFM, 4.1.10, 1) and 4.2.2] to compute the local monodromies of the middle convolution $MC_\chi((f_*\overline{\mathbb{Q}}_\ell/\overline{\mathbb{Q}}_\ell)[1])$. QED

Armed with this result, we return to the case

$$n = 1,\ \mathrm{order}(\chi) \geq 3,\ e \geq 4 \text{ prime to } p,\ \chi^e = \mathbb{1}.$$

If $p = 2$, then G_{geom} contains a pseudoreflection of order $2\times\mathrm{order}(\chi) \geq 6$. By Mitchell's Theorem 5.3.4, G_{geom} is not a primitive irreducible finite subgroup of $GL(e-2)$ for any $e \geq 4$.

Suppose now that p is odd. We once again apply Mitchell's Theorem 5.3.4. As χ has $\mathrm{order}(\chi) \geq 3$, $\chi\chi_2$ is never trivial, and never of order 2. We have

$\mathrm{order}(\chi\chi_2) = 3$ if and only if $\mathrm{order}(\chi) = 6$,

$\mathrm{order}(\chi\chi_2) = 4$ if and only if $\mathrm{order}(\chi) = 4$,

$\mathrm{order}(\chi\chi_2) = 5$ if and only if $\mathrm{order}(\chi) = 10$.

By Mitchell's Theorem 5.3.4, G_{geom} cannot be a primitive irreducible finite subgroup of $GL(e-2)$ except possibly when χ has order 6, 4, or 10, and, at the same time, we have $e - 2 \leq 4$. Since $\chi^e = \mathbb{1}$ by hypothesis, the only cases in which G_{geom} could be finite are

$\mathrm{order}(\chi) = 4$, $e = 4$, $p \geq 3$,

$\mathrm{order}(\chi) = 6$, $e = 6$, $p \geq 5$.

We have seen in Theorem 5.3.9 that G_{geom} is in fact finite in these cases. This concludes the proof of the $n = 1$ case.

(5.6) Proof of Theorem 5.5.2 in the case $n \geq 2$

(5.6.1) Exactly as in the proof of Theorem 5.2.2, we now pass from the n=1 case to the general case, by an induction on n, based on the "added variable trick", i.e., the observation that, for any integer $e \geq 1$, if k is a field in which e is invertible, and if an n-variable polynomial

$$f(x_1,\ldots, x_n)$$

lies in $\mathcal{SD}(n, e)(k)$, then the polynomial in n+1 variables

$$f(x_1,\ldots, x_n) - y^n$$

lies in $\mathcal{SD}(n+1, e)(k)$. We denote by

$$\mathrm{AddedVar} : \mathcal{SD}(n, e) \to \mathcal{SD}(n+1, e)$$

$$f(x_1,\dots, x_n) \mapsto f(x_1,\dots, x_n) - y^e$$

this closed immersion.

(5.6.2) What is different in the case where $\chi^e = \mathbb{1}$, is the decomposition of the lisse sheaf $\mathrm{AddedVar}^*\mathcal{M}(n+1, e, \chi)$ on $\mathcal{SD}(n, e)$. The problem is that now $\overline{\chi}$ is among the nontrivial characters ρ of order dividing e, which indexed the previous decomposition, cf. Theorem 5.4.2. Not surprisingly, the nature of the component indexed by $\overline{\chi}$ is quite special.

(5.6.3) To explain it, we make a brief digression. Over $\mathcal{SD}(n, e)$, we have the product $\mathbb{A}^n \times \mathcal{SD}(n, e)$, with coordinates (v, f). This product space contains the corresponding "universal" hypersurface

$$\mathrm{AffHyp}(n, e) : f(v) = 0 \text{ in } \mathbb{A}^n \times \mathcal{SD}(n, e).$$

Let us denote by

$$\pi(n,e) : \mathrm{AffHyp}(n, e) \to \mathcal{SD}(n, e)$$

the projection in this family, whose fibre over a point f_0 in $\mathcal{SD}(n, e)$ is just the affine hypersurface in $\mathbb{A}^n$ defined by the vanishing of f_0 Of interest to us below will be the sheaf $R^{n-1}\pi(n,e)_!\overline{\mathbb{Q}}_\ell$ on $\mathcal{SD}(n, e)$. By the definition of $\mathcal{SD}(n, e)$, the universal affine hypersurface over it is the complement of the smooth hyperplane section $X_0 = 0$ in the smooth hypersurface in $\mathbb{P}^n$ of equation $F(X_0, X_1,\dots, X_n) = 0$, where

$$F(X_0, X_1,\dots, X_n) := (X_0)^e f(X_1/X_0,\dots, X_n/X_0)$$

is the homogenization of f. Let us denote by

$$\overline{\pi}(n,e) : \mathrm{ProjHyp}(n, e) \to \mathcal{SD}(n, e)$$

the corresponding universal family of smooth, degree e hypersurfaces in $\mathbb{P}^n$ which are transverse to the hyperplane $X_0 = 0$. Let us also denote by

$$\overline{\pi}(n-1,e) : \mathrm{ProjHyp}(n-1, e) \to \mathcal{SD}(n, e)$$

the family of smooth, degree e hypersurfaces in $\mathbb{P}^{n-1}$ given by intersecting with the hyperplane section $X_0 = 0$. So the fibre over a point f_0 in $\mathcal{SD}(n, e)$ is the projective hypersurface defined by the vanishing of the part of f_0 which is homogeneous of highest degree e. [If n = 1, these fibres are empty.] The sheaves $R^i\overline{\pi}(n,e)_!\overline{\mathbb{Q}}_\ell = R^i\overline{\pi}(n,e)_*\overline{\mathbb{Q}}_\ell$ and $R^i\overline{\pi}(n-1,e)_!\overline{\mathbb{Q}}_\ell = R^i\overline{\pi}(n-1,e)_*\overline{\mathbb{Q}}_\ell$ are lisse on $\mathcal{SD}(n, e)$, and pure of weight i. We have natural restriction maps

$$R^i\overline{\pi}(n,e)_!\overline{\mathbb{Q}}_\ell \to R^i\overline{\pi}(n-1,e)_!\overline{\mathbb{Q}}_\ell.$$

(5.6.4) For $n \geq 2$, we define

$$\mathrm{Prim}^{n-1}\overline{\pi}(n,e)_!\overline{\mathbb{Q}}_\ell$$

$$:= \text{Kernel of } R^{n-1}\overline{\pi}(n,e)_!\overline{\mathbb{Q}}_\ell \to R^{n-1}\overline{\pi}(n-1,e)_!\overline{\mathbb{Q}}_\ell,$$

and

$$\mathrm{Prim}^{n-2}\overline{\pi}(n-1,e)_!\overline{\mathbb{Q}}_\ell$$
$$:= \text{Coker of } R^{n-2}\overline{\pi}(n,e)_!\overline{\mathbb{Q}}_\ell \to R^{n-2}\overline{\pi}(n-1,e)_!\overline{\mathbb{Q}}_\ell.$$

These are the usual sheaves of middle-dimensional primitive cohomology of the smooth hypersurface fibres.

Lemma 5.6.5 For $n \geq 2$, $e \geq 3$, any field k in which e is invertible, and any prime ℓ invertible in k, we have the following results concerning the sheaves $R^i\pi(n,e)_!\overline{\mathbb{Q}}_\ell$ on $\mathcal{SD}(n, e)$.

1) $R^i\pi(n,e)_!\overline{\mathbb{Q}}_\ell = 0$ unless $i = n-1$ or $i = 2n-2$.

2) $R^{2n-2}\pi(n,e)_!\overline{\mathbb{Q}}_\ell = \overline{\mathbb{Q}}_\ell(1-n)$.

3) $R^{n-1}\pi(n,e)_!\overline{\mathbb{Q}}_\ell$ sits in a short exact sequence

$$0 \to \mathrm{Prim}^{n-2}\overline{\pi}(n-1,e)_!\overline{\mathbb{Q}}_\ell \to R^{n-1}\pi(n,e)_!\overline{\mathbb{Q}}_\ell$$
$$\to \mathrm{Prim}^{n-1}\overline{\pi}(n,e)_!\overline{\mathbb{Q}}_\ell \to 0.$$

4) $R^{n-1}\pi(n,e)_!\overline{\mathbb{Q}}_\ell$ is lisse on $\mathcal{SD}(n, e)$, mixed of weights n-1 and n-2, and we have

$$\mathrm{Gr}^{n-1}(R^{n-1}\pi(n,e)_!\overline{\mathbb{Q}}_\ell) \cong \mathrm{Prim}^{n-1}\overline{\pi}(n,e)_!\overline{\mathbb{Q}}_\ell.$$

proof Assertion 4) results from Assertion 3) and the fact [De-Weil I, 1.6] that $R^i\overline{\pi}(n,e)_!\overline{\mathbb{Q}}_\ell$ and $R^i\overline{\pi}(n-1,e)_!\overline{\mathbb{Q}}_\ell$ are both pure of weight i. Assertions 1), 2), and 3) are standard, and result from the excision sequence attached to

$$\mathrm{AffHyp}(n, e) \subset \mathrm{ProjHyp}(n, e) \supset \mathrm{ProjHyp}(n-1, e),$$

and the known cohomological structure of smooth projective hypersurfaces. QED

Theorem 5.6.6 Let $n \geq 2$ and $e \geq 3$ be integers. Let k be a finite field of characteristic p, in which e is invertible, and which contains the e'th roots of unity. Let ℓ be a prime with $\ell \neq p$, and let

$$\chi : k^\times \to \overline{\mathbb{Q}}_\ell^\times$$

be a nontrivial multiplicative character. Suppose that $\chi^e = \mathbb{1}$. For each nontrivial multiplicative character ρ of $k^\times$ with $\rho^e = \mathbb{1}$ but with $\rho \neq \mathbb{1}$, consider the Jacobi sum

$$J(\chi, \rho) := \Sigma_{u \text{ in } k^\times} \chi(1-u)\rho(u).$$

[Thus $-J(\chi, \overline{\chi}) = \chi(-1)$, while the others are pure of weight 1.]

For a lisse sheaf $\mathcal{G}$ on a reasonable scheme X, denote by $\mathcal{G}^{ss}$ its semisimplification as a representation of $\pi_1(X)$.

1) We have an isomorphism of lisse sheaves on $\mathcal{SD}(n, e)_k$

$$(\mathrm{AddedVar}^*\mathcal{M}(n+1, e, \chi))^{ss}$$
$$\cong (R^{n-1}\pi_!\overline{\mathbb{Q}}_\ell((n-1)/2)\otimes(\chi(-1))^{\deg}$$
$$\oplus_{\rho\neq\mathbb{1},\ \overline{\chi},\ \rho^e=\mathbb{1}} \mathcal{M}(n, e, \chi\rho)(1/2)\otimes(-J(\chi,\rho))^{\deg})^{ss}.$$

2) We have an isomorphism of lisse sheaves on $\mathcal{SD}(n, e)_{\overline{k}}$

$$(\mathrm{AddedVar}^*\mathcal{M}(n+1, e, \chi))^{ss}$$
$$\cong (R^{n-1}\pi_!\overline{\mathbb{Q}}_\ell \oplus_{\rho\neq \mathbb{1}, \overline{\chi}, \rho^e=\mathbb{1}} \mathcal{M}(n, e, \chi\rho))^{ss}.$$

3) We have an isomorphism of lisse sheaves on $\mathcal{SD}(n, e)_{\overline{k}}$

$$\mathrm{AddedVar}^*(\mathrm{Gr}^0\mathcal{M}(n+1, e, \chi))$$
$$\cong \mathrm{Prim}^{n-1}\overline{\pi}_{n!}\overline{\mathbb{Q}}_\ell \oplus_{\rho\neq \mathbb{1}, \overline{\chi}, \rho^e=\mathbb{1}} \mathrm{Gr}^0\mathcal{M}(n, e, \chi\rho).$$

proof The proof is quite similar to that of Theorem 5.4.2. Exactly as there, we show that we have a direct sum decomposition

$$\mathrm{AddedVar}^*\mathcal{M}(n+1, e, \chi)(-(n+1)/2)$$
$$= \oplus_{\rho^e=\mathbb{1}} R^{n+1}pr_{2!}(\mathcal{L}_{\chi(f(v)-y)}\otimes\mathcal{L}_{\rho(y)}),$$

we infer that each term

$$R^{n+1}pr_{2!}(\mathcal{L}_{\chi(f(v)-y)}\otimes\mathcal{L}_{\rho(y)})$$

is lisse on $\mathcal{SD}(n, e)$, and we infer that for $i \neq n+1$, we have

$$\oplus_{\rho^e=\mathbb{1}} R^ipr_{2!}(\mathcal{L}_{\chi(f(v)-y)}\otimes\mathcal{L}_{\rho(y)}) = 0, \text{ for } i \neq n+1.$$

Exactly as there, this vanishing allows us to compute the trace function of each summand in terms of character sums. For $\rho = \mathbb{1}$, the trace function vanishes, so we have

$$R^{n+1}pr_{2!}(\mathcal{L}_{\chi(f(v)-y)}) = 0.$$

For $\rho^e = \mathbb{1}$, but $\rho \neq \mathbb{1}, \overline{\chi}$, we find that the two lisse sheaves on $\mathcal{SD}(n, e)$

$$R^{n+1}pr_{2!}(\mathcal{L}_{\chi(f(v)-y)}\otimes\mathcal{L}_{\rho(y)})$$

and

$$\mathcal{M}(n, e, \chi\rho)(-n/2)\otimes(-J(\chi,\rho))^{\deg}$$

have the same trace function, and hence by Chebotarev have isomorphic semisimplications as representations of $\pi_1^{arith}(\mathcal{SD}(n\ e))$.

For $\rho = \overline{\chi}$, the trace function of $R^{n+1}pr_{2!}(\mathcal{L}_{\chi(f(v)-y)}\otimes\mathcal{L}_{\overline{\chi}(y)})$ is given as follows. Take E/k a finite extension and f in $\mathcal{SD}(n, e)(E)$. The trace of $\mathrm{Frob}_{E,f}$ is

$$(-1)^{n+1}\Sigma_{(x,\ y) \text{ in } \mathbb{A}^{n+1}(E)}\ \chi_E(f(x)-y)\overline{\chi}_E(y).$$

The trace function of $\mathrm{Frob}_{E,f}$ on $R^{n-1}\pi_!\overline{\mathbb{Q}}_\ell(-1)$ is readily calculated, using the above lemma, to be

$$(\#E)(-1)^{n-1}(\#\{f = 0 \text{ in } \mathbb{A}^n(E)\} - (\#E)^{n-1}).$$

So the fact that

$$R^{n+1}pr_{2!}(\mathcal{L}_{\chi(f(v)-y)}\otimes\mathcal{L}_{\overline{\chi}(y)})$$

and

$$R^{n-1}\pi_!\overline{\mathbb{Q}}_\ell(-1)\otimes(\chi(-1))^{\deg}$$

have the same trace function comes down to the following elementary lemma.

Lemma 5.6.7 For any nontrivial multiplicative character χ of any finite field E, any integer n ≥ 1, and for any n-variable polynomial f with coefficients in E, we have the identity

$$\Sigma_{(x,\ y)\ \text{in}\ \mathbb{A}^{n+1}(E)}\ \chi(f(x) - y)\overline{\chi}(y)$$
$$= \chi(-1)(\#E)\#\{f = 0 \text{ in } \mathbb{A}^n(E)\} - \chi(-1)(\#E)^n.$$

proof Consider the first sum. First fix a point x in $\mathbb{A}^n(E)$ where f(x) ≠ 0, say f(x) = α, with α ≠ 0. Then the sum over y is

$$\Sigma_{y \text{ in } E}\ \chi(\alpha - y)\overline{\chi}(y) = \Sigma_{y \text{ in } E - \{0,\ \alpha\}}\ \chi(\alpha - y)\overline{\chi}(y)$$
$$= \Sigma_{y \text{ in } E - \{0,\ \alpha\}}\ \chi((\alpha - y)/y)$$
$$= \Sigma_{u \text{ in } \mathbb{P}^1(E) - \{0,\ -1,\ \infty\}}\ \chi(u)$$
$$= -\chi(-1).$$

Now fix a point x in $\mathbb{A}^n(E)$ where f(x) = 0. Then the sum over y is

$$\Sigma_{y \text{ in } E}\ \chi(-y)\overline{\chi}(y) = \Sigma_{y \text{ in } E - \{0\}}\ \chi(-y)\overline{\chi}(y)$$
$$= \chi(-1)(\#E - 1).$$

So the first sum is

$$-\chi(-1)\#\{f \neq 0 \text{ in } \mathbb{A}^n(E)\} + \chi(-1)(\#E - 1)\#\{f = 0 \text{ in } \mathbb{A}^n(E)\}$$
$$= -\chi(-1)\#\mathbb{A}^n(E) + \chi(-1)(\#E)\#\{f = 0 \text{ in } \mathbb{A}^n(E)\}.$$ QED

With this lemma, we see that Assertion 1) holds. Assertion 2) results trivially, since π_1^{geom} is a normal subgroup of π_1^{arith}. To prove Assertion 3), apply Gr^0 to the isomorphism in Assertion 1). This shows that both sides in 3) have the same geometric semisimpification. But each summand separately is known to be geometrically irreducible, cf. [De-Weil II, 4.4.1 and 4.4.9] for the geometric irreducibility of $\mathrm{Prim}^{n-1}\overline{\pi}(n,e)_!\overline{\mathbb{Q}}_\ell$. Now recall that both sides of 3) are pure, hence are geometrically semisimple. QED for Theorem 5.6.6

(5.6.8) For n=1, the situation is a bit different. In this case, we have

$$\mathrm{AffHyp}(1,\ e) = \mathrm{ProjHyp}(1,\ e).$$

The maps

$$\pi(1,e) : \mathrm{AffHyp}(n,\ e) \to \mathcal{SD}(n,\ e),$$
$$\overline{\pi}(1,e) : \mathrm{ProjHyp}(1,\ e) \to \mathcal{SD}(1,\ e),$$

coincide, and are finite etale of degree e. We have a surjective Trace morphism

$$\text{Trace: } R^0\overline{\pi}(n,e)_!\overline{\mathbb{Q}}_\ell \to \overline{\mathbb{Q}}_\ell.$$

We define

$$\mathrm{Prim}^0\overline{\pi}(n,e)_!\overline{\mathbb{Q}}_\ell$$
$$:= \text{Kernel of Trace: } R^0\overline{\pi}(n,e)_!\overline{\mathbb{Q}}_\ell \to \overline{\mathbb{Q}}_\ell.$$

For ease of later reference, we summarize this discussion in the following lemma.

Lemma 5.6.9 For n = 1, e ≥ 3, any field k in which e is invertible, and any prime ℓ invertible in k, we have the following results concerning the sheaves $R^i\pi(1,e)_!\overline{\mathbb{Q}}_\ell = R^i\overline{\pi}(1,e)_!\overline{\mathbb{Q}}_\ell$ on $\mathcal{SD}(1, e)$.

1) $R^i\pi(1,e)_!\overline{\mathbb{Q}}_\ell = 0$ unless $i = 0$.

2) $R^0\pi(n,e)_!\overline{\mathbb{Q}}_\ell$ is lisse of rank e and pure of weight 0.

3) $R^0\pi(1,e)_!\overline{\mathbb{Q}}_\ell$ sits in a short exact sequence

$$0 \to \mathrm{Prim}^0\overline{\pi}(1,e)_!\overline{\mathbb{Q}}_\ell \to R^0\pi(1,e)_!\overline{\mathbb{Q}}_\ell \to \overline{\mathbb{Q}}_\ell \to 0.$$

In the case n=1, the previous Theorem 5.6.6 becomes the following.

Theorem 5.6.10 Let $n = 1$ and $e \geq 3$ be integers. Let k be a finite field of characteristic p, in which e is invertible, and which contains the e'th roots of unity. Let ℓ be a prime with $\ell \neq p$, and let

$$\chi : k^\times \to \overline{\mathbb{Q}}_\ell^{\times}$$

be a nontrivial multiplicative character. Suppose that $\chi^e = \mathbb{1}$. For each nontrivial multiplicative character ρ of $k^\times$ with $\rho^e = \mathbb{1}$ but with $\rho \neq \mathbb{1}$, consider the Jacobi sum

$$J(\chi, \rho) := \Sigma_{u \text{ in } k^\times} \chi(1-u)\rho(u).$$

[Thus $-J(\chi, \overline{\chi}) = \chi(-1)$, while the others are pure of weight 1.]

For a lisse sheaf $\mathcal{G}$ on a reasonable scheme X, denote by $\mathcal{G}^{ss}$ its semisimplification as a representation of $\pi_1(X)$.

1) We have an isomorphism of lisse sheaves on $\mathcal{SD}(n, e)_k$

$$(\mathrm{AddedVar}^*\mathcal{M}(2, e, \chi))^{ss}$$
$$\cong (\mathrm{Prim}^0\overline{\pi}(1,e)_!\overline{\mathbb{Q}}_\ell \otimes (\chi(-1))^{\deg}$$
$$\oplus_{\rho \neq \mathbb{1}, \overline{\chi}, \rho^e = \mathbb{1}} \mathcal{M}(1, e, \chi\rho)(1/2) \otimes (-J(\chi,\rho))^{\deg})^{ss}.$$

2) We have an isomorphism of lisse sheaves on $\mathcal{SD}(n, e)_{\overline{k}}$

$$(\mathrm{AddedVar}^*\mathcal{M}(n+1, e, \chi))^{ss}$$
$$\cong (\mathrm{Prim}^0\overline{\pi}(1,e)_!\overline{\mathbb{Q}}_\ell \oplus_{\rho \neq \mathbb{1}, \overline{\chi}, \rho^e = \mathbb{1}} \mathcal{M}(1, e, \chi\rho))^{ss}.$$

3) We have an isomorphism of lisse sheaves on $\mathcal{SD}(n, e)_{\overline{k}}$

$$\mathrm{AddedVar}^*(\mathrm{Gr}^0\mathcal{M}(n+1, e, \chi))$$
$$\cong \mathrm{Prim}^0\overline{\pi}(1,e)_!\overline{\mathbb{Q}}_\ell \oplus_{\rho \neq \mathbb{1}, \overline{\chi}, \rho^e = \mathbb{1}} \mathrm{Gr}^0\mathcal{M}(1, e, \chi\rho).$$

proof The proof is identical to that of Theorem 5.6.6 up through the computation of the trace function of $R^{n+1}pr_{2!}(\mathcal{L}_{\chi(f(v) - y)} \otimes \mathcal{L}_{\overline{\chi}(y)})$ in the discussion of the case when $\rho = \overline{\chi}$. At that point, we continue as follows.

The trace function of $\mathrm{Frob}_{E,f}$ on $\mathrm{Prim}^0\overline{\pi}(1,e)_!\overline{\mathbb{Q}}_\ell(-1)$ is readily calculated, using the above Lemma 5.6.9, to be

$$(\#E)(\#\{f = 0 \text{ in } \mathbb{A}^n(E)\} - 1).$$

The fact that

$$R^2pr_{2!}(\mathcal{L}_{\chi(f(v) - y)} \otimes \mathcal{L}_{\overline{\chi}(y)})$$

and

$$\mathrm{Prim}^0\overline{\pi}(1,e)_!\overline{\mathbb{Q}}_\ell(-1) \otimes (\chi(-1))^{\deg}$$

have the same trace function comes down to the n=1 case of Lemma 5.6.7. QED

(5.6.11) With Theorem 5.6.10 established, we can proceed by induction to treat the case $n \geq 1$.

(5.6.12) Let us first treat the case $n = 2$, since here we will find two last exceptional cases. Fix a value of $e \geq 3$ prime to p, and a nontrivial χ with $\chi^e = \mathbb{1}$. If $e = 3$, then χ has order 3, and, as we will see below, $\mathrm{Gr}^0\mathcal{M}(2, 3, \chi_3)$ has finite G_{geom}. If $e = 4$ and χ has order 2, we will see below that $\mathrm{Gr}^0\mathcal{M}(2, 4, \chi_2)$ has finite G_{geom}. If e $= 4$ and χ has order 4, then one of the summands in $\mathrm{AddedVar}^*\mathrm{Gr}^0\mathcal{M}(2, 4, \chi_4)$ is $\mathrm{Gr}^0\mathcal{M}(1, 4, (\chi_4)^2 = \chi_2)$, which we have seen has infinite G_{geom}. If $e=5$, then χ has order 5, and one of the summands of $\mathrm{AddedVar}^*\mathrm{Gr}^0\mathcal{M}(2, 5, \chi_5)$ is $\mathrm{Gr}^0\mathcal{M}(1, 5, (\chi_5)^4 = \overline{\chi}_5)$, which has infinite G_{geom}. If $e = 6$, and χ has order 2, then one of the summands is $\mathrm{Gr}^0\mathcal{M}(1, 6, (\chi_2)(\chi_6) = \chi_3)$, which has infinite G_{geom}. If $e = 6$, and χ has order 3, then one of the summands is $\mathrm{Gr}^0\mathcal{M}(1, 6, (\chi_3)(\chi_6) = \chi_2)$, which has infinite G_{geom}. If $e = 6$, and χ has order 6, one of the summands is

$$\mathrm{Gr}^0\mathcal{M}(1, 6, (\chi_6)(\chi_2\overline{\chi}_6) = \chi_2),$$

which has infinite G_{geom}. Suppose now $e \geq 7$. Then one of the summands is $\mathrm{Gr}^0\mathcal{M}(1, e, \chi(\text{some } \rho \neq \overline{\chi}, \mathbb{1} \text{ of order dividing } e))$. This has infinite G_{geom}, because the only exceptions in the n=1 case had $e \leq 6$.

(5.6.13) To complete the discussion of the n=2 case, we prove that the two possibly exceptional cases we found do in fact have finite G_{geom}.

Theorem 5.6.14

1) $\mathrm{Gr}^0\mathcal{M}(2, 3, \chi_3)$ has finite G_{geom} in characteristic $p \neq 3$.

2) $\mathrm{Gr}^0\mathcal{M}(2, 4, \chi_2)$ have finite G_{geom} in characteristic $p \neq 2$, with G_{geom} the Weyl group $W(E_7)$ in its 7-dimensional reflection representation.

proof 1) For a finite field k in which 3 is invertible and which contains the cube roots of unity, χ_3 a cubic multiplicative character of k, f(x, y) in $\mathcal{SD}(2, 3)(k)$, the stalk of $\mathcal{M}(2, 3, \chi_3)(-1)$ is

$$H_c^2(\mathbb{A}^2\otimes_k\overline{k}, \mathcal{L}_{\chi_3(f(x, y))})$$

$=$ the χ_3-isotypical component of

$$H_c^2((\text{the affine cubic surface } X_f : w^3 = f(x, y))\otimes_k\overline{k}, \overline{\mathbb{Q}}_\ell),$$

for the action of $\mu_3(k)$ which moves w alone. This affine surface X_f has a natural compactification to a projective smooth cubic surface

Z_f in $\mathbb{P}^3$, with homogeneous equation

$$W^3 = F(X, Y, Z) := Z^3 f(X/Z, Y/Z).$$

The stalk of $\mathrm{Gr}^0\mathcal{M}(2, 3, \chi_3)(-1)$ at f is the χ_3-isotypical component of

$$(\mathrm{Prim}^2(Z_f\otimes_k\overline{k}, \overline{\mathbb{Q}}_\ell)).$$

So by Chebotarev the χ_3-isotypical component of the lisse sheaf $\mathrm{Prim}^2(Z_{\mathrm{univ}}/\mathcal{SD}(2, 3))$ on $\mathcal{SD}(2, 3)$ is $\mathrm{Gr}^0\mathcal{M}(2, 3, \chi_3)(-1)$, and we have a direct sum decomposition

$$\mathrm{Prim}^2(Z_{\mathrm{univ}}/\mathcal{SD}(2, 3)) \text{ on } \mathcal{SD}(2, 3)$$
$$\cong \mathrm{Gr}^0\mathcal{M}(2, 3, \chi_3)(-1) \oplus \mathrm{Gr}^0\mathcal{M}(2, 3, \overline{\chi}_3)(-1).$$

So it suffices to show that the lisse sheaf $\mathrm{Prim}^2(Z_{\mathrm{univ}}/\mathcal{SD}(2, 3))$ has finite G_{geom}. But the Prim^2 for the universal family of all smooth cubic hypersurfaces in $\mathbb{P}^3$, of which our family $Z_{\mathrm{univ}}/\mathcal{SD}(2, 3)$ is a subfamily, is well known to have finite G_{geom}, corresponding to the fact that H^2 of a smooth cubic surface is entirely algebraic, cf. [Ka-Sar-RMFEM, 11.4.9]. [Alternate end of proof: reduce to the complex case, as in Theorem 5.4.9, and use the fact that Prim^2 of a smooth cubic surface is entirely of type (1, 1).]

2) We now turn to the case of $\mathrm{Gr}^0\mathcal{M}(2, 4, \chi_2)$. At a point f(x, y) in $\mathcal{SD}(2, 4)(k)$, the stalk of $\mathcal{M}(2, 4, \chi_2)(-1)$ is

$$H_c^2(\mathbb{A}^2\otimes_k\overline{k}, \mathcal{L}_{\chi_2(f(x, y))})$$
$$= H_c^2((\text{the affine surface } X_f : t^2 = f(x, y))\otimes_k\overline{k}, \overline{\mathbb{Q}}_\ell),$$

which is itself the part of

$$H_c^2((\text{the affine surface } Y_f : w^4 = f(x, y))\otimes_k\overline{k}, \overline{\mathbb{Q}}_\ell)$$

which is invariant under the action of the group $\Gamma := \pm 1$ on w alone. Then the stalk of $\mathrm{Gr}^0\mathcal{M}(2, 4, \chi_2)(-1)$ at f is the part of Prim^2 of the smooth quartic hypersurface Z_f in $\mathbb{P}^3$

$$Z_f : W^4 = F(X, Y, Z) := Z^4 f(X/Z, Y/Z)$$

which is invariant under the action of Γ on W. The quotient surface

$$S_f := Z_f/\Gamma$$

is itself smooth, and has the a priori description as the double cover of $\mathbb{P}^2$ ramified along the smooth quartic curve C_f given by

$$C_f\colon F(X, Y, Z) = 0 \text{ in } \mathbb{P}^2.$$

Thus the stalk of $\mathrm{Gr}^0\mathcal{M}(2, 4, \chi_2)(-1)$ at f is $\mathrm{Prim}^2(S_f) = \mathrm{Prim}^2(Z_f)^\Gamma$. Such a double cover S_f is precisely a "Del Pezzo surface of degree 2", cf. [Manin, page 119]. Over $\overline{k}$, S_f is the blow up of $\mathbb{P}^2$ at seven points, no three of which are colinear and no six of which lie on a conic. [The degree two map to $\mathbb{P}^2$ which exhibits it as the double cover ramified along the smooth quartic curve C_f is given by the $\mathbb{P}^2$ of cubics passing through the seven points.] Because S_f is a blow up of $\mathbb{P}^2$ at seven points, its Prim^2 is entirely algebraic. Therefore a power of every $\mathrm{Frob}_{k,f}$ on $\mathrm{Prim}^2(S_f)(1) = \mathrm{Gr}^0\mathcal{M}(2, 4, \chi_2)_f$ is the identity. Therefore [Ka-ESDE 8.14.3.1] the group G_{arith} for $\mathrm{Gr}^0\mathcal{M}(2, 4, \chi_2)$ is finite, and hence its subgroup G_{geom} is finite.

Here is an alternate proof of the finiteness of G_{geom}. We must show that in the universal family of the quartic surfaces Z_f constructed above, say $Z_{univ}/\mathcal{SD}(2, 4)$, the Γ invariants in the lisse sheaf $\mathrm{Prim}^2(Z_{univ}/\mathcal{SD}(2, 4))$ have finite G_{geom}. To prove this, we first reduce to the complex case, and use the Hodge theoretic criterion [Ka-ASDE, 4.4.2]. It suffices to show that on any complex K3 surface of the shape

$$W^4 = F(X, Y, Z),$$

with F a homogeneous quartic which defines a smooth curve in $\mathbb{P}^2$, the Γ-invariant part of its Prim^2 is entirely of type (1, 1). To see this, note that the Γ-invariants form a sub-Hodge structure of Prim^2. So it suffices to show that there are no nonzero Γ-invariants in $H^{2,0}$. But $H^{2,0}$ is the one-dimensional space spanned by the holomorphic 2-form $\omega := dxdy/w^3$ (in affine coordinates w, x, y on $w^4 = f(x, y)$), and ω is visibly anti-invariant under the ±1 action.

It remains to show that, in any characteristic $p \neq 2$, G_{geom} for $\mathrm{Gr}^0\mathcal{M}(2, 4, \chi_2)$ is the Weyl group of E_7 in its 7-dimensional reflection representation. For this, we return to the Del Pezzo point of view. For the universal family $S_{univ}/\mathcal{SD}(2, 4)$ of S_f's, we have

$$\mathrm{Gr}^0\mathcal{M}(2, 4, \chi_2) \cong \mathrm{Prim}^2(S_{univ}/\mathcal{SD}(2, 4))(1).$$

Consider a Del Pezzo surface X of degree 2 over an algebraically closed field, i.e., $\mathbb{P}^2$ blown up at seven points, no three of which are colinear and no six of which lie on a conic. Consider its Neron-Severi group NS(X). One knows [Manin, 23.9] that NS(X) is $\mathbb{Z}$-free of rank 8, and that for any prime ℓ invertible on X we have

$$NS(X)\otimes\overline{\mathbb{Q}}_\ell = H^2(X,\overline{\mathbb{Q}}_\ell)(1).$$

Moreover, in the orthogonal complement (for the intersection pairing) of the canonical class in NS(X), call it $NS(X)^\perp$, the elements of self-intersection -2 form a root system of type E_7, and we have

$$NS(X)^\perp\otimes\overline{\mathbb{Q}}_\ell = \mathrm{Prim}^2(X,\overline{\mathbb{Q}}_\ell)(1).$$

In any characteristic $p \neq 2$, $S_{univ}/\mathcal{SD}(2, 4)$ is a family of Del

Pezzo surfaces of degree 2, so the group G_{geom}, and indeed G_{arith}, is a priori a subgroup of the Weyl group $W(E_7)$, in its 7-dimensional reflection representation.

Since $W(E_7)$ is finite, the normal subgroup $G_{geom} \subset G_{arith}$ is of finite index, and the quotient is a finite cyclic group. If over k, G_{geom} has index n in $G_{arith/k}$, then over the extension k_n/k of degree n, G_{geom} is equal to G_{arith/k_n}. So at the expense of replacing k by a finite extension, we reduce to the case when $G_{geom} = G_{arith}$.

Thus G_{geom} is a subgroup of $W(E_7)$ with fourth moment 3, cf. the first paragraph of the proof of 5.5.2, hence (by 2.6.3) is primitive (thought of as a subgroup of $GL(7) := GL(7, \overline{\mathbb{Q}}_\ell)$). If we knew that G_{geom} contained a reflection, we could conclude, by invoking Theorem 2.6.9, that G_{geom} is $W(E_7)$.

To show that G_{geom} contains a reflection, it suffices to show that, over any sufficiently large field k of odd characteristic, G_{arith} contains a reflection. So it suffices to exhibit a single Del Pezzo surface X/k of degree 2, such that Frobenius $Frob_k$ acts on NS(X) by a reflection, and then to write X/k as S_f for some f in $\mathcal{SD}(2, 4)(k)$. Then $Frob_{k,f}$ in G_{arith} is the desired reflection. [Since Frobenius preserves the canonical class, if it acts on NS(X) as a reflection, it also acts on $NS(X)^\perp$ as a reflection.] To get such an X/k, simply blow up $\mathbb{P}^2$ over k in

5 k-points and one closed point of degree 2.

This trick for getting a reflection in G_{geom} is inspired by [Erne, section 3]. [The reader should be warned that this otherwise delightful paper contains an error on page 21: if one blows up $\mathbb{P}^2$ over k in

3 k-points and one closed point of degree 4,

one gets a Frobenius whose projection mod ±1 to $Sp(6, \mathbb{F}_2)$ lies in the conjugacy class 4C in ATLAS notation, not in the class 4A as asserted.] QED

(5.6.15) We now turn to the remaining cases. We must show that G_{geom} for $Gr^0\mathcal{M}(n+1, e, \chi)$ is never finite, for

$$n+1 \geq 3,\ e \geq 3 \text{ prime to } p,\ \chi \text{ nontrivial},\ \chi^e = \mathbb{1}.$$

We have already shown that $AddedVar^*Gr^0\mathcal{M}(n+1, e, \chi)$ contains as a direct factor $Prim^{n-1}\overline{\pi}(n,e)_!\overline{\mathbb{Q}}_\ell$, the $Prim^{n-1}$ along the fibres for the universal family of projective smooth hypersurfaces of degree e in $\mathbb{P}^n$ which are transverse to the hyperplane $X_0 = 0$. This family is the restriction to a dense open set of the parameter space of the universal family of projective smooth hypersurfaces of degree e in

$\mathbb{P}^n$ (with no imposed transversality to the hyperplane $X_0 = 0$), so its Prim^{n-1} has the same G_{geom} as for the entire universal family. As explained in [Ka-Sar-RMFEM, 11.4.9], $\mathrm{Prim}^{n-1}\overline{\pi}(n,e)_!\overline{\mathbb{Q}}_\ell$ for $n \geq 2$ and $e \geq 3$ has infinite (indeed, as large as possible, but that does not concern us here) G_{geom} except in one single case, namely $(n = 3, e = 3)$, the case of cubic surfaces. Thus we find that for

$$n+1 \geq 3,\ e \geq 3 \text{ prime to } p,\ \chi \text{ nontrivial},\ \chi^e = \mathbb{1},$$

G_{geom} for $\mathrm{Gr}^0\mathcal{M}(n+1, e, \chi)$ is infinite except possibly for the single case $\mathrm{Gr}^0\mathcal{M}(4, 3, \chi)$. To treat this last case, we use the fact that $\mathrm{AddedVar}^*\mathrm{Gr}^0\mathcal{M}(4, 3, \chi)$ contains as a direct factor $\mathrm{Gr}^0\mathcal{M}(3, 3, \chi^2)$, which we have already shown to have infinite G_{geom}. QED

Chapter 6: Middle additive convolution

(6.1) Middle convolution and its effect on local monodromy

(6.1.1) Let k be an algebraically closed field, ℓ a prime invertible in k, and K and L objects in $D^b_c(\mathbb{A}^1, \overline{\mathbb{Q}}_\ell)$. Recall [Ka-RLS, 2.5] that the ! convolution $K*_!L$ and the $*$ convolution $K*_*L$ are defined as objects in $D^b_c(\mathbb{A}^1, \overline{\mathbb{Q}}_\ell)$ as follows. On $\mathbb{A}^2$ with its two projections to $\mathbb{A}^1$, one forms the external tensor product $K\boxtimes L := (pr_1{}^*K)\otimes(pr_2{}^*L)$. Then one uses the sum map

$$\mathrm{sum}: \mathbb{A}^2 \to \mathbb{A}^1,$$

to define

$$K*_!L := R\mathrm{sum}_!(K\boxtimes L),$$
$$K*_*L := R\mathrm{sum}_*(K\boxtimes L).$$

(6.1.2) Suppose now that K and L are both perverse. There is no reason for either $K*_!L$ or $K*_*L$ to be perverse. We say that the perverse object K has $\mathcal{P}$ if, for every perverse object L, both $K*_!L$ and $K*_*L$ are perverse. If K has $\mathcal{P}$, we define the middle additive convolution $K*_{mid}L$ to be the image, in the abelian category of perverse sheaves on $\mathbb{A}^1$, of the canonical "forget supports" map:

$$K*_{mid}L := \mathrm{Image}(R\mathrm{sum}_!(K\boxtimes L) \to R\mathrm{sum}_*(K\boxtimes L)).$$

One knows [Ka-RSL, 2.6.17] that if both K and L have $\mathcal{P}$, then their middle convolution $K*_{mid}L$ has $\mathcal{P}$. On the other hand, if K has $\mathcal{P}$ but L is constant (e.g., if L is $\overline{\mathbb{Q}}_\ell[1]$), then all three of $K*_!L$, $K*_*L$, and $K*_{mid}L$ are constant, and one knows that no nonzero constant perverse sheaf has $\mathcal{P}$.

(6.1.3) Suppose henceforth that k has positive characteristic p, and fix a nontrivial additive character ψ of the prime field $\mathbb{F}_p$. Then Fourier Transform is an autoequivalence $K \mapsto FT_\psi(K)$ of the abelian category of perserve sheaves on $\mathbb{A}^1$. One knows [Ka-RLS, 2.10.3] that a perverse object K on $\mathbb{A}^1$ has $\mathcal{P}$ if and only if $FT_\psi(K)$ is a middle extension, i.e., of the form $(j_*\mathcal{N})[1]$ for $j : U \subset \mathbb{A}^1$ the inclusion of a dense open set and for $\mathcal{N}$ a lisse $\overline{\mathbb{Q}}_\ell$-sheaf on U. Moreover, for K and L both perverse objects with $\mathcal{P}$, if we pick a common dense open set $j : U \subset \mathbb{A}^1$ such that $FT_\psi(K) = (j_*\mathcal{N})[1]$ and $FT_\psi(L) = (j_*\mathcal{M})[1]$, with $\mathcal{N}$ and $\mathcal{M}$ lisse $\overline{\mathbb{Q}}_\ell$-sheaves on U, then we have [Ka-RLS, 2.10.8]

$$FT_\psi(K*_{mid}L) = (j_*(\mathcal{N}\otimes\mathcal{M}))[1],$$

and [Ka-RLS, 2.10.1]

$$FT_\psi(K*_!L) = ((j_*\mathcal{N})\otimes(j_*\mathcal{M}))[1],$$

(6.1.4) There is a simple classification of perverse irreducibles on $\mathbb{A}^1$. If K is perverse irreducible, then K is either itself an irreducible middle extension (a middle extension $(j_*\mathcal{N})[1]$ as above, with $\mathcal{N}$ irreducible as a lisse sheaf on U), or is punctual, δ_α for some α in k. Applying this classification to the perverse irreducible object $FT_\psi(K)$, and using the result of the previous paragraph, we see that a perverse irreducible K has $\mathcal{P}$ unless K is $\mathcal{L}_{\psi(\alpha x)}[1]$ for some α in k (i.e., unless $FT_\psi(K)$ is a δ_α).

(6.1.5) Consider now a perverse object K which is semisimple, i.e., a direct sum of perverse irreducibles. Clearly such a K has $\mathcal{P}$ if and only if each of its irreducible constituents has $\mathcal{P}$, i.e., if and only if none of its irreducible constituents is $\mathcal{L}_{\psi(\alpha x)}[1]$ for some α in k.

Lemma 6.1.6 Suppose K and L are each perverse, semisimple, and have $\mathcal{P}$. Then their middle convolution $K*_{mid}L$ is perverse, semisimple, and has $\mathcal{P}$.

proof On the Fourier Transform side, this is the assertion that given lisse sheaves $\mathcal{N}$ and $\mathcal{M}$ on U which are both completely reducible as $\overline{\mathbb{Q}}_\ell$-representations of $\pi_1(U)$, $\mathcal{N}\otimes\mathcal{M}$ is also completely reducible as a $\overline{\mathbb{Q}}_\ell$-representation of $\pi_1(U)$. But one knows that over a field of characteristic zero, the tensor product of completely reducible finite-dimensional representations of any group G is again completely reducible, cf. [Chev, page 88]. QED

(6.1.7) Let us say that a perverse semisimple K is non-punctual if none of its irreducible constituents is punctual, i.e., if it is a direct sum of irreducible middle extensions. It has $\mathcal{P}$ if and only if none of the summands is $\mathcal{L}_{\psi(\alpha x)}[1]$ for any α in k. Thus the class of perverse semisimple K which are non-punctual and which have $\mathcal{P}$ is stable by Fourier Transform. Any perverse semisimple K which is non-punctual is a middle extension, $K = (j_*\mathcal{K})[1]$. If in addition K has $\mathcal{P}$, then $\mathcal{K}$ is a "Fourier sheaf" in the sense of [Ka-ESDE, 7.3.5], indeed $\mathcal{K}$ is a direct sum of irreducible Fourier sheaves.

(6.1.8) It is **not true** in general that if K and L are each perverse, semisimple, have $\mathcal{P}$, and are non-punctual, then their middle convolution $K*_{mid}L$ (which by the above lemma is perverse semisimple and has $\mathcal{P}$) is also non-punctual. For instance, take K and L to be inverse Kummer sheaves $\mathcal{L}_\chi[1]$ and $\mathcal{L}_{\overline{\chi}}[1]$, for χ a nontrivial multiplicative character of a finite subfield of k. Then up to a Tate twist $K*_{mid}L$ is δ_0.

Lemma 6.1.9 Suppose K and L are both perverse, irreducible, non-punctual, and have $\mathcal{P}$. Then $K*_{mid}L$ (which by the above lemma is perverse semisimple and has $\mathcal{P}$) is non-punctual if and only if there exists no isomorphism

$$L \cong [x \mapsto x+\alpha]^*[-1]^*DK$$

for any α in $\mathbb{A}^1$.

proof Write $FT_\psi(K)$ as $(j_*\mathcal{N})[1]$ and write $FT_\psi(L)$ as $(j_*\mathcal{M})[1]$, for some irreducibles $\mathcal{N}$ and $\mathcal{M}$ on a common dense open set $U \subset \mathbb{A}^1$. We know that

$$FT_\psi(K*_{mid}L) = (j_*(\mathcal{N}\otimes\mathcal{M}))[1].$$

Suppose that $K*_{mid}L$ fails to be non-punctual. This means that $(j_*(\mathcal{N}\otimes\mathcal{M}))[1]$ has some $\mathcal{L}_{\psi(\alpha x)}[1]$ as a constituent, i.e., that for some α in k, $\mathcal{L}_{\psi(\alpha x)}\otimes\mathcal{N}\otimes\mathcal{M}$ has $\overline{\mathbb{Q}}_\ell$ as a constituent. Since $\mathcal{M}$ and $\mathcal{N}$ are irreducible, $\mathcal{M}$ and $\mathcal{L}_{\psi(\alpha x)}\otimes\mathcal{N}$ are both irreducible. If $\mathcal{L}_{\psi(\alpha x)}\otimes\mathcal{N}\otimes\mathcal{M}$ has $\overline{\mathbb{Q}}_\ell$ as a constituent, then projection onto that constituent is a nonzero pairing of irreducible lisse sheaves on U,

$$\mathcal{L}_{\psi(\alpha x)}\otimes\mathcal{N} \times \mathcal{M} \to \overline{\mathbb{Q}}_\ell,$$

i.e., it is a nonzero map of sheaves

$$\mathcal{M} \to (\mathcal{L}_{\psi(\alpha x)}\otimes\mathcal{N})^\vee.$$

Since source and target are irreducible, such a nonzero map is an isomorphism. Conversely, if $\mathcal{M} \cong (\mathcal{L}_{\psi(\alpha x)}\otimes\mathcal{N})^\vee$, then $\mathcal{L}_{\psi(\alpha x)}\otimes\mathcal{N}\otimes\mathcal{M}$ $= \mathrm{End}(\mathcal{M}_\lambda)$ visibly has $\overline{\mathbb{Q}}_\ell$ as a constituent. Thus $K*_{mid}L$ fails to be non-punctual if and only if there exists an isomorphism

$$\mathcal{M} \cong (\mathcal{L}_{\psi(\alpha x)}\otimes\mathcal{N})^\vee,$$

i.e., an isomorphism

$$(j_*\mathcal{M})[1] \cong \mathcal{L}_{\psi(-\alpha x)}\otimes(j_*\mathcal{N}^\vee)[1],$$

i.e., an isomorphism

$$FT_\psi(L) \cong \mathcal{L}_{\psi(-\alpha x)}\otimes(FT_\psi([-1]^*DK)),$$

i.e., an isomorphism

$$FT_\psi(L) \cong FT_\psi([x \mapsto x+\alpha]^*[-1]^*DK),$$

i.e., an isomorphism

$$L \cong [x \mapsto x+\alpha]^*[-1]^*DK. \text{ QED}$$

(6.1.10) In order to apply the above result, we need criteria for showing that L is not isomorphic to any additive translate of $[-1]^*DK$. We will give two such criteria, both based on the local monodromies at finite distance of K and L. The first criterion is based entirely on the number and locations of the finite singularities.

Theorem 6.1.11 Suppose K is perverse, semisimple, and has $\mathcal{P}$. Suppose L is perverse irreducible and non-punctual, i.e., suppose L is an irreducible middle extension, say $L = (j_*\mathcal{L})[1]$ for some irreducible $\mathcal{L}$ on U. Denote by s the number of finite singularities of L, i.e., the number of points in $\mathbb{A}^1$ at which the sheaf $j_*\mathcal{L}$ on $\mathbb{A}^1$ fails to be lisse. For each λ in $k^\times$, consider the automorphism $x \mapsto \lambda x$ of $\mathbb{A}^1$, and define

$$\mathrm{MultTrans}_\lambda(L) := [x \mapsto \lambda x]_\star L = [x \mapsto \lambda^{-1}x]^\star L.$$

Then we have the following results.

1) Suppose $s \geq 1$. For each λ in $k^\times$, $\mathrm{MultTrans}_\lambda(L)$ is perverse irreducible, non-punctual, and has $\mathcal{P}$.

2) Suppose $s \geq 2$. Denote by $m(K)$ the number of distinct isomorphism classes which occur among those non-punctual irreducible constituents of K which have precisely s finite singularities. For all but at most $s(s-1)m(K)$ values of λ in $k^\times$, the middle convolution $K \star_{mid} \mathrm{MultTrans}_\lambda(L)$ is (perverse, semisimple, has $\mathcal{P}$, and is) non-punctual.

proof Suppose $s \geq 1$. To prove 1), we argue as follows. Since $x \mapsto \lambda x$ is an automorphism of $\mathbb{A}^1$, and L is perverse irreducible and non-punctual, so is $\mathrm{MultTrans}_\lambda(L)$. Since $\mathrm{MultTrans}_\lambda(L)$ has finite singularities, it is not $\mathcal{L}_{\psi(\alpha x)}[1]$ for any α, and so $\mathrm{MultTrans}_\lambda(L)$ has $\mathcal{P}$. This proves 1). By Lemma 6.1.7 we know that, for every λ in $k^\times$, $K \star_{mid} \mathrm{MultTrans}_\lambda(L)$ is perverse, semisimple, and has $\mathcal{P}$.

Suppose now $s \geq 2$. To prove 2), we argue as follows. We must show that for all but at most $s(s-1)m(K)$ λ in $k^\times$, $K \star_{mid} \mathrm{MultTrans}_\lambda(L)$ is non-punctual. Since K is semisimple, we reduce immediately to the case when K is perverse irreducible and has $\mathcal{P}$.

If K is punctual, say δ_α, $K \star_{mid} \mathrm{MultTrans}_\lambda(L)$ is non-punctual for every λ in $k^\times$, since (any flavor of) convolution with δ_α is just additive translation by α, and L and all its multiplicative translates $\mathrm{MultTrans}_\lambda(L)$ are non-punctual. Thus there are no exceptional λ, as asserted.

Suppose now that K is an irreducible middle extension that has $\mathcal{P}$, say $K = (j_\star \mathcal{K})[1]$ with $\mathcal{K}$ irreducible on U. By the previous result, $K \star_{mid} \mathrm{MultTrans}_\lambda(L)$ fails to be non-punctual if and only if there exists an isomorphism

$$\mathrm{MultTrans}_\lambda(L) \cong [x \mapsto x+\alpha]^\star[-1]^\star DK$$

for some α in $\mathbb{A}^1$.

We must show that there are at most $s(s-1)$ values of λ in $k^\times$ such that for some α in k we have such an isomorphism. For this, we argue as follows. By assumption, K is an irreducible middle extension $(j_\star \mathcal{K})[1]$. Then DK is, up to a Tate twist, the irreducible middle extension $(j_\star \mathcal{K}^\vee)[1]$.

Suppose first that K is lisse on all of $\mathbb{A}^1$. Then for every α in k, $[x \mapsto x+\alpha]^\star[-1]^\star DK$ is lisse on $\mathbb{A}^1$. But $\mathrm{MultTrans}_\lambda(L)$ has at least two finite singularities, so such an isomorphism exists for no λ in $k^\times$.

Suppose next that K has $r \geq 1$ finite singularities, say $\beta_1,\ldots, \beta_r$. Then DK has $\beta_1,\ldots, \beta_r$ as its finite singularities, and

$$[x \mapsto x+\alpha]^*[-1]^*DK$$

has as its finite singularities the r points

$$-\alpha - \beta_1, -\alpha - \beta_2,\ldots, -\alpha - \beta_r.$$

On the other hand, L had $s \geq 2$ finite singularities, say

$$\gamma_1,\ldots, \gamma_s.$$

Then $\mathrm{MultTrans}_\lambda(L)$ has s finite singularities

$$\lambda\gamma_1,\ldots, \lambda\gamma_s.$$

If $r \neq s$, then there can be no isomorphism

$$\mathrm{MultTrans}_\lambda(L) \cong [x \mapsto x+\alpha]^*[-1]^*DK,$$

simply because the two sides have different numbers of finite singularities.

If $r= s$, and if such an isomorphism exists, then for some permutation σ of $\{1,\ldots, s\}$, we have

$$\lambda\gamma_i = -\alpha - \beta_{\sigma(i)} \text{ for } i =1 \text{ to } s.$$

Subtracting the first two such equations (remember $s \geq 2$), we get

$$\lambda(\gamma_1 - \gamma_2) = \beta_{\sigma(2)} - \beta_{\sigma(1)}.$$

So we may solve for λ:

$$\lambda = (\beta_{\sigma(2)} - \beta_{\sigma(1)})/(\gamma_1 - \gamma_2).$$

Thus for each of the $s(s-1)$ possible values of $(\sigma(1), \sigma(2))$, we get a possible λ. For any other λ in $k^\times$, there can exist no such isomorphism. QED

Corollary 6.1.12 Suppose K is perverse, semisimple, non-punctual, and has $\mathcal{P}$. Denote by $r \geq 0$ the number of finite singularities of K; if $r \geq 1$, denote them $\beta_1,\ldots, \beta_r$. Suppose L is perverse irreducible and non-punctual, i.e.,suppose L is an irreducible middle extension, say $L = (j_*\mathcal{L})[1]$ for some irreducible $\mathcal{L}$ on U. Suppose L has $s \geq 2$ finite singularities $\gamma_1,\ldots, \gamma_s$.

1) If $r < s$, then for every λ in $k^\times$, $K*_{\mathrm{mid}}\mathrm{MultTrans}_\lambda(L)$ is (perverse, semisimple, has $\mathcal{P}$, and is) non-punctual.

2) If $r \geq s$, then $K*_{\mathrm{mid}}\mathrm{MultTrans}_\lambda(L)$ is (perverse, semisimple, has $\mathcal{P}$, and is) non-punctual, provided that λ in $k^\times$ is not any of the $r(r-1)$ ratios

$$(\beta_{i_0} - \beta_{i_1})/(\gamma_1 - \gamma_2), \text{ with } i_0 \neq i_1.$$

proof In case 1), $m(K) = 0$. In case 2, any irreducible constituent of K with s finite singularities has its finite singularities among the β_i, so the excluded λ for that irreducible constituent are among the named ratios. QED

(6.1.13) For ease of later reference, we state the following elementary lemma, whose proof is left to the reader.

Lemma 6.1.14 Let $r \geq 1$ and $s \geq 1$ be integers, and k a field. Let λ be an element of $k^\times$. Let $\beta_1, ..., \beta_r$ be r distinct elements of k. Let $\gamma_1, ..., \gamma_s$ be s distinct elements of k.
1) If $r = 1$ or if $s = 1$, the rs numbers $\beta_i + \lambda\gamma_j$, $i = 1$ to r, $j = 1$ to s, are all distinct in k.
2) If $r \geq 2$ and $s \geq 2$, the rs numbers $\beta_i + \lambda\gamma_j$, $i = 1$ to r, $j = 1$ to s, are all distinct in k, provided that λ in $k^\times$ is not any of the $r(r-1)(s)(s-1)$ ratios

$$(\beta_{i_0} - \beta_{i_1})/(\gamma_{j_0} - \gamma_{j_1}), \text{ with } i_0 \neq i_1 \text{ and } j_0 \neq j_1.$$

(6.1.15) We now give a criterion for the non-punctuality of $K*_{\text{mid}}L$ based on the local monodromy at the finite singularities. It has the advantage of sometimes applying when L has only one finite singularity. In the ensuing discussion, we will "compare" the local monodromy of one lisse sheaf, say $\mathcal{K}$, on an open dense set of $\mathbb{A}^1$ at one finite singularity, say β, with the local monodromy of another lisse sheaf, say $\mathcal{L}$, on an open dense set of $\mathbb{A}^1$ at another finite singularity, say γ. We use additive translation by γ, resp. β, to view $I(\gamma)$-representations, resp. $I(\beta)$-representations as $I(0)$-representations. Then we view both the local monodromy of $\mathcal{K}$ at β and the local monodromy of $\mathcal{L}$ at γ as $I(0)$-representations, and it is as such that we compare them.

Theorem 6.1.16 Suppose K is perverse, semisimple, non-punctual, and has $\mathcal{P}$. Thus K is $(j_*\mathcal{K})[1]$ for some semisimple $\mathcal{K}$ on U. Suppose L is perverse irreducible and non-punctual, i.e., suppose L is an irreducible middle extension, say $L = (j_*\mathcal{L})[1]$ for some irreducible $\mathcal{L}$ on U. Suppose there exists a finite singularity γ of $\mathcal{L}$ such that the local monodromy of $\mathcal{L}$ at γ is not a direct factor of the local monodromy of $\mathcal{K}^\vee$ at any finite singularity β of $\mathcal{K}$. Then $K*_{\text{mid}}L$ is (perverse, semisimple, has $\mathcal{P}$, and is) non-punctual. Moreover, for every λ in $k^\times$, $K*_{\text{mid}}\text{MultTrans}_\lambda(L)$ is (perverse, semisimple, has $\mathcal{P}$, and is) non-punctual.

proof Since L has a finite singularity and is perverse irreducible, it has $\mathcal{P}$. We know that $K*_{\text{mid}}L$ is non-punctual unless there exists an irreducible constituent K_i of K, an α in $\mathbb{A}^1$, and an isomorphism

$$L \cong [x \mapsto x+\alpha]^*[-1]^*DK_i.$$

If such an isomorphism exists, then the local monodromy of $\mathcal{L}$ at γ is isomorphic to the local monodromy of $\mathcal{K}_i^\vee$ at $\beta := -\gamma - \alpha$, which is a direct summand of the local monodromy of $\mathcal{K}^\vee$ at $-\gamma - \alpha$, contradiction. Therefore $K*_{\text{mid}}L$ is (perverse, semisimple, has $\mathcal{P}$, and is) non-punctual. For any λ in $k^\times$, $\text{MultTrans}_\lambda(L)$ has the same local monodromy at $\lambda\gamma$ as L did at γ, so satisfies the same hypotheses. QED

(6.1.17) We now recall some relations between the local monodromies at finite distance of K, L, and $K*_{mid}\mathrm{MultTrans}_\lambda(L)$, for sufficiently general λ. The key point is that, for sufficiently general λ, each of these three is perverse, semisimple, non-punctual, and has $\mathcal{P}$. In particular, each is a middle extension, and the Fourier Transform of each is a middle extension. This allows us to use the fundamental results of Laumon, which relate the local monodromies at finite distance of a middle extension to the $I(\infty)$-representation of its Fourier Transform, cf. [Ka-ESDE, 7.4.2].

Theorem 6.1.18 Suppose K is perverse, semisimple, non-punctual, and has $\mathcal{P}$. Write $K = (j_*\mathcal{K})[1]$. Denote by $r \geq 0$ the number of finite singularities of $\mathcal{K}$; if $r \geq 1$, denote them $\beta_1,\ldots, \beta_r$. Suppose L is perverse irreducible and non-punctual, $L = (j_*\mathcal{L})[1]$. Suppose $\mathcal{L}$ has $s \geq 1$ finite singularities $\gamma_1,\ldots, \gamma_s$. Suppose that either

a) $s = 1$, and Theorem 6.1.11 applies,

or

b) $s \geq 2$.

Suppose further that either K or L has all ∞-slopes ≤ 1 (e.g., this holds if either K or L is tame at ∞). Then we have the following results.

1) If $s=1$, take any λ in $k^\times$. If $s \geq 2$, take any λ in $k^\times$ which is not any of the $r(r-1)(s)(s-1)$ ratios

$$(\beta_{i_0} - \beta_{i_1})/(\gamma_{j_0} - \gamma_{j_1}), \text{ with } i_0 \neq i_1 \text{ and } j_0 \neq j_1.$$

Then $K*_{mid}\mathrm{MultTrans}_\lambda(L)$ (which is perverse, semisimple has $\mathcal{P}$, and is non-punctual, by the results above) has precisely rs finite singularities, at the points $\beta_i + \lambda\gamma_j$, $i = 1$ to r, $j = 1$ to s.

2) Fix any λ in $k^\times$ not excluded in 1) above. Write

$$\mathrm{MultTrans}_\lambda(L) = (j_*\mathcal{L}_\lambda)[1],$$

$$K*_{mid}\mathrm{MultTrans}_\lambda(L) = (j_*\mathcal{Q}_\lambda)[1].$$

Denote by $\mathcal{K}(\beta_i)$ the $I(\beta_i)$-representation given by $\mathcal{K}$, by $\mathcal{L}(\gamma_j)$ the $I(\gamma_j)$-representation given by $\mathcal{L}$, and by $\mathcal{Q}_\lambda(\beta_i + \lambda\gamma_j)$ the $I(\beta_i + \lambda\gamma_j)$-representation given by $\mathcal{Q}_\lambda$. Denote by $FT_\psi\mathrm{loc}(s, \infty)$ Laumon's local Fourier Transform functors, cf. [Lau-TF, 2.4.2.3], [Ka-ESDE, 7.4.1]. Then we have an isomorphism of $I(\infty)$-representations

$$FT_\psi\mathrm{loc}(\beta_i + \lambda\gamma_j, \infty)(\mathcal{Q}_\lambda(\beta_i + \lambda\gamma_j)/\mathcal{Q}_\lambda(\beta_i + \lambda\gamma_j)^{I(\beta_i + \lambda\gamma_j)}) \cong$$

$$FT_\psi\mathrm{loc}(\beta_i, \infty)(\mathcal{K}(\beta_i)/\mathcal{K}(\beta_i)^{I(\beta_i)})\otimes FT_\psi\mathrm{loc}(\lambda\gamma_j, \infty)(\mathcal{L}_\lambda(\lambda\gamma_j)/\mathcal{L}_\lambda(\lambda\gamma_j)^{I(\lambda\gamma_j)}).$$

3) Suppose in addition that $\mathcal{K}$ is tamely ramified at the finite singularity β_i, and that $\mathcal{L}$ is tamely ramified at the finite singularity γ_j. Use additive translation by $\beta_i + \lambda\gamma_j$, β_i, and γ_j respectively to view

$I(\beta_i + \lambda\gamma_j)$-representations, $I(\beta_i)$-representations, and $I(\gamma_j)$-representations respectively all as $I(0)$-representations. Then we have an isomorphism of of $I(0)$-representations

$$\mathcal{Q}_\lambda(\beta_i + \lambda\gamma_j)/\mathcal{Q}_\lambda(\beta_i + \lambda\gamma_j)^{I(\beta_i + \lambda\gamma_j)} \text{ moved to } 0$$
$$\cong (\mathcal{K}(\beta_i)/\mathcal{K}(\beta_i)^{I(\beta_i)})\otimes(\mathcal{L}(\gamma_j)/\mathcal{L}(\gamma_j)^{I(\gamma_j)}) \text{ moved to } 0.$$

proof This is a simple consequence of Laumon's determination of the monodromy of a Fourier Transform, cf. [Ka-ESDE, 7.4.1 and 7.4.2], [Lau-TF, 2.3.3.1 (iii) and 2.4.2.3], and [Ka-RLS, proof of 3.0.4, 3.3.5, and 3.3.6]. Let us write

$$FT_\psi(K) = (j_*\mathcal{N})[1],$$
$$\mathrm{MultTrans}_\lambda(L) = (j_*\mathcal{L}_\lambda)[1],$$
$$FT_\psi(L) = (j_*\mathcal{M})[1],$$
$$FT_\psi(\mathrm{MultTrans}_\lambda(L)) = \mathrm{MultTrans}_{1/\lambda}FT_\psi(L) = (j_*\mathcal{M}_\lambda)[1],$$
$$FT_\psi(K*_{mid}\mathrm{MultTrans}_\lambda(L)) = (j_*\mathcal{R}_\lambda)[1].$$

Then as $I(\infty)$-representations, we have

$$\mathcal{N}(\infty)$$
$$\cong FT_\psi loc(\infty, \infty)(\mathcal{K}(\infty)) \bigoplus \oplus_i \mathcal{L}_{\psi(\beta_i t)}\otimes FT_\psi loc(\beta_i, \infty)(\mathcal{K}(\beta_i)/\mathcal{K}(\beta_i)^{I(\beta_i)}),$$

and

$$\mathcal{M}_\lambda(\infty)$$
$$\cong FT_\psi loc(\infty, \infty)(\mathcal{L}_\lambda(\infty)) \bigoplus$$
$$\oplus_j \mathcal{L}_{\psi(\lambda\gamma_j t)}\otimes FT_\psi loc(\lambda\gamma_j, \infty)(\mathcal{L}_\lambda(\lambda\gamma_j)/\mathcal{L}_\lambda(\lambda\gamma_j)^{I(\lambda\gamma_j)}).$$

By the fundamental relation [Ka-RLS, 2.10.8], we have

$$\mathcal{R}_\lambda(\infty) \cong \mathcal{N}(\infty)\otimes\mathcal{M}_\lambda(\infty).$$

Recall that each $FT_\psi loc(s, \infty)$ for s in $\mathbb{A}^1$ takes values in $I(\infty)$-representations of slope < 1, and is an equivalence

$$I(s)\text{-representations} \cong I(\infty)\text{-representations with all slopes} < 1.$$

Moreover, for variable s in $\mathbb{A}^1$, the various $FT_\psi loc(s, \infty)$ are carried into each other by additive translation.

Recall that $FT_\psi loc(\infty, \infty)$ takes values in $I(\infty)$-representations with all slopes > 1, and vanishes precisely on those $I(\infty)$-representations with all slopes ≤ 1. Thus for K perverse semisimple with $\mathcal{P}$ and non-punctual, we recover the finite singularities of K as those β such that $\mathcal{N}(\infty)\otimes\mathcal{L}_{\psi(-\beta t)}$ has a nonzero "slope < 1" part, we recover

$$FT_\psi loc(\infty, \infty)(\mathcal{K}(\infty))$$

as the "slope > 1" part of $\mathcal{N}(\infty)$, and we recover

$$\oplus_i \mathcal{L}_{\psi(\beta_i t)}\otimes FT_\psi loc(\beta_i, \infty)(\mathcal{K}(\beta_i)/\mathcal{K}(\beta_i)^{I(\beta_i)})$$

as the "slope ≤ 1" part of $\mathcal{N}(\infty)$.

The hypothesis that either K or L has all ∞-slopes ≤ 1 means that either $FT_\psi loc(\infty, \infty)(\mathcal{K}(\infty))$ or $FT_\psi loc(\infty, \infty)(\mathcal{L}_\lambda(\infty))$ vanishes, i.e., that

$$FT_\psi loc(\infty, \infty)(\mathcal{K}(\infty)) \otimes FT_\psi loc(\infty, \infty)(\mathcal{L}_\lambda(\infty)) = 0.$$

Recall also [Ka-GKM,1.3] the crude behaviour of ∞-slopes under tensor product:

$$(\text{all slopes} > 1) \otimes (\text{all slopes} \le 1) = (\text{all slopes} > 1),$$
$$(\text{all slopes} \le 1) \otimes (\text{all slopes} \le 1) = (\text{all slopes} \le 1),$$
$$(\text{all slopes} < 1) \otimes (\text{all slopes} < 1) = (\text{all slopes} < 1).$$

So when we compute $\mathcal{R}_\lambda(\infty) \cong \mathcal{N}(\infty) \otimes \mathcal{M}_\lambda(\infty)$ by expanding out both tensorees, the one term whose slopes were in doubt, namely

$$FT_\psi loc(\infty, \infty)(\mathcal{K}(\infty)) \otimes FT_\psi loc(\infty, \infty)(\mathcal{L}_\lambda(\infty)),$$

conveniently vanishes. We find

$$\mathcal{R}_\lambda(\infty) \cong (\text{all slopes} > 1) \bigoplus \oplus_{i,j} \mathcal{L}_{\psi((\beta_i + \lambda\gamma_j)t)} \otimes V_{i,j},$$

with $V_{i,j}$ the $I(\infty)$-representation with all slopes < 1 given by

$$FT_\psi loc(\beta_i, \infty)(\mathcal{K}(\beta_i)/\mathcal{K}(\beta_i)^{I(\beta_i)}) \otimes FT_\psi loc(\lambda\gamma_j, \infty)(\mathcal{L}_\lambda(\lambda\gamma_j)/\mathcal{L}_\lambda(\lambda\gamma_j)^{I(\lambda\gamma_j)}).$$

We have already seen that the rs points $\beta_i + \lambda\gamma_j$ are all distinct. This decomposition shows that the finite singularities are precisely these rs points. Moreover, at each of the rs points $\beta_i + \lambda\gamma_j$ we have the relation

$$FT_\psi loc(\beta_i + \lambda\gamma_j, \infty)(\mathcal{Q}_\lambda(\beta_i + \lambda\gamma_j)/\mathcal{Q}_\lambda(\beta_i + \lambda\gamma_j)^{I(\beta_i + \lambda\gamma_j)}) \cong$$
$$FT_\psi loc(\beta_i, \infty)(\mathcal{K}(\beta_i)/\mathcal{K}(\beta_i)^{I(\beta_i)}) \otimes FT_\psi loc(\lambda\gamma_j, \infty)(\mathcal{L}_\lambda(\lambda\gamma_j)/\mathcal{L}_\lambda(\lambda\gamma_j)^{I(\lambda\gamma_j)}).$$

Furthermore, if, for some i and j, $\mathcal{K}$ is tame at β_i and $\mathcal{L}$ is tame at β_j, then the explicit description of $FT_\psi loc(s, \infty)$ on tame $I(s)$-representations, namely

$$FT_\psi loc(s, \infty)(M(s)) = [x-s \mapsto 1/(x - s)]^* M(s)$$

when $M(s)$ is tame, shows that

$$\mathcal{Q}_\lambda(\beta_i + \lambda\gamma_j)/\mathcal{Q}_\lambda(\beta_i + \lambda\gamma_j)^{I(\beta_i + \lambda\gamma_j)} \text{ moved to } 0$$
$$\cong (\mathcal{K}(\beta_i)/\mathcal{K}(\beta_i)^{I(\beta_i)}) \otimes (\mathcal{L}(\gamma_j)/\mathcal{L}(\gamma_j)^{I(\gamma_j)}) \text{ moved to } 0,$$

as required. QED

Unipotent Pseudoreflection Input Corollary 6.1.19 Hypotheses as in Theorem 6.1.18, suppose in addition that the local monodromy of $\mathcal{K}$ at the finite singularity β_i is a unipotent pseudoreflection. Fix any λ in $k^\times$. If $s \ge 2$, suppose λ is not any of the $r(r-1)(s)(s-1)$ ratios

$$(\beta_{i_0} - \beta_{i_1})/(\gamma_{j_0} - \gamma_{j_1}), \text{ with } i_0 \ne i_1 \text{ and } j_0 \ne j_1,$$

and write

$$K *_{mid} MultTrans_\lambda(L) = (j_* \mathcal{Q}_\lambda)[1].$$

Use additive translation by $\beta_i + \lambda\gamma_j$, β_i and γ_j respectively to view $I(\beta_i + \lambda\gamma_j)$-representations, $I(\beta_i)$-representations, and $I(\gamma_j)$-

representations respectively all as I(0)-representations. Then for any j, the local monodromy of $\mathcal{Q}_\lambda$ at $\beta_i + \lambda\gamma_j$ is related to the local monodromy of $\mathcal{L}$ at γ_j by an isomorphism of I(0)-representations

$$\mathcal{Q}_\lambda(\beta_i + \lambda\gamma_j)/\mathcal{Q}_\lambda(\beta_i + \lambda\gamma_j)^{I(\beta_i + \lambda\gamma_j)} \text{ moved to } 0$$
$$\cong \mathcal{L}(\gamma_j)/\mathcal{L}(\gamma_j)^{I(\gamma_j)} \text{ moved to } 0.$$

proof This is a special case of part 2) of the theorem, according to which for every i and j we have an isomorphism of $I(\infty)$-representations

$$FT_\psi loc(\beta_i + \lambda\gamma_j, \infty)(\mathcal{Q}_\lambda(\beta_i + \lambda\gamma_j)/\mathcal{Q}_\lambda(\beta_i + \lambda\gamma_j)^{I(\beta_i + \lambda\gamma_j)}) \cong$$
$$FT_\psi loc(\beta_i, \infty)(\mathcal{K}(\beta_i)/\mathcal{K}(\beta_i)^{I(\beta_i)}) \otimes FT_\psi loc(\lambda\gamma_j, \infty)(\mathcal{L}_\lambda(\lambda\gamma_j)/\mathcal{L}_\lambda(\lambda\gamma_j)^{I(\lambda\gamma_j)}).$$

If the local monodromy of $\mathcal{K}$ at the finite singularity β_i is a unipotent pseudoreflection, then

$$FT_\psi loc(\beta_i, \infty)(\mathcal{K}(\beta_i)/\mathcal{K}(\beta_i)^{I(\beta_i)}) = FT_\psi loc(\beta_i, \infty)(\overline{\mathbb{Q}}_\ell) = \overline{\mathbb{Q}}_\ell,$$

so we find an isomorphism of $I(\infty)$-representations

$$FT_\psi loc(\beta_i + \lambda\gamma_j, \infty)(\mathcal{Q}_\lambda(\beta_i + \lambda\gamma_j)/\mathcal{Q}_\lambda(\beta_i + \lambda\gamma_j)^{I(\beta_i + \lambda\gamma_j)}) \cong$$
$$FT_\psi loc(\lambda\gamma_j, \infty)(\mathcal{L}_\lambda(\lambda\gamma_j)/\mathcal{L}_\lambda(\lambda\gamma_j)^{I(\lambda\gamma_j)}).$$

By means of the translation convention to view

$$\mathcal{Q}_\lambda(\beta_i + \lambda\gamma_j)/\mathcal{Q}_\lambda(\beta_i + \lambda\gamma_j)^{I(\beta_i + \lambda\gamma_j)}$$

and

$$\mathcal{L}_\lambda(\lambda\gamma_j)/\mathcal{L}_\lambda(\lambda\gamma_j)^{I(\lambda\gamma_j)}$$

as I(0)-representations, we can rewrite this

$$FT_\psi loc(0, \infty)(\mathcal{Q}_\lambda(\beta_i + \lambda\gamma_j)/\mathcal{Q}_\lambda(\beta_i + \lambda\gamma_j)^{I(\beta_i + \lambda\gamma_j)} \text{ moved to } 0) \cong$$
$$FT_\psi loc(0, \infty)(\mathcal{L}_\lambda(\lambda\gamma_j)/\mathcal{L}_\lambda(\lambda\gamma_j)^{I(\lambda\gamma_j)} \text{ moved to } 0).$$

But $FT_\psi loc(0, \infty)$ is an equivalence of categories

$$I(0)\text{-representations} \cong I(\infty)\text{-representations with all slopes} < 1,$$

so we infer the existence of an I(0)-isomorphism

$$\mathcal{Q}_\lambda(\beta_i + \lambda\gamma_j)/\mathcal{Q}_\lambda(\beta_i + \lambda\gamma_j)^{I(\beta_i + \lambda\gamma_j)} \text{ moved to } 0$$
$$\cong \mathcal{L}_\lambda(\lambda\gamma_j)/\mathcal{L}_\lambda(\lambda\gamma_j)^{I(\lambda\gamma_j)} \text{ moved to } 0.$$

Meanwhile, multiplicative translation by λ carries $\mathcal{L}_\lambda(\lambda\gamma_j)/\mathcal{L}_\lambda(\lambda\gamma_j)^{I(\lambda\gamma_j)}$ to $\mathcal{L}(\gamma_j)/\mathcal{L}(\gamma_j)^{I(\gamma_j)}$, so we have

$$\mathcal{L}_\lambda(\lambda\gamma_j)/\mathcal{L}_\lambda(\lambda\gamma_j)^{I(\lambda\gamma_j)} \text{ moved to } 0 \cong \mathcal{L}(\gamma_j)/\mathcal{L}(\gamma_j)^{I(\gamma_j)} \text{ moved to } 0.$$

QED

Unipotent Pseudoreflection Input Corollary bis 6.1.20 Hypotheses as in Theorem 6.1.18, suppose in addition that the local monodromy of $\mathcal{L}$ at the finite singularity γ_j is a unipotent pseudoreflection. Fix any λ in $k^\times$. If $s \geq 2$, suppose λ is not any of the $r(r-1)(s)(s-1)$ ratios

$$(\beta_{i_0} - \beta_{i_1})/(\gamma_{j_0} - \gamma_{j_1}), \text{ with } i_0 \neq i_1 \text{ and } j_0 \neq j_1,$$

and write

$$K*_{mid}\mathrm{MultTrans}_\lambda(L) = (j_*\mathcal{Q}_\lambda)[1].$$

Use additive translation by $\beta_i + \lambda\gamma_j$, β_i, and γ_j respectively to view $I(\beta_i + \lambda\gamma_j)$-representations, $I(\beta_i)$-representations, and $I(\gamma_j)$-representations respectively all as $I(0)$-representations. Then for any j, the local monodromy of $\mathcal{Q}_\lambda$ at $\beta_i + \lambda\gamma_j$ is related to the local monodromy of $\mathcal{L}$ at γ_j by an isomorphism of $I(0)$-representations

$$\mathcal{Q}_\lambda(\beta_i + \lambda\gamma_j)/\mathcal{Q}_\lambda(\beta_i + \lambda\gamma_j)^{I(\beta_i + \lambda\gamma_j)} \text{ moved to } 0$$
$$\cong \mathcal{K}(\beta_i)/\mathcal{K}(\beta_i)^{I(\beta_i)} \text{ moved to } 0.$$

proof This is again a special case of part 2) of the theorem. Its proof is essentially identical to that of the previous version, with $\mathcal{K}$ and $\mathcal{L}$ interchanged. QED

Pseudoreflection Output Corollary 6.1.21 Hypotheses as in Theorem 6.1.18, suppose in addition that the local monodromy of $\mathcal{K}$ at the finite singularity β_i is a tame pseudoreflection of determinant χ_i, and that the local monodromy of $\mathcal{L}$ at the finite singularity γ_j is a tame pseudoreflection of determinant ρ_j. Fix any λ in $k^\times$. If $s \geq 2$, suppose λ is not any of the r(r-1)(s)(s-1) ratios

$$(\beta_{i_0} - \beta_{i_1})/(\gamma_{j_0} - \gamma_{j_1}), \text{ with } i_0 \neq i_1 \text{ and } j_0 \neq j_1.$$

Write

$$K*_{mid}\mathrm{MultTrans}_\lambda(L) = (j_*\mathcal{Q}_\lambda)[1].$$

Then the local monodromy of $\mathcal{Q}_\lambda$ at $\beta_i + \lambda\gamma_j$ is a tame pseudoreflection of determinant $\chi_i\rho_j$.

proof This is a special case of part 3) of the theorem. QED

Non-semisimplicity Corollary 6.1.22 Hypotheses as in Theorem 6.1.18, fix a finite singularity β_i of $\mathcal{K}$, and a finite singularity γ_j of $\mathcal{L}$. Suppose that either a) or b) holds.

a) $\mathcal{K}(\beta_i)/\mathcal{K}(\beta_i)^{I(\beta_i)}$ is a non-semisimple representation of $I(\beta_i)$.

b) $\mathcal{L}(\gamma_j)/\mathcal{L}(\gamma_j)^{I(\gamma_j)}$ is a non-semisimple representation of $I(\gamma_j)$.

Fix any λ in $k^\times$. If $s \geq 2$, suppose λ is not any of the r(r-1)(s)(s-1) ratios

$$(\beta_{i_0} - \beta_{i_1})/(\gamma_{j_0} - \gamma_{j_1}), \text{ with } i_0 \neq i_1 \text{ and } j_0 \neq j_1.$$

Write

$$K*_{mid}\mathrm{MultTrans}_\lambda(L) = (j_*\mathcal{Q}_\lambda)[1].$$

Then $\mathcal{Q}_\lambda(\beta_i + \lambda\gamma_j)/\mathcal{Q}_\lambda(\beta_i + \lambda\gamma_j)^{I(\beta_i + \lambda\gamma_j)}$ is a non-semisimple representation of $I(\beta_i + \lambda\gamma_j)$.

proof by Theorem 6.1.18, part 2), we have an isomorphism of $I(\infty)$-representations

$$\mathrm{FT}_\psi\mathrm{loc}(\beta_i + \lambda\gamma_j, \infty)(\mathcal{Q}_\lambda(\beta_i + \lambda\gamma_j)/\mathcal{Q}_\lambda(\beta_i + \lambda\gamma_j)^{I(\beta_i + \lambda\gamma_j)}) \cong$$
$$\mathrm{FT}_\psi\mathrm{loc}(\beta_i, \infty)(\mathcal{K}(\beta_i)/\mathcal{K}(\beta_i)^{I(\beta_i)})\otimes\mathrm{FT}_\psi\mathrm{loc}(\lambda\gamma_j, \infty)(\mathcal{L}_\lambda(\lambda\gamma_j)/\mathcal{L}_\lambda(\lambda\gamma_j)^{I(\lambda\gamma_j)}).$$

To make use of this, recall that for each s in $\mathbb{A}^1$, $\mathrm{FT}_\psi\mathrm{loc}(s, \infty)$ is an equivalence of the categories of finite-dimensional, continuous $\overline{\mathbb{Q}}_\ell$-representations

$$I(s)\text{-representations} \cong I(\infty)\text{-representations with all slopes} < 1.$$

As such, it carries direct sums to direct sums, irreducibles to irreducibles, semisimple objects to semisimple objects, and non-semisimple objects to non-semisimple objects. So in order to show that $\mathcal{Q}_\lambda(\beta_i + \lambda\gamma_j)/\mathcal{Q}_\lambda(\beta_i + \lambda\gamma_j)^{I(\beta_i + \lambda\gamma_j)}$ is non-semisimple, it suffices to show that

$$\mathrm{FT}_\psi\mathrm{loc}(\beta_i + \lambda\gamma_j, \infty)(\mathcal{Q}_\lambda(\beta_i + \lambda\gamma_j)/\mathcal{Q}_\lambda(\beta_i + \lambda\gamma_j)^{I(\beta_i + \lambda\gamma_j)})$$

is non-semisimple. Since at least one of $\mathcal{K}(\beta_i)/\mathcal{K}(\beta_i)^{I(\beta_i)}$ or $\mathcal{L}_\lambda(\lambda\gamma_j)/\mathcal{L}_\lambda(\lambda\gamma_j)^{I(\lambda\gamma_j)}$ is non-semisimple, at least one of the tensorees in

$$\mathrm{FT}_\psi\mathrm{loc}(\beta_i, \infty)(\mathcal{K}(\beta_i)/\mathcal{K}(\beta_i)^{I(\beta_i)})\otimes\mathrm{FT}_\psi\mathrm{loc}(\lambda\gamma_j, \infty)(\mathcal{L}_\lambda(\lambda\gamma_j)/\mathcal{L}_\lambda(\lambda\gamma_j)^{I(\lambda\gamma_j)}).$$

is non-semisimple. So it suffices to show the following well known lemma, applied to $\Gamma = I(\infty)$, $V = \mathrm{FT}_\psi\mathrm{loc}(\beta_i, \infty)(\mathcal{K}(\beta_i)/\mathcal{K}(\beta_i)^{I(\beta_i)})$ and $W = \mathrm{FT}_\psi\mathrm{loc}(\lambda\gamma_j, \infty)(\mathcal{L}_\lambda(\lambda\gamma_j)/\mathcal{L}_\lambda(\lambda\gamma_j)^{I(\lambda\gamma_j)})$. We include its proof for ease of reference.

Lemma 6.1.23 Let Γ be a topological group, V and W two nonzero finite-dimensional continuous $\overline{\mathbb{Q}}_\ell$-representations of Γ. Denote by Γ_{discr} the discrete group underlying Γ. Then we have the following results.

1) A finite-dimensional continuous $\overline{\mathbb{Q}}_\ell$-representation M of Γ is semisimple as a continuous $\overline{\mathbb{Q}}_\ell$-representation of Γ if and only if it is semisimple as a $\overline{\mathbb{Q}}_\ell$-representations of Γ_{discr}.

2) $V\otimes W$ is semisimple if and only if both V and W are semisimple.

proof 1) If M is semisimple as a continuous $\overline{\mathbb{Q}}_\ell$-representation of Γ, it is the direct sum of irreducible $\overline{\mathbb{Q}}_\ell$-representation of Γ, which happen to be continuous, so it is semisimple as a $\overline{\mathbb{Q}}_\ell$-representation of Γ_{discr}. Conversely, if M is semisimple as a $\overline{\mathbb{Q}}_\ell$-representations of Γ_{discr}, then M admits a basis in which Γ_{discr} acts block-diagonally, with each block irreducible. But the fact that M is a continuous representation of Γ is independent of a particular choice of basis, so each of the irreducible diagonal blocks is itself a continuous representation of Γ.

2) By part 1), applied to V, W, and $V\otimes W$, we can forget the topology, i.e., it suffices to treat the case when Γ is a discrete group. In this case, we may replace Γ by its Zariski closure, call it G, in $GL(V\oplus W)$.

If both V and W are semisimple, so is $V\oplus W$, and hence G is reductive (because we are in characteristic zero, and G has a faithful completely reducible representation, namely $V \oplus W$). And in characteristic zero, any finite-dimensional representation of a reductive group, e.g., $V \otimes W$, is completely reducible.

Suppose that at least one of V or W is not semisimple. Then $V \oplus W$ is not semisimple, and hence G is not reductive. Therefore its unipotent radical $\mathcal{R}_u(G)$ is nonzero. To show that $V \otimes W$ is not semisimple, it suffices to show that $\mathcal{R}_u(G)$ acts nontrivially on $V \otimes W$. For this, we argue as follows. Since $V \oplus W$ is a faithful representation of G, it is a faithful representation of $\mathcal{R}_u(G)$. Therefore $\mathcal{R}_u(G)$ acts nontrivially at least one of V or W (otherwise it would act trivially on $V \oplus W$, hence would be the trivial subgroup of G). Since every element of $\mathcal{R}_u(G)$ acts unipotently, we can pick an element γ in $\mathcal{R}_u(G)$ which acts nontrivially on at least one of V or W, and unipotently on both V and W. This element γ acts on $V \otimes W$ as the tensor product

$$(\gamma|V) \otimes (\gamma|W)$$

of two unipotent automorphisms. Such a tensor product is the identity if and only if each tensoree is the identity. Therefore γ acts nontrivially on $V \otimes W$, and hence $\mathcal{R}_u(G)$ acts nontrivially on $V \otimes W$. QED

(6.2) Interlude: some galois theory in one variable

(6.2.1) We continue to work over an algebraically closed field k in which the prime ℓ is invertible. Let C/k be a proper, smooth, connected curve of genus g, and let $D = \Sigma_i a_i P_i$ be an effective divisor on C, of degree $d := \Sigma_i a_i$. We suppose that

$$d := \deg(D) \geq 2g + 3.$$

Denote by L(D) the Riemann-Roch space

$$L(D) = H^0(C, I^{-1}(D)).$$

We view L(D) as the k-points of an affine space of dimension d+1-g over k, and we view functions f in L(D) as morphisms from C-D to $\mathbb{A}^1$. Let us recall from [Ka-TLFM] the following theorem.

Theorem 6.2.2 ([Ka-TLFM, 2.2.6 and 2.4.2]) Let k be an algebraically closed field, and let C/k be a projective, smooth connected curve, of genus denoted g. Fix an effective divisor D on C of degree $d \geq 2g+3$. Fix a finite subset S of C - D. Then in L(D) viewed as the k-points of an affine space of dimension d+1-g, there is a dense open set U such that any f in U has the following properties:

1) the divisor of poles of f is D, and f is Lefschetz on C-D, i.e., if we view f as a finite flat map of degree d from C - D to $\mathbb{A}^1$, then all but finitely many of the fibres of f over $\mathbb{A}^1$ consist of d distinct points, and the remaining fibres consist of d-1 distinct points, d-2 of which occur with multiplicity 1, and one which occurs with multiplicity 2.

2) f separates the points of S, i.e., $f(s_1) = f(s_2)$ if and only if $s_1 = s_2$,

and f is finite etale in a neighborhood of each fibre $f^{-1}f(s)$. Put another way, there are #S fibres over $\mathbb{A}^1$ which each have d points and which each contain a single point of S.

(6.2.3) Any f in L(D) with divisor of poles D defines a finite flat map of degree d from C - D to $\mathbb{A}^1$. If f is Lefschetz on C-D, f has at most finitely many cricital points (zeroes of the differential df) in C - D, whose images under f are the finitely many critical values of f in $\mathbb{A}^1$. Thus f by restriction defines a finite etale map of degree d

$$\begin{array}{l} C - D - f^{-1}(\mathrm{CritValues}(f)) \\ f \downarrow \\ \mathbb{A}^1 - \mathrm{CritValues}(f). \end{array}$$

Theorem 6.2.4 Let k be an algebraically closed field, and let C/k be a projective, smooth connected curve, of genus denoted g. Fix an effective divisor $D = \Sigma_i a_i P_i$ on C of degree $d \geq 2g+3$. Suppose in addition that each a_i invertible in k. Fix any f in L(D) whose divisor of poles is D, and which is Lefschetz on C-D. Then we have the following results.

1) For any ℓ invertible in k, the middle extension sheaf $f_*\overline{\mathbb{Q}}_\ell/\overline{\mathbb{Q}}_\ell$ on $\mathbb{A}^1$ is irreducible.

2) The restriction to $\mathbb{A}^1$ - CritValues(f) of $f_*\overline{\mathbb{Q}}_\ell$ is a lisse sheaf, whose geometric monodromy group G_{geom} is the full symmetric group S_d, in its standard d-dimensional permutation representation. Local monodromy at each finite singularity (i.e., at each critical value of f in $\mathbb{A}^1$) is a reflection, the action of a transposition in S_d.

3) The restriction to $\mathbb{A}^1$ - CritValues(f) of $f_*\overline{\mathbb{Q}}_\ell/\overline{\mathbb{Q}}_\ell$ is a lisse sheaf, whose geometric monodromy group G_{geom} is the full symmetric group S_d, in its (d-1)-dimensional "deleted permutation representation". Local monodromy at each finite singularity (i.e., at each critical value of f in $\mathbb{A}^1$) is a reflection, the action of a transposition in S_d.

4) The sheaf $f_*\overline{\mathbb{Q}}_\ell$ has at least one finite singularity. If 2 is invertible in k, $f_*\overline{\mathbb{Q}}_\ell$ has at least two finite singularities.

proof The sheaf $f_*\overline{\mathbb{Q}}_\ell$ is a middle extension, cf. [Ka-TLFM, proof of 3.3.1], which contains $\overline{\mathbb{Q}}_\ell$ as subsheaf. Consider the trace morphism

$$\mathrm{Trace} : f_*\overline{\mathbb{Q}}_\ell \to \overline{\mathbb{Q}}_\ell.$$

The map (1/d)Trace splits the inclusion of $\overline{\mathbb{Q}}_\ell$ into $f_*\overline{\mathbb{Q}}_\ell$. So we have a direct sum decomposition

$$f_*\overline{\mathbb{Q}}_\ell = \overline{\mathbb{Q}}_\ell \oplus \mathrm{Ker}(\mathrm{Trace}) \cong \overline{\mathbb{Q}}_\ell \oplus f_*\overline{\mathbb{Q}}_\ell/\overline{\mathbb{Q}}_\ell$$

Thus $f_*\overline{\mathbb{Q}}_\ell/\overline{\mathbb{Q}}_\ell$ is a middle extension, being a direct factor of a middle extension.

The restriction to $\mathbb{A}^1$ - CritValues(f) of $f_*\overline{\mathbb{Q}}_\ell$ (respectively of $f_*\overline{\mathbb{Q}}_\ell/\overline{\mathbb{Q}}_\ell$) is lisse, and its G_{geom} is a transitive subgroup, say Γ, of S_d in its standard d-dimensional permutation representation (resp. in its (d-1)-dimensional "deleted permutation representation"), simply because f is finite etale of degree d over CritValues(f) with connected source space C - D - {zeroes of df in C-D}. If we show that $\Gamma = S_d$, then $f_*\overline{\mathbb{Q}}_\ell/\overline{\mathbb{Q}}_\ell$ is an irreducible middle extension on $\mathbb{A}^1$, simply because the deleted permutation representation of S_d is irreducible.

To show that $\Gamma = S_d$, it suffices to show that Γ is generated by transpositions. The assumption that each a_i is invertible in k means that all poles of f are of order invertible in k, and this in turn guarantees that $f_*\overline{\mathbb{Q}}_\ell$ is tamely ramified at ∞. Therefore Γ is generated by all the local monodromies at all the finite singularities of $f_*\overline{\mathbb{Q}}_\ell$. Because f is Lefschetz on C-D, at each finite singularity γ, the fibre consists of d-2 points of multiplicity 1, and one point of multiplicity 2. Thus each element of the inertia group $I(\gamma)$ has at least d-2 fixed points, so is either the identity or is a transposition. [Since the rank of $f_*\overline{\mathbb{Q}}_\ell$ drops by one at γ, and $f_*\overline{\mathbb{Q}}_\ell$ is a middle extension, the local monodromy at γ cannot be trivial. But we don't know yet that there exist any finite singularities γ.] Thus Γ is a transitive subgroup of S_d, generated by a possibly empty collection of transpositions. Therefore $\Gamma = S_d$. This in turn shows that $f_*\overline{\mathbb{Q}}_\ell$ has at least one finite singularity γ, otherwise Γ would be the trivial group.

The finite singularities of $f_*\overline{\mathbb{Q}}_\ell$, i.e., the critical values of f in $\mathbb{A}^1$, are the images under f of the zeroes of the differential df in C-D. Because f has divisor of poles D and all the a_i are prime to p, df has poles at each point of D. So the zeroes of df all lie in C-D. Because f is Lefschetz on C-D, f maps the zeroes of df bijectively to the finite singularities of $f_*\overline{\mathbb{Q}}_\ell$. Hence the number of finite singularities of $f_*\overline{\mathbb{Q}}_\ell$ is the number of distinct zeroes of df. When 2 is invertible in k, the fact that f is Lefschetz on C - D means precisely that df has only simple zeroes in C - D, and that f separates these zeroes. So when 2 is invertible, the number of finite singularities of $f_*\overline{\mathbb{Q}}_\ell$ is the degree of the divisor of zeroes of df. The divisor of df has degree 2g-2, so

$$\deg(\text{divisor of zeroes of df}) = 2g - 2 + \deg(\text{divisor of poles of df}).$$

Now f has polar divisor $D = \Sigma_i a_i P_i$ with all a_i invertible in k, so the divisor of poles of df is $\Sigma_i (a_i + 1)P_i$, whose degree is

$$\deg(D) + \deg(D^{red}) \geq \deg(D) + 1 \geq 2g + 4.$$

Thus we find

$$\deg(\text{divisor of zeroes of df}) \geq 4g + 2.$$

So when 2 is invertible in k, the number of finite singularities of $f_*\overline{\mathbb{Q}}_\ell$ is at least two.

Here is an alternate proof of this last result. If 2 is invertible in k, then local monodromy at each finite singularity, being of order 2, is tame, and so $f_\star \overline{\mathbb{Q}}_\ell/\overline{\mathbb{Q}}_\ell$, which has rank $d-1 \geq 2g + 2 \geq 2$, is an irreducible representation of dimension $d-1 > 1$ of the group

$$\pi_1{}^{\mathrm{tame}}(\mathbb{A}^1 - \text{finite singularites of } f_\star \overline{\mathbb{Q}}_\ell/\overline{\mathbb{Q}}_\ell).$$

Therefore this group cannot be abelian. But $\pi_1{}^{\mathrm{tame}}(\mathbb{G}_m)$ is abelian. So there are at least two finite singularites of $f_\star \overline{\mathbb{Q}}_\ell/\overline{\mathbb{Q}}_\ell$ if 2 is invertible in k. QED

Remark 6.2.5 Here is an example to show that in conclusion 4), we must exclude characteristic 2 if we want to insure that $f_\star \overline{\mathbb{Q}}_\ell$ has at least two finite singularities, cf. [Ka-TLFM, 2.5.4]. Take an integer $n \geq 1$, take C to be $\mathbb{P}^1$, and take D to be $(2n+1)\infty$. In characteristic 2, the function $f = x^2 + x^{2n+1}$ on $\mathbb{A}^1$ is Lefschetz: its unique singular fibre as a finite flat degree $2n+1$ map of $\mathbb{A}^1$ to itself is the fibre over the origin, which consists of 2n points. Thus f is finite etale over $\mathbb{G}_m$, and so $f_\star \overline{\mathbb{Q}}_\ell$ is lisse on $\mathbb{G}_m$.

(6.2.6) What happens to the above corollary if we drop the hypothesis that in the divisor $D = \Sigma_i a_i P_i$, each a_i invertible in k? The first "problem" we have to deal with is that, for any f in L(D), if the characteristic p of k divides a_i, then the differential df has a pole of order at most a_i at P_i.

Lemma 6.2.7 Let k be an algebraically closed field of characteristic $p > 0$, and let C/k be a projective, smooth connected curve, of genus denoted g. Fix an effective divisor $D = \Sigma_i a_i P_i$ on C of degree $d \geq 2g+2$. Define integers b_i by

$$b_i := a_i, \text{ if } p \text{ divides } a_i,$$
$$b_i := a_i + 1, \text{ if } p \text{ does not divide } a_i.$$

The set V consisting of those functions f in L(D) such that the divisor of poles of f is D, and such that the divisor of poles of df is $\Sigma_i b_i P_i$, form a dense open set in L(D).

proof Fix a uniformizing parameter z_i at P_i, for each P_i which occurs in D. The formal expansion of f at P_i has the form

$$f = \Sigma_{1 \leq n \leq a_i} C(n,i,f)/(z_i)^n + (\text{holomorphic at } P_i)$$

with coefficients C(n,i,f) in k. For each point P_i and for each integer n with $1 \leq n \leq a_i$, the map $f \mapsto C(n,i,f)$ is a linear form on L(D). A function f in L(D) has divisor of poles D if and only if, for each P_i, we have nonvanishing of the most polar term:

$$C(a_i, i, f) \neq 0.$$

The divisor of poles of df is $\Sigma_i\ b_iP_i$ if and only if, in addition, for each P_i whose a_i is divisible by p, we have nonvanishing of the next most polar term:

$$C(a_i -1, i, f) \neq 0.$$

Thus the functions f in L(D) which lie in V are precisely the functions on which all of the linear forms

$$f \mapsto C(a_i, i, f), \text{ for each i},$$
$$f \mapsto C(a_i -1, i, f), \text{ for those i with } a_i \text{ divisible by p},$$

are nonzero. So it suffices to see that each of these linear forms is not identically zero on L(D), for then V is the complement in L(D) of a finite union of hyperplanes.

The kernel in L(D) of $f \mapsto C(a_i, i, f)$ is $L(D - P_i)$. As $d \geq 2g +2$, we have $\deg(D - P_i) = d - 1 \geq 2g + 1$, so by Riemann-Roch, we get

$$\ell(D - P_i) = \deg(D - P_i) + 1 - g = \ell(D) - 1.$$

Thus the kernel in L(D) of $f \mapsto C(a_i, i, f)$ is a proper subspace, and hence the linear form $f \mapsto C(a_i, i, f)$ on L(D) is nonzero.

For i with a_i divisible by p, we have $a_i \geq 2$, and the kernel in $L(D - P_i)$ of $f \mapsto C(a_i - 1, i, f)$ is $L(D - 2P_i)$. Again by Riemann-Roch, we have

$$\ell(D - 2P_i) = \deg(D - 2P_i) + 1 - g = \ell(D - P_i) - 1.$$

Thus the kernel in $L(D - P_i)$ of $f \mapsto C(a_i - 1, i, f)$ is a proper subspace, and so the linear form $f \mapsto C(a_i - 1, i, f)$ is nonzero on $L(D - P_i)$, and hence it is nonzero on the larger space L(D). QED

(6.2.8) The following lemma, of a similar flavor, will be useful later.

Lemma 6.2.9 Hypotheses and notations as in Lemma 6.2.7, let $T \subset \mathbb{A}^1(k)$ be a finite set of points. Inside the dense open set $V \subset L(D)$ of the previous lemma, the subset $V_{etale/T} \subset V$ consisting of those functions f in V such that f as a finite flat map of C - D to $\mathbb{A}^1$ of degree deg(D) is finite etale over each t in T (i.e., such that for each t in T, $f^{-1}(t)$ consists of deg(D) distinct points in C - D) is an open dense set of V.

proof We first show that $V_{etale/T}$ is nonempty. Take any function f in V. Then df is nonzero (because its polar divisor is nonzero), so f has finitely many critical points in C - D. Then f lies in $V_{etale/T}$ if and only the finite set CritValues(f) := f({CritPoints(f)}) is disjoint from T. Additively translating f by a constant λ does not alter the set of critical points, but translates the set of critical values by that same constant λ. Therefore $f + \lambda$ lies in $V_{etale/T}$ for all but the finitely many λ of the form t - c with t in T and with c in CritValues(f).

To see that $V_{etale/T}$ is open in V, we first reduce to the case when T is a single point t (since $V_{etale/T} = \cap_{t \text{ in } T} V_{etale/t}$). Over $\mathbb{A}^1 \times V$, we have a finite flat map of degree deg(D)

$$\pi : (C - D)\times V \to \mathbb{A}^1 \times V,$$
$$(x, f) \mapsto (f(x), f).$$

Pulling back to the closed subspace $t \times V$ of the base, we get a Cartesian diagram

$$\begin{array}{ccc} Z & \subset & (C - D)\times V \\ \downarrow \pi_t & & \downarrow \pi \\ V & \subset & \mathbb{A}^1 \times V. \end{array}$$

Zariski locally on V, V is Spec(A) and Z is Spec(B) for B some A-algebra which is a free A-module of rank deg(D), say with A-basis $\{e_i\}_i$, and $V_{etale/T}$ is the open set of Spec(A) where the discriminant $\det(\mathrm{Trace}_{B/A}(e_ie_j))$ of B as A-algebra is invertible. QED

(6.2.10) The following theorem is joint work with E. Rains.

Theorem 6.2.11 (joint with Rains) Let k be an algebraically closed field of positive characteristic p, and let C/k be a projective, smooth connected curve, of genus denoted g. Fix an effective divisor $D = \Sigma_i a_i P_i$ on C of degree $d \ge 2g+3$. Define integers b_i by

$$b_i := a_i, \text{ if } p \text{ divides } a_i,$$
$$b_i := a_i + 1, \text{ if } p \text{ does not divide } a_i.$$

Let f be a function in L(D) such that the divisor of poles of f is D, the divisor of poles of df is $\Sigma_i b_i P_i$, and such that f is Lefschetz on C - D. Suppose we are not in the following exceptional case:

$p = 2$, $D = \Sigma_i 2P_i$, there exists a function g in $L(\Sigma_i P_i)$ and there exist scalars α, β in $k^\times$ and γ in k such that $f = \alpha g^2 + \beta g + \gamma$.

Then we have the following results.

1) For any ℓ invertible in k, the middle extension sheaf $f_*\overline{\mathbb{Q}}_\ell/\overline{\mathbb{Q}}_\ell$ on $\mathbb{A}^1$ is irreducible.

2) The restriction to $\mathbb{A}^1$ - CritValues(f) of $f_*\overline{\mathbb{Q}}_\ell$ is a lisse sheaf, whose geometric monodromy group G_{geom} is the full symmetric group S_d, in its standard d-dimensional permutation representation. Local monodromy at each finite singularity (i.e., at each critical value of f in $\mathbb{A}^1$) is a reflection, the action of a transposition in S_d.

3) The restriction to $\mathbb{A}^1$ - CritValues(f) of $f_*\overline{\mathbb{Q}}_\ell/\overline{\mathbb{Q}}_\ell$ is a lisse sheaf, whose geometric monodromy group G_{geom} is the full symmetric group S_d, in its (d-1)-dimensional "deleted permutation representation". Local monodromy at each finite singularity (i.e., at each critical value of f in $\mathbb{A}^1$) is a reflection, the action of a transposition in S_d.

4) The sheaf $f_*\overline{\mathbb{Q}}_\ell$ has at least one finite singularity. If 2 is invertible in k, and we are not in the case

$$p = 3,\ g = 0,\ D = 3P,$$

then the sheaf $f_{\star}\overline{\mathbb{Q}}_{\ell}$ has at least two finite singularities.

(6.2.12) Before beginning the proof of the theorem, let us clarify the exceptional case, and give two "genericity" corollaries of the theorem.

Lemma 6.2.13 Let k be an algebraically closed field of characteristic 2, and let C/k be a projective, smooth connected curve, of genus denoted g. Fix an effective divisor D of the form $D = \Sigma_i 2P_i$ of degree $d \geq 2g+3$. For each P_i, pick a uniformizing parameter z_i at P_i. Given f in L(D), write its formal expansion at P_i in the form

$$f = C(2,i,f)/(z_i)^2 + C(1,i,f)/z_i + (\text{holomorphic at } P_i)$$

with coefficients C(2,i,f) and C(1,i,f) in k. Denote by Z the closed subset of L(D) defined by the equations

$$C(2,i,f)C(1,j,f)^2 = C(2,j,f)C(1,i,f)^2 \text{ for all } i \neq j,$$

i.e., Z is the locus where the two vectors $\{C(2,i,f)\}_i$ and $\{C(1,i,f)^2\}_i$ are linearly dependent. Then Z is a proper closed subset of L(D). For any g in $L(\Sigma_i P_i)$, and for any scalars α, β, γ in k, the function $\alpha g^2 + \beta g + \gamma$ lies in Z.

proof Given g in $L(\Sigma_i P_i)$, write its formal expansion at P_i in the form

$$g = C(1,i,g)/z_i + (\text{holomorphic at } P_i).$$

Since we are in characteristic 2, the expansion of g^2 at P_i is

$$g^2 = C(1,i,g)^2/(z_i)^2 + (\text{holomorphic at } P_i).$$

Thus the expansion of $\alpha g^2 + \beta g + \gamma$ at P_i is

$$\alpha g^2 + \beta g + \gamma = \alpha C(1,i,g)^2/(z_i)^2 + \beta C(1,i,g)/z_i + (\text{holomorphic at } P_i).$$

Thus for the function $f := \alpha g^2 + \beta g + \gamma$, we have

$$C(2,i,f) = \alpha C(1,i,g)^2,$$
$$C(1,i,f) = \beta C(1,i,g).$$

So the two vectors $\{C(2,i,f)\}_i$ and $\{C(1,i,f)^2\}_i$ are both scalar multiples of the vector $\{C(1,i,g)^2\}_i$, hence are linearly dependent. Thus the function $f := g^2 + \alpha g + \beta$ lies in the closed set Z.

To see that Z is a proper closed subset, pick two indices $i \neq j$. This is possible because $\deg(\Sigma_i 2P_i) \geq 2g + 3 \geq 3$, so there are at least two distinct points P_i. Pick a function f in $L(D - P_j)$ whose polar divisor is $D - P_j$. This is possible by Riemann-Roch, because $D - P_j$ has degree $\geq 2g+2 \geq 2g$. Such an f has a double pole at P_i and a simple pole at P_j, so it has

$$C(2,i,f) \neq 0,\ C(1,j,f) \neq 0,\ C(2,j,f) = 0,$$

and hence does not satisfy the equation

$$C(2,i,f)C(1,j,f)^2 = C(2,j,f)C(1,i,f)^2$$

for the chosen i and j. QED

Corollary 6.2.14 Let k be an algebraically closed field of characteristic 2, and let C/k be a projective, smooth connected curve, of genus denoted g. Fix an effective divisor D of the form $D = \Sigma_i 2P_i$ of degree $d \geq 2g+3$. Let f be a function in L(D) such that the divisor of poles of f is D, the divisor of poles of df is $\Sigma_i b_i P_i$, f is Lefschetz on C - D, and f does not lie in the proper closed set Z. Then all the conclusions of Theorem 6.2.11 hold.

proof Immediate from Theorem 6.2.11 and Lemma 6.2.13 above. QED

Corollary 6.2.15 Let k be an algebraically closed field of positive characteristic p, and let C/k be a projective, smooth connected curve, of genus denoted g. Fix an effective divisor $D = \Sigma_i a_i P_i$ on C of degree $d \geq 2g+3$. There exists a dense open set in L(D) such that for any f in this dense open set, all the conclusions of Theorem 6.2.11 hold.

proof For f and df to have imposed divisors of poles, and for f to be Lefschetz on C-D, are conditions which all hold on a dense open set U of L(D). If p = 2 and $D = \Sigma_i 2P_i$, intersect this U with the complement of Z. QED

(6.3) Proof of Theorem 6.2.11

(6.3.1) In the case when all the a_i were prime to p, the proof relied on the group theoretic fact that a transitive subgroup Γ of the symmetric group S_d generated by transpositions is all of S_d. In the general case, the proof relies on the fact that a primitive (in the sense of permuation groups) subgroup of S_d, $d \geq 3$, which contains a transposition, is all of S_d. Recall that a subgroup Γ of S_d, $d \geq 3$, is said to be primitive if the only partitions of the set {1, 2,..., d} into disjoint nonempty subsets $\mathcal{P}_\alpha$ which are permuted among themselves by Γ are the one set partition and the d set partition. The following lemma is well known, cf. [Serre-TGT, page 40, proof of part 2 of Lemma 4.4.4].

Lemma 6.3.2 A primitive subgroup Γ of S_d, $d \geq 3$, which contains a transposition, is all of S_d.

proof Here is a simple proof, which I learned from Eric Rains. Because $d \geq 3$, the trivial group is not primitive. If Γ is primitive, it is transitive (otherwise its orbits form a partition which violates the primitivity of Γ). Denote by H the subgroup of Γ generated by all the transpositions in Γ. Then H is a normal subgroup of Γ which is non-trivial. H acts transitively (otherwise its orbits form a partition which violates the primitivity of Γ). Thus H is a transitive subgroup of S_d generated by transpositions, hence is all of S_d. So a fortiori Γ is S_d. QED

(6.3.3) The following criterion for primitivity is standard.

Lemma 6.3.4 A transitive subgroup Γ of S_d, $d \geq 3$, is primitive if and only if the stablizer in Γ of a point in $\{1, 2,..., d\}$ is a maximal subgroup of Γ.

proof Suppose first Stab(1) is a maximal subgroup of Γ. Let $\{\mathcal{P}_\alpha\}$ be any partition of $\{1, 2,..., d\}$ into $r > 1$ disjoint nonempty subsets $\mathcal{P}_\alpha$ which are permuted among themselves by Γ. We must show $r = d$. To see this, let $\mathcal{P}_1$ contain the element 1. Let Stab($\mathcal{P}_1$) denote the stabilizer in Γ of $\mathcal{P}_1$. Then we have inclusions

$$\mathrm{Stab}(1) \subset \mathrm{Stab}(\mathcal{P}_1) \subset \Gamma.$$

Because Γ is transitive, Γ transitively permutes the $\mathcal{P}_\alpha$. So the index of $\mathrm{Stab}(\mathcal{P}_1) \subset \Gamma$ is r, the number of partition sets. As $r > 1$, Stab($\mathcal{P}_1$) is a proper subgroup of Γ. As Stab(1) is a maximal subgroup of Γ, Stab(1) = Stab($\mathcal{P}_1$). But Stab(1) has index d in Γ (because Γ is transitive), so $r = d$.

If Stab(1) is not a maximal subgroup of Γ, let H be a strictly intermediate subgroup. View $\{1, 2,..., d\}$ as the homogeneous space Γ/Stab(1). Then the fibres of the natural map of Γ-spaces

$$\Gamma/\mathrm{Stab}(1) \to \Gamma/H$$

are a partition of $\{1, 2,..., d\}$ which violates primitivity. QED

(6.3.5) Now let us turn to the proof of Theorem 6.2.11. Thus k is an algebraically closed field of positive characteristic p, C/k is a projective, smooth connected curve, of genus denoted g, and D is an effective divisor $D = \Sigma_i a_i P_i$ on C of degree $d \geq 2g+3$. The integers b_i are defined by

$b_i := a_i$, if p divides a_i,

$b_i := a_i + 1$, if p does not divide a_i.

Finally, f is a function in L(D) whose divisor of poles of f is D, which is Lefschetz on C - D, and such that the divisor of poles of df is $\Sigma_i b_i P_i$.

(6.3.6) Exactly as in the proof of the prime-to-p version, Theorem 6.2.4, we see that $f_\star\overline{\mathbb{Q}}_\ell$ is a middle extension, and that we have a direct sum decomposition $f_\star\overline{\mathbb{Q}}_\ell \cong \overline{\mathbb{Q}}_\ell \oplus f_\star\overline{\mathbb{Q}}_\ell/\overline{\mathbb{Q}}_\ell$. We further see that the restriction to $\mathbb{A}^1$ - CritValues(f) of $f_\star\overline{\mathbb{Q}}_\ell$ (resp. of $f_\star\overline{\mathbb{Q}}_\ell/\overline{\mathbb{Q}}_\ell$) is lisse, and its G_{geom} is a transitive subgroup, say Γ, of S_d in its standard d-dimensional permutation representation (resp. in its (d-1)-dimensional "deleted permutation representation").

(6.3.7) Still as in the proof of Theorem 6.2.4, we see that at each finite singularity γ, the fibre consists of d-2 points of multiplicity 1, and one point of multiplicity 2. Thus each element of the inertia group $I(\gamma)$ has at least d-2 fixed points, so is either the identity or is the transposition which interchanges the two roots which have coalesced at γ. Since the rank of $f_*\overline{\mathbb{Q}}_\ell$ drops by one at γ, and $f_*\overline{\mathbb{Q}}_\ell$ is a middle extension, the local monodromy at γ cannot be trivial, so there are elements of $I(\gamma)$ which act as transpositions.

(6.3.8) We next estimate from below the number of finite singularities of $f_*\overline{\mathbb{Q}}_\ell$, i.e., the number of critical values of f. The critical values of f in $\mathbb{A}^1$ are the images under f of the zeroes of the differential df in C-D. Because df has divisor of poles $\Sigma_i b_i P_i$, df has poles at each point of D. So the zeroes of df all lie in C-D. Because f is Lefschetz on C-D, f maps the zeroes of df bijectively to the finite singularities of $f_*\overline{\mathbb{Q}}_\ell$. Hence the number of finite singularities of $f_*\overline{\mathbb{Q}}_\ell$ is the number of distinct zeroes of df. The total number of zeroes of df, counting multiplicity, is readily calculated, just using the fact that the divisor of df has degree 2g-2, while the polar divisor of df is $\Sigma_i b_i P_i$. So we get

(6.3.8.1)
$$\begin{aligned} &\deg(\text{divisor of zeroes of df}) \\ &= 2g - 2 + \deg(\text{divisor of poles of df}) \\ &= 2g - 2 + \Sigma_i b_i \geq 2g - 2 + \Sigma_i a_i \\ &= 2g - 2 + \deg(D) \geq 2g - 2 + 2g + 3 \\ &\geq 4g + 1 \geq 1. \end{aligned}$$

So in any characteristic, df has zeroes: $f_*\overline{\mathbb{Q}}_\ell$ has finite singularities Since the local monodromy at each finite singularity is a transposition, the group Γ contains transpositions.

(6.3.9) When 2 is invertible in k, and f is Lefschetz on C - D, df has only simple zeroes in C - D, so df has precisely $2g - 2 + \Sigma_i b_i$ distinct zeroes, and there are $2g - 2 + \Sigma_i b_i$ finite singularities. We now show that in odd characteristic p, we have $2g - 2 + \Sigma_i b_i \geq 2$, except in the case when p = 3, g = 0, and D = 3P. Indeed, we can have the equality

$$2g - 2 + \Sigma_i b_i = 1$$

if and only if in the above chain of inequalities 6.3.8.1, every inequality is an equality, i.e., if and only if

$$g = 0,\ \deg(D) = 2g + 3,\ a_i = b_i \text{ for all i},$$

i.e.,

$$g = 0,\ \deg(D) = 3,\ p \text{ divides } a_i \text{ for all i}.$$

Since each a_i is divisible by the odd characteristic p, we have $a_i \geq 3$ for all i. But deg(D) = 3, so D = 3P, and p = 3.

(6.3.10) It remains only to show that the transitive group Γ is primitive. For this, we think of the function f first as a finite flat degree d, generically etale, map from C to $\mathbb{P}^1$, then as an inclusion of function fields $k(\mathbb{P}^1) \subset k(C)$ which makes $k(C)/k(\mathbb{P}^1)$ a finite separable extension of degree d. Let us denote by $K/k(\mathbb{P}^1)$ the galois closure of $k(C)/k(\mathbb{P}^1)$. Then Γ, the monodromy group G_{geom} of $f_*\overline{\mathbb{Q}}_\ell$ on $\mathbb{A}^1$ - CritValues(f), is just the galois group of $K/k(\mathbb{P}^1)$, in its standard d-dimensional permutation representation. The intermediate field $k(C)/k(\mathbb{P}^1)$ inside $K/k(\mathbb{P}^1)$ corresponds to the subgroup of Γ which, viewing Γ inside S_d, is the stabilizer of one point. Now Γ is primitive if and only if this subgroup of Γ is maximal, i.e., if and only if there exist no intermediate fields L,

$$k(\mathbb{P}^1) \subset L \subset k(C),$$

with $k(\mathbb{P}^1) \neq L \neq k(C)$.

(6.3.11) We argue by contradiction. Suppose there exists such an intermediate field. Let us define

$$d_1 := \deg(L/k(\mathbb{P}^1)),\ d_2 := \deg(k(C)/L).$$

Thus d_1 and d_2 are both at least 2, and $d_1d_2 = d$. The field L is the function field of some proper smooth connected curve C_1/k, whose genus we denote g_1. The inclusions of function fields above gives us finite flat k-morphisms

$$g : C \to C_1 \text{ of degree } d_2,$$

$$h : C_1 \to \mathbb{P}^1 \text{ of degree } d_1$$

whose composite is $f : C \to \mathbb{P}^1$. Denote by $D_1 \subset C_1$ the effective divisor of degree d_1 which is the scheme-theoretic inverse image of ∞ under the morphism h. In other words, view h as a rational function on C_1, and denote by D_1 the polar divisor of h. Write

$$D_1 := \Sigma_j e_jQ_j.$$

(6.3.12) We first claim that the map $h : C_1 \to \mathbb{P}^1$ is finite etale over $\mathbb{A}^1$. Since g is finite and flat of degree d_1, it suffices to show that the fibre of h over every k-valued point α of $\mathbb{A}^1$ contains d_1 points. To see this, we argue as follows. Since f is $h\circ g$, $D := f^{-1}(\infty) = g^{-1}(h^{-1}(\infty)) = g^{-1}(D_1)$, so g and h induce finite flat maps

$$g : C - D \to C_1 - D_1,$$

$$h: C_1 - D_1 \to \mathbb{A}^1,$$

whose composite is

$$f: C - D \to \mathbb{A}^1.$$

Because f is Lefschetz on C - D, the fibre $f^{-1}(\alpha)$ contains at least d-1 points. Because g is finite and flat of degree d_2, for every k-valued point β of $C_1 - D_1$, $g^{-1}(\beta)$ contains at most d_2 points. So if $h^{-1}(\alpha)$ contains at most d_1 -1 points, then $g^{-1}(h^{-1}(\alpha))$ contains at most

$$(d_1 - 1)d_2$$

points. But $g^{-1}(h^{-1}(\alpha))$ is $f^{-1}(\alpha)$, which contains at least d - 1 points. Therefore we have the inequalities

$$(d_1 - 1)d_2 \geq \# f^{-1}(\alpha) \geq d - 1.$$

But $d_1 d_2 = d$, so we get

$$d - d_2 \geq d - 1,$$

which is impossible, because $d_2 \geq 2$ by hypothesis.

(6.3.13) In terms of the polar divisor $D_1 := \Sigma_j e_j Q_j$ of h, we define integers f_j by

$$f_j := e_j, \text{ if p divides } e_j,$$
$$f_j := e_j + 1, \text{ if p does not divide } e_j.$$

(6.3.14) We claim that the divisor of poles of the differential dh is $\Sigma_j f_j Q_j$. Let us temporarily admit this claim, and use it to complete the proof of the theorem. Because h is finite etale over $\mathbb{A}^1$, dh has neither zero nor pole in $C_1 - D_1$. So the divisor of dh is $- \Sigma_j f_j Q_j$. The divisor of dh has degree $2g_1 - 2$. Thus we have the equality

$$2g_1 - 2 = - \Sigma_j f_j.$$

This can only happen if $g_1 = 0$ and $\Sigma_j f_j = 2$. So we have

$$2 = \Sigma_j f_j \geq \Sigma_j e_j = \deg(D_1) = d_1.$$

Since $d_1 \geq 2$, we have $\Sigma_j f_j = \Sigma_j e_j$, i.e., every e_j is divisible by p. But $\Sigma_j e_j = 2$, so there is only one e_j, and p = 2. Thus C_1 is $\mathbb{P}^1$, and D_1 is 2Q, for some point Q in $\mathbb{P}^1$. Applying an automorphism of $\mathbb{P}^1$, we may assume that Q is the point ∞ in $\mathbb{P}^1$. Then $C_1 - D_1$ is $\mathbb{A}^1$, and h is a quadratic polynomial $\alpha x^2 + \beta x + \gamma$, which makes $\mathbb{A}^1$ finite etale over itself of degree two. Therefore $\alpha \neq 0$ (because h has degree two) and $\beta \neq 0$ (otherwise dh vanishes). In this case, $f = h \circ g$ is $\alpha g^2 + \beta g + \gamma$. Denote by $D_2 \subset C$ the divisor of poles of g. Since $f = h \circ g$, and h has divisor of poles 2∞, we see that D, the divisor of poles of f, is related to D_2 by

$$D = 2D_2.$$

Since p = 2, the divisor of poles of the differential df is D. This in turn implies that g has only simple poles, i.e., that $D = \Sigma_i 2P_i$. Indeed, if at some point P g has a pole of order $n \geq 1$, then in terms of a uniformizing parameter z at P we have

$$g = 1/z^n + (1/z^{n-1})(\text{holomorphic at P}).$$

Because p = 2, $f = \alpha g^2 + \beta g + \gamma$ has

$$df = \beta dg,$$

and hence df has a pole of order at most n+1 at P. But f has a pole of order 2n at P, so by hypothesis (remember p = 2) df also has a pole of order 2n at P. Therefore $2n \leq n + 1$, possible only when n =1.

(6.3.15) This shows that Γ is indeed primitive, except in the excluded exceptional case:

$p = 2$, $D = \Sigma_i 2P_i$, there exists a function g in $L(\Sigma_i P_i)$ and there exist scalars α, β in $k^\times$ and γ in k such that $f = \alpha g^2 + \beta g + \gamma$.

(6.3.16) It remains to prove that the divisor of poles of the differential dh is $\Sigma_j f_j Q_j$. At a point Q_j in D_1 with multiplicity e_j prime to p, h has a pole of order e_j, and dh has a pole of order $f_j := 1 + e_j$. At a point Q in D_1 with multiplicity e divisible by p, we must show that dh has a pole of the same order e. Pick a point P in D_2 lying over Q. Let z be a uniformizing parameter at P, and let w be a uniformizing parameter at Q. Denote by $m \geq 1$ the order of zero of $g - g(P)$ at the point P. Then $f = h \circ g$ has a pole of order em at P.
(6.3.17) Rescaling the uniformizing parameter z if necessary, the formal expansion of g at P has the form

$$g - g(P) = z^m(1 + az^1 + ...).$$

Rescaling w if necessary, the formal expansion of h at Q is

$$h = 1/w^e - b/w^{e-1} + ... = (1/w^e)(1 - bw + ...).$$

Equivalently, the formal expansion of 1/h at P is

$$1/h = w^e(1 + bw + ...).$$

We must show that $b \neq 0$.
(6.3.18) The formal expansion of $1/f = 1/(h \circ g) = (1/h) \circ g$ at P is

$$1/f = z^{me}(1 + az + ...)^e(1 + bz^m + ...).$$

Since p divides e, we have

$$(1 + az + ...)^e = 1 + (z^p) = 1 + (z^2).$$

Thus

$$1/f = z^{me}(1 + bz^m + ...)(1 + (z^2)).$$

So the expansion of f at P is

$$f = z^{-me}(1 - bz^m + ...)(1 + (z^2)).$$

But by hypothesis on f, its expansion at P is

$$f = \alpha/z^{em} + \beta/z^{em-1} + ...,$$

with $\alpha\beta \neq 0$. Comparing these two expansions, we infer that $m = 1$, otherwise f has no $1/z^{em-1}$ term, and we infer that $b = -\beta \neq 0$. QED

(6.4) Interpretation in terms of Swan conductors

(6.4.1) In Theorem 6.2.11, an essential hypothesis on the function f is that at any point P where f has a pole whose order e is not prime to p, the differential df has a pole of the same order e at P, or, equivalently, $\mathrm{ord}_P(df/f) = 0$. In this section, we give the interpretation of this condition in terms of Swan conductors.
(6.4.2) Let k be an algebraically closed field, C/k a projective, smooth connected curve, $D = \Sigma_i a_i P_i$ an effective divisor on X, and f in L(D) a function with divisor of poles D. View f as a finite flat map

$$f : C \to \mathbb{P}^1$$

of degree d. Suppose that f is finite etale over a dense open set

$U = \mathbb{P}^1 - T$ of $\mathbb{P}^1$. Put $V := C - f^{-1}(T)$.

(6.4.3) For any ℓ invertible in k, and for any lisse $\overline{\mathbb{Q}}_\ell$-sheaf $\mathcal{G}$ on U, the Euler Poincaré formula [Ray] asserts that

(6.4.3.1) $\chi_c(U, \mathcal{G}) = \chi_c(U, \overline{\mathbb{Q}}_\ell)\mathrm{rank}(\mathcal{G}) - \Sigma_{t \text{ in } T} \mathrm{Swan}_t(\mathcal{G})$.

If we begin with a lisse $\overline{\mathbb{Q}}_\ell$-sheaf $\mathcal{H}$ on V, and take for $\mathcal{G}$ the direct image $f_*\mathcal{H}$, then

(6.4.3.2) $$\chi_c(V, \mathcal{H}) = \chi_c(U, f_*\mathcal{H}),$$

and the Euler Poincaré formula becomes

(6.4.3.3) $\chi_c(V, \mathcal{H}) = \chi_c(U, \overline{\mathbb{Q}}_\ell)\times d\times \mathrm{rank}(\mathcal{H}) - \Sigma_{t \text{ in } T} \mathrm{Swan}_t(f_*\mathcal{H})$.

Fix one point t in T, and denote by $x_1,..., x_n$ the points of C lying over t. As representation of I(t), $\mathcal{G}(t)$ is $(f_*\mathcal{H})(t)$, which is the direct sum

(6.4.3.4) $$(f_*\mathcal{H})(t) = \oplus_i \mathrm{Ind}_{I(x_i)}{}^{I(t)} \mathcal{H}(x_i).$$

Denote by K the function field of $\mathbb{P}^1$ over k, and by L the function field of C over k. Denote by K_t and L_{x_i} their completions at the indicated points, and by

$$f_{x_i} : \mathrm{Spec}(L_{x_i}) \to \mathrm{Spec}(K_t)$$

the map induced on (the spectra of) these completions. Geometrically, we have

$$(f_*\mathcal{H})(t) = \oplus_i (f_{x_i})_*\mathcal{H}(x_i).$$

Thus we have

(6.4.3.5) $\mathrm{Swan}_t(f_*\mathcal{H}) = \Sigma_{x \mapsto t} \mathrm{Swan}_t((f_x)_*\mathcal{H}(x))$.

So we have

(6.4.3.6) $\chi_c(V, \mathcal{H})$

$$= \chi_c(U, \overline{\mathbb{Q}}_\ell)\times d\times \mathrm{rank}(\mathcal{H}) - \Sigma_{t \text{ in } T} \Sigma_{x \mapsto t} \mathrm{Swan}_t((f_x)_*\mathcal{H}(x)).$$

Take for $\mathcal{H}$ the constant sheaf $\overline{\mathbb{Q}}_\ell$ on V. We get the Hurwitz formula

(6.4.3.7) $\chi_c(V, \overline{\mathbb{Q}}_\ell) = \chi_c(U, \overline{\mathbb{Q}}_\ell)d - \Sigma_{t \text{ in } T} \Sigma_{x \mapsto t} \mathrm{Swan}_t((f_x)_*\overline{\mathbb{Q}}_\ell)$.

Lemma 6.4.4 If $x \mapsto t$ and t is 0 or ∞, we have the formula

$$\mathrm{Swan}_t((f_x)_*\overline{\mathbb{Q}}_\ell) - 1 = \mathrm{ord}_x(df/f).$$

For $x \mapsto t$ with t in $\mathbb{G}_m$, we have

$$\mathrm{Swan}_t((f_x)_*\overline{\mathbb{Q}}_\ell) - 1 = \mathrm{ord}_x(df/(f - t)).$$

proof It suffices to treat the case when $t = 0$. [By additive translation, we reduced the case of t in $\mathbb{A}^1$ to this case. By replacing f by 1/f, we reduce the case $t = \infty$ to this case.] In this case, f has a zero of order $e \geq 1$ at x, and L_x/K_t is a finite separable extension of degree e. In terms of a uniformizing parameter Π in L_x, the expansion of f at x is

$$f = \Sigma_{n \geq e} \alpha_n\Pi^n, \ \alpha_e \neq 0,$$

and K_t is the subfield $k((f))$ of $k((\Pi))$, with uniformizing parameter f.

We next apply to L_x/K_t the following general facts. Let F be a nonarchimedean local field with algebraically closed residue field. For

E/F any finite separable extension, of degree denoted e, let M/F be a finite galois extension of F which contains E. Denote by G the galois group of M/F, and by $H \subset G$ the subgroup which fixes H. Let W be a $\overline{\mathbb{Q}}_\ell$-finite-dimensional G-module. If we view W as a lisse $\overline{\mathbb{Q}}_\ell$-sheaf on Spec(F), we may speak of its Swan conductor Swan(W), which is related to the Artin conductor $\mathrm{Artin}_G(W)$ of W as G-module by

$$\mathrm{Artin}_G(W) = \dim(W/W^G) + \mathrm{Swan}(W).$$

For $W := \mathrm{Ind}_H{}^G(\overline{\mathbb{Q}}_\ell)$, $\dim(W^G) = \langle \mathrm{Ind}_H{}^G(\overline{\mathbb{Q}}_\ell), \overline{\mathbb{Q}}_\ell \rangle_G = 1$, by Frobenius reciprocity, so

$$\mathrm{Artin}_G(\mathrm{Ind}_H{}^G(\overline{\mathbb{Q}}_\ell)) = e - 1 + \mathrm{Swan}(\mathrm{Ind}_H{}^G(\overline{\mathbb{Q}}_\ell)).$$

On the other hand, one knows [Serre-CL, VI, §2, Corollaire on page 109, applied to ψ := the trivial character of H] that, denoting by $\Delta_{E/F}$ the discriminant of E/F, one has

$$\mathrm{Artin}_G(\mathrm{Ind}_H{}^G(\overline{\mathbb{Q}}_\ell)) = \mathrm{ord}_F(\Delta_{E/F}).$$

Denote by $\mathcal{D}_{E/F}$ the different of E/F. Since E/F is fully ramified (remember that F has algebraically closed residue field), and $\Delta_{E/F} = \mathrm{Norm}_{E/F}(\mathcal{D}_{E/F})$, we have

$$\mathrm{Artin}_G(\mathrm{Ind}_H{}^G(\overline{\mathbb{Q}}_\ell)) = \mathrm{ord}_E(\mathcal{D}_{E/F}).$$

So we have

$$\mathrm{Swan}(\mathrm{Ind}_H{}^G(\overline{\mathbb{Q}}_\ell)) - 1 = \mathrm{ord}_E(\mathcal{D}_{E/F}) - e.$$

The different $\mathcal{D}_{E/F}$ is the annihilator in $\mathcal{O}_E$ of $\Omega^1{}_{\mathcal{O}_E/\mathcal{O}_F}$. So in equal characteristic, if we fix uniformizing parameters π of F and Π of E, $\mathcal{D}_{E/F}$ is the ideal generated by $d\pi/d\Pi$ in $\mathcal{O}_E$. Thus $\mathrm{ord}_E(\mathcal{D}_{E/F}) - e$ is the order of zero or pole of $d\pi/\pi$ as as section of $\Omega^1{}_{\mathcal{O}_E/k} := (\mathcal{O}_E)d\Pi$:

$$\mathrm{Swan}(\mathrm{Ind}_H{}^G(\overline{\mathbb{Q}}_\ell)) - 1 = \mathrm{ord}_E(d\pi/\pi).$$

Applying this in the situation L_x/K_t with the uniformizer π taken to be f, we find the asserted formula. QED

Corollary 6.4.5 Suppose f has a zero or pole at x, say

$$\mathrm{ord}_x(f) = e \neq 0.$$

Then we have the following results.

1) If e is invertible in k, then $\mathrm{Swan}_{f(x)}((f_x)_\star\overline{\mathbb{Q}}_\ell) = 0$.

2) If k has positive characteristic p and p divides e, pick a uniformizing parameter Π at x, and write the formal expansion of f at x,

$$f = \Pi^e\Sigma_{n \geq 0}\, \alpha_n\Pi^n,$$

with $\alpha_0 \neq 0$. Then we have

$$\mathrm{Swan}_{f(x)}((f_x)_\star\overline{\mathbb{Q}}_\ell)$$
$$= \text{the least prime-to-p integer } m \text{ with } \alpha_m \neq 0.$$

In particular, when p divides e, we have

$$\mathrm{Swan}_{f(x)}((f_x)_*\overline{\mathbb{Q}}_\ell) = 1 \text{ if and only if } \alpha_1 \neq 0.$$

(6.4.6) So the hypothesis of Theorem 6.2.11 concerning the poles of df could be restated as saying that at each pole x of f, we have the inequality

$$\mathrm{Swan}_\infty((f_x)_*\overline{\mathbb{Q}}_\ell) \le 1.$$

Let us say that an f satisfying this condition has Swan-minimal poles. To end the proof of Theorem 6.2.11, we showed by a power series computation that if f has Swan-minimal poles and f is a composition $h\circ g$, then h has Swan-minimal poles. This can also be seen as a special case of a general fact about the behaviour of Swan conductors under finite direct image. Suppose we have a proper smooth curve Y/k, together with maps $g : X \to Y$ and $h: Y \to \mathbb{P}^1$ such that $f = h\circ g$. For any pole y of h, and any $\overline{\mathbb{Q}}_\ell$-representation $\mathcal{G}(y)$ of the inertia group I(y), we have the direct image formula

(6.4.6.1) $\mathrm{Swan}_\infty((h_y)_*\mathcal{G}(y))$

$$= \mathrm{Swan}_y(\mathcal{G}(y)) + \mathrm{rank}(\mathcal{G}(y))\mathrm{Swan}_\infty((h_y)_*\overline{\mathbb{Q}}_\ell),$$

cf. [Ka-TLFM, 1.6.4.1]. Now take a pole x of f, with image $y = g(x)$ in Y, and take for $\mathcal{G}(y)$ the direct image $(g_x)_*\overline{\mathbb{Q}}_\ell$. Then

$$(h_y)_*\mathcal{G}(y) = (h_y)_*(g_x)_*\overline{\mathbb{Q}}_\ell = (f_x)_*\overline{\mathbb{Q}}_\ell,$$

so the formula gives

(6.4.6.2) $\mathrm{Swan}_\infty((f_x)_*\overline{\mathbb{Q}}_\ell)$

$$= \mathrm{Swan}_y((g_x)_*\overline{\mathbb{Q}}_\ell) + \mathrm{rank}((g_x)_*\overline{\mathbb{Q}}_\ell)\mathrm{Swan}_\infty((h_y)_*\overline{\mathbb{Q}}_\ell).$$

Since Swan conductors are non-negative integers, and $\mathrm{rank}((g_x)_*\overline{\mathbb{Q}}_\ell)$ is at least one, we have an a priori inequality

(6.4.6.3) $$\mathrm{Swan}_\infty((f_x)_*\overline{\mathbb{Q}}_\ell) \ge \mathrm{Swan}_\infty((h_y)_*\overline{\mathbb{Q}}_\ell).$$

In particular, if f has Swan-minimal poles, so does h.

(6.4.7) As a consequence of the above discussion, we extract the following simple corollary.

Corollary 6.4.8 If f has Swan-minimal poles, then the middle extension sheaves $f_*\overline{\mathbb{Q}}_\ell/\overline{\mathbb{Q}}_\ell$ and $f_*\overline{\mathbb{Q}}_\ell$ on $\mathbb{A}^1$ both have all $I(\infty)$-breaks ≤ 1.

proof Indeed, $f_*\overline{\mathbb{Q}}_\ell/\overline{\mathbb{Q}}_\ell$ is a direct summand of $f_*\overline{\mathbb{Q}}_\ell$, so it suffices to treat $f_*\overline{\mathbb{Q}}_\ell$. The $I(\infty)$-representation $f_*\overline{\mathbb{Q}}_\ell$ is the direct sum, over the poles x of f, of the $I(\infty)$-representations $(f_x)_*\overline{\mathbb{Q}}_\ell$. As we have seen above, each of these has $\mathrm{Swan}_\infty((f_x)_*\overline{\mathbb{Q}}_\ell) \le 1$. But the Swan conductor of an $I(\infty)$-representation is the sum of its $I(\infty)$-breaks, each of which is a non-negative rational number. Therefore each $(f_x)_*\overline{\mathbb{Q}}_\ell$ has all its $I(\infty)$-breaks ≤ 1, and hence their direct sum, $(f_x)_*\overline{\mathbb{Q}}_\ell$ as $I(\infty)$-representation, has all its $I(\infty)$-breaks ≤ 1. QED

(6.5) Middle convolution and purity

(6.5.1) In this section, we work on $\mathbb{A}^1$ over a finite field k, of characteristic p. We fix a prime $\ell \neq p$, and an embedding ι of $\overline{\mathbb{Q}}_\ell$ into $\mathbb{C}$. For ease of later reference, we begin by stating a basic consequence for ! convolution of Deligne's Weil II results.

Lemma 6.5.2 Let K (resp. L) be an object in $D^b{}_c(\mathbb{A}^1, \overline{\mathbb{Q}}_\ell)$ which is ι-mixed of weight $\leq a$ (resp. $\leq b$). Then their ! convolution $K*_!L$ is ι-mixed of weight $\leq a + b$.

proof The external tensor product $K\boxtimes L := (pr_1{}^*K)\otimes(pr_2{}^*L)$ on $\mathbb{A}^2$ is ι-mixed of weight $\leq a + b$, so by Deligne's Weil II result [De-Weil II, 3.3.1], $K*_!L := Rsum_!(K\boxtimes L)$ is ι-mixed of weight $\leq a + b$. QED.

(6.5.3) We now turn to our main target.

Theorem 6.5.4 Let K and L be perverse sheaves on $\mathbb{A}^1/k$. Suppose that K and L are both ι-pure of weight zero. Suppose that K has $\mathcal{P}$ geometrically (i.e., after extension of scalars to $\overline{k}$). Then we have the following results.

1) The ! convolution $K*_!L$ is ι-mixed of weight ≤ 0.

2) The middle convolution $K*_{mid}L$ is ι-pure of weight zero.

3) The canonical surjective map of perverse sheaves

$$K*_!L \to K*_{mid}L$$

has a kernel which is ι-mixed of weight ≤ -1, and induces an isomorphism of perverse sheaves on $\mathbb{A}^1/k$

$$Gr_W{}^0(K*_!L) \cong K*_{mid}L.$$

proof Assertion 1) results from Lemma 6.5.2 above

To prove 2), note that, as K and L are ι-pure of weight 0, their Verdier duals DK and DL are ι-mixed of weight ≤ 0. The Verdier dual of $K*_*L$ is $DK*_!DL$, which by 1) is ι-mixed of weight ≤ 0. Therefore $K*_*L$ is ι-mixed of weight ≥ 0. Because K has $\mathcal{P}$, both $K*_!L$ and $K*_*L$ are perverse, and $K*_{mid}L$ is, as a perverse sheaf, both a quotient of $K*_!L$, so ι-mixed of weight ≤ 0, and a subobject of $K*_*L$, so ι-mixed of weight ≥ 0. Thus $K*_{mid}L$ is ι-pure of weight 0.

To prove 3), we first reduce to the case where, in addition, L is geometrically isotypical. For this, we exploit the fact that L, being perverse and ι-pure of weight 0, is geometrically semisimple. Consider its isotypical decomposition over $\overline{k}$. The finitely many nonzero isotypical components are permuted by $Gal(\overline{k}/k)$, so, at the expense of replacing k by a finite extension, we may assume that each isotypical component is $Gal(\overline{k}/k)$-stable, and hence that the isomorphism class of the geometric irreducible underlying each isotypical component is $Gal(\overline{k}/k)$-stable. Then over k, L is a direct sum of pieces, each of which is perverse, ι-pure of weight 0, and geometrically isotypical. Since 3) is compatible with direct sums, it suffices to prove 3) in the case when L is perverse, ι-pure of weight 0, and geometrically isotypical.

If the geometric irreducible underlying L has $\mathcal{P}$, then L has $\mathcal{P}$

geometrically, and we argue as follows. Fix a nontrivial $\overline{\mathbb{Q}}_\ell^\times$-valued additive character ψ of the prime field $\mathbb{F}_p$. We will exploit the fact that Fourier Transform FT_ψ is an autoequivalence of the category of perverse sheaves on $\mathbb{A}^1/k$, which increases weights by one.

Since K and L both have $\mathcal{P}$, we can write $FT_\psi(K)$ as $(j_*\mathcal{N})[1]$ and write $FT_\psi(L)$ as $(j_*\mathcal{M})[1]$, for some lisse $\overline{\mathbb{Q}}_\ell$-sheaves $\mathcal{N}$ and $\mathcal{M}$ on a common dense open set $U \subset \mathbb{A}^1/k$, inclusion denoted $j: U \to \mathbb{A}^1$, cf. [Ka-RLS, 2.10.3]. We know [Ka-RLS, 2.10.8 and 2.10.1] that

$$FT_\psi(K*_{mid}L) = (j_*(\mathcal{N}\otimes\mathcal{M}))[1],$$
$$FT_\psi(K*_!L) = ((j_*\mathcal{N})\otimes(j_*\mathcal{M}))[1].$$

Because K and L are both ι-pure of weight zero, $FT_\psi(K)$ and $FT_\psi(L)$ are both ι-pure of weight one, or equivalently, $\mathcal{N}$ and $\mathcal{M}$ are both ι-pure of weight zero. Thus $\mathcal{N}\otimes\mathcal{M}$ is also ι-pure of weight zero, and hence $j_*(\mathcal{N}\otimes\mathcal{M})$ is ι-mixed of weight ≤ 0. Similarly $(j_*\mathcal{N})\otimes(j_*\mathcal{M})$ is ι-mixed of weight ≤ 0. We have a natural injective map of usual sheaves on $\mathbb{A}^1/k$

$$(j_*\mathcal{N})\otimes(j_*\mathcal{M}) \to j_*(\mathcal{N}\otimes\mathcal{M})$$

which sits in a short exact sequence of usual sheaves on $\mathbb{A}^1/k$,

$$0 \to (j_*\mathcal{N})\otimes(j_*\mathcal{M}) \to j_*(\mathcal{N}\otimes\mathcal{M}) \to j_*(\mathcal{N}\otimes\mathcal{M})/((j_*\mathcal{N})\otimes(j_*\mathcal{M})) \to 0.$$

The cokernel sheaf $j_*(\mathcal{N}\otimes\mathcal{M})/((j_*\mathcal{N})\otimes(j_*\mathcal{M}))$ is punctual, and it is ι-mixed of weight ≤ 0, being a quotient of $j_*(\mathcal{N}\otimes\mathcal{M})$. This exact sequence of usual sheaves gives a short exact sequence of perverse sheaves on $\mathbb{A}^1/k$,

$$0 \to j_*(\mathcal{N}\otimes\mathcal{M})/((j_*\mathcal{N})\otimes(j_*\mathcal{M}))$$
$$\to ((j_*\mathcal{N})\otimes(j_*\mathcal{M}))[1] \to (j_*(\mathcal{N}\otimes\mathcal{M}))[1] \to 0.$$

In this short exact sequence the first term is ι-mixed of weight ≤ 0. Applying $FT_{\overline{\psi}}(1)$, which is the inverse to FT_ψ and which decreases weights by one, we get a short exact sequence

$$0 \to (\iota\text{-mixed of weight} \le -1) \to K*_!L \to K*_{mid}L \to 0,$$

in which the map $K*_!L \to K*_{mid}L$ is the natural one. We already know that $K*_{mid}L$ is ι-pure of weight 0. So applying the exact functor $Gr_W{}^0$ concludes the proof of 3) in the case when the geometric irreducible underlying L has $\mathcal{P}$.

If the geometric irreducible underlying L does not have $\mathcal{P}$, that irreducible must be $\mathcal{L}_{\psi(\alpha x)}[1]$ for some α in k. Indeed, the only perverse irreducibles over $\overline{k}$ which do not have $\mathcal{P}$ are the $\mathcal{L}_{\psi(\alpha x)}[1]$, for α in $\overline{k}$; the requirement that its isomorphism class be $Gal(\overline{k}/k)$-stable forces α to lie in k. Then L is

$$L \cong \mathcal{L}_{\psi(\alpha x)}[1]\otimes\mathcal{A}(1/2)$$

for $\mathcal{A}$ some geometrically constant sheaf on $\mathbb{A}^1/k$ which is ι-pure of weight 0. So it suffices to treat the case when L is $\mathcal{L}_{\psi(\alpha x)}(1/2)[1]$. In this case, using $\mathcal{L}_{\psi(\alpha(x+y))} \cong \mathcal{L}_{\psi(\alpha x)} \otimes \mathcal{L}_{\psi(\alpha y)}$, we find

$$K*_!\mathcal{L}_{\psi(\alpha x)}(1/2)[1] = \mathcal{L}_{\psi(\alpha x)}(1/2)[1] \otimes R\Gamma_c(\mathbb{A}^1\otimes_k\bar{k}, K\otimes\mathcal{L}_{\psi(-\alpha x)}),$$

$$K*_*\mathcal{L}_{\psi(\alpha x)}(1/2)[1] = \mathcal{L}_{\psi(\alpha x)}(1/2)[1] \otimes R\Gamma(\mathbb{A}^1\otimes_k\bar{k}, K\otimes\mathcal{L}_{\psi(-\alpha x)}).$$

The "forget supports" map

$$K*_!\mathcal{L}_{\psi(\alpha x)}(1/2)[1] \to K*_*\mathcal{L}_{\psi(\alpha x)}(1/2)[1]$$

is (the tensor product with $\mathcal{L}_{\psi(\alpha x)}(1/2)[1]$ of) the "forget supports" map

$$R\Gamma_c(\mathbb{A}^1\otimes_k\bar{k}, K\otimes\mathcal{L}_{\psi(-\alpha x)}) \to R\Gamma(\mathbb{A}^1\otimes_k\bar{k}, K\otimes\mathcal{L}_{\psi(-\alpha x)}).$$

Notice that $K\otimes\mathcal{L}_{\psi(-\alpha x)}$ is perverse, ι-pure of weight zero, and geometrically has $\mathcal{P}$. So assertion 3) now results from the following lemma, applied to $K\otimes\mathcal{L}_{\psi(-\alpha x)}$.

Lemma 6.5.5. Suppose K on $\mathbb{A}^1/k$ is perverse, ι-pure of weight 0, and geometrically has $\mathcal{P}$. Then we have the following results.

1) $H_c^i(\mathbb{A}^1\otimes_k\bar{k}, K) = 0$ for $i \neq 0$,

2) $H_c^0(\mathbb{A}^1\otimes_k\bar{k}, K)$ is ι-mixed of weight ≤ 0,

3) $H^i(\mathbb{A}^1\otimes_k\bar{k}, K) = 0$ for $i \neq 0$,

4) $H^0(\mathbb{A}^1\otimes_k\bar{k}, K)$ is ι-mixed of weight ≥ 0,

5) the kernel of the natural "forget supports" map

$$H_c^0(\mathbb{A}^1\otimes_k\bar{k}, K) \to H^0(\mathbb{A}^1\otimes_k\bar{k}, K)$$

is ι-mixed of weight ≤ -1, and this map induces an isomorphism

$$\mathrm{Gr}_W^0(H_c^0(\mathbb{A}^1\otimes_k\bar{k}, K)) \cong \mathrm{Image}(H_c^0(\mathbb{A}^1\otimes_k\bar{k}, K) \to H^0(\mathbb{A}^1\otimes_k\bar{k}, K)).$$

proof Assertion 1) is the perversity (remember K has $\mathcal{P}$ geometrically) of

$$K*_!\bar{\mathbb{Q}}_\ell[1] = \bar{\mathbb{Q}}_\ell[1]\otimes R\Gamma_c(\mathbb{A}^1\otimes_k\bar{k}, K).$$

Assertion 2) results from [De-Weil II, 3.3.1]. Assertions 3) and 4) for K are the Poincaré duals of assertions 1) and 2) respectively for DK.

To prove 5), we break K geometrically (i.e., over $\bar{k}$) into isotypical components (remember that K, being ι-pure, is geometrically semisimple, cf. [BBD, 5.3.8]). Each isotypical component is either a middle extension, or is punctual. Lumping together the isotypical components of each type, we get an arithmetic (i.e., over k) decomposition of K as a direct sum

$$K = K_{\mathrm{pct}} \oplus K_{\mathrm{midext}}$$

of a punctual perverse sheaf K_{pct} and of a middle extension perverse sheaf on $\mathbb{A}^1/k$, both of which are ι-pure of weight 0.

If K is punctual, then the "forget supports" map

$$H_c^0(\mathbb{A}^1\otimes_k\bar{k}, K) \to H^0(\mathbb{A}^1\otimes_k\bar{k}, K)$$

is an isomorphism, and both its source and target are ι-pure of weight 0.

If K is a middle extension, say $K = \mathcal{G}[1]$ for some middle extension sheaf on $\mathbb{A}^1$, we argue as follows. Denote by

$$j : \mathbb{A}^1 \subset \mathbb{P}^1$$

the inclusion. On $\mathbb{P}^1$ we have the three perverse sheaves $j_!K = j_!\mathcal{G}[1]$, $Rj_*K = Rj_*\mathcal{G}[1]$, and $j_{!*}K = j_*\mathcal{G}[1]$. The object $j_{!*}K = j_*\mathcal{G}[1]$ is ι-pure of weight 0 on $\mathbb{P}^1$.

The short exact sequence of usual sheaves on $\mathbb{P}^1$

$$0 \to j_!\mathcal{G} \to j_*\mathcal{G} \to j_*\mathcal{G}/j_!\mathcal{G} \cong \mathcal{G}^{I(\infty)} \text{ as pct. sheaf at } \infty \to 0$$

gives a short exact sequence of perverse sheaves on $\mathbb{P}^1$

$$0 \to j_*\mathcal{G}/j_!\mathcal{G} \to j_!\mathcal{G}[1] \to j_*\mathcal{G}[1] \to 0.$$

Since the middle extension $\mathcal{G}[1]$ is ι-pure of weight 0 on $\mathbb{A}^1$, $\mathcal{G}$ is the extension from a dense open set $U \subset \mathbb{A}^1$ of a lisse sheaf $\mathcal{G}|U$ which is ι-pure of weight -1 on U. Then $j_*\mathcal{G}$ is ι-mixed of weight ≤ -1, cf. [De-Weil II, 1.8.1], as is its quotient $j_*\mathcal{G}/j_!\mathcal{G} \cong \mathcal{G}^{I(\infty)}$. The long exact cohomology sequence on $\mathbb{P}^1\otimes_k\bar{k}$ gives a right exact sequence

$$(\text{weight} \le -1) \to H^0_c(\mathbb{A}^1\otimes_k\bar{k}, K) \to H^0(\mathbb{P}^1\otimes_k\bar{k}, j_{!*}K) \to 0.$$

The third term $H^0(\mathbb{P}^1\otimes_k\bar{k}, j_{!*}K)$ is ι-pure of weight zero. Thus we have

$$\mathrm{Gr}_W{}^0(H^0{}_c(\mathbb{A}^1\otimes_k\bar{k}, K)) = H^0(\mathbb{P}^1\otimes_k\bar{k}, j_{!*}K).$$

We also have a short exact sequence of perverse sheaves on $\mathbb{P}^1$

$$0\to j_*\mathcal{G}[1] \to Rj_*\mathcal{G}[1] \to H^1(I(\infty), \mathcal{G}(\infty)) \text{ as pct. sheaf at } \infty \to 0,$$

which gives a left exact sequence

$$0 \to H^0(\mathbb{P}^1\otimes_k\bar{k}, j_{!*}K) \to H^0(\mathbb{A}^1\otimes_k\bar{k}, K) \to H^1(I(\infty), \mathcal{G}(\infty)).$$

Thus we have

$$\mathrm{Gr}_W{}^0(H^0{}_c(\mathbb{A}^1\otimes_k\bar{k}, K)) = H^0(\mathbb{P}^1\otimes_k\bar{k}, j_{!*}K)$$
$$= \mathrm{Image}(H^0{}_c(\mathbb{A}^1\otimes_k\bar{k}, K) \to H^0(\mathbb{A}^1\otimes_k\bar{k}, K)),$$

as required. QED

This concludes the proof of Theorem 6.5.4.

Corollary 6.5.6 Fix an integer $r \ge 2$. Let $K_1, K_2,\ldots,K_r$ be perverse sheaves on $\mathbb{A}^1/k$. Suppose that each K_i is ι-pure of weight zero, and has $\mathcal{P}$ geometrically. Then we have the following results.

1) The ! convolution $K_1*_!K_2*_!\ldots*_!K_r$ is ι-mixed of weight ≤ 0.

2) The middle convolution $K_1*_{mid}K_2*_{mid}\cdots*_{mid}K_r$ has $\mathcal{P}$, and is ι-pure of weight zero.

3) The canonical surjective map of perverse sheaves

$$K_1*_!K_2*_!\ldots*_!K_r \to K_1*_{mid}K_2*_{mid}\cdots*_{mid}K_r$$

has a kernel which is ι-mixed of weight ≤ -1, and induces an isomorphism of perverse sheaves on $\mathbb{A}^1/k$

$$\mathrm{Gr}_W{}^0(K_1 *_! K_2 *_! \ldots *_! K_r) \cong K_1 *_{mid} K_2 *_{mid} \cdots *_{mid} K_r.$$

proof The external tensor product $K_1 \boxtimes K_2 \boxtimes \ldots \boxtimes K_r :=$ $(pr_1{}^*K_1) \otimes (pr_2{}^*K_2) \otimes \ldots \otimes (pr_r{}^*K_r)$ on $\mathbb{A}^r$ is ι-mixed of weight ≤ 0, so by Deligne's Weil II result [De-Weil II, 3.3.1],

$$K_1 *_! K_2 *_! \ldots *_! K_r := \mathrm{Rsum}_!(K_1 \boxtimes K_2 \boxtimes \ldots \boxtimes K_r)$$

is ι-mixed of weight ≤ 0. This proves assertion 1). From [Ka-RLS, 2.6.5 and 2.6.17], we know that $*_{mid}$ is an associative product

$$\text{(perverse sheaves with } \mathcal{P}) \times \text{(perverse sheaves with } \mathcal{P})$$
$$\to \text{(perverse sheaves with } \mathcal{P}).$$

So assertion 2) results from the previous Theorem 6.5.4 by an obvious induction. For assertion 3), we argue as follows. If $r = 2$, we invoke the previous Theorem 6.5.4. If $r \ge 3$, we may assume by induction that the canonical surjective map of perverse sheaves

$$K_1 *_! K_2 *_! \ldots *_! K_{r-1} \to K_1 *_{mid} K_2 *_{mid} \cdots *_{mid} K_{r-1}$$

sits in a short exact sequence of perverse sheaves

$$0 \to (\text{weight} \le -1) \to K_1 *_! K_2 *_! \ldots *_! K_{r-1}$$
$$\to K_1 *_{mid} K_2 *_{mid} \cdots *_{mid} K_{r-1} \to 0.$$

Because K_r has $\mathcal{P}$, the distinguished triangle we get from this by forming the ! convolution with K_r has all its terms perverse, so we get a short exact sequence

$$0 \to (\text{weight} \le -1) *_! K_r \to K_1 *_! K_2 *_! \ldots *_! K_r$$
$$\to (K_1 *_{mid} K_2 *_{mid} \cdots *_{mid} K_{r-1}) *_! K_r \to 0$$

of perverse sheaves. The first term

$$\mathrm{Ker}_1 := (\text{weight} \le -1) *_! K_r$$

is mixed of weight ≤ -1, by Lemma 6.5.2.

By the previous Theorem 6.5.4 applied to the two objects $K_1 *_{mid} K_2 *_{mid} \cdots *_{mid} K_{r-1}$ and K_r, the canonical surjective map of perverse sheaves

$$(K_1 *_{mid} K_2 *_{mid} \cdots *_{mid} K_{r-1}) *_! K_r \to K_1 *_{mid} K_2 *_{mid} \cdots *_{mid} K_r$$

sits in a short exact sequence of perverse sheaves

$$0 \to (\text{weight} \le -1) := \mathrm{Ker}_2 \to (K_1 *_{mid} K_2 *_{mid} \cdots *_{mid} K_{r-1}) *_! K_r$$
$$\to K_1 *_{mid} K_2 *_{mid} \cdots *_{mid} K_r \to 0.$$

Combining these two short exact sequences, we get a short exact sequence

$$0 \to \mathrm{Ker}_3 \to K_1 *_! K_2 *_! \ldots *_! K_r \to K_1 *_{mid} K_2 *_{mid} \cdots *_{mid} K_r \to 0,$$

in which Ker_3 is ι-mixed of weight ≤ -1, being an extension of Ker_2 by Ker_1. Applying the exact functor $\mathrm{Gr}_W{}^0$, we get

$$\mathrm{Gr}_W{}^0(K_1 *_! K_2 *_! \ldots *_! K_r) \cong K_1 *_{mid} K_2 *_{mid} \cdots *_{mid} K_r. \qquad \text{QED}$$

(6.6) Application to the monodromy of multiplicative character sums in several variables

(6.6.1) Let us recall the general context. We have a finite field k of characteristic p, a prime number ℓ invertible in k, and a nontrivial multiplicative character $\chi : k^\times \to \overline{\mathbb{Q}}_\ell{}^\times$ of $k^\times$, extended by zero to all of k. We fix integers $e \geq 3$ and $n \geq 1$. We denote by $\mathcal{P}(n, e)$ the space of polynomial functions on $\mathbb{A}^n$ of degree $\leq e$. Recall from 5.1.10 that an n-variable polynomial f in $\mathcal{P}(n,e)(\overline{k})$ is a strong Deligne polynomial if it satisfies the following three conditions D0), D1), and D2).

D0) The closed subscheme $f=0$ in $\mathbb{A}^n$ is smooth of codimension one.
D1) When we write $f = \Sigma_{i\leq e} F_i$ as a sum of homogeneous forms, F_e is nonzero, and, in the case $n \geq 2$, the closed subscheme of $\mathbb{P}^{n-1}$ defined by the vanishing of F_e is smooth of codimension one.
D2) The integer e is prime to p.

(6.6.2) For a fixed integer e which is prime to p, the strong Deligne polynomials form a dense open set $\mathcal{SD}(n,e)$ of $\mathcal{P}(n,e)$.

(6.6.3) Recall further from Corollary 5.1.14 that if e is prime to p, there is a lisse $\overline{\mathbb{Q}}_\ell$-sheaf $\mathcal{M}(n,e,\chi)|\mathcal{SD}(n,e)$ on $\mathcal{SD}(n, e)$, whose trace function is given as follows, cf. 5.1.15: for E/k any finite extension, and for any f in $\mathcal{SD}(n,e)(E)$, we have

$$\mathrm{Trace}(\mathrm{Frob}_{E,f} \mid \mathcal{M}(n,e,\chi))$$
$$= (-1)^n(\# E)^{-n/2}\Sigma_{v \text{ in } \mathbb{A}^n(E)} \chi_E(f(v)).$$

The sheaf $\mathcal{M}(n,e,\chi)|\mathcal{SD}(n,e)$ has rank $(e - 1)^n$. If χ^e is nontrivial, this sheaf is pure of weight 0. If χ^e is trivial, this sheaf is mixed of weights 0 and -1; its Gr^{-1} is lisse of rank $(1/e)((e-1)^n - (-1)^n)$, and its Gr^0 is lisse of rank $N(n, e) := (1/e)((e-1)^{n+1} - (-1)^{n+1})$.

(6.6.4) In the previous chapter, we proved general theorems concerning the geometric monodromy group G_{geom} for the sheaf $\mathcal{M}(n,e,\chi)|\mathcal{SD}(n,e)$ when χ^e is nontrivial, and for the sheaf $\mathrm{Gr}^0(\mathcal{M}(n,e,\chi)|\mathcal{SD}(n,e))$ when χ^e is trivial. When χ has order 2 (i.e., k has odd characteristic, and χ is the quadratic character χ_2) and n is even, we showed that G_{geom} is either SO or O, except in two exceptional cases, namely ($n=2$, $e=3$, $p \geq 5$) and ($n=2$, $e=4$, $p \geq 3$), when G_{geom} is finite. We will now show that the SO case does not occur, by showing that G_{geom} contains a reflection, whenever χ is the quadratic character χ_2 and n is even. This result is a special case of the following one, which tells us that, in any odd characteristic, G_{geom} always contains a specific sort of pseudoreflection.

Theorem 6.6.5 Suppose k is a finite field of odd characteristic p. Let $n \geq 1$ and $e \geq 3$ be integers, and let χ be a nontrivial multiplicative character of $k^\times$. Suppose that e is prime to p. Then G_{geom} for the sheaf $\mathrm{Gr}^0(\mathcal{M}(n,e,\chi)|\mathcal{SD}(n,e))$ contains a (necessarily tame) pseudoreflection of determinant $\chi(\chi_2)^n$. [Recall (5.1.14) that if χ^e is nontrivial, then $\mathrm{Gr}^0(\mathcal{M}(n,e,\chi)|\mathcal{SD}(n,e))$ is just $\mathcal{M}(n,e,\chi)|\mathcal{SD}(n,e)$ itself.]

(6.7) Proof of Theorem 6.6.5, and applications

(6.7.1) The result is geometric, so we may freely replace the finite field k by any finite extension of itself. Over such an extension, we will exhibit a one-parameter family of strong Deligne polynomials such that after pullback of the sheaf in question to this one-parameter family, one of its local monodromies is a pseudoreflection of the asserted type.

(6.7.2) Replacing k by a finite extension if necessary, we can find a one-variable monic polynomial f(x) over k of degree e which is Lefschetz on $\mathbb{A}^1$, i.e., such that its derivative f'(x) has e-1 distinct zeroes, say $\alpha_1,\ldots,\alpha_{e-1}$, and which has e-1 distinct critical values $\beta_i := f(\alpha_i)$. At the expense of further enlarging k, we may assume that all the critical points α_i and, a fortiori, all the critical values $\beta_i := f(\alpha_i)$, lie in k. Now pick a sequence of finite extensions

$$k = k_1 \subset k_2 \subset \ldots \subset k_n$$

such that $k_i \neq k_j$ for $i \neq j$, and pick scalars $\lambda_i \in k_i$ i =1 to n, so that $\lambda_1 \neq 0$, and so that, for $i \geq 2$, $\lambda_i \notin k_{i-1}$. We claim that the one-parameter family (with parameter t) of polynomials in n variables given by

$$t \mapsto t - \Sigma_{j=1 \text{ to } n} \lambda_j f(x_j)$$

is a family of strong Deligne polynomials, over the parameter space

$$\mathbb{A}^1 - S,$$

for

$$S := \{\text{the } (e-1)^n \text{ distinct sums } \Sigma_j \lambda_j\beta_j\}.$$

The critical points of $\Sigma_{j=1 \text{ to } n} \lambda_j f(x_j)$ are the $(e-1)^n$ points in $\mathbb{A}^n$ each of whose coordinates is one of the e-1 critical points α_i of f. Its critical values are the points $\Sigma_j \lambda_j\beta_j$, where each β_j is one of the e-1 critical values of f. There are $(e-1)^n$ distinct such sums $\Sigma_j \lambda_j\beta_j$, simply because the critical values β_i of f all lie in k, while the numbers λ_i are linearly independent over k. Thus for t_0 not in S, the hypersurface

$$t_0 - \Sigma_{j=1 \text{ to } n} \lambda_j f(x_j) = 0$$

in $\mathbb{A}^n$ is smooth of codimension one. Moreover, it is a strong Deligne polynomial, as its leading form is $\Sigma_i \lambda_i(x_i)^e$, and e is prime to p.

(6.7.3) We now take k_n as our new ground field. Let us denote by $\mathcal{M}_{1par}$ the ("one-parameter") pullback of $\mathcal{M}(n, e, \chi)$ to $\mathbb{A}^1 - S$. Thus $\mathcal{M}_{1par}$ is a lisse sheaf on $\mathbb{A}^1 - S$ whose trace function is given as follows: for any finite extension of k_n, and for any t in $\mathbb{A}^1(E) - S$, we have

$$\mathrm{Trace}(\mathrm{Frob}_{E,t} | \mathcal{M}_{1\mathrm{par}})$$
$$= (-1)^n(\#E)^{-n/2}\Sigma_{v \text{ in } \mathbb{A}^n(E)} \chi_E(t - \Sigma \lambda_i f(v_i)).$$

We further expand the inner sum:

$$\Sigma_{v \text{ in } \mathbb{A}^n(E)} \chi_E(t - \Sigma \lambda_i f(v_i))$$
$$= \Sigma_{w \text{ in } \mathbb{A}^n(E)} \chi_E(t - \Sigma w_i)\Pi_{i=1 \text{ to } n}\#\{v_i \text{ in } E \text{ with } \lambda_i f(v_i) = w_i\}.$$

(6.7.4) If we fix E and view this sum as a function of t in E, we recognize it as the multiple additive convolution on E of the n+1 $\overline{\mathbb{Q}}_\ell$-valued functions on E given by

$$x \mapsto \chi_E(x)$$

and the n functions

$$t \mapsto \#(y \text{ in } E \text{ with } \lambda_i f(y) = x\}.$$

The first function is the trace function of $j_!\mathcal{L}_\chi$, the remain ones are the trace functions of the sheaves $(\lambda_i f)_\star \overline{\mathbb{Q}}_\ell$, for i=1 to n.

(6.7.5) Taking account of the factor $(-1)^n(\#E)^{-n/2}$, we find

Key Identity 6.7.6 The restriction to $\mathbb{A}^1 - S$ of the multiple ! convolution

$$(j_\star \mathcal{L}_\chi[1](1/2))\star_!((\lambda_1 f)_\star \overline{\mathbb{Q}}_\ell[1](1/2))\star_!...\star_!((\lambda_n f)_\star \overline{\mathbb{Q}}_\ell[1](1/2))$$

has the same trace function as $\mathcal{M}_{1\mathrm{par}}[1](1/2)$.

(6.7.7) Up to this point, the multiple ! convolution only makes sense as a derived category object. We will now see that it is in fact perverse. Denote by L the perverse sheaf

$$L := (f_\star \overline{\mathbb{Q}}_\ell/\overline{\mathbb{Q}}_\ell)[1](1/2)$$

on $\mathbb{A}^1$. Thanks to Theorem 6.2.4, L is perverse and geometrically irreducible, lisse of rank e-1 on $\mathbb{A}^1$ - CritValues(f), with local monodromy a tame reflection at each of its e-1 finite singularities, and with G_{geom} the full symmetric group S_e. We note for later use that L is tame at ∞ (because e is prime to p). Also, L has $\mathcal{P}$, since it is geometrically irreducible and has finite singularities. For each λ_i, we have

$$\mathrm{Mult}_{\lambda_i}(L) = ((\lambda_i f_\star)\overline{\mathbb{Q}}_\ell/\overline{\mathbb{Q}}_\ell)[1](1/2),$$
$$(\lambda_1 f)_\star \overline{\mathbb{Q}}_\ell[1](1/2) = \mathrm{Mult}_{\lambda_i}(L) \oplus \overline{\mathbb{Q}}_\ell[1](1/2).$$

Lemma 6.7.8 The multiple ! convolution

$$(j_\star \mathcal{L}_\chi[1](1/2))\star_!((\lambda_1 f)_\star \overline{\mathbb{Q}}_\ell[1](1/2))\star_!...\star_!((\lambda_n f)_\star \overline{\mathbb{Q}}_\ell[1](1/2))$$

is perverse. It is equal to the perverse sheaf

$$(j_\star \mathcal{L}_\chi[1](1/2))\star_!\mathrm{Mult}_{\lambda_1}(L)\star_!...\star_!\mathrm{Mult}_{\lambda_n}(L).$$

proof The perverse sheaf $(j_\star \mathcal{L}_\chi[1](1/2))$ has $\mathcal{P}$ geometrically [Ka-RLS, 2.9.1], and one knows that

$$H_c(\mathbb{A}^1 \otimes_k \overline{k}, j_!\mathcal{L}_\chi) = 0.$$

So we have

$$(j_*\mathcal{L}_\chi[1](1/2))*_!\overline{\mathbb{Q}}_\ell[1](1/2) = 0.$$

For any perverse M, the object $(j_*\mathcal{L}_\chi[1](1/2))*_!M$ is perverse, and (commutativity of ! convolution) satisfies

$$((j_*\mathcal{L}_\chi[1](1/2))*_!M)*_!(\overline{\mathbb{Q}}_\ell[1](1/2)) = 0.$$

Thus we find that

$$(j_*\mathcal{L}_\chi[1](1/2))*_!((\lambda_1 f)_*\overline{\mathbb{Q}}_\ell[1](1/2))$$
$$= (j_*\mathcal{L}_\chi[1](1/2))*_!(\mathrm{Mult}_{\lambda_1}(L) \oplus \overline{\mathbb{Q}}_\ell[1](1/2))$$
$$= (j_*\mathcal{L}_\chi[1](1/2))*_!\mathrm{Mult}_{\lambda_1}(L)$$

is perverse. We now proceed by induction on n. Because each $\mathrm{Mult}_{\lambda_i}(L)$ has $\mathcal{P}$, we can take for M successively the perverse sheaves $\mathrm{Mult}_{\lambda_1}(L)$, $\mathrm{Mult}_{\lambda_1}(L)*_!\mathrm{Mult}_{\lambda_2}(L)$, et cetera, and we find the asserted equality. QED

Combining this lemma with the Key Identity 6.7.6 we find

Corollary 6.7.9 The restriction to $\mathbb{A}^1 - S$ of the perverse sheaf

$$(j_*\mathcal{L}_\chi[1](1/2))*_!\mathrm{Mult}_{\lambda_1}(L)*_!\ldots*_!\mathrm{Mult}_{\lambda_n}(L).$$

has the same trace function as $\mathcal{M}_{1\mathrm{par}}[1](1/2)$.

(6.7.10) There exists a dense open set $U \subset \mathbb{A}^1 - S$ on which the perverse sheaf $(j_*\mathcal{L}_\chi[1](1/2))*_!\mathrm{Mult}_{\lambda_1}(L)*_!\ldots*_!\mathrm{Mult}_{\lambda_n}(L)$ is of the form $\mathcal{N}(1/2)[1]$, for some lisse sheaf $\mathcal{N}$ on U. Because the perverse sheaf in question is ι-mixed of weight ≤ 0, $\mathcal{N}$ is ι-mixed of weight ≤ 0. The lisse sheaves $\mathcal{N}$ and $\mathcal{M}_{1\mathrm{par}}|U$ have the same trace function. Both $\mathcal{N}$ and $\mathcal{M}_{1\mathrm{par}}|U$ are ι-mixed, of weight ≤ 0. Since both are lisse on U, we know a priori [De-Weil II, 3.4.1 (ii)] that their weight filtrations are by lisse subsheaves on U. This being the case, $\mathrm{Gr}^0(\mathcal{N})$ and of $\mathrm{Gr}^0(\mathcal{M}_{1\mathrm{par}}|U)$ are both lisse sheaves on U, whose trace functions we can recover point by point, by just keeping those Frobenius eigenvalues of $\mathcal{N}$ and of $\mathcal{M}_{1\mathrm{par}}$ respectively which, via ι, lie on the unit circle. Therefore $\mathrm{Gr}^0(\mathcal{N})$ and $\mathrm{Gr}^0(\mathcal{M}_{1\mathrm{par}}|U)$ have the same trace function. By Chebotarev, $\mathrm{Gr}^0(\mathcal{N})$ and $\mathrm{Gr}^0(\mathcal{M}_{1\mathrm{par}}|U)$ have isomorphic semisimplification as representations of $\pi_1^{\mathrm{arith}}(U)$, so a fortiori they have isomorphic semisimplification as representations of $\pi_1^{\mathrm{geom}}(U)$. But by a fundamental result of Deligne [De-Weil II,3.4.1 (iii)], any lisse sheaf which is ι-pure on U is semisimple as a representation of $\pi_1^{\mathrm{geom}}(U)$. Thus $\mathrm{Gr}^0(\mathcal{N})$ and $\mathrm{Gr}^0(\mathcal{M}_{1\mathrm{par}}|U)$ are geometrically isomorphic on U.

(6.7.11) Recall that $\mathcal{M}_{1\mathrm{par}}|U$ is the pullback of $\mathcal{M}(n, e, \chi)|\mathcal{SD}(n, e)$ to U. So in order to show that G_{geom} for $\mathrm{Gr}^0(\mathcal{M}(n, e, \chi)|\mathcal{SD}(n, e))$

contains a pseudoreflection of determinant $\chi(\chi_2)^n$, it suffices to show that G_{geom} for its pullback $Gr^0(\mathcal{M}_{1par}|U)$ contains such a pseudoreflection. Since $Gr^0(\mathcal{N})$ and $Gr^0(\mathcal{M}_{1par}|U)$ are geometrically isomorphic on U, it suffices to show that G_{geom} for $Gr^0(\mathcal{N})$ contains such a pseudoreflection. Now $Gr^0(\mathcal{N})(1/2)[1]$ is the restriction to U of the perverse sheaf

$$Gr^0((j_*\mathcal{L}_\chi[1](1/2))*_!Mult_{\lambda_1}(L)*_!...*_!Mult_{\lambda_n}(L))$$
$$= (j_*\mathcal{L}_\chi[1](1/2))*_{mid}Mult_{\lambda_1}(L)*_{mid}...*_{mid}Mult_{\lambda_n}(L).$$

We will show that this perverse sheaf is a middle extension which is lisse on $\mathbb{A}^1$ - S, and at each point of S has local monodromy a (necessarily tame) pseudoreflection of determinant $\chi(\chi_2)^n$.

(6.7.12) To do this, we apply Corollary 6.1.12, Theorem 6.1.18, and Pseudoreflection Output Corollary 6.1.21 to analyze the perverse sheaf

$$(j_*\mathcal{L}_\chi[1](1/2))*_{mid}Mult_{\lambda_1}(L)*_{mid}...*_{mid}Mult_{\lambda_n}(L)$$

and its local monodromy at finite distance on $\mathbb{A}^1$ over $\bar{k}$. In applying those results, we take the "L" in those results to be our L. Recall that L is perverse and geometrically irreducible, lisse of rank e-1 on $\mathbb{A}^1$ - CritValues(f), with local monodromy a tame reflection at each of its e-1 finite singularities. Recall that L has $\mathcal{P}$ (because it is geometrically irreducible and has finite singularities). Recall also from Corollary 6.4.5 that L has all its $I(\infty)$-slopes ≤ 1. In fact, since e is prime to p, L is tame at ∞. We successively take the "K" in those results to be

$$j_*\mathcal{L}_\chi[1](1/2),$$
$$(j_*\mathcal{L}_\chi[1](1/2))*_{mid}Mult_{\lambda_1}(L),$$
$$.$$
$$.$$
$$.$$
$$(j_*\mathcal{L}_\chi[1](1/2))*_{mid}Mult_{\lambda_1}(L)*_{mid}...*_{mid}Mult_{\lambda_{n-1}}(L).$$

The initial K, namely $j_*\mathcal{L}_\chi[1](1/2)$, is perverse and geometrically irreducible (and hence geometrically semisimple), non-punctual, and has $\mathcal{P}$. It is lisse on $\mathbb{G}_m$, and its local monodromy at 0 is a tame pseudoreflection of determinant χ.

(6.7.13) By our successive choice of the λ_i, we see by induction on n that

$$(j_*\mathcal{L}_\chi[1](1/2))*_{mid}Mult_{\lambda_1}(L)*_{mid}...*_{mid}Mult_{\lambda_n}(L)$$

is geometrically semisimple, non-punctual, has $\mathcal{P}$, is lisse outside the $(e-1)^n$ points of S, and at each point of S has local monodromy a tame pseudoreflection of determinant $\chi(\chi_2)^n$. This concludes the proof of Theorem 6.6.5. QED

(6.7.14) In the special case when χ is χ_2 and n is even, we get the existence of a reflection in G_{geom}.

Corollary 6.7.15 Suppose k is a finite field of odd characteristic p. Let $n \geq 1$ and $e \geq 3$ be integers, and let χ_2 be the quadratic character of $k^\times$. Suppose that e is prime to p, and that n is even. Then G_{geom} for the sheaf

$\mathcal{M}(n,e,\chi_2)|\mathcal{SD}(n,e)$, if e is odd,

$Gr^0(\mathcal{M}(n,e,\chi_2)|\mathcal{SD}(n,e))$, if e is even,

contains a reflection.

(6.7.16) Once we have a reflection, we can sharpen Theorems 5.2.2 and 5.5.2 in the case $\chi = \chi_2$, n even.

Theorem 6.7.17 Suppose k is a finite field of characteristic odd p. Let $n \geq 1$ and $e \geq 3$ be integers, and let χ_2 be the quadratic character of $k^\times$. Suppose further that e is prime to p, and that n is even. Then we have the following results.

1) If e is odd (i.e., if $(\chi_2)^e \neq \mathbb{1}$), then G_{geom} for $\mathcal{M}(n,e,\chi)|\mathcal{SD}(n,e)$ is $O((e-1)^n)$, except in the case ($n = 2$, $e = 3$, $p \geq 5$), in which case G_{geom} is a finite primitive subgroup of $O(4)$ with fourth moment 3, which contains a reflection.

2) If e is even (i.e., if $(\chi_2)^e = \mathbb{1}$), then G_{geom} is $O(N(n, e))$, except in the case ($n = 2$, $e = 4$, $p \geq 3$), in which case G_{geom} is the Weyl group of E_7 in its reflection representation.

Remark 6.7.18 We can also use the full strength of Theorem 6.6.5, combined with the fundamental results from group theory recalled in Chapter 2, to give a very short proof of most of Theorems 5.2.2 and 5.5.2. We say "most of" because we omit characteristic 2, and we omit, in $n \leq 2$ variables, some low degrees. Here is the precise statement of what we get this "soft" way.

Theorem 6.7.19 Suppose k is a finite field of odd characteristic p. Let $n \geq 1$ and $e \geq 3$ be integers, and let χ be a nontrivial multiplicative character of $k^\times$. Suppose that e is prime to p. Consider the lisse sheaf

$\mathcal{M}(n,e,\chi)|\mathcal{SD}(n,e) = Gr^0(\mathcal{M}(n,e,\chi)|\mathcal{SD}(n,e))$, if χ^e is nontrivial,

$Gr^0(\mathcal{M}(n,e,\chi)|\mathcal{SD}(n,e))$, if χ^e is trivial,

and denote by N its rank. Thus

$N = (e-1)^n$, if $\chi^e \neq \mathbb{1}$,

$N = (1/e)((e-1)^{n+1} - (-1)^{n+1})$, if $\chi^e = \mathbb{1}$.

Then we have the following results for the group G_{geom} attached to this sheaf.

1) If χ does not have order 2, and if any of the following conditions a), b), or c) holds, then G_{geom} contains SL(N):

a) N > 4,

b) N > 2, and $\chi(\chi_2)^n$ has order > 3,

c) $\chi(\chi_2)^n$ has order ≥ 6.

2) If χ has order 2, and if n is odd, then G_{geom} = Sp(N).

3) If χ has order 2, if n is even, and if N > 8, then G_{geom} = O(N).

proof Repeating verbatim the first page of the proofs of Theorems 5.2.2 and 5.5.2, we reduce to a small number of possibilities for G_{geom}.

If χ does not have order 2, then G_{geom} either contains SL(N), or is a finite primitive subgroup of GL(N) which, thanks to Theorem 6.6.5, contains a pseudoreflection of determinant $\chi(\chi_2)^n$. As χ does not have order 2, the order of $\chi(\chi_2)^n$ is at least 3. For N ≥ 2, there are no such finite primitive groups if any of conditions a), b), or c) hold. This is immediate from the r=1 case of Theorem 2.6.7. And if N = 1, a case which can only arise if n=1, e=3, and $\chi^e = \mathbb{1}$, there is nothing to prove.

If χ has order 2 and n is odd, then G_{geom} is either Sp(N) or is finite. But it cannot be finite, as it contains a unipotent pseudoreflection.

If χ has order 2 and n is even, then G_{geom} is either SO(N) or O(N) or is a finite primitive subgroup of O(N) with fourth moment $M_4 \leq 3$ which, thanks to Theorem 6.6.5, contains a reflection. So the SO(N) case does not arise. For N > 8, the finite case cannot arise either. Indeed, by Mitchell's Theorem 2.6.8, a finite primitive subgroup of GL(N) which contains a reflection has image in PGL(N) the symmetric group S_{N+1} in its deleted permutation representation, and any such group has fourth moment $M_4 > 3$ (in fact, equal to 4, but we do not need this extra precision). QED

(6.7.20) When χ does not have order 2, we can also determine G_{geom} exactly, so long as it is not finite. For any integer a ≥ 1, let us denote by $GL_a(N)$ the Zariski-closed subgroup of GL(N) defined as

$$GL_a(N) := \{A \text{ in } GL(N) \text{ with } \det(A)^a = 1\}.$$

Theorem 6.7.21 Suppose k is a finite field of odd characteristic p. Let n ≥ 1 and e ≥ 3 be integers, and let χ be a nontrivial multiplicative character of $k^\times$, whose order is not 2. Suppose that e is prime to p. Consider the lisse sheaf

$\mathcal{M}(n,e,\chi)|\mathcal{SD}(n,e) = Gr^0(\mathcal{M}(n,e,\chi)|\mathcal{SD}(n,e))$, if χ^e is nontrivial,

$Gr^0(\mathcal{M}(n,e,\chi)|\mathcal{SD}(n,e))$, if χ^e is trivial,

and denote by N its rank. Thus

$N = (e-1)^n$, if $\chi^e \neq \mathbb{1}$,

$N = (1/e)((e-1)^{n+1} - (-1)^{n+1})$, if $\chi^e = \mathbb{1}$.

Suppose that the group G_{geom} attached to this sheaf contains SL(N), e.g., suppose that any of the following conditions a), b), or c) holds:

a) $N > 4$,

b) $N > 2$, and $\chi(\chi_2)^n$ has order > 3,

c) $\chi(\chi_2)^n$ has order ≥ 6,

cf. Theorem 6.7.19.

Define integers a and b as follows:

a := the order of the character $\chi(\chi_2)^n$,

b := the number of roots of unity in $\mathbb{Q}(\chi) = \mathbb{Q}(\chi(\chi_2)^n)$.

[Thus b = a if a is even, and b = 2a if a is odd.]

Then we have the following results.

1) We have the inclusions

$$GL_a(N) \subset G_{geom} \subset GL_b(N).$$

2) If b = a, then $G_{geom} = GL_a(N)$.

3) Suppose that n is odd, b = 2a, and $\chi^e = \mathbb{1}$. Then $G_{geom} = GL_a(N)$.

4) Suppose that n is odd, b = 2a, $\chi^e \neq \mathbb{1}$, and e is odd. Then $G_{geom} = GL_a(N)$.

5) Suppose that n is odd, b = 2a, $\chi^e \neq \mathbb{1}$, and e is even. Then $G_{geom} = GL_b(N)$.

6) Suppose that n is even, b = 2a, and $\chi^e = \mathbb{1}$. Then $G_{geom} = GL_a(N)$.

7) Suppose that n is even, b = 2a, and $\chi^e \neq \mathbb{1}$. Then $G_{geom} = GL_b(N)$.

proof 1) The key point is that the characteristic polynomials of all Frobenii on the lisse sheaves

$\mathcal{M}(n,e,\chi)(-n/2)|\mathcal{SD}(n,e)$, for any nontrivial χ,

$(Gr^0(\mathcal{M}(n,e,\chi))(-n/2)|\mathcal{SD}(n,e))$, if χ^e is trivial,

have coefficients in the ring of integers $\mathcal{O}(\chi)$ of the subfield $\mathbb{Q}(\chi)$ of $\overline{\mathbb{Q}}_\ell$. Moreover, as we vary the prime $\ell \neq$ char(k) and the embedding of the abstract field $\mathbb{Q}(\chi)$ into $\overline{\mathbb{Q}}_\ell$, we obtain a compatible system of lisse sheaves on the space $\mathcal{SD}(n, e)$. [By a compatible system, we mean the following. Let k be a finite field of characteristic p, X/k a lisse, geometrically connected k-scheme, and let K be a number field. We denote by Λ the set of all pairs

(a prime number $\ell \neq p$, a field embedding $\lambda : K \subset \overline{\mathbb{Q}}_\ell$).

Suppose we are given for each λ in Λ a lisse $\overline{\mathbb{Q}}_\ell$-sheaf $\mathcal{G}_\lambda$ on X. We say the collection $\{\mathcal{G}_\lambda\}_\lambda$ forms a compatible system if for every finite extension E/k, and for every E-valued point f in X(E), the characteristic polynomials, for variable λ in Λ,

$$\det(1 - T\mathrm{Frob}_{E,f} \mid \mathcal{G}_\lambda)$$

all lie in K[T] and are all equal in K[T]. If this is the case, then all the coefficients lie in $\mathcal{O}_K[1/p]$, and the highest degree coefficient, $\det(\mathrm{Frob}_{E,f} \mid \mathcal{G}_\lambda)$, is a unit in $\mathcal{O}_K[1/p]$.]

This is obvious for the lisse sheaf $\mathcal{M}(n,e,\chi)(-n/2)|\mathcal{S}\mathcal{D}(n,e)$, since by Corollary 5.1.15, for any finite extension E/k, and any f in $\mathcal{S}\mathcal{D}(n,e)(E)$, we have

$$\mathrm{Trace}(\mathrm{Frob}_{E,f} \mid \mathcal{M}(n,e,\chi)(-n/2)) = (-1)^n \Sigma_{v \text{ in } \mathbb{A}^n(E)} \chi_E(f(v)),$$

and the groups $H_c^i((\mathbb{A}^n \otimes_k \overline{E})[1/f], \mathcal{L}_{\chi(f)}) = H_c^i(\mathbb{A}^n \otimes_k \overline{E}, \mathcal{L}_{\chi(f)})$ vanish for $i \neq n$, cf. [Ka-ENSMCS, 5.1]. Thus we have

$$\det(1 - T\mathrm{Frob}_{E,f} \mid \mathcal{M}(n,e,\chi)(-n/2))$$
$$= \det(1 - T\mathrm{Frob}_E \mid H_c^n((\mathbb{A}^n[1/f]) \otimes \overline{E}, \mathcal{L}_{\chi(f)})),$$

and we have the vanishing

$$H_c^i((\mathbb{A}^n[1/F]) \otimes \overline{E}, \mathcal{L}_{\chi(f)}) = 0 \text{ for } i \neq n.$$

Thus the characteristic polynomial of $\mathrm{Frob}_{e,f}$ on $\mathcal{M}(n,e,\chi)(-n/2)$ is equal to the L-function of $\mathbb{A}^n[1/f]/E$ with coefficients in $\mathcal{L}_{\chi(f)}$: we have the formula

$$L(\mathbb{A}^n[1/f]/E, \mathcal{L}_{\chi(f)})(T)^{(-1)^{n+1}} = \det(1 - T\mathrm{Frob}_{E,f} \mid \mathcal{M}(n,e,\chi)(-n/2)).$$

When $\chi^e = \mathbb{1}$, we need the explicit cohomological description of the characteristic polynomials of $\mathrm{Frob}_{E,f}$ on $\mathrm{Gr}^0(\mathcal{M}(n,e,\chi))(-n/2)$ as follows, cf. [Ka-ENSMCS, Theorem 18]. Attached to the strong Deligne polynomial f in n variables over E of degree e, we have the homogeneous form F in n+1 variables over E of the same degree e, obtained by homogenizing f, say

$$F(X_0,\ldots, X_n) := (X_0)^e f(X_1/X_0,\ldots, X_n/X_0).$$

In the projective space $\mathbb{P}^n$ with homogeneous coordinates the X_i, we have the affine open set $\mathbb{P}^n[1/F]$. Because $\chi^e = \mathbb{1}$, we may form the lisse sheaf $\mathcal{L}_{\chi(F)}$ on $\mathbb{P}^n[1/F]$. We have

$$\det(1 - T\mathrm{Frob}_{E,f} \mid \mathrm{Gr}^0(\mathcal{M}(n,e,\chi))(-n/2))$$
$$= \det(1 - T\mathrm{Frob}_E \mid H_c^n((\mathbb{P}^n[1/F]) \otimes \overline{E}, \mathcal{L}_{\chi(F)})),$$

and we have the vanishing

$$H_c^i((\mathbb{P}^n[1/F]) \otimes \overline{E}, \mathcal{L}_{\chi(F)}) = 0 \text{ for } i \neq n.$$

Thus the characteristic polynomial of $\mathrm{Frob}_{e,f}$ on $\mathrm{Gr}^0(\mathcal{M}(n,e,\chi))(-n/2)$ is equal to the L-function of $\mathbb{P}^n[1/F]/E$ with coefficients in $\mathcal{L}_{\chi(F)}$: we have the formula

$$L(\mathbb{P}^n[1/F]/E, \mathcal{L}_{\chi(F)})(T)^{(-1)^{n+1}}$$
$$= \det(1 - T\mathrm{Frob}_{E,f} \mid \mathrm{Gr}^0(\mathcal{M}(n,e,\chi))(-n/2)).$$

We next apply to these sheaves the following general lemmas.

Lemma 6.7.22 Let k be a finite field in which ℓ is invertible, X/k a lisse, geometrically connected k-scheme, and $\mathcal{G}$ a lisse $\overline{\mathbb{Q}}_\ell$-sheaf on X. Let $K \subset \overline{\mathbb{Q}}_\ell$ be a subfield which contains all traces $\mathrm{Trace}(\mathrm{Frob}_{E,x} \mid \mathcal{G})$, for all finite extensions E/k and all points x in X(E). If the group $\mu(K)$ of all roots of unity in K is finite, say of order b, then $\det(\mathcal{G})^{\otimes b}$ is geometrically constant, i.e., G_{geom} for $\mathcal{G}$ lies in $GL_b(\mathrm{rank}(\mathcal{G}))$.

proof This is proven, but not stated (!) in the argument of [Ka-ACT, first half of proof of 5.2bis]. QED

Lemma 6.7.23 Let k be a finite field in which ℓ is invertible, X/k a lisse, geometrically connected k-scheme, and K a number field. Let $\{\mathcal{G}_\lambda\}_\lambda$ be a compatible system of lisse sheaves of rank one on X. Fix an element γ in $\pi_1{}^{geom}(X)$. Then $\gamma|\mathcal{G}_\lambda$ is a root of unity in K, independent of λ.

proof The assertion is geometric, so we may replace k by a finite extension of itself if need be and assume that X(k) is nonempty. Pick a rational point f in X(k), define

$$\alpha := \mathrm{Frob}_{k,f} \mid \mathcal{G}_\lambda.$$

Thus α is a unit in $\mathcal{O}_K[1/p]$, independent of λ. Replacing each $\mathcal{G}_\lambda$ by $\mathcal{G}_\lambda \otimes (1/\alpha)^{deg}$, we reduce to the case where, in addition,

$$\mathrm{Frob}_{k,f} \mid \mathcal{G}_\lambda = 1.$$

That $\gamma|\mathcal{G}_\lambda$ is a root of unity results from the fact that each $\mathcal{G}_\lambda$ separately is geometrically of finite order, cf. [De-Weil II, 1.3.4]. Thus each $\mathcal{G}_\lambda$ is arithmetically of finite order. Comparing two different λ, there is a finite quotient Γ of $\pi_1{}^{arith}(X)$ through which both factor. But in any such finite quotient, every element is, by Chebotarev, the image of a Frobenius class. So for any element γ in Γ, $\gamma|\mathcal{G}_\lambda$ lies in K, independent of λ. QED

Corollary 6.7.24 Hypotheses as in the lemma above, for any element β in the Weil group W(X) := {elements in $\pi_1{}^{arith}(X)$ with integral degree}, $\beta|\mathcal{G}_\lambda$ lies in K, and is independent of λ.

proof Denote by k_n/k the extension of degree n. For n large, both $X(k_n)$ and $X(k_{n+1})$ are nonempty. Pick f in $X(k_n)$, g in $X(k_{n+1})$. Pick any choice of Frobenius elements $A = \mathrm{Frob}_{k_n,f}$ and $B = \mathrm{Frob}_{k_{n+1},g}$ in $\pi_1{}^{arith}(X)$. Then if β in W(X) has degree d, either d=0 and β lies in

$\pi_1{}^{geom}(X)$, a case covered by the lemma, or $\gamma := \beta(BA^{-1})^d$ lies in $\pi_1{}^{geom}(X)$. Thus

$$\beta|\mathcal{G}_\lambda = (\gamma|\mathcal{G}_\lambda)(B|\mathcal{G}_\lambda)^{-d}(A|\mathcal{G}_\lambda)^d. \qquad \text{QED}$$

Lemma 6.7.25 Let k be a finite field in which ℓ is invertible, X/k a lisse, geometrically connected k-scheme, and K a number field. Let $\{\mathcal{G}_\lambda\}_\lambda$ and $\{\mathcal{H}_\lambda\}_\lambda$ be two compatible systems of lisse sheaves on X. Let a be a strictly positive integer. Suppose we know that, for each λ, $\det(\mathcal{G}_\lambda)^{\otimes a}$ and $\det(\mathcal{H}_\lambda)^{\otimes a}$ each have order dividing 2 as characters of $\pi_1{}^{geom}(X)$. Suppose that for some λ_0 with ℓ_0 odd, we have a congruence

$$\det(\mathrm{Frob}_{E,f} \mid \mathcal{G}_{\lambda_0}) \equiv \det(\mathrm{Frob}_{E,f} \mid \mathcal{H}_{\lambda_0}) \bmod \lambda_0,$$

for every finite extension E/k, and for every E-valued point f in X(E). Then for every γ in $\pi_1{}^{geom}(X)$, and every pair λ_1 and λ_2 of λ's, we have

$$\det(\gamma \mid \mathcal{G}_{\lambda_1})^a = \det(\gamma \mid \mathcal{H}_{\lambda_2})^a.$$

proof By the previous lemma, applied separately to $\{\det(\mathcal{G}_\lambda)^{\otimes a}\}_\lambda$ and to $\{\det(\mathcal{H}_\lambda)^{\otimes a}\}_\lambda$, it suffices to prove that for every γ in $\pi_1{}^{geom}(X)$, we have

$$\det(\gamma \mid \mathcal{G}_{\lambda_0})^a = \det(\gamma \mid \mathcal{H}_{\lambda_0})^a.$$

Since both sides are ± 1, it suffices to prove the congruence

$$\det(\gamma \mid \mathcal{G}_{\lambda_0})^a \equiv \det(\gamma \mid \mathcal{H}_{\lambda_0})^a \bmod \lambda_0.$$

In fact, this congruence holds for every γ in $\pi_1{}^{arith}(X)$. Indeed, both $\det(\mathcal{G}_{\lambda_0})^{\otimes a}$ and $\det(\mathcal{H}_{\lambda_0})^{\otimes a}$ mod λ_0 are characters with values in the finite field $\mathbb{F}_{\lambda_0}$, so they factor through a common finite quotient Γ of $\pi_1{}^{arith}(X)$. In any such finite quotient, every element is, by Chebotarev, the image of a Frobenius class. QED

With these preliminaries out of the way, we now return to the proof of part 1) of Theorem 6.7.21. We now have an a priori inclusion

$$G_{geom} \subset GL_b(N).$$

On the other hand, G_{geom} contains $SL(N)$, by hypothesis, and by Theorem 6.6.5, G_{geom} contains a pseudoreflection of determinant $\chi(\chi_2)^n$. Thus we have an a priori inclusion

$$GL_a(N) \subset G_{geom}.$$

This concludes the proof of 1).

Assertion 2), "mise pour memoire", results trivially from 1).

In assertions 3), 4), and 5), n is odd and b=2a. This means that

$\chi\chi_2$ is nontrivial of odd order a. Thus $\chi = \chi_a\chi_2$, with χ_a nontrivial of odd order a.

To prove 3), we argue as follows. Because $\chi^e = \mathbb{1}$ here, we are dealing with the Gr^0 sheaves. For any finite extension E/k, and any f in $\mathcal{SD}(n,e)(E)$, we have

$$L(\mathbb{P}^n[1/F]/E, \mathcal{L}_{\chi(F)})(T)^{(-1)^{n+1}}$$
$$= \det(1 - TFrob_{E,f} \mid Gr^0(\mathcal{M}(n,e,\chi))(-n/2)).$$

We wish to show that $\det(Gr^0(\mathcal{M}(n,e,\chi))(-n/2))^{\otimes a}$, which a priori as a character of $\pi_1{}^{geom}$ has order dividing 2, is in fact geometrically trivial. We will show this by using the fact that we can distinguish 1 from -1 by reducing modulo any odd prime. We use the "change of χ, change of ℓ" congruence argument of [Ka-ACT, pages 206-207], i.e., we successively apply Lemma 6.7.25, to show that we have a geometric isomorphism

$$\det(Gr^0(\mathcal{M}(n,e,\chi_a\chi_2))(-n/2))^{\otimes a}$$
$$\cong \det(Gr^0(\mathcal{M}(n,e,\chi_2))(-n/2))^{\otimes a}.$$

Since n is odd, $Gr^0(\mathcal{M}(n,e,\chi_2))$ is symplectically self dual, so it has trivial determinant.

To prove 4), we use a similar argument. We again use the "change of χ_a, change of ℓ" congruence argument of [Ka-ACT, pages 206-207] to show that we have a geometric isomorphism

$$\det(\mathcal{M}(n,e,\chi_a\chi_2)(-n/2))^{\otimes a}$$
$$\cong \det(\mathcal{M}(n,e,\chi_2)(-n/2))^{\otimes a}.$$

Since ne is odd, $\mathcal{M}(n,e,\chi_2)$ is symplectically self dual, so it has trivial determinant.

To prove 5), there are two ways we can proceed. The first is the more geometric argument, but it requires $n \geq 3$. We again use the "change of χ_a, change of ℓ" congruence argument of [Ka-ACT, pages 206-207] to show that we have a geometric isomorphism

$$\det(\mathcal{M}(n,e,\chi_a\chi_2)(-n/2))^{\otimes a}$$
$$\cong \det((\mathcal{M}(n,e,\chi_2)(-n/2))^{\otimes a}.$$

However, this time e is even, hence $(\chi_2)^e = \mathbb{1}$, and $\mathcal{M}(n,e,\chi_2)(-n/2)$ is no longer pure of weight n, but is rather mixed of weights n and n-1. Recall that we denoted by F the homogenization of f. Let us denote by f_e the leading form of f. Then [Ka-ENSMCS, Theorem 18] we have a short exact sequence

$$0 \to H_c{}^{n-1}((\mathbb{P}^{n-1}[1/f_e])\otimes\overline{E}, \mathcal{L}_{\chi_2(f_e)}) \to H_c{}^n((\mathbb{A}^n[1/f])\otimes\overline{E}, \mathcal{L}_{\chi_2(f)})$$
$$\to H_c{}^n((\mathbb{P}^n[1/F])\otimes\overline{E}, \mathcal{L}_{\chi_2(F)}) \to 0,$$

and we have the vanishings

$$H_c^i((\mathbb{P}^{n-1}[1/f_e])\otimes\overline{E}, \mathcal{L}_{\chi_2(f_e)}) = 0 \text{ for } i \neq n-1,$$

$$H_c^i((\mathbb{A}^n[1/f])\otimes\overline{E}, \mathcal{L}_{\chi_2(f)}) = 0 \text{ for } i \neq n,$$

$$H_c^i((\mathbb{P}^n[1/F])\otimes\overline{E}, \mathcal{L}_{\chi_2(F)}) = 0 \text{ for } i \neq n.$$

There exists a lisse sheaf $\mathcal{G}(n, \chi_2)$ on $\mathcal{SD}(n,e)$ with

$$\det(1 - T\mathrm{Frob}_{E,f} \mid \mathcal{G}(n, \chi_2))$$
$$= \det(1 - T\mathrm{Frob}_E \mid H_c^{n-1}((\mathbb{P}^{n-1}[1/f_e])\otimes\overline{E}, \mathcal{L}_{\chi_2(f_e)})),$$

which sits in a short exact sequence of lisse sheaves on $\mathcal{SD}(n,e)$,

$$0 \to \mathcal{G}(n, \chi_2) \to \mathcal{M}(n,e,\chi_2)(-n/2) \to \mathrm{Gr}^0(\mathcal{M}(n,e,\chi_2))(-n/2) \to 0.$$

Thus we have an isomorphism

$$\det(\mathcal{M}(n,e,\chi_2)(-n/2))^{\otimes a}$$
$$\cong (\det(\mathcal{G}(n, \chi_2))^{\otimes a})\otimes\det(\mathrm{Gr}^0(\mathcal{M}(n,e,\chi_2))(-n/2))^{\otimes a}.$$

Since n is odd, $\mathrm{Gr}^0(\mathcal{M}(n,e,\chi_2))$ is symplectically self dual, so it has trivial determinant. Thus we have a geometric isomorphism

$$\det(\mathcal{M}(n,e,\chi_2)(-n/2))^{\otimes a} \cong \det(\mathcal{G}(n, \chi_2))^{\otimes a}.$$

We must show that $\det(\mathcal{G}(n, \chi_2))^{\otimes a}$ is not geometrically trivial.

To do this, we will exhibit a pullback of $\det(\mathcal{G}(n, \chi_2))^{\otimes a}$ which is not geometrically trivial. Consider the map

$$\mathrm{incl} : \mathcal{SD}(n-1, e) \to \mathcal{SD}(n,e),$$
$$g \mapsto 1 + G,$$

where G is the homogenization of g. [The reader will easily verify that if g lies in $\mathcal{SD}(n-1, e)$ and if e is prime to p, then 1 + G lies in $\mathcal{SD}(n, e)$.] The two lisse sheaves

$$\mathrm{incl}^*(\mathcal{G}(n, \chi_2)) \text{ and } \mathrm{Gr}^0(\mathcal{M}(n-1, e, \chi_2)((1-n)/2))$$

on $\mathcal{SD}(n-1, e)$ have the same characteristic polynomials of Frobenius everywhere, so they have isomorphic arithmetic semisimplifications. As $\mathrm{Gr}^0(\mathcal{M}(n-1, e, \chi_2))$ is geometrically and hence arithmetically irreducible, we in fact have an isomorphism

$$\mathrm{incl}^*(\mathcal{G}(n, \chi_2)) \cong \mathrm{Gr}^0(\mathcal{M}(n-1, e, \chi_2)((1-n)/2)),$$

and hence a geometric isomorphism

$$\mathrm{incl}^*(\mathcal{G}(n, \chi_2)) \cong \mathrm{Gr}^0(\mathcal{M}(n-1, e, \chi_2)).$$

So it suffices to show that $\det(\mathrm{Gr}^0(\mathcal{M}(n-1, e, \chi_2)))^{\otimes a}$ is not geometrically trivial.

Because n is odd and $n \geq 3$, $n-1$ is even, and $n-1 \geq 2$. We know that $\mathrm{Gr}^0(\mathcal{M}(n-1, e, \chi_2))$ is orthogonally self dual. By Theorem 6.6.5, its G_{geom} contains reflections. Thus

$$\det(\mathrm{Gr}^0(\mathcal{M}(n-1, e, \chi_2)))^{\otimes a}$$

is not geometrically trivial, as required. This concludes the first proof of 5), valid when $n \geq 3$.

Here is a second proof of 5), valid always. Since $\chi^e \neq \mathbb{1}$, we are dealing with the lisse sheaf $\mathcal{M}(n,e,\chi)|\mathcal{SD}(n,e)$, whose trace function is given by

$$\mathrm{Trace}(\mathrm{Frob}_{E,f} \mid \mathcal{M}(n,e,\chi)(-n/2)) = (-1)^n \Sigma_{v \text{ in } \mathbb{A}^n(E)} \chi_E(f(v)).$$

If we multiply f by a nonzero scalar λ in $E^\times$, we get

$$\mathrm{Trace}(\mathrm{Frob}_{E,\lambda f} \mid \mathcal{M}(n,e,\chi)(-n/2)) = (-1)^n \Sigma_{v \text{ in } \mathbb{A}^n(E)} \chi_E(\lambda f(v))$$
$$= \chi_E(\lambda)\mathrm{Trace}(\mathrm{Frob}_{E,f} \mid \mathcal{M}(n,e,\chi)(-n/2)).$$

Indeed, we have an isomorphism of $\mathrm{Gal}(\overline{E}/E)$-representations

$$H_c{}^n(\mathbb{A}^n \otimes \overline{E}, \mathcal{L}_{\chi(\lambda f)}) \cong (\chi_E(\lambda))^{\deg} \otimes H_c{}^n(\mathbb{A}^n \otimes \overline{E}, \mathcal{L}_{\chi(f)}).$$

Thus for $N := \mathrm{rank}(\mathcal{M}(n,e,\chi)) = (e - 1)^n$, we have

$$\det(\mathrm{Frob}_{E,\lambda f} \mid \mathcal{M}(n,e,\chi)) = (\chi_E(\lambda))^N \det(\mathrm{Frob}_{E,f} \mid \mathcal{M}(n,e,\chi)).$$

So we obtain $(\chi_E(\lambda))^N$ as a ratio of determinants of Frobenii on $\mathcal{M}(n,e,\chi)$ at two E-valued points, f and λf, of $\mathcal{SD}(n,e)$. Therefore $(\chi_E(\lambda))^N$ is a value taken by the character $\det(\mathcal{M}(n,e,\chi))$ on the geometric fundamental group $\pi_1{}^{geom}(\mathcal{SD}(n,e))$. Namely, it is the value taken at any element of $\pi_1{}^{geom}(\mathcal{SD}(n,e))$ of the form AB^{-1} for any element A in the conjugacy class $\mathrm{Frob}_{E,\lambda f}$ and any element B in the conjugacy class $\mathrm{Frob}_{E,f}$. [Although A and B each lie in $\pi_1{}^{arith}(\mathcal{SD}(n,e))$, they have the same degree, namely $\deg(E/k)$, so their "ratio" AB^{-1} has degree 0, and hence lies in $\pi_1{}^{geom}(\mathcal{SD}(n,e))$.]

Because e is even, the quantity $N = (e - 1)^n$ is odd. Since χ had even order $b = 2a$, χ^N has even order. Thus for any finite extension E/k, the set of all values $\{(\chi_E(\lambda))^N\}$, as λ runs over $E^\times$, forms a group of even order. Thus $\det(\mathcal{M}(n,e,\chi))$ as a character of $\pi_1{}^{geom}(\mathcal{SD}(n,e))$ has even order. So we cannot have $G_{geom} = GL_a(N)$. The only remaining possibility is the asserted one, namely that $G_{geom} = GL_b(N)$. This concludes the second proof of 5).

In assertions 6) and 7), n is even, and $b = 2a$. This means that $\chi = \chi_a$ is a character of odd order a.

To prove 6), we argue as follows. Because $\chi^e = \mathbb{1}$, we have

$$L(\mathbb{P}^n[1/F]/E, \mathcal{L}_{\chi_a(F)})(T)^{(-1)^{n+1}}$$
$$= \det(1 - T\mathrm{Frob}_{E,f} \mid \mathrm{Gr}^0(\mathcal{M}(n,e,\chi_a))(-n/2)).$$

We wish in this case to prove that $\det(\mathrm{Gr}^0(\mathcal{M}(n, e, \chi)))^{\otimes a}$ is geometrically trivial, knowing a priori that geometrically it has order dividing 2. We again use the "change of χ_a, change of ℓ" congruence argument of [Ka-ACT, pages 206-207], but this time we stop when we have stripped away all but one odd prime power from the order of χ (remember χ has odd order). Thus we have a geometric isomorphism

$$\det(\mathrm{Gr}^0(\mathcal{M}(n,e,\chi))(-n/2))^{\otimes a}$$
$$\cong \det(\mathrm{Gr}^0(\mathcal{M}(n,e,\rho))(-n/2))^{\otimes a},$$

for some nontrivial character ρ whose prime power order ℓ^ν divides a. The key point observation is that modulo any prime λ of $\mathcal{O}(\chi)$ which lies over ℓ, ρ becomes trivial. Thus for every finite extension E/k, and for every f in $\mathcal{SD}(n, e)$, we get a mod λ congruence

$$1/\mathrm{Zeta}(\mathbb{P}^n[1/F]/E)(T)$$
$$\equiv \det(1 - T\mathrm{Frob}_{E,f} \mid \mathrm{Gr}^0(\mathcal{M}(n,e,\rho))(-n/2)) \bmod \lambda.$$

Because F defines a smooth hypersurface in $\mathbb{P}^n$, its cohomological structure is particularly simple, and we readily calculate

$$1/\mathrm{Zeta}(\mathbb{P}^n[1/F]/E)(T) = \mathrm{Zeta}((F=0 \text{ in } \mathbb{P}^n)/E)(T)/\mathrm{Zeta}(\mathbb{P}^n/E)(T)$$
$$= \det(1 - T\mathrm{Frob}_E \mid \mathrm{Prim}^{n-1}((F=0 \text{ in } \mathbb{P}^n)\otimes\overline{E}, \overline{\mathbb{Q}}_\ell))(1 - T(\#E)^n).$$

Thus we have the mod λ congruence

$$\det(1 - T\mathrm{Frob}_{E,f} \mid \mathrm{Gr}^0(\mathcal{M}(n,e,\rho))(-n/2))$$
$$\equiv \det(1 - T\mathrm{Frob}_E \mid \mathrm{Prim}^{n-1}((F=0 \text{ in } \mathbb{P}^n)\otimes\overline{E}, \overline{\mathbb{Q}}_\ell))(1 - T(\#E)^n).$$

The cohomology groups $\mathrm{Prim}^{n-1}((F=0 \text{ in } \mathbb{P}^n)\otimes\overline{E}, \overline{\mathbb{Q}}_\ell)$ fit together to form a lisse sheaf Prim(n-1, d) on $\mathcal{SD}(n, e)$ with

$$\det(1 - T\mathrm{Frob}_{E,f} \mid \mathrm{Prim}(n-1, d))$$
$$= \det(1 - T\mathrm{Frob}_E \mid \mathrm{Prim}^{n-1}((F=0 \text{ in } \mathbb{P}^n)\otimes\overline{E}, \overline{\mathbb{Q}}_\ell)).$$

Thus we have a congruence

$$\det(\mathrm{Prim}(n-1, d))\otimes((\#k)^n)^{\deg} \equiv \det(\mathrm{Gr}^0(\mathcal{M}(n,e,\rho))(-n/2)) \bmod \lambda.$$

Since k contains the a'th and hence the ℓ^ν'th roots of unity, we have $\#k \equiv 1 \bmod \lambda$, so we have

$$\det(\mathrm{Prim}(n-1, d)) \equiv \det(\mathrm{Gr}^0(\mathcal{M}(n,e,\rho))) \bmod \lambda.$$

As n is even, n-1 is odd, so Prim(n-1, d)((n-1)/2) is symplectically self dual, and hence has trivial determinant. Therefore $\det(\mathrm{Gr}^0(\mathcal{M}(n,e,\rho)))$ mod λ is geometrically trivial. So a fortiori $\det(\mathrm{Gr}^0(\mathcal{M}(n, e, \chi)))^{\otimes a}$ mod λ is geometrically trivial. But as $\det(\mathrm{Gr}^0(\mathcal{M}(n, e, \chi)))^{\otimes a}$ geometrically has order 1 or 2, and λ lies over an odd prime ℓ, we see that $\det(\mathrm{Gr}^0(\mathcal{M}(n, e, \chi)))^{\otimes a}$ is geometrically trivial, as required.

The proof of 7) proceeds along similar lines. This time, $\chi^e \neq \mathbb{1}$, and we are dealing with $\mathcal{M}(n, e, \chi)$. We have

$$L(\mathbb{P}^n[1/F]/E, \mathcal{L}_{\chi(F)})(T)^{(-1)^{n+1}}$$
$$= \det(1 - T\mathrm{Frob}_{E,f} \mid \mathcal{M}(n,e,\chi)(-n/2)).$$

We wish in this case to prove that $\det(\mathrm{Gr}^0(\mathcal{M}(n, e, \chi)))^{\otimes a}$ is geometrically nontrivial, knowing a priori that geometrically it has order dividing 2. Exactly as in the proof of 6), we get a geometric isomorphism

$$\det(\mathcal{M}(n,e,\chi)(-n/2))^{\otimes a}$$
$$\cong \det(\mathcal{M}(n,e,\rho)(-n/2))^{\otimes a},$$

for some nontrivial character ρ whose prime power order ℓ^{ν} divides a. Again as in 6) above, we fix a prime λ of $\mathcal{O}(\chi)$ lying over ℓ. For every finite extension E/k, and for every f in $\mathcal{SD}(n, e)$, we get a mod λ congruence

$$1/\mathrm{Zeta}(\mathbb{A}^n[1/f]/E)(T)$$
$$\equiv \det(1 - T\mathrm{Frob}_{E,f} \mid \mathcal{M}(n,e,\rho)(-n/2)) \bmod \lambda.$$

In terms of the homogenization F and the leading form f_e of f, we have

$$\mathbb{A}^n[1/f] = \mathbb{P}^n[1/F] - \mathbb{P}^{n-1}[1/f_e].$$

Thus we have

$$1/\mathrm{Zeta}(\mathbb{A}^n[1/f]/E)(T)$$
$$= \mathrm{Zeta}(\mathbb{P}^{n-1}[1/f_e]/E)(T)/\mathrm{Zeta}(\mathbb{P}^n[1/F]/E)(T)$$
$$= \det(1 - T\mathrm{Frob}_E \mid \mathrm{Prim}^{n-1}((F=0 \text{ in } \mathbb{P}^n)\otimes\overline{E}, \overline{\mathbb{Q}}_\ell))(1 - T(\#E)^n)$$
$$\times(\det(1 - T\mathrm{Frob}_E \mid \mathrm{Prim}^{n-2}((f_e=0 \text{ in } \mathbb{P}^{n-1})\otimes\overline{E}, \overline{\mathbb{Q}}_\ell))(1 - T(\#E)^{n-1}))^{-1}.$$

Thus we obtain a congruence mod λ

$$\det(1 - T\mathrm{Frob}_{E,f} \mid \mathcal{M}(n,e,\rho)(-n/2))$$
$$\equiv \det(1 - T\mathrm{Frob}_E \mid \mathrm{Prim}^{n-1}((F=0 \text{ in } \mathbb{P}^n)\otimes\overline{E}, \overline{\mathbb{Q}}_\ell))(1 - T(\#E)^n)$$
$$\times(\det(1 - T\mathrm{Frob}_E \mid \mathrm{Prim}^{n-2}((f_e=0 \text{ in } \mathbb{P}^{n-1})\otimes\overline{E}, \overline{\mathbb{Q}}_\ell))(1 - T(\#E)^{n-1}))^{-1}.$$

Now $\#E \equiv 1 \bmod \ell$, so we may rewrite this as the congruence

$$\det(1 - T\mathrm{Frob}_{E,f} \mid \mathcal{M}(n,e,\rho))$$
$$\times\det(1 - T\mathrm{Frob}_E \mid \mathrm{Prim}^{n-2}((f_e=0 \text{ in } \mathbb{P}^{n-1})\otimes\overline{E}, \overline{\mathbb{Q}}_\ell))$$
$$\equiv \det(1 - T\mathrm{Frob}_E \mid \mathrm{Prim}^{n-1}((F=0 \text{ in } \mathbb{P}^n)\otimes\overline{E}, \overline{\mathbb{Q}}_\ell)) \bmod \lambda.$$

The cohomology groups $\mathrm{Prim}^{n-2}((f_e=0 \text{ in } \mathbb{P}^{n-1})\otimes\overline{E}, \overline{\mathbb{Q}}_\ell)$ fit together to form a lissse sheaf Prim(n-2,e) on $\mathcal{SD}(n, e)$ with

$$\det(1 - T\mathrm{Frob}_{E,f} \mid \mathrm{Prim}(n-2, d))$$
$$= \det(1 - T\mathrm{Frob}_E \mid \mathrm{Prim}^{n-2}((f_e=0 \text{ in } \mathbb{P}^{n-1})\otimes\overline{E}, \overline{\mathbb{Q}}_\ell)).$$

Just as in the proof of 6) above, the cohomology groups

$$\mathrm{Prim}^{n-1}((F=0 \text{ in } \mathbb{P}^n)\otimes\overline{E}, \overline{\mathbb{Q}}_\ell)$$

fit together to form a lisse sheaf Prim(n-1,e) on $\mathcal{SD}(n, e)$ with

$$\det(1 - T\mathrm{Frob}_{E,f} \mid \mathrm{Prim}(n-1, d))$$
$$= \det(1 - T\mathrm{Frob}_E \mid \mathrm{Prim}^{n-1}((F=0 \text{ in } \mathbb{P}^n)\otimes\overline{E}, \overline{\mathbb{Q}}_\ell)).$$

Thus we have a congruence

$$\det(\mathcal{M}(n,e,\rho))\otimes\det(\mathrm{Prim}(n-2,e)) \equiv \det(\mathrm{Prim}(n-1,e)) \bmod \lambda,$$

and hence a congruence

$$(\det(\mathcal{M}(n,e,\rho))^{\otimes a})\otimes\det(\mathrm{Prim}(n-2,e))^{\otimes a}$$
$$\equiv (\det(\mathrm{Prim}(n-1,e)))^{\otimes a} \bmod \lambda.$$

As n is even, n-1 is odd, so Prim(n-1,e)((n-1)/2) is symplectically self dual, and hence has trivial determinant. By the same token, Prim(n-2,e)((n-1)/2) is orthogonally self dual, so has determinant of order 1 or 2. We already know that $\det(\mathcal{M}(n,e,\rho))^{\otimes a}$ is geometrically of order 1 or 2. Therefore from the above congruence we get a geometric isomorphism

$$\det(\mathcal{M}(n,e,\rho))^{\otimes a} \cong \det(\mathrm{Prim}(n-2,e))^{\otimes a} = \det(\mathrm{Prim}(n-2,e)),$$

the last equality because a is odd. So it remains to show that the determinant of the lisse sheaf Prim(n-2,e) is geometrically nontrivial. For this, it suffices to show that after some pullback, the determinant is geometrically nontrivial.

We first pull back from $\mathcal{SD}(n, e)$ to the space $\mathcal{H}_{n,e}$ of forms of degree e in n variables which define smooth hypersurfaces in $\mathbb{P}^{n-1}$, by the map

$$\mathcal{H}_{n,e} \to \mathcal{SD}(n, e),$$
$$G \mapsto 1 + G.$$

In the case n=2, Abel's theorem tells us that G_{geom} for this pullback of Prim(0, e) is the symmetric group S_e in its deleted permutation representation, cf. [Ka-LAMM, 2.4.3]. For $n \geq 4$ even, and $e \geq 3$, we further pull back Prim(n-2,e) to a Lefschetz pencil of degree e hypersurface sections of $\mathbb{P}^{n-1}$. Then the Picard-Lefschetz formula shows that G_{geom} for this pullback of Prim(n-2,e) contains a reflection. Hence det(Prim(n-2,e)) is geometrically nontrivial, as required. QED

Remark 6.7.26 It is instructive to compare part 5) of the above theorem, in the case n=1, with [Ka-ACT, 5.4, 2)]. Thus χ is $\chi_2\chi_a$ with χ_a nontrivial of ddd order a, the degree e is even and prime to p, and $\chi^e \neq \mathbb{1}$. Above, we computed G_{geom} for the sums $\Sigma_x\ \chi(f(x))$ as f varies over all degree e polynomials with nonzero discriminant, and showed it to be $GL_{2a}(e-1)$. In [Ka-ACT, 5.4, 2)], we computed G_{geom} for the two-parameter subfamily with parameters (α, β) obtained by choosing an f_0 of degree e whose second derivative f_0'' is not identically zero, and looking only at f's of the form $f(x) + \alpha x + \beta$. We showed that this smaller family had G_{geom} the smaller group $GL_a(e-1)$. If we instead look at the three-parameter family $\lambda f_0(x) + \alpha x + \beta$, with parameters (λ, α, β), the argument given in the second proof of 5) above shows that for this family G_{geom} is once again the larger group $GL_{2a}(e-1)$.

(6.8) Application to the monodromy of additive character sums in several variables

(6.8.1) Let us recall the general context. We have a finite field k of characteristic p, a prime number ℓ invertible in k, and a nontrivial additive character $\chi : k \to \overline{\mathbb{Q}}_\ell^\times$ of k. We fix integers $e \geq 3$ and $n \geq 1$. We denote by $\mathcal{P}(n, e)$ the space of polynomial functions on $\mathbb{A}^n$ of degree $\leq e$. Recall from 3.5.8 that an n-variable polynomial f in $\mathcal{P}(n,e)(\overline{k})$ is a Deligne polynomial if it satisfies the following two conditions D1), and D2).

D1) When we write $f = \Sigma_{i \leq e} F_i$ as a sum of homogeneous forms, F_e is nonzero, and, in the case $n \geq 2$, the closed subscheme of $\mathbb{P}^{n-1}$ defined by the vanishing of F_e is smooth of codimension one.

D2) The integer e is prime to p.

(6.8.2) For a fixed integer e which is prime to p, the Deligne polynomials form a dense open set $\mathcal{D}(n,e)$ of $\mathcal{P}(n,e)$.

(6.8.3) Recall further from Corollaries 3.5.11-12 that if e is prime to p, there is a lisse $\overline{\mathbb{Q}}_\ell$-sheaf $\mathcal{M}(n,e,\psi)|\mathcal{D}(n,e)$ on $\mathcal{D}(n, e)$, whose trace function is given as follows: for E/k any finite extension, and for any f in $\mathcal{D}(n,e)(E)$, we have

$$\mathrm{Trace}(\mathrm{Frob}_{E,f} \mid \mathcal{M}(n,e,\psi))$$
$$= (-1)^n(\#E)^{-n/2}\Sigma_{v \text{ in } \mathbb{A}^n(E)} \psi_E(f(v)).$$

The sheaf $\mathcal{M}(n,e,\psi)|\mathcal{D}(n,e)$ has rank $(e - 1)^n$, and is pure of weight 0.

(6.8.4) In Theorem 3.9.2, we proved the following general theorem concerning the geometric monodromy group G_{geom} for the sheaf $\mathcal{M}(n,e,\psi)|\mathcal{D}(n,e)$.

Theorem 6.8.5 (= 3.9.2 restated) Let k be a finite field, $p := \mathrm{char}(k)$, $\ell \neq p$ and ψ a nontrivial additive $\overline{\mathbb{Q}}_\ell^\times$-valued of k. Fix $n \geq 1$, $e \geq 3$, with e prime to p. Denote by $\mathcal{D}(n,e)$ the space of Deligne polynomials, and denote by $\mathcal{M}(n,e,\psi)|\mathcal{D}(n,e)$ the lisse, geometrically irreducible, and pure of weight zero $\overline{\mathbb{Q}}_\ell$-sheaf of rank $(e-1)^n$ on $\mathcal{D}(n,e)$ whose trace function is given by

$$\mathrm{Trace}(\mathrm{Frob}_{E,f} \mid \mathcal{M}(n,e,\psi)) = (-1)^n(\#E)^{-n/2}\Sigma_{v \text{ in } \mathbb{A}^n(E)} \psi_E(f(v)).$$

Suppose that any of the following five conditions holds:

a) $p \geq 7$,

b) $n \geq 3$,

c) $p = 5$ and $e \geq 4$,

d) $p = 3$ and $e \geq 7$,

e) $p= 2$ and $e \geq 7$.

Then we have the following results concerning the group G_{geom} for the lisse sheaf $\mathcal{M}(n,e,\psi)|\mathcal{D}(n,e)$.

1) If $p \neq 2$, G_{geom} contains $SL((e-1)^n)$

2) If $p = 2$ and n is odd, $G_{geom} = Sp((e-1)^n)$.

3) If $p = 2$ and n is even, G_{geom} is either $SO((e-1)^n)$ or $O((e-1)^n)$.

(6.8.6) We then considered "small" families of Deligne polynomials, as follows. We fixed an integer d prime to p, a Deligne polynomial

$$F \text{ in } \mathcal{D}(n,d)(k),$$

and an integer $e < d$. For any finite extension E/k, and for any f in $\mathcal{P}(n,e)(E)$, the sum $F + f$ is again a Deligne polynomial of degree d. So we have a closed immersion

$$\mathcal{P}(n,e) \to \mathcal{D}(n,d),$$
$$f \mapsto F+f.$$

The restriction of the lisse sheaf $\mathcal{M}(n,d)|\mathcal{D}(n,d)$ to $\mathcal{P}(n,e)$ by means of this closed immersion gives a lisse sheaf

$$\mathcal{M}(n,e,\psi,F) \text{ on } \mathcal{P}(n,e)$$

of rank $(d-1)^n$, whose trace function is given by

$$\mathrm{Trace}(\mathrm{Frob}_{E,f} \mid \mathcal{M}(n,e,\psi,F))$$
$$= (-1)^n(\#E)^{-n/2}\Sigma_{v \text{ in } \mathbb{A}^n(E)} \psi_E(F(v) + f(v)).$$

Concerning these sheaves, we proved the following theorem.

Theorem 6.8.7 (= 3.9.6 restated) Let k be a finite field, $p := \mathrm{char}(k)$, $\ell \neq p$ and ψ a nontrivial additive $\overline{\mathbb{Q}}_\ell^\times$-valued of k. Fix integers

$$n \geq 1,\ d > e \geq 3$$

with d prime to p. Fix a k-rational Deligne polynomial F in n variables of degree d,

$$F \text{ in } \mathcal{D}(n,d)(k).$$

Suppose that any of the following five conditions holds:

a) $p \geq 7$,

b) $n \geq 3$,

c) $p = 5$ and $e \geq 4$,

d) $p = 3$ and $e \geq 7$,

e) $p= 2$ and $e \geq 7$.

Then we have the following results concerning the group G_{geom} for the lisse sheaf $\mathcal{M}(n,e,\psi,F)$ on $\mathcal{P}(n,e)$.

1) If $p \neq 2$, G_{geom} contains $SL((d-1)^n)$

2) If $p = 2$ and n is odd, $G_{geom} = Sp((d-1)^n)$.

3) If $p = 2$ and n is even, G_{geom} is either $SO((d-1)^n)$ or $O((d-1)^n)$.

(6.8.8) We then turned to the case of strongly odd Deligne polynomials, those in which every monomial which occurs has odd degree. For each odd e prime to p, the strongly odd Deligne polynomials of degree e form a closed subspace

$$\mathcal{D}(n,e,\mathrm{odd}) \subset \mathcal{D}(n,e)$$

of all Deligne polynomials of the given degree e. The restriction of $\mathcal{M}(n,e,\psi)$ to $\mathcal{D}(n,e,\mathrm{odd})$ is self dual, symplectically if n is odd and orthogonally if n is even, cf. Theorem 3.10.6. Concerning these sheaves, we proved the following theorem.

Theorem 6.8.9 (= 3.10.7 restated) Let k be a finite field of characteristic p, ℓ a prime with $\ell \neq p$, and ψ a nontrivial additive $\overline{\mathbb{Q}}_\ell^\times$-valued of k. Fix integers

$n \geq 1$, $e \geq 3$,

with e prime to p and odd. Suppose that any of the following five conditions holds:

a) $p \geq 7$,

b) $p \neq 3$ and $n \geq 3$,

c) $p = 5$ and $e \geq 7$,

d $p = 3$ and $e \geq 7$,

e) $p= 2$ and $e \geq 7$.

Then we have the following results concerning the group G_{geom} for the lisse sheaf $\mathcal{M}(n,e,\psi)|\mathcal{D}(n,e,odd)$.

1) If n is odd, $G_{geom} = Sp((e-1)^n)$.

2) If n is even, G_{geom} is either $SO((e-1)^n)$ or $O((e-1)^n)$.

(6.8.10) We will now explain how to sharpen, in many cases, these results on the monodromy of additive character sums and determine G_{geom} exactly.

(6.8.11) We begin with the sheaves $\mathcal{M}(n,e,\psi)$ on $\mathcal{D}(n, e)$, and the sheaves $\mathcal{M}(n, e, \psi, F)$ on $\mathcal{P}(n, e)$. Recall that for any integer $a \geq 1$, we denote by $GL_a(N) \subset GL(N)$ the subgroup

$$GL_a(N) = \{A \text{ in } GL(N) \text{ such that } \det(A)^a = 1\}.$$

Lemma 6.8.12 For $e \geq 3$ prime to p, and $n \geq 1$, the group G_{geom} for $\mathcal{M}(n,e,\psi)$ on $\mathcal{D}(n, e)$ lies in

$$GL_{2p}((e-1)^n), \text{ if p is odd,}$$

$$GL_2((e-1)^n), \text{ if } p = 2.$$

proof The Tate-twisted sheaf $\mathcal{M}(n, e, \psi)(-n/2)$ has all its traces in the field $\mathbb{Q}(\zeta_p)$ of p'th roots of unity, cf. Corollary 3.5.12. As the sheaves $\mathcal{M}(n, e, \psi)$ and $\mathcal{M}(n, e, \psi)(-n/2)$ have the same G_{geom}, the result is now a special case of Lemma 6.7.22. QED

Lemma 6.8.13 Suppose $e \geq 3$ is prime to p. If in addition e-1 is prime to p,then for any $n \geq 1$, the group G_{geom} for $\mathcal{M}(n,e,\psi)$ on $\mathcal{D}(n, e)$ contains an element whose determinant is a primitive p'th root of unity in $\overline{\mathbb{Q}}_\ell$.

proof Take any Deligne polynomial f in $\mathcal{D}(n, e)(k)$, and take any element α in k such that $\psi(\alpha) \neq 1$. As $Frob_k$ modules, we have

$$H_c^n(\mathbb{A}^n\otimes\overline{k}, \mathcal{L}_{\psi(f+\alpha)}) \cong H_c^n(\mathbb{A}^n\otimes\overline{k}, \mathcal{L}_{\psi(f)})\otimes (\psi(\alpha))^{\deg}.$$

Comparing determinants of Frobenius, we find

$$\det(Frob_{k,\ f+\alpha} \mid \mathcal{M}(n, e, \psi))$$
$$= \det(Frob_{k,\ f} \mid \mathcal{M}(n, e, \psi))\times(\psi(\alpha))^{(e-1)^n}.$$

Because e-1 is prime to p, $(\psi(\alpha))^{(e-1)^n}$ is again a primitive p'th root of unity, and the above equation shows it is the determinant of all elements in $\pi_1{}^{geom}$ of the form

(any choice of $\mathrm{Frob}_{k,\ f+\alpha}$)(any choice of $\mathrm{Frob}_{k,\ f}$)$^{-1}$. QED

Lemma 6.8.14 Suppose $d > e \geq 3$, with d prime to p. Let $n \geq 1$. Let F in $\mathcal{D}(n, d)$ be a Deligne polynomial of degree d.

1) The group G_{geom} for $\mathcal{M}(n,e,\psi,F)$ on $\mathcal{P}(n, e)$ lies in $GL_p((d-1)^n)$.

2) If in addition d-1 is prime to p, then G_{geom} contains an element whose determinant is a primitive p'th root of unity in $\overline{\mathbb{Q}}_\ell$.

proof For the first assertion, note that $\mathcal{M}(n,e,\psi,F)|\mathcal{P}(n, e)$ is a pullback of $\mathcal{M}(n,d,\psi)|\mathcal{D}(n,d)$, whose larger G_{geom} lies in

$$GL_{2p}((d-1)^n),\text{ if p is odd,}$$
$$GL_2((d-1)^n),\text{ if } p = 2$$

(by Lemma 6.8.12, applied with e := d). So for p=2, there is nothing more to prove. If p is odd, then $\det(\mathcal{M}(n,e,\psi,F))^{\otimes p} \mid \mathcal{P}(n, e)\otimes\overline{k}$ is a lisse sheaf of rank one on the affine space $\mathcal{P}(n, e)\otimes\overline{k}$ of order dividing 2, i.e., an element of the cohomology group

$$H^1(\mathcal{P}(n, e)\otimes\overline{k}, \mathbb{Z}/2\mathbb{Z}) = 0.$$

The second assertion is proven exactly as the previous lemma, using F and F+α. QED

Theorem 6.8.15 Suppose $d > e \geq 3$, with d prime to p. Let $n \geq 1$. Let F in $\mathcal{D}(n, d)$ be a Deligne polynomial of degree d. Suppose that d-1 is prime to p, and suppose that any of the following four conditions holds:

a) $p \geq 7$,

b) $n \geq 3$,

c) $p = 5$ and $e \geq 4$,

d) $p = 3$ and $e \geq 7$,

Then G_{geom} for the lisse sheaf $\mathcal{M}(n,e,\psi,F)|\mathcal{P}(n,e)$ is $GL_p((d-1)^n)$.

proof Since d and d-1 are prime to p, p must be odd. From Theorem 6.8.7, we know G_{geom} contains $SL((d-1)^n)$. The previous lemma shows that G_{geom} lies in $GL_p((d-1)^n)$, and contains elements whose determinants have order p. QED

(6.8.16) To go further, we now make use of the results of the previous section giving the monodromy of multiplicative character sums.

(6.8.17) For $n \geq 2$, and $e \geq 1$, denote by

$$\mathrm{Homog}(n,e) \subset \mathcal{P}(n,e)$$

the linear space of homogeneous forms of degree d in n variables, and by

$$\mathrm{NSHomog}(n,e) \subset \mathrm{Homog}(n,e)$$

the dense open set consisting of nonsingular (NS) forms, i.e., those forms F of degree e such that the closed subscheme X_F of $\mathbb{P}^{n-1}$ defined by the vanishing of F is smooth of codimension one.

(6.8.18) Over NSHomog(n,e), we have the universal family of smooth degree e hypersurfaces in $\mathbb{P}^{n-1}$, say

$$\pi : \mathcal{X} \to \mathrm{NSHomog}(n,e),$$
$$\pi^{-1}(F) = X_F.$$

Here $\mathcal{X}$ is the closed subscheme of $\mathbb{P}^{n-1} \times \mathrm{NSHomog}(n,e)$ with coordinates (x, F) where F(x) = 0, and π is the second projection.

(6.8.19) On NSHomog(n,e), we have the lisse sheaf

$$R^{n-2}\pi_*\overline{\mathbb{Q}}_\ell.$$

For even n, $R^{n-2}\pi_*\overline{\mathbb{Q}}_\ell$ contains as a direct factor the geometrically constant sheaf $H^{n-2}(\mathbb{P}^{n-1}\otimes\overline{k}, \overline{\mathbb{Q}}_\ell) \cong \overline{\mathbb{Q}}_\ell((2-n)/2)$, spanned by the (n-2)/2 power of the class L of a hyperplane. We define the lisse sheaf $\mathrm{Prim}^{n-2}(e)$ on NSHomog(n,e) as follows:

$\mathrm{Prim}^{n-2}(e)$

:= $R^{n-2}\pi_*\overline{\mathbb{Q}}_\ell$, for n odd,

:= the orthogonal, for cup product, of the constant sheaf $H^{n-2}(\mathbb{P}^{n-1}\otimes\overline{k}, \overline{\mathbb{Q}}_\ell)$ in $R^{n-2}\pi_*\overline{\mathbb{Q}}_\ell$, for n even.

Theorem 6.8.19 The cup product pairing

$$\mathrm{Prim}^{n-2}(e) \times \mathrm{Prim}^{n-2}(e) \to \overline{\mathbb{Q}}_\ell(2-n)$$

is a perfect pairing. It is symplectic for n odd, and orthogonal for n even. For $n \geq 2$ even, the group G_{geom} for $\mathrm{Prim}^{n-2}(e)$ contains a reflection.

proof For n odd, $\mathrm{Prim}^{n-2}(e)$ is just $R^{n-2}\pi_*\overline{\mathbb{Q}}_\ell$, and the assertion is an instance of Poincaré duality. For n even, the nondegeneracy of the pairing results from the fact that $L^{(n-2)/2}$ has nonzero self-intersection on any X_F. That G_{geom} contains a reflection for $n \geq 4$ even results by pulling back to a Lefschetz pencil and using the Picard-Lefschetz formula. If n = 2, it results from Abel's theorem, i.e., that G_{geom} for $\mathrm{Prim}^0(e)$ is the symmetric group S_e in its deleted permutation representation. QED

(6.8.20) We also have the universal family of complements

$$\rho : \mathcal{U} \to \mathrm{NSHomog}(n,e),$$
$$\rho^{-1}(F) = \mathbb{P}^{n-1} - X_F := \mathbb{P}^{n-1}[1/F].$$

Here $\mathcal{U}$ is the complement in $\mathbb{P}^{n-1}\times\mathrm{NSHomog}(n,e)$, and ρ is the second projection. For χ a nontrivial multiplicative character with $\chi^e = \mathbb{1}$, we have the lisse sheaf $\mathcal{L}_{\chi(F(x))}$ on $\mathcal{U} = \{(x, F) \mid F(x) \neq 0\}$. According to [Ka-ENSMCS, Theorem 18, applied fibre by fibre], we have

$$R^i\rho_!\mathcal{L}_{\chi(F(x))} = 0 \text{ for } i \neq n-1,$$

and each stalk of $R^{n-1}\rho_!\mathcal{L}_{\chi(F(x))}$ is pure of weight n-1 and has dimension $(1/e)((e-1)^n - (-1)^n)$.

Lemma 6.8.21 For χ a nontrivial multiplicative character with $\chi^e = \mathbb{1}$, the sheaf $R^{n-1}\rho_!\mathcal{L}_{\chi(F(x))}$ on NSHomog(n,e) is lisse of rank

$$(1/e)((e-1)^n - (-1)^n),$$

and pure of weight 0.

proof Since the morphism ρ is affine and smooth of relative dimension n-1, $R^{n-1}\rho_!\mathcal{L}_{\chi(F(x))}$ is a sheaf "of perverse origin", and the constancy of its rank means precisely that it is lisse, cf. [Ka-SMD, Proposition 11]. QED

Lemma 6.8.22 For χ a nontrivial multiplicative character with $\chi^e = \mathbb{1}$, and $\overline{\chi}$ the conjugate character $1/\chi$, the cup product pairing

$$R^{n-1}\rho_!\mathcal{L}_{\chi(F(x))} \times R^{n-1}\rho_!\mathcal{L}_{\overline{\chi}(F(x))} \to R^{2n-2}\rho_!\overline{\mathbb{Q}}_\ell \cong \overline{\mathbb{Q}}_\ell(1-n)$$

is a perfect pairing. If e is even and χ is χ_2, the quadratic character, this pairing is symplectic if n is odd, and orthogonal if n is even.

proof To see that this pairing of lisse sheaves is a perfect pairing, it suffices to check at a single point F, where it follows from usual Poincaré duality and the fact that the natural "forget supports" map induces an isomorphism

$$H_c^{n-1}(\mathbb{P}^{n-1}[1/F]\otimes\overline{k}, \mathcal{L}_{\chi(F)}) \cong H^{n-1}(\mathbb{P}^{n-1}[1/F]\otimes\overline{k}, \mathcal{L}_{\chi(F)}).$$

That this pairing has the asserted symmetry property when e is even and χ is χ_2 results from standard sign properties of cup product. QED

(6.8.23) The sheaves $R^{n-1}\rho_!\mathcal{L}_{\chi(F(x))}$ on NSHomog(n,e) are closely related to the sheaves $Gr^0(\mathcal{M}(n-1, e, \chi))((1-n)/2)$ on $\mathcal{SD}(n-1, e)$.

Lemma 6.8.24 Under the morphism

$$i : \mathcal{SD}(n-1, e) \to \text{NSHomog}(n,e)$$

defined by homogenization,

$$f(x_1,\dots, x_{n-1}) \mapsto F(x_0,\dots, x_{n-1}) := (x_0)^e f(\dots, x_i/x_0,\dots),$$

we have

$$i^*R^{n-1}\rho_!\mathcal{L}_{\chi(F(x))} \cong Gr^0(\mathcal{M}(n-1, e, \chi))((1-n)/2).$$

proof That the two have the same characteristic polynomials of Frobenius is given by [Ka-ENSMCS, Theorem 18 and preceding paragraph]. Therefore the two have isomorphic semisimplifications. The target is geometrically irreducible, cf. 5.1.5 and 1.15.6, hence is arithmetically irreducible. Therefore the two sides are both arithmetically irreducible, and isomorphic. QED

Corollary 6.8.25 Suppose $e \geq 3$ is even, and $\chi = \chi_2$. If $n \geq 3$ is odd, then G_{geom} for $R^{n-1}\rho_!\mathcal{L}_{\chi_2(F(x))}$ contains a reflection.

proof Since e is prime to p, $p \neq 2$. After pullback, the group G_{geom} only gets smaller. After pullback by i, we get the sheaf $Gr^0(\mathcal{M}(n-1, e, \chi))((1-n)/2)$, whose G_{geom} contains a reflection, thanks to Theorem 6.6.5. QED

(6.8.26) Now we recall from Chapter 3 the following result.

Theorem 6.8.27 (= 3.6.4 restated) Suppose $e \geq 3$ is prime to p. Suppose that E/k is a finite extension which contains the e'th roots of unity. Suppose $n \geq 2$. Let F in NSHomog(n,e)(E) be a nonsingular form of degree e in n variables. For any finite extension L/E, we have the identity

$$(-1)^n\Sigma_{v \text{ in } \mathbb{A}^n(L)}\ \psi_L(F(v))$$

$$= \mathrm{Trace}(\mathrm{Frob}_L, \mathrm{Prim}^{n-2}(X_F\otimes_E\overline{E}, \overline{\mathbb{Q}}_\ell)(-1))$$

$$+ \Sigma_{\chi \neq \mathbb{1}, \chi^e = \mathbb{1}}(-g(\psi_E, \overline{\chi}_E))\mathrm{Trace}(\mathrm{Frob}_L, H_c^{n-1}(\mathbb{P}^{n-1}[1/F]\otimes_E\overline{E}, \mathcal{L}_{\chi(F)})).$$

Corollary 6.8.28 Suppose that $e \geq 3$ is prime to p, and that $n \geq 2$. Then we have a geometric isomorphism of lisse sheaves on NSHomog(n, e)

$$\mathcal{M}(n, e, \psi)|\mathrm{NSHomog}(n, e)$$

$$\cong \mathrm{Prim}^{n-2}(e) \oplus \oplus_{\chi^e = \mathbb{1}, \chi \neq \mathbb{1}} R^{n-1}\rho_!\mathcal{L}_{\chi(F(x))}.$$

proof The assertion is geometric, so we may extend scalars, and reduce to the case when k contains the e'th roots of unity. Then the lisse sheaves on NSHomog(n, e)

$$\mathcal{M}(n, e, \psi)(-n/2)|\mathrm{NSHomog}(n, e)$$

and

$$\mathrm{Prim}^{n-2}(e)(-1) \oplus \oplus_{\chi^e = \mathbb{1}, \chi \neq \mathbb{1}} R^{n-1}\rho_!\mathcal{L}_{\chi(F(x))}\otimes(-g(\psi, \overline{\chi}))^{\deg}$$

have the same trace function on all Frobenii $\mathrm{Frob}_{E,F}$. By Chebotarev, their arithmetic semisimplifications are isomorphic, and hence their geometric semisimplifications are isomorphic. But these lisse sheaves are all pure of weight n, so are already geometrically semisimple. QED

Theorem 6.8.29 Suppose that $e \geq 3$ is prime to p, and that $n \geq 2$. If ne is even, then G_{geom} for $\mathcal{M}(n, e, \psi)|\mathrm{NSHomog}(n, e)$ contains an element of determinant -1.

proof Suppose first that $n \geq 2$ is even. Consider the geometric isomorphism

$$\mathcal{M}(n, e, \psi)|\mathrm{NSHomog}(n, e)$$

$$\cong \mathrm{Prim}^{n-2}(e) \oplus \oplus_{\chi^e = \mathbb{1}, \chi \neq \mathbb{1}} R^{n-1}\rho_!\mathcal{L}_{\chi(F(x))}.$$

If e is odd, the characters χ which occur fall into pairs of conjugates, and for each such pair the corresponding summand

$$R^{n-1}\rho_!\mathcal{L}_{\chi(F(x))} \oplus R^{n-1}\rho_!\mathcal{L}_{\overline{\chi}(F(x))}$$

is the direct sum of a lisse sheaf and its dual. Such a direct sum has trivial determinant. If e is even, there is one leftover term after the pairing up of conjugates, namely

$$R^{n-1}\rho_!\mathcal{L}_{\chi_2(F(x))}.$$

But as n is even, n-1 is odd, and hence this leftover term is symplectically self dual, and hence has trivial determinant. Thus when n is even, we have a geometric isomorphism

$$\mathcal{M}(n, e, \psi)|\mathrm{NSHomog}(n, e)$$
$$\cong \mathrm{Prim}^{n-2}(e) \oplus (\text{a lisse sheaf with trivial determinant}).$$

There exist elements γ in $\pi_1{}^{\mathrm{geom}}(\mathrm{NSHomog}(n, e))$ which act on $\mathrm{Prim}^{n-2}(e)$ with determinant -1. [Indeed, as n is even, $\mathrm{Prim}^{n-2}(e)$ is orthogonally self dual, and its G_{geom} contains a reflection. Hence its determinant, which a priori has geometric order dividing 2, is geometrically nontrivial.] Any such element γ then acts on

$$\mathcal{M}(n, e, \psi)|\mathrm{NSHomog}(n, e)$$
$$\cong \mathrm{Prim}^{n-2}(e) \oplus (\text{a lisse sheaf with trivial determinant})$$

with determinant -1.

Suppose next that $n \geq 3$ is odd, but that e is even. Then $p \neq 2$, since e is prime to p. Consider the geometric isomorphism

$$\mathcal{M}(n, e, \psi)|\mathrm{NSHomog}(n, e)$$
$$\cong \mathrm{Prim}^{n-2}(e) \oplus \oplus_{\chi^e = 1, \chi \neq 1} R^{n-1}\rho_!\mathcal{L}_{\chi(F(x))}.$$

Now $\mathrm{Prim}^{n-2}(e)$ is symplectically self dual, so has trivial determinant, as does the direct sum

$$R^{n-1}\rho_!\mathcal{L}_{\chi(F(x))} \oplus R^{n-1}\rho_!\mathcal{L}_{\overline{\chi}(F(x))}$$

corresponding to a pair of conjugate characters. Thus we have a geometric isomorphism

$$\mathcal{M}(n, e, \psi)|\mathrm{NSHomog}(n, e)$$
$$\cong R^{n-1}\rho_!\mathcal{L}_{\chi_2(F(x))} \oplus (\text{a lisse sheaf with trivial determinant}).$$

But now there exist elements γ in $\pi_1{}^{\mathrm{geom}}(\mathrm{NSHomog}(n, e))$ which act on $R^{n-1}\rho_!\mathcal{L}_{\chi_2(F(x))}$ with determinant -1, because this is an orthogonally self dual sheaf whose G_{geom} contains a reflection. Any such element γ then acts on $\mathcal{M}(n, e, \psi)|\mathrm{NSHomog}(n, e)$ with determinant -1. QED

(6.8.30) We now record two corollaries, based on the inclusions

$$\mathrm{NSHomog}(n, e) \subset \mathcal{D}(n,e),$$

and, if e is odd,

$$\mathrm{NSHomog}(n, e) \subset \mathcal{D}(n,e,\mathrm{odd}).$$

Corollary 6.8.31 Suppose that $e \geq 3$ is prime to p, and that $n \geq 2$. If ne is even, then G_{geom} for $\mathcal{M}(n, e, \psi)$ contains an element of determinant -1.

Corollary 6.8.32 Suppose that $n \geq 2$ is even, and that $e \geq 3$ is prime to p and odd. Then G_{geom} for the lisse sheaf $\mathcal{M}(n,e,\psi)|\mathcal{D}(n,e,odd)$ contains an element of determinant -1.

(6.8.33) We now put together the information we have assembled.

Theorem 6.8.34 Let $e \geq 3$ and $n \geq 1$. Suppose that $e(e-1)$ is prime to p and that ne is even. If any of the following four conditions holds:

a) $p \geq 7$,
b) $n \geq 3$,
c) $p = 5$ and $e \geq 4$,
d) $p = 3$ and $e \geq 7$,

then G_{geom} for $\mathcal{M}(n, e, \psi)|\mathcal{D}(n, e)$ is $GL_{2p}((e-1)^n)$.

proof. Since $e(e-1)$ is prime to p, we have $p \neq 2$. By Theorem 6.8.5, G_{geom} contains $SL((e-1)^n)$. By Lemma 6.8.12, $G_{geom} \subset GL_{2p}((e-1)^n)$. Thus we have

$$SL((e-1)^n) \subset G_{geom} \subset GL_{2p}((e-1)^n).$$

So it suffices to show that G_{geom} contains an element whose determinant is a primitive 2p'th root of unity. By Lemma 6.8.13, G_{geom} contains an element α whose determinant is a primitive p'th root of unity ζ_p. If $n \geq 2$, then by Corollary 6.8.31, G_{geom} contains an element β whose determinant is -1, and $\alpha\beta$ is the desired element.

Suppose now n=1. Then we need another mechanism to produce an element β in G_{geom} whose determinant is -1. We once again restrict to the n=1 analogue of NSHomog(n, e), which by abuse of notation we call NSHomog(1, e), namely the one-parameter family $a \mapsto ax^e$, parameterized by a in $\mathbb{G}_m$. Over any field containing the e'th roots of unity, we have

$$\begin{aligned} -\Sigma_x \psi(ax^e) &= \Sigma_y \psi(ay) \Sigma_{\chi \text{ with } \chi^e = \mathbb{1}} \chi(y) \\ &= -\Sigma_{\chi \neq \mathbb{1} \text{ with } \chi^e = \mathbb{1}} \Sigma_y \psi(ay)\chi(y) \\ &= \Sigma_{\chi \neq \mathbb{1} \text{ with } \chi^e = \mathbb{1}} \overline{\chi}(a)(-\Sigma_y \psi(y)\chi(y)). \end{aligned}$$

In other words, the lisse sheaves on NSHomog(1, e) = $\mathbb{G}_m$, both pure of weight 1,

$$\mathcal{M}(1,e,\psi)(-1/2)|\text{NSHomog}(1, e)$$

and

$$\oplus_{\chi \neq \mathbb{1} \text{ with } \chi^e = \mathbb{1}} \mathcal{L}_{\overline{\chi}} \otimes (-g(\psi, \chi))^{deg},$$

have the same trace function. Hence we have a geometric isomorphism

$$\mathcal{M}(1,e,\psi) \mid \mathrm{NSHomog}(1, e) \cong \oplus_{\chi \neq \mathbb{1} \text{ with } \chi^e = \mathbb{1}} \mathcal{L}_\chi,$$

whose determinant, e being even, is $\mathcal{L}_{\chi_2}$. Thus already G_{geom} for $\mathcal{M}(1,e,\psi)|\mathrm{NSHomog}(1, e)$ contains an element β of determinant -1, so a fortiori the larger group G_{geom} for $\mathcal{M}(1,e,\psi)|\mathcal{D}(1,e)$ contains such an element β. QED

Theorem 6.8.35 Let $e \geq 3$ and $n \geq 2$. Suppose that e is odd and prime to p, and suppose that $n \geq 2$ is even. Suppose that any of the following five conditions holds:

a) $p \geq 7$,
b) $p \neq 3$ and $n \geq 3$,
c) $p = 5$ and $e \geq 7$,
d) $p = 3$ and $e \geq 7$,
e) $p= 2$ and $e \geq 7$.

Then G_{geom} for $\mathcal{M}(n, e, \psi)|\mathcal{D}(n, e, \text{odd})$ is $O((e-1)^n)$.

proof By Theorem 6.8.9, G_{geom} is either $SO((e-1)^n)$ or $O((e-1)^n)$. But by Corollary 6.8.32, G_{geom} contains an element of determinant -1, which rules out the SO possibility. QED

Appendix A6: Swan-minimal poles

(A6.1) Swan conductors of direct images

(A6.1.1) Let L be a complete, discretely valued field of positive characteristic p, whose residue field k is algebraically closed. In terms of any choice of uniformizing parameter z, L is the Laurent series field k((z)). Let f in $L^\times$ have a Swan-minimal pole (cf. 6.4.6), i.e., $\mathrm{ord}_L(f) = -e$, for some $e > 0$, and $\mathrm{ord}_L(df/f) \le 0$. Then L is a separable, degree e extension of the subfield K := k((1/x)), via x := f. We view K as the ∞-adic completion of the function field $k(\mathbb{G}_m)$ of $\mathbb{G}_m$:= Spec(k[x,1/x]). We view f as a finite etale map from Spec(L) to Spec(K), and form the lisse sheaf $f_*\overline{\mathbb{Q}}_\ell$ of rank e on Spec(K). In down to earth galois-theoretic terms, $f_*\overline{\mathbb{Q}}_\ell$ is the e-dimensional representation ρ_f of

$$I(\infty) := \mathrm{Gal}(\overline{K}/K)$$

given as follows. Pick any finite galois extension M/K which contains L, with

$$G := \mathrm{Gal}(M/K),\ H := \mathrm{Gal}(M/L).$$

Then ρ_f factors through the quotient G of $I(\infty)$, and as a $\overline{\mathbb{Q}}_\ell$-representation of G we have

$$\rho_f = \mathrm{Ind}_H{}^G(\mathbb{1}).$$

(A6.1.2) We wish to compute the $I(\infty)$-breaks of $f_*\overline{\mathbb{Q}}_\ell$, i.e., the upper numbering breaks of $\rho_f = \mathrm{Ind}_H{}^G(\mathbb{1})$. We already know by Lemma 6.4.4 that the sum of the $I(\infty)$-breaks, $\mathrm{Swan}_\infty(f_*\overline{\mathbb{Q}}_\ell)$, is given by

$$\mathrm{Swan}_\infty(f_*\overline{\mathbb{Q}}_\ell) = 1 + \mathrm{ord}_L(df/f).$$

(A6.1.3) If e is prime to p, then $\mathrm{ord}_L(df/f) = -1$, so $\mathrm{Swan}_\infty(f_*\overline{\mathbb{Q}}_\ell) = 0$, and hence all the $I(\infty)$-breaks of $f_*\overline{\mathbb{Q}}_\ell$ are zero. In this case, 1/f is the e'th power of a uniformizing parameter of L, and L/K is the tame extension of K of degree e. So L/K is cyclic of degree e, and we have the more precise information that $f_*\overline{\mathbb{Q}}_\ell$ is the direct sum of all the characters of $I(\infty)$ of order dividing e.

(A6.1.4) In the case when p divides e, then by assumption we have $\mathrm{ord}_L(df/f) = 0$, and hence $\mathrm{Swan}_\infty(f_*\overline{\mathbb{Q}}_\ell) = 1$. Now $f_*\overline{\mathbb{Q}}_\ell$ is a semisimple $I(\infty)$-representation (because it factors through a finite quotient). Since it has $\mathrm{Swan}_\infty = 1$, it must [Ka-GKM, 1.11] be of the form

(rank r, all breaks 1/r)
$\oplus$ ($\oplus$ of tame characters of finite order).

How do we compute r, and how do we determine which tame characters occur?

Theorem A6.1.5 Suppose f in $L^\times$ has a Swan-minimal pole of order $e = e_0 q$, with e_0 prime to p and with $q = p^n$ for some $n \geq 1$. Then $f_* \overline{\mathbb{Q}}_\ell$ as $I(\infty)$-representation is the direct sum

(rank $r = e_0(q-1)$, all breaks $1/r$)

$\oplus$ ($\oplus$ all characters of order dividing e_0).

proof Pick a uniformizing parameter z of L, and expand f. We get

$$f = \alpha z^{-e}(1 + a_1 z + \text{higher terms}),$$

for some nonzero scalar α in k. By Swan-minimality, a_1 is nonzero in k. By rescaling the parameter, i.e., using $(a_1/e_0)z$ as parameter, we may assume that $a_1 = e_0$:

$$f = \alpha z^{-e}(1 + e_0 z + \text{higher terms}).$$

Because e is divisible by e_0, and e_0 is prime to p, f/α has a unique e_0'th root f_0 in L of the form

$$f_0 = z^{-q}(1 + z + \text{higher terms}).$$

Then $f = \alpha(f_0)^{e_0}$ is the composition of f_0 followed by the e_0'th power endomorphism $[e_0]$ of $\mathbb{G}_m$ given by $x \mapsto x^{e_0}$, followed by the multiplicative translation automorphism $\mathrm{MultTrans}_\alpha : x \mapsto \alpha x$ of $\mathbb{G}_m$:

$$f = [\mathrm{MultTrans}_\alpha][e_0]\circ f_0.$$

Thus

$$f_* \overline{\mathbb{Q}}_\ell = [\mathrm{MultTrans}_\alpha]_*[e_0]_*(f_{0*}\overline{\mathbb{Q}}_\ell).$$

In terms of fields, we have

$$L \supset k((1/f_0)) \supset k\,(((1/f_0)^{e_0})) = k((\alpha/f)) = k((1/f)) = K.$$

It suffices to prove that

$$f_{0*}\overline{\mathbb{Q}}_\ell \cong (\text{rank } r = (q-1), \text{ all breaks } 1/r) \oplus \mathbb{1},$$

because for any e_0 prime to p, and for any r, we have $I(\infty)$-isomorphisms

$[e_0]_*$(rank r, all breaks $1/r$) = (rank $e_0 r$, all breaks $1/e_0 r$),

$[e_0]_*(\mathbb{1})$ = $\oplus$ all characters of order dividing e_0,

$[\mathrm{MultTrans}_\alpha]_*$(rank r, all breaks $1/r$) = (rank r, all breaks $1/r$),

$[\mathrm{MultTrans}_\alpha]_*(\mathcal{L}_\chi) = \mathcal{L}_\chi$, for any tame character χ of $I(\infty)$.

Thus we are reduced to the case when f has a Swan-minimal pole of p-power order q, and has $a_1 = -1$:

$$f = z^{-q}(1 + z + \text{higher terms})$$
$$= z^{-q}(1 + z + z^2\Gamma),$$

with Γ in k[[z]].

We claim that there exists a parameter Z of L of the form

$$Z = z + \text{higher terms} = z(1 + z\Delta),$$

for some Δ in k[[z]], such that

$$f = Z^{-q}(1 + Z).$$

We want to find Δ in $k[[z]]$ such that

$$z^{-q}(1 + z + z^2\Gamma) = z^{-q}(1 + z\Delta)^{-q}(1 + z + z^2\Delta),$$

i.e.,

$$(1 + z\Delta)^q(1 + z + z^2\Gamma) = 1 + z + z^2\Delta,$$

i.e.,

$$(1 + z^q\Delta^q)(1 + z + z^2\Gamma) = 1 + z + z^2\Delta,$$

i.e.,

$$1 + z + z^2\Gamma + z^q(1 + z + z^2\Gamma)\Delta^q = 1 + z + z^2\Delta,$$

i.e.,

$$z^2\Gamma + z^q(1 + z + z^2\Gamma)\Delta^q = z^2\Delta,$$

i.e.,

$$\Gamma - \Delta + z^{q-2}(1 + z + z^2\Gamma)\Delta^q = 0.$$

In other words, the desired Δ is a root in $k[[z]]$ of a polynomial equation with coefficients in $k[[z]]$, of the form

$$F(X) := AX^q - X + \Gamma = 0.$$

We have $dF/dX = 1$ identically. If $q > 2$, then $X = \Gamma$ is a solution mod z, so by Hensel's lemma there is a unique solution Δ which has the same constant term as Γ. If $q = 2$, then we must first choose in k a solution, call it γ, of the equation $X^2 - X + \Gamma(0)$. Then by Hensel's lemma there is a unique solution Δ with constant term γ. So in all cases we have constructed the desired Δ.

In terms of the parameter Z, our f is

$$f = Z^{-q} + Z^{1-q}.$$

We now deal with this explicit f by means of a global argument. We work on $\mathbb{A}^1 := \mathrm{Spec}(k[X])$, and consider the polynomial

$$\varphi(X) := X^{q-1}(X + 1) = X^q + X^{q-1}.$$

We view φ as a finite flat map of $\mathbb{A}^1 - \{0, -1\}$ to $\mathbb{G}_m$. It extends to a finite flat map $\overline{\varphi}$ of $\mathbb{P}^1$ to itself, which is fully ramified over ∞. In terms of the uniformizing parameter $Z := 1/X$ at ∞ on the source, the map φ becomes our f, namely $Z^{-q} + Z^{1-q}$. Thus $f_*\overline{\mathbb{Q}}_\ell$ as $I(\infty)$-representation is obtained globally as follows. We consider the map

$$\varphi : \mathbb{A}^1 - \{0, -1\} \to \mathbb{G}_m,\ X \mapsto X^{q-1}(X + 1).$$

Then $f_*\overline{\mathbb{Q}}_\ell$ is the $I(\infty)$-representation attached to the sheaf $\varphi_*\overline{\mathbb{Q}}_\ell$ on $\mathbb{G}_m$.

It remains to analyze the sheaf $\varphi_*\overline{\mathbb{Q}}_\ell$ on $\mathbb{G}_m$. Because we are in characteristic p, the map φ makes $\mathbb{A}^1 - \{0, -1\}$ a finite etale covering of $\mathbb{G}_m$. So the sheaf $\varphi_*\overline{\mathbb{Q}}_\ell$ on $\mathbb{A}^1$ is lisse on $\mathbb{G}_m$. Over 0, there are two points in the fibre, 0 with ramification index $q-1$, and -1 with ramification index 1. Thus as $I(0)$-representation, $\varphi_*\overline{\mathbb{Q}}_\ell$ is tame, isomorphic to

$$\mathbb{1} \oplus (\oplus \text{ all characters of order dividing } q-1).$$

As already noted in A6.1.4 above, $\varphi_*\overline{\mathbb{Q}}_\ell$ as $I(\infty)$-representation, or equivalently $f_* \overline{\mathbb{Q}}_\ell$, is of the form

(rank r, all breaks 1/r)

$\oplus$ ($\oplus$ of tame characters of finite order).

We now analyze which tame characters can occur here. By Frobenius reciprocity, $f_* \overline{\mathbb{Q}}_\ell = \mathrm{Ind}_H{}^G(\mathbb{1})$ contains the trivial representation once. So we have only to show that no nontrivial tame character χ of finite order of $I(\infty)$ occurs in $\varphi_* \overline{\mathbb{Q}}_\ell$.

We will prove this as follows. Any nontrivial tame character of finite order of $I(\infty)$ is the restriction to $I(\infty)$ of a Kummer sheaf $\mathcal{L}_\chi$ attached to a nontrivial tame character χ of $\pi_1(\mathbb{G}_m)$. So what we must show is that for any nontrivial tame character χ of $\pi_1(\mathbb{G}_m)$, if we denote by $j : \mathbb{G}_m \to \mathbb{P}^1$ the inclusion, then the sheaf

$$\mathcal{G}(\chi) := j_*(\mathcal{L}_\chi \otimes \varphi_* \overline{\mathbb{Q}}_\ell)$$

on $\mathbb{P}^1$ has vanishing stalk at ∞.

To analyze the dimensions of the stalks of $\mathcal{G}(\chi)$ at 0 and ∞, we use the Euler Poincaré formula, which gives

$$\begin{aligned}
&\chi(\mathbb{P}^1, \mathcal{G}(\chi))\\
&= \dim(\mathcal{G}(\chi)_0) + \dim(\mathcal{G}(\chi)_\infty) + \chi(\mathbb{G}_m, j^*\mathcal{G}(\chi))\\
&= \dim(\mathcal{G}(\chi)_0) + \dim(\mathcal{G}(\chi)_\infty) - 1,
\end{aligned}$$

the last equality because $j^*\mathcal{G}(\chi) = \mathcal{L}_\chi \otimes j^*\varphi_* \overline{\mathbb{Q}}_\ell$ is lisse on $\mathbb{G}_m$, tame at 0, and has Swan conductor 1 at ∞.

We now express this Euler characteristic "upstairs". Thus we consider the diagram

$$\begin{array}{ccccc}
 & \mathbb{A}^1 - \{0, -1\} & \xrightarrow{k} & \mathbb{P}^1 & \\
\varphi & \downarrow & & \downarrow & \overline{\varphi} \\
 & \mathbb{G}_m & \xrightarrow[j]{} & \mathbb{P}^1 &
\end{array}$$

By the projection formula, $\mathcal{L}_\chi \otimes \varphi_* \overline{\mathbb{Q}}_\ell = \varphi_* \mathcal{L}_{\chi(\varphi)}$, and hence

$$\mathcal{G}(\chi) := j_* \varphi_* \mathcal{L}_{\chi(\varphi)} = \overline{\varphi}_* k_* \mathcal{L}_{\chi(\varphi)}.$$

Thus we have

$$\begin{aligned}
\chi(\mathbb{P}^1, \mathcal{G}(\chi)) &= \chi(\mathbb{P}^1, \overline{\varphi}_* k_* \mathcal{L}_{\chi(\varphi)})\\
&= \chi(\mathbb{P}^1, k_* \mathcal{L}_{\chi(\varphi)}).
\end{aligned}$$

The lisse sheaf $\mathcal{L}_{\chi(\varphi)} = \mathcal{L}_{\chi^{q-1}(x)} \otimes \mathcal{L}_{\chi(x+1)}$ on $\mathbb{A}^1 - \{0, -1\}$ is lisse of rank one, and nontrivially ramified at both $x = -1$, because χ is nontrivial by assumption, and at $x = \infty$ (because χ^q is nontrivial if χ is). It is ramified at $x=0$ if and only if $\chi^{q-1} \neq \mathbb{1}$. It is everywhere tame.

We first consider the case when $\chi^{q-1} \neq \mathbb{1}$. Then we have

$$\chi(\mathbb{P}^1, k_*\mathcal{L}_{\chi(\varphi)}) = \chi_c(\mathbb{A}^1 - \{0, -1\}, \mathcal{L}_{\chi(\varphi)})$$
$$= \chi_c(\mathbb{A}^1 - \{0, -1\}, \overline{\mathbb{Q}}_\ell) = 1 - 2 = -1.$$

Comparing with our earlier formula

$$\dim(\mathcal{G}(\chi)_0) + \dim(\mathcal{G}(\chi)_\infty) - 1$$

for this same Euler characteristic, we see that $\mathcal{G}(\chi)_\infty = 0$ in this case.

Now we consider the case when $\chi \neq \mathbb{1}$, but $\chi^{q-1} = \mathbb{1}$. In this case, $k_*\mathcal{L}_{\chi(\varphi)}$ is lisse at x=0, so we have

$$\chi(\mathbb{P}^1, k_*\mathcal{L}_{\chi(\varphi)}) = \chi_c(\mathbb{A}^1 - \{-1\}, \mathcal{L}_{\chi(\varphi)}) = 0.$$

Comparing with the earlier formula, we see that

$$\dim(\mathcal{G}(\chi)_0) + \dim(\mathcal{G}(\chi)_\infty) = 1.$$

But we have $\dim(\mathcal{G}(\chi)_0) = 1$ when $\chi^{q-1} = \mathbb{1}$, and hence $\mathcal{G}(\chi)_\infty = 0$ in this case as well. QED

(A6.2) An application to Swan conductors of pullbacks

(A6.2.1) In the setup of the previous section, suppose we are given a two-dimensional $\overline{\mathbb{Q}}_\ell$-representation M of $I(\infty) := \mathrm{Gal}(\overline{K}/K)$, both of whose $I(\infty)$-breaks are 1/2. The pullback f^*M is the restriction of M to the open subgroup $I_L(\infty) := \mathrm{Gal}(\overline{K}/L)$. We wish to calculate the Swan conductor of f^*M.

Lemma A6.2.2 Suppose f in $L^\times$ has a Swan-minimal pole of order e. Let M be a two-dimensional $\overline{\mathbb{Q}}_\ell$-representation M of $I(\infty) := \mathrm{Gal}(\overline{K}/K)$, both of whose $I(\infty)$-breaks are 1/2.

1) If e is prime to p, then f^*M has both $I_L(\infty)$-breaks e/2, and

$$\mathrm{Swan}(f^*M) = e.$$

2) If $e = e_0 q$ with $q = p^n$, $n \geq 1$, and e_0 prime to p, then we have the following results.

2a) If $e_0(q-1) \geq 3$, then

$$\mathrm{Swan}(f^*M) = e - 2.$$

2b) If $e = p = 3$, we have the inequality

$$\mathrm{Swan}(f^*M) \leq e - 2,$$

and for all but precisely one value of μ in $k^\times$, we have

$$\mathrm{Swan}((\mu f)^*M) = e - 2 = 1.$$

For the unique exceptional μ, call it μ_0, we have

$$\mathrm{Swan}((\mu_0 f)^*M) = e - 3 = 0.$$

2c) If $e = p = 2$, then we have the equality

$$\mathrm{Swan}(f^*M) = 1.$$

proof In case 1), L/K is the tame extension of degree e, and the effect of pullback is to multiply each break, and the Swan conductor, by the degree e, cf. [Ka-GKM, 1.13 and 1.13.1].

In case 2), we argue as follows. For virtual representations of virtual dimension zero, the Swan conductor is invariant under induction, cf. [De-Const, 5.5.2 and 5.6.1], [Ka-TLFM, proof of 1.6.4.1]. So we have

$$\begin{aligned}\mathrm{Swan}(f^*M) &= \mathrm{Swan}(f^*M - 2\overline{\mathbb{Q}}_\ell) = \mathrm{Swan}(f_*(f^*M - 2\overline{\mathbb{Q}}_\ell))\\ &= \mathrm{Swan}(M\otimes f_*\overline{\mathbb{Q}}_\ell) - 2\mathrm{Swan}(f_*\overline{\mathbb{Q}}_\ell)\\ &= \mathrm{Swan}(M\otimes f_*\overline{\mathbb{Q}}_\ell) - 2,\end{aligned}$$

the last equality because f has a Swan-minimal pole of order divisible by p, cf. Corollary 6.4.5. We have seen in Theorem A6.1.5 that $f_*\overline{\mathbb{Q}}_\ell$ as $I(\infty)$-representation has the shape

$$(\text{rank } r = e_0(q-1), \text{ all breaks } 1/r)$$
$$\oplus\ (\oplus \text{ all characters of order dividing } e_0).$$

In case 2a), when $e_0(q-1) \geq 3$, all the breaks of $f_*\overline{\mathbb{Q}}_\ell$ are $< 1/2$. As both breaks of M are 1/2, all the 2e breaks of $M\otimes f_*\overline{\mathbb{Q}}_\ell$ are 1/2. So in case 2a),we have $\mathrm{Swan}(M\otimes f_*\overline{\mathbb{Q}}_\ell) = e$, as required.

If we are not in case 2a), then $e_0(q-1) \leq 2$. This happens only in two cases, ($q = p = 3$, $e_0 = 1$) and ($q = p = 2$, $e_0 = 1$).

In case 2b), $e = p = 3$, and $f_*\overline{\mathbb{Q}}_\ell$ has the shape

$$\overline{\mathbb{Q}}_\ell \oplus (\text{rank } 2, \text{ both breaks } 1/2).$$

In this case $f_*\overline{\mathbb{Q}}_\ell$ has all its breaks $\leq 1/2$, so $M\otimes f_*\overline{\mathbb{Q}}_\ell$ has all its 2e breaks $\leq 1/2$. Thus we certainly have $\mathrm{Swan}(M\otimes f_*\overline{\mathbb{Q}}_\ell) \leq e$, and so we have the asserted inequality.

To go further in case 2b), we use the fact that we are not in characteristic 2. Then every tame $\overline{\mathbb{Q}}_\ell^\times$-valued character of $I(\infty)$ has a tame square root. Since det(M) is tame (its unique break is $\leq 1/2$, so 0), we can replace M by $M\otimes\mathcal{L}_\chi$ for some tame character χ of $I(\infty)$ and reduce to the case where det(M) is $\mathcal{L}_{\chi_2}$, for χ_2 the unique character of $I(\infty)$ of order 2. In this case, denoting by $[2] : \mathbb{G}_m \to \mathbb{G}_m$ the squaring map on $\mathbb{G}_m$, we have

$$M \cong [2]_*\mathcal{L}_{\psi(\alpha x)},$$

for ψ a nontrivial additive character of the prime field $\mathbb{F}_p$, and some α in $k^\times$. Indeed, to see this recall first [Ka-GKM, 5.6.1] that $[2]_*\mathcal{L}_{\psi(2x)}$ is the $I(\infty)$-representation attached to the Kloosterman sheaf $\mathrm{Kl}(\psi;\mathbb{1}, \chi_2)$, and hence that

$$\det([2]_*\mathcal{L}_{\psi(2x)}) = \mathcal{L}_{\chi_2}.$$

Then apply [Ka-ESDE, 8.6.3] to conclude that M is a multiplicative translate of $[2]_*\mathcal{L}_{\psi(2x)}$. A similar analysis of $f_*\overline{\mathbb{Q}}_\ell/\overline{\mathbb{Q}}_\ell$ shows that for some tame character ρ of $I(\infty)$, and for some β in $k^\times$, we have

$$f_*\overline{\mathbb{Q}}_\ell/\overline{\mathbb{Q}}_\ell \cong [2]_*(\mathcal{L}_\rho\otimes\mathcal{L}_{\psi(\beta x)}).$$

For λ in $k^\times$, we have

$$(\lambda^{-2}f)_*\overline{\mathbb{Q}}_\ell/\overline{\mathbb{Q}}_\ell \cong [x \mapsto \lambda^{-2}x]_*(f_*\overline{\mathbb{Q}}_\ell/\overline{\mathbb{Q}}_\ell)$$
$$\cong [x \mapsto \lambda^{-2}x]_*[2]_*(\mathcal{L}_\rho\otimes\mathcal{L}_{\psi(\beta x)})$$
$$\cong [2]_*[x \mapsto \lambda^{-1}x]_*(\mathcal{L}_\rho\otimes\mathcal{L}_{\psi(\beta x)})$$
$$\cong [2]_*(\mathcal{L}_\rho\otimes\mathcal{L}_{\psi(\lambda\beta x)}).$$

[We use the translation invariance of the sheaves $\overline{\mathbb{Q}}_\ell$ and $\mathcal{L}_\rho$, in the first and fourth isomorphisms respectively.] Taking pullback by the squaring map, we find

$$[2]^*M \cong \mathcal{L}_{\psi(\alpha x)} \oplus \mathcal{L}_{\psi(-\alpha x)},$$
$$[2]^*((\lambda^{-2}f)_*\overline{\mathbb{Q}}_\ell/\overline{\mathbb{Q}}_\ell) \cong \mathcal{L}_\rho\otimes\mathcal{L}_{\psi(\lambda\beta x)} \oplus \mathcal{L}_\rho\otimes\mathcal{L}_{\psi(-\lambda\beta x)}.$$

So we have

$$[2]^*(M\otimes((\lambda^{-2}f)_*\overline{\mathbb{Q}}_\ell/\overline{\mathbb{Q}}_\ell))$$
$$\cong \mathcal{L}_\rho\otimes(\mathcal{L}_{\psi((\alpha+\lambda\beta)x)} \oplus \mathcal{L}_{\psi((-\alpha+\lambda\beta)x)} \oplus \mathcal{L}_{\psi((\alpha-\lambda\beta)x)} \oplus \mathcal{L}_{\psi((-\alpha-\lambda\beta)x)},$$

which has all four breaks 1 unless λ is $\pm\alpha/\beta$, i.e., unless $\lambda^{-2} = (\beta/\alpha)^2$, in which case there are two breaks 1 and two breaks 0. So except for this single excluded value of λ^{-2}, $M\otimes((\lambda^{-2}f)_*\overline{\mathbb{Q}}_\ell/\overline{\mathbb{Q}}_\ell)$ itself has all four breaks 1/2, and we have

$$\mathrm{Swan}(M\otimes(\lambda^{-2}f)_*\overline{\mathbb{Q}}_\ell) = \mathrm{Swan}(M) + \mathrm{Swan}(M\otimes(((\lambda^{-2}f)_*\overline{\mathbb{Q}}_\ell)/\overline{\mathbb{Q}}_\ell))$$
$$= 1 + 2 = 3 = e,$$

as required. For the single exceptional value $\lambda_0{}^{-2}$ of λ^{-2}, we get

$$\mathrm{Swan}(M\otimes(\lambda_0{}^{-2}f)_*\overline{\mathbb{Q}}_\ell) = 1 + 1 = 2,$$

in which case we have the rather remarkable fact that $(\lambda_0{}^{-2}f)^*(M)$ is tame. Then $\mu_0 := \lambda_0{}^{-2}$ is the unique excluded value of α in 2b).

We now turn to case 2c), when $e = p = 2$. In this case, we have

$$f_*\overline{\mathbb{Q}}_\ell \cong \overline{\mathbb{Q}}_\ell \oplus \mathcal{L}_{\psi(x)},$$

with slopes 0 and 1. So in this case, $M\otimes f_*\overline{\mathbb{Q}}_\ell$ has two slopes 1/2 and two slopes 1, so we get

$$\mathrm{Swan}(M\otimes f_*\overline{\mathbb{Q}}_\ell) = 3,$$

and hence

$$\mathrm{Swan}(f^*M) = \mathrm{Swan}(M\otimes f_*\overline{\mathbb{Q}}_\ell) - 2 = 1. \text{ QED}$$

(A6.3) Interpretation in terms of canonical extensions

(A6.3.1) We continue to view K as the ∞-adic completion of the function field $k(\mathbb{G}_m)$. Then the separable extension L/K, viewed as a finite etale connected covering of Spec(K), has a "canonical extension" [Ka-LG] to a finite etale connected covering $\pi : E \to \mathbb{G}_m$ This canonical extension is characterized by two properties. The first is that it extends L/K in the sense that Spec(L)/Spec(K) is obtained from $E/\mathbb{G}_m$ by base change via the natural map $\mathrm{Spec}(K) \to \mathbb{G}_m$ corresponding to the inclusion of rings $k[x, 1/x] \subset K = k((1/x))$. The second is that $E/\mathbb{G}_m$ is a "special" finite etale covering; this means that the geometric monodromy group G_{geom} for the lisse sheaf $\pi_*\overline{\mathbb{Q}}_\ell$ on $\mathbb{G}_m$ has a unique p-Sylow subgroup, or equivalently that some p-Sylow subgroup is a normal subgroup. By the first requirement, we recover $f_*\overline{\mathbb{Q}}_\ell$ as the $I(\infty)$-representation attached to $\pi_*\overline{\mathbb{Q}}_\ell$.

Theorem A6.3.2 Suppose f in $L^\times$ has a Swan-minimal pole of order $e = e_0 q$, with e_0 prime to p and with $q = p^n$ for some $n \geq 1$. Then the canonical extension of L/K is the pullback, by a multiplicative translate on the base $\mathbb{G}_m$, of the finite etale covering

$$\pi_e : \mathbb{A}^1 - \{0, -1\} \to \mathbb{G}_m,$$
$$X \mapsto (X^{q-1}(1+X))^{e_0}.$$

For any nontrivial additive character ψ of $\mathbb{F}_q$, we have an isomorphism of lisse sheaves on $\mathbb{G}_m$,

$$\pi_{e*}\overline{\mathbb{Q}}_\ell \cong [e_0]_*(\mathbb{1}) \oplus [e_0(q-1)]_*\mathcal{L}_\psi.$$

proof We have seen in the proof of Theorem A6.1.5 that the pullback by some multiplicative translate on the base $\mathbb{G}_m$ of the finite etale covering π_e is an extension of L/K. It remains to see that π_e is a special covering, and that $\pi_{e*}\overline{\mathbb{Q}}_\ell$ is as asserted. It suffices to show this last statement, that

$$\pi_{e*}\overline{\mathbb{Q}}_\ell \cong [e_0]_*(\mathbb{1}) \oplus [e_0(q-1)]_*\mathcal{L}_\psi.$$

For if this holds, then the covering is indeed special. Indeed, after pullback by $[e_0(q-1)]^*$, i.e., after restriction to a normal subgroup of $\pi_1(\mathbb{G}_m)$ of prime-to-p index $e_0(q-1)$, we obtain a lisse sheaf whose geometric monodromy group is visibly a p-group, namely

$$[e_0(q-1)]^*(\pi_{e*}\overline{\mathbb{Q}}_\ell)$$
$$\cong [e_0(q-1)]^*([e_0]_*(\mathbb{1})) \oplus [e_0(q-1)]^*[e_0(q-1)]_*\mathcal{L}_\psi$$
$$\cong (e_0 \text{ copies of } \mathbb{1}) \oplus (\oplus_{\zeta \text{ in } \mu_{e_0(q-1)}(k)} \mathcal{L}_{\psi_\zeta}),$$

where we denote by ψ_ζ the additive character $x \mapsto \psi(\zeta x)$.

To show that

$$\pi_{e*}\overline{\mathbb{Q}}_\ell \cong [e_0]_*(\mathbb{1}) \oplus [e_0(q-1)]_*\mathcal{L}_\psi,$$

we rewrite it as

$$\pi_{e*}\overline{\mathbb{Q}}_\ell \cong [e_0]_*(\mathbb{1} \oplus [q-1]_*\mathcal{L}_\psi).$$

We then use the tautological fact that

$$\pi_e = [e_0] \circ \pi_q,$$

to reduce to showing that

$$\pi_{q*}\overline{\mathbb{Q}}_\ell \cong \mathbb{1} \oplus [q-1]_*\mathcal{L}_\psi.$$

For this, we argue as follows. We have a direct sum decomposition

$$\pi_{q*}\overline{\mathbb{Q}}_\ell \cong \overline{\mathbb{Q}}_\ell \oplus (\pi_{q*}\overline{\mathbb{Q}}_\ell)/\overline{\mathbb{Q}}_\ell.$$

On the other hand, we have seen in the proof of Theorem A6.1.5 that we have

$$\pi_{q*}\overline{\mathbb{Q}}_\ell \mid I(0) \cong \overline{\mathbb{Q}}_\ell \oplus [q-1]_*\overline{\mathbb{Q}}_\ell$$
$$\cong \overline{\mathbb{Q}}_\ell \oplus (\oplus \text{ all characters of order dividing } q-1),$$

and we have

$$\pi_{q*}\overline{\mathbb{Q}}_\ell \mid I(\infty) \cong \overline{\mathbb{Q}}_\ell \oplus(\text{rank } r = q-1, \text{ all breaks } 1/r).$$

Therefore $(\pi_{q*}\overline{\mathbb{Q}}_\ell)/\overline{\mathbb{Q}}_\ell$ is a lisse sheaf on $\mathbb{G}_m$ whose $I(0)$-representation is

$$\oplus \text{ all characters of order dividing } q-1,$$

and whose $I(\infty)$-representation is totally wild, of Swan conductor one. Since the covering π_q already exists over the prime field $\mathbb{F}_p$, it exists over $\mathbb{F}_q$, and hence $(\pi_{q*}\overline{\mathbb{Q}}_\ell)/\overline{\mathbb{Q}}_\ell$ exists as a lisse sheaf on $\mathbb{G}_m/\mathbb{F}_q$. We now apply [Ka-GKM, 8.7.1], according to which our lisse sheaf on $\mathbb{G}_m/\mathbb{F}_q$ is geometrically isomorphic to a multiplicative translate, by an $\mathbb{F}_q$-point of $\mathbb{G}_m$, of the Kloosterman sheaf

$$\mathrm{Kl}(\psi, \chi_1, \ldots, \chi_{q-1}),$$

with the χ_i all the q-1 multiplicative characters of $\mathbb{F}_q^\times$. By [Ka-GKM, 5.6.1 or 5.6.2], we have a geometric isomorphism

$$[q-1]_*\mathcal{L}_{\overline{\psi}} \cong \mathrm{Kl}(\psi, \chi_1, \ldots, \chi_{q-1}).$$

On the other hand, all the sheaves $\mathcal{L}_\psi$ for ψ a nontrivial additive character of $\mathbb{F}_q$, are transitively permuted by the translation action of $\mathbb{G}_m(\mathbb{F}_q) = \mu_{q-1}(k)$, which is the galois group of the Kummer covering $[q-1]$. Therefore the isomorphism class of the direct image $[q-1]_*\mathcal{L}_\psi$ is independent of the choice of nontrivial additive character ψ of $\mathbb{F}_q$. So we have

$$[q-1]_*\mathcal{L}_\psi \cong \mathrm{Kl}(\psi, \chi_1, \ldots, \chi_{q-1}).$$

So we now have a geometric isomorphism between the sheaf $(\pi_{q*}\overline{\mathbb{Q}}_\ell)/\overline{\mathbb{Q}}_\ell$ on $\mathbb{G}_m/\mathbb{F}_q$ and some $\mathbb{G}_m(\mathbb{F}_q)$-translate of $[q-1]_*\mathcal{L}_\psi$. Both these sheaves are geometrically irreducible (because they are $I(\infty)$-irreducible), and have finite arithmetic monodromy. So there exists a root of unity α in $\overline{\mathbb{Q}}_\ell^\times$, a choice of b in $\mathbb{F}_q^\times$, and an isomorphism of lisse sheaves on $\mathbb{G}_m/\mathbb{F}_q$,

$$[x \mapsto bx]^*((\pi_{q*}\overline{\mathbb{Q}}_\ell)/\overline{\mathbb{Q}}_\ell) \cong \alpha^{\deg}\otimes[q-1]_*\mathcal{L}_\psi.$$

We must show that $b = 1$ in $\mathbb{F}_q$, and that $\alpha = 1$ in $\overline{\mathbb{Q}}_\ell^\times$. We do this by comparing the trace functions of both sides of the isomorphism above at $\mathbb{F}_q$-valued points of $\mathbb{G}_m$. At a point t in $\mathbb{G}_m(\mathbb{F}_q)$, the trace of $\alpha^{\deg}\otimes[q-1]_*\mathcal{L}_\psi$ is given by

$$\mathrm{Trace}(\mathrm{Frob}_{\mathbb{F}_q,t} \mid \alpha^{\deg}\otimes[q-1]_*\mathcal{L}_\psi)$$
$$= \alpha\Sigma_{x \text{ in } \mathbb{F}_q^\times \text{ with } x^{q-1} = t}\ \psi(x).$$

But for x in $\mathbb{F}_q^\times$, $x^{q-1} = 1$. Therefore we have

$$\mathrm{Trace}(\mathrm{Frob}_{\mathbb{F}_q,t} \mid \alpha^{\deg}\otimes[q-1]_*\mathcal{L}_\psi) = 0, \text{ if } t \neq 1.$$

If $t = 1$, then we have

$$\mathrm{Trace}(\mathrm{Frob}_{\mathbb{F}_q,1} \mid \alpha^{\deg}\otimes[q-1]_*\mathcal{L}_\psi)$$
$$= \alpha\Sigma_{x \text{ in } \mathbb{F}_q^\times} \psi(x)$$
$$= -\alpha.$$

In other words, we have

$$\mathrm{Trace}(\mathrm{Frob}_{\mathbb{F}_q,t} \mid \alpha^{\deg}\otimes[q-1]_*\mathcal{L}_\psi) = -\alpha\delta_{1,t}.$$

Therefore the trace function of $[x \mapsto bx]^*(\pi_{q*}\overline{\mathbb{Q}}_\ell)/\overline{\mathbb{Q}}_\ell)$ at $\mathbb{F}_q$-valued points of $\mathbb{G}_m$ is supported at $t=1$, where it takes the value $-\alpha$. In other words, the trace function of $(\pi_{q*}\overline{\mathbb{Q}}_\ell)/\overline{\mathbb{Q}}_\ell$ at $\mathbb{F}_q$-valued points of $\mathbb{G}_m$ is supported at $t=b$, where it takes the value $-\alpha$. So we need only show that

$$\mathrm{Trace}(\mathrm{Frob}_{\mathbb{F}_q,1} \mid (\pi_{q*}\overline{\mathbb{Q}}_\ell)/\overline{\mathbb{Q}}_\ell) = -1.$$

Then by comparing supports we see that $b = 1$, and comparing values we see that $\alpha = 1$.

To show this, we simply compute. We have

$$\mathrm{Trace}(\mathrm{Frob}_{\mathbb{F}_q,1} \mid (\pi_{q*}\overline{\mathbb{Q}}_\ell)/\overline{\mathbb{Q}}_\ell)$$
$$= -1 + \mathrm{Trace}(\mathrm{Frob}_{\mathbb{F}_q,1} \mid \pi_{q*}\overline{\mathbb{Q}}_\ell)$$
$$= -1 + \#\{x \text{ in } \mathbb{F}_q - \{0, -1\} \text{ with } \pi_q(x) = 1\}$$
$$= -1 + \#\{x \text{ in } \mathbb{F}_q - \{0, -1\} \text{ with } x^{q-1}(x+1) = 1\}$$
$$= -1 + \#\{x \text{ in } \mathbb{F}_q - \{0, -1\} \text{ with } x + 1 = 1\}$$
$$= -1,$$

the next to last inequality because for x in $\mathbb{F}_q^\times$, $x^{q-1} = 1$. [This same calculation shows that for t in $\mathbb{F}_q^\times$, $t \neq 1$, we have

$$\mathrm{Trace}(\mathrm{Frob}_{\mathbb{F}_q,t} \mid (\pi_{q*}\overline{\mathbb{Q}}_\ell)/\overline{\mathbb{Q}}_\ell)$$
$$= -1 + \#\{x \text{ in } \mathbb{F}_q - \{0, -1\} \text{ with } x + 1 = t\} = 0,$$

giving an a priori proof that the trace function of $(\pi_{q*}\overline{\mathbb{Q}}_\ell)/\overline{\mathbb{Q}}_\ell$ at $\mathbb{F}_q$-valued points of $\mathbb{G}_m$ is supported at $t=1$, where it takes the value -1.] QED

(A6.3.3) For the purposes of the next section, it is convenient to restate the above theorem in slightly different form.

Theorem A6.3.4 (= 6.3.2 bis) Suppose f in $L^\times$ has a Swan-minimal pole of order $e = e_0q$, with e_0 prime to p and with $q = p^n$ for some $n \geq 1$. Then the canonical extension of L/K is the pullback, by a multiplicative translate on the base $\mathbb{G}_m$, of the finite etale covering

$$\pi_e : \mathbb{A}^1 - \{0, 1\} \to \mathbb{G}_m,$$
$$X \mapsto (X^{q-1}(1 - X))^{e_0}.$$

For any nontrivial additive character ψ of $\mathbb{F}_q$, we have an isomorphism of lisse sheaves on $\mathbb{G}_m$,

$$\pi_{e*}\overline{\mathbb{Q}}_\ell \cong [e_0]_*(\mathbb{1}) \oplus [e_0(q-1)]_*\mathcal{L}_\psi.$$

proof This is simply the previous theorem, after the automorphism $X \mapsto -X$ of the source $\mathbb{A}^1$. QED

(A6.4) Belyi polynomials, non-canonical extensions, and hypergeometric sheaves

(A6.4.1) We continue to work over an algebraically closed field k of positive characteristic p. The polynomial $(X^{q-1}(1 - X))^{e_0}$ which describes the canonical extension in Theorem A6.3.4 above is a special case of a Belyi polynomial $X^a(1 - X)^b$.

Lemma A6.4.2 Let k be an algebraically closed field of positive characteristic p. Let a and b be positive integers, both prime to p, whose sum is divisible by p, say

$$a + b = e = e_0 q$$

with e_0 prime to p, and with $q = p^n$ for some $n \geq 1$. Consider the Belyi polynomial

$$B_{a,b}(X) := X^a(1 - X)^b.$$

Then $B_{a,b}$ makes $\mathbb{A}^1 - \{0, 1\}$ a finite etale covering of $\mathbb{G}_m$ of degree e.

proof The map $B_{a,b}$ makes $\mathbb{A}^1$ a finite flat covering of itself of degree e. We readily compute

$$(d/dX)(B_{a,b}(X)) = (a -(a+b)X)X^{a-1}(1 - X)^{b-1}$$
$$= aX^{a-1}(1 - X)^{b-1}.$$

Thus the only possible critical points of $B_{a,b}$ in $\mathbb{A}^1$ are $X=0$ and $X=1$, both of which map to 0 under $B_{a,b}$. QED

Lemma A6.4.3 Hypotheses as in the lemma above, the finite etale covering

$$B_{a,b} : \mathbb{A}^1 - \{0, 1\} \to \mathbb{G}_m$$

is a canonical extension if and only if (a, b) is either $(e_0, e_0(q-1))$ or $(e_0(q-1), e_0)$).

proof Indeed, $B_{a,b}(X)$ has a Swan-minimal pole of order e at ∞ on the source, and so, by Theorem A6.3.4 above, its canonical extension from ∞ is (a multiplicative translate on the target of) the covering $B_{e_0(q-1),e_0}$, whose local monodromy at 0 is the direct sum of all characters of order dividing e_0, together with all characters of order dividing $e_0(q-1)$. We recover e_0 as the number of distinct characters which occur twice at 0, and we recover $e_0(q-1)$ as the largest order of a character on the list. But the covering $B_{a,b}$ has local monodromy at 0 the direct sum of all the characters of order dividing a, together with all the characters of order dividing b. On that list, there are precisely gcd(a, b) characters which occur twice, and the highest order of a character on the list is max(a, b). Thus if the $B_{a,b}$ covering is canonical, then gcd(a b) = e_0, and max(a, b) = $e_0(q-1)$. This in turn implies that (a, b) is either $(e_0, e_0(q-1))$ or $(e_0(q-1), e_0))$. Both of these are canonical, the second by Theorem A6.3.4, and the first by applying the automorphism $X \mapsto 1 - X$ of the source $\mathbb{A}^1$. QED

(A6.4.4) We now turn to the question of identifying the lisse sheaf $B_{a,b*}\overline{\mathbb{Q}}_\ell$ on $\mathbb{G}_m$ provided by a Belyi polynomial. The answer involves the hypergeometric sheaves of [Ka-ESDE, 8.4].

Theorem A6.4.5 Hypotheses and notations as in Lemma A6.4.2 above, suppose in addition that a and b are relatively prime. Fix a finite subfield k_0 of k which contains the roots of unity of order abe_0, and fix a nontrivial additive character ψ of k_0. Denote

$$\mathrm{NTChar}(a) := \{\text{nontrivial characters } \chi \text{ of } k_0{}^\times \text{ with } \chi^a = \mathbb{1}\},$$

and similarly for NTChar(b) and NTChar(e_0). Then there exists a unique λ in $k^\times$ such that there exists a geometric isomorphism of lisse sheaves on $\mathbb{G}_m$

$$B_{a,b*}\overline{\mathbb{Q}}_\ell \cong \overline{\mathbb{Q}}_\ell \oplus \mathcal{H}_\lambda(!, \psi; \mathbb{1}, \mathrm{NTChar}(a), \mathrm{NTChar}(b); \mathrm{NTChar}(e_0)).$$

proof The sheaf $B_{a,b*}\overline{\mathbb{Q}}_\ell$ is lisse on $\mathbb{G}_m$, and visibly has finite monodromy, so is completely reducible. It is tamely ramified at 0. Its I(0)-representation is the direct sum

$$\mathbb{1} \oplus (\oplus_{\chi \text{ in NTChar}(a)} \mathcal{L}_\chi) \oplus \mathbb{1} \oplus (\oplus_{\rho \text{ in NTChar}(b)} \mathcal{L}_\rho).$$

The function $B_{a,b}$ has a Swan-minimal pole of order e at ∞, so by Theorem A6.1.5, the I(∞)-representation attached to $B_{a,b*}\overline{\mathbb{Q}}_\ell$ is

$$(\text{rank } r = e_0(q-1), \text{ all breaks } 1/r)$$
$$\oplus\ \mathbb{1} \oplus \oplus_{\Lambda \text{ in NTChar}(e_0)} \mathcal{L}_\Lambda .$$

In particular, $B_{a,b*}\overline{\mathbb{Q}}_\ell$ is tame at 0, and its Swan_∞ is one.

Now use the fact that $B_{a,b*}\overline{\mathbb{Q}}_\ell$ is completely reducible as a lisse sheaf on $\mathbb{G}_m$ to write it as a direct sum

$$\oplus_i \mathcal{G}_i$$

of geometrically irreducible lisse sheaves on $\mathbb{G}_m$. From the equations

$$0 = \mathrm{Swan}_0(B_{a,b*}\overline{\mathbb{Q}}_\ell) = \Sigma_i \mathrm{Swan}_0(\mathcal{G}_i),$$
$$1 = \mathrm{Swan}_\infty(B_{a,b*}\overline{\mathbb{Q}}_\ell) = \Sigma_i \mathrm{Swan}_\infty(\mathcal{G}_i),$$

we see that all the $\mathcal{G}_i$ are tame at 0, and that all but one is tame at ∞, with the exceptional one, say $\mathcal{G}_1$, having $\mathrm{Swan}_\infty(\mathcal{G}_1) = 1$. Now any $\mathcal{G}_i$ with $i \geq 2$ is a geometrically irreducible lisse sheaf on $\mathbb{G}_m$ which is everywhere tame and with finite monodromy, so is a Kummer sheaf $\mathcal{L}_\sigma$ for some multiplicative character σ of some finite subfield of k. But if $\mathcal{L}_\sigma$ is direct summand of $B_{a,b*}\overline{\mathbb{Q}}_\ell$, then $\mathcal{L}_\sigma$ is a direct summand of both the I(0) and the I(∞)-representations attached to $B_{a,b*}\overline{\mathbb{Q}}_\ell$. Since we have assumed that a and b are relatively prime, the three numbers (a, b, a+b = e_0q) are pairwise relatively prime, and hence the only possible σ is $\mathbb{1}$, and it can only occur once.

Therefore we have

$$B_{a,b*}\overline{\mathbb{Q}}_\ell \cong \mathcal{G}_1 \oplus \overline{\mathbb{Q}}_\ell.$$

In this decomposition, $\mathcal{G}_1$ is a geometrically irreducible lisse sheaf on $\mathbb{G}_m$ whose I(0)-representation is

$$\mathbb{1} \oplus (\oplus_{\chi \text{ in NTChar}(a)} \mathcal{L}_\chi) \oplus (\oplus_{\rho \text{ in NTChar}(b)} \mathcal{L}_\rho)$$

and whose I(∞)-representation is

$$(\text{rank } r = e_0(q-1), \text{ all breaks } 1/r) \oplus (\oplus_{\Lambda \text{ in NTChar}(e_0)} \mathcal{L}_\Lambda).$$

According to [Ka-ESDE, 8.5.3.1] applied to $\mathcal{G}_1$, there is a unique λ in $k^\times$ and a geometric isomorphism

$$\mathcal{G}_1 \cong \mathcal{H}_\lambda(!, \psi; \mathbb{1}, \text{NTChar}(a), \text{NTChar}(b); \text{NTChar}(e_0)).$$ QED

Corollary A6.4.6 Hypotheses and notations as in the theorem above, suppose in addition that $e_0 = 1$, i.e., that a+b = q. Then $B_{a,b*}\overline{\mathbb{Q}}_\ell/\overline{\mathbb{Q}}_\ell$ is geometrically isomorphic to a unique multiplicative translation of the Kloosterman sheaf Kl(ψ, $\mathbb{1}$, NTChar(a), NTChar(b)).

proof This is the special case $e_0 = 1$ of the theorem: a hypergeometric sheaf with no tame characters at ∞ is a Kloosterman sheaf. QED

(A6.4.7) To analyze a Belyi polynomial $B_{A,B}$ with A and B prime to p but not necessarily relatively prime, and with A+B divisible by p, we apply the following corollary, with d := gcd(A, B), a := A/d, b := B/d.

Corollary A6.4.8 Hypotheses and notations as in the theorem above, let d be a positive integer which is prime to p. Consider the Belyi polynomial $B_{da,db}(X) = (B_{a,b}(X))^d$. Then there exists an isomorphism of lisse sheaves on $\mathbb{G}_m$

$$B_{da,db*}\overline{\mathbb{Q}}_\ell \cong [d]_*(\overline{\mathbb{Q}}_\ell) \oplus [d]_*(B_{a,b*}\overline{\mathbb{Q}}_\ell/\overline{\mathbb{Q}}_\ell).$$

proof Since $B_{da,db} = [d]\circ B_{a,b}$, this is immediate from the theorem. QED

Remark A6.4.9 For any hypergeometric sheaf $\mathcal{H}$(!, ψ; χ's; ρ's) and for any prime-to-p integer d, $[d]_*\mathcal{H}$ is, geometrically, another hypergeometric sheaf, whose characters at 0 are all the d'th roots of the χ's, and whose characters at ∞ are all the d'th roots of the ρ's, cf. [Ka-ESDE, 8.9.1]. Since $B_{a,b*}\overline{\mathbb{Q}}_\ell/\overline{\mathbb{Q}}_\ell$ is itself an explicit hypergeometric sheaf, the second term $[d]_*(B_{a,b*}\overline{\mathbb{Q}}_\ell/\overline{\mathbb{Q}}_\ell)$ is an equally explicit hypergeometric sheaf.

Chapter 7: Pullbacks to curves from $\mathbb{A}^1$

(7.1) The general pullback setting

(7.1.1) In this chapter, we will consider the following general "pullback to a curve" situation. We work over a finite field k. We fix a prime number ℓ invertible in k, and a field embedding $\overline{\mathbb{Q}}_\ell \subset \mathbb{C}$. On $\mathbb{A}^1/k$, we are given a geometrically irreducible, geometrically nonconstant perverse sheaf K, which is ι-pure of weight 0. We assume that K is non-punctual, and that, geometrically, K has $\mathcal{P}$. Thus K is of the form $\mathcal{G}[1]$, for $\mathcal{G}$ a geometrically irreducible middle extension sheaf on $\mathbb{A}^1$ which is not geometrically of the form $\mathcal{L}_{\psi(\alpha x)}$ for any α in k. We denote by $S \subset \mathbb{A}^1$ the set of finite singularities of $\mathcal{G}$, and we denote by $j : \mathbb{A}^1 - S \to \mathbb{A}^1$ the inclusion. Then $\mathcal{G} \cong j_*j^*\mathcal{G}$, and $j^*\mathcal{G}$ is lisse, geometrically irreducible ι-pure of weight -1 on $\mathbb{A}^1 - S$. Because $\mathcal{G}$ is geometrically nonconstant, we have $H_c^i(\mathbb{A}^1\otimes_k\overline{k}, \mathcal{G}) = 0$ for $i \neq 1$, i.e., we have

$$H_c^i(\mathbb{A}^1\otimes_k\overline{k}, K) = 0 \text{ for } i \neq 0.$$

(7.1.2) We now consider a proper, smooth, geometrically connected curve C/k, of genus denoted g, together with an effective divisor D on C with

$$\deg(D) \geq 2g + 3.$$

We view the Riemann Roch space L(D) as a space of $\mathbb{A}^1$-valued functions on the open curve C-D, with

$$\tau : L(D) \to \mathrm{Hom}_{k\text{-schemes}}(C\text{-}D, \mathbb{A}^1)$$

the natural evaluation map. As we have noted in 1.1.9, this space of functions is 4-separating. Indeed, it is d-separating, for $d := \deg(D) - (2g-1)$.

(7.1.3) We take as "standard input" (cf. 1.15.4) the following data:

the integer $m = 1$,

the perverse sheaf K on $\mathbb{A}^1/k$,

the affine k-scheme $V := C\text{-}D$,

the k-morphism $h : V \to \mathbb{A}^1$ given by $h = 0$,

the perverse sheaf $L := \overline{\mathbb{Q}}_\ell(1/2)[1]$ on C-D,

the integer $d := \deg(D) - (2g-1)$,

the space of functions $(L(D), \tau)$ on C-D.

[This is indeed standard input: the condition that

$$H^*_c((V\times\mathbb{A}^m)\otimes\overline{k}, \mathrm{pr}_1{}^*L\otimes\mathrm{pr}_2{}^*K)$$

be concentrated in degree $\leq m$ is trivially satisfied.]

(7.1.4) We then form the perverse sheaf

$$M = \mathrm{Twist}(L = \overline{\mathbb{Q}}_\ell(1/2)[1], K, \mathcal{F} = L(D), h = 0)$$

on the space L(D). Let us make this perverse sheaf explicit.

(7.1.5) For each finite extension field E/k inside $\overline{k}$, and for each f in $L(D)\otimes_k E$, we form the pullback sheaf $f^*\mathcal{G}$ on $(C-D)_E$, and compute its compact cohomology

$$H_c^*((C - D)_E\otimes_E\overline{k}, f^*\mathcal{G}).$$

By definition, we have

$$\mathcal{H}^{i-\ell(D)}(M)_{(E,\ f)} := H_c^{i+1}((C - D)_E\otimes_E\overline{k}, f^*\mathcal{G})(1/2).$$

Since the sheaf $f^*\mathcal{G}$ has no nonzero punctual sections we have

$$H_c^0((C - D)_E\otimes_E\overline{k}, f^*\mathcal{G}) = 0, \text{ for every f.}$$

Since C - D is a curve, we have

$$H_c^{i\geq 3}((C - D)_E\otimes_E\overline{k}, f^*\mathcal{G}) = 0, \text{ for every f.}$$

Thus the only possibly nonvanishing cohomology sheaves of M are $\mathcal{H}^{-\ell(D)}(M)$ and $\mathcal{H}^{1-\ell(D)}(M)$. Since M is perverse on L(D), $\mathcal{H}^{1-\ell(D)}(M)$ is generically zero. So on any dense open set $U \subset L(D)$ on which M is lisse (:= has lisse cohomology sheaves), we have $\mathcal{H}^{1-\ell(D)}(M)|U = 0$.

(7.1.6) In favorable cases, we can make explicit a dense open set on which M is lisse. Recall that S is the set of finite singularities of $\mathcal{G}$. Denote by

$$U_{D,S} \subset L(D)$$

the dense open set whose $\overline{k}$-valued points are those f in $L(D)(\overline{k})$ with the following two properties:

the divisor of poles of f is D,
f is finite etale over over S.

Lemma 7.1.7 Suppose that either $\mathcal{G}$ is tamely ramified at ∞, or that the divisor D is "prime to p", in the sense that when we write D over $\overline{k}$ as $\Sigma_i a_iP_i$, each a_i is prime to p. Then the perverse sheaf M is lisse on $U_{D,S}$.

proof Over the space $U := U_{D,S}$, we have the constant curve C_U, with coordinates (x, f), which contains the constant relative divisor $D^{red}{}_U$, (those points (x, f) where x is some P_i) and also the relative divisor $f^{-1}(S)$ (those points (x, f) where f(x) lies in S). These divisors are disjoint, and each is finite etale over $U_{D,S}$, the first of degree $\deg(D^{red})$, the second of degree $\deg(D)\#S(\overline{k})$. On the open relative curve $(C - D^{red})_U$, we have the sheaf $f^*\mathcal{G}$. Denote by

$$\alpha : (C - D^{red})_U - f^{-1}(S) \to (C - D^{red})_U$$

and

$$\beta : f^{-1}(S) \to (C - D^{red})_U$$

the inclusions. Denote the various structural morphisms by

$$\pi : (C - D^{red})_U \to U,$$
$$\pi_0 : (C - D^{red})_U - f^{-1}(S) \to U,$$

and

$$\rho : f^{-1}(S) \to U.$$

It is tautological that $R\pi_!(f^*\mathcal{G})$ is essentially M|U, up to a shift and a Tate twist. For each i, we have

$$\mathcal{H}^{i-\ell(D)}(M)|U = R^i\pi_!(f^*\mathcal{G})(1/2).$$

We wish to show that all the $R^i\pi_!(f^*\mathcal{G})$ are lisse on U. Consider the excision sequence for α and β:

$$\ldots \to R^i\pi_{0!}(\alpha^*f^*\mathcal{G}) \to R^i\pi_!(f^*\mathcal{G}) \to R^i\rho_!(\beta^*f^*\mathcal{G}) \to \ldots .$$

The sheaf $\beta^*f^*\mathcal{G}$ is lisse on $f^{-1}(S)$, and ρ is finite etale, so the terms $R^i\rho_!(\beta^*f^*\mathcal{G})$ are all lisse on U (and vanish for $i \neq 0$). So it suffices to show that all the $R^i\pi_{0!}(\alpha^*f^*\mathcal{G})$ are lisse on U. The sheaf $\alpha^*f^*\mathcal{G}$ is lisse on the open relative curve $(C - D^{red})_U - f^{-1}(S)$, and this open curve is the complement in a proper smooth curve C_U of a divisor which is finite etale over U. By Deligne's semicontinuity theorem [Lau-SC, 2.1.1 and 2.1.2], it suffices to show that for variable f in $U(\bar{k})$, the sum of the Swan conductors of $f^*\mathcal{G}$ at all the $\bar{k}$-valued points of $D \amalg f^{-1}(S)$ is constant. To see this, we argue as follows.

Since f is finite etale of degree deg(D) over S, we have

$$\Sigma_{x \text{ in } f^{-1}(S)(\bar{k})} \mathrm{Swan}_x(f^*\mathcal{G}) = \deg(D)\Sigma_{s \text{ in } S(\bar{k})} \mathrm{Swan}_s(\mathcal{G}),$$

which is independent of f in U. It is to insure that

$$\Sigma_{P_i \text{ in } D(\bar{k})} \mathrm{Swan}_{P_i}(f^*\mathcal{G})$$

is independent of f in U that we made the hypothesis that either $\mathcal{G}$ is tame at ∞, or D is prime to p. Suppose first that $\mathcal{G}$ is tame at ∞. Then $f^*\mathcal{G}$ is tame at each point of D, and this sum vanishes. If D is prime to p, then at P_i, $f^*\mathcal{G}$ as representation of the inertia group $I(P_i)$ is the pullback of $\mathcal{G}$ as $I(\infty)$-representation by a degree a_i Kummer covering, and $\mathrm{Swan}_{P_i}(f^*\mathcal{G}) = a_i\mathrm{Swan}_\infty(\mathcal{G})$. So we get

$$\Sigma_{P_i \text{ in } D(\bar{k})} \mathrm{Swan}_{P_i}(f^*\mathcal{G}) = \deg(D)\mathrm{Swan}_\infty(\mathcal{G}),$$

a formula which is also valid in the case when $\mathcal{G}$ is tame at ∞, which shows that this sum is also independent of f in U. QED

(7.1.8) In the general case, we do not know a simple, explicit description of a dense open set U on which M is lisse. What we can describe explicitly are dense open sets U for which $\mathcal{H}^{1-\ell(D)}(M)|U = 0$, i.e., explicit dense open U such that for f in $U(\bar{k})$, we have

$$H_c^2((C - D)\otimes_k\bar{k}, f^*\mathcal{G}) = 0.$$

(7.1.9) Write D over $\overline{k}$ as $\Sigma_i a_i P_i$. Define integers b_i by

$b_i := a_i$, if p divides a_i,

$b_i := a_i + 1$, if p does not divide a_i.

Recall (from Theorem 6.2.2, Lemma 6.2.7, and Corollary 6.2.15) that there is a dense open set $U_1 \subset L(D)$ such that for f in $U_1(\overline{k})$, f has divisor of poles D, df has divisor of poles $\Sigma b_i P_i$, f is Lefschetz on C - D, $f_\star \overline{\mathbb{Q}}_\ell / \overline{\mathbb{Q}}_\ell$ is an irreducible middle extension sheaf on $\mathbb{A}^1$ with G_{geom} the full symmetric group $S_{\deg(D)}$ in its deleted permutation representation, $f_\star \overline{\mathbb{Q}}_\ell / \overline{\mathbb{Q}}_\ell$ has at least one finite singularity, and its local monodromy at each finite singularity is a reflection.

(7.1.10) Denote by $U_2 \subset L(D)$ the dense open set consisting of those f in U_1 which are finite etale over the set S of finite singularities of $\mathcal{G}$, i.e., U_2 is $U_1 \cap U_{D,S}$

Lemma 7.1.11 Suppose that any of the following four conditions holds:

a) $\mathrm{rank}(\mathcal{G}) \neq \deg(D) - 1$,

b) $\mathcal{G}$ is not, geometrically, orthogonally self dual,

c) $\mathcal{G}$ is lisse on $\mathbb{A}^1$,

d) At some finite singularity s of $\mathcal{G}$, local monodromy is not a reflection (i.e., there exists an element in the inertia group I(s) which acts nontrivially, but not as a reflection).

Then for any f in $U_1(\overline{k})$, we have

$$H_c^2((C - D)\otimes_k \overline{k}, f^\star \mathcal{G}) = 0.$$

proof 1) This results from the projection formula, and the interpretation of H_c^2 on a curve in terms of coinvariants under π_1. For f in $L(D)(\overline{k})$ nonconstant, the Leray spectral sequence for f gives

$$\begin{aligned} H_c^2((C - D)\otimes_k \overline{k}, f^\star \mathcal{G}) &= H_c^2(\mathbb{A}^1\otimes_k \overline{k}, f_\star f^\star \mathcal{G}) \\ &= H_c^2(\mathbb{A}^1\otimes_k \overline{k}, \mathcal{G}\otimes f_\star \overline{\mathbb{Q}}_\ell) \\ &= H_c^2(\mathbb{A}^1\otimes_k \overline{k}, \mathcal{G}\otimes (f_\star \overline{\mathbb{Q}}_\ell/\overline{\mathbb{Q}}_\ell)) \oplus H_c^2(\mathbb{A}^1\otimes_k \overline{k}, \mathcal{G}). \end{aligned}$$

The second summand vanishes because $\mathcal{G}$ is a geometrically irreducible middle extension which is geometrically nonconstant. For f in $U_1(\overline{k})$, $(f_\star \overline{\mathbb{Q}}_\ell/\overline{\mathbb{Q}}_\ell)$ is geometrically irreducible and self dual, so, $\mathcal{G}$ being itself geometrically irreducible, the first summand vanishes unless there exists a geometric isomorphism

$$\mathcal{G} \cong (f_\star \overline{\mathbb{Q}}_\ell/\overline{\mathbb{Q}}_\ell)$$

on $(C - D)_{\overline{k}}$. There can be no such isomorphism if the two sides have different ranks, or different duality types, or different sets of finite singularities, or different local monodromies at some common singularity. QED

Lemma 7.1.12

1)For any f in $U_2(\overline{k})$, we have $H_c^2((C - D)\otimes_k \overline{k}, f^\star \mathcal{G}) = 0$.

2) If $\mathcal{G}$ is geometrically Lie-irreducible on $\mathbb{A}^1 - S$ and $\mathrm{rank}(\mathcal{G}) \geq 2$, then for any nonconstant f in $L(D)(\overline{k})$, we have $H_c^2((C - D)\otimes_k \overline{k}, f^*\mathcal{G}) = 0$.

proof 1) Either S is empty, and we apply case c) of the previous Lemma 7.1.11, or S is nonempty, in which case $(f_*\overline{\mathbb{Q}}_\ell/\overline{\mathbb{Q}}_\ell)$ but not $\mathcal{G}$ is lisse at points of S.

2) The pullback $f^*\mathcal{G}$, restricted to $C - D - f^{-1}(S)$, is geometrically irreducible of rank ≥ 2. By the birational invariance of H_c^2, we have the asserted vanishing. QED

(7.1.13) Now we have, in the general case, a dense open set U_2 with $\mathcal{H}^{1-\ell(D)}(M)|U = 0$. Notice that U_2 lies in the dense open set $U_{D,S}$ which "works" under the special hypotheses of Lemma 7.1.7 above. On the other hand, $\mathcal{H}^{-\ell(D)}(M)$ is a sheaf of perverse origin on L(D), so it is lisse precisely on the dense open set U_{max} consisting of those points where it has maximal rank [Ka-SMD, Proposition 12]. So we can assert that M is lisse on the dense open set $U_2 \cap U_{max}$, but the fact is that in the general case we do not know an explicit description of U_{max}, nor even an explicit description of a dense open set which lies in U_{max}.

(7.1.14) Let $U \subset U_{D,S}$ be a dense open set on which M is lisse. Then $M|U$ is $\mathcal{M}(1/2)[\ell(D)]$, for $\mathcal{M}$ the lisse sheaf $R^1\pi_!(f^*\mathcal{G})|U$. The sheaf $\mathcal{M}$ is ι-mixed of weight ≤ 0. Our next task is to understand concretely the sheaf $\mathrm{Gr}^0(\mathcal{M})$ and its direct sum decomposition

$$\mathrm{Gr}^0(\mathcal{M}) = \mathrm{Gr}^0(\mathcal{M})_{ncst} \oplus \mathrm{Gr}^0(\mathcal{M})_{cst},$$

and to estimate from below the rank of $\mathrm{Gr}^0(\mathcal{M})_{ncst}$.

(7.1.15) To begin with, let us factor the morphism

$$\pi : (C - D^{red})_U \to U,$$

through the tautological map

$$f : (C - D^{red})_U \to \mathbb{A}^1{}_U,$$
$$(x, f) \mapsto (f(x), f).$$

Denote by

$$pr_2 : \mathbb{A}^1{}_U \to U$$

the structural map. Using the Leray spectral sequence for the finite map f, we have, for every i,

$$R^i\pi_!(f^*\mathcal{G})|U = R^ipr_{2!}(f_*f^*\mathcal{G})|U = R^ipr_{2!}(\mathcal{G}\otimes f_*\overline{\mathbb{Q}}_\ell)|U$$
$$= R^ipr_{2!}(\mathcal{G}\otimes(f_*\overline{\mathbb{Q}}_\ell/\overline{\mathbb{Q}}_\ell))|U \oplus R^ipr_{2!}(\mathcal{G})|U$$
$$= R^ipr_{2!}(\mathcal{G}\otimes(f_*\overline{\mathbb{Q}}_\ell/\overline{\mathbb{Q}}_\ell))|U$$
$$\oplus \text{ (the constant sheaf } H_c^i(\mathbb{A}^1\otimes_k\overline{k}, \mathcal{G}) \text{ on } U).$$

Since $R^1\pi_!(f^*\mathcal{G})|U$ is lisse and the other $R^i\pi_!(f^*\mathcal{G})|U$ vanish, we see that $R^1pr_{2!}(\mathcal{G}\otimes(f_*\overline{\mathbb{Q}}_\ell/\overline{\mathbb{Q}}_\ell))|U$ is lisse, and the other $R^ipr_{2!}(\mathcal{G}\otimes(f_*\overline{\mathbb{Q}}_\ell/\overline{\mathbb{Q}}_\ell))|U$ vanish. Similarly $H_c{}^i(\mathbb{A}^1\otimes_k\overline{k}, \mathcal{G})$ vanishes for $i \neq 1$.

(7.1.16) Let us define a lisse sheaf $\mathcal{M}_{ncst}$ on U by

$$\mathcal{M}_{ncst} := R^1pr_{2!}(\mathcal{G}\otimes(f_*\overline{\mathbb{Q}}_\ell/\overline{\mathbb{Q}}_\ell))|U.$$

Thus we have a direct sum decomposition of $\mathcal{M}$,

$$\mathcal{M} = \mathcal{M}_{ncst} \oplus (\text{the constant sheaf } H_c{}^1(\mathbb{A}^1\otimes_k\overline{k}, \mathcal{G})).$$

Passing to Gr^0, we find a direct sum decomposition

$$Gr^0(\mathcal{M}) = Gr^0(\mathcal{M}_{ncst}) \oplus (\text{the constant sheaf } H_c{}^1(\mathbb{A}^1\otimes_k\overline{k}, \mathcal{G})_{wt=0}).$$

Lemma 7.1.17 We have an isomorphism $Gr^0(\mathcal{M})_{ncst} \cong Gr^0(\mathcal{M}_{ncst})$ of lisse sheaves on U.

proof We also have the decomposition

$$Gr^0(\mathcal{M}) = Gr^0(\mathcal{M})_{ncst} \oplus Gr^0(\mathcal{M})_{cst}.$$

In this decomposition, we know from 2.1.1.5 that $Gr^0(\mathcal{M})_{cst}$ is the constant sheaf $H_c{}^1(\mathbb{A}^1\otimes_k\overline{k}, \mathcal{G})_{wt=0}$. Comparing the two decompositions, we find that $Gr^0(\mathcal{M}_{ncst})$ and $Gr^0(\mathcal{M})_{ncst}$ have the same trace function. By Chebotarev, it follows that $Gr^0(\mathcal{M}_{ncst})$ and $Gr^0(\mathcal{M})_{ncst}$ have isomorphic $\pi_1(U)$-semisimplifications. But from the general theory, we know that $Gr^0(\mathcal{M})_{ncst}$ is $\pi_1{}^{geom}(U)$-irreducible, and hence $\pi_1(U)$-irreducible. Therefore $Gr^0(\mathcal{M}_{ncst})$ and $Gr^0(\mathcal{M})_{ncst}$ are isomorphic as $\pi_1(U)$-representations. QED

Lemma 7.1.18 Denoting by g the genus of C, we have the inequality

$$\mathrm{rank}(Gr^0(\mathcal{M})_{ncst}) \geq (2g - 1)\mathrm{rank}(\mathcal{G}) + (\deg(D) - 1)\#S(\overline{k}) - \mathrm{Swan}_\infty(\mathcal{G}).$$

If either $\mathcal{G}$ is tame at ∞ or D is prime to p, then we have the stronger inequality

$$\mathrm{rank}(Gr^0(\mathcal{M})_{ncst}) \geq (2g - 1)\mathrm{rank}(\mathcal{G}) + (\deg(D) - 1)\#S(\overline{k}) + (\deg(D) - 1)\mathrm{Swan}_\infty(\mathcal{G}).$$

proof Take an f in $U_{D,S}(\overline{k})$ at which M is lisse and at which the analogous object Twist(L = $\overline{\mathbb{Q}}_\ell(1/2)[1]$, DK, $\mathcal{F}$ = L(D), h = 0) made with the dual DK of K is also lisse. The stalk at f of $Gr^0(\mathcal{M})_{ncst}$ is the quotient vector space

$$H_c{}^1((C - D)\otimes_k\overline{k}, f^*\mathcal{G})_{wt=0}/H_c{}^1(\mathbb{A}^1\otimes_k\overline{k}, \mathcal{G})_{wt=0}.$$

Because $\mathcal{G}$ is a middle extension on $\mathbb{A}^1$ which is ι-pure of weight -1, and f is finite etale over the finite singularities S of $\mathcal{G}$, the sheaf $f^*\mathcal{G}$ on C - D is again a middle extension, still pure of weight -1. So we can rewrite the pure parts in terms of "parabolic cohomology" as follows. Denote by

$$j_D : C - D \to C$$

and

$$j_\infty : \mathbb{A}^1 \to \mathbb{P}^1$$

the inclusions. Then we have [De-Weil II, 1.8.1 and 3.2.3]

$$H_c^1((C - D)\otimes_k \bar{k}, f^*\mathcal{G})_{wt=0} = H^1(C\otimes_k \bar{k}, j_{D*}f^*\mathcal{G}),$$

$$H_c^1(\mathbb{A}^1\otimes_k \bar{k}, \mathcal{G})_{wt=0} = H^1(\mathbb{P}^1\otimes_k \bar{k}, j_{\infty *}\mathcal{G}).$$

By the birational invariance of H_c^2, we have

$$0 = H_c^2((C - D)\otimes_k \bar{k}, f^*\mathcal{G}) = H^2(C\otimes_k \bar{k}, j_{D*}f^*\mathcal{G}),$$

$$0 = H_c^2(\mathbb{A}^1\otimes_k \bar{k}, \mathcal{G}) = H^2(\mathbb{P}^1\otimes_k \bar{k}, j_{\infty *}\mathcal{G}).$$

Applying this last vanishing to the dual middle extension, we find by Poincaré duality that

$$0 = H^0((C - D)\otimes_k \bar{k}, f^*\mathcal{G}) = H^0(C\otimes_k \bar{k}, j_{D*}f^*\mathcal{G}),$$

$$0 = H^0(\mathbb{A}^1\otimes_k \bar{k}, \mathcal{G}) = H^0(\mathbb{P}^1\otimes_k \bar{k}, j_{\infty *}\mathcal{G}).$$

So we have the dimension formulas

$$\dim(H_c^1((C - D)\otimes_k \bar{k}, f^*\mathcal{G})_{wt=0}) = -\chi(C\otimes_k \bar{k}, j_{D*}f^*\mathcal{G}),$$

$$\dim(H_c^1(\mathbb{A}^1\otimes_k \bar{k}, \mathcal{G})_{wt=0}) = -\chi(\mathbb{P}^1\otimes_k \bar{k}, j_{\infty *}\mathcal{G}).$$

Thus we find the rank formula

$$\mathrm{rank}(\mathrm{Gr}^0(\mathcal{M})_{ncst}) = -\chi(C\otimes_k \bar{k}, j_{D*}f^*\mathcal{G}) + \chi(\mathbb{P}^1\otimes_k \bar{k}, j_{\infty *}\mathcal{G}).$$

To make use of this, let us recall a convenient form of the Euler Poincaré formula. For a constructible sheaf $\mathcal{H}$ on $C\otimes_k \bar{k}$, of generic rank denoted $\mathrm{rank}(\mathcal{H})$, and lisse outside the finite set $\mathrm{Sing}(\mathcal{H})$, we have

$$\chi(C\otimes_k \bar{k}, \mathcal{H}) = \chi(C\otimes_k \bar{k})\mathrm{rank}(\mathcal{H}) - \Sigma_{x \text{ in } \mathrm{Sing}(\mathcal{H})} \mathrm{TotalDrop}_x(\mathcal{H}),$$

where

$$\mathrm{TotalDrop}_x(\mathcal{H}) := \mathrm{Swan}_x(\mathcal{H}) + \mathrm{rank}(\mathcal{H}) - \dim(\mathcal{H}_x).$$

We apply this both to $j_{D*}f^*\mathcal{G}$ on $C\otimes_k \bar{k}$, and to $j_{\infty *}\mathcal{G}$ on $(\mathbb{P}^1\otimes_k \bar{k})$. Because f is finite etale of degree deg(D) over S, we have

$$-\chi(C\otimes_k \bar{k}, j_{D*}f^*\mathcal{G})$$
$$= (2g - 2)\mathrm{rank}(\mathcal{G}) + \Sigma_{x \text{ in } f^{-1}(S)} \mathrm{TotalDrop}_x(j_{D*}f^*\mathcal{G})$$
$$+ \Sigma_{P \text{ in } D(\bar{k})} \mathrm{TotalDrop}_P(j_{D*}f^*\mathcal{G})$$
$$= (2g - 2)\mathrm{rank}(\mathcal{G}) + \deg(D)\Sigma_{s \text{ in } S(\bar{k})} \mathrm{TotalDrop}_s(\mathcal{G})$$
$$+ \Sigma_{P \text{ in } D(\bar{k})} \mathrm{TotalDrop}_P(j_{D*}f^*\mathcal{G})$$

$$\geq (2g - 2)\mathrm{rank}(\mathcal{G}) + \deg(D)\Sigma_{s \text{ in } S(\bar{k})} \mathrm{TotalDrop}_s(\mathcal{G}) + \Sigma_{P \text{ in } D(\bar{k})} \mathrm{Swan}_P(f^*\mathcal{G}).$$

We also have

$$\chi(\mathbb{P}^1\otimes_k\bar{k}, j_{\infty *}\mathcal{G}) = 2\mathrm{rank}(\mathcal{G}) - \Sigma_{s \text{ in } S(\bar{k})} \mathrm{TotalDrop}_s(\mathcal{G}) - \mathrm{TotalDrop}_\infty(j_{\infty *}\mathcal{G})$$
$$\geq 2\mathrm{rank}(\mathcal{G}) - \Sigma_{s \text{ in } S(\bar{k})} \mathrm{TotalDrop}_s(\mathcal{G}) - \mathrm{rank}(\mathcal{G}) - \mathrm{Swan}_\infty(\mathcal{G})$$
$$= \mathrm{rank}(\mathcal{G}) - \Sigma_{s \text{ in } S(\bar{k})} \mathrm{TotalDrop}_s(\mathcal{G}) - \mathrm{Swan}_\infty(\mathcal{G}).$$

Adding these up, we get

$$-\chi(C\otimes_k\bar{k}, j_{D*}f^*\mathcal{G}) + \chi(\mathbb{P}^1\otimes_k\bar{k}, j_{\infty *}\mathcal{G})$$
$$\geq (2g - 1)\mathrm{rank}(\mathcal{G}) + (\deg(D) - 1)\Sigma_{s \text{ in } S(\bar{k})} \mathrm{TotalDrop}_s(\mathcal{G}) + \Sigma_{P \text{ in } D(\bar{k})} \mathrm{Swan}_P(f^*\mathcal{G}) - \mathrm{Swan}_\infty(\mathcal{G}).$$

For each s in $S(\bar{k})$, $\mathcal{G}$ is a middle extension which is not lisse at s, so we have the trivial inequality

$$\mathrm{TotalDrop}_s(\mathcal{G}) \geq 1.$$

So we have

$$-\chi(C\otimes_k\bar{k}, j_{D*}f^*\mathcal{G}) + \chi(\mathbb{P}^1\otimes_k\bar{k}, j_{\infty *}\mathcal{G})$$
$$\geq (2g - 1)\mathrm{rank}(\mathcal{G}) + (\deg(D) - 1)\#S(\bar{k}) + \Sigma_{P \text{ in } D(\bar{k})} \mathrm{Swan}_P(f^*\mathcal{G}) - \mathrm{Swan}_\infty(\mathcal{G}).$$

To get the first asserted inequality, simply throw away the terms

$$\Sigma_{P \text{ in } D(\bar{k})} \mathrm{Swan}_P(f^*\mathcal{G}).$$

If either $\mathcal{G}$ is tame at ∞ or D is prime to p, we have

$$\Sigma_{P \text{ in } D(\bar{k})} \mathrm{Swan}_P(f^*\mathcal{G}) = \deg(D)\mathrm{Swan}_\infty(\mathcal{G}),$$

which gives the second asserted inequality. QED

Corollary 7.1.19 If either $\mathcal{G}$ is tame at ∞ or D is prime to p, then we have the inequality

$$\mathrm{rank}(\mathrm{Gr}^0(\mathcal{M})_{\mathrm{ncst}}) \geq (\deg(D) + 2g - 2)\mathrm{rank}(\mathcal{G}).$$

proof From the middle of the proof of the last lemma, we extract the inequality

$$\mathrm{rank}(\mathrm{Gr}^0(\mathcal{M})_{\mathrm{ncst}}) \geq (2g - 1)\mathrm{rank}(\mathcal{G}) + (\deg(D) - 1)\Sigma_{s \text{ in } S(\bar{k})} \mathrm{TotalDrop}_s(\mathcal{G}) + \Sigma_{P \text{ in } D(\bar{k})} \mathrm{Swan}_P(f^*\mathcal{G}) - \mathrm{Swan}_\infty(\mathcal{G})$$
$$= (2g - 1)\mathrm{rank}(\mathcal{G}) + (\deg(D) - 1)(\Sigma_{s \text{ in } S(\bar{k})} \mathrm{TotalDrop}_s(\mathcal{G}) + \mathrm{Swan}_\infty(\mathcal{G})).$$

From the Euler Poincaré formula for $\mathcal{G}$ on $\mathbb{A}^1$, we see that

$$\Sigma_{s \text{ in } S(\bar{k})} \mathrm{TotalDrop}_s(\mathcal{G}) + \mathrm{Swan}_\infty(\mathcal{G})$$
$$= \mathrm{rank}(\mathcal{G}) - \chi(\mathbb{A}^1\otimes_k\bar{k}, \mathcal{G})$$
$$= \mathrm{rank}(\mathcal{G}) + h_c{}^1(\mathbb{A}^1\otimes_k\bar{k}, \mathcal{G})$$
$$\geq \mathrm{rank}(\mathcal{G}).$$

So we get

$\mathrm{rank}(\mathrm{Gr}^0(\mathcal{M})_{ncst})$
$\geq (2g - 1)\mathrm{rank}(\mathcal{G}) + (\deg(D) - 1)\mathrm{rank}(\mathcal{G})$
$\geq (\deg(D) + 2g - 2)\mathrm{rank}(\mathcal{G})$. QED

(7.2) General results on G_{geom} for pullbacks

Theorem 7.2.1 (uses the truth of the Larsen Eighth Moment Conjecture) Suppose that $\deg(D) \geq 2g + 7$. Denote

$$N := \mathrm{rank}(\mathrm{Gr}^0(\mathcal{M})_{ncst}).$$

Then we have the following results concerning the group G_{geom} for $(\mathrm{Gr}^0(\mathcal{M})_{ncst}$.

1) If $\mathcal{G}|\mathbb{A}^1 - S$ is not geometrically self dual, and if $N \geq 8$, then $G_{geom} \supset SL(N)$.

2) If $\mathcal{G}|\mathbb{A}^1 - S$ is, geometrically, orthogonally self dual, and if $N \geq 8$, then $G_{geom} = Sp(N)$.

3) If $\mathcal{G}|\mathbb{A}^1 - S$ is, geometrically, symplectically self dual, and if $N \geq 8$, then G_{geom} is either $SO(N)$ or $O(N)$.

proof This is a special case of Theorem 2.5.2. QED

Remark 7.2.2 In the above theorem, suppose in addition that either $\mathcal{G}$ is tame at ∞ or D is prime to p. Then the hypotheses on the size of N are nearly always satisfied. Indeed, by Corollary 7.1.19, we have the inequality

$$N \geq (\deg(D) + 2g - 2)\mathrm{rank}(\mathcal{G}) \geq (4g + 5)\mathrm{rank}(\mathcal{G}).$$

So we always have $N \geq 4$. If $\mathcal{G}|\mathbb{A}^1 - S$ is, geometrically, symplectically self dual, then $\mathrm{rank}(\mathcal{G})$ is even, so we have $N \geq 10$. If $\mathcal{G}|\mathbb{A}^1 - S$ is, geometrically, orthogonally self dual, we have $N \geq 8$ except, possibly, in the case ($\mathrm{rank}(\mathcal{G}) = 1$, $g = 0$, and $\deg(D) \leq 9$).

Theorem 7.2.3 Suppose that

1) $\deg(D) \geq 2g+3$,

2) $\mathcal{G}|\mathbb{A}^1 - S$ is, geometrically, symplectically self dual, and the local monodromy at some point s in $S(\bar{k})$ is a unipotent pseudoreflection.

Then G_{geom} for $\mathrm{Gr}^0(\mathcal{M})_{ncst}$ contains a reflection. If in addition

$$N := \mathrm{rank}(\mathrm{Gr}^0(\mathcal{M})_{ncst}) \geq 9,$$

then G_{geom} is $O(N)$. If N is 7 or 8, then G_{geom} is either $O(N)$ or it is the Weyl group of E_N in its reflection representation.

proof Once we prove that G_{geom} contains a reflection, we argue as follows. Since $\deg(D) \geq 2g + 3$, $L(D)$ is 4-separating. As $N \geq 4$, the Higher Moment Theorem 1.20.2 and its Corollary 1.20.3 show that G_{geom} lies in $O(N)$, and has fourth moment 3. By purity, we know that G_{geom} is semisimple. The result is then an instance of Theorem 2.6.9, for $N \geq 9$, or of Theorem 2.6.11, for $N=7$ or $N=8$.

It remains to prove that G_{geom} for $Gr^0(\mathcal{M})_{ncst}$ contains a reflection. To do this, it suffices to show that after pullback to some curve, G_{geom} for the pullback contains a reflection. In fact, we will obtain a reflection as a local monodromy of such a pullback.

Recall that $\mathcal{M}$ lives on a dense open set $U \subset L(D)$ such that $M|U$ is lisse. Write D over $\overline{k}$ as $\Sigma_i a_i P_i$. Define integers b_i by

$$b_i := a_i, \text{ if p divides } a_i,$$
$$b_i := a_i + 1, \text{ if p does not divide } a_i.$$

Recall from 7.1.9 that there is a dense open set $U_1 \subset L(D)$ such that for f in $U_1(\overline{k})$, f has divisor of poles D, df has divisor of poles $\Sigma b_i P_i$ (i.e., f has Swan-minimal poles, cf. 6.4.6), f is Lefschetz on C - D, $f_\star \overline{\mathbb{Q}}_\ell/\overline{\mathbb{Q}}_\ell$ is an irreducible middle extension sheaf on $\mathbb{A}^1$ with G_{geom} the full symmetric group $S_{deg(D)}$ in its deleted permutation representation, $f_\star \overline{\mathbb{Q}}_\ell/\overline{\mathbb{Q}}_\ell$ has at least one finite singularity, and its local monodromy at each finite singularity is a reflection.

Pick a function f in $U_1(\overline{k})$ such that -f lies in $U(\overline{k})$. Because U_1 and U are open in L(D), for all but finitely many λ in $\overline{k}^\times$, λf lies in $U_1(\overline{k})$, and $-\lambda f$ lies in $U(\overline{k})$.

The idea is to fix a sufficiently general λ in $U(\overline{k})$, and then to consider the $\overline{k}$-map

$$\mathbb{A}^1 \to L(D)$$

given by

$$t \mapsto t - \lambda f.$$

Since $-\lambda f$ lies in $U(\overline{k})$, $t - \lambda f$ lies in U for all but finitely many t. So if we exclude a finite closed subscheme Z of $\mathbb{A}^1/\overline{k}$, we get a map

$$\mathbb{A}^1 - Z \to U,$$
$$t \mapsto t - \lambda f.$$

We will show that after we pull back $Gr^0(\mathcal{M})_{ncst}$ to $\mathbb{A}^1 - Z$, its local monodromy at some point of Z is a reflection.

Consider the perverse sheaf

$$L := (f_\star \overline{\mathbb{Q}}_\ell/\overline{\mathbb{Q}}_\ell)[1].$$

Thus L on $\mathbb{A}^1/\overline{k}$ is perverse, irreducible, non-punctual, has $\mathcal{P}$, each of its local monodromies at finite distance is a reflection, and, by Corollary 6.4.5, all its ∞-slopes are ≤ 1. The perverse sheaf $K = \mathcal{G}[1]$ on $\mathbb{A}^1/\overline{k}$ is perverse, irreducible, non-punctual, and has $\mathcal{P}$. By assumption, the local monodromy of K at some finite singularity is a unipotent pseudoreflection. We next apply Theorem 6.1.18 and its Corollary 6.1.19 to this situation. We conclude that for all but finitely many λ in $\overline{k}^\times$, the middle convolution

$$K \star_{mid} \mathrm{MultTrans}_\lambda(L)$$

is perverse, semisimple, has $\mathcal{P}$, and is non-punctual, say

$$K*_{mid}\mathrm{MultTrans}_\lambda(L) = \mathcal{Q}_\lambda[1],$$

and the local monodromy of $\mathcal{Q}_\lambda$ at some finite singularity is a reflection. In particular, we can choose λ in $\bar{k}^\times$ such that this holds, and such that λf lies in $U_1(\bar{k})$, and $-\lambda f$ lies in $U(\bar{k})$. Fix such a choice of λ. At the expense of enlarging the ground field k, we may assume that f lies in L(D)(k), and that λ lies in $k^\times$.

Choose a finite closed subscheme Z of $\mathbb{A}^1$ such that $\mathbb{A}^1 - Z$ maps to U and such that both of the perverse sheaves (remember K has $\mathcal{P}$)

$$K*_!\mathrm{MultTrans}_\lambda(L) \text{ and } K*_{mid}\mathrm{MultTrans}_\lambda(L)$$

are lisse on $\mathbb{A}^1 - Z$. To conclude the proof, we will now show that for this choice of λ and Z, we have an isomorphism

$$\mathcal{Q}_\lambda \mid \mathbb{A}^1 - Z \cong \mathrm{Gr}^0(\mathcal{M})_{ncst} \mid \mathbb{A}^1 - Z$$

of lisse sheaves on $\mathbb{A}^1 - Z$. For this, we argue as follows. It is tautologous that the perverse sheaves

$$\mathcal{M}_{ncst}[1] \mid \mathbb{A}^1 - Z$$

and

$$K*_!\mathrm{MultTrans}_\lambda(L) \mid \mathbb{A}^1 - Z$$

have the same trace function. Both are lisse sheaves on $\mathbb{A}^1 - Z$, placed in degree -1. Write

$$K*_!\mathrm{MultTrans}_\lambda(L) \mid \mathbb{A}^1 - Z = \mathcal{Q}_{!\lambda}[1].$$

Then both $\mathcal{M}_{ncst} \mid \mathbb{A}^1 - Z$ and $\mathcal{Q}_{!\lambda}$ are lisse on $\mathbb{A}^1 - Z$, ι-mixed of weight ≤ 0, and have the same trace function. Therefore

$$\mathrm{Gr}^0(\mathcal{M}_{ncst}) \mid \mathbb{A}^1 - Z \text{ and } \mathrm{Gr}^0(\mathcal{Q}_{!\lambda})$$

have the same trace function. So by Chebotarev their arithmetic semisimplifications are isomorphic. Both of these are lisse and ι-pure of weight zero, so both are geometrically, and a fortiori arithmetically, semisimple. Therefore we have an isomorphism of lisse sheaves on $\mathbb{A}^1 - Z$,

$$\mathrm{Gr}^0(\mathcal{M}_{ncst}) \mid \mathbb{A}^1 - Z \cong \mathrm{Gr}^0(\mathcal{Q}_{!\lambda}).$$

We have already seen in Lemma 7.1.17 that

$$\mathrm{Gr}^0(\mathcal{M}_{ncst}) \cong \mathrm{Gr}^0(\mathcal{M})_{ncst}.$$

And by Theorem 6.4.5, applied to K and to $\mathrm{MultTrans}_\lambda(L)(1/2)$, we have

$$\mathrm{Gr}^0(K*_!\mathrm{MultTrans}_\lambda(L)(1/2)) \cong K*_{mid}\mathrm{MultTrans}_\lambda(L)(1/2).$$

Passing to $\mathbb{A}^1 - Z$, this isomorphism gives

$$\mathrm{Gr}^0(\mathcal{Q}_{!\lambda}) \cong \mathcal{Q}_\lambda.$$

Putting this all together, we get the desired isomorphism

$$\mathcal{Q}_\lambda \mid \mathbb{A}^1 - Z \cong \mathrm{Gr}^0(\mathcal{M})_{ncst} \mid \mathbb{A}^1 - Z. \qquad \text{QED}$$

We record for later use the following corollary.

Corollary 7.2.4 Suppose that

1) $\deg(D) \geq 2g+3$,

2) $\mathcal{G}|\mathbb{A}^1 - S$ is, geometrically, symplectically self dual, and the local monodromy at some point s in $S(\bar{k})$ is a unipotent pseudoreflection.

Then G_{geom} for $Gr^0(\mathcal{M})$ contains a reflection.

proof Indeed, we have a direct sum decomposition

$$Gr^0(\mathcal{M}) = Gr^0(\mathcal{M})_{ncst} \oplus Gr^0(\mathcal{M})_{cst},$$

in which the second summand is geometrically constant. So if an element γ in $\pi_1{}^{geom}(U)$ acts as a reflection $\rho(\gamma)$ on $Gr^0(\mathcal{M})_{ncst}$, this same γ acts as the reflection $\rho(\gamma) \oplus$ (identity) on $Gr^0(\mathcal{M})$. QED

Theorem 7.2.5 (uses the truth of the Larsen Eighth Moment Conjecture) Suppose that the following four conditions hold.

1) $\deg(D) \geq 2g + 7$.

2) $\mathcal{G}|\mathbb{A}^1 - S$ is, geometrically, symplectically self dual, and there exists a point s in $S(\bar{k})$ at which the local monodromy is tame, with $\mathcal{G}(s)/\mathcal{G}(s)^{I(s)}$ of odd dimension.

3) Either $p \neq 2$, or the local monodromy of $\mathcal{G}$ at s is unipotent.

4) $N := \mathrm{rank}(Gr^0(\mathcal{M})_{ncst}) \geq 8$.

Then G_{geom} for $Gr^0(\mathcal{M})_{ncst}$ is $O(N)$.

proof From Theorem 7.2.1, we see that G_{geom} is either $SO(N)$ or $O(N)$. Restricting $\mathcal{M}_{ncst}$ to a curve $\mathbb{A}^1 - Z$ as in the proof of Theorem 7.2.3, it suffices to show that, in the notations of that proof, the convolution sheaf $\mathcal{Q}_\lambda$ has some local monodromy of determinant -1.

Suppose first $p \neq 2$. In this case, the result is immediate from Theorem 6.1.18, part 3), applied to $\mathcal{G}[1]$ and to $(f_*\bar{\mathbb{Q}}_\ell/\bar{\mathbb{Q}}_\ell)[1]$. This result tells us that if γ is a finite singularity of $f_*\bar{\mathbb{Q}}_\ell/\bar{\mathbb{Q}}_\ell$, then $s + \lambda\gamma$ is finite singularity of $\mathcal{G}$, and we have an isomorphism of $I(0)$-representations

$$\mathcal{Q}_\lambda(s + \lambda\gamma)/\mathcal{Q}_\lambda(s + \lambda\gamma)^{I(s + \lambda\gamma_j)} \text{ moved to } 0$$

$$\cong (\mathcal{G}(s)/\mathcal{G}(s)^{I(s)} \text{ moved to } 0)\otimes(\text{the quadratic character } \mathcal{L}_{\chi_2}).$$

As $\mathcal{G}$ is, geometrically, symplectically self dual, $\det(\mathcal{G})$ is geometrically trivial, so trivial on $I(s)$. Visibly $\det(\mathcal{G}(s)^{I(s)})$ is trivial on $I(s)$. Therefore the $I(0)$-representation

$$(\mathcal{G}(s)/\mathcal{G}(s)^{I(s)} \text{ moved to } 0)\otimes(\text{the quadratic character } \mathcal{L}_{\chi_2})$$

has determinant $(\mathcal{L}_{\chi_2})^{\otimes n}$, for $n := \dim(\mathcal{G}(s)/\mathcal{G}(s)^{I(s)})$.

Since n is odd by hypothesis, we have $(\mathcal{L}_{\chi_2})^{\otimes n} = \mathcal{L}_{\chi_2}$. The above isomorphism then shows that $\mathcal{Q}_\lambda(s + \lambda\gamma)/\mathcal{Q}_\lambda(s + \lambda\gamma)^{I(s + \lambda\gamma_j)}$

has determinant $\mathcal{L}_{\chi_2}$ moved to $s + \lambda\gamma$ as $I(s + \lambda\gamma)$-representation. As $\mathcal{Q}_\lambda(s + \lambda\gamma)^{I(s + \lambda\gamma_j)}$ has trivial determinant as $I(s + \lambda\gamma)$-representation, we see that $\mathcal{Q}_\lambda$ as $I(s + \lambda\gamma)$-representation has determinant $\mathcal{L}_{\chi_2}$ moved to $s + \lambda\gamma$.

Suppose now that $p = 2$, and that the local monodromy of $\mathcal{G}$ at s is unipotent. In this case, we have, by Theorem 6.1.18, part 2), an isomorphism of $I(\infty)$-representations

$$FT_\psi loc(s + \lambda\gamma, \infty)(\mathcal{Q}_\lambda(s + \lambda\gamma)/\mathcal{Q}_\lambda(s+ \lambda\gamma)^{I(s + \lambda\gamma)}) \cong$$
$$FT_\psi loc(s, \infty)(\mathcal{G}(s)/\mathcal{G}(s)^{I(s)})\otimes FT_\psi loc(\lambda\gamma, \infty)(\text{a quadratic character } \rho),$$

for ρ some character of order 2 of $I(\lambda\gamma)$.

Because $\mathcal{G}$ is unipotent at s, and hence tame at s, the $I(\infty)$-representation

$$FT_\psi loc(s, \infty)(\mathcal{G}(s)/\mathcal{G}(s)^{I(s)}) \cong [x - s \mapsto 1/(x-s)]^*(\mathcal{G}(s)/\mathcal{G}(s)^{I(s)})$$

is unipotent, of odd dimension $n := \dim(\mathcal{G}(s)/\mathcal{G}(s)^{I(s)})$. Thus

$$FT_\psi loc(s + \lambda\gamma, \infty)(\mathcal{Q}_\lambda(s + \lambda\gamma)/\mathcal{Q}_\lambda(s+ \lambda\gamma)^{I(s + \lambda\gamma)})$$

is an odd-dimensional successive extension of $FT_\psi loc(\lambda\gamma, \infty)(\rho)$ by itself. Therefore $\mathcal{Q}_\lambda(s + \lambda\gamma)/\mathcal{Q}_\lambda(s+ \lambda\gamma)^{I(s + \lambda\gamma)}$ is an odd-dimensional successive extension of ρ (translated to $s + \lambda\gamma$) by itself, and we conclude as above that $\det(\mathcal{Q}_\lambda)$ is nontrivial of order 2 as $I(s + \lambda\gamma)$-representation. QED

(7.2.6) In a similar vein, we have the following result.

Theorem 7.2.7 Suppose that the following two conditions hold.
1) $\deg(D) \geq 2g+3$.
2) For some point s in $S(\overline{k})$, $\mathcal{G}(s)/\mathcal{G}(s)^{I(s)}$ is not a semisimple representation of $I(s)$.

Then we have the following results for the group G_{geom} for $Gr^0(\mathcal{M})_{ncst}$.

1) If $\mathcal{G}|\mathbb{A}^1 - S$ is, not geometrically self dual, then
$$N := \mathrm{rank}(Gr^0(\mathcal{M})_{ncst}) \geq 2,$$
and $G_{geom} \supset SL(N)$.

2)If $\mathcal{G}|\mathbb{A}^1 - S$ is, geometrically, symplectically self dual, then
$$N := \mathrm{rank}(Gr^0(\mathcal{M})_{ncst}) \geq 3,$$
and G_{geom} is $SO(N)$ or $O(N)$.

3)If $\mathcal{G}|\mathbb{A}^1 - S$ is, geometrically, orthogonally self dual, then
$$N := \mathrm{rank}(Gr^0(\mathcal{M})_{ncst}) \geq 2,$$
and $G_{geom} = Sp(N)$.

proof We restrict to an $\mathbb{A}^1 - Z$ as in the proof ofTheorem 7.2.3, using Non-semisimplicity Corollary 6.1.22 to see that $\mathcal{Q}_\lambda$ has a non-semisimple local monodromy representation at some point. As this local monodromy representation is not semisimple, it cannot have finite image, and it must have dimension at least 2. Therefore we have $N \geq 2$. If $N = 2$ in case 3), we use the fact that G_{geom} is, by purity, a semisimple subgroup of Sp(2), which is not finite, so must be Sp(2). In the remaining cases, we use Larsen's alternative (Theorem 2.2.2) via the Higher Moment Theorem 1.20.2. Since the finite case has been ruled out, we get all the conclusions, except the assertion that $N \geq 3$ in case 2). But G_{geom} is a semisimple group, and neither SO(2) nor O(2) is semisimple. QED

Corollary 7.2.8 Suppose that the following three conditions hold.
1) $\deg(D) \geq 2g + 3$.
2) $\mathcal{G}|\mathbb{A}^1 - S$ is, geometrically, symplectically self dual, and there exists a point s in $S(\bar{k})$ at which the local monodromy is tame, with $\mathcal{G}(s)/\mathcal{G}(s)^{I(s)}$ of odd dimension and not semisimple as an I(s)-representation.
3) Either $p \neq 2$, or the local monodromy of $\mathcal{G}$ at s is unipotent.
Then $N \geq 3$, and G_{geom} for $Gr^0(\mathcal{M})_{ncst}$ is O(N).
proof The previous theorem shows that G_{geom} is either SO(N) or O(N). The argument used in proving Theorem 7.2.5 shows that G_{geom} contains an element of determinant -1. QED

(7.3) Application to pullback families of elliptic curves and of their symmetric powers

(7.3.1) In this section, we further specialize the perverse sheaf K on $\mathbb{A}^1/k$ with which we have been working. Let us write $\mathbb{A}^1$ as Spec(k[t]). We begin with an elliptic curve E_t over the function field k(t). Denote by $j : \mathbb{A}^1 - S \subset \mathbb{A}^1$ the inclusion of an open dense set over which $E_t/k(t)$ extends to an elliptic curve $\pi : \mathcal{E}_t \rightarrow \mathbb{A}^1 - S$. On $\mathbb{A}^1 - S$, we form the lisse, rank two sheaf $R^1\pi_*\overline{\mathbb{Q}}_\ell(1)$. It is pure of weight -1, and symplectically self dual toward $\overline{\mathbb{Q}}_\ell(1)$. [We remark in passing that in any characteristic $p \geq 5$, $\mathcal{G}$ is automatically everywhere tamely ramified [Ka-TLFM,7.6.2 and 7.5.1], in particular at ∞.]

(7.3.2) We assume that $E_t/k(t)$ has nonconstant j invariant, and that it has **multiplicative reduction** at some point s of $S(\bar{k})$. This means precisely [SGA 7, Expose IX, 3.5] that the local monodromy of $R^1\pi_*\overline{\mathbb{Q}}_\ell(1)$ at s is a unipotent pseudoreflection, i.e., it is Unip(2), a single unipotent Jordan block of size 2. The group G_{geom} for $R^1\pi_*\overline{\mathbb{Q}}_\ell(1)$ is SL(2). [Indeed, it is a semisimple (by purity, cf. [De-Weil II, 1.3.9 and 3.4.1 (iii)]) subgroup of SL(2) = Sp(2), so is either finite or is SL(2). As it contains a unipotent pseudoreflection, it must be SL(2).]

(7.3.3) We form the middle extension sheaf $\mathcal{G} := j_*R^1\pi_*\overline{\mathbb{Q}}_\ell(1)$ on $\mathbb{A}^1$, and the perverse sheaf $K := \mathcal{G}[1]$ on $\mathbb{A}^1$. Thus K is arithmetically self dual as a perverse sheaf, geometrically irreducible, pure of weight zero, non-punctual, and geometrically has $\mathcal{P}$. It is everywhere tamely ramified if $p \geq 5$.

(7.3.4) More generally, for each integer $n \geq 1$, we form the lisse sheaf

$$\mathrm{Symm}^n(R^1\pi_*\overline{\mathbb{Q}}_\ell)((n+1)/2)$$

on U. It is lisse of rank $n + 1$, and pure of weight -1. If n is odd (resp. even), it is symplectically (resp. orthogonally) self dual toward $\overline{\mathbb{Q}}_\ell(1)$. It is geometrically irreducible, because all the symmetric powers of the standard representation of SL(2) are irreducible. Its local monodromy at s is $\mathrm{Symm}^n(\mathrm{Unip}(2)) = \mathrm{Unip}(n+1)$, a single unipotent Jordan block of size $n+1$. We define the middle extension sheaf

$$\mathcal{G}_n := j_*\mathrm{Symm}^n(R^1\pi_*\overline{\mathbb{Q}}_\ell)((n+1)/2)$$

on $\mathbb{A}^1$, and the perverse sheaf

$$K_n := \mathcal{G}_n[1] \text{ on } \mathbb{A}^1.$$

Thus each K_n is arithmetically self dual as a perverse sheaf, geometrically irreducible, pure of weight zero, non-punctual, and geometrically has $\mathcal{P}$. For $n=1$, we recover $\mathcal{G}$ and K.

For each $n \geq 1$, we have

(7.3.4.1) $$\mathcal{G}_n(s)/\mathcal{G}_n(s)^{I(s)} = \mathrm{Unip}(n+1)/\text{invariants} \cong \mathrm{Unip}(n).$$

(7.3.5) Denote by

$$j_\infty : \mathbb{A}^1 \to \mathbb{P}^1$$

the inclusion. In the terminology of elliptic curves over function fields, the (unitarized) L-function of $E_t/k(t)$ is the reversed characteristic polynomial

$$L(E_t/k(t), T) := \det(1 - T\mathrm{Frob}_k \mid H^1(\mathbb{P}^1\otimes_k\overline{k}, j_{\infty*}\mathcal{G})).$$

And for each $n \geq 1$, the (unitarized) L-function of its n'th symmetric power is the reversed characteristic polynomial

$$L(\mathrm{Symm}^n(E_t/k(t)), T) := \det(1 - T\mathrm{Frob}_k \mid H^1(\mathbb{P}^1\otimes_k\overline{k}, j_{\infty*}\mathcal{G}_n)).$$

(7.3.6) We now bring to bear our proper, smooth, geometrically connected curve C/k, of genus denoted g, together with its effective divisor D of degree

$$\deg(D) \geq 2g + 3.$$

Recall that $U_{D,S} \subset L(D)$ is the dense open set whose $\overline{k}$-valued points consist of those f in $L(D)(\overline{k})$ whose divisor of poles is D, and which are finite etale over S.

(7.3.7) For f any nonconstant function in k(C), we can form the pullback by f of the elliptic curve $E_t/k(t)$, which we denote $E_f/k(C)$. Concretely, $E_t/k(t)$ can be given by a generalized Weierstrass equation

$$E_t : y^2 + a_1(t)y + a_3(t)xy = x^3 + a_2(t)x^2 + a_4(t)x + a_6(t),$$

with the $a_i(t)$ rational functions of t, holomorphic on $\mathbb{A}^1$ -S such that the discriminant Δ(the $a_i(t)$'s) is invertible on $\mathbb{A}^1$ - S. Then $E_f/k(C)$ is the elliptic curve over k(C) obtained by the substitution $t \mapsto f$ in the rational functions $a_i(t)$:

$$E_f : y^2 + a_1(f)y + a_3(f)xy = x^3 + a_2(f)x^2 + a_4(f)x + a_6(f).$$

(7.3.8) If we take f in $U_{D,S}(k)$, the f defines a morphism from C - D to $\mathbb{A}^1$, and a morphism from C - D - $f^{-1}(S)$ to $\mathbb{A}^1$ - S. For such an f, $E_f/k(C)$ spreads out to an elliptic curve

$$\pi_f : \mathcal{E}_f \to C - D - f^{-1}(S).$$

Denote by

$$j_f : C - D - f^{-1}(S) \to C - D$$

and

$$j_D : C- D \to C$$

the inclusions. Because f in $U_{D,S}(k)$ is finite etale over S, we have

$$j_{f\star}R^1\pi_{f\star}\overline{\mathbb{Q}}_\ell(1) = f^\star\mathcal{G},$$

and, for each $n \geq 1$,

$$j_{f\star}\mathrm{Symm}^n(R^1\pi_{f\star}\overline{\mathbb{Q}}_\ell)((n+1)/2) = f^\star\mathcal{G}_n.$$

So the unitarized L-functions of $E_f/k(C)$ and its symmetric powers are given by

$$L(E_f/k(C), T) = \det(1 - T\mathrm{Frob}_k \mid H^1(C\otimes_k\overline{k}, j_{D\star}f^\star\mathcal{G})),$$

and, for each $n \geq 1$,

$$L(\mathrm{Symm}^n(E_f/k(C)), T) = \det(1 - T\mathrm{Frob}_k \mid H^1(C\otimes_k\overline{k}, j_{D\star}f^\star\mathcal{G}_n)).$$

Similarly if we take k_d/k the extension field of degree d, and f in $U_{D,S}(k_d)$, we get, for each $n \geq 1$,

$$L(\mathrm{Symm}^n(E_f/k_d(C)), T) = \det(1 - T\mathrm{Frob}_{k_d} \mid H^1(C\otimes_k\overline{k}, j_{D\star}f^\star\mathcal{G}_n)).$$

(7.3.9) Now view the cohomology group $H^1(C\otimes_k\overline{k}, j_{D\star}f^\star\mathcal{G}_n)$ as the Gr^0 of the cohomology group $H_c{}^1((C - D)\otimes_k\overline{k}, f^\star\mathcal{G}_n)$. Similarly, view $H^1(\mathbb{P}^1\otimes_k\overline{k}, j_{\infty\star}\mathcal{G}_n)$ as the Gr^0 of the subgroup (via $f^\star$)

$$H_c{}^1(\mathbb{A}^1\otimes_k\overline{k}, \mathcal{G}_n) \subset H_c{}^1((C - D)\otimes_k\overline{k}, f^\star\mathcal{G}_n).$$

So we have a Frob_{k_d}-equivariant inclusion

$$H^1(\mathbb{P}^1\otimes_k\overline{k}, j_{\infty\star}\mathcal{G}_n) \subset H^1(C\otimes_k\overline{k}, j_{D\star}f^\star\mathcal{G}_n),$$

and hence a divisibility of L-functions for each $n \geq 1$:

$$L(\mathrm{Symm}^n(E_t/k_d(t)), T) \mid L(\mathrm{Symm}^n(E_f/k_d(C)), T).$$

(7.3.10) The quotient

$$L(\mathrm{Symm}^n(E_f/k_d(C)), T)/L(\mathrm{Symm}^n(E_t/k_d(t)), T)$$

is a polynomial which one might call the "new part" of the L-function of $\mathrm{Symm}^n(E_f/k_d(C))$. Of course, this notion of "new part" is not intrinsic to the elliptic curve $E_f/k_d(C)$, but rather depends on viewing it as the pullback by f of $E_t/k(t)$.

(7.3.11) Let us make explicit how these "new parts" of L-functions are captured by the perverse sheaves

$$M_n := M = \mathrm{Twist}(L = \overline{\mathbb{Q}}_\ell(1/2)[1], K_n, \mathcal{F} = L(D), h = 0).$$

If either $\mathcal{G}$ is tame at ∞ (in which case all the $\mathcal{G}_n$ are tame at ∞, and which is automatic if $p \geq 5$), or if D is prime to p, then, by Lemma 7.1.7, the perverse sheaves M_n are all lisse on the dense open set $U_{D,S}$. In general, we pick for each n a dense open set $U_n \subset U_{D,S}$ on which M_n is lisse.

(7.3.12) On the open set U_n, M_n is $\mathcal{M}_n(1/2)[\ell(D)]$, with $\mathcal{M}_n$ lisse and ι-mixed of weight ≤ 0, given stalkwise by

$$\mathcal{M}_{n,f} = H_c^1((C - D)\otimes_k \overline{k}, f^*\mathcal{G}_n).$$

We have

$$\mathrm{Gr}^0(\mathcal{M}_n)_f = H^1(C\otimes_k \overline{k}, j_{D*}f^*\mathcal{G}_n),$$

$$\mathrm{Gr}^0(\mathcal{M}_n)_{cst} = \text{the constant sheaf } H^1(\mathbb{P}^1\otimes_k \overline{k}, j_{\infty *}\mathcal{G}_n),$$

$$(\mathrm{Gr}^0(\mathcal{M}_n)_{ncst})_f = H^1(C\otimes_k \overline{k}, j_{D*}f^*\mathcal{G}_n)/H^1(\mathbb{P}^1\otimes_k \overline{k}, j_{\infty *}\mathcal{G}_n).$$

Thus for k_d/k the extension field of degree d, and for f in $U_n(k_d)$, we have

$$\det(1 - T\mathrm{Frob}_{k_d,f} \mid \mathrm{Gr}^0(\mathcal{M}_n)_{ncst})$$

$$= L(\mathrm{Symm}^n(E_f/k_d(C)), T)/L(\mathrm{Symm}^n(E_t/k_d(t)), T).$$

(7.3.13) We now turn to the determination of the group G_{geom} for $\mathrm{Gr}^0(\mathcal{M}_n)_{ncst}$. For $n \geq 2$, the answer is wonderfully simple.

Theorem 7.3.14 Suppose $\deg(D) \geq 2g + 3$. For $n \geq 2$, and $N := \mathrm{rank}(\mathrm{Gr}^0(\mathcal{M}_n)_{ncst})$, the group G_{geom} for $\mathrm{Gr}^0(\mathcal{M}_n)_{ncst}$ is $O(N)$ if n is odd, and is $Sp(N)$ if n is even.

proof For $n \geq 2$,

$$\mathcal{G}_n(s)/\mathcal{G}_n(s)^{I(s)} = \mathrm{Unip}(n+1)/\text{invariants} \cong \mathrm{Unip}(n)$$

is not semisimple, and has dimension n. If n is even, then $\mathcal{G}_n$ is, geometrically, orthogonally self dual, and the result is a special case of Theorem 7.2.7, part 3). If n is odd, then $\mathcal{G}_n$ is, geometrically, symplectically self dual, and the result is a special case of Corollary 7.2.8. QED

(7.3.15) The case n=1 is more complicated, because G_{geom} can be finite in certain cases. Nonetheless, the "general case" behaves well.

Theorem 7.3.16 Suppose $\deg(D) \geq 2g + 3$. Suppose

$$N := \mathrm{rank}(\mathrm{Gr}^0(\mathcal{M}_1)_{ncst}) \geq 9.$$

Then the group G_{geom} for $Gr^0(\mathcal{M}_1)_{ncst}$ is $O(N)$.
proof This is a special case of Theorem 7.2.3. QED

(7.3.17) We also have the following short list of possibilities when N is 7 or 8.
Theorem 7.3.18 Suppose $\deg(D) \geq 2g + 3$. Suppose $N := \mathrm{rank}(Gr^0(\mathcal{M}_1)_{ncst})$ is 7 or 8. Then the group G_{geom} for $Gr^0(\mathcal{M}_1)_{ncst}$ is either $O(N)$ or it is the Weyl group of E_N in its reflection representation.
proof This is a special case of Theorem 7.2.3. QED

(7.4) Cautionary examples

(7.4.1) Here are five examples to show that the hypothesis $N \geq 9$ in the theorem above is essential. These examples are all based on the general fact that if we pick polynomials $a_1(t)$, $a_2(t)$, $a_3(t)$, $a_4(t)$, $a_6(t)$ in $k[t]$ with $\deg(a_i) \leq i$, and with nonconstant j invariant, then the ellliptic surface over $\mathbb{P}^1$ whose generic fibre is the Weierstrass equation

$$y^2 + a_1(t)xy + a_3(t)y = x^3 + a_2(t)x^2 + a_4(t)x + a_6(t).$$

is a rational surface, cf. [Shioda-MWL, 10.14 on page 237].
(7.4.2) In each example, we specify an elliptic curve $E_t/k(t)$ given as a Weierstrass equation with coefficients in $k[t]$, we form the middle extension sheaf $\mathcal{G} = \mathcal{G}_1$ on $\mathbb{A}^1$ attached to it, we denote by $S \subset \mathbb{A}^1$ the finite singularities of $\mathcal{G}$, we take C to be $\mathbb{P}^1$, we take D to be the divisor $d\infty$ for an integer $d \geq 3$, and we write $M_{1,d}$ for the perverse sheaf on L(D) attached to the situation. In each of the five examples, the sheaf $\mathcal{G}$ is in fact the H_c^1 along the fibres of the given affine Weierstrass equation over $\mathbb{A}^1$, thanks to Corollary 7.5.5 of the Appendix to this chapter.
(7.4.3) When we are in characteristic $p \geq 5$, the sheaf $\mathcal{G}$ is automatically everywhere tamely ramified, in particular at ∞. So for $p \geq 5$, $M_{1,d}$ is lisse on $U_{D,S}$, the space of polynomials of degree d which are finite etale over each $\bar{k}$-valued point of S. In case p is 2 or 3, $\mathcal{G}$ need not be tame at ∞, and will not be in our examples. But if we take only d prime to p, then it remains true, by Lemma 7.1.7, that $M_{1,d}$ is lisse on $U_{D,S}$.
Example 7.4.4 In this first example, take $p \geq 5$, and take as $E_t/k(t)$ the elliptic curve

$$y^2 = 4x^3 - 3x - t.$$

Its discriminant is $3^3 - 27t^2 = 27(1 - t^2)$. At $t = \pm 1$, we have multiplicative reduction. At ∞, we have bad, but potentially good (after taking the sixth root of t), reduction. Since $\mathcal{G}$ is tame at ∞, the Euler Poincaré formula gives

$$\chi(\mathbb{A}^1\otimes_k\bar{k}, \mathcal{G}) = \chi(\mathbb{A}^1\otimes_k\bar{k} - \{\pm 1\}, \mathcal{G}) + 2$$
$$= (1 - 2)\mathrm{rank}(\mathcal{G}) + 2 = 0.$$

Therefore $H_c^1(\mathbb{A}^1\otimes_k\bar{k}, \mathcal{G}) = 0$, and hence its Gr^0 quotient vanishes:

$$H^1(\mathbb{P}^1\otimes_k\bar{k}, j_{\infty\star}\mathcal{G}) = 0.$$

Each $M_{1,d}$ is lisse on $U_{D,S=\{\pm 1\}}$, the space of polynomials $f(t)$ of degree d such that $f(t)^2 - 1$ has 2d distinct zeroes over $\bar{k}$. For such an f, E_f has multiplicative reduction at the 2d zeroes of $f(t)^2 - 1$. It has good reduction at ∞ if 6 divides d; otherwise it has bad but potentially good reduction at ∞. By the vanishing $H^1(\mathbb{P}^1\otimes_k\bar{k}, j_{\infty\star}\mathcal{G}) = 0$ explained above, we have

$$\mathrm{Gr}^0(\mathcal{M}_{1,d})_{cst} = 0, \ \mathrm{Gr}^0(\mathcal{M}_{1,d}) \cong \mathrm{Gr}^0(\mathcal{M}_{1,d})_{ncst}.$$

Using the Euler Poincaré formula, we readily compute

$$N_d := \text{rank of } \mathrm{Gr}^0(\mathcal{M}_{1,d}) = \text{rank of } \mathrm{Gr}^0(\mathcal{M}_{1,d})_{ncst}$$
$$= -[(1 - 2d)2 + 2d] = 2d-2, \text{ if 6 does not divide d,}$$
$$= -[(2-2d)2 + 2d] = 2d-4, \text{ if 6 divides d.}$$

So we have $N_d \geq 9$ precisely for $d \geq 7$. For d = 5 or 6, we have $N_d = 8$.

In fact, for $d \leq 6$, G_{geom} for $\mathrm{Gr}^0(\mathcal{M}_{1,d}) \cong \mathrm{Gr}^0(\mathcal{M}_{1,d})_{ncst}$ is finite. As noted in 7.4.1 above, for any nonconstant polynomial f of degree ≤ 6, the ellliptic surface X over $\mathbb{P}^1$ whose generic fibre is E_f is a rational surface. Hence all of its cohomology is algebraic, and therefore the action of Frobenius on $H^2(X\otimes_k\bar{k},.\bar{\mathbb{Q}}_\ell(1))$ is of finite order. But $H^1(\mathbb{P}^1\otimes_k\bar{k}, j_{\infty\star}f^{\star}\mathcal{G})$ is a subquotient, by Proposition 7.5.2 of the Appendix to this chapter, of $H^2(X\otimes_k\bar{k},.\bar{\mathbb{Q}}_\ell(1))$, so the action of Frobenius on $H^1(\mathbb{P}^1\otimes_k\bar{k}, j_{\infty\star}f^{\star}\mathcal{G})$ is also of finite order. So for $d \leq 6$, every Froebenius on $\mathrm{Gr}^0(\mathcal{M}_{1,d})$ is of finite order, and hence [Ka-ESDE, 8.14.3] G_{geom} is finite.

Example 7.4.5 In this second example, take $p \geq 5$, and take as $E_t/k(t)$ the elliptic curve

$$y^2 = 4x^3 - 3tx - 1.$$

Its discriminant is $(3t)^3 - 27 = 27(t^3 - 1)$. At t in $\mu_3(\bar{k})$, we have multiplicative reduction. At ∞, we have bad, but potentially good (after taking the fourth root of t), reduction. Since $\mathcal{G}$ is tame at ∞, the Euler Poincaré formula gives

$$\chi(\mathbb{A}^1\otimes_k\bar{k}, \mathcal{G}) = \chi(\mathbb{A}^1\otimes_k\bar{k} - \mu_3(\bar{k}), \mathcal{G}) + 3$$
$$= (1 - 3)\mathrm{rank}(\mathcal{G}) + 3 = -1.$$

Therefore $h_c^1(\mathbb{A}^1\otimes_k\bar{k}, \mathcal{G}) = 1$. In this case, we have

$$H_c^1(\mathbb{A}^1\otimes_k\bar{k}, \mathcal{G}) = H^1(\mathbb{P}^1\otimes_k\bar{k}, j_{\infty\star}\mathcal{G}),$$

so we have

$$\dim H^1(\mathbb{P}^1\otimes_k\bar{k}, j_{\infty*}\mathcal{G}) = 1.$$

Each $M_{1,d}$ is lisse on $U_{D,S=\mu_3}$, the space of polynomials $f(t)$ of degree d such that $f(t)^3 - 1$ has $3d$ distinct zeroes over $\bar{k}$. For such an f, E_f has multiplicative reduction at the $3d$ zeroes of $f(t)^3 - 1$. It has good reduction at ∞ if 4 divides d; otherwise it has bad but potentially good reduction at ∞. By the calculation $h^1(\mathbb{P}^1\otimes_k\bar{k}, j_{\infty*}\mathcal{G}) = 1$ explained above, we have

$$\mathrm{Gr}^0(\mathcal{M}_{1,d})_{cst} = \text{the rank one constant sheaf } h^1(\mathbb{P}^1\otimes_k\bar{k}, j_{\infty*}\mathcal{G}),$$

$$\mathrm{rank}\mathrm{Gr}^0(\mathcal{M}_{1,d}) = 1 + \mathrm{rank}\mathrm{Gr}^0(\mathcal{M}_{1,d})_{ncst}.$$

Using the Euler Poincaré formula, we readily compute the rank N_d of $\mathrm{Gr}^0(\mathcal{M}_{1,d})_{ncst}$:

$$\begin{aligned} N_d + 1 &:= \text{rank of } \mathrm{Gr}^0(\mathcal{M}_{1,d}) = 1 + \text{rank of } \mathrm{Gr}^0(\mathcal{M}_{1,d})_{ncst} \\ &= -[(1 - 3d)2 + 3d] = 3d-2, \text{ if 4 does not divide } d, \\ &= -[(2-3d)2 + 3d] = 3d-4, \text{ if 4 divides } d. \end{aligned}$$

Thus we have

$$\begin{aligned} N_d &= 3d - 3, \text{ if 4 does not divide } d, \\ &= 3d - 5, \text{ if 4 divides } d. \end{aligned}$$

So we have $N_d \geq 9$ precisely for $d \geq 5$. For $d = 4$, we have $N_d = 7$. For $d=3$, we have $N_d = 6$.

In fact, for $d \leq 4$, G_{geom} for $\mathrm{Gr}^0(\mathcal{M}_{1,d})$ (and hence for its direct summand $\mathrm{Gr}^0(\mathcal{M}_{1,d})_{ncst}$) is finite. Exactly as in the first example, this finiteness results from the fact that for any nonconstant polynomial f of degree ≤ 4, the ellliptic surface X over $\mathbb{P}^1$ whose generic fibre is E_f is a rational surface.

Example 7.4.6 In this third example, again take $p \geq 5$, but take as $E_t/k(t)$ the elliptic curve

$$y^2 = 4x^3 - 3tx - t.$$

Its discriminant is $(3t)^3 - 27t^2 = 27t^2(t - 1)$. At $t=1$, we have multiplicative reduction. At $t=0$, we have bad, but potentially good (after taking the sixth root of t), reduction. At ∞, we have bad, but potentially good (after taking the fourth root of t), reduction. Since $\mathcal{G}$ is tame at ∞, we see from the Euler Poincaré formula that

$$\begin{aligned} \chi(\mathbb{A}^1\otimes_k\bar{k}, \mathcal{G}) &= \chi(\mathbb{A}^1\otimes_k\bar{k} - \{0,1\}, \mathcal{G}) + 1 \\ &= (1 - 2)\mathrm{rank}(\mathcal{G}) + 1 = -1. \end{aligned}$$

Therefore $h_c{}^1(\mathbb{A}^1\otimes_k\bar{k}, \mathcal{G}) = 1$. In this case, we have

$$H_c{}^1(\mathbb{A}^1\otimes_k\bar{k}, \mathcal{G}) = H^1(\mathbb{P}^1\otimes_k\bar{k}, j_{\infty*}\mathcal{G}),$$

so we have

$$\dim H^1(\mathbb{P}^1\otimes_k\bar{k}, j_{\infty*}\mathcal{G}) = 1.$$

Each $M_{1,d}$ is lisse on $U_{D,S=\{0,1\}}$, the space of polynomials $f(t)$ of degree d such that $f(t)(f(t) - 1)$ has 2d distinct zeroes over $\bar{k}$. For such an f, E_f has multiplicative reduction at the d zeroes of $f(t) - 1$, and bad but potentially good reduction at the d zeroes of $f(t)$. It has good reduction at ∞ if 4 divides d, otherwise it has bad but potentially good reduction at ∞. By the calculation $h^1(\mathbb{P}^1\otimes_k\bar{k}, j_{\infty\star}\mathcal{G}) = 1$ explained above, we have

$$\mathrm{Gr}^0(\mathcal{M}_{1,d})_{cst} = \text{the rank one constant sheaf } h^1(\mathbb{P}^1\otimes_k\bar{k}, j_{\infty\star}\mathcal{G}),$$
$$\mathrm{rankGr}^0(\mathcal{M}_{1,d}) = 1 + \mathrm{rankGr}^0(\mathcal{M}_{1,d})_{ncst}.$$

Using the Euler Poincaré formula, we readily compute the rank N_d of $\mathrm{Gr}^0(\mathcal{M}_{1,d})_{ncst}$:

$$N_d + 1 := \text{rank of } \mathrm{Gr}^0(\mathcal{M}_{1,d}) = 1 + \text{rank of } \mathrm{Gr}^0(\mathcal{M}_{1,d})_{ncst}$$
$$= -[(1 - 2d)2 + d] = 3d-2, \text{ if 4 does not divide d},$$
$$= -[(2-2d)2 + d] = 3d-4, \text{ if 4 divides d}.$$

Thus we have

$$N_d = 3d - 3, \text{ if 4 does not divide d},$$
$$= 3d - 5, \text{ if 4 divides d}.$$

So we have $N_d \geq 9$ precisely for $d \geq 5$. For $d = 4$, we have $N_d = 7$. For $d=3$, we have $N_d = 6$.

Once again, for $d \leq 4$, G_{geom} for $\mathrm{Gr}^0(\mathcal{M}_{1,d})$ (and hence for its direct summand $\mathrm{Gr}^0(\mathcal{M}_{1,d})_{ncst}$) is finite. Exactly as in the first example, this finiteness results from the fact that for any nonconstant polynomial f of degree ≤ 4, the ellliptic surface X over $\mathbb{P}^1$ whose generic fibre is E_f is a rational surface.

Example 7.4.7 Take $p = 3$, and take as $E_t/k(t)$ the elliptic curve

$$y^2 = x^3 + x^2 + t.$$

This curve has good reduction over $\mathbb{G}_m$, and multiplicative reduction at 0.

We claim that $\mathrm{Swan}_\infty(\mathcal{G}) = 1$. To see this, consider the locus

$$\mathcal{E}^{aff} : y^2 = x^3 + x^2 + t, \ t \neq 0,$$

in $\mathbb{A}^2\times\mathbb{G}_m$, and denote by

$$\pi : \mathcal{E}^{aff} \to \mathbb{G}_m,$$
$$(x, y, t) \mapsto t,$$

the structural map. The only nonvanishing $R^i\pi_!\mathbb{Q}_\ell$ are $R^1\pi_!\mathbb{Q}_\ell = \mathcal{G}$, and $R^2\pi_!\mathbb{Q}_\ell = \mathbb{Q}_\ell(-1)$. Both are lisse. From the Leray spectral sequence, we get

$$\chi_c(\mathcal{E}^{aff}\otimes_k\bar{k}, \mathbb{Q}_\ell)$$
$$= -\chi_c(\mathbb{G}_m\otimes_k\bar{k}, R^1\pi_!\mathbb{Q}_\ell) + \chi_c(\mathbb{G}_m\otimes_k\bar{k}, R^2\pi_!\mathbb{Q}_\ell)$$
$$= -\chi_c(\mathbb{G}_m\otimes_k\bar{k}, \mathcal{G}) + \chi_c(\mathbb{G}_m\otimes_k\bar{k}, \mathbb{Q}_\ell(-1)).$$

Since $\mathcal{G}$ is lisse on $\mathbb{G}_m$ and tame at 0, the first term is $\mathrm{Swan}_\infty(\mathcal{G})$. The second term vanishes. So we must show that

$$\chi_c(\mathcal{E}^{\mathrm{aff}} \otimes_k \bar{k}, \mathbb{Q}_\ell) = 1.$$

We can view $\mathcal{E}^{\mathrm{aff}}$ as the open set of $\mathbb{A}^2$ where $y^2 - x^3 - x^2$ is invertible (solve for t). So we have

$$\begin{aligned}&\chi_c(\mathcal{E}^{\mathrm{aff}} \otimes_k \bar{k}, \mathbb{Q}_\ell)\\ &= \chi_c(\mathbb{A}^2 \otimes_k \bar{k}, \mathbb{Q}_\ell) - \chi_c((y^2 = x^3 + x^2 \text{ in } \mathbb{A}^2) \otimes_k \bar{k}, \mathbb{Q}_\ell)\\ &= 1 - \chi_c((y^2 = x^3 + x^2 \text{ in } \mathbb{A}^2) \otimes_k \bar{k}, \mathbb{Q}_\ell).\end{aligned}$$

To see that the affine curve $y^2 = x^3 + x^2$ has Euler characteristic zero, cut it up as the disjoint union of the origin (0, 0) and (putting $Y := y/x$) of the open curve ($Y^2 = x + 1$, $x \neq 0$), which is itself the complement of two points $Y = \pm 1$ in the curve $Y^2 = x + 1$, which is an $\mathbb{A}^1$. So our curve is "point + $\mathbb{A}^1$ - 2 points", so has Euler characteristic zero.

The Euler Poincaré formula gives

$$\chi_c(\mathbb{A}^1 \otimes_k \bar{k}, \mathcal{G}) = 1 + \chi_c(\mathbb{G}_m \otimes_k \bar{k}, \mathcal{G}) = 1 - \mathrm{Swan}_\infty(\mathcal{G}) = 0.$$

Therefore $H_c^1(\mathbb{A}^1 \otimes_k \bar{k}, \mathcal{G}) = 0$, and hence its Gr^0 quotient vanishes:

$$H^1(\mathbb{P}^1 \otimes_k \bar{k}, j_{\infty\star}\mathcal{G}) = 0.$$

Take $d \geq 3$ prime to 3. For f a polynomial of degree d with d distinct roots, $f^\star\mathcal{G}$ is a middle extension on $\mathbb{A}^1$, lisse on $\mathbb{A}^1 - f^{-1}(0)$. It has unipotent local monodromy at each zero of f, and its Swan conductor at ∞ is d. By the vanishing $H^1(\mathbb{P}^1 \otimes_k \bar{k}, j_{\infty\star}\mathcal{G}) = 0$ explained above, we have

$$\mathrm{Gr}^0(\mathcal{M}_{1,d})_{\mathrm{cst}} = 0,\ \mathrm{Gr}^0(\mathcal{M}_{1,d}) \cong \mathrm{Gr}^0(\mathcal{M}_{1,d})_{\mathrm{ncst}}.$$

Using the Euler Poincaré formula, we readily compute

$$\begin{aligned}N_d &:= \text{rank of } \mathrm{Gr}^0(\mathcal{M}_{1,d}) = \text{rank of } \mathrm{Gr}^0(\mathcal{M}_{1,d})_{\mathrm{ncst}}\\ &= -[(1 - d)2 + d - d] = 2d - 2.\end{aligned}$$

So for d prime to 3, we have $N_d \geq 9$ precisely for $d \geq 7$. For $d = 5$, we have $N_d = 8$, and for $d = 4$ we have $N_d = 6$.

Once again, for d = 4 or 5, G_{geom} for $\mathrm{Gr}^0(\mathcal{M}_{1,d}) = \mathrm{Gr}^0((\mathcal{M}_{1,d})_{\mathrm{ncst}})$ is finite. Exactly as in the first example, this finiteness results from the fact that for any nonconstant polynomial f of degree ≤ 6, the ellliptic surface X over $\mathbb{P}^1$ whose generic fibre is E_f is a rational surface.

Example 7.4.8 Take p = 2, and take as $E_t/k(t)$ the elliptic curve

$$y^2 + xy = x^3 + t.$$

This curve has good reduction over $\mathbb{G}_m$, and multiplicative reduction at 0.

We claim that $\mathrm{Swan}_\infty(\mathcal{G}) = 1$. The proof is almost identical to the proof of this same fact in the characteristic 3 example just above, and is left to the reader.

The Euler Poincaré formula gives

$$\chi_c(\mathbb{A}^1\otimes_k\bar{k}, \mathcal{G}) = 1 + \chi_c(\mathbb{G}_m\otimes_k\bar{k}, \mathcal{G}) = 1 - \mathrm{Swan}_\infty(\mathcal{G}) = 0.$$

Therefore $H_c{}^1(\mathbb{A}^1\otimes_k\bar{k}, \mathcal{G}) = 0$, and hence its Gr^0 quotient vanishes:

$$H^1(\mathbb{P}^1\otimes_k\bar{k}, j_{\infty\star}\mathcal{G}) = 0.$$

Take $d \geq 3$ prime to 2. For f a polynomial of degree d with d distinct roots, $f^\star\mathcal{G}$ is a middle extension on $\mathbb{A}^1$, lisse on $\mathbb{A}^1 - f^{-1}(0)$. It has unipotent local monodromy at each zero of f, and its Swan conductor at ∞ is d. By the vanishing $H^1(\mathbb{P}^1\otimes_k\bar{k}, j_{\infty\star}\mathcal{G}) = 0$ explained above, we have

$$\mathrm{Gr}^0(\mathcal{M}_{1,d})_{\mathrm{cst}} = 0,\ \mathrm{Gr}^0(\mathcal{M}_{1,d}) \cong \mathrm{Gr}^0(\mathcal{M}_{1,d})_{\mathrm{ncst}}.$$

Using the Euler Poincaré formula, we readily compute

$$N_d := \text{rank of } \mathrm{Gr}^0(\mathcal{M}_{1,d}) = \text{rank of } \mathrm{Gr}^0(\mathcal{M}_{1,d})_{\mathrm{ncst}}$$
$$= -[(1 - d)2 + d - d] = 2d - 2.$$

So for d prime to 2, we have $N_d \geq 9$ precisely for $d \geq 7$. For $d = 5$, we have $N_d = 8$, and for $d = 3$ we have $N_d = 4$.

Once again, for $d = 3$ or 5, G_{geom} for $\mathrm{Gr}^0(\mathcal{M}_{1,d}) = \mathrm{Gr}^0((\mathcal{M}_{1,d})_{\mathrm{ncst}})$ is finite. Exactly as in the first example, this finiteness results from the fact that for any nonconstant polynomial f of degree ≤ 6, the ellliptic surface X over $\mathbb{P}^1$ whose generic fibre is E_f is a rational surface.

(7.5) Appendix: Degeneration of Leray spectral sequences

(7.5.1) The following result is certainly well known to the experts, but I do not know a convenient reference.

Proposition 7.5.2 Let k be a finite field, C/k a proper, smooth, geometrically connected curve, X/k proper smooth, and geometrically connected of dimension $n \geq 2$, and $f : X \to C$ a proper morphism which over a dense open set U of C is smooth of relative dimension n-1, with geometrically connected fibres. Fix a prime number ℓ which is invertible in k. We have the following results.

1) The Leray spectral sequence for f,

$$E_2{}^{p,q} = H^p(C\otimes_k\bar{k}, R^qf_\star\mathbb{Q}_\ell) \Rightarrow H^{p+q}(X\otimes_k\bar{k}, \mathbb{Q}_\ell),$$

degenerates at E_2.

2) For $j : U \to C$ the inclusion of any dense open set over which f is smooth, and any integer i, the adjunction map

$$R^if_\star\mathbb{Q}_\ell \to j_\star j^\star R^if_\star\mathbb{Q}_\ell$$

is surjective, and sits in a short exact sequence

$$0 \to \mathrm{Pct}^i \to R^if_\star\mathbb{Q}_\ell \to j_\star j^\star R^if_\star\mathbb{Q}_\ell \to 0,$$

with Pct^i a punctual sheaf, supported in C - U.

3) The sheaf Pct^i vanishes if and only if the sheaf Pct^{2n-i} vanishes: the adjunction map

$$R^if_*\mathbb{Q}_\ell \to j_*j^*R^if_*\mathbb{Q}_\ell$$

is an isomorphism if and only if the adjunction map

$$R^{2n-i}f_*\mathbb{Q}_\ell \to j_*j^*R^{2n-i}f_*\mathbb{Q}_\ell$$

is an isomorphism.

4) For $j : U \to C$ the inclusion of any dense open set over which f is smooth, we have

$$R^if_*\mathbb{Q}_\ell \cong j_*j^*R^if_*\mathbb{Q}_\ell \text{ for } i \le 1.$$

proof Because $C\otimes_k\bar{k}$ has cohomological dimension 2, the only possibly nonvanishing $E_2{}^{p,q}$ have p either 0, 1, or 2. So the only possibly nonvanishing differentials in the spectral sequence are

$$d_2 : H^0(C\otimes_k\bar{k}, R^qf_*\mathbb{Q}_\ell) \to H^2(C\otimes_k\bar{k}, R^{q-1}f_*\mathbb{Q}_\ell).$$

The source is mixed of weight $\le q$, while the target, which we can rewrite as $H_c{}^2(U\otimes_k\bar{k}, R^{q-1}f_*\mathbb{Q}_\ell\,|U)$, is pure of weight q+1, and so d_2 must vanish, as it is $\mathrm{Gal}(\bar{k}/k)$-equivariant.

Assertion 2) results from the local invariant cycle theorem [De-Weil II, 3.6.1], according to which the canonical map

$$R^if_*\mathbb{Q}_\ell \to j_*j^*R^if_*\mathbb{Q}_\ell$$

is surjective for every i. So we have a short exact sequence on C,

$$0 \to \mathrm{Pct}^i \to R^if_*\mathbb{Q}_\ell \to j_*j^*R^if_*\mathbb{Q}_\ell \to 0,$$

for each i, where Pct^i is a punctual sheaf.

To prove 3), we argue as follows. Because Pct_i is punctual, we have the equivalence

$$\mathrm{Pct}^i = 0 \text{ if and only if } H^0(C\otimes_k\bar{k}, \mathrm{Pct}^i) = 0.$$

Taking cohomology on $C\otimes_k\bar{k}$, we find a short exact sequence of H^0's,

(∗)

$$0 \to H^0(C\otimes_k\bar{k}, \mathrm{Pct}^i) \to H^0(C\otimes_k\bar{k}, R^if_*\mathbb{Q}_\ell) \to H^0(C\otimes_k\bar{k}, j_*j^*R^if_*\mathbb{Q}_\ell) \to 0,$$

and isomorphisms,

(∗∗)
$$H^1(C\otimes_k\bar{k}, R^if_*\mathbb{Q}_\ell) \cong H^1(C\otimes_k\bar{k}, j_*j^*R^if_*\mathbb{Q}_\ell),$$
$$H^2(C\otimes_k\bar{k}, R^if_*\mathbb{Q}_\ell) \cong H^2(C\otimes_k\bar{k}, j_*j^*R^if_*\mathbb{Q}_\ell).$$

Now by Poincaré duality on C and on the smooth fibres of f, we have

(∗∗∗) the groups

$$H^p(C\otimes_k\bar{k}, j_*j^*R^qf_*\mathbb{Q}_\ell) \text{ and } H^{2-p}(C\otimes_k\bar{k}, j_*j^*R^{2n-2-q}f_*\mathbb{Q}_\ell)(n)$$

are $\mathbb{Q}_\ell$-dual, and hence have equal dimensions.

On the other hand, by Poincaré duality on X, we have

(****) the groups

$$H^i(X\otimes_k\bar{k}, \mathbb{Q}_\ell) \text{ and } H^{2n-i}(X\otimes_k\bar{k}, \mathbb{Q}_\ell)(n)$$

are $\mathbb{Q}_\ell$-dual, and hence have equal dimensions.

Equating their dimensions, using the degeneration proven in part 1) and (*) and (**) above, we find

$$\begin{aligned} &h^0(C\otimes_k\bar{k}, \mathrm{Pct}^i) + h^0(C\otimes_k\bar{k}, j_*j^*R^if_*\mathbb{Q}_\ell) \\ &\qquad + h^1(C\otimes_k\bar{k}, j_*j^*R^{i-1}f_*\mathbb{Q}_\ell) + h^2(C\otimes_k\bar{k}, j_*j^*R^{i-2}f_*\mathbb{Q}_\ell) \\ &= h^0(C\otimes_k\bar{k}, \mathrm{Pct}^{2n-i}) + h^0(C\otimes_k\bar{k}, j_*j^*R^{2n-i}f_*\mathbb{Q}_\ell) \\ &\qquad + h^1(C\otimes_k\bar{k}, j_*j^*R^{2n-i-1}f_*\mathbb{Q}_\ell) + h^2(C\otimes_k\bar{k}, j_*j^*R^{2n-i-2}f_*\mathbb{Q}_\ell). \end{aligned}$$

Cancelling using (***), we are left with

$$h^0(C\otimes_k\bar{k}, \mathrm{Pct}^i) = h^0(C\otimes_k\bar{k}, \mathrm{Pct}^{2n-i}),$$

which proves 3). For 4), we need only use 3) and remark that the sheaves $R^{2n-1}f_*\mathbb{Q}_\ell$ and $R^{2n}f_*\mathbb{Q}_\ell$ both vanish. This holds by proper base change, and the fact that f, being flat (an integral connected scheme over a regular scheme of dimension one is either flat or lies entirely over a single closed point), has all its fibres of the same dimension, here n-1. QED

Remark 7.5.3 The case i=2 of part 3) tells us that we have an isomorphism

$$R^2f_*\mathbb{Q}_\ell \cong j_*j^*R^2f_*\mathbb{Q}_\ell,$$

if and only if all the fibres of f are geometrically irreducible. Notice that having geometrically irreducible fibres is antithetical to having normal crosing divisors as special fibres. And even if we start with a situation in which all the fibres of f are geometrically irreducible, we will destroy this property if we blow up X at a closed point.

Corollary 7.5.4 Let k be a finite field, C/k a proper, smooth, geometrically connected curve, and D an effective divisor on C. Pick functions a_1, a_2, a_3, a_4, a_6 in L(D), and consider the locus in $\mathbb{P}^2\times(C - D)$ defined by the homogenized generalized Weierstrass equation

$$Y^2Z + a_1XYZ + a_3YZ^2 = X^3 + a_2X^2Z + a_4XZ^2 + a_6Z^3.$$

View this locus as a relative curve $f_0 : X_0 \to C - D$. Suppose that the discriminant $\Delta(a_1, a_2, a_3, a_4, a_6)$ is not identically zero in L(12D), and that at each zero δ of Δ in $(C - D)(\bar{k})$, one of the two following conditions holds.

a) Δ has a simple zero at δ.

b) Each of a_3, a_4, a_6 has a zero at δ, and a_6 has a simple zero at δ.

Then the total space X_0 is smooth over k, and for any prime ℓ invertible in k, the sheaves $R^0f_{0*}\mathbb{Q}_\ell$ and $R^1f_{0*}\mathbb{Q}_\ell$ on C - D are middle extensions, lisse on the open set C - D - Δ of C - D where Δ is invertible. The sheaf $R^0 f_{0*}\mathbb{Q}_\ell$ is the constant sheaf $\mathbb{Q}_\ell$.

proof Over $C - D - \Delta$, f_0 is a relative elliptic curve, so smooth over a smooth curve, so its total space is smooth over k. Fix a $\overline{k}$-valued zero δ of Δ in $C - D$. The fibre of f_0 over δ is smooth outside a single point, say x_0. Also X_0 is smooth at each point which is smooth in its fibre. That X_0 is smooth over k, or equivalently regular (k being a perfect field) at x_0 if either a) or b) holds is standard [Sil-ATEC, Lemma 9.5].

By resolution for surfaces, we can compactify $f_0 : X_0 \to C - D$ to a proper morphism $f : X \to C$ with X a proper smooth geometrically connected surface over k. Because f is proper and smooth with geometrically connected fibres over $C - D - \Delta$, all the $R^i f_*\mathbb{Q}_\ell$ are lisse on $C - D - \Delta$, and $R^0 f_*\mathbb{Q}_\ell \mid C - D - \Delta$ is the constant sheaf $\mathbb{Q}_\ell$. By Proposition 7.5.2, part 4), $R^0 f_*\mathbb{Q}_\ell$ and $R^1 f_*\mathbb{Q}_\ell$ are middle extensions on C, and hence their restrictions to $C - D$ are middle extension on $C - D$. That $R^0 f_{0*}\mathbb{Q}_\ell = \mathbb{Q}_\ell$ results from its being a middle extension, and the previously noted fact that $R^0 f_*\mathbb{Q}_\ell \mid C - D - \Delta$ is the constant sheaf $\mathbb{Q}_\ell$. QED

Corollary 7.5.5 Hypotheses and notations as in the previous corollary, consider the locus in $\mathbb{A}^2 \times (C - D)$ defined by the generalized Weierstrass equation

$$y^2 + a_1xy + a_3y = x^3 + a_2x^2 + a_4x + a_6.$$

View this locus as a relative affine curve $\pi : X_0{}^{\mathrm{aff}} \to C - D$. For any prime ℓ invertible in k, the natural map is an isomorphism

$$R^1\pi_!\mathbb{Q}_\ell \cong R^1 f_{0*}\mathbb{Q}_\ell,$$

and $R^1\pi_!\mathbb{Q}_\ell$ is a middle extension.

proof We obtain $\pi : X_0{}^{\mathrm{aff}} \to C - D$ from $f_0 : X_0 \to C - D$ by deleteting the zero section. That $R^1\pi_!\mathbb{Q}_\ell \to R^1 f_{0*}\mathbb{Q}_\ell$ is an isomorphism can be checked fibre by fibre: removing a single point from a connected proper curve does not change its $H_c{}^1$. We have proven in the above corollary that $R^1 f_{0*}\mathbb{Q}_\ell$ is a middle extension. QED

Chapter 8: One variable twists on curves

(8.1) Twist sheaves in the sense of [Ka-TLFM]

(8.1.1) In this chapter, we will use the general machine we have developed to discuss briefly the twist sheaves to which [Ka-TLFM] was devoted.

(8.1.2) We work over a finite field k. We fix a prime number ℓ invertible in k, and a field embedding $\overline{\mathbb{Q}}_\ell \subset \mathbb{C}$. We fix a nontrivial multiplicative character

$$\chi : k^\times \to \overline{\mathbb{Q}}_\ell^\times.$$

On $\mathbb{G}_m/k$, we have the corresponding lisse, rank one Kummer sheaf $\mathcal{L}_\chi$, whose trace function is given as follows: for E/k a finite extension field, and for α in $\mathbb{G}_m(E) = E^\times$, we have

$$\mathrm{Trace}(\mathrm{Frob}_{E,\alpha} \mid \mathcal{L}_\chi) = \chi(\mathrm{Norm}_{E/k}(\alpha)).$$

We denote by $j : \mathbb{G}_m \to \mathbb{A}^1$ the inclusion. On $\mathbb{A}^1$, we form the perverse sheaf

$$K := j_*\mathcal{L}_\chi(1/2)[1],$$

which is geometrically irreducible, and pure of weight zero. It is geometrically self dual only for χ of order two.

(8.1.3) We also fix a proper, smooth, geometrically connected curve C/k, of genus denoted g, together with an effective divisor D on C with

$$\deg(D) \geq 2g + 3.$$

We also fix a reduced closed subscheme S of C - D with $\dim(S) \leq 0$, i.e., S is a finite, possibly empty, set of closed points of C - D.

(8.1.4) The Riemann Roch space L(D) will occur in what follows both as a space of $\mathbb{A}^1$-valued functions on the open curve C-D, with

$$\tau : L(D) \to \mathrm{Hom}_{k\text{-schemes}}(C\text{-}D, \mathbb{A}^1)$$

the natural evaluation map, and as a space of $\mathbb{A}^1$-valued functions on the open curve C-D - S, again with τ the natural evaluation map. In either context, this space of functions is 4-separating. Indeed, it is d-separating, for d:= deg(D) - (2g-1).

(8.1.5) We suppose given on C - D - S a geometrically irreducible perverse sheaf L, which is ι-pure of weight zero and non-punctual. In other words, L is $\mathcal{G}[1]$ for $\mathcal{G}$ an irreducible middle extension sheaf on C - D - S, which on some dense open set is lisse, ι-pure of weight -1, and not geometrically constant. When S is nonempty, we denote by

$$j_S : C - D - S \to C - D$$

the inclusion. Then $j_{S*!}L := (j_{S*}\mathcal{G})[1]$ is a geometrically irreducible perverse sheaf on C - D, which is ι-pure of weight zero, non-punctual, and geometrically nonconstant. We have

$$H_c^i((C - D - S)\otimes_k \bar{k}, L) = 0 \text{ for } i \neq 0,$$

$$H_c^i((C - D)\otimes_k \bar{k}, j_{S*!}L) = 0 \text{ for } i \neq 0.$$

(8.1.6) We denote by $\mathrm{Sing}(L)_{\mathrm{finite}} \subset C - D - S$ the finite set of closed points of C - D - S at which L, or equivalently $\mathcal{G}$, is not lisse. And we denote by $\mathrm{Sing}(j_{S*}\mathcal{G})_{\mathrm{finite}}$ the finite set of closed points of C - D at which $j_{S*}\mathcal{G}$ is not lisse. So we have trivial inclusions

$$\mathrm{Sing}(L)_{\mathrm{finite}} \subset \mathrm{Sing}(j_{S*}\mathcal{G})_{\mathrm{finite}} \subset S \cup \mathrm{Sing}(L)_{\mathrm{finite}}.$$

(8.1.7) Recall [Ka-TLFM, 5.0.6] that we denote by

$$Fct(C, \deg(D), D, S) \subset L(D)$$

the dense open set whose $\bar{k}$-points consist of those functions f in $L(D)(\bar{k})$ whose divisor of poles is D, and which, over $\bar{k}$, have deg(D) distinct zeroes, all of which lie in C - D - S.

(8.1.8) There are now two slightly different "standard inputs", cf. 1.15.4, we are interested in. The first is

the integer m = 1,
the perverse sheaf $K := j_*\mathcal{L}_\chi(1/2)[1]$ on $\mathbb{A}^1/k$,
the affine k-scheme V := C - D - S,
the k-morphism $h : V \to \mathbb{A}^1$ given by h = 0,
the perverse sheaf $L := \mathcal{G}[1]$ on C - D - S,
the integer d:= deg(D) - (2g-1),
the space of functions $(L(D), \tau)$ on C - D - S.

With this input, we form the perverse sheaf on L(D)

$$M_S := \mathrm{Twist}(L, K, \mathcal{F} = L(D) \text{ on } C - D - S, h = 0).$$

(8.1.9) If we take S to be empty, we get the second standard input of interest, namely

the integer m = 1,
the perverse sheaf $K := j_*\mathcal{L}_\chi(1/2)[1]$ on $\mathbb{A}^1/k$,
the affine k-scheme V := C - D,
the k-morphism $h : V \to \mathbb{A}^1$ given by h = 0,
the perverse sheaf $j_{S*!}L := j_{S*}\mathcal{G}[1]$ on C - D,
the integer d:= deg(D) - (2g-1),
the space of functions $(L(D), \tau)$ on C - D.

With this input, we form the perverse sheaf on L(D)

$$M_\emptyset := \mathrm{Twist}(j_{S*!}L, K, \mathcal{F} = L(D) \text{ on } C - D, h = 0).$$

There is a natural excision morphism

$$M_S \to M_\emptyset$$

of perverse sheaves on L(D).

Lemma 8.1.10 The excision map induces an isomorphism

$$\mathrm{Gr}^0(M_S((\ell(D) - 1)/2)) \cong \mathrm{Gr}^0(M_\emptyset((\ell(D) - 1)/2)).$$

proof We can also think of M_S as formed on C - D using $j_{S!}L$:

$$M_S := \mathrm{Twist}(j_{S!}L, K, \mathcal{F} = L(D) \text{ on } C - D, h = 0).$$

Since $H_c{}^*(\mathbb{A}\otimes\bar{k}, K) = 0$ for our K, we can apply Corollary 1.4.5 to the short exact sequence

$$0 \to i_S{}^*j_{S*}\mathcal{G} \to j_{S!}L \to j_{S*!}L \to 0.$$

We get a short exact sequence of perverse sheaves on L(D),

$$0 \to \mathrm{Ker} \to M_S \to M_\emptyset \to 0,$$

with

$$\mathrm{Ker} := \mathrm{Twist}(\, i_S{}^*j_{S*}\mathcal{G}, K, \mathcal{F} = L(D) \text{ on } C - D, h = 0).$$

Twisting, we get an exact sequence

$$0 \to \mathrm{Ker}((\ell(D) - 1)/2) \to M_S((\ell(D) - 1)/2) \to M_\emptyset((\ell(D) - 1)/2) \to 0.$$

Because $\mathcal{G}$ is mixed of weight ≤ -1, $i_S{}^*j_{S*}\mathcal{G}$ is a punctual perverse sheaf on C - D which is ι-mixed of weight ≤ -1. By Corollary 1.15.11, $\mathrm{Ker}((\ell(D) - 1)/2)$ is ι-mixed of weight ≤ -1, and hence its Gr^0 vanishes. So applying the exact functor [BBD, 5.3.5] Gr^0 to this exact sequence gives the asserted result.

Lemma 8.1.11 1) The perverse sheaves M_S and $M_\emptyset$ on L(D) are each lisse on $Fct(C, \deg(D), D, \mathrm{SUSing}(L)_{\mathrm{finite}})$.

2) On this open set, $\mathrm{Gr}^0(\mathcal{M}_S) \cong \mathrm{Gr}^0(\mathcal{M}_\emptyset)$ is the lisse sheaf $\mathrm{Twist}_{\chi,C,D}(j_{S*}\mathcal{G})$ of [Ka-TLFM, 5.2.2.1].

3) We have the inequality

$$\mathrm{rank}(\mathrm{Gr}^0(\mathcal{M}_\emptyset)) \ge (2g - 2 + \deg(D))\mathrm{rank}(\mathcal{G}) + \#\mathrm{Sing}(j_{S*}\mathcal{G})_{\mathrm{finite}}(\bar{k}).$$

proof 1) That each is lisse on $Fct(C, \deg(D), D, \mathrm{SUSing}(L)_{\mathrm{finite}})$ is proven by first showing that, on this open set, each is of the form

(a single sheaf)$[\ell(D)]$,

and then by using the Euler Poincaré formula to show that the stalks of this sheaf have constant rank on the open set, cf. [Ka-TLFM, 5.2.1]. The sheaf in question being of perverse origin, it is lisse if its stalks have constant rank, cf. [Ka-SMD, Proposition 11].

2) That $\mathrm{Gr}^0(\mathcal{M}_\emptyset)$ is the lisse sheaf $\mathrm{Twist}_{\chi,C,D}(j_{S*}\mathcal{G})$ on the space $Fct(C, \deg(D), D, \mathrm{SUSing}(L)_{\mathrm{finite}})$ is essentially a tautology.

3) The inequality for its rank is proven in [Ka-TLFM, 5.2.1, part 5)]. QED

Lemma 8.1.12 We have $\mathrm{Gr}^0(\mathcal{M}_\emptyset) \cong \mathrm{Gr}^0(\mathcal{M}_\emptyset)_{\mathrm{ncst}}$.

proof Since $H_c{}^*(\mathbb{A}\otimes\bar{k}, K) = 0$, Kunneth gives

$$H_c{}^*((V\times\mathbb{A}^1)\otimes\bar{k}, \mathrm{pr}_1{}^*L\otimes\mathrm{pr}_2{}^*K) = 0.$$

By Corollary 1.20.3, part 3), this implies that both $\mathrm{Gr}^0(M_\emptyset)_{\mathrm{cst}}$ and $\mathrm{Gr}^0(\mathcal{M}_\emptyset)_{\mathrm{cst}}$ vanish. QED

(8.2) Monodromy of twist sheaves in the sense of [Ka-TLFM]

Theorem 8.2.1 (uses the truth of the Larsen Eighth Moment Conjecture in Guralnick-Tiep form) Denote by N the rank of $\mathrm{Gr}^0(\mathcal{M}_\emptyset) = \mathrm{Twist}_{\chi,C,D}(j_{S\star}\mathcal{G})$. Suppose that $\deg(D) \geq 2g + 7$. Then we have the following results.

1) If either χ is not of order 2, or K is not geometrically self dual, then G_{geom} for $\mathrm{Gr}^0(\mathcal{M}_\emptyset)$ contains SL(N).

2) If χ has order two, and if $\mathcal{G}| C - D - S - \mathrm{Sing}(L)$ is, geometrically, symplectically self dual, then G_{geom} for $\mathrm{Gr}^0(\mathcal{M})_{ncst}$ is either SO(N) or O(N).

3) If $N \geq 8$, if χ has order two, and if $\mathcal{G}| C - D - S - \mathrm{Sing}(L)$ is, geometrically, orthogonally self dual, then G_{geom} for $\mathrm{Gr}^0(\mathcal{M})_{ncst}$ is Sp(N).

proof The hypothesis that $\deg(D) \geq 2g + 7$ insures that the space of functions L(D) is 8-separating. The inequality

$$\mathrm{rank}(\mathrm{Gr}^0(\mathcal{M}_\emptyset)) \geq (2g - 2 + \deg(D))\mathrm{rank}(\mathcal{G}) + \#\mathrm{Sing}(j_{S\star}\mathcal{G})_{finite}(\bar{k}),$$

proven above in Lemma 8.1.11, part 3), together with the hypothesis that $\deg(D) \geq 2g + 7$ shows that $N \geq 5$, and that $N \geq 10$ if $\mathrm{rank}(\mathcal{G}) \geq 2$. So in case 1), we simply apply Larsen's Alternative and the Guralnick-Tiep Theorem 2.5.4. In case 2), we have $\mathrm{rank}(\mathcal{G}) \geq 2$ because $\mathcal{G}$ is symplectically self dual, so $\mathrm{rank}(\mathcal{G})$ is even. Therefore $N \geq 10$ in case 2). In case 3), we have $N \geq 8$ by hypothesis. So cases 2) and 3) result from Theorem 2.5.2. QED

Theorem 8.2.2 Suppose the following four conditions hold.

1) $\deg(D) \geq 2g + 3$.

2) χ has order two.

3) $\mathcal{G}| C - D - S - \mathrm{Sing}(L)$ is, geometrically, symplectically self dual.

4) There exists a point β in $\mathrm{Sing}(j_{S\star}\mathcal{G})_{finite}(\bar{k})$ at which the local monodromy is tame, with $\mathcal{G}(\beta)/\mathcal{G}(\beta)^{I(\beta)}$ of odd dimension.

Then we have the following results concerning the group G_{geom} for $\mathrm{Gr}^0(\mathcal{M}_\emptyset) = \mathrm{Twist}_{\chi,C,D}(j_{S\star}\mathcal{G})$.

1) G_{geom} contains an element of determinant -1.

2) If the local monodromy at β is a unipotent pseudoreflection, then G_{geom} contains a reflection.

proof Replacing $\mathcal{G}$ on C - D - S by $j_{S\star}\mathcal{G}$ on C - D, we reduce to the case when S is empty. For any f in $Fct(C, \deg(D), D, \mathrm{Sing}(L)_{finite})(\bar{k})$, the construction $\lambda \mapsto \lambda - f$ defines a map

$$\rho : U := \mathbb{A}^1 - \{\mathrm{CritValues}(f) \cup f(\mathrm{Sing}(L)_{finite})\} \to Fct(C, \deg(D), D, \mathrm{Sing}(L)_{finite}).$$

Since f is a finite map $f : C - D \to \mathbb{A}^1$, $f_\star L := f_\star\mathcal{G}[1]$ is perverse on $\mathbb{A}^1$.

According to [Ka-TLFM, 5.3.5], we have a geometric isomorphism of lisse sheaves on $U := \mathbb{A}^1 - \{\mathrm{CritValues}(f) \cup f(\mathrm{Sing}(L)_{\mathrm{finite}})\}$,

$$\rho^*\mathrm{Twist}_{\chi,C,D}(\mathcal{G}) \cong (j_*\mathcal{L}_\chi[1]) *_{\mathrm{mid}} (f_*\mathcal{G}[1]) \mid U.$$

We will show that for a well chosen f, G_{geom} for this pullback to a curve already contains an element of the required kind (either having determinant -1, or being a reflection) as a local monodromy.

Choose the function f so that it satisfies the following extra condition: for each β in $\mathrm{Sing}(L)_{\mathrm{finite}}(\bar{k})$, the fibre $f^{-1}(f(\beta))$ over $f(\beta)$ consists of deg(D) distinct $\bar{k}$-points, of which only β lies in $\mathrm{Sing}(L)_{\mathrm{finite}}$. This is possible, by [Ka-TLFM, 2.2.6, part 2), applied with $S := \mathrm{Sing}(L)_{\mathrm{finite}}(\bar{k})$]. Then by the Irreducible Induction Criterion [Ka-TLFM, 3.3.1], the perverse sheaf $f_*L := f_*\mathcal{G}[1]$ is geometrically irreducible on $\mathbb{A}^1$. As f_*L is non-punctual and has generic rank $\deg(D)\mathrm{rank}(\mathcal{G}) > 1$, it certainly has $\mathcal{P}$.

At the point $f(\beta)$, use f to identify the inertia group $I(\beta)$ with the inertia group $I(f(\beta))$. With this identification, we have an $I(f(\beta))$-isomorphism

$$(f_*\mathcal{G})(f(\beta)) \cong \mathcal{G}(\beta) \oplus (\text{a trivial } I(f(\beta))\text{-representation}),$$

simply because f is finite etale over $f(\beta)$, and β is the only singularity of $\mathcal{G}$ in the fibre $f^{-1}(f(\beta))$ over $f(\beta)$.

Now apply Theorem 6.1.18, part 3), to $K = j_*\mathcal{L}_\chi(1/2)[1]$ and to f_*L, at the finite singularity 0 of K, and at the finite singularity $f(\beta)$ of f_*L. We find that for the lisse sheaf on U

$$\mathcal{H} := \rho^*\mathrm{Twist}_{\chi,C,D}(\mathcal{G}),$$

we have an $I(f(\beta))$-isomorphism

$$\mathcal{H}(f(\beta))/\mathcal{H}(f(\beta))^{I(f(\beta))} \cong (\mathcal{G}(\beta)/\mathcal{G}(\beta)^{I(\beta)}) \otimes \mathcal{L}_{\chi(x-\beta)}.$$

Suppose first that $\mathcal{G}(\beta)/\mathcal{G}(\beta)^{I(\beta)}$ has odd dimension. Just as in the proof of Theorem 7.2.5), the fact that $\mathcal{G}$ is symplectic shows that $I(\beta)$ acts trivially on $\det(\mathcal{G}(\beta)/\mathcal{G}(\beta)^{I(\beta)})$, and hence acts by $\mathcal{L}_{\chi(x-\beta)}$ on $\det((\mathcal{G}(\beta)/\mathcal{G}(\beta)^{I(\beta)}) \otimes \mathcal{L}_{\chi(x-\beta)})$. So we find that $I(f(\beta))$ acts on $\mathcal{H}(f(\beta))/\mathcal{H}(f(\beta))^{I(f(\beta))}$ and hence also on $\mathcal{H}(f(\beta))$ with determinant $\mathcal{L}_{\chi(x-f(\beta))}$. Thus we get elements of determinant -1 in the inertia group at $f(\beta)$.

If the local monodromy at β is a unipotent pseudoreflection, i.e., if $\mathcal{G}(\beta)/\mathcal{G}(\beta)^{I(\beta)}$ has dimension one, then the above discussion shows that local monodromy at $f(\beta)$ is a reflection. QED

Corollary 8.2.3 Suppose that the following four conditions hold.
1) $\deg(D) \geq 2g + 3$.
2) χ has order two.
3) $\mathcal{G} \mid C - D - S - \mathrm{Sing}(L)$ is, geometrically, symplectically self dual.
4) There exists a point β in $\mathrm{Sing}(j_{S*}\mathcal{G})_{\mathrm{finite}}(\bar{k})$ at which the local

monodromy is a unipotent pseudoreflection.
Then we have the following results concerning the group G_{geom} for $Gr^0(\mathcal{M}_\varnothing) = Twist_{\chi,C,D}(j_{S*}\mathcal{G})$.

1) If $N := rank(Gr^0(\mathcal{M}_1)) \geq 9$, then G_{geom} is $O(N)$.

2) If N is 7 or 8, then G_{geom} is either $O(N)$ or it is the Weyl group of E_N in its reflection representation.

proof By the previous result, G_{geom} contains a reflection. Now repeat the first paragraph of the proof of Theorem 7.2.3. QED

(8.2.4) We also have the following result, analogous to Theorem 7.2.7.

Theorem 8.2.5 Denote by N the rank of $Gr^0(\mathcal{M}_\varnothing) = Twist_{\chi,C,D}(j_{S*}\mathcal{G})$. Suppose that

1) $\deg(D) \geq 2g+3$,

2) for some point s in $S(\bar{k})$, $\mathcal{G}(s)/\mathcal{G}(s)^{I(s)}$ is not a semisimple representation of $I(s)$.

Then we have the following results for the group G_{geom} for $Gr^0(\mathcal{M}_\varnothing) = Twist_{\chi,C,D}(j_{S*}\mathcal{G})$.

1) If either χ is not of order 2, or K is not geometrically self dual, then $N \geq 2$, and $G_{geom} \supset SL(N)$.

2) If χ has order two, and if $\mathcal{G}| C - D - S - Sing(L)$ is, geometrically, symplectically self dual, then $N \geq 3$, and G_{geom} is $SO(N)$ or $O(N)$.

3) If χ has order two, and if $\mathcal{G}| C - D - S - Sing(L)$ is, geometrically, orthogonally self dual, then $N \geq 2$, and $G_{geom} = Sp(N)$.

proof The proof is entirely analogous to the proof of Theorem 7.2.7, and is left to the reader. QED

Chapter 9: Weierstrass sheaves as inputs

(9.1) Weierstrass sheaves

(9.1.1) In this chapter, we will study various several parameter families of elliptic curves, which will serve as "inputs" in later chapters. What these families have in common is that all of them are given as families of Weierstrass equations.

(9.1.2) We begin with some general results.

Proposition 9.1.3 Let k be a field, and S/k a separated k-scheme of finite type which is locally a complete intersection, everywhere of some dimension d. Pick functions a_1, a_2, a_3, a_4, a_6 in $\Gamma(S, \mathcal{O}_S)$, and consider the locus $\mathcal{E}^{aff}$ in $\mathbb{A}^2\times S$ defined by the generalized Weierstrass equation

$$y^2 + a_1xy + a_3y = x^3 + a_2x^2 + a_4x + a_6.$$

View $\mathcal{E}^{aff}$ as a relative affine curve $\pi : \mathcal{E}^{aff} \to S$. Then for any prime number ℓ invertible in k, $R^1\pi_!\overline{\mathbb{Q}}_\ell[d]$ is a perverse sheaf on S.

proof Because $\mathcal{E}^{aff}$ is defined in $\mathbb{A}^2\times S$ by one equation, and S itself is locally a complete intersection, we see that $\mathcal{E}^{aff}$ itself is locally a complete intersection, everywhere of dimension d+1. Therefore [Ka-SMD, Corollary 6] $\overline{\mathbb{Q}}_\ell[d+1]$ is perverse on $\mathcal{E}^{aff}$. Now $\mathcal{E}^{aff}$ is a finite flat double cover of the x-line, by the map

$$\rho : \mathcal{E}^{aff} \to \mathbb{A}^1\times S,$$
$$(x, y, s) \mapsto (x, s).$$

Since ρ is finite, $\rho_*\overline{\mathbb{Q}}_\ell[d+1] = \rho_!\mathbb{Q}_\ell[d+1]$ is perverse on $\mathbb{A}^1\times S$. Because ρ is finite and flat of degree 2, we have the Trace map

$$\text{Trace} : \rho_*\overline{\mathbb{Q}}_\ell = \rho_*\rho^*\overline{\mathbb{Q}}_\ell \to \overline{\mathbb{Q}}_\ell.$$

We also have the natural adjunction map

$$\overline{\mathbb{Q}}_\ell \to \rho_*\rho^*\overline{\mathbb{Q}}_\ell.$$

The composite

$$\overline{\mathbb{Q}}_\ell \to \rho_*\rho^*\overline{\mathbb{Q}}_\ell \to \overline{\mathbb{Q}}_\ell$$

is multiplication by 2, cf. [SGA 4, Expose XVIII, Thm. 2.9, Var4, I]. So we have a direct sum decomposition

$$\rho_*\overline{\mathbb{Q}}_\ell = \overline{\mathbb{Q}}_\ell \oplus (\mathcal{K} := \text{Ker}(\text{Trace} : \rho_*\overline{\mathbb{Q}}_\ell \to \overline{\mathbb{Q}}_\ell)).$$

Thus $\mathcal{K}[d+1]$ is a direct summand of $\rho_*\overline{\mathbb{Q}}_\ell[d+1]$, hence is itself perverse on $\mathbb{A}^1\times S$.

We make use of this perversity as follows. We factor π as

$$\mathcal{E}^{\mathrm{aff}} \xrightarrow{\rho} \mathbb{A}^1\times S \xrightarrow{pr_2} S.$$

We will show that

$$R^1(pr_2)_!\mathcal{K} \cong R^1\pi_!\overline{\mathbb{Q}}_\ell,$$

$$R^i(pr_2)_!\mathcal{K} = 0 \text{ for } i \neq 1.$$

Once we have shown this, then we have

$$R(pr_2)_!\mathcal{K}[d+1] = R^1(pr_2)_!\mathcal{K}[d] = R^1\pi_!\overline{\mathbb{Q}}_\ell[d].$$

This shows that $R(pr_2)_!\mathcal{K}[d+1]$ is semiperverse on S, since it is a single sheaf, concentrated in dimension $d = \dim(S)$. To show that $R(pr_2)_!\mathcal{K}[d+1]$ is perverse, we must show that its dual $R(pr_2)_*(D(\mathcal{K}[d+1]))$ is semiperverse. But $D(\mathcal{K}[d+1])$ is perverse, being the dual of the perverse object $\mathcal{K}[d+1]$, and pr_2 is an affine morphism, so the semiperversity results from Artin's theorem, cf. [SGA 4, Expose XIV, 3.1] and [BBD, 4.1.1].

We next calculate the $R^i(pr_2)_!\mathcal{K}$. Because ρ is finite, the Leray spectral sequence gives

$$R^i\pi_!\overline{\mathbb{Q}}_\ell = R^i(pr_2)_!(\rho_*\overline{\mathbb{Q}}_\ell) = R^i(pr_2)_!\overline{\mathbb{Q}}_\ell \oplus R^i(pr_2)_!\mathcal{K}.$$

The cohomology of $\mathbb{A}^1$ is known: we have

$$R^2(pr_2)_!\overline{\mathbb{Q}}_\ell = \overline{\mathbb{Q}}_\ell(-1),$$

$$R^i(pr_2)_!\overline{\mathbb{Q}}_\ell = 0 \text{ for } i \neq 2.$$

The vanishing of $R^1(pr_2)_!\overline{\mathbb{Q}}_\ell$ shows that $R^1(pr_2)_!\mathcal{K} \cong R^1\pi_!\overline{\mathbb{Q}}_\ell$. The only possibly nonvanishing $R^i(pr_2)_!\mathcal{K}$ have i in {0, 1, 2}. To show that $R^i(pr_2)_!\mathcal{K} = 0$ for $i = 0$ and 2, we argue as follows. The geometric fibres of $\mathcal{E}^{\mathrm{aff}}/S$ are geometrically irreducible affine plane curves, so $R^0\pi_!\overline{\mathbb{Q}}_\ell = 0$ [Ka-Sar-RMFEM, 10.1.5, applied fibre by fibre] and $R^2\pi_!\overline{\mathbb{Q}}_\ell = \overline{\mathbb{Q}}_\ell(-1)$. Looking fibre by fibre and comparing dimensions, we get the vanishing of $R^i(pr_2)_!\mathcal{K}$ for $i = 0$ and 2. QED

Proposition 9.1.4 Hypotheses and notations as in Proposition 9.1.3 above, suppose in addition that k is a finite field, and that S/k is smooth and geometrically connected. Suppose further that the discriminant Δ of our Weierstrass equation

$$y^2 + a_1xy + a_3y = x^3 + a_2x^2 + a_4x + a_6$$

is nonzero in $\Gamma(S, \mathcal{O}_S)$, and that its j invariant is nonconstant. Then we have the following results.

1) The perverse sheaf $R^1\pi_!\overline{\mathbb{Q}}_\ell[d]((d+1)/2)$ on S is mixed of weight ≤ 0.

2) $\mathrm{Gr}^0(R^1\pi_!\overline{\mathbb{Q}}_\ell[d]((d+1)/2))$ is geometrically irreducible, geometrically nonconstant, and autodual. Its restriction to $S[1/\Delta]$ is $\mathcal{W}(d/2)[d]$, for

$\mathcal{W}$ the lisse sheaf $R^1\pi_!\overline{\mathbb{Q}}_\ell(1/2)|S[1/\Delta]$, which is pure of weight zero and symplectically self dual, with G_{geom} = SL(2).

proof That $R^1\pi_!\overline{\mathbb{Q}}_\ell[d]((d+1)/2)$ is mixed of weight ≤ 0, or equivalently that $R^1\pi_!\overline{\mathbb{Q}}_\ell$ on S is mixed of weight ≤ 1, is a special case of Deligne's theorem [De-Weil II, 3.3.1]. In the case at hand, it is due to Hasse. On the open set $S[1/\Delta]$ where Δ is invertible, $R^1\pi_!\overline{\mathbb{Q}}_\ell$ is lisse of rank 2 and (by [Hasse]) pure of weight one. At finite field-valued points of S where Δ vanishes, the stalk of $R^1\pi_!\overline{\mathbb{Q}}_\ell$ is the $H_c{}^1$ of a singular Weierstrass cubic, which either vanishes ("additive case") or is one-dimensional ("multiplicative case"); in the latter case, the Frobenius eigenvalue is ±1. The hypothesis that j is nonconstant implies [De-Weil II, 3.5.5] that G_{geom} for $R^1\pi_!\overline{\mathbb{Q}}_\ell \mid S[1/\Delta]$ is SL(2). This in turn implies that on $S[1/\Delta]$, the perverse sheaf $R^1\pi_!\overline{\mathbb{Q}}_\ell[d]((d+1)/2)$ is geometrically irreducible, and is its own Gr^0.

To show that $Gr^0(R^1\pi_!\overline{\mathbb{Q}}_\ell[d]((d+1)/2))$ is geometrically irreducible on S, we apply the criterion of Second Corollary 1.8.3, part 3). Putting $M := R^1\pi_!\overline{\mathbb{Q}}_\ell[d]((d+1)/d)$, we must show that there exists a real $\varepsilon > 0$ such that for variable finite extensions E/k, we have

$$\Sigma_{x \text{ in } S(E)} |M(E, x)|^2 = 1 + O((\#E)^{-\varepsilon/2}).$$

Thanks to the Othogonality Theorem 1.7.2, part 3), the geometric irreducibility of $R^1\pi_!\overline{\mathbb{Q}}_\ell \mid S[1/\Delta]$ tells that we have

$$\Sigma_{x \text{ in } S[1/\Delta](E)} |M(E, x)|^2 = 1 + O((\#E)^{-1/2}).$$

The terms where $\Delta = 0$ make a negligible contribution. Indeed, each term $|M(E, x)|^2$ for x an E-valued point of the locus $\Delta = 0$ is, as noted above, either 0 or $(\#E)^{-d-1}$, and the number of terms over which we sum is $O((\#E)^{d-1})$.

That the geometrically irreducible object M is self dual results (by Third Corollary 1.8.4, part 3)) from the fact that via any ι, its trace function takes real values.

That $Gr^0(R^1\pi_!\overline{\mathbb{Q}}_\ell[d]((d+1)/2))$ is not geometrically constant is obvious from the fact that already on the dense open set $S[1/\Delta]$ it is not geometrically constant, indeed its G_{geom} is SL(2). QED

Vanishing Proposition 9.1.5 Let k be a field, d ≥ 1 an integer, $\mathbb{A}^d/k$ the affine space $\mathrm{Spec}(k[t_1,..., t_d])$. Choose a_1, a_2, a_3, and a_4 arbitrarily in the subring $k[t_1,..., t_{d-1}]$ of polynomials which do not involve t_d. Choose $a_6 := t_d$. Over $\mathbb{A}^d/k$, consider the Weierstrass cubic $\mathcal{E}^{aff}$

$$y^2 + a_1xy + a_3y = x^3 + a_2x^2 + a_4x + a_6.$$

Then for any prime ℓ invertible in k, we have

$$H_c{}^*(\mathbb{A}^d \otimes_k \bar{k}, R^1\pi_!\bar{\mathbb{Q}}_\ell) = 0,$$

$$H_c{}^*(\mathbb{A}^d \otimes_k \bar{k}, R^1\pi_!\bar{\mathbb{Q}}_\ell[d]) = 0.$$

proof The two statements are trivially equivalent. We will prove the first. Consider the Leray spectral sequence for π,

$$E_2{}^{a,b} := H_c{}^a(\mathbb{A}^d \otimes_k \bar{k}, R^b\pi_!\bar{\mathbb{Q}}_\ell) \Rightarrow H_c{}^{a+b}(\mathcal{E}^{aff} \otimes_k \bar{k}, \bar{\mathbb{Q}}_\ell).$$

On the one hand, we have

$$R^i\pi_!\bar{\mathbb{Q}}_\ell = 0 \text{ for i not in } \{1, 2\},$$

$$R^2\pi_!\bar{\mathbb{Q}}_\ell = \bar{\mathbb{Q}}_\ell(-1).$$

So $E_2{}^{a,b}$ vanishes unless b is 1 or 2. The only nonvanishing $E_2{}^{a,2}$ term is $E_2{}^{2d,2}$. The terms $E_2{}^{a,1}$ vanish for a outside [0, 2d]. From this pattern of vanishing

```
                                    *
*    *    *    *    ...             *
```

we see that the spectral sequence degenerates at E_2, and that

$$H_c{}^a(\mathbb{A}^d \otimes_k \bar{k}, R^1\pi_!\bar{\mathbb{Q}}_\ell) \cong H_c{}^{a+1}(\mathcal{E}^{aff} \otimes_k \bar{k}, \bar{\mathbb{Q}}_\ell), \text{ for } 0 \le a \le 2d.$$

Because a_6 is t_d, and the other a_i do not involve t_d, we can use the Weierstrass equation to "solve for a_6", i.e., we have

$$\mathcal{E}^{aff} \cong \mathbb{A}^{d+1}, \text{ coordinates } x, y, t_1,\ldots, t_{d-1}.$$

Therefore we have

$$H_c{}^i(\mathcal{E}^{aff} \otimes_k \bar{k}, \bar{\mathbb{Q}}_\ell) = 0 \text{ for } i \neq 2d+2.$$

Hence we have $H_c{}^*(\mathbb{A}^d \otimes_k \bar{k}, R^1\pi_!\bar{\mathbb{Q}}_\ell) = 0$. QED

(9.2) The situation when 2 is invertible

(9.2.1) In this section, we work over a $\mathbb{Z}[1/2]$-scheme S, i.e., a scheme S on which 2 is invertible. This allows us to complete the square in a Weierstrass cubic, and to reduce to the case $a_1 = a_3 = 0$. More precisely, pick functions a_1, a_2, a_3, a_4, a_6 in $\Gamma(S, \mathcal{O}_S)$, and consider the Weierstrass cubic $\mathcal{E}^{aff}$ defined by

$$y^2 + a_1xy + a_3y = x^3 + a_2x^2 + a_4x + a_6.$$

The quantity $\tilde{y} := y + (a_1x + a_3)/2$ satisfies the equation

$$\tilde{y}^2 = x^3 + \tilde{a}_2x^2 + \tilde{a}_4x + \tilde{a}_6,$$

where

$$\tilde{a}_2 := a_2 + (a_1)^2/4,$$

$$\tilde{a}_4 := a_4 + (a_1a_3)/2,$$

$$\tilde{a}_6 := a_6 + (a_3)^2/4.$$

Twisting Lemma 9.2.2 Let S be a $\mathbb{Z}[1/2]$-scheme, a_2, a_4, a_6 functions in $\Gamma(S, \mathcal{O}_S)$. Consider the locus $\mathcal{E}^{aff}$ in $\mathbb{A}^2\times S$ defined by the Weierstrass cubic

$$y^2 = x^3 + a_2x^2 + a_4x + a_6.$$

View $\mathcal{E}^{aff}$ as a relative affine curve $\pi : \mathcal{E}^{aff} \to S$.

Fix an invertible function f on S, and denote by $\mathcal{E}_f{}^{aff}$ the locus in $\mathbb{A}^2\times S$ defined by the Weierstrass equation

$$y^2 = x^3 + fa_2x^2 + f^2a_4x + f^3a_6.$$

View $\mathcal{E}_f{}^{aff}$ as a relative affine curve $\pi_f : \mathcal{E}^{aff} \to S$.
For any prime number ℓ invertible on S, we have a canonical isomorphism of sheaves on S,

$$R^1(\pi_f)_!\overline{\mathbb{Q}}_\ell \cong \mathcal{L}_{\chi_2(f)}\otimes R^1\pi_!\overline{\mathbb{Q}}_\ell,$$

where χ_2 denotes the quadratic character, and $\mathcal{L}_{\chi_2}$ the corresponding Kummer sheaf.

proof Denote by $\rho : \mathcal{E}^{aff} \to \mathbb{A}^1\times S$ the projection onto the x-line. Because 2 is invertible, and $\mathcal{E}^{aff}$ has equation of the form $y^2 = f(x)$, we have

$$\rho_*\overline{\mathbb{Q}}_\ell = \overline{\mathbb{Q}}_\ell \oplus \mathcal{L}_{\chi_2(f(x))}$$
$$= \overline{\mathbb{Q}}_\ell \oplus \mathcal{L}_{\chi_2(x^3 + a_2x^2 + a_4x + a_6)}.$$

Just as in the proof of Proposition 9.1.3, the vanishing of $R^1(pr_2)_!\overline{\mathbb{Q}}_\ell$ gives

$$R^1\pi_!\overline{\mathbb{Q}}_\ell \cong R^1(pr_2)_!\mathcal{L}_{\chi_2(x^3 + a_2x^2 + a_4x + a_6)}.$$

Applying this same formula to π_f, we get

$$R^1(\pi_f)_!\overline{\mathbb{Q}}_\ell \cong R^1(pr_2)_!\mathcal{L}_{\chi_2(x^3 + fa_2x^2 + f^2a_4x + f^3a_6)}.$$

Now apply the S-automorphism $(x, s) \mapsto (xf, s)$ of $\mathbb{A}^1\times S$ to rewrite the left hand side as

$$R^1(pr_2)_!\mathcal{L}_{\chi_2(x^3 + fa_2x^2 + f^2a_4x + f^3a_6)}$$
$$\cong R^1(pr_2)_!\mathcal{L}_{\chi_2((f^3)(x^3 + a_2x^2 + a_4x + a_6))}$$
$$\cong R^1(pr_2)_!\mathcal{L}_{\chi_2((f)(x^3 + a_2x^2 + a_4x + a_6))}$$
(because χ_2 is quadratic)
$$\cong \mathcal{L}_{\chi_2(f)}\otimes R^1(pr_2)_!\mathcal{L}_{\chi_2(x^3 + a_2x^2 + a_4x + a_6)}$$
(by the projection formula)
$$\cong \mathcal{L}_{\chi_2(f)}\otimes R^1\pi_!\overline{\mathbb{Q}}_\ell.$$
QED

(9.3) Theorems of geometric irreducibility in odd characteristic

(9.3.1) In this section, we work over a field k in which 2 is invertible. We will give a number of families of Weierstrass cubics over affine spaces, $\pi : \mathcal{E}^{aff} \to \mathbb{A}^d$, for which the perverse sheaf $R^1\pi_!\overline{\mathbb{Q}}_\ell[d]$ on $\mathbb{A}^d$ is geometrically irreducible.

(9.3.2) There are three simple principles we will apply, given as 9.3.3-5 below. The first two are standard. We include proofs for ease of reference.

External tensor product Lemma 9.3.3 Let k be a field in which a prime ℓ is invertible. Let S and T be geometrically irreducible nonempty separated k-schemes of finite type. Suppose K and L are $\overline{\mathbb{Q}}_\ell$-perverse sheaves on S and T respectively. Suppose that K is geometrically irreducible on S, and that L is geometrically irreducible on T. Then their external tensor product

$$K \boxtimes L := (pr_1{}^*K) \otimes (pr_2{}^*L)$$

is geometrically irreducible on $S \times_k T$. If K and L are both self dual, then $K \boxtimes L$ is self dual.

proof We first reduce to the case when k is algebraically closed, S is the support of K, T is the support of L, and S and T are both reduced. Then [BBD, 4.3.2 and 4.3.3] we can find affine open sets $U \subset S$ and $V \subset T$, inclusions denoted $j : U \to S$, $k : V \to T$, with U/k and V/k smooth and connected, and irreducible lisse sheaves $\mathcal{G}$ and $\mathcal{H}$ on U and V respectively, such that $K = j_{!*}\mathcal{G}[\dim U]$, $L = k_{!*}\mathcal{H}[\dim V]$. Their external tensor product $K \boxtimes L$ is the middle extension $(j \times k)_{!*}((\mathcal{G} \boxtimes \mathcal{H})[\dim(U \times V)])$. So it suffices [BBD, 4.3.1 and 4.3.3] to see that $\mathcal{G} \boxtimes \mathcal{H}$ is geometrically irreducible on $U \times V$, i.e., that if we pick base points u in U and v in V, the action ρ of $\pi_1(U \times V, u \times v)$ on $\mathcal{G} \boxtimes \mathcal{H}$ is irreducible. But there is a canonical homomorphism

$$\alpha : \pi_1(U, u) \times \pi_1(V, v) \to \pi_1(U \times V, u \times v).$$

The composite $\rho \circ \alpha$ is the external tensor product of the representations $\mathcal{G}$ of $\pi_1(U, u)$ and $\mathcal{H}$ of $\pi_1(V, v)$. Since the external tersor product of finite-dimensional irreducible $\overline{\mathbb{Q}}_\ell$-representations of two groups G and H is an irreducible representation of $G \times H$, $\rho \circ \alpha$ is irreducible, and a fortiori ρ itself is irreducible.

That the external tensor product of self dual perverse sheaves is self dual results from [BBD, 4.2.7 (b)]. QED

Local Nature Lemma 9.3.4 Let k be a field in which a prime ℓ is invertible. Let S be a geometrically irreducible separated k-scheme of finite type, which is the union of finitely many Zariski open sets U_i. Suppose K is a $\overline{\mathbb{Q}}_\ell$-perverse sheaf on S, whose support is S. Then the following conditions are equivalent:

1) K is geometrically irreducible on S,

2) $K|U_i$ is geometrically irreducible on U_i, for each i.

proof We reduce immediately to the case when k is algebraically closed. Since the support of K is S, there exists an affine open set

$V \subset S$ such that $K|V$ is $\mathcal{G}[d]$, for a lisse $\overline{\mathbb{Q}}_\ell$-sheaf $\mathcal{G}$ on V. At the expense of shrinking V, we may further assume that $V \subset U_i$ for each i. Now K is geometrically irreducible on S if and only if $\mathcal{G}$ is geometrically irreducible as lisse sheaf on V and if, denoting by $j : V \to S$ the inclusion, the canonical adjunction map of perverse sheaves on S

$$K \to Rj_\star j^\star K = Rj_\star(\mathcal{G}[d])$$

is injective, and maps K isomorphically to the perverse subsheaf $j_{!\star}(\mathcal{G}[d])$ of $Rj_\star(\mathcal{G}[d])$.

On the other hand, $K|U_i$ is geometrically irreducible on U_i if and only if $\mathcal{G}$ is geometrically irreducible as lisse sheaf on V and if, denoting by $j_i : V \to U_i$ the inclusion, the canonical adjunction map of perverse sheaves on U_i

$$K|U_i \to Rj_{i\star}j^\star K = Rj_{i\star}(\mathcal{G}[d])$$

is injective, and maps $K|U_i$ isomorphically to the perverse subsheaf $j_{i!\star}(\mathcal{G}[d])$ of $Rj_{i\star}(\mathcal{G}[d])$. By transitivity [BBD, 2.1.7], if we denote by $k_i : U_i \to S$ the inclusion, we have

$$j_{!\star}(\mathcal{G}[d]) = k_{i!\star}j_{i!\star}(\mathcal{G}[d]).$$

Restricting to U_i, this gives

$$(j_{!\star}(\mathcal{G}[d]))|\, U_i = j_{i!\star}(\mathcal{G}[d]).$$

If either 1) or 2) holds, then $\mathcal{G}$ is geometrically irreducible as lisse sheaf on V. If 1) holds, the above equality says that $K|U_i$ is the middle extension $j_{i!\star}(\mathcal{G}[d])$, and hence 2) holds. If 2) holds, both K and $j_{!\star}(\mathcal{G}[d])$ are perverse subsheaves of the perverse sheaf $Rj_\star(\mathcal{G}[d])$ which agree on each U_i, and hence coincide. Thus 1) holds. QED

Missing Points Lemma 9.3.5 Let k be an algebraically closed field in which a prime ℓ is invertible. Let S/k be an affine, smooth, connected k-scheme, of dimension $d \geq 1$. Let $\mathcal{G}$ be a constructible $\overline{\mathbb{Q}}_\ell$-sheaf on S. Suppose that the following three conditions hold.

1) $\mathcal{G}[d]$ is a perverse sheaf on S.

2) $H_c{}^\star(S, \mathcal{G}[d]) = 0$.

3) There exists a finite subset $Z \subset S(k)$ such that $\mathcal{G}|Z = 0$ and such that $\mathcal{G}[d]|S - Z$ is irreducible and self dual as perverse sheaf on $S - Z$.

Then $\mathcal{G}[d]$ is irreducible and self dual as a perverse sheaf on S.

proof Denote by $j : S - Z \to S$ the inclusion. By 3), $j^\star\mathcal{G}[d]$ is perverse irreducible and self dual on $S - Z$. So its middle extension $j_{!\star}(j^\star\mathcal{G}[d])$ is perverse irreducible and self dual on S. So it suffices to show that $\mathcal{G}[d] \cong j_{!\star}(j^\star\mathcal{G}[d])$. But $\mathcal{G}|Z = 0$ by 3), i.e., $\mathcal{G} \cong j_!(j^\star\mathcal{G})$. So it suffices to show that $j_!(j^\star\mathcal{G}[d]) \cong j_{!\star}(j^\star\mathcal{G}[d])$. Since $j^\star\mathcal{G}[d]$ is self dual, it suffices by duality to show that $j_{!\star}(j^\star\mathcal{G}[d]) \cong Rj_\star(j^\star\mathcal{G}[d])$. Since Z is a finite set of k-points, we obtain $j_{!*}(j^*\mathcal{G}[d])$ by first forming $Rj_*(j^*\mathcal{G}[d])$ and then by throwing away all cohomology sheaves in degree ≥ 0. So we have $j_{!*}(j^*\mathcal{G}[d]) \cong Rj_*(j^*\mathcal{G}[d])$ if and only if the object $Rj_*(j^*\mathcal{G}[d])$ has no nonzero cohomology sheaves in degree ≥ 0, if and only if the object $Rj_*j^*\mathcal{G}$ has no nonzero cohomology sheaves in degree $\geq d$. Now above degree zero, $Rj_*j^*\mathcal{G}$ is concentrated on Z. So we must show that $(Rj_*j^*\mathcal{G})|Z$ is concentrated in degree $\leq d$. As Z is finite, it is the same to show that $H^i(Z, (Rj_*j^*\mathcal{G})|Z) = 0$ for $i \geq d$.

To see this, we argue as follows. We have a distinguished triangle

$$\dots j_!j^*\mathcal{G} \to Rj_*j^*\mathcal{G} \to (Rj_*j^*\mathcal{G})|Z \to \dots .$$

By 2), we have $H_c{}^*(S, \mathcal{G}[d]) = 0$, a vanishing we rewrite as

$$H_c{}^*(S, j_!(j^*\mathcal{G}[d])) = 0.$$

Since $j^*\mathcal{G}[d]$ is self dual, by 3), the dual of this vanishing is

$$H^*(S, Rj_*(j^*\mathcal{G}[d])) = 0,$$

i.e.,

$$H^*(S, Rj_*j^*\mathcal{G}) = 0.$$

Looking back at the distinguished triangle, we see that the coboundary map is an isomorphism

$$H^i(Z, (Rj_*j^*\mathcal{G})|Z) \cong H^{i+1}(S, j_!j^*\mathcal{G}).$$

Because S is affine of dimension d, and $j_!j^*\mathcal{G}$ is a single constructible sheaf, we have

$$H^{i+1}(S, j_!j^*\mathcal{G}) = 0 \text{ for } i \geq d.$$

Therefore we find

$$H^i(Z, (Rj_*j^*\mathcal{G})|Z) = 0 \text{ for } i \geq d,$$

as required. QED

Theorem 9.3.6 Let k be a field in which 2 is invertible, $\mathbb{A}^2 = \mathrm{Spec}(k[s, t])$, and $\pi : \mathcal{E}^{aff} \to \mathbb{A}^2$ the relative affine curve defined by the Weierstrass equation

$$y^2 = x^3 + sx^2 + s^3t.$$

Then the perverse sheaf $R^1\pi_!\overline{\mathbb{Q}}_\ell[2]$ on $\mathbb{A}^2$ is geometrically irreducible and geometrically self dual, and $H_c{}^*(\mathbb{A}^2\otimes_k\overline{k}, R^1\pi_!\overline{\mathbb{Q}}) = 0$.

proof Immediate reduction to the case when k is algebraically closed. Over each point of the locus $s = 0$, we have fibre $y^2 = x^3$. Looking fibre by fibre, we see that $R^1\pi_!\overline{\mathbb{Q}}_\ell \mid (s = 0)$ vanishes. On the locus where s is invertible, our curve is the s-twist of the curve

$$\mathcal{E}_1{}^{aff} : y^2 = x^3 + x^2 + t,$$

whose structural morphism we denote $\pi_1 : \mathcal{E}_1{}^{\mathrm{aff}} \to \mathbb{A}^1$.

By Twisting Lemma 9.2.2, $R^1\pi_!\overline{\mathbb{Q}}_\ell$ on $\mathbb{A}^2[1/s] \cong \mathbb{G}_m \times \mathbb{A}^1$ is the external tensor product

$$R^1\pi_!\overline{\mathbb{Q}}_\ell \cong \mathcal{L}_{\chi_2(s)} \boxtimes R^1\pi_{1!}\overline{\mathbb{Q}}_\ell.$$

Denote by $j : \mathbb{G}_m \to \mathbb{A}^1$ the inclusion. Then $j_*\mathcal{L}_{\chi_2(s)}$ vanishes at $s=0$. So if we extend by zero from $\mathbb{A}^2[1/s]$ to $\mathbb{A}^2$, we get an isomorphism on $\mathbb{A}^2$,

$$R^1\pi_!\overline{\mathbb{Q}}_\ell \cong j_*\mathcal{L}_{\chi_2(s)} \boxtimes R^1\pi_{1!}\overline{\mathbb{Q}}_\ell,$$

i.e.,

$$R^1\pi_!\overline{\mathbb{Q}}_\ell[2] \cong j_{!*}(\mathcal{L}_{\chi_2(s)}[1]) \boxtimes R^1\pi_{1!}\overline{\mathbb{Q}}_\ell[1].$$

The asserted vanishing $H_c{}^*(\mathbb{A}^2 \otimes_k \overline{k}, R^1\pi_!\overline{\mathbb{Q}}) = 0$ is obvious from this isomorphism, the Kunneth formula, and the vanishing of $H_c{}^*(\mathbb{A}^1 \otimes_k \overline{k}, j_*\mathcal{L}_{\chi_2})$.

It remains to establish the irreducibility and self duality of $R^1\pi_!\overline{\mathbb{Q}}_\ell[2] \cong j_{!*}(\mathcal{L}_{\chi_2(s)}[1]) \boxtimes R^1\pi_{1!}\overline{\mathbb{Q}}_\ell[1]$. The first factor $j_{!*}(\mathcal{L}_{\chi_2(s)}[1])$ is perverse irreducible and self dual. So it suffices, by Lemma 9.3.3, to show that $R^1\pi_{1!}\overline{\mathbb{Q}}_\ell[1]$ is perverse irreducible and self dual. The curve $y^2 = x^3 + x^2 + t$ has nonconstant j, and its $\Delta = -16(4t + 27t^2)$ has only simple zeroes in any characteristic not two. So $R^1\pi_{1!}\overline{\mathbb{Q}}_\ell$ is a middle extension, by Corollary 7.5.5. Its restriction to $\mathbb{A}^1[1/\Delta]$ is geometrically irreducible, because the j invariant is not constant. Therefore $R^1\pi_{1!}\overline{\mathbb{Q}}_\ell[1]$ is the middle extension from $\mathbb{A}^1[1/\Delta]$ of a perverse irreducible which is self dual, so is itself perverse irreducible and self dual. QED

Corollary 9.3.7 Hypotheses as in Theorem 9.3.6 above, suppose in addition that k is a finite field. Then the geometrically irreducible perverse sheaf $R^1\pi_!\overline{\mathbb{Q}}_\ell[2](3/2)$ on $\mathbb{A}^2$ is self dual.

proof Immediate from Proposition 9.1.4, part 2). QED

Theorem 9.3.8 Let k be a field in which 2 is invertible. Over $\mathbb{A}^2 = \mathrm{Spec}(k[a_2, a_6])$, consider the relative affine curve

$$\pi : \mathcal{E}^{\mathrm{aff}} \to \mathbb{A}^2$$

defined by the Weierstrass equation

$$y^2 = x^3 + a_2x^2 + a_6.$$

Then the perverse sheaf $R^1\pi_!\overline{\mathbb{Q}}_\ell[2]$ on $\mathbb{A}^2$ is geometrically irreducible and geometrically self dual, and $H_c{}^*(\mathbb{A}^2 \otimes_k \overline{k}, R^1\pi_!\overline{\mathbb{Q}}) = 0$.

proof Immediate reduction to the case when k is algebraically closed. The asserted vanishing $H_c{}^*(\mathbb{A}^2\otimes_k\bar{k}, R^1\pi_!\bar{\mathbb{Q}}) = 0$ is a special case of Proposition 9.1.5.

We next show that over the open set $\mathbb{A}^2[1/a_2]$ where a_2 is invertible, $R^1\pi_!\bar{\mathbb{Q}}_\ell[2] \mid \mathbb{A}^2[1/a_2]$ is geometrically irreducible. Indeed, the change of variables $s = a_2$, $t = a_6/(a_2)^3$ gives us the family

$$y^2 = x^3 + sx^2 + s^3t$$

over the open set of the (s, t) plane where s is invertible. So by Theorem 9.3.6, we find that $R^1\pi_!\bar{\mathbb{Q}}_\ell[2] \mid \mathbb{A}^2[1/a_2]$ is geometrically irreducible and self dual. Since the j invariant is nonconstant, $R^1\pi_!\bar{\mathbb{Q}}_\ell[2] \mid \mathbb{A}^2[1/\Delta]$ is geometrically irreducible and self dual. The union of the the two open sets $\mathbb{A}^2[1/a_2]$ and $\mathbb{A}^2[1/\Delta]$ is $\mathbb{A}^2 - (0,0)$. Indeed, Δ is

$$\Delta = -16a_6(4(a_2)^3 + 27a_6),$$

so if Δ and a_2 both vanish, then a_6 also vanishes. Thus $R^1\pi_!\bar{\mathbb{Q}}_\ell[2]$ is irreducible and self dual as perverse sheaf on $\mathbb{A}^2 - (0,0)$.

We now observe that $R^1\pi_!\bar{\mathbb{Q}}_\ell$ vanishes at (0,0) (by inspection). By Proposition 9.1.3, $R^1\pi_!\bar{\mathbb{Q}}_\ell[2]$ is perverse on $\mathbb{A}^2$. By Proposition 9.1.5, we have $H_c{}^*(\mathbb{A}^2, R^1\pi_!\bar{\mathbb{Q}}_\ell) = 0$. So we may invoke the Missing Points Lemma 9.3.5, with $S = \mathbb{A}^2$, $Z = (0, 0)$, and $\mathcal{G} = R^1\pi_!\bar{\mathbb{Q}}_\ell$. QED

Corollary 9.3.9 Hypotheses as in Theorem 9.3.8 above, suppose in addition that k is a finite field. Then the geometrically irreducible perverse sheaf $R^1\pi_!\bar{\mathbb{Q}}_\ell[2](3/2)$ on $\mathbb{A}^2$ is self dual.
proof Immediate from Proposition 9.1.4, part 2). QED

Theorem 9.3.10 Let k be a field in which 6 is invertible, $\mathbb{A}^2 = \mathrm{Spec}(k[s, t])$, and $\pi : \mathcal{E}^{\mathrm{aff}} \to \mathbb{A}^2$ the relative affine curve defined by the Weierstrass equation

$$y^2 = 4x^3 - s^2tx - s^3t.$$

Then the perverse sheaf $R^1\pi_!\bar{\mathbb{Q}}_\ell[2]$ on $\mathbb{A}^2$ is geometrically irreducible and geometrically self dual, and $H_c{}^*(\mathbb{A}^2\otimes_k\bar{k}, R^1\pi_!\bar{\mathbb{Q}}) = 0$.
proof The proof is very similar to that of Theorem 9.3.6 Just as we did there, we first reduce to the case when k is algebraically closed. Looking fibre by fibre, we see that $R^1\pi_!\bar{\mathbb{Q}}_\ell$ vanishes on the locus s=0. On the open set where s is invertible, our curve is the s-twist of the curve

$$\mathcal{E}_1{}^{\mathrm{aff}} : y^2 = 4x^3 - tx - t,$$

whose structural morphism we denote $\pi_1 : \mathcal{E}_1{}^{\mathrm{aff}} \to \mathbb{A}^1$.
We find

$$R^1\pi_!\overline{\mathbb{Q}}_\ell[2] \cong j_{!*}(\mathcal{L}_{\chi_2(s)}[1])\boxtimes R^1\pi_{1!}\overline{\mathbb{Q}}_\ell[1],$$

which makes clear the asserted vanishing $H_c{}^*(\mathbb{A}^2\otimes_k\overline{k}, R^1\pi_!\overline{\mathbb{Q}}) = 0$.

It remains to show that $R^1\pi_{1!}\overline{\mathbb{Q}}_\ell[1]$ is perverse irreducible and self dual on $\mathbb{A}^1$. Since the j invariant $j = 1728t/(t - 27)$ of $\mathcal{E}_1{}^{\mathrm{aff}}$ is nonconstant, $R^1\pi_{1!}\overline{\mathbb{Q}}_\ell \mid \mathbb{A}^1[1/\Delta]$ is lisse, irreducible, and self dual. As $\Delta = t^2(t - 27)$, we see from Corollary 7.5.5 that $R^1\pi_{1!}\overline{\mathbb{Q}}_\ell$ is a middle extension, and hence that $R^1\pi_{1!}\overline{\mathbb{Q}}_\ell[1]$ is perverse irreducible and self dual on $\mathbb{A}^1$. QED

Corollary 9.3.11 Hypotheses as in Theorem 9.3.10 above, suppose in addition that k is a finite field. Then the geometrically irreducible perverse sheaf $R^1\pi_!\overline{\mathbb{Q}}_\ell[2](3/2)$ on $\mathbb{A}^2$ is self dual.
proof Immediate from Proposition 9.1.4, part 2). QED

Theorem 9.3.12 Let k be a field in which 6 is invertible. Over $\mathbb{A}^2 = \mathrm{Spec}(k[g_2, g_3])$, consider the relative affine curve $\pi : \mathcal{E}^{\mathrm{aff}} \to \mathbb{A}^2$ defined by the Weierstrass equation

$$y^2 = 4x^3 - g_2x - g_3.$$

Then the perverse sheaf $R^1\pi_!\overline{\mathbb{Q}}_\ell[2]$ on $\mathbb{A}^2$ is geometrically irreducible and geometrically self dual, and $H_c{}^*(\mathbb{A}^2\otimes_k\overline{k}, R^1\pi_!\overline{\mathbb{Q}}) = 0$.
proof The vanishing $H_c{}^*(\mathbb{A}^2\otimes_k\overline{k}, R^1\pi_!\overline{\mathbb{Q}}) = 0$ is a special case of Proposition 9.1.5. On the open set where g_2g_3 is invertible, the change of variables

$$s = g_3/g_2,\ t = (g_2)^3/(g_3)^2,$$

i.e.,

$$g_2 = s^2t,\ g_3 = s^3t,$$

makes the open set $\mathbb{A}^2[1/g_2g_3]$ of the (g_2, g_3)-plane isomorphic to the open set $\mathbb{A}^2[1/st]$ of the (s, t)-plane, and carries our family

$$y^2 = 4x^3 - g_2x - g_3$$

to the twist family

$$y^2 = 4x^3 - s^2tx - s^3t.$$

By the previous Theorem 9.3.10, $R^1\pi_!\overline{\mathbb{Q}}_\ell[2] \mid \mathbb{A}^2[1/g_2g_3]$ is geometrically irreducible and self dual.

Since the j invariant $j = 1728(g_2)^3/((g_2)^3 - 27(g_3)^2)$ is nonconstant, $R^1\pi_!\overline{\mathbb{Q}}_\ell[2] \mid \mathbb{A}^2[1/\Delta]$ is also geometrically irreducible and self dual. The union of the two open sets $\mathbb{A}^2[1/g_2g_3]$ and $\mathbb{A}^2[1/\Delta]$

is $\mathbb{A}^2 - (0,0)$. Indeed, as $\Delta = (g_2)^3 - 27(g_3)^2$, if both Δ and g_2g_3 vanish, then both g_2 and g_3 vanish. Therefore the restriction of $R^1\pi_!\overline{\mathbb{Q}}_\ell[2]$ to $\mathbb{A}^2 - (0,0)$ is geometrically irreducible and geometrically self dual. By inspection, $R^1\pi_!\overline{\mathbb{Q}}_\ell$ vanishes at (0,0). As already proven, we have $H_c{}^*(\mathbb{A}^2\otimes_k\overline{k}, R^1\pi_!\overline{\mathbb{Q}}_\ell) = 0$. So we have only to apply the Missing Points Lemma 9.3.5 to the situation $S = \mathbb{A}^2$, $Z = (0,0)$, $\mathcal{G} = R^1\pi_!\overline{\mathbb{Q}}_\ell$. QED

Corollary 9.3.13 Hypotheses as in Theorem 9.3.12 above, suppose in addition that k is a finite field. Then the geometrically irreducible perverse sheaf $R^1\pi_!\overline{\mathbb{Q}}_\ell[2](3/2)$ on $\mathbb{A}^2$ is self dual.
proof Immediate from Proposition 9.1.4, part 2). QED

(9.3.14) Strangely enough, the last two results 9.3.12-13 remain true in characteristic 3, although their proofs are necessarily quite different, since these are now families of curves whose j invariant $j = 1728 = 0$ is constant.
Theorem 9.3.15 Let k be a field of characteristic 3, $\mathbb{A}^1 = \mathrm{Spec}(k[t])$, and $\pi : \mathcal{E}^{\mathrm{aff}} \to \mathbb{A}^1$ the relative affine curve defined by the Weierstrass equation

$$y^2 = 4x^3 - tx - t.$$

Then the perverse sheaf $R^1\pi_!\overline{\mathbb{Q}}_\ell[1]$ on $\mathbb{A}^1$ is geometrically irreducible and geometrically self dual.
proof Immediate reduction to the case when k is algebraically closed. Since k has characteristic 3, $\Delta = t^3$. So $R^1\pi_!\overline{\mathbb{Q}}_\ell|\mathbb{G}_m$ is a lisse sheaf of rank two which is self dual.

By Corollary 7.5.5, $R^1\pi_!\overline{\mathbb{Q}}_\ell$ is a middle extension from $\mathbb{G}_m$, hence is self dual. What we must show is that the lisse sheaf $\mathcal{G} := R^1\pi_!\overline{\mathbb{Q}}_\ell|\mathbb{G}_m$ is geometrically irreducible on $\mathbb{G}_m$. For this, we argue as follows. Because our family has constant j invariant $j = 1728 = 0$, $\mathcal{G}$ has finite geometric monodromy and hence is a semisimple representation of $\pi_1(\mathbb{G}_m)$. Denote by

$$[2]: \mathbb{G}_m \to \mathbb{G}_m$$
$$t \mapsto t^2$$

the squaring map. It makes $\mathbb{G}_m$ a Galois finite etale covering of itself with Galois group ± 1. We will show that $[2]^*\mathcal{G}$ is the direct sum of two distinct characters α and β of $\pi_1(\mathbb{G}_m)$, which are different but which are interchanged by the nontrivial galois automorphism $[-1]$ of the upper $\mathbb{G}_m$. Once we have shown this, then Mackey theory tells us that $[2]_*\alpha = [2]_*\beta$ is irreducible. So we can compute the inner product (in the realm of semisimple finite-dimensional $\overline{\mathbb{Q}}_\ell$-representations of $\pi_1(\mathbb{G}_m)$)

$$\begin{aligned}\langle \mathcal{G}, [2]_*\alpha\rangle &= (1/2)(\langle \mathcal{G}, [2]_*\alpha\rangle + \langle \mathcal{G}, [2]_*\beta\rangle)\\ &= (1/2)\langle \mathcal{G}, [2]_*[2]^*\mathcal{G}\rangle\\ &= (1/2)\langle [2]^*\mathcal{G}, [2]^*\mathcal{G}\rangle\\ &= (1/2)\langle \alpha + \beta, \alpha + \beta\rangle\\ &= (1/2)(1 + 1) = 1.\end{aligned}$$

Since $\mathcal{G}$ has rank 2, we must have $\mathcal{G} = [2]_*\alpha$, which as noted above is irreducible.

To analyze $[2]^*\mathcal{G}$, we view it as $R^1\pi_{[2]!}\overline{\mathbb{Q}}_\ell$ for the curve $\pi_{[2]} : \mathcal{E}_{[2]}{}^{\mathrm{aff}} \to \mathbb{G}_m$ defined by

$$y^2 = 4x^3 - t^2x - t^2.$$

This curve over $\mathbb{G}_m$ is the t-twist of the curve $\pi_{[2t]} : \mathcal{E}_{[2t]}{}^{\mathrm{aff}} \to \mathbb{G}_m$ defined by

$$y^2 = 4x^3 - x - 1/t.$$

We recall that $4 = 1$, and write this last equation as

$$x^3 - x = y^2 + 1/t.$$

Let us denote by ψ and $\overline{\psi}$ the two nontrivial additive characters of $\mathbb{F}_3$, and by $\mathcal{L}_\psi$ and $\mathcal{L}_{\overline{\psi}}$ the corresponding Artin-Schreier sheaves on $\mathbb{A}^1$. Denote by ρ the projection of $\mathcal{E}_{[2t]}{}^{\mathrm{aff}}$ onto the y-line over $\mathbb{G}_m$. Then we have

$$\begin{aligned}\rho_*\overline{\mathbb{Q}}_\ell &= \overline{\mathbb{Q}}_\ell \oplus \mathcal{L}_{\psi(y^2 + 1/t)} \oplus \mathcal{L}_{\overline{\psi}(y^2 + 1/t)}\\ &= \overline{\mathbb{Q}}_\ell \oplus \mathcal{L}_{\psi(y^2 + 1/t)} \oplus \mathcal{L}_{\psi(-y^2 - 1/t)}\\ &\cong \overline{\mathbb{Q}}_\ell \oplus \mathcal{L}_{\psi(1/t)}\otimes\mathcal{L}_{\psi(y^2)} \oplus \mathcal{L}_{\psi(-1/t)}\otimes\mathcal{L}_{\psi(-y^2)}.\end{aligned}$$

Applying $pr_2 : \mathbb{A}^1 \times \mathbb{G}_m \to \mathbb{G}_m$, we get

$$\begin{aligned}R^1\pi_{[2t]!}\overline{\mathbb{Q}}_\ell &= R^1pr_{2!}(\rho_*\overline{\mathbb{Q}}_\ell)\\ &\cong R^1pr_{2!}(\mathcal{L}_{\psi(1/t)}\otimes\mathcal{L}_{\psi(y^2)}) \oplus R^1pr_{2!}(\mathcal{L}_{\psi(-1/t)}\otimes\mathcal{L}_{\psi(-y^2)})\\ &\cong \mathcal{L}_{\psi(1/t)}\otimes R^1pr_{2!}(\mathcal{L}_{\psi(y^2)}) \oplus \mathcal{L}_{\psi(-1/t)})\otimes R^1pr_{2!}(\mathcal{L}_{\psi(-y^2)}).\end{aligned}$$

The sheaves $R^1pr_{2!}(\mathcal{L}_{\psi(y^2)})$ and $R^1pr_{2!}(\mathcal{L}_{\psi(-y^2)})$ on $\mathbb{G}_m$ are isomorphic (by $y \mapsto iy$) and constant (Kunneth formula), of rank one, so geometrically they are isomorphic to the constant sheaf. Thus we find

$$R^1\pi_{[2t]!}\overline{\mathbb{Q}}_\ell \cong \mathcal{L}_{\psi(1/t)} \oplus \mathcal{L}_{\psi(-1/t)}.$$

So by the twisting lemma, we have

$$\begin{aligned}[2]^*\mathcal{G} &\cong \mathcal{L}_{\chi_2(t)}\otimes R^1\pi_{[2t]!}\overline{\mathbb{Q}}_\ell\\ &\cong \mathcal{L}_{\chi_2(t)}\otimes\mathcal{L}_{\psi(1/t)} \oplus \mathcal{L}_{\chi_2(t)}\otimes\mathcal{L}_{\psi(-1/t)}.\end{aligned}$$

The sheaves $\mathcal{L}_{\chi_2(t)}$ and $\mathcal{L}_{\chi_2(-t)}$ are geometrically isomorphic. On the other hand, the sheaves $\mathcal{L}_{\psi(1/t)}$ and $\mathcal{L}_{\psi(-1/t)}$ are not geometrically isomorphic, as they are inverses and both nontrivial of order 3. So putting

$$\alpha := \mathcal{L}_{\chi_2(t)} \otimes \mathcal{L}_{\psi(1/t)},$$

we have

$$[2]^*\mathcal{G} = \alpha + \beta,\ \alpha \neq \beta,\ \beta = [-1]^*\alpha,$$

as required. QED

Corollary 9.3.16 Hypotheses as in Theorem 9.3.15 above, suppose in addition that k is a finite field. Then the geometrically irreducible perverse sheaf $R^1\pi_!\overline{\mathbb{Q}}_\ell[1](1)$ on $\mathbb{A}^1$ is self dual.

proof Immediate from Proposition 9.1.4, part 2). QED

Theorem 9.3.17 Let k be a field of characteristic 3, $\mathbb{A}^2 = \mathrm{Spec}(k[s, t])$, and $\pi : \mathcal{E}^{\mathrm{aff}} \to \mathbb{A}^2$ the relative affine curve defined by the Weierstrass equation

$$y^2 = 4x^3 - s^2tx - s^3t.$$

Then the perverse sheaf $R^1\pi_!\overline{\mathbb{Q}}_\ell[2]$ on $\mathbb{A}^2$ is geometrically irreducible and geometrically self dual, and $H_c{}^*(\mathbb{A}^2\otimes_k\overline{k}, R^1\pi_!\overline{\mathbb{Q}}) = 0$.

proof Repeat the first ten lines of the proof of Theorem 9.3.10, and then invoke Theorem 9.3.15 above. QED

Corollary 9.3.18 Hypotheses as in Theorem 9.3.17 above, suppose in addition that k is a finite field. Then the geometrically irreducible perverse sheaf $R^1\pi_!\overline{\mathbb{Q}}_\ell[2](3/2)$ on $\mathbb{A}^2$ is self dual.

proof Immediate from Proposition 9.1.4, part 2). QED

Theorem 9.3.19 Let k be a field of characteristic 3, $\mathbb{A}^2 = \mathrm{Spec}(k[g_2, g_3])$. Consider the relative affine curve $\pi : \mathcal{E}^{\mathrm{aff}} \to \mathbb{A}^2$ defined by the Weierstrass equation

$$y^2 = 4x^3 - g_2x - g_3.$$

Then the perverse sheaf $R^1\pi_!\overline{\mathbb{Q}}_\ell[2]$ on $\mathbb{A}^2$ is geometrically irreducible and geometrically self dual, and $H_c{}^*(\mathbb{A}^2\otimes_k\overline{k}, R^1\pi_!\overline{\mathbb{Q}}) = 0$.

proof Immediate reduction to the case when k is algebraically closed. The vanishing $H_c{}^*(\mathbb{A}^2\otimes_k\overline{k}, R^1\pi_!\overline{\mathbb{Q}}) = 0$ is a special case of Proposition 9.1.5.

Here Δ is $(g_2)^3$. On the open set $\mathbb{A}^2[1/g_2]$, $R^1\pi_!\overline{\mathbb{Q}}_\ell$ is lisse of rank two and self dual. Pull back by the map

$$\varphi : \mathbb{G}_m \to \mathbb{A}^2[1/g_2],$$
$$t \mapsto (t, t).$$

By Theorem 9.3.15, $\varphi^*(R^1\pi_!\overline{\mathbb{Q}}_\ell)$ is irreducible, and hence $R^1\pi_!\overline{\mathbb{Q}}_\ell \mid \mathbb{A}^2[1/g_2]$ is irreducible.

Denote by

$$j : \mathbb{A}^2[1/g_2] \to \mathbb{A}^2$$

the inclusion. It remains only to see that $R^1\pi_!\overline{\mathbb{Q}}_\ell[2]$ is its own middle extension $j_{!*}j^*R^1\pi_!\overline{\mathbb{Q}}_\ell[2]$ from $\mathbb{A}^2[1/g_2]$. Because we are in characteristic 3, over the locus $g_2 = 0$, every geometric fibre is isomorphic to $y^2 = x^3$, and so $R^1\pi_!\overline{\mathbb{Q}}_\ell[2]$ vanishes on this locus, i.e., we have

$$R^1\pi_!\overline{\mathbb{Q}}_\ell[2] = j_!j^*R^1\pi_!\overline{\mathbb{Q}}_\ell[2].$$

We will show that this holds outside a finite set $Z \subset \mathbb{A}^2(k)$ of the form $0 \times Z_3$. Once we have shown this, then we have only to apply Missing Points Lemma 9.3.5.

For this, it suffices to show that

$$j_!(j^*R^1\pi_!\overline{\mathbb{Q}}_\ell[2]) \mid \mathbb{A}^2 - Z \cong Rj_*(j^*R^1\pi_!\overline{\mathbb{Q}}_\ell[2]) \mid \mathbb{A}^2 - Z.$$

This will be a consequence of Deligne's generic base change theorem [De-Fin, 1.9 (ii)]. We view $j : \mathbb{A}^2[1/g_2] \to \mathbb{A}^2$ as a morphism of $\mathbb{A}^1 := \mathrm{Spec}(k[g_3])$ schemes. Over a dense open set

$$U = \mathbb{A}^1 - Z_3 \subset \mathbb{A}^1 := \mathrm{Spec}(k[g_3]),$$

the formation of $Rj_*(j^*R^1\pi_!\overline{\mathbb{Q}}_\ell)$ commutes with arbitrary change of base on U.

Let us spell out what this means. We may shrink U, and assume 0 lies in Z_3. For any b in $U(k) \subset k^\times$, the restriction of $Rj_*(j^*R^1\pi_!\overline{\mathbb{Q}}_\ell)$ to the line $g_2 \mapsto (g_2, b)$ in $\mathbb{A}^2$ is the following. We consider the one-parameter family

$$\mathcal{E}_b{}^{\mathrm{aff}} : y^2 = x^3 + g_2x^2 + b$$

over the g_2-line, with structural morphism $\pi_b : \mathcal{E}_b{}^{\mathrm{aff}} \to \mathbb{A}^1$. In characteristic 3, the discriminant is $(g_2)^3$. Let us denote by

$$j_b : \mathbb{A}^1 - \{0\} \to \mathbb{A}^1$$

the inclusion. Then $j_b{}^*R^1\pi_{b!}\overline{\mathbb{Q}}_\ell$ is lisse on $\mathbb{G}_m$, and we have

$$Rj_{b*}(j_b{}^*R^1\pi_{b!}\overline{\mathbb{Q}}_\ell) \cong Rj_*(j^*R^1\pi_!\overline{\mathbb{Q}}_\ell) \mid (\text{the line } g_2 \mapsto (g_2, b)).$$

How do we make use of this? Because $b \neq 0$ and k is algebraically closed, we can write b as β^3. Then our b-frozen family, written in terms of $\tilde{x} := x+\beta$ and y, has equation

$$y^2 = (\tilde{x})^3 + g_2(\tilde{x} - \beta) = (\tilde{x})^3 + g_2\tilde{x} - \beta g_2.$$

Here we may apply Corollary 7.5.5 to conclude that $R^1\pi_{b!}\overline{\mathbb{Q}}_\ell$ on $\mathbb{A}^1$ is a middle extension across $g_2 = 0$. Looking at the fibre over 0, we see that $R^1\pi_{b!}\overline{\mathbb{Q}}_\ell$ vanishes at 0. Since it is the middle extension across zero of the lisse sheaf $j_b{}^*R^1\pi_{b!}\overline{\mathbb{Q}}_\ell$, we have

$$j_{b!}(j_b{}^*R^1\pi_{b!}\overline{\mathbb{Q}}_\ell) \cong Rj_{b*}(j_b{}^*R^1\pi_{b!}\overline{\mathbb{Q}}_\ell).$$

Therefore the object

$$Rj_*(j^*R^1\pi_!\overline{\mathbb{Q}}_\ell) \mid (\text{the line } g_2 \mapsto (g_2, b))$$

vanishes at the point (0, b), so long as b does not lie in Z_3. Thus if we take for Z the finite set $0 \times Z_3$, we have

$$j_!(j^*R^1\pi_!\overline{\mathbb{Q}}_\ell[2]) \mid \mathbb{A}^2 - Z \cong Rj_*(j^*R^1\pi_!\overline{\mathbb{Q}}_\ell[2]) \mid \mathbb{A}^2 - Z,$$

as required. QED

Corollary 9.3.20 Hypotheses as in Theorem 9.3.19 above, suppose in addition that k is a finite field. Then the geometrically irreducible perverse sheaf $R^1\pi_!\overline{\mathbb{Q}}_\ell[2](3/2)$ on $\mathbb{A}^2$ is self dual.

proof Immediate from Proposition 9.1.4, part 2). QED

Remark 9.3.21 Here is another method of showing that for k a field of characteristic $p > 2$, and $\pi : \mathcal{E}^{\mathrm{aff}} \to \mathbb{A}^2 = \mathrm{Spec}(k[g_2, g_3])$ defined by the Weierstrass equation

$$y^2 = 4x^3 - g_2x - g_3,$$

has $R^1\pi_!\overline{\mathbb{Q}}_\ell|\mathbb{A}^2[1/\Delta]$ geometrically irreducible. Rather than reduce to the case when k is algebraically closed, we will instead reduce to the case when k is $\mathbb{F}_p$. This is legitimate because for any field K (here taken to be $\mathbb{F}_p$), for any geometrically connected K-scheme X (here taken to be $\mathbb{A}^2[1/\Delta]$), for any algebraically closed overfield L/K, and for any geometric point x of $X\otimes_K L$, with image point x' in $X\otimes_K\overline{K}$, the natural homomorphism

$$\pi_1(X\otimes_K L, x) \to \pi_1(X\otimes_K\overline{K}, x')$$

is surjective, cf. [Ka-LG, 1.2.2] and [EGA IV 4, 5.21]. Denote by $pr_2 : \mathbb{A}^1\times\mathbb{A}^2 \to \mathbb{A}^2$ the projection of the x-line. Because 2 is invertible, we have (cf. the proof of Lemma 9.2.2)

$$R^1\pi_!\overline{\mathbb{Q}}_\ell = R^1pr_{2!}\mathcal{L}_{\chi_2}(4x^3 - g_2x - g_3),$$

and

$$R^ipr_{2!}\mathcal{L}_{\chi_2}(4x^3 - g_2x - g_3) = 0, \text{ for } i \neq 1.$$

Thus we may view $\mathbb{A}^2$ as the space $\mathcal{F}$ of polynomial functions of degree at most one on $\mathbb{A}^1$, and the perverse sheaf $R^1\pi_!\overline{\mathbb{Q}}_\ell(1)[2]$ on $\mathbb{A}^2$ as the object M = Twist(L, K, $\mathcal{F}$, h) attached to the standard input (cf. 1.15.4) over $k = \mathbb{F}_p$ given by

the integer m = 1,

the perverse sheaf $K = j_*\mathcal{L}_{\chi_2}(1/2)[1]$ on $\mathbb{A}^m/k$,

the affine k-scheme of finite type $V = \mathbb{A}^1/k$,

the k-morphism $h : V \to \mathbb{A}^m$ given by $h(x) = 4x^3$,

the perverse sheaf $L = \overline{\mathbb{Q}}_\ell(1/2)[1]$ on V/k,

the integer d > 2

the space of functions $(\mathcal{F}, \tau)$ on V given by polynomials of degree at most one.

Because $H_c^*(\mathbb{A}^1 \otimes_k \bar{k}, j_*\mathcal{L}_{\chi_2}(1/2)[1]) = 0$, we see from Lemma 1.15.5, part 3), that $Gr^0(M(1/2)) = Gr^0(M(1/2))_{ncst}$. By the Standard Input Theorem 1.15.6, part 2), $Gr^0(M(1/2)) = Gr^0(M(1/2))_{ncst}$ is geometrically irreducible on $\mathbb{A}^2$. Therefore $Gr^0(M(1/2)) \mid \mathbb{A}^2[1/\Delta]$ is either geometrically irreducible, or it is zero. But the object $M(1/2) \mid \mathbb{A}^2[1/\Delta]$ is already pure of weight zero and nonzero: it is the lisse rank two sheaf $R^1\pi_!\bar{\mathbb{Q}}_\ell(3/2)|\mathbb{A}^2[1/\Delta]$, placed in degree -2. Thus

$$R^1\pi_!\bar{\mathbb{Q}}_\ell(3/2)|\mathbb{A}^2[1/\Delta] = Gr^0(M(1/2)) \mid \mathbb{A}^2[1/\Delta].$$

Hence the lisse sheaf $R^1\pi_!\bar{\mathbb{Q}}_\ell|\mathbb{A}^2[1/\Delta]$ is geometrically irreducible. QED

(9.4) Geometric Irreducibility in even characteristic

Twisting Lemma 9.4.1 Let S be an $\mathbb{F}_2$-scheme, a_2, a_4, a_6 functions in $\Gamma(S, \mathcal{O}_S)$. Consider the locus $\mathcal{E}^{aff}$ in $\mathbb{A}^2 \times S$ defined by the Weierstrass cubic

$$y^2 + xy = x^3 + a_2x^2 + a_4x + a_6.$$

View $\mathcal{E}^{aff}$ as a relative affine curve $\pi : \mathcal{E}^{aff} \to S$.

Fix a function f on S, and denote by $\mathcal{E}_f^{aff}$ the locus in $\mathbb{A}^2 \times S$ defined by the Weierstrass equation

$$y^2 + xy = x^3 + (a_2 + f)x^2 + a_4x + a_6.$$

View $\mathcal{E}_f^{aff}$ as a relative affine curve $\pi_f : \mathcal{E}^{aff} \to S$.

For any prime number ℓ invertible on S, we have a canonical isomorphism of sheaves on S,

$$R^1(\pi_f)_!\bar{\mathbb{Q}}_\ell \cong \mathcal{L}_{\psi(f)} \otimes R^1\pi_!\bar{\mathbb{Q}}_\ell,$$

where ψ denotes the nontrivial additive character of $\mathbb{F}_2$, and $\mathcal{L}_\psi$ the corresponding Artin-Schreier sheaf.

proof Denote by $\rho : \mathcal{E}^{aff} \to \mathbb{A}^1 \times S$ the projection onto the x-line. This finite flat double covering is ramified over x = 0, so if we denote by

$$j : \mathbb{G}_m \times S \to \mathbb{A}^1 \times S$$

the inclusion, we have

$$\rho_*\bar{\mathbb{Q}}_\ell = \bar{\mathbb{Q}}_\ell \oplus (\mathcal{K} := \mathrm{Ker}(\mathrm{Trace} : \rho_*\bar{\mathbb{Q}}_\ell \to \bar{\mathbb{Q}}_\ell)),$$

with $\mathcal{K}$ supported in $\mathbb{G}_m \times S$. Thus $\mathcal{K} = j_!j^*\mathcal{K}$, and we can compute $j^*\mathcal{K}$ as follows. Over $\mathbb{G}_m \times S$, we take new coordinates $(x, \tilde{y} := y/x)$. In terms of these, our equation becomes an Artin-Schreier equation

$$(\tilde{y})^2 + \tilde{y} = x + a_2 + a_4/x + a_6/x^2.$$

So we find

$$j^*\mathcal{K} = \mathcal{L}_{\psi(x + a_2 + a_4/x + a_6/x^2)}.$$

So we find that, denoting now by $pr_{2,\mathbb{G}_m}$ the projection

$$pr_{2,\mathbb{G}_m} : \mathbb{G}_m \times S \to S,$$

we have

$$R^1(\pi)_!\overline{\mathbb{Q}}_\ell \cong R^1(pr_{2,\mathbb{G}_m})_!\mathcal{L}_{\psi(x + a_2 + a_4/x + a_6/x^2)}.$$

Repeating the same argument with π_f instead of π, we find

$$\begin{aligned} &R^1(\pi_f)_!\overline{\mathbb{Q}}_\ell \\ &\cong R^1(pr_{2,\mathbb{G}_m})_!\mathcal{L}_{\psi(x + a_2 + f + a_4/x + a_6/x^2)} \\ &\cong R^1(pr_{2,\mathbb{G}_m})_!(\mathcal{L}_{\psi(f)}\otimes\mathcal{L}_{\psi(x + a_2 + a_4/x + a_6/x^2)}) \\ &\cong \mathcal{L}_{\psi(f)}\otimes R^1(pr_{2,\mathbb{G}_m})_!\mathcal{L}_{\psi(x + a_2 + a_4/x + a_6/x^2)} \\ &\cong \mathcal{L}_{\psi(f)}\otimes R^1\pi_!\overline{\mathbb{Q}}_\ell. \end{aligned}$$

QED

Theorem 9.4.2 Let k be a field of characteristic 2, $\mathbb{A}^2 := \mathrm{Spec}(k[a_2, a_6])$. Consider the relative affine curve

$$\pi : \mathcal{E}^{\mathrm{aff}} \to \mathbb{A}^2$$

defined by the Weierstrass equation

$$y^2 + xy = x^3 + a_2x^2 + a_6.$$

Then the perverse sheaf $R^1\pi_!\overline{\mathbb{Q}}_\ell[2]$ on $\mathbb{A}^2$ is geometrically irreducible and geometrically self dual, and $H_c^*(\mathbb{A}^2\otimes_k\overline{k}, R^1\pi_!\overline{\mathbb{Q}}) = 0$.

proof The vanishing $H_c^*(\mathbb{A}^2\otimes_k\overline{k}, R^1\pi_!\overline{\mathbb{Q}}) = 0$ is a special case of Propostion 9.1.5. Denote by $\pi_{0,6} : \mathcal{E}_{0,6}{}^{\mathrm{aff}} \to \mathbb{A}^1 := \mathrm{Spec}(k[a_6])$ the relative affine curve defined by putting $a_2 = 0$:

$$y^2 + xy = x^3 + a_6.$$

In this family, $\Delta = a_6$ and $j = 1/a_6$.

By the previous Lemma 9.4.1, $R^1\pi_!\overline{\mathbb{Q}}_\ell$ on $\mathbb{A}^2$ is the external tensor

$$R^1\pi_!\overline{\mathbb{Q}}_\ell \cong \mathcal{L}_{\psi(a_2)}\boxtimes R^1\pi_{0,6!}\overline{\mathbb{Q}}_\ell.$$

The sheaf $\mathcal{L}_{\psi(a_2)}$ is lisse of rank one on $\mathbb{A}^1$. So it remains to show that $R^1\pi_{0,6!}\overline{\mathbb{Q}}_\ell$ is a geometrically irreducible and geometrically self dual middle extension. It is lisse on $\mathbb{G}_m$ (because $\Delta = a_6$). It is geometrically irreducible and geometrically self dual on $\mathbb{G}_m$ because the j invariant $j = 1/a_6$ is nonconstant. It is a middle extension across $a_6 = 0$ by Corollary 7.5.5. QED

Corollary 9.4.3 Hypotheses as in Theorem 9.4.2 above, suppose in addition that k is a finite field. Then the geometrically irreducible perverse sheaf $R^1\pi_!\overline{\mathbb{Q}}_\ell[2](3/2)$ on $\mathbb{A}^2$ is self dual.

proof Immediate from Proposition 9.1.4, part 2). QED

Theorem 9.4.4 Let k be a field of characteristic 2, $\mathbb{A}^2 := \mathrm{Spec}(k[a_2, a_4])$. Consider the relative affine curve

$$\pi : \mathcal{E}^{\mathrm{aff}} \to \mathbb{A}^2$$

defined by the Weierstrass equation

$$y^2 + xy = x^3 + a_2x^2 + a_4x.$$

Then the perverse sheaf $R^1\pi_!\overline{\mathbb{Q}}_\ell[2]$ on $\mathbb{A}^2$ is geometrically irreducible and geometrically self dual, and $H_c{}^*(\mathbb{A}^2\otimes_k\overline{k}, R^1\pi_!\overline{\mathbb{Q}}) = 0$.

proof The vanishing $H_c{}^*(\mathbb{A}^2\otimes_k\overline{k}, R^1\pi_!\overline{\mathbb{Q}}) = 0$ is a special case of Proposition 9.1.5. Denote by $\pi_{0,4} : \mathcal{E}_{0,4}{}^{\mathrm{aff}} \to \mathbb{A}^1 := \mathrm{Spec}(k[a_4])$ the relative affine curve defined by putting $a_2 = 0$:

$$y^2 + xy = x^3 + a_4x.$$

In this family, $\Delta = (a_4)^2$ and $j = 1/(a_4)^2$.

Just as in the preceding theorem, we have

$$R^1\pi_!\overline{\mathbb{Q}}_\ell \cong \mathcal{L}_{\psi(a_2)}\boxtimes R^1\pi_{0,4!}\overline{\mathbb{Q}}_\ell.$$

We must show that $R^1\pi_{0,4!}\overline{\mathbb{Q}}_\ell$ is a geometrically irreducible and geometrically self dual middle extension. It is lisse on $\mathbb{G}_m$ (because $\Delta = (a_4)^2$). It is geometrically irreducible and geometrically self dual on $\mathbb{G}_m$ because the j invariant $j = 1/(a_4)^2$ is nonconstant. It is a middle extension across $t = 0$ because it is the absolute Frobenius pullback of the middle extension sheaf $R^1\pi_{0,6!}\overline{\mathbb{Q}}_\ell$ on $\mathbb{A}^1$ which occurred in the previous theorem. Indeed, on $\mathbb{A}^1 := \mathrm{Spec}(k[t])$, we have $\mathrm{Frob}_2{}^*(R^1\pi_{0,6!}\overline{\mathbb{Q}}_\ell)$ is $R^1\pi_{0,6,\mathrm{Fr}!}\overline{\mathbb{Q}}_\ell$ for the relative affine curve $\pi_{0,6,\mathrm{Fr}} : \mathcal{E}_{0,6,\mathrm{Fr}}{}^{\mathrm{aff}} \to \mathbb{A}^1$ defined by the Weierstrass equation

$$y^2 + xy = x^3 + t^2.$$

So here we have

$$\begin{aligned} R^1(\pi_{0,6,\mathrm{Fr}})_!\overline{\mathbb{Q}}_\ell &\cong R^1(pr_{2,\mathbb{G}_m})_!\mathcal{L}_{\psi(x + (t/x)^2)} \\ &\cong R^1(pr_{2,\mathbb{G}_m})_!\mathcal{L}_{\psi(x + t/x)} \\ &:= R^1\pi_{0,4!}\overline{\mathbb{Q}}_\ell. \end{aligned}$$

Thus $R^1\pi_{0,4!}\overline{\mathbb{Q}}_\ell \cong \mathrm{Frob}_2{}^*(R^1\pi_{0,6!}\overline{\mathbb{Q}}_\ell)$ is itself a middle extension. QED

Remark 9.4.5 The middle extension sheaf $R^1\pi_{0,4!}\overline{\mathbb{Q}}_\ell$ on $\mathbb{A}^1$ is the middle extension of the Kloosterman sheaf $Kl_2(\psi)$ [Ka-GKM, 11.0.1] on $\mathbb{G}_m$.

Corollary 9.4.6 Hypotheses as in Theorem 9.4.4 above, suppose in addition that k is a finite field. Then the geometrically irreducible perverse sheaf $R^1\pi_!\overline{\mathbb{Q}}_\ell[2](3/2)$ on $\mathbb{A}^2$ is self dual.

proof Immediate from Proposition 9.1.5, part 2). QED

(9.4.7) To end this section, we given a case where j remains constant, j = 0 = 1728, in our family.

Theorem 9.4.8 Let k be a field of characteristic 2, $\mathbb{A}^2 := \mathrm{Spec}(k[a_4, a_6])$. Consider the relative affine curve $\pi : \mathcal{E}^{aff} \to \mathbb{A}^2$ defined by the Weierstrass equation

$$y^2 + y = x^3 + a_4x + a_6.$$

Then the perverse sheaf $R^1\pi_!\overline{\mathbb{Q}}_\ell[2]$ on $\mathbb{A}^2$ is geometrically irreducible and geometrically self dual, and $H_c{}^*(\mathbb{A}^2\otimes_k\overline{k}, R^1\pi_!\overline{\mathbb{Q}}) = 0$.

proof The vanishing $H_c{}^*(\mathbb{A}^2\otimes_k\overline{k}, R^1\pi_!\overline{\mathbb{Q}}) = 0$ is a special case of Proposition 9.1.5. In this family, $\Delta = 1$. So $R^1\pi_!\overline{\mathbb{Q}}_\ell$ is lisse of rank two on all of $\mathbb{A}^2$, and geometrically self dual. To see that it is geometrically irreducible, pull back by the map

$$\varphi : \mathbb{A}^1 \to \mathbb{A}^2,$$
$$t \mapsto (t, 0),$$

obtaining the curve $\pi_{0,4} : \mathcal{E}_{0,4}{}^{aff} \to \mathbb{A}^1 := \mathrm{Spec}(k[t])$ defined by

$$y^2 + y = x^3 + tx.$$

Then $\varphi^{*1}\pi_!\overline{\mathbb{Q}}_\ell$ is $R^1\pi_{0,4!}\overline{\mathbb{Q}}_\ell$, which in turn is

$$R^1\pi_{0,4!}\overline{\mathbb{Q}}_\ell = R^1(pr_2 : \mathbb{A}^1\times\mathbb{A}^1 \to \mathbb{A}^1)_!\mathcal{L}_{\psi(x^3 + tx)}.$$

It is geometrically irreducible on $\mathbb{A}^1$, because

$$R^1\pi_{0,4!}\overline{\mathbb{Q}}_\ell[1] = R^1(pr_2 : \mathbb{A}^1\times\mathbb{A}^1 \to \mathbb{A}^1)_!\mathcal{L}_{\psi(x^3 + a_4x)}[1]$$

is the Fourier Transform on $\mathbb{A}^1$ of the perverse, geometrically irreducible object $\mathcal{L}_{\psi(x^3)}$ on $\mathbb{A}^1$. QED

Remark 9.4.9 It is also useful to keep in mind the description (verification left to the reader) of $R^1\pi_!\overline{\mathbb{Q}}_\ell$ as the external tensor product

$$R^1\pi_!\overline{\mathbb{Q}}_\ell \cong (R^1\pi_{0,4!}\overline{\mathbb{Q}}_\ell)\boxtimes\mathcal{L}_{\psi(a_6)}.$$

Corollary 9.4.10 Hypotheses as inTheorem 9.4.8 above, suppose in addition that k is a finite field. Then the geometrically irreducible perverse sheaf $R^1\pi_!\overline{\mathbb{Q}}_\ell[2](3/2)$ on $\mathbb{A}^2$ is self dual.

proof Immediate from Proposition 9.1.5, part 2). QED

Chapter 10: Weierstrass families

(10.1) Universal Weierstrass families in arbitrary characteristic

(10.1.1) We first give the motivation. Over any field K, or indeed Zariski locally over (the spec of) any ring R, an elliptic curve E/R can always be given by a Weierstrass equation

$$E_{a's} : y^2 + a_1xy + a_3y = x^3 + a_2x^2 + a_4x + a_6,$$

with coefficients a_i in R, subject to the condition that the discriminant Δ be invertible in R. For any unit λ in R, the map

$$(x, y) \mapsto (X, Y) := (\lambda^2x, \lambda^3y)$$

is an isomorphism from the curve

$$y^2 + a_1xy + a_3y = x^3 + a_2x^2 + a_4x + a_6$$

to the curve

$$Y^2 + \lambda a_1XY + \lambda^2a_3Y = X^3 + \lambda^2a_2X^2 + \lambda^4a_4X + \lambda^6a_6.$$

(10.1.2) Recall [Sil-ATEC, page 364] the precise formula for Δ. Given quantities a_1, a_2, a_3, a_4, a_6 in any ring R, one defines quantities b_2, b_4, b_6, and b_8 in R by the formulas

$$b_2 := (a_1)^2 + 4a_2,$$

$$b_4 := a_1a_3 + 2a_4,$$

$$b_6 := (a_3)^2 + 4a_6,$$

$$b_8 := (a_1)^2a_6 + 4a_2a_6 - a_1a_3a_4 + a_2(a_3)^2 - (a_2)^4.$$

Then one defines the quantity $\Delta := \Delta(a_1, a_2, a_3, a_4, a_6)$ in R by

$$\Delta := -(b_2)^2b_8 -8(b_4)^3 - 27(b_6)^2 + 9b_2b_4b_6.$$

Notice that if one gives a_k weight k, then b_k is isobaric of weight k, and Δ is isobaric of weight 12, as a universal $\mathbb{Z}$-polynomial in the a_i's.

(10.1.3) We now specialize to the case when K is a function field in one variable over a finite field. Thus we work over a finite field k of characteristic p, in which a prime ℓ is invertible. We fix a projective, smooth, geometrically connected curve C/k, of genus denoted g. We also fix on C an effective divisor D. We assume that

$$\deg(D) \geq 2g + 3.$$

We denote

$$V := C - D.$$

Suppose we are given functions f_k in the Riemann Roch spaces L(kD), for k = 1, 2, 3, 4, 6. Then we form the Weierstrass equation

$$E_{f's} : y^2 + f_1xy + f_3y = x^3 + f_2x^2 + f_4x + f_6.$$

We view it as a relative affine curve over V, with structural morphism

$$\pi_{f's} : E_{f's} \to V.$$

Its discriminant $\Delta := \Delta(f_1, f_2, f_3, f_4, f_6)$ then lies in L(12D). If Δ is nonzero, our Weierstrass equation defines an elliptic curve over K. And every elliptic curve over K, indeed every Weierstrass equation over K with nonvanishing Δ, is obtained this way for some choice of effective divisor D with $\deg(D) \geq 2g + 3$.

(10.1.4) For a fixed divisor D, we are interested in the variation of the L-function of $E_{f's}/k(C)$ as $(f_1, f_2, f_3, f_4, f_6)$ varies in

$$L(D)\times L(2D)\times L(3D)\times L(4D)\times L(6D).$$

Similarly, for each finite extension field k_d/k, we are interested in the variation of the L-function of $E_{f's}/k_d(C)$ as the f's vary in

$$(L(D)\times L(2D)\times L(3D)\times L(4D)\times L(6D))\otimes_k k_d.$$

For this to be a reasonable question, we must restrict to a dense open set of $L(D)\times L(2D)\times L(3D)\times L(4D)\times L(6D)$ (viewed as an affine space over k) over which the L-function is a polynomial of constant degree.

(10.1.5) We denote by

$$GW_{I_1 fd}(C, D) \subset L(D)\times L(2D)\times L(3D)\times L(4D)\times L(6D)$$

the open set whose $\bar{k}$-valued points are the f's over $\bar{k}$ such that the function Δ(f's) in L(12D) has divisor of poles 12D and has 12deg(D) distinct zeroes "at finite distance", i.e., in $(C - D)(\bar{k})$. We will see below that this open set is nonempty, and hence dense. Here GW is intended to evoke "general Weierstrass". The subscript $I_1 fd$ is intended to evoke "Neron type I_1 at finite distance", a property enjoyed by the $E_{f's}$ with f's in $GW_{I_1 fd}(C, D)(\bar{k})$ simply because their Δ's have only simple zeroes at finite distance. In particular, such an $E_{f's}$ has multiplicative reduction, with local monodromy a unipotent pseudoreflection, at each zero of Δ at finite distance. Because $E_{f's}$ has at least one point of multiplicative reduction, its j invariant is nonconstant.

Lemma 10.1.6 The open set $GW_{I_1 fd}(C, D)$ is nonempty.

proof We argue in each characteristic separately. In characteristic 2, we consider a curve of the form

$$y^2 + xy + f_3 y = x^3 + 1,$$

i.e., we take the vector of f's to be $(1, 0, f_3, 0, 1)$. Here Δ is equal to

$$(f_3)^4 + (f_3)^3 + 1.$$

The polynomial $t^4 + t^3 + 1$ has distinct roots over $\bar{k}$. We have only to take f_3 in $L(3D)\otimes_k \bar{k}$ which has divisor of poles 3D and which, as a finite flat map of degree 3deg(D) from C - D to $\mathbb{A}^1$, is finite etale over each of the 4 zeroes of $t^4 + t^3 + 1$. Such choices of f_3 exist, thanks to Lemma 6.2.9, applied to the divisor 3D.

In characteristic 3, we consider a curve of the form

$$y^2 = x^3 + x^2 + f_4x + 1,$$

i.e., we take the vector of f's to be $(0, 1, 0, f_4, 1)$. Here Δ is equal to

$$(f_4)^3 - (f_4)^2 + 1.$$

The polynomial $t^3 - t^2 + 1$ has distinct roots over $\overline{k}$. We have only to take f_4 in $L(4D)\otimes_k\overline{k}$ which has divisor of poles 4D and which, as a finite flat map of degree 4deg(D) from C - D to $\mathbb{A}^1$, is finite etale over each of the 3 zeroes of $t^3 - t^2 + 1$. Such choices of f_4 exist, thanks to Lemma 6.2.9, applied to the divisor 4D.

In characteristic $p \geq 5$, we consider a curve of the form

$$y^2 = x^3 - (3/4)x - f_6/4$$

(which we could rewite in the more familiar form

$$(2y)^2 = 4x^3 - 3x - f_6),$$

i.e., we take the vector of f's to be $(0, 0, 0, -3/4, -f_6/4)$. Here Δ is equal to

$$-27((f_6)^2 - 1).$$

The polynomial $27(t^2 - 1)$ has distinct roots in $\overline{k}$. We have only to take f_6 in $L(6D)\otimes_k\overline{k}$ which has divisor of poles 6D and which, as a finite flat map of degree 6deg(D) from C - D to $\mathbb{A}^1$, is finite etale over each of the 2 zeroes of $27(t^2 - 1)$. Such choices of f_6 exist, thanks to Lemma 6.2.9, applied to the divisor 6D.

In characteristic $p > 5$, we could instead consider a curve of the form

$$y^2 = x^3 - (3/4)f_4x - 1/4$$

(which we could rewite in the more familiar form

$$(2y)^2 = 4x^3 - 3f_4x - 1),$$

i.e., we take the vector of f's to be $(0, 0, 0, -3f_4/4, -1/4)$. Here Δ is equal to

$$27((f_4)^3 - 1).$$

The polynomial $27(t^3 - 1)$ has distinct roots in $\overline{k}$. We have only to take f_4 in $L(4D)\otimes_k\overline{k}$ which has divisor of poles 4D and which, as a finite flat map of degree 4deg(D) from C - D to $\mathbb{A}^1$, is finite etale over each of the 3 zeroes of $27(t^3 - 1)$. Such choices of f_4 exist, thanks to Lemma 6.2.9, applied to the divisor 4D. QED

Lemma 10.1.7 Given f's in $GW_{I_1fd}(C, D)(\overline{k})$, indeed given f's in $L(D)\times L(2D)\times L(3D)\times L(4D)\times L(6D)$ whose Δ has divisor of poles 12D, $E_{f's}/\overline{k}(C)$ has good reduction "at infinity", i.e., at each point P of $D(\overline{k})$.

proof Pick a point P in $D(\overline{k})$, say of multiplicity $r \geq 1$ in D. In terms of a uniformizing parameter π at P, the formal expansion of f_k at P is

$$f_k = \pi^{-kr}F_k,$$

with F_k holomorphic at P. By the change of variables

$$(x, y) \mapsto (\pi^{2r}x, \pi^{3r}y),$$

the curve $E_{f's}$ over the P-adic completion K_P of K becomes isomorphic to the curve $E_{F's}$ over that field. By isobaricity, we have

$$\Delta(F's) = \pi^{12r}\Delta(f's).$$

Since $\Delta(f's)$ has a pole of order 12r at P, $\Delta(F's)$ is invertible at P. As the F's are holomorphic at P, we see that $E_{F's}$ has good reduction at P. QED

Lemma 10.1.8 Let k_d/k be a finite extension, f's in $GW_{I_1 fd}(C, D)(k_d)$, and $\pi_{f's} : E_{f's} \to (C - D)\otimes_k k_d$ the structural morphism. Denote by

$$j : V := C - D \to C$$

the inclusion. Then we have the following results.

1) The sheaf $R^1\pi_{f's!}\overline{\mathbb{Q}}_\ell$ on $V\otimes_k k_d$ is a geometrically irreducible middle extension of generic rank two.

2) This sheaf is lisse on $(V\otimes_k k_d)[1/\Delta(f's)]$, pure of weight one, and its local monodromy at each zero of $\Delta(f's)$ in $V(\overline{k})$ is a unipotent pseudoreflection.

3) The sheaf $j_*R^1\pi_{f's!}\overline{\mathbb{Q}}_\ell$ on $C\otimes_k k_d$ is lisse at each point at infinity.

4) The cohomology groups $H_c^i(V\otimes_k \overline{k}, R^1\pi_{f's!}\overline{\mathbb{Q}}_\ell)$ vanish for $i \neq 1$.

5) The cohomology group $H_c^1(V\otimes_k \overline{k}, R^1\pi_{f's!}\overline{\mathbb{Q}}_\ell)$ is mixed of weight ≤ 2, and we have the formula

$$\dim H_c^1(V\otimes_k \overline{k}, R^1\pi_{f's!}\overline{\mathbb{Q}}_\ell) = 12\deg(D) - 2\chi_c(V\otimes_k \overline{k}, \overline{\mathbb{Q}}_\ell).$$

6) The cohomology groups $H_c^i(C\otimes_k \overline{k}, j_*R^1\pi_{f's!}\overline{\mathbb{Q}}_\ell)$ vanish for $i \neq 1$.

7) The cohomology group $H_c^1(C\otimes_k \overline{k}, j_*R^1\pi_{f's!}\overline{\mathbb{Q}}_\ell)$ is pure of weight two, it is the weight two quotient of $H_c^1(V\otimes_k \overline{k}, R^1\pi_{f's!}\overline{\mathbb{Q}}_\ell)$, and we have the formula

$$\begin{aligned}\dim H_c^1(C\otimes_k \overline{k}, j_*R^1\pi_{f's!}\overline{\mathbb{Q}}_\ell) &= 12\deg(D) - 2\#D(\overline{k}) - 2\chi_c(V\otimes_k \overline{k}, \overline{\mathbb{Q}}_\ell) \\ &= 4g - 4 + 12\deg(D).\end{aligned}$$

proof 1) That $R^1\pi_{f's!}\overline{\mathbb{Q}}_\ell$ on $V\otimes_k k_d$ is a middle extension of generic rank two is a special case of Corollary 7.5.5, because all the zeroes of Δ are simple. It is geometrically irreducible because the j invariant of $E_{f's}$ is nonconstant. Assertion 2) is "mise pour memoire". Assertion 3) results from the previous Lemma 10.1.7. Assertions 4) and 6) result from 1). Assertions 5) and 7) result from 1), from Deligne's main theorem in Weil II [De-Weil II, 3.3.1] for the mixedness and purity, and from the Euler Poincaré formula, whose straightforward application is left to the reader. QED

Corollary 10.1.9 Denote by N the common dimension

$$N := \dim H_c^1(C\otimes_k\overline{k}, j_*R^1\pi_{f's!}\overline{\mathbb{Q}}_\ell).$$

We have the inequality

$$N \geq 28g + 32.$$

proof Indeed, we have the exact formula $N = 4g - 4 + 12\deg(D)$ (by Lemma 10.1.8, part 7)), and the inequality $\deg(D) \geq 2g + 3$. QED

Corollary 10.1.10 For any finite extension k_d/k, and for any point f's in $GW_{I_1fd}(C, D)(k_d)$, the unitarized L-function of $E_{f's}/k_d(C)$ is given by

$$L(E_{f's}/k_d(C), T) = \det(1 - T\mathrm{Frob}_{k_d} \mid H^1(C\otimes_k\overline{k}, j_*R^1\pi_{f's!}\overline{\mathbb{Q}}_\ell(1))).$$

proof This results from the fact that $j_*R^1\pi_{f,g!}\overline{\mathbb{Q}}_\ell$ is a middle extension from the open set $V[1/(\Delta(f's))]\otimes_k\overline{k}$ where it is lisse, and the vanishing given by Lemma 10.1.8, part 6). QED

(10.1.11) Our next task is to capture these L-functions in the context of suitable perverse sheaves $M = \mathrm{Twist}(L, K, \mathcal{F}, h)$ attached to suitable "standard input", cf. 1.15.4. Over

$$\mathbb{A}^5 := \mathrm{Spec}(k[a_1, a_2, a_3, a_4, a_6]),$$

we have the universal Weierstrass curve

$$E_{a's} : y^2 + a_1xy + a_3y = x^3 + a_2x^2 + a_4x + a_6,$$

whose structural morphism we denote

$$\pi_{a's} : E_{a's} \to \mathbb{A}^5.$$

We take

the integer $m = 5$,

the perverse sheaf $K = R^1\pi_{a's!}\overline{\mathbb{Q}}_\ell(3)[5]$ on $\mathbb{A}^5/k = \mathrm{Spec}(k[a's])$,

the affine k-scheme V/k given by $V := C - D$,

the k-morphism $h : V \to \mathbb{A}^5$ given by $h = 0$,

the perverse sheaf $L = \overline{\mathbb{Q}}_\ell(1/2)[1]$ on V/k,

the integer $d = 1 - 2g + \deg(D)$,

the space of $\mathbb{A}^5$-valued functions $(\mathcal{F}, \tau)$ on V given by the finite-dimensional k-vector space

$$\mathcal{F} = L(D)\times L(2D)\times L(3D)\times L(4D)\times L(6D),$$

and the k-linear map

$$\tau : \mathcal{F} \to \mathrm{Hom}_{k\text{-schemes}}(V, \mathbb{A}^5),$$
$$(f_1, f_2, f_3, f_4, f_6) \mapsto$$
$$\text{the map } v \mapsto (f_1(v), f_2(v), f_3(v), f_4(v), f_6(v)).$$

It results from Propostions 9.1.4 and 9.1.5 that this is standard input, and, using Kunneth, that $H^*_c((V\times\mathbb{A}^m)\otimes\overline{k}, pr_1{}^*L\otimes pr_2{}^*K) = 0$.

Lemma 10.1.12 The perverse sheaf M is lisse on the dense open set $GW_{I_1 fd}(C, D)$.

proof On this dense open set, there is only one nonvanishing cohomology sheaf (by Lemma 10.1.8, part 4)), namely $\mathcal{H}^{-\dim(\mathcal{F})}(M)$, which is tautologically a sheaf of perverse origin on the space $\mathcal{F}$. The stalks of this sheaf are

$$f's \mapsto H_c^1(V\otimes_k \overline{k}, R^1\pi_{f's!}\overline{\mathbb{Q}}_\ell)(3),$$

which have constant rank on $GW_{I_1 fd}(C, D)$, by Lemma 10.1.8, part 5). By [Ka-SMD, Proposition 11], $\mathcal{H}^{-\dim(\mathcal{F})}(M)$ is lisse on $GW_{I_1 fd}(C, D)$. QED

(10.1.13) Thus on the open set $GW_{I_1 fd}(C, D)$, M is

$$\mathcal{M}(5/2)[\ell(D) + \ell(2D) + \ell(3D) + \ell(4D) + \ell(6D)],$$

with $\mathcal{M}$ the lisse sheaf, mixed of weight ≤ 0, given stalkwise by

$$\mathcal{M}_{f's} := H_c^1(V\otimes_k \overline{k}, R^1\pi_{f's!}\overline{\mathbb{Q}}_\ell)(1).$$

The sheaf $Gr^0(\mathcal{M})$ on $GW_{I_1 fd}(C, D)$ is lisse of rank N, orthogonally self dual, and pure of weight zero. We have

$$Gr^0(\mathcal{M})_{f's} = H_c^1(C\otimes_k \overline{k}, j_\star R^1\pi_{f's!}\overline{\mathbb{Q}}_\ell)(1).$$

Because of the vanishing $H^\star{}_c((V\times\mathbb{A}^m)\otimes\overline{k}, pr_1{}^\star L\otimes pr_2{}^\star K) = 0$ noted in 10.1.11 above, we have (by Corollary 1.20.3, part 3))

$$Gr^0(\mathcal{M})_{cst} = 0,\ Gr^0(\mathcal{M}) = Gr^0(\mathcal{M})_{ncst}. \qquad \text{QED}$$

Theorem 10.1.14 Let k be a finite field, in which a prime ℓ is invertible. Fix a projective, smooth, geometrically connected curve C/k, of genus denoted g. Fix on C an effective divisor D of degree $\ge 2g + 3$. Then the group G_{geom} for the lisse sheaf $Gr^0(\mathcal{M})$ on the dense open set $GW_{I_1 fd}(C, D)$ is O(N), for $N := 4g - 4 + 12\deg(D)$ the rank of $Gr^0(\mathcal{M})$.

proof As noted in Corollary 10.1.9 above, we have $N \ge 28g + 32$, so certainly $N \ge 9$. By the argument given in the first paragraph of the proof of Theorem 7.2.3, it suffices to show that G_{geom} contains a reflection.

To show this, we argue in each characteristic separately, exactly as we did in proving Lemma 10.1.6. The idea, in all cases, is to invoke Corollary 7.2.4 suitably. We will spell out the details in the characteristic 2 case, and just indicate the changes to be made in treating the case of other characteristic.

In characteristic 2, we consider the pullback of $\mathcal{M}$ to the inverse image of $GW_{I_1 fd}(C, D)$ in L(3D) by the map

$$i : L(3D) \to L(D)\times L(2D)\times L(3D)\times L(4D)\times L(6D),$$
$$f_3 \mapsto (1, 0, f_3, 0, 1).$$

Then $i^{-1}(GW_{I_1 fd}(C, D))$ is the dense open set $U \subset L(3D)$ whose $\overline{k}$-valued points are those f_3 in $L(3D)\otimes_k \overline{k}$ which have divisor of poles 3D and which, as finite flat maps of degree 3deg(D) from C - D to $\mathbb{A}^1$, are finite etale over each of the 4 zeroes of $X^4 + X^3 + 1$. The lisse sheaf $i^*\mathcal{M}$ on U is precisely the $\mathcal{M}$ of 7.1.14 attached to the perverse sheaf M constructed in 7.1.4 attached to the the following situation.

Over $\mathbb{A}^1 := \mathrm{Spec}(k[t])$, we have the curve

$$E_t : y^2 + xy + ty = x^3 + 1,$$

with structural morphism

$$\pi_t : E_t \to \mathbb{A}^1.$$

Its Δ is $t^4 + t^3 + 1$, which has simple zeroes. We denote by $S \subset \mathbb{A}^1$ the zero set of $t^4 + t^3 + 1$:

$$S := \{t \mid t^4 + t^3 + 1 = 0\} \subset \mathbb{A}^1.$$

At each of these zeroes, E_t has multiplicative reduction. On $\mathbb{A}^1$, we form the sheaf

$$\mathcal{G} := R^1\pi_{t!}\overline{\mathbb{Q}}_\ell(1).$$

We see from Corollary 7.5.5 that $\mathcal{G}$ is a middle extension. Its restriction to $\mathbb{A}^1$ - S is lisse of rank 2, geometrically irreducible (because j is nonconstant), pure of weight -1, and symplectically self dual toward $\overline{\mathbb{Q}}_\ell(1)$. At each point in $S(\overline{k})$, the local monodromy of $\mathcal{G}$ is a unipotent pseudoreflection. At ∞, E_t has local monodromy of order 3 (because E_t acquires good reduction at ∞ after passing to $k(t^{1/3})$), and hence $\mathcal{G}$ is tame at ∞. We put

$$K := \mathcal{G}[1].$$

We use this K on $\mathbb{A}^1$, we take our curve C and the divisor 3D to be the (C, D) used in the construction of 7.1.2-4. This construction gives us a perverse sheaf on L(3D) which, up to a Tate twist and a shift, is the pullback i^*M. Our open set $U \subset L(3D)$ is precisely the set $U_{D,S}$ of 7.1.6, and our $i^*\mathcal{M}$ on U is precisely the lisse sheaf on $U_{D,S}$ denoted $\mathcal{M}$ in 7.1.14. We then apply Corollary 7.2.4, which tells us that G_{geom} for $Gr^0(i^*\mathcal{M}) = i^*Gr^0(\mathcal{M})$ contains a reflection. Therefore the larger group G_{geom} for $Gr^0(\mathcal{M})$ contains a reflection.

In characteristic 3, we use the analogous argument, but for the map

$$i : L(4D) \to L(D)\times L(2D)\times L(3D)\times L(4D)\times L(6D),$$
$$f_4 \mapsto (0, 1, 0, f_4, 1),$$

and the initial family

$$y^2 = x^3 + x^2 + tx + 1.$$

In characteristic $p \geq 5$, we have two choices of how to give the analogous argument. We can use the map

$$i : L(4D) \to L(D)\times L(2D)\times L(3D)\times L(4D)\times L(6D),$$
$$f_4 \mapsto (0, 0, 0, -3f_4/4, -1/4),$$

and the initial one one-parameter family

$$y^2 = x^3 - (3/4)tx - 1/4.$$

Or we can use the map

$$i : L(6D) \to L(D)\times L(2D)\times L(3D)\times L(4D)\times L(6D),$$
$$f_6 \mapsto (0, 0, 0, -3/4, -f_6/4),$$

and the initial one-parameter family

$$y^2 = x^3 - (3/4)x - t/4.$$

In all cases, we apply Corollary 7.2.4, which tells us that G_{geom} for $Gr^0(i^*\mathcal{M}) = i^*Gr^0(\mathcal{M})$ contains a reflection, and we conclude that the larger group G_{geom} for $Gr^0(\mathcal{M})$ contains a reflection. QED

(10.2) Usual Weierstrass families in characteristic $p \geq 5$

(10.2.1) In this section, we work over a finite field k with characteristic $p \geq 5$. We fix a projective, smooth, geometrically connected curve C/k, of genus denoted g. We also fix on C two effective divisors D_2 and D_3. We assume that

$$\deg(D_2) \geq 2g + 3,\ \deg(D_3) \geq 2g + 3.$$

From the two divisors D_2 and D_3, we form the auxiliary divisor

$$D_{max} := \mathrm{Max}(3D_2, 2D_3).$$

Concretely, if $D_2 = \Sigma n_P P$ and $D_3 = \Sigma m_P P$, then

$$D_{max} := \Sigma\ \mathrm{Max}(3n_P, 2m_P)P.$$

We denote

$$V := C - (D_2 \cup D_3) = V - D_{max}.$$

Given functions f_2 and f_3 in the Riemann Roch spaces $L(D_2)$ and $L(D_3)$ respectively, we form the Weierstrass equation

$$E_{f's} : y^2 = 4x^3 - f_2x - f_3,$$

which we view as a relative affine curve over V, with structural morphism

$$\pi_{f's} : E_{f's} \to V.$$

Its discriminant $\Delta(f_2, f_3)$ is given by

$$\Delta(f_2, f_3) = (f_2)^3 - 27(f_3)^2,$$

and lies a priori in $L(D_{max})$.

(10.2.2) For fixed divisors D_2 and D_3 as above, we are interested in the variation of the L-function of $E_{f's}/k(C)$ as (f_2, f_3) varies in $L(D_2)\times L(D_3)$. Similarly, for each finite extension field k_d/k, we are interested in the variation of the L-function of $E_{f's}/k_d(C)$ as the f's vary in $(L(D_2)\times L(D_3))\otimes_k k_d$.

(10.2.3) For this to be a reasonable question, we must restrict to a dense open set of $L(D_2)\times L(D_3)$ (viewed as an affine space over k) over which the L-function is a polynomial of constant degree.

(10.2.4) We denote by

$$W_{I_1fd}(C, D_2, D_3) \subset L(D_2)\times L(D_3)$$

the open set whose $\bar{k}$-valued points are the f's over $\bar{k}$ such that the function $\Delta(f_2, f_3)$ in $L(D_{max})$ has divisor of poles D_{max}, and has $\deg(D_{max})$ distinct zeroes "at finite distance", i.e., in $V(\bar{k})$. We will see below that this open set is nonempty, and hence dense. Here W is intended to evoke "Weierstrass". The subscript I_1fd is intended to evoke "Neron type I_1 at finite distance", a property enjoyed by the $E_{f's}$ with f's in $W_{I_1fd}(C, D_2, D_3)(\bar{k})$ simply because their Δ's have only simple zeroes at finite distance. In particular, such an $E_{f's}$ has multiplicative reduction, with local monodromy a unipotent pseudoreflection, at each zero of Δ at finite distance. Because $E_{f's}$ has at least one point of multiplicative reduction, its j invariant is nonconstant.

Lemma 10.2.5 The open set $W_{I_1fd}(C, D_2, D_3)$ is nonempty.

proof First, let us show that the open set where $\Delta(f_2, f_3)$ in $L(D_{max})$ has divisor of poles D_{max} is nonempty. For this, take any f_2 in $L(D_2)$ with divisor of poles D_2, and any f_3 in $L(D_3)$ with divisor of poles D_3. For all but finitely many values of λ in $\bar{k}$, Δ for the pair $(f_2, \lambda f_3)$,

$$\Delta(f_2, \lambda f_3) = (f_2)^3 - 27(g_3)^2,$$

visibly has divisor of poles D_{max}.

Now take (f_2, f_3) in $(L(D_2)\times L(D_3))\otimes_k \bar{k}$ such that f_2 and f_3 have divisor of poles D_2 and D_3 respectively, such that both f_2 and f_3 have Swan-minimal poles, cf. Lemma 6.2.7 and 6.4.6, and such that $\Delta(f_2, f_3)$ has divisor of poles D_{max}. For any scalars (a, b) in $\mathbb{A}^2(\bar{k})$, the pair $(f_2 + a, f_3 + b)$ has the same property, that $\Delta(f_2 + a, f_3 + b)$ has divisor of poles D_{max}. We will show that for some (a, b) in $\mathbb{A}^2(\bar{k})$, the pair $(f_2 + a, f_3 + b)$ lies in $W_{I_1fd}(C, D_2, D_3)(\bar{k})$.

To see this, we argue as follows. We work over $\bar{k}$. In $V\times\mathbb{A}^2$, with coordinates (v, a, b), consider the hypersurface Z of equation

$$(a + f_2(v))^3 - 27(b + f_3(v))^2 = 0.$$

By means of the automorphism of $V\times\mathbb{A}^2$ given by

$$(v, a, b) \mapsto (v, a - f_2(v), b - g_3(v)),$$

Z receives isomorphically the product

$$V\times C \cong V,$$

for C the curve in $\mathbb{A}^2$ of equation

$$a^3 - 27b^2 = 0.$$

This curve is smooth outside (0, 0), and C - (0, 0) is $\mathbb{G}_m$, by the map

$$\mathbb{G}_m \to C - (0, 0),$$
$$t \mapsto (3t^2, t^3).$$

Thus our original hypersurface Z is smooth outside the graph Γ_φ of the map

$$\varphi : V \to \mathbb{A}^2,$$
$$v \mapsto \varphi(v) := (-f_2(v), -f_3(v)).$$

Thus Z - $\Gamma$is a geometrically connected smooth surface, isomorphic to $V \times \mathbb{G}_m$.

Consider the projection

$$\pi : Z \to \mathbb{A}^2$$

which is the composite of the inclusion

$$Z \subset V \times \mathbb{A}^2$$

with the projection of $V \times \mathbb{A}^2$ onto $\mathbb{A}^2$. The scheme-theoretic fibre of π over a $\bar{k}$-valued point (a, b) of $\mathbb{A}^2$ is the closed subscheme of $V \otimes_k \bar{k}$ defined by $\Delta(a + f_2(v), b + f_3(v)) = 0$. A priori, this is a finite subscheme of $V \otimes_k \bar{k}$ which is finite and flat over $\bar{k}$ of degree $\deg(D_{max})$. We must show that for some choice of (a, b), this fibre is etale over $\bar{k}$. For this, it suffices to show that the map π is etale on a dense open set of V, say on V - W for some closed subscheme W of dimension at most one. Then $\pi(W)$ lies in some proper closed subscheme T of $\mathbb{A}^2$, the map π is etale over $\mathbb{A}^2$ - T, and any $\bar{k}$-valued point (a, b) of $\mathbb{A}^2$ - T does the job.

To see that π is generically etale on Z, we look at π restricted to Z - Γ. Via the isomorphism above

$$V \times \mathbb{G}_m \cong Z - \Gamma,$$
$$(v, t) \mapsto (v, 3t^2 - f_2(v), t^3 - f_3(v)),$$

the map π becomes the map

$$\tilde{\pi} : V \times \mathbb{G}_m \to \mathbb{A}^2,$$
$$(v, t) \mapsto (3t^2 - f_2(v), t^3 - f_3(v)).$$

Zariski locally on V, we may choose an $\mathcal{O}_V$ basis ∂ of the tangent bundle <u>Der</u>(V/k). Using it, the Jacobian matrix $d\tilde{\pi}$ is

$$\begin{matrix} 6t & 3t^2 \\ -(\partial f_2)(v) & -(\partial f_3)(v), \end{matrix}$$

and the Jacobian determinant at (v, t) is

$$-6t(\partial f_3)(v) + 3t^2(\partial f_2)(v).$$

Since t lies in $\mathbb{G}_m$, this vanishes if and only if

$$-2(\partial f_3)(v) + t(\partial f_2)(v) = 0,$$

i.e., if and only if the two vectors $(\partial f_2, \partial f_3)(v)$ and (t, -2) are orthogonal. Thus either the map φ is etale on a dense open set of the source, or for every v in V(k) and for every t in $\bar{k}^\times$, we have the perpendicularity relation

$$\mathrm{grad}(f_2, f_3)(v) \perp (t, -2).$$

This second possibility cannot hold. Indeed, since f_2 and f_3 both have Swan-minimal poles, neither df_2 nor df_3 vanishes identically, and hence for some v in $V(\bar{k})$, we have

$$\mathrm{grad}(f_2, f_3)(v) \neq (0, 0).$$

For this fixed v, the orthogonality

$$\mathrm{grad}(f_2, f_3)(v) \perp (t, -2)$$

can hold for at most one value of t, since $(t_1, -2)$ and $(t_2, -2)$ are linearly independent for $t_1 \neq t_2$ in $\bar{k}$.

Therefore the map φ is etale on a dense open set of the source Z. QED.

Lemma 10.2.6 Write D_{max} over $\bar{k}$ as $\Sigma_P c_P P$. Forany point (f_2, f_3) in $W_{I_1 fd}(C, D_2, D_3)(\bar{k})$, or more generally for any point (f_2, f_3) in $(L(D_2)\times L(D_3))(\bar{k})$ such that $\Delta(f_2, f_3)$ has divisor of poles D_{max}, the reduction type of $E_{f's}$ at a point P of $D_{max}(\bar{k})$ is determined by the integer c_P as follows. If $c_P \equiv 0 \bmod 12$, then $E_{f's}$ has good reduction at P, otherwise it has potentially good but additive reduction at P.

proof Fix a point P in $D_{max}(\bar{k})$. Write $c := c_P$, the multiplicity of P in D_{max}. Denote by n (resp. m) the multiplicity with which P occurs in D_2 (resp. in D_3). Thus $c = \mathrm{Max}(3n, 2m)$. Pick a uniformizing parameter z at P. Thus we have, over the P-adic completion K_P of the function field $K := \bar{k}(C)$, an elliptic curve

$$y^2 = 4x^3 - f_2 x - f_3$$

with $\mathrm{ord}_P(f_2) \geq -n$, $\mathrm{ord}_P(f_3) \geq -m$, and $\mathrm{ord}_P(\Delta(f_2, f_3)) = -c$.

Suppose first that $c \equiv 0 \bmod 12$, say $c = 12d$. Because $c = \mathrm{Max}(3n, 2m)$, we find

$$\mathrm{ord}_P(z^{4d}f_2) \geq 4d - n = (1/3)(c - 3n) \geq 0,$$

$$\mathrm{ord}_P(z^{6d}f_3) \geq 6d - m = (1/2)(c - 2m) \geq 0,$$

$$\mathrm{ord}_P(\Delta(z^{4d}f_2, z^{6d}f_3)) = \mathrm{ord}_P(z^{12d}\Delta(f_2, f_3)) = 0.$$

Thus the K_P-isomorphic curve

$$y^2 = 4x^3 - z^{4d}f_2 x - z^{6d}f_3$$

has good reduction at P, as asserted.

This same argument shows that even when c is not divisible by 12, if we pass to the degree 12, fully ramified extension L_P of K_P obtained by adjoining $z^{1/12}$, then our curve acquires good reduction at the unique place over P. Thus our curve always has potentially good reduction at P. If $c = \mathrm{ord}_P(\Delta)$ is not divisible by 12, our curve cannot have good reduction at P. But a curve with potentially good reduction has either good reduction or additive reduction. QED

Lemma 10.2.7 Let k_d/k be a finite extension, (f_2, f_3) a point in $W_{I_1 fd}(C, D_2, D_3)(k_d)$, $\pi_{f's} : E_{f's} \to V\otimes_k k_d$ the structural morphism. Denote by

$$j : V := C - D_{max} \to C$$

the inclusion. Then we have the following results.

1) The sheaf $R^1\pi_{f's!}\overline{\mathbb{Q}}_\ell$ on $V\otimes_k k_d$ is a geometrically irreducible middle extension of generic rank two.

2) This sheaf is lisse on $(V\otimes_k k_d)[1/\Delta(f_2, f_3)]$, pure of weight one, and its local monodromy at each zero of $\Delta(f_2, f_3)$ in $V(\overline{k})$ is a unipotent pseudoreflection.

3) The sheaf $j_\star R^1\pi_{f's!}\overline{\mathbb{Q}}_\ell$ on $C\otimes_k k_d$ is lisse at each point P at infinity with $c_P \equiv 0 \bmod 12$, and it vanishes at all other points at ∞.

4) The cohomology groups $H_c^i(V\otimes_k\overline{k}, R^1\pi_{f's!}\overline{\mathbb{Q}}_\ell)$ vanish for $i \neq 1$.

5) The cohomology group $H_c^1(V\otimes_k\overline{k}, R^1\pi_{f's!}\overline{\mathbb{Q}}_\ell)$ is mixed of weight ≤ 2, and we have the formula

$$\dim H_c^1(V\otimes_k\overline{k}, R^1\pi_{f's!}\overline{\mathbb{Q}}_\ell) = \deg(D_{max}) - 2\chi_c(V\otimes_k\overline{k}, \overline{\mathbb{Q}}_\ell).$$

6) The cohomology groups $H_c^i(C\otimes_k\overline{k}, j_\star R^1\pi_{f's!}\overline{\mathbb{Q}}_\ell)$ vanish for $i \neq 1$.

7) The cohomology group $H_c^1(C\otimes_k\overline{k}, j_\star R^1\pi_{f's!}\overline{\mathbb{Q}}_\ell)$ is pure of weight two, it is the weight two quotient of $H_c^1(V\otimes_k\overline{k}, R^1\pi_{f's!}\overline{\mathbb{Q}}_\ell)$, and we have the formula

$$\begin{aligned}
&\dim H_c^1(C\otimes_k\overline{k}, j_\star R^1\pi_{f's!}\overline{\mathbb{Q}}_\ell)\\
&\quad= \deg(D_{max}) - 2\#\{P \text{ in } D_{max}(\overline{k}) \text{ with } c_P \equiv 0 \bmod 12\}\\
&\qquad\qquad -2\chi_c(V\otimes_k\overline{k}, \overline{\mathbb{Q}}_\ell)\\
&\quad= 4g - 4 + 2\#D_{max}(\overline{k}) + \deg(D_{max})\\
&\qquad\qquad - 2\#\{P \text{ in } D_{max}(\overline{k}) \text{ with } c_P \equiv 0 \bmod 12\}\\
&\quad= 4g - 4 + \deg(D_{max})\\
&\qquad\qquad + 2\#\{P \text{ in } D_{max}(\overline{k}) \text{ with } c_P \not\equiv 0 \bmod 12\}.
\end{aligned}$$

proof Because we are in characteristic $p \geq 5$, the sheaf $R^1\pi_{f's!}\overline{\mathbb{Q}}_\ell$ is automatically everywhere tame. Having made this observation, the proof is essentially identical to the proof of Lemma 10.1.8, and is left to the reader. QED

Corollary 10.2.8 Denote by N the common dimension

$$N := \dim H_c^1(C\otimes_k\overline{k}, j_\star R^1\pi_{f's!}\overline{\mathbb{Q}}_\ell).$$

We have the inequalities

$$N \geq 4g - 4 + \deg(D_{max}) \geq 4g - 4 + 3(2g + 3) = 10g + 5.$$

proof Immediate from Lemma 10.2.7, part 7) above, the definition of D_{max}, and the inequality $\deg(D_2) \geq 2g + 3$. QED

Corollary 10.2.9 For any finite extension k_d/k, and for any point f's in $W_{I_1 fd}(C, D_2, D_3)(k_d)$, the unitarized L-function of $E_{f's}/k_d(C)$ is given by

$$L(E_{f's}/k_d(C), T) = \det(1 - T\mathrm{Frob}_{k_d} \mid H^1(C\otimes_k\bar{k}, j_\star R^1\pi_{f's!}\bar{\mathbb{Q}}_\ell(1))).$$

proof This results from the fact that $j_\star R^1\pi_{f,g!}\bar{\mathbb{Q}}_\ell$ is a middle extension from the open set $V[1/\Delta(f_2, f_3)]\otimes_k\bar{k}$ where it is lisse, and the vanishing given by Lemma 10.2.7, part 6). QED

(10.2.10) Our next task is to capture these L-functions in the context of suitable perverse sheaves M = Twist(L, K, $\mathcal{F}$, h) attached to suitable "standard input", cf. 1.15.4. Over

$$\mathbb{A}^2 := \mathrm{Spec}(k[g_2, g_3]),$$

we have the Weierstrass curve

$$E_{g's} : y^2 = 4x^3 - g_2x - g_3,$$

whose structural morphism we denote

$$\pi_{g's} : E_{g's} \to \mathbb{A}^2.$$

We take

the integer m = 2,

the perverse sheaf $K = R^1\pi_{g's!}\bar{\mathbb{Q}}_\ell(3/2)[2]$ on $\mathbb{A}^2 = \mathrm{Spec}(k[g_2,g_3])$,

the affine k-scheme V/k given by $V := C - D_{max}$,

the k-morphism $h : V \to \mathbb{A}^2$ given by h = 0,

the perverse sheaf $L = \bar{\mathbb{Q}}_\ell(1/2)[1]$ on V/k,

the integer $d = 1 - 2g + \mathrm{Min}(\deg(D_2), \deg(D_3))$,

the space of $\mathbb{A}^2$-valued functions $(\mathcal{F}, \tau)$ on V given by the finite-dimensional k-vector space

$$\mathcal{F} = L(D_2)\times L(D_3),$$

and the k-linear map

$$\tau : \mathcal{F} \to \mathrm{Hom}_{k\text{-schemes}}(V, \mathbb{A}^2),$$
$$(f_2, f_3) \mapsto$$
$$\text{the map } v \mapsto (f_2(v), f_3(v)).$$

It results from Propositions 9.1.4 and 9.1.5 that this is standard input, and, using Kunneth, that $H^*_c((V\times\mathbb{A}^m)\otimes\bar{k}, pr_1{}^*L\otimes pr_2{}^*K) = 0$.

Lemma 10.2.11 The perverse sheaf M is lisse on the dense open set $W_{I_1 fd}(C, D_2, D_3)$.

proof On this dense open set, there is only one nonvanishing cohomology sheaf (by Lemma 10.2.7, part 4)), namely $\mathcal{H}^{-\dim(\mathcal{F})}(M)$, which is tautologically a sheaf of perverse origin on the space $\mathcal{F}$. The stalks of this sheaf are

$$f's \mapsto H_c{}^1(V\otimes_k\bar{k}, R^1\pi_{f's!}\bar{\mathbb{Q}}_\ell)(3/2),$$

which have constant rank on $W_{I_1 fd}(C, D)$ (by Lemma 10.2.7, part 5)). By [Ka-SMD, Proposition 11], $\mathcal{H}^{-\dim(\mathcal{F})}(M)$ is lisse on $W_{I_1 fd}(C, D_2, D_3)$. QED

(10.2.12) Thus on the open set $W_{I_1 fd}(C, D_2, D_3)$, M is

$$\mathcal{M}(1)[\ell(D_2) + \ell(D_3)],$$

with $\mathcal{M}$ the lisse sheaf, mixed of weight ≤ 0, given stalkwise by

$$\mathcal{M}_{f's} := H_c^1(V\otimes_k \bar{k}, R^1\pi_{f's!}\bar{\mathbb{Q}}_\ell)(1).$$

The sheaf $Gr^0(\mathcal{M})$ on $W_{I_1 fd}(C, D_2, D_3)$ is lisse of rank N, orthogonally self dual, and pure of weight zero. We have

$$Gr^0(\mathcal{M})_{f's} = H_c^1(C\otimes_k \bar{k}, j_\star R^1\pi_{f's!}\bar{\mathbb{Q}}_\ell)(1).$$

Because of the vanishing $H^*_c((V\times\mathbb{A}^m)\otimes\bar{k}, pr_1{}^*L\otimes pr_2{}^*K) = 0$ noted in 10.2.10 above, we have (by Corollary 1.20.3, part3))

$$Gr^0(\mathcal{M})_{cst} = 0, \ Gr^0(\mathcal{M}) = Gr^0(\mathcal{M})_{ncst}.$$

Theorem 10.2.13 Let k be a finite field of characteristic $p \ge 5$, in which a prime ℓ is invertible. Fix a projective, smooth, geometrically connected curve C/k, of genus denoted g. Fix on C effective divisors D_2 and D_3, both of degree $\ge 2g + 3$. Define $D_{max} := \mathrm{Max}(3D_2, 2D_3)$. Write D_{max} over $\bar{k}$ as $\Sigma c_P P$. Denote by

$$N := 4g - 4 + \deg(D_{max}) + 2\#\{P \text{ in } D_{max}(\bar{k}) \text{ with } c_P \not\equiv 0 \bmod 12\}$$

the rank of the lisse sheaf $Gr^0(\mathcal{M})$ on the dense open set $W_{I_1 fd}(C, D_2, D_3)$. Suppose that

$$N \ge 9.$$

Suppose also that

$$\text{either } 3D_2 \ge 2D_3, \text{ or } 2D_3 \ge 3D_2,$$

i.e., suppose that $D_{max} = 2D_2$ or that $D_{max} = 2D_3$. Then the group G_{geom} for the lisse sheaf $Gr^0(\mathcal{M})$ on $W_{I_1 fd}(C, D_2, D_3)$ is O(N).

proof The argument is similar to that used in proving Theorem 10.1.14. By assumption, we have $N \ge 9$. By the argument given in the first paragraph of the proof of Theorem 7.2.3, it suffices to show that G_{geom} contains a reflection.

Suppose first we are in the case $3D_2 \ge 2D_3$. In this case, we consider the pullback of $\mathcal{M}$ to the inverse image of $W_{I_1 fd}(C, D_2, D_3)$ in $L(D_2)$ by the map

$$i : L(D_2) \to L(D_2)\times L(D_3),$$
$$f_2 \mapsto (3f_2, 1).$$

Here $i^{-1}(W_{I_1 fd}(C, D_2, D_3))$ is the dense open set in $L(D_2)$ whose $\bar{k}$-

valued points are those f_2 with divisor of poles D_2 and which, as finite flat maps from $C - D_{max} = C - D_2 \to \mathbb{A}^1$, are finite etale over the cube roots of unity.

The lisse sheaf $i^*\mathcal{M}$ on U is precisely the $\mathcal{M}$ of 7.1.14 attached to the perverse sheaf M constructed in 7.1.2-4, attached to the following situation.

Over $\mathbb{A}^1 := \mathrm{Spec}(k[t])$, we have the curve

$$E_t : y^2 = 4x^3 - 3tx - 1,$$

with structural morphism

$$\pi_t : E_t \to \mathbb{A}^1.$$

Its Δ is $27(t^3 - 1)$, which has simple zeroes $S = \mu_3$. At each of these zeroes, E_t has multiplicative reduction. On $\mathbb{A}^1$, we form the sheaf

$$\mathcal{G} := R^1\pi_{t!}\overline{\mathbb{Q}}_\ell(1).$$

We see from Corollary 7.5.5 that $\mathcal{G}$ is a middle extension. Its restriction to $\mathbb{A}^1[1/\Delta]$ is lisse of rank 2, geometrically irreducible (because j is nonconstant), pure of weight -1, and symplectically self dual toward $\overline{\mathbb{Q}}_\ell(1)$. At each point in $S(\overline{k}) := \mu_3(\overline{k})$, the local monodromy of $\mathcal{G}$ is a unipotent pseudoreflection. At ∞, E_t has local monodromy of order 4 (because E_t acquires good reduction at ∞ after passing to $k(t^{1/4})$), hence $\mathcal{G}$ is tame at ∞. We put

$$K := \mathcal{G}[1].$$

We use this K on $\mathbb{A}^1$, and we take our curve C and the divisor D_2 to be the (C, D) used in the construction of 7.1.2-4 This construction gives us a perverse sheaf on $L(D_2)$ which, up to a Tate twist and a shift, is the pullback i^*M. Our open set $U \subset L(D_2)$ is precisely the set $U_{D,S}$ of 7.1.6 and our $i^*\mathcal{M}$ on U is precisely the lisse sheaf on $U_{D,S}$ denoted $\mathcal{M}$ in 7.1.14. We then apply Corollary 7.2.4, which tells us that G_{geom} for $Gr^0(i^*\mathcal{M}) = i^*Gr^0(\mathcal{M})$ contains a reflection. Therefore the larger group G_{geom} for $Gr^0(\mathcal{M})$ contains a reflection.

Suppose now we are in the case $2D_3 \geq 3D_2$. In this case, we consider the pullback of $\mathcal{M}$ to the inverse image of $W_{I_1fd}(C, D_2, D_3)$ in $L(D_3)$ by the map

$$i : L(D_3) \to L(D_2)\times L(D_3),$$

$$f_3 \mapsto (3, f_3).$$

Here $i^{-1}(W_{I_1fd}(C, D_2, D_3))$ is the dense open set in $L(D_3)$ whose $\overline{k}$-valued points are those f_3 with divisor of poles D_3 and which, as finite flat maps from $C - D_{max} = C - D_3 \to \mathbb{A}^1$, are finite etale over the square roots of unity.

The lisse sheaf $i^*\mathcal{M}$ on U is precisely the $\mathcal{M}$ of 7.1.14 attached to the perverse sheaf M constructed in 7.1.2-4, attached to the following situation.

Over $\mathbb{A}^1 := \mathrm{Spec}(k[t])$, we have the curve

$$E_t : y^2 = 4x^3 - 3x - t,$$

with structural morphism

$$\pi_t : E_t \to \mathbb{A}^1.$$

Its Δ is $27(1 - t^2)$, which has simple zeroes $S = \mu_2$. At each of these zeroes, E_t has multiplicative reduction. On $\mathbb{A}^1$, we form the sheaf

$$\mathcal{G} := R^1\pi_{t!}\overline{\mathbb{Q}}_\ell(1).$$

We see from Corollary 7.5.5 that $\mathcal{G}$ is a middle extension. Its restriction to $\mathbb{A}^1[1/\Delta]$ is lisse of rank 2, geometrically irreducible (because j is nonconstant), pure of weight -1, and symplectically self dual toward $\overline{\mathbb{Q}}_\ell(1)$. At each point in $S(\overline{k}) := \mu_2(\overline{k})$, the local monodromy of $\mathcal{G}$ is a unipotent pseudoreflection. At ∞, E_t has local monodromy of order 6 (because E_t acquires good reduction at ∞ after passing to $k(t^{1/6})$), and hence $\mathcal{G}$ is tame at ∞. We put

$$K := \mathcal{G}[1].$$

We use this K on $\mathbb{A}^1$, and we take our curve C and the divisor D_3 to be the (C, D) used in the construction of 7.1.2-4. This construction gives us a perverse sheaf on $L(D_3)$ which, up to a Tate twist and a shift, is the pullback i^*M. Our open set $U \subset L(D_3)$ is precisely the set $U_{D,S}$ of 7.1.6, and our $i^*\mathcal{M}$ on U is precisely the lisse sheaf on $U_{D,S}$ denoted $\mathcal{M}$ in 7.1.14. We then apply Corollary 7.2.4, which tells us that G_{geom} for $Gr^0(i^*\mathcal{M}) = i^*Gr^0(\mathcal{M})$ contains a reflection. Therefore the larger group G_{geom} for $Gr^0(\mathcal{M})$ contains a reflection. QED

Remark 10.2.14 We expect that the above theorem will remain true if we drop the hypothesis that either $3D_2 \geq 2D_3$ or $2D_3 \geq 3D_2$. In this more general case, we have only the following less satisfactory result.

Theorem 10.2.15 Let k be a finite field of characteristic $p \geq 5$, in which a prime ℓ is invertible. Fix a projective, smooth, geometrically connected curve C/k, of genus denoted g. Fix on C effective divisors D_2 and D_3, both of degree $\geq 2g + 3$. Define $D_{max} := \mathrm{Max}(3D_2, 2D_3)$. Denote by

$$N := 4g - 4 + \deg(D_{max}) \\ + 2\#\{P \text{ in } D_{max}(\overline{k}) \text{ with } cp \not\equiv 0 \bmod 12\}$$

the rank of the lisse sheaf $Gr^0(\mathcal{M})$ on the dense open set

$W_{I_1 fd}(C, D_2, D_3)$. Suppose that at least one of the following conditions holds:

$$4g - 4 + 3\deg(D_2) \geq 10,$$
$$4g - 4 + 2\deg(D_3) \geq 9.$$

Then

$$N \geq 9,$$

and the group G_{geom} for the lisse sheaf $Gr^0(\mathcal{M})$ on $W_{I_1 fd}(C, D_2, D_3)$ is either $SO(N)$ or $O(N)$. If in addition N is odd, then G_{geom} is $O(N)$.

proof That $N \geq 9$ is immediate, since

$$\begin{aligned} N &\geq 4g - 4 + \deg(D_{max}) \\ &\geq 4g - 4 + \mathrm{Max}(\deg(3D_2), \deg(2D_3)) \\ &= \mathrm{Max}(4g - 4 + 3\deg(D_2), 4g - 4 + 2\deg(D_3)) \geq 9. \end{aligned}$$

Since both D_2 and D_3 have degree $\geq 2g + 3$, the space $L(D_2)\times L(D_3)$ of $\mathbb{A}^2$-valued functions on $C - D_{max}$ is at least 4-separating. As $N \geq 4$, the Higher Moment Theorem 1.20.2 and 2.1.1 show that G_{geom} lies in $O(N)$, and has fourth moment 3. By purity, we know that G_{geom} is semisimple. By Larsen's Alternative 2.2.2, G_{geom} is either finite, or $SO(N)$, or $O(N)$.

We next show that the group G_{geom} contains the scalar -1. The argument is similar to that given in [Ka-TLFM, proof of 5.5.2, part 3)]. At the expense of extending scalars from k to a finite extension, we may pick a k-valued point (f_2, f_3) of $W_{I_1 fd}(C, D_2, D_3)$. Then the pullback of $Gr^0(\mathcal{M})$ to the one-parameter family

$$i : \mathbb{G}_m \to W_{I_1 fd}(C, D_2, D_3),$$
$$t \mapsto (t^2 f_2, t^3 f_3),$$

which amounts to looking at the L-functions of all quadratic twists of the single curve $E_{f's}/k(C) : y^2 = 4x^3 - f_2 x - f_3$, is the tensor product

$$\begin{aligned} &i^* Gr^0(\mathcal{M}) \\ &\cong \mathcal{L}_{\chi_2(t)} \otimes (\text{the constant sheaf } H^1(C\otimes_k \bar{k}, j_\star R^1\pi_{f's!}\bar{\mathbb{Q}}_\ell)(1)), \end{aligned}$$

whose G_{geom} is the subgroup $\{\pm 1\}$ of $O(N)$. Since the G_{geom} for $i^*Gr^0(\mathcal{M})$ is a subgroup of G_{geom} for $Gr^0(\mathcal{M})$, we see that G_{geom} for $Gr^0(\mathcal{M})$ contains the scalar -1.

We now show that G_{geom} is not finite. Once we have this result, it then follows from Larsen's criterion that G_{geom} must be either $SO(N)$, or $O(N)$. Since G_{geom} contains -1, the $SO(N)$ case is not possible if N is odd.

To show that G_{geom} is not finite, we make use of the semicontinuity results of [Ka-SMD, Corollary 10]. We apply these results to the sheaf of perverse origin $\mathcal{H}^{-\dim(\mathcal{F})}(M)$ on the space $L(D_2)\times L(D_3)$. This allows us to compare the G_{geom} of its (lisse) restriction $\mathcal{M}(1)$ to the dense open set $U := W_{I_1fd}(C, D_2, D_3)$, to the G_{geom} of its its (lisse) pullback $\mathcal{M}_2(1)$ to a dense open set U_2 of $L(D_2)$, by the map

$$i_2 : L(D_2) \to L(D_2)\times L(D_3),$$
$$f_2 \mapsto (3f_2, 1),$$

if

$$4g - 4 + 3\deg(D_2) \geq 10,$$

or to the G_{geom} of its lisse pullback $\mathcal{M}_3(1)$ to a dense open set U_3 of $L(D_3)$, by the map

$$i_3 : L(D_3) \to L(D_2)\times L(D_3),$$
$$f_2 \mapsto (3, f_3),$$

if

$$4g - 4 + 2\deg(D_3) \geq 9.$$

In considering this plethora of data, there are a total of fifteen lisse sheaves on three different spaces we need to look at. On the space $W_{I_1fd}(C, D_2, D_3)$, we have the five lisse sheaves

$$Gr^{-1}(\mathcal{M}),\ \mathcal{M},\ Gr^0(\mathcal{M}) = Gr^0(\mathcal{M})_{ncst},\ Gr^0(\mathcal{M})_{cst} = 0.$$

The first three sit in a short exact sequence

$$0 \to Gr^{-1}(\mathcal{M}) \to \mathcal{M} \to Gr^0(\mathcal{M}) \to 0.$$

We denote by

$$G_{-1} : = \text{the group } G_{geom} \text{ for } Gr^{-1}(\mathcal{M}),$$
$$G_{big} := \text{the group } G_{geom} \text{ for } \mathcal{M},$$
$$G_0 := \text{the group } G_{geom} \text{ for } Gr^0(\mathcal{M}).$$

Thus

$$G_{big}/\mathcal{R}_u(G_{big}) \cong G_0 \times G_{-1}.$$

On the space U_2, we have the five lisse sheaves

$$Gr^{-1}(\mathcal{M}_2),\ \mathcal{M}_2,\ Gr^0(\mathcal{M}_2) = Gr^0(\mathcal{M}_2)_{ncst} \oplus Gr^0(\mathcal{M}_2)_{cst}.$$

The first three sit in a short exact sequence

$$0 \to Gr^{-1}(\mathcal{M}_2) \to \mathcal{M}_2 \to Gr^0(\mathcal{M}_2) \to 0.$$

We denote by

$$G_{-1,2} : = \text{the group } G_{geom} \text{ for } Gr^{-1}(\mathcal{M}_2),$$
$$G_{big,2} := \text{the group } G_{geom} \text{ for } \mathcal{M}_2,$$
$$G_{0,2} := \text{the group } G_{geom} \text{ for } Gr^0(\mathcal{M}_2),$$
$$G_{0,2,ncst} \cong G_{0,2} := \text{the group } G_{geom} \text{ for } Gr^0(\mathcal{M}_2)_{ncst}.$$

Thus

$$G_{big,2}/\mathcal{R}_u(G_{big,2}) \cong G_{0,2} \times G_{-1,2} \cong G_{0,2,ncst} \times G_{-1,2}.$$

The lisse sheaf $\mathcal{M}_2$ is precisely the one constructed in the proof of Theorem 10.2.13 in the case when $3D_2 \geq 2D_3$, where we considered the pullbacks of the curve $y^2 = 4x^3 - 3tx - 1$ over $k(t)$ by functions f_2 in $L(D_2)$. The geometrically constant sheaf $Gr^0(\mathcal{M}_2)_{cst}$ has rank one.

On the space U_3, we have the five lisse sheaves

$$Gr^{-1}(\mathcal{M}_3),\ \mathcal{M}_3,\ Gr^0(\mathcal{M}_3) = Gr^0(\mathcal{M}_3)_{ncst},\ Gr^0(\mathcal{M}_3)_{cst} = 0.$$

The first three sit in a short exact sequence

$$0 \to Gr^{-1}(\mathcal{M}_3) \to \mathcal{M}_3 \to Gr^0(\mathcal{M}_3) \to 0.$$

We denote by

$$G_{-1,3} := \text{the group } G_{geom} \text{ for } Gr^{-1}(\mathcal{M}_3),$$
$$G_{big,3} := \text{the group } G_{geom} \text{ for } \mathcal{M}_3,$$
$$G_{0,3} := \text{the group } G_{geom} \text{ for } Gr^0(\mathcal{M}_3).$$

Thus

$$G_{big,3}/\mathcal{R}_u(G_{big,3}) \cong G_{0,3} \times G_{-1,3}.$$

The lisse sheaf $\mathcal{M}_3$ is precisely the one constructed in the proof of Theorem 10.2.13 in the case when $2D_3 \geq 3D_2$, where we considered the pullbacks of the curve $y^2 = 4x^3 - 3x - t$ over $k(t)$ by functions f_3 in $L(D_3)$.

We now show that G_{geom} for $Gr^0(\mathcal{M})$, i.e., the group G_0, is not finite. To do this, we introduce the following notation. For any linear algebraic group G over $\overline{\mathbb{Q}}_\ell$, we denote by $r(G)$ its "rank", in the sense

$$r(G) = \text{maximum dimension of an algbraic torus } (\mathbb{G}_m)^n \text{ in } G.$$

The semicontinuity result [Ka-SMD, Corollary 10 (2c)] gives us the inequalities

$$r(G_{big}) \geq r(G_{big,2}),$$
$$r(G_{big}) \geq r(G_{big,3}).$$

On the other hand, we have the tautologous equalities

$$r(G_{big}) = r(G_0) + r(G_{-1}),$$
$$r(G_{big,2}) = r(G_{0,2}) + r(G_{-1,2}) = r(G_{0,2,ncst}) + r(G_{-1,2}),$$
$$r(G_{big,3}) = r(G_{0,3}) + r(G_{-1,3}).$$

So we have the inequalities

$$r(G_0) + r(G_{-1}) \geq r(G_{0,2,ncst}) + r(G_{-1,2}) \geq r(G_{0,2,ncst}),$$
$$r(G_0) + r(G_{-1}) \geq r(G_{0,3}) + r(G_{-1,3}) \geq r(G_{0,3}).$$

Now suppose that G_{geom} for $Gr^0(\mathcal{M})$, i.e., the group G_0, is finite. We argue by contradiction. If G_0 is finite, then $r(G_0) = 0$, so we get inequalities

$$r(G_{-1}) \geq r(G_{0,2,ncst}),$$
$$r(G_{-1}) \geq r(G_{0,3}).$$

Let us denote by N_2 the rank of $Gr^0(\mathcal{M}_2)_{ncst}$, and by N_3 the rank of $Gr^0(\mathcal{M}_3)$. We have

$$N_2 + 1 = \text{rank of } Gr^0(\mathcal{M}_2) \geq 4g - 4 + \deg(3D_2),$$

$$N_3 = \text{rank of } Gr^0(\mathcal{M}_3) \geq 4g - 4 + \deg(2D_3).$$

From Theorem 7.3.16, we have the following results concerning the groups $G_{0,2,ncst}$ and $G_{0,3}$:

$$G_{0,2,ncst} = O(N_2), \text{ if } N_2 \geq 9,$$

$$G_{0,3} = O(N_3), \text{ if } N_3 \geq 9.$$

Now the orthogonal group $O(n)$ has rank $[n/2]$, the integral part of $n/2$. So if G_0 is finite, we get the following inequalities:

$$r(G_{-1}) \geq r(G_{0,2,ncst}) = [N_2/2] \geq [(4g - 5 + \deg(3D_2))/2],$$
$$\text{if } 4g - 4 + 3\deg(D_2) \geq 10,$$

and

$$r(G_{-1}) \geq r(G_{0,3}) = [N_3/2] \geq [(4g - 4 + \deg(2D_3))/2],$$
$$\text{if } 4g - 4 + 2\deg(D_3) \geq 9.$$

We now need to estimate the rank of G_{-1}. For this, we need to understand the structure of $Gr^{-1}(\mathcal{M})$. Over $\bar{k}$, write the divisors D_2 and D_3 as

$$D_2 = \Sigma\, a_P P,\ D_3 = \Sigma b_P P.$$

Then

$$D_{max} = \Sigma\, c_P P,\ c_P = \text{Max}(3a_P, 2b_P).$$

Then $Gr^{-1}(\mathcal{M})$ is the direct sum, indexed by those points P of $D_{max}(\bar{k})$ with $c_P \equiv 0 \bmod 12$, of lisse sheaves of rank 2, which are the H^1's along the fibres of families of elliptic curves over the base, cf. Lemmas 10.2.6 and 10.2.7. Furthermore, except at those points where in addition $3a_P = 2b_P$, we have a family with constant j invariant ($j = 1728$ if $3a_P > 2b_P$, and $j = 0$ if $2b_P > 3a_P$). Let us denote by

$$S := \{\text{points P in } D_{max}(\bar{k}) \text{ with } 3a_P = 2b_P \equiv 0 \bmod 12\}.$$

Thus we have

$$Gr^{-1}(\mathcal{M})$$
$$\cong \oplus_{s \text{ in } S} (\text{lisse sheaves of rank 2, with } G_{geom} \subset SL(2))$$
$$\oplus(\text{some lisse sheaves of rank 2 whose } G_{geom} \text{ is finite}).$$

Thus we have an inclusion

$$G_{-1} \subset (SL(2))^{\#S} \times (\text{a finite group}),$$

and so an inequality

$$r(G_{-1}) \leq \#S.$$

On the other hand, from the very definition of S we see that any point P in S lies in both D_2 and in D_3, that $a_P \equiv 0 \bmod 4$, and that $b_P \equiv 0 \bmod 6$. Thus we have the two inequalities

$$\#S \leq \deg(D_2)/4,\ \#S \leq \deg(D_3)/6.$$

Putting this all together, we find the inequalities

$$\deg(D_2)/4 \geq \#S \geq r(G_{-1}) \geq [N_2/2] \geq [(4g - 5 + \deg(3D_2))/2],$$
$$\text{if } 4g - 4 + 3\deg(D_2) \geq 10,$$
$$\deg(D_3)/6 \geq \#S \geq r(G_{-1}) \geq [N_3/2] \geq [(4g - 4 + \deg(2D_3))/2],$$
$$\text{if } 4g - 4 + 2\deg(D_3) \geq 9.$$

In the first case, we have

$$\deg(D_2)/4 \geq [(4g - 5 + \deg(3D_2))/2]$$
$$\geq (4g - 6 + \deg(3D_2))/2$$
$$= 2g - 3 + (3/2)\deg(D_2),$$

i.e., we have

$$0 \geq 2g - 3 + (5/4)\deg(D_2) > 2g - 3 + \deg(D_2) \geq 4g,$$

the last inequality because $\deg(D_2) \geq 2g + 3$ by hypothesis. The resulting inequality $0 > 4g$ is the desired contradiction.

In the second case, we have

$$\deg(D_3)/6 \geq [(4g - 4 + \deg(2D_3))/2] = 2g - 2 + \deg(D_3),$$

i.e., we have

$$0 \geq 2g - 2 + (5/6)\deg(D_3) \geq 2g - 2 + (5/6)(2g + 3)$$
$$\geq (11/3)g + 1/2,$$

again a contradiction. QED

Chapter 11: FJTwist families and variants

(11.1) (FJ, twist) families in characteristic $p \geq 5$

(11.1.1) We first give the motivation. Let K be a field, E/K and E'/K two elliptic curves over K with the same j invariant: $j(E/K) = j(E'/K)$ in K. Suppose further that $j(E/K)(j(E/K) - 1728)$ is invertible in K. Then E'/K is a quadratic twist of E/K. Our main interest will be in the case when 6 is invertible in K. But first let us review the cases of characteristics 2 and 3.

(11.1.2) If K has characteristic 2, then E is ordinary, and so can be written in the form

$$y^2 + xy = x^3 + a_2x^2 + a_6.$$

Here, $j = 1/a_6$, so a_6 is determined by j, and the effect of a_2 is to perform a quadratic twist using the Artin-Schreier sheaf $\mathcal{L}_\psi$, cf. Twisting Lemma 9.4.1.

(11.1.3) If K has characteristic 3, then E is ordinary, and so can be written in the form

$$y^2 = x^3 + a_2x^2 + a_6$$

with a_2 invertible. Here $j = -(a_2)^3/a_6$. This curve is the quadratic twist by a_2 of the curve $y^2 = x^3 + x^2 + a_6/(a_2)^3$.

(11.1.4) We henceforth assume that 6 is invertible in K. In this case, we denote by FJ the following fractional linear expression in j:

$$FJ := j/(j - 1728).$$

In terms of the Weierstrass cubic

$$y^2 = 4x^3 - g_2x - g_3,$$

we have

$$j = 1728(g_2)^3/((g_2)^3 - 27(g_3)^2),$$

$$FJ = (g_2)^3/27(g_3)^2,$$

$$j = 1728FJ/(FJ - 1).$$

Given a value j in K, $j \neq 0$, $j \neq 1728$, put $t := FJ := j/(j - 1728)$. Thus t lies in $K - \{0, 1\}$. The curve

$$E_{1,t} : y^2 = 4x^3 - 3tx - t$$

has $j(E_{1,t}/K) = j$, or, equivalently, has $FJ(E_{1,t}/K) = t$. We think of this $E_{1,t}/K$ as the "reference" elliptic curve with given j, $j \neq 0$, $j \neq 1728$. Any E'/K with the same j is a quadratic twist of the reference curve E_t/K, so is of the form

$$E' \cong E_{s,t} : y^2 = 4x^3 - 3s^2tx - s^3t,$$

for some s in $K^\times$. We call t the FJ parameter of E', and we call s a twist parameter for E'.

(11.1.5) Thus as t varies over K - {0, 1}, and s varies over $K^\times$, the curves

$$E_{s,t} : y^2 = 4x^3 - 3s^2tx - s^3t$$

exhaust the K-isomorphism classes of those elliptic curves over K whose j is neither 0 nor 1728. [Of course each isomorphism class it attained infinitely often, since multiplying s by a nonzero square keeps us in the same isomorphism class.]

(11.1.6) We now specialize to the case when K is a function field in one variable over a finite field. Thus we work over a finite field k of characteristic $p \geq 5$, in which a prime ℓ is invertible. We fix a projective, smooth, geometrically connected curve C/k, of genus denoted g. We also fix on C two effective divisors D_0 and D_1. We assume that

$$\deg(D_i) \geq 2g + 3,$$

for i=0, 1.

(11.1.7) Given a function f in the Riemann Roch space $L(D_0)$ and a function g in the Riemann Roch space $L(D_1)$, we consider the Weierstrass equation

$$E_{f,g} : y^2 = 4x^3 - 3f^2gx - f^3g.$$

Provided that $g \neq 1$ and $f \neq 0$, this is an elliptic curve over k(C). As we allow the efffective divisors D_0 and D_1 to grow, we sweep out (with repetition) all the elliptic curves over k(C) whose j is neither 0 nor 1728.

(11.1.8) For fixed efffective divisors D_0 and D_1, we are interested in the variation of the L-function of $E_{f,g}/k(C)$ as (f, g) vary in $L(D_0) \times L(D_1)$. Similarly, for each finite extension field k_d/k, we are interested in the variation of the L-function of $E_{f,g}/k_d(C)$ as (f, g) vary in $L(D_0)\otimes_k k_d \times L(D_1)\otimes_k k_d$. For this to be a reasonable question, we must restrict to a dense open set of $L(D_0) \times L(D_1)$ (viewed as an affine space over k) over which the L-function is a polynomial of constant degree.

(11.1.9) We denote

$$V := C - (D_0 \cup D_1).$$

We define

$$\mathrm{FJTwist}(D_0, D_1) \subset L(D_0) \times L(D_1)$$

to be the dense open set whose $\bar{k}$-valued points are the pairs (f, g) with f in $L(D_0)\otimes_k \bar{k}$, g in $L(D_1)\otimes_k \bar{k}$, such that

1) f has divisor of poles D_0,

2) g has divisor of poles D_1,

3) fg(g-1) has $\deg(D_0) + 2\deg(D_1)$ distinct zeroes in $V(\bar{k})$.

[Equivalently, f has $\deg(D_0)$ distinct zeroes, which all lie in $V(\bar{k})$, g has $\deg(D_1)$ distinct zeroes, which all lie in $V(\bar{k})$, g-1 has $\deg(D_1)$

distinct zeroes, which all lie in $V(\overline{k})$, and these three zero sets are pairwise disjoint.]

(11.1.10) For any $d \geq 1$ and for any (f, g) in $\mathrm{FJTwist}(D_0, D_1)(k_d)$, $E_{f,g}$ is an elliptic curve over $k_d(C)$. It has good reduction over $(V\otimes_k k_d)[1/(fg(g-1))]$, it has multiplicative reduction at the $\deg(D_1)$ zeroes of $g-1$, and it has additive reduction at the $\deg(D_0) + \deg(D_1)$ zeroes of fg. Its reduction type at the $\overline{k}$-valued points of $D_0 \cup D_1$ depends only on the divisors D_0 and D_1: it is independent of the particular choice of (d, f, g) such that (f, g) in $\mathrm{FJTwist}(D_0, D_1)(k_d)$. Denote by $\pi_{f,g} : \mathcal{E}_{f,g}{}^{\mathrm{aff}} \to V\otimes_k k_d$ the relative affine curve of equation

$$y^2 = 4x^3 - 3f^2gx - f^3g.$$

Its $R^1\pi_{f,g!}\overline{\mathbb{Q}}_\ell$ on $V\otimes_k k_d$ has a tensor product decomposition as follows. On $\mathbb{A}^1 = \mathrm{Spec}(k[t])$, we have

$$\pi_{1,t} : \mathcal{E}_{1,t}{}^{\mathrm{aff}} \to \mathbb{A}^1,$$

the relative affine curve of equation

$$y^2 = 4x^3 - 3tx - t.$$

Its $R^1\pi_{1,t!}\overline{\mathbb{Q}}_\ell$ on $\mathbb{A}^1$ is a middle extension sheaf (by Corollary 7.5.5), which is lisse outside the two points $\{0, 1\}$. Because the map g is finite etale over both these points, the pullback $g^*R^1\pi_{1,t!}\overline{\mathbb{Q}}_\ell$ is a middle extension on $V\otimes_k k_d$. Similarly, the extension by zero from $\mathbb{G}_m$ of $\mathcal{L}_{\chi_2}$ is a middle extension on $\mathbb{A}^1$, lisse outside 0, while f is finite etale over 0, so the pullback $f^*(\mathcal{L}_{\chi_2}) := \mathcal{L}_{\chi_2(f)}$ is a middle extension. We have an isomorphism on $V\otimes_k k_d$

$$R^1\pi_{f,g!}\overline{\mathbb{Q}}_\ell \cong \mathcal{L}_{\chi_2(f)}\otimes g^*R^1\pi_{1,t!}\overline{\mathbb{Q}}_\ell.$$

(11.1.11) Notice that the two tensor factors, $\mathcal{L}_{\chi_2(f)}$ and $g^*R^1\pi_{1,t!}\overline{\mathbb{Q}}_\ell$, have disjoint ramification. As each is a middle extension, so is their tensor product. Since $\mathcal{E}_{f,g}$ has nonconstant j invariant, the middle extension $R^1\pi_{f,g!}\overline{\mathbb{Q}}_\ell$ is geometrically irreducible. Thus we have:

Lemma 11.1.12 For (f, g) in $\mathrm{FJTwist}(D_0, D_1)(k_d)$, $R^1\pi_{f,g!}\overline{\mathbb{Q}}_\ell$ on $V\otimes_k k_d$ is a geometrically irreducible middle extension of generic rank two, which is lisse on $(V\otimes_k k_d)[1/(fg(g-1))]$, and which has the tensor product structure

$$R^1\pi_{f,g!}\overline{\mathbb{Q}}_\ell \cong \mathcal{L}_{\chi_2(f)}\otimes g^*R^1\pi_{1,t!}\overline{\mathbb{Q}}_\ell.$$

(11.1.13) Denote by $j : V \to C$ the inclusion. The unitarized L-function of $E_{f,g}/k_d(C)$ is given by

$$L(E_{f,g}/k_d(C), T) = \det(1 - T\mathrm{Frob}_{k_d} \mid H^1(C\otimes_k \bar{k}, j_*R^1\pi_{f,g!}\bar{\mathbb{Q}}_\ell(1))).$$

(11.1.14) Because both $\mathcal{L}_{\chi_2}$ and $R^1\pi_{1,t!}\bar{\mathbb{Q}}_\ell$ are tamely ramified at ∞ on $\mathbb{A}^1$, for each point P in $(C - V)(\bar{k}) = (D_0 \cup D_1)(\bar{k})$, the I(P)-representation attached to $R^1\pi_{f,g!}\bar{\mathbb{Q}}_\ell \cong \mathcal{L}_{\chi_2(f)}\otimes g^*R^1\pi_{1,t!}\bar{\mathbb{Q}}_\ell$ is tame, and depends only on the the data $(\mathrm{ord}_P(f), \mathrm{ord}_P(g))$. For (f, g) in $\mathrm{FJTwist}(D_0, D_1)(\bar{k})$, this data is just $(-\mathrm{ord}_P(D_0), -\mathrm{ord}_P(D_1))$. Here is the explicit recipe.

Lemma 11.1.15 Over $\bar{k}$, write

$$D_0 = \Sigma a_P P,\ D_1 = \Sigma b_P P.$$

For any point P in $(D_0 \cup D_1)(\bar{k})$, the 2-dimensional I(P)-representation attached to $R^1\pi_{f,g!}\bar{\mathbb{Q}}_\ell \cong \mathcal{L}_{\chi_2(f)}\otimes g^*R^1\pi_{1,t!}\bar{\mathbb{Q}}_\ell$ is tame, and given as follows. Pick a topological generator γ_P of the tame quotient $I(P)^{\mathrm{tame}}$ of I(P). Then $(\gamma_P)^4$ acts trivially, and the action of γ_P is given by the following table.

a_P	b_P	eigenvalues of γ_P
even	0 mod 4	1, 1
odd	0 mod 4	-1, -1
even	2 mod 4	-1, -1
odd	2 mod 4	1, 1
arbitrary	1 or 3 mod 4	i, -i.

proof We first analyze the 2-dimensional I(P)-representation attached to $g^*R^1\pi_{1,t!}\bar{\mathbb{Q}}_\ell$, i.e., to the curve

$$y^2 = 4x^3 - 3gx - g.$$

At a point P in D_0 but not in D_1, the function g is invertible at P, as is g(g-1), and so this curve has good reduction at P, and γ_P acts trivially. Similarly, if P lies in D_1 but $b_P \equiv 0 \bmod 4$, this curve again has good reduction at P, and γ_P acts trivially. This good reduction when $b_P \equiv 0 \bmod 4$ shows that we always attain good reduction after adjoining the fourth root of a uniformizing parameter at P, and hence that $(\gamma_P)^4$ acts trivially in all cases. If P lies in D_1 but $b_P \equiv 2 \bmod 4$, then this curve is a quadratic twist at P of a curve with good reduction, and γ_P acts as the scalar -1. Finally, if $b_P \equiv \pm 1 \bmod 4$, then neither this curve nor its quadratic twist has good reduction at P, and so γ_P acts with eigenvalues $\pm i$. Tensoring with $\mathcal{L}_{\chi_2(f)}$, we find the asserted table. QED

Corollary 11.1.16 For P in $(D_0 \cup D_1)(\bar{k})$, we have the formula

$$2 - \dim(R^1\pi_{f,g!}\bar{\mathbb{Q}}_\ell)^{I(P)} \quad = 0 \text{ if } 2a_P + b_P \equiv 0 \bmod 4,$$
$$= 2, \text{ if not.}$$

proof Immediate from Lemma 11.1.15. QED

Lemma 11.1.17 Denote by $j : V \subset C$ the inclusion. Then for (f, g) in $\mathrm{FJTwist}(D_0, D_1)(\bar{k})$, we have

1) $H_c^i(V\otimes_k\bar{k}, R^1\pi_{f,g!}\bar{\mathbb{Q}}_\ell) = 0$ for $i \neq 1$,

2) $\dim H_c^1(V\otimes_k\bar{k}, R^1\pi_{f,g!}\bar{\mathbb{Q}}_\ell) = -\chi_c(V\otimes_k\bar{k}, R^1\pi_{f,g!}\bar{\mathbb{Q}}_\ell)$
$= 4g - 4 + 2\deg(D_0) + 3\deg(D_1) + 2\#((D_0 \cup D_1)(\bar{k}))$,

3) $H_c^i(C\otimes_k\bar{k}, j_*R^1\pi_{f,g!}\bar{\mathbb{Q}}_\ell) = 0$ for $i \neq 1$,

4) $\dim H_c^1(C\otimes_k\bar{k}, j_*R^1\pi_{f,g!}\bar{\mathbb{Q}}_\ell) = -\chi_c(C\otimes_k\bar{k}, j_*R^1\pi_{f,g!}\bar{\mathbb{Q}}_\ell)$

$$= 4g - 4 + 2\deg(D_0) + 3\deg(D_1) + \Sigma_{P \text{ in } (D_0\cup D_1)(\bar{k})} (2 - \dim(R^1\pi_{f,g!}\bar{\mathbb{Q}}_\ell)^{I(P)})$$

$$= 4g - 4 + 2\deg(D_0) + 3\deg(D_1) + 2\#\{P \text{ in}(D_0 \cup D_1)(\bar{k}) \text{ with } 2a_P + b_P \not\equiv 0 \bmod 4\}$$

$$\geq 4g - 4 + 2\deg(D_0) + 3\deg(D_1).$$

proof Assertions 1) and 3) result from the fact that the coefficient sheaf is a geometrically irreducible middle extension which is not geometrically trivial. Assertion 2) is a straightforward application of the Euler Poincaré formula, because the coefficient sheaf is everywhere tame, and we know the rank of each of its stalks, namely 2 on $(V[1/(fg(g-1))])\otimes_k\bar{k}$, one at the $\deg(D_1)$ zeroes of g-1, and zero at the $\deg(D_0) + \deg(D_1)$ zeroes of fg. Thus we have

$$\chi_c(V\otimes_k\bar{k}, R^1\pi_{f,g!}\bar{\mathbb{Q}}_\ell)$$
$$= 2\chi_c(V[1/(fg(g-1))]\otimes_k\bar{k}) + \deg(D_1)$$
$$= 2(2 - 2g - \#((D_0 \cup D_1)(\bar{k})) - \deg(D_0) - 2\deg(D_1)) + \deg(D_1)$$
$$= 4 - 4g - 2\#((D_0 \cup D_1)(\bar{k})) - 2\deg(D_0) - 3\deg(D_1).$$

To prove 4), we argue as follows. By 3), we have the first equality. Because $R^1\pi_{f,g!}\bar{\mathbb{Q}}_\ell$ is everywhere tame, we have

$$-\chi_c(C\otimes_k\bar{k}, j_*R^1\pi_{f,g!}\bar{\mathbb{Q}}_\ell)$$
$$= -\chi_c(V\otimes_k\bar{k}, R^1\pi_{f,g!}\bar{\mathbb{Q}}_\ell) - \Sigma_{P \text{ in } (D_0\cup D_1)(\bar{k})} \dim(R^1\pi_{f,g!}\bar{\mathbb{Q}}_\ell)^{I(P)}$$
$$= 4g - 4 + 2\deg(D_0) + 3\deg(D_1) + 2\#((D_0 \cup D_1)(\bar{k})) - \Sigma_{P \text{ in } (D_0\cup D_1)(\bar{k})} \dim(R^1\pi_{f,g!}\bar{\mathbb{Q}}_\ell)^{I(P)}$$

$= 4g - 4 + 2\deg(D_0) + 3\deg(D_1)$

$+ \Sigma_{P \text{ in } (D_0 \cup D_1)(\bar{k})} (2 - \dim(R^1\pi_{f,g!}\bar{\mathbb{Q}}_\ell)^{I(P)})$

$= 4g - 4 + 2\deg(D_0) + 3\deg(D_1)$

$+ 2\#\{P \text{ in}(D_0 \cup D_1)(\bar{k}) \text{ with } 2a_P + b_P \not\equiv 0 \bmod 4\}$

$\geq 4g - 4 + 2\deg(D_0) + 3\deg(D_1)$,

the penultimate inequality by the previous Corollary 11.1.16. QED

Corollary 11.1.18 Denote by N the common dimension

$$N := \dim H_c^1(C\otimes_k\bar{k}, j_*R^1\pi_{f,g!}\bar{\mathbb{Q}}_\ell)$$

for (f, g) in $\mathrm{FJTwist}(D_0, D_1)(\bar{k})$. We have the inequality

$$N \geq 14g + 11.$$

proof Indeed, we have by assumption the inequalities

$$\deg(D_i) \geq 2g + 3,$$

for i = 0, 1. The previous result gives

$$N \geq 4g - 4 + 2\deg(D_0) + 3\deg(D_1),$$

so we find

$$N \geq 4g - 4 + 2(2g + 3) + 3(2g + 3) = 14g + 11.$$

Corollary 11.1.19 For any finite extension k_d/k, and for any (f, g) in $\mathrm{FJTwist}(D_0, D_1)(k_d)$, the unitarized L-function of $E_{f,g}/k_d(C)$ is given by

$$L(E_{f,g}/k_d(C), T) = \det(1 - T\mathrm{Frob}_{k_d} \mid H^1(C\otimes_k\bar{k}, j_*R^1\pi_{f,g!}\bar{\mathbb{Q}}_\ell(1))).$$

proof This results from the fact that $j_*R^1\pi_{f,g!}\bar{\mathbb{Q}}_\ell$ is a middle extension from the open set $(V[1/(fg(g-1))])\otimes_k\bar{k}$ where it is lisse, and the vanishing given by Lemma 11.1.17, part, 3). QED

(11.1.20) Our next task is to capture these L-functions in the context of suitable perverse sheaves M = Twist(L, K, $\mathcal{F}$, h) attached to suitable "standard input", cf. 1.15.4. We take

the integer m = 2,

the perverse sheaf $K = R^1\pi_{s,t!}\bar{\mathbb{Q}}_\ell(3/2)[2]$

$\cong \mathcal{L}_{\chi_2(s)}(1/2)[1]\boxtimes R^1\pi_{1,t!}\bar{\mathbb{Q}}_\ell(1)[1]$ on $\mathbb{A}^2/k = \mathrm{Spec}(k[s,t])$,

the affine k-scheme V/k given by $V := C - (D_0 \cup D_1)$,

the k-morphism $h : V \to \mathbb{A}^2$ given by h = 0,

the perverse sheaf $L = \bar{\mathbb{Q}}_\ell(1/2)[1]$ on V/k,

the integer $d = 1 - 2g + \min(\deg(D_0), \deg(D_1))$,

the space of $\mathbb{A}^2$-valued functions ($\mathcal{F}$, τ) on V given by the finite-dimensional k-vector space $\mathcal{F} = L(D_0)\times L(D_1)$ and the k-linear map

$$\tau : \mathcal{F} \to \mathrm{Hom}_{k\text{-schemes}}(V, \mathbb{A}^2),$$
$$(f, g) \mapsto \text{the map } v \mapsto (f(v), g(v)).$$

It results from Theorem 9.3.10 that this is standard input, and, using Kunneth, that $H^*_c((V\times\mathbb{A}^m)\otimes\bar{k}, pr_1{}^*L\otimes pr_2{}^*K) = 0$.

Lemma 11.1.21 The perverse sheaf M is lisse on the dense open set FJTwist(D_0, D_1).

proof On this dense open set, there is only one nonvanishing cohomology sheaf, namely $\mathcal{H}^{-\dim(\mathcal{F})}(M)$, which is tautologically a sheaf of perverse origin on the space $\mathcal{F}$, cf. Lemma 11.1.17, part 1). The stalks of this sheaf are

$$(f,g) \mapsto H_c{}^1(V\otimes_k\bar{k}, R^1\pi_{f,g!}\bar{\mathbb{Q}}_\ell)(2),$$

which have constant rank on FJTwist(D_0, D_1), cf. Lemma 11.1.17, part 2). By [Ka-SMD, Proposition 11], $\mathcal{H}^{-\dim(\mathcal{F})}(M)$ is lisse on FJTwist(D_0, D_1). QED

(11.1.22) Thus on the open set FJTwist(D_0, D_1), M is $\mathcal{M}(1)[\ell(D_0) + \ell(D_1)]$, with $\mathcal{M}$ the lisse sheaf, mixed of weight ≤ 0, given stalkwise by

$$\mathcal{M}_{f,g} := H_c{}^1(V\otimes_k\bar{k}, R^1\pi_{f,g!}\bar{\mathbb{Q}}_\ell)(1).$$

The sheaf $\mathrm{Gr}^0(\mathcal{M})$ on FJTwist(D_0, D_1) is lisse of rank

$$N = 4g - 4 + 2\deg(D_0) + 3\deg(D_1)$$
$$+ 2\#\{P \text{ in}(D_0 \cup D_1)(\bar{k}) \text{ with } 2a_P + b_P \not\equiv 0 \bmod 4\},$$

orthogonally self dual, and pure of weight zero. We have

$$\mathrm{Gr}^0(\mathcal{M})_{f,g} = H_c{}^1(C\otimes_k\bar{k}, j_*R^1\pi_{f,g!}\bar{\mathbb{Q}}_\ell)(1).$$

Because of the vanishing $H^*_c((V\times\mathbb{A}^m)\otimes\bar{k}, pr_1{}^*L\otimes pr_2{}^*K) = 0$ noted in 11.1.20 above, we have (by 1.20.3, part 3))

$$\mathrm{Gr}^0(\mathcal{M})_{cst} = 0,\ \mathrm{Gr}^0(\mathcal{M}) = \mathrm{Gr}^0(\mathcal{M})_{ncst}.$$

Theorem 11.1.23 Let k be a finite field of characteristic $p \ge 5$, in which a prime ℓ is invertible. Fix a projective, smooth, geometrically connected curve C/k, of genus denoted g. Fix on C two effective divisors D_0 and D_1, both of degree $\ge 2g + 3$. Then the group G_{geom} for the lisse sheaf $\mathrm{Gr}^0(\mathcal{M})$ on the dense open set FJTwist(D_0, D_1) is O(N), for

$$N = 4g - 4 + 2\deg(D_0) + 3\deg(D_1)$$
$$+ 2\#\{P \text{ in}(D_0 \cup D_1)(\bar{k}) \text{ with } 2a_P + b_P \not\equiv 0 \bmod 4\},$$

the rank of $\mathrm{Gr}^0(\mathcal{M})$.

proof As noted above in 11.1.18, we have $N \ge 14g + 11$, so certainly $N \ge 9$. By the argument given in the first paragraph of the proof of Theorem 7.2.3, it suffices to show that G_{geom} contains a reflection.

For this, we argue as follows. At the expense of replacing k by a finite extension of itself, we may suppose that there exists a k-rational point (f_0, g_0) in the dense open set FJTwist(D_0, D_1). We then freeze the function g_0. Thus g_0 has divisor of poles D_1, and both g_0 and g_0-1 have $\deg(D_1)$ distinct zeroes over $\bar{k}$, all located in V. We now look at all pairs (f, g_0) which lie in FJTwist(D_0, D_1), i.e., we look at all f with divisor of poles D_0, such that f has $\deg(D_0)$ distinct zeroes over $\bar{k}$, all of which lie in $V[1/g_0(g_0-1)]$.

Write the open curve $V[1/g_0(g_0-1)]$ as $C - D_0 - S_0$, with S_0 a reduced, finite closed subcheme of $C - D_0$. Write V as $C - D_0 - S$, with S either empty or a reduced, finite closed subcheme of $C - D_0$. And also write $V[1/g_0(g_0-1)]$ as $V - S_{01}$, with S_{01} the reduced, finite closed subcheme of V defined by the vanishing of $g_0(g_0 - 1)$. Thus

$$V[1/g_0(g_0-1)] = C - D_0 - S_0,$$
$$V = C - D_0 - S.$$

In the notation of [Ka-TLFM, 5.0.6] and 8.1.7, a pair (f, g_0) lies in FJTwist(D_0, D_1) if and only if f lies in the dense open subset of $L(D_0)$ given by

$$Fct(C, \deg(D_0), D_0, S_0) \subset L(D_0).$$

We will show that after we restrict $\mathcal{M}$ to

$$Fct(C, \deg(D_0), D_0, S_0)\times\{g_0\} \subset \text{FJTwist}(D_0, D_1),$$

thus obtaining a lisse sheaf $\mathcal{M}_{restr}$ on $Fct(C, \deg(D_0), D_0, S_0)$, the group G_{geom} for $Gr^0(\mathcal{M}_{restr})$ contains a reflection (and hence the group G_{geom} for $Gr^0(\mathcal{M})$ contains a reflection). We will do this by relating $\mathcal{M}_{restr}$ to the twist sheaves of [Ka-TLFM] and Chapter 8.

Consider the sheaf

$$\mathcal{G} := g_0{}^*R^1\pi_{1,t!}\bar{\mathbb{Q}}_\ell(1) \text{ on } C - D_1.$$

The sheaf $R^1\pi_{1,t!}\bar{\mathbb{Q}}_\ell$ is a geometrically irreducible middle extension on $\mathbb{A}^1$ of generic rank two, lisse outside 0 and 1, vanishing at 0, and with unipotent local monodromy at 1. Since g_0 as a finite flat map of $C - D_1$ to $\mathbb{A}^1$ of degree $\deg(D_1)$ is finite etale over both 0 and 1, $\mathcal{G}$ is a geometrically irreducible (FJ = g_0 is nonconstant) middle extension of generic rank two on $C - D_1$. It is lisse outside of S_{01}, it has unipotent local monodromy at the $\deg(D_1)$ zeroes of $g_0 - 1$, and it vanishes at the $\deg(D_1)$ zeroes of g_0.

Consider the inclusions

$$j_S : V \subset C - D_0,$$
$$j_1 : V \subset C - D_1.$$

Then

(11.1.23.1) $$L_1 := (j_1)^*\mathcal{G}[1] = (g_0 : V \to \mathbb{A}^1)^*R^1\pi_{1,t!}\bar{\mathbb{Q}}_\ell(1)[1]$$

is a perverse and geometrically irreducible middle extension on V, and

$$L_0 := j_{S!*}L_1 = j_{S*}(j_1)^*\mathcal{G}[1]$$

is a perverse and geometrically irreducible middle extension on $C - D_0$.

Now consider the following two standard inputs.

The first is

the integer $m = 1$,

the perverse sheaf $K := j_*\mathcal{L}_\chi(1/2)[1]$ on $\mathbb{A}^1/k$,

the affine k-scheme $V_0 := C - D_0$,

the k-morphism $h : V_0 \to \mathbb{A}^1$ given by $h = 0$,

the perverse sheaf $L_0 := (j_S)_*(j_1)^*\mathcal{G}[1]$ on $C - D_0$,

the integer $d := \deg(D) - (2g-1)$,

the space of functions $(L(D_0), \tau)$ on $C - D_0$.

With this input, we form the perverse sheaf on $L(D_0)$

$$M_0 := \mathrm{Twist}(L_0, K, \mathcal{F} = L(D_0) \text{ on } C - D_0, h = 0).$$

The second standard input is

the integer $m = 1$,

the perverse sheaf $K := j_*\mathcal{L}_\chi(1/2)[1]$ on $\mathbb{A}^1/k$,

the affine k-scheme $V = C - D_0 - D_1$,

the k-morphism $h : V \to \mathbb{A}^1$ given by $h = 0$,

the perverse sheaf $L_1 := (j_1)^*\mathcal{G}[1] = (\mathcal{G}|V)[1]$,

the integer $d := \deg(D) - (2g-1)$,

the space of functions $(L(D_0), \tau)$ on V.

With this input, we form the perverse sheaf on $L(D_0)$

$$M_1 := \mathrm{Twist}(L_1, K, \mathcal{F} = L(D_0) \text{ on } V, h = 0).$$

Key Lemma 11.1.24 The perverse sheaf

$$M_1 := \mathrm{Twist}(L_1, K, \mathcal{F} = L(D_0) \text{ on } V, h = 0)$$

on $L(D_0) \cong L(D_0)\times\{g_0\}$ is, up to a shift and a Tate twist, the restriction of the perverse sheaf M on $L(D_0)\times L(D_1)$ to $L(D_0)\times\{g_0\}$. More precisely, we have

$$M \mid L(D_0)\times\{g_0\} \cong M_1(1/2)[\ell(D_1)].$$

proof This is a tautology, thanks to proper base change, the fact that

$$R^1\pi_{f,g!}\overline{\mathbb{Q}}_\ell \cong \mathcal{L}_{\chi_2(f)}\otimes g^*R^1\pi_{1,t!}\overline{\mathbb{Q}}_\ell,$$

cf. 11.1.12, and the observation, made in 11.1.23.1 above that

$$L_1 := (j_1)^*\mathcal{G}[1] = (g_0 : V \to \mathbb{A}^1)^*R^1\pi_{1,t!}\overline{\mathbb{Q}}_\ell(1)[1]. \text{ QED}$$

We now bring to bear Lemmas 8.1.10 and 8.1.11, applied to the perverse sheaves M_0 and M_1 on $L(D_0)$. [In the notations of those lemmas, our D_0 is its D, our M_1 is its $M_\varnothing$, and our M_0 is its M_S.] Thus both M_1 and M_0 are lisse on $\mathit{Fct}(C, \deg(D_0), D, S_0)$, and we have

$$\mathrm{Gr}^0(M_0((\ell(D_0) - 1)/2)) \cong \mathrm{Gr}^0(M_1((\ell(D_0) - 1)/2)).$$

We introduce lisse sheaves $\mathcal{M}_0$ and $\mathcal{M}_1$ on *Fct*(C, deg(D_0), D, S_0) by

$$M_0|Fct(C, \deg(D_0), D, S_0) = \mathcal{M}_0(1/2)[\ell(D_0)],$$
$$M_1|Fct(C, \deg(D_0), D, S_0) = \mathcal{M}_1(1/2)[\ell(D_0)].$$

Restricting to FJTwist(D_0, D_1), we obtain an isomorphism of lisse sheaves on *Fct*(C, deg(D_0), D, S_0),

$$\mathcal{M}_{restr} := \mathcal{M} \mid Fct(C, \deg(D_0), D, S_0)\times\{g_0\} \cong \mathcal{M}_1,$$

cf. Key Lemma 11.1.24. Passing to Gr^0, we obtain

$$\mathrm{Gr}^0(\mathcal{M}_{restr}) \cong \mathrm{Gr}_0(\mathcal{M}_1).$$

By Lemma 8.1.10, we have

$$\mathrm{Gr}_0(\mathcal{M}_1) \cong \mathrm{Gr}_0(\mathcal{M}_0) \cong \mathrm{Twist}_{\chi_2,C,D_0}((j_S)_*(j_1)^*\mathcal{G}).$$

The local monodromy of $(j_S)_*(j_1)^*\mathcal{G}$ at each zero of $g_0 - 1$ is a unipotent pseudoreflection. By Theorem 8.2.2, conclusion 2), G_{geom} for $\mathrm{Gr}_0(\mathcal{M}_0) \cong \mathrm{Gr}^0(\mathcal{M}_{restr})$ contains a reflection. QED

(11.2) (j^{-1}, twist) families in characteristic 3

(11.2.1) In this section, we consider the analogue in characteristic 3 of the (FJ, twist) families we considered when 6 was invertible. The theory is quite similar to that in higher characteristic, but we must now take into account phenomena of wild ramification, which are themselves reflections of the fact that in charactertistic 2 or 3, the two special j values 0 and 1728 coalesce.

(11.2.2) Over a field K of characteristic 3, an elliptic curve with nonzero j is ordinary, and can be written in the form

$$E_{s,t} : y^2 = x^3 + sx^2 + s^3t,$$

with both s and t invertible. Here $j = 1/t$, and $E_{s,t}$ is the quadratic twist by s of the curve

$$E_{1,t} : y^2 = x^3 + x^2 + t.$$

As s and t each run over $K^\times$, the curves $E_{s,t}$ exhaust, with infinite repetition, the isomorphism classes of elliptic curves E/K with $j \neq 0$.

(11.2.3) We now specialize to the case when K is a function field in one variable over a finite field of characteristic 3. Thus we work over a finite field k of characteristic 3, in which a prime ℓ is invertible. We fix a projective, smooth, geometrically connected curve C/k, of genus denoted g. We also fix on C two effective divisors D_0 and D_1. We assume that

$$\deg(D_i) \geq 2g + 3,$$

for i=0, 1.

(11.2.4) Given a function f in the Riemann Roch space $L(D_0)$ and a function g in the Riemann Roch space $L(D_1)$, we consider the

Weierstrass equation

$$E_{f,g} : y^2 = x^3 + fx^2 + f^3g.$$

Provided that $fg \neq 0$, this is an elliptic curve over $k(C)$, with $j = 1/g$. As we allow the effective divisors D_0 and D_1 to grow, we sweep out (with repetition) all the elliptic curves over $k(C)$ with $j \neq 0$.

(11.2.5) We are interested in the variation with (f, g) in $L(D_0)\times L(D_1)$ of the L-function of $E_{f,g}/k(C)$. Similarly, for each finite extension field k_d/k, we are interested in the variation with (f, g) in $L(D_0)\otimes_k k_d \times L(D_1)\otimes_k k_d$ of the L-function of $E_{f,g}/k_d(C)$. For this to be a reasonable question, we must restrict to a dense open set of $L(D_0)\times L(D_1)$ (viewed as an affine space over k) over which the L-function is a polynomial of constant degree.

(11.2.6) We define

$$V := C - (D_0 \cup D_1).$$

Our next task is to define a dense open set

$$j^{-1}\mathrm{Twist}(D_0, D_1) \subset L(D_0)\times L(D_1)$$

over which the L-function is a polynomial of constant degree. We will do this in two steps. We will first define a preliminary dense open set

$$j^{-1}\mathrm{Twist}(D_0, D_1)_{prelim} \subset L(D_0)\times L(D_1),$$

and then we will define $j^{-1}\mathrm{Twist}(D_0, D_1)$ as a dense open set

$$j^{-1}\mathrm{Twist}(D_0, D_1) \subset j^{-1}\mathrm{Twist}(D_0, D_1)_{prelim}.$$

(11.2.7) We define

$$j^{-1}\mathrm{Twist}(D_0, D_1)_{prelim} \subset L(D_0)\times L(D_1)$$

to be the dense open set whose $\bar{k}$-valued points are the pairs (f, g) with f in $L(D_0)\otimes_k \bar{k}$, g in $L(D_1)\otimes_k \bar{k}$, such that the following four conditions hold:

1) f has divisor of poles D_0,

2) g has divisor of poles D_1,

3) g has Swan-minimal poles, cf. 6.4.6,

4) fg has $\deg(D_0) + \deg(D_1)$ distinct zeroes in $V(\bar{k})$ (i.e., f has $\deg(D_0)$ distinct zeroes, which all lie in $V(\bar{k})$, g has $\deg(D_1)$ distinct zeroes, which all lie in $V(\bar{k})$, and these two zero sets are disjoint).

(11.2.8) Take any finite extension k_d/k and any k_d-valued point (f, g) in $j^{-1}\mathrm{Twist}(D_0, D_1)_{prelim}$. Denote by

$$\pi_{f,g} : \mathcal{E}_{f,g}{}^{aff} \to V$$

the relative affine curve of equation

$$E_{f,g} : y^2 = x^3 + fx^2 + f^3g.$$

Its $R^1\pi_{f,g!}\bar{\mathbb{Q}}_\ell$ on $V\otimes_k k_d$ has a tensor product decomposition as follows. On $\mathbb{A}^1 = \mathrm{Spec}(k[t])$, we have

$$\pi_{1,t} : \mathcal{E}_{1,t}{}^{\mathrm{aff}} \to \mathbb{A}^1,$$

the relative affine curve of equation

$$E_{1,t} : y^2 = x^3 + x^2 + t.$$

Its $R^1\pi_{1,t!}\overline{\mathbb{Q}}_\ell$ on $\mathbb{A}^1$ is an irreducible middle extension sheaf, cf. 7.4.7 and 7.5.5. It is lisse on $\mathbb{G}_m$ and its local monodromy at the origin is a unipotent pseudoreflection. Because the map g is finite etale over 0, the pullback $g^*R^1\pi_{1,t!}\overline{\mathbb{Q}}_\ell$ is a middle extension on $V\otimes_k k_d$. Similarly, $f^*(\mathcal{L}_{\chi_2}) := \mathcal{L}_{\chi_2(f)}$ is a middle extension. We have an isomorphism on $V\otimes_k k_d$

$$R^1\pi_{f,g!}\overline{\mathbb{Q}}_\ell \cong \mathcal{L}_{\chi_2(f)}\otimes g^*R^1\pi_{1,t!}\overline{\mathbb{Q}}_\ell.$$

As the two tensor factors are middle extensions with disjoint ramification, $R^1\pi_{f,g!}\overline{\mathbb{Q}}_\ell$ is a middle extension. Since $\mathcal{E}_{f,g}$ has nonconstant j invariant, the middle extension $R^1\pi_{f,g!}\overline{\mathbb{Q}}_\ell$ is geometrically irreducible. Thus we have

Lemma 11.2.9 For (f, g) in $j^{-1}\mathrm{Twist}(D_0, D_1)_{\mathrm{prelim}}(k_d)$, $R^1\pi_{f,g!}\overline{\mathbb{Q}}_\ell$ on $V\otimes_k k_d$ is a geometrically irreducible middle extension of generic rank two, which is lisse on $(V\otimes_k k_d)[1/fg]$, and which has the tensor product structure

$$R^1\pi_{f,g!}\overline{\mathbb{Q}}_\ell \cong \mathcal{L}_{\chi_2(f)}\otimes g^*R^1\pi_{1,t!}\overline{\mathbb{Q}}_\ell.$$

(11.2.10) Denote by $j : V \to C$ the inclusion. The unitarized L-function of $E_{f,g}/k_d(C)$ is given by

$$L(E_{f,g}/k_d(C), T) = \det(1 - T\mathrm{Frob}_{k_d} \mid H^1(C\otimes_k \overline{k}, j_\star R^1\pi_{f,g!}\overline{\mathbb{Q}}_\ell(1))).$$

We can recover $H^1(C\otimes_k \overline{k}, j_\star R^1\pi_{f,g!}\overline{\mathbb{Q}}_\ell(1))$ as the weight zero quotient of

$$H_c{}^1(V\otimes_k \overline{k}, R^1\pi_{f,g!}\overline{\mathbb{Q}}_\ell(1))$$
$$= H_c{}^1(V\otimes_k \overline{k}, \mathcal{L}_{\chi_2(f)}\otimes g^*R^1\pi_{1,t!}\overline{\mathbb{Q}}_\ell(1)).$$

Lemma 11.2.11 For (f, g) in $j^{-1}\mathrm{Twist}(D_0, D_1)_{\mathrm{prelim}}(\overline{k})$, we have

1) $H_c{}^i(V\otimes_k \overline{k}, R^1\pi_{f,g!}\overline{\mathbb{Q}}_\ell) = 0$ for $i \neq 1$,

2) $\dim H_c{}^1(V\otimes_k \overline{k}, R^1\pi_{f,g!}\overline{\mathbb{Q}}_\ell) = -\chi_c(V\otimes_k \overline{k}, R^1\pi_{f,g!}\overline{\mathbb{Q}}_\ell)$
$= -2\chi_c((V\otimes_k \overline{k})[1/fg], \overline{\mathbb{Q}}_\ell)$
$-\deg(D_1) + \Sigma_{\text{poles P of g}} \mathrm{Swan}_P(g^*R^1\pi_{1,t!}\overline{\mathbb{Q}}_\ell).$

proof Assertion 1) results from the fact that the coefficient sheaf is a geometrically irreducible middle extension which is not geometrically trivial. Assertion 2) is a straightforward application of the Euler Poincaré formula. Indeed, the sheaf

$$R^1\pi_{f,g!}\overline{\mathbb{Q}}_\ell \cong \mathcal{L}_{\chi_2(f)}\otimes g^*R^1\pi_{1,t!}\overline{\mathbb{Q}}_\ell$$

is lisse where fg is invertible. At the $\deg(D_0)$ zeroes of f, it is tame, and its stalk vanishes. At the $\deg(D_1)$ zeroes of g, it is tame, and local monodromy is a unipotent pseudoreflection. Thus we have

$$\chi_c(V\otimes_k\overline{k},\ R^1\pi_{f,g!}\overline{\mathbb{Q}}_\ell)$$
$$= \chi_c((V\otimes_k\overline{k})[1/fg],\ \mathcal{L}_{\chi_2(f)}\otimes g^*R^1\pi_{1,t!}\overline{\mathbb{Q}}_\ell) + \deg(D_1).$$

The lisse rank one sheaf $\mathcal{L}_{\chi_2(f)}$ on $(V\otimes_k\overline{k})[1/fg]$ is everywhere tame. Therefore the the lisse sheaf $\mathcal{L}_{\chi_2(f)}\otimes g^*R^1\pi_{1,t!}\overline{\mathbb{Q}}_\ell)$ on $(V\otimes_k\overline{k})[1/fg]$ is tame except possibly at the poles of g, and at any pole P of g, we have

$$\mathrm{Swan}_P(\mathcal{L}_{\chi_2(f)}\otimes g^*R^1\pi_{1,t!}\overline{\mathbb{Q}}_\ell) = \mathrm{Swan}_P(g^*R^1\pi_{1,t!}\overline{\mathbb{Q}}_\ell).$$

So the Euler Poincaré formula gives

$$\chi_c((V\otimes_k\overline{k})[1/fg],\ \mathcal{L}_{\chi_2(f)}\otimes g^*R^1\pi_{1,t!}\overline{\mathbb{Q}}_\ell)$$
$$= 2\chi_c((V\otimes_k\overline{k})[1/fg],\ \overline{\mathbb{Q}}_\ell) - \Sigma_{\text{poles P of g}}\ \mathrm{Swan}_P(g^*R^1\pi_{1,t!}\overline{\mathbb{Q}}_\ell).$$

Putting this all together, we find assertion 2). QED

(11.2.12) The cohomology groups $H_c^1(V\otimes_k\overline{k},\ R^1\pi_{f,g!}\overline{\mathbb{Q}}_\ell(1))$ can be all captured by a suitable perverse sheaf $M = \mathrm{Twist}(L, K, \mathcal{F}, h)$ attached to suitable "standard input", cf. 1.15.4 We take

the integer $m = 2$,

the perverse sheaf $K = R^1\pi_{s,t!}\overline{\mathbb{Q}}_\ell(3/2)[2]$

$$\cong \mathcal{L}_{\chi_2(s)}(1/2)[1]\boxtimes R^1\pi_{1,t!}\overline{\mathbb{Q}}_\ell(1)[1] \text{ on } \mathbb{A}^2/k = \mathrm{Spec}(k[s,t]),$$

the affine k-scheme V/k given by $V := C - (D_0 \cup D_1)$,

the k-morphism $h : V \to \mathbb{A}^2$ given by $h = 0$,

the perverse sheaf $L = \overline{\mathbb{Q}}_\ell(1/2)[1]$ on V/k,

the integer $d = 1 - 2g + \min(\deg(D_0), \deg(D_1))$,

the space of $\mathbb{A}^2$-valued functions $(\mathcal{F}, \tau)$ on V given by the finite-dimensional k-vector space $\mathcal{F} = L(D_0)\times L(D_1)$ and the k-linear map

$$\tau : \mathcal{F} \to \mathrm{Hom}_{k\text{-schemes}}(V, \mathbb{A}^2),$$
$$(f, g) \mapsto \text{the map } v \mapsto (f(v), g(v)).$$

It results from Theorem 9.3.6 that this is standard input, and, using Kunneth, that $H^*_c((V\times\mathbb{A}^m)\otimes\overline{k},\ pr_1{}^*L\otimes pr_2{}^*K) = 0$.

Lemma 11.2.13 On the dense open set $j^{-1}\mathrm{Twist}(D_0, D_1)_{\mathrm{prelim}}$, the perverse sheaf M has only one nonvanishing cohomology sheaf, namely $\mathcal{H}^{-\dim(\mathcal{F})}(M)$, whose stalks are

$$(f,g) \mapsto H_c^1(V\otimes_k\bar{k}, R^1\pi_{f,g!}\bar{\mathbb{Q}}_\ell)(2).$$

proof This is just a translation of the previous Lemma 11.2.11. QED

(11.2.14) Thus on the open set $j^{-1}\mathrm{Twist}(D_0, D_1)_{\mathrm{prelim}}$, M is the object $\mathcal{M}(1)[\ell(D_0) + \ell(D_1)]$, with $\mathcal{M} := \mathcal{H}^{-\dim(\mathcal{F})}(M)(-1)$ the sheaf of perverse origin, mixed of weight ≤ 0, given stalkwise by

$$\mathcal{M}_{f,g} := H_c^1(V\otimes_k\bar{k}, R^1\pi_{f,g!}\bar{\mathbb{Q}}_\ell)(1).$$

We now invoke the fact [Ka-SMD, Prop. 12] that for a sheaf of perverse origin $\mathcal{M}$ on a smooth connected k-scheme, here $j^{-1}\mathrm{Twist}(D_0, D_1)_{\mathrm{prelim}}$, the set of points where its stalk has maximum rank is a dense open set,

$$U_{\max} \subset j^{-1}\mathrm{Twist}(D_0, D_1)_{\mathrm{prelim}},$$

and $\mathcal{M}$ is lisse on $U_{\max}$.

(11.2.15) We now define the dense open set $j^{-1}\mathrm{Twist}(D_0, D_1)$ to be

$$j^{-1}\mathrm{Twist}(D_0, D_1) := U_{\max} \subset j^{-1}\mathrm{Twist}(D_0, D_1)_{\mathrm{prelim}}.$$

Thanks to the dimension formula of Lemma 11.2.11, part 2) above, we can also characterize $U_{\max}$ as the set of points (f, g) in $j^{-1}\mathrm{Twist}(D_0, D_1)_{\mathrm{prelim}}$ at which

$$\Sigma_{\text{poles P of g}} \mathrm{Swan}_P(g^*R^1\pi_{1,t!}\bar{\mathbb{Q}}_\ell)$$

attains it maximum value.

Lemma 11.2.16 Let M be a two-dimensional continuous $\bar{\mathbb{Q}}_\ell$-representation of $I(\infty)$ with trivial determinant. Then both $I(\infty)$-breaks of M are Swan(M)/2. In particular, if Swan(M) > 0, then M is totally wild, and $M^{P(\infty)} = M^{I(\infty)} = 0$.

proof If M is any irreducible $\bar{\mathbb{Q}}_\ell$-representation of $I(\infty)$, then the break-decomposition [Ka-GKM, 1.1, 1.2] shows that M has a single $I(\infty)$ slope, Swan(M)/dim(M), with multiplicity dim(M).

If our two-dimensional M is reducible, then, because it has trivial determinant, it is an extension of a linear character ρ of $I(\infty)$ by the inverse character ρ^{-1}. Both ρ and ρ^{-1} have the same Swan conductor, say x, so M has a single break x, with multiplicity 2. Since $\mathrm{Swan}(M) = \mathrm{Swan}(\rho) + \mathrm{Swan}(\rho^{-1})$, we find that x = Swan(M)/2.

If Swan(M) > 0, then all breaks of M are nonzero, and hence [Ka-GKM, 1.9] $M^{P(\infty)} = 0$. The inclusion $M^{I(\infty)} \subset M^{P(\infty)}$ gives the final conclusion. QED

Lemma 11.2.17 Write D_1 over $\bar{k}$ as $\Sigma a_P P$. Define integers c_P, one for each P in D_1, by

$$c_P = a_P, \text{ if } a_P \text{ is prime to } 3,$$
$$= a_P - 2, \text{ if } 3|a_P.$$

Suppose g in $L(D_1)$ has divisor of poles D_1 and has Swan-minimal poles. Then we have the following results.

1) We have the inequality

$$\Sigma_{\text{poles P of g}} \mathrm{Swan}_P(g^*R^1\pi_{1,t!}\overline{\mathbb{Q}}_\ell) \le \Sigma_P c_P.$$

2) If $a_P \neq 3$ for all P, then we have the equality

$$\Sigma_{\text{poles P of g}} \mathrm{Swan}_P(g^*R^1\pi_{1,t!}\overline{\mathbb{Q}}_\ell) = \Sigma_P c_P.$$

3) For all but at most $\#\{P \text{ in } D_1 \text{ with } a_P = 3\}$ values of λ in $\overline{k}^\times$, we have the equality

$$\Sigma_{\text{poles P of g}} \mathrm{Swan}_P((\lambda g)^*R^1\pi_{1,t!}\overline{\mathbb{Q}}_\ell) = \Sigma_P c_P.$$

proof The sheaf $R^1\pi_{1,t!}\overline{\mathbb{Q}}_\ell$ is lisse of rank two on $\mathbb{G}_m$, has geometrically trivial determinant $\overline{\mathbb{Q}}_\ell(-1)$, and has $\mathrm{Swan}_\infty = 1$. Hence its $I(\infty)$-representation, call it M, has trivial determinant, rank two, and Swan conductor 1. By Lemma 11.2.16, both $I(\infty)$-breaks are 1/2. It now follows from Lemma A6.2.2 that at each pole P of g, we have

$$\mathrm{Swan}_P((\lambda g)^*R^1\pi_{1,t!}\overline{\mathbb{Q}}_\ell) = c_P, \text{ for every } \lambda \text{ in } \overline{k}^\times, \text{ if } a_P \neq 3,$$

$$\mathrm{Swan}_P((\lambda g)^*R^1\pi_{1,t!}\overline{\mathbb{Q}}_\ell) = c_P, \text{ for all but one } \lambda \text{ in } \overline{k}^\times, \text{ if } a_P = 3.$$

So there are at most as many exceptional λ in $\overline{k}^\times$ as there are triple poles of g. QED

Corollary 11.2.18 A point (f, g) in $j^{-1}\mathrm{Twist}(D_0, D_1)_{\mathrm{prelim}}(\overline{k})$ lies in $j^{-1}\mathrm{Twist}(D_0, D_1)$ if and only if

$$\Sigma_{\text{poles P of g}} \mathrm{Swan}_P(g^*R^1\pi_{1,t!}\overline{\mathbb{Q}}_\ell) = \Sigma_P c_P.$$

Lemma 11.2.19 Notations as in Lemma 11.2.17 above, write D_0 over $\overline{k}$ as $\Sigma_Q b_Q Q$. Let (f, g) be a $\overline{k}$-valued point of $j^{-1}\mathrm{Twist}(D_0, D_1)$, and denote by $j : V \to C$ the inclusion. We have the following results.

1) $H_c^i(V\otimes_k\overline{k}, R^1\pi_{f,g!}\overline{\mathbb{Q}}_\ell) = 0$ for $i \neq 1$.

2) $\dim H_c^1(V\otimes_k\overline{k}, R^1\pi_{f,g!}\overline{\mathbb{Q}}_\ell)$

$$= -2\chi_c((V\otimes_k\overline{k})[1/fg], \overline{\mathbb{Q}}_\ell) - \deg(D_1) + \Sigma_P c_P.$$

3) $H^i(C\otimes_k\overline{k}, j_\star R^1\pi_{f,g!}\overline{\mathbb{Q}}_\ell(1)) = 0$ for $i \neq 1$.

4) $h^1(C\otimes_k\overline{k}, j_\star R^1\pi_{f,g!}\overline{\mathbb{Q}}_\ell(1)) = -\chi_c(C\otimes_k\overline{k}, j_\star R^1\pi_{f,g!}\overline{\mathbb{Q}}_\ell(1))$

$$= -2\chi_c((V\otimes_k\overline{k})[1/fg], \overline{\mathbb{Q}}_\ell) - 2\#\{P \text{ in } D_1 \text{ with } 3|a_P\}$$
$$- 2\#\{Q \text{ in } (D_0 - D_0\cap D_1)(\overline{k}) \text{ with } b_Q \text{ even}\}.$$

5) $h^1(C\otimes_k\overline{k}, j_\star R^1\pi_{f,g!}\overline{\mathbb{Q}}_\ell(1)) \ge 12g + 8$, with equality if and only if every a_P is divisible by 3, every b_Q for Q in $(D_0 - D_0\cap D_1)(\overline{k})$ is even, and $\deg(D_0) = \deg(D_1) = 2g + 3$.

proof We have the vanishing asserted in 1) and 3) because $R^1\pi_{f,g!}\overline{\mathbb{Q}}_\ell(1)$ is a geometrically irreducible middle extension of generic rank two. Assertion 2) is immediate from Lemma 11.2.11 and Corollary 11.2.18. To prove 4), we argue as follows. For brevity, denote by $\mathcal{F}$ the lisse sheaf on $(V\otimes_k\bar{k})[1/fg]$ given by

$$\mathcal{F} := R^1\pi_{f,g!}\bar{\mathbb{Q}}_\ell(1)|(V\otimes_k\bar{k})[1/fg] = \mathcal{L}_{\chi_2(f)}\otimes g^*R^1\pi_{1,t!}\bar{\mathbb{Q}}_\ell.$$

By the Euler Poincaré formula, we have

$$\begin{aligned}
&H_c^1(C\otimes_k\bar{k}, j_*R^1\pi_{f,g!}\bar{\mathbb{Q}}_\ell) := H_c^1(C\otimes_k\bar{k}, j_*\mathcal{F})\\
&= -\chi_c(C\otimes_k\bar{k}, j_*\mathcal{F})\\
&= -\chi_c^1((V\otimes_k\bar{k})[1/fg], \mathcal{F}) - \Sigma_{x \text{ in } C\otimes_k\bar{k} - (V\otimes_k\bar{k})[1/fg]} \dim(\mathcal{F}^{I(x)})\\
&= -2\chi_c((V\otimes_k\bar{k})[1/fg], \bar{\mathbb{Q}}_\ell) + \Sigma_P c_P\\
&\qquad - \Sigma_{x \text{ in } C\otimes_k\bar{k} - (V\otimes_k\bar{k})[1/fg]} \dim(\mathcal{F}^{I(x)}).
\end{aligned}$$

To compute the $\mathcal{F}^{I(x)}$ terms, we must examine more closely the sheaf $\mathcal{F} := \mathcal{L}_{\chi_2(f)}\otimes g^*R^1\pi_{1,t!}\bar{\mathbb{Q}}_\ell$. Recall that $R^1\pi_{1,t!}\bar{\mathbb{Q}}_\ell$ is lisse on $\mathbb{G}_m$, its local monodromy at 0 is a unipotent pseudoreflection, and its $I(\infty)$ representation is totally wild, with both breaks 1/2. At a zero x of f, g has neither zero nor pole, so $g^*R^1\pi_{1,t!}\bar{\mathbb{Q}}_\ell$ is lisse at x, while f has a simple zero, so $\mathcal{F}^{I(x)} = 0$. At each of the $\deg(D_1)$ zeroes x of g, f is invertible, and $g^*R^1\pi_{1,t!}\bar{\mathbb{Q}}_\ell$ has local monodromy a unipotent pseudoreflection, so $\dim(\mathcal{F}^{I(x)}) = 1$. At any pole x of g, $g^*R^1\pi_{1,t!}\bar{\mathbb{Q}}_\ell$ has a nonzero Swan conductor. As it is a two-dimensional $I(x)$-representation with trivial determinant, it must be totally wild at x, cf. Lemma 11.2.16. But $\mathcal{L}_{\chi_2(f)}$ is everyhwere tame, so $\mathcal{F}$ is totally wild at x, and so $\mathcal{F}^{I(x)} = 0$. Finally, at a pole x of f which is not a pole of g, g is invertible, so $g^*R^1\pi_{1,t!}\bar{\mathbb{Q}}_\ell$ is lisse at x, and so $\dim(\mathcal{F}^{I(x)}) = 2\dim((\mathcal{L}_{\chi_2(f)})^{I(x)})$ is 0, if f has an odd order pole at x, and it is 2 is f has an even order pole. Thus we find

$$\begin{aligned}
&h^1(C\otimes_k\bar{k}, j_*R^1\pi_{f,g!}\bar{\mathbb{Q}}_\ell(1))\\
&= -2\chi_c((V\otimes_k\bar{k})[1/fg], \bar{\mathbb{Q}}_\ell) + \Sigma_P c_P\\
&\qquad - \deg(D_1) - 2\#\{Q \text{ in } (D_0 - D_0\cap D_1)(\bar{k}) \text{ with } b_Q \text{ even}\}.
\end{aligned}$$

To conclude the proof of 4), we simply note that by definition of the integers c_P, we have

$$\Sigma_P c_P - \deg(D_1) = \Sigma_P (c_P - a_P) = -2\#\{P \text{ in } D_1 \text{ with } 3|a_P\}.$$

To prove 5), we use 4) and compute

$$\begin{aligned}
&h^1(C\otimes_k\bar{k}, j_*R^1\pi_{f,g!}\bar{\mathbb{Q}}_\ell(1))\\
&= -2\chi_c((V\otimes_k\bar{k})[1/fg], \bar{\mathbb{Q}}_\ell)\\
&\qquad -2\#\{P \text{ in } D_1 \text{ with } 3|a_P\}\\
&\qquad - 2\#\{Q \text{ in } (D_0 - D_0\cap D_1)(\bar{k}) \text{ with } b_Q \text{ even}\}
\end{aligned}$$

$= -2(2 - 2g - \#((D_0 \cup D_1)(\bar{k})) - \deg(D_0) - \deg(D_1))$
$-2\#\{P \text{ in } D_1 \text{ with } 3|a_P\}$
$- 2\#\{Q \text{ in } (D_0 - D_0 \cap D_1)(\bar{k}) \text{ with } b_Q \text{ even}\}$
$= -4 + 4g + 2\#((D_0 \cup D_1)(\bar{k})) + 2\deg(D_0) + 2\deg(D_1)$
$-2\#\{P \text{ in } D_1 \text{ with } 3|a_P\}$
$- 2\#\{Q \text{ in } (D_0 - D_0 \cap D_1)(\bar{k}) \text{ with } b_Q \text{ even}\}$
$\geq -4 + 4g + 2\deg(D_0) + 2\deg(D_1)$
$\geq -4 + 4g + 2(2g + 3) + 2(2g + 3) \geq 12g + 8.$

To get the last two equalities, we use first the inequality

$$\#((D_0 \cup D_1)(\bar{k}))$$
$$\geq \#\{P \text{ in } D_1 \text{ with } 3|a_P\} + \#\{Q \text{ in } (D_0 - D_0 \cap D_1)(\bar{k}) \text{ with } b_Q \text{ even}\},$$

and then the inequalities

$$\deg(D_i) \geq 2g + 3, \text{ for } i = 0, 1.$$ QED

Corollary 11.2.20 Denote by N the common dimension

$$N := \dim H_c^1(C\otimes_k \bar{k}, j_\star R^1\pi_{f,g!}\bar{\mathbb{Q}}_\ell)$$

for (f, g) in $j^{-1}\mathrm{Twist}(D_0, D_1)(\bar{k})$.

1) We have the inequality

$$N \geq 8.$$

2) We have $N \geq 9$ except in the case when $g = 0$, $D_1 = 3P$ for some rational point P, and D_0 is either 3P or P + 2Q for some second rational point Q.

proof This is immediate from Lemma 11.2.19, part 5). QED

(11.2.21) Our situation now is this. On the space $j^{-1}\mathrm{Twist}(D_0, D_1)$, we have the lisse sheaf $\mathcal{M}$, which is mixed of weight ≤ 0. The lisse sheaf $\mathrm{Gr}^0(\mathcal{M})$ is orthogonally self dual, and captures the L-function; for any finite extension k_d/k, and any point (f, g) in $j^{-1}\mathrm{Twist}(D_0, D_1)(k_d)$, we have

$$\mathrm{Gr}^0(\mathcal{M})_{(f,g)} = H_c^1(C\otimes_k \bar{k}, j_\star R^1\pi_{f,g!}\bar{\mathbb{Q}}_\ell(1)),$$

and the unitarized L-function of $E_{f,g}/k_d(C)$ is given by

$$L(E_{f,g}/k_d(C), T) = \det(1 - T\mathrm{Frob}_{k_d} \mid H^1(C\otimes_k \bar{k}, j_\star R^1\pi_{f,g!}\bar{\mathbb{Q}}_\ell(1))).$$

Theorem 11.2.22 Let k be a finite field of characteristic p = 3, ℓ a prime not 3. Fix a projective, smooth, geometrically connected curve C/k, of genus denoted g. Fix on C two effective divisors D_0 and D_1, both of degree $\geq 2g + 3$. Then the group G_{geom} for the lisse sheaf $\mathrm{Gr}^0(\mathcal{M})$ on the dense open set $j^{-1}\mathrm{Twist}(D_0, D_1)$ is O(N), for $N \geq 12g + 8$ the rank of $\mathrm{Gr}^0(\mathcal{M})$.

proof By a slight variation on the proof of Theorem 11.1.23 which we leave to the reader, we prove that G_{geom} contains a reflection. If $N \geq 9$, then we are done, by the argument given in the first paragraph of the proof of Theorem 7.2.3.

It remains to treat the case $N = 8$. Here we have either $O(8)$ or the Weyl group $W(E_8)$, by Theorem 2.6.11. We will give an ad hoc argument to show that when $N = 8$, G_{geom} cannot be finite, and hence must be $O(8)$.

If we have $N = 8$, then by Corollary 11.2.20 we have $g = 0$, D_1 is $3P$ for some rational point P, and D_0 is either $3P$ or $P + 2Q$ for some second rational point Q. Thus the curve C is $\mathbb{P}^1$, and an automorphism of $C = \mathbb{P}^1$ carries P to ∞ and Q, if present, to 0. Thus D_1 is $3[\infty]$, and D_0 is either $3[\infty]$ ("the first case") or $[\infty] + 2[0]$ ("the second case"). Let us denote by $Gr^0(\mathcal{M}_1)$ and $Gr^0(\mathcal{M}_2)$ the $Gr^0(\mathcal{M})$ sheaves attached to the two cases. In fact, these two cases are equivalent, as follows. In terms of the coordinate t on $\mathbb{A}^1 \subset \mathbb{P}^1$, the map

$$f \mapsto f/t^2$$

is an isomorphism

$$L(3[\infty]) \cong L([\infty] + 2[0]).$$

Given (f, g) in $j^{-1}\mathrm{Twist}(3[\infty], 3[\infty])$ with $f(0)$ invertible, the point $(f/t^2, g)$ lies in $j^{-1}\mathrm{Twist}(3[\infty], ([\infty] + 2[0])$, and this map

$$(f, g) \mapsto (f/t^2, g)$$

defines an isomorphism

$$\alpha : \{\text{the dense open set of } j^{-1}\mathrm{Twist}(3[\infty], 3[\infty]) \text{ with } f(0) \text{ invertible}\}$$
$$\cong j^{-1}\mathrm{Twist}(3[\infty], ([\infty] + 2[0]).$$

Given a finite extension k_d/k, the characteristic polynomial of Frobenius at a point (f, g) in $j^{-1}\mathrm{Twist}(3[\infty], 3[\infty])(k_d)$ with $f(0)$ invertible is the unitarized L-function of the elliptic curve $E_{f,g}$ over $k_d(t)$, while the characteristic polynomial of Frobenius at the corresponding point $(f/t^2, g)$ in $j^{-1}\mathrm{Twist}(3[\infty], ([\infty] + 2[0])(k_d)$ is the unitarized L-function of the elliptic curve $E_{f/t^2,g}$ over $k_d(t)$. But the curve $E_{f,g}$ is the quadratic twist by f of the curve $E_{1,g}$, and hence, as f and f/t^2 have ratio a square, the curves $E_{f,g}$ and $E_{f/t^2,g}$ over $k_d(t)$ are isomorphic, and hence have the same unitarized L-functions. Therefore we have an isomorphism of lisse sheaves

$$\alpha^*Gr^0(\mathcal{M}_2) \cong Gr^0(\mathcal{M}_1)| \{\text{the open set where } f(0) \text{ is invertible}\}.$$

So it suffices to treat the case $D_0 = D_1 = 3[\infty]$.

This case starts life over the prime field $\mathbb{F}_3$. The polynomials

$$g(t) := t^3 + t^2 - 1,$$
$$f(t) := t^3 - t^2 + 1,$$

are distinct irreducibles over $\mathbb{F}_3$ (each is a cubic with no zero in $\mathbb{F}_3$), and g has a Swan-minimal pole at ∞. So for at least one choice of $\varepsilon = \pm 1$ in $\mathbb{F}_3{}^\times$, the point $(f, \varepsilon g)$ lies in $j^{-1}\mathrm{Twist}(3[\infty], 3[\infty])(\mathbb{F}_3)$.

We argue by contradiction. Suppose $\mathrm{Gr}^0(\mathcal{M})$ has finite G_{geom}. Then it has finite G_{arith}. [Indeed, G_{arith} lies in $O(N)$ and normalizes the irreducible finite subgroup G_{geom} of $O(N)$. Put $n := \#\mathrm{Aut}_{gp}(G_{geom})$. Then for every element γ of G_{arith}, γ^n commutes with G_{geom}, so γ^n is a scalar in $O(N)$, so $\gamma^{2n} = 1$, so G_{arith} is an algebraic group with $\mathrm{Lie}(G_{arith})$ killed by $2n$, so $\mathrm{Lie}(G_{arith}) = 0$, so G_{arith} is finite.] So every eigenvalue of $\mathrm{Frob}_{\mathbb{F}_3,(f,\varepsilon g)}$ on $\mathrm{Gr}^0(\mathcal{M})_{(f,\varepsilon g)}$ is a root of unity. Equivalently, every eigenvalue of $\mathrm{Frob}_{\mathbb{F}_3}$ on

$$H^1(\mathbb{P}^1\otimes_{\mathbb{F}_3}\overline{\mathbb{F}}_3, j_\star R^1\pi_{f,\varepsilon g!}\overline{\mathbb{Q}}_\ell)$$

is of the form $3\times$(a root of unity).The sheaf $R^1\pi_{f,\varepsilon g!}\overline{\mathbb{Q}}_\ell) \cong \mathcal{L}_{\chi_2(f)}\otimes(\varepsilon g)^\star R^1\pi_{1,t!}\overline{\mathbb{Q}}_\ell$ is totally wild at ∞, so we have

$$H^1(\mathbb{P}^1\otimes_{\mathbb{F}_3}\overline{\mathbb{F}}_3, j_\star R^1\pi_{f,\varepsilon g!}\overline{\mathbb{Q}}_\ell) \cong H_c{}^1(\mathbb{A}^1\otimes_{\mathbb{F}_3}\overline{\mathbb{F}}_3, R^1\pi_{f,\varepsilon g!}\overline{\mathbb{Q}}_\ell).$$

Hence the ordinary integer

$$\mathrm{Trace}(\mathrm{Frob}_{\mathbb{F}_3} | H_c{}^1(\mathbb{A}^1\otimes_{\mathbb{F}_3}\overline{\mathbb{F}}_3, R^1\pi_{f,\varepsilon g!}\overline{\mathbb{Q}}_\ell))$$

is divisible by 3. The groups $H_c{}^i(\mathbb{A}^1\otimes_{\mathbb{F}_3}\overline{\mathbb{F}}_3, R^1\pi_{f,\varepsilon g!}\overline{\mathbb{Q}}_\ell)$ vanish for $i \neq 1$ (because $R^1\pi_{f,\varepsilon g!}\overline{\mathbb{Q}}_\ell$ is an irreducible middle extension of generic rank two). For any t in $\mathbb{A}^1(\mathbb{F}_3)$, we have $f(t) \neq 0$ (because f is irreducible), and so

$$\begin{aligned}&\mathrm{Trace}(\mathrm{Frob}_{\mathbb{F}_3,t} \mid R^1\pi_{f,\varepsilon g!}\overline{\mathbb{Q}}_\ell)\\ &\quad= -\Sigma_{x \text{ in } \mathbb{F}_3} \chi_2(x^3 + f(t)x^2 + f(t)^3\varepsilon g(t))\\ &\quad= -\chi_2(f(t))\Sigma_{x \text{ in } \mathbb{F}_3} \chi_2(x^3 + x^2 + \varepsilon g(t)),\end{aligned}$$

the last equality by replacing x by $f(t)x$ in the summation over x. So by the Lefschetz Trace Formula, we have

$$\begin{aligned}&\mathrm{Trace}(\mathrm{Frob}_{\mathbb{F}_3} | H_c{}^1(\mathbb{A}^1\otimes_{\mathbb{F}_3}\overline{\mathbb{F}}_3, R^1\pi_{f,\varepsilon g!}\overline{\mathbb{Q}}_\ell))\\ &\quad= \Sigma_{t \text{ in } \mathbb{F}_3} \chi_2(f(t))\Sigma_{x \text{ in } \mathbb{F}_3} \chi_2(x^3 + x^2 + \varepsilon g(t)).\end{aligned}$$

We now achieve the desired contradiction, by showing that the integer

$$\Sigma_{t \text{ in } \mathbb{F}_3} \chi_2(f(t))\Sigma_{x \text{ in } \mathbb{F}_3} \chi_2(x^3 + x^2 + \varepsilon g(t))$$

is not divisible by 3. Indeed, for α in $\mathbb{F}_3$, we have the congruence

$$\chi_2(\alpha) \equiv \alpha \bmod (3).$$

So we have congruences mod (3),

$$\Sigma_{t \text{ in } \mathbb{F}_3} \chi_2(f(t))\Sigma_{x \text{ in } \mathbb{F}_3} \chi_2(x^3 + x^2 + \varepsilon g(t))$$
$$\equiv \Sigma_{x \text{ in } \mathbb{F}_3} \Sigma_{t \text{ in } \mathbb{F}_3} f(t)(x^3 + x^2 + \varepsilon g(t)).$$

Now the function $x^3 + x^2$ takes the value 0 at x=0 and x=2, and the value 2 at x=1. So summing over the three x values separately we have an equality in $\mathbb{F}_3$,

$$\Sigma_{x \text{ in } \mathbb{F}_3} \Sigma_{t \text{ in } \mathbb{F}_3} f(t)(x^3 + x^2 + \varepsilon g(t)).$$
$$= 2\Sigma_{t \text{ in } \mathbb{F}_3} f(t)(0 + \varepsilon g(t)) + \Sigma_{t \text{ in } \mathbb{F}_3} f(t)(2 + \varepsilon g(t))$$
$$= 2\Sigma_{t \text{ in } \mathbb{F}_3} f(t)\varepsilon g(t) + 2\Sigma_{t \text{ in } \mathbb{F}_3} f(t) + \Sigma_{t \text{ in } \mathbb{F}_3} f(t)\varepsilon g(t)$$
$$= 2\Sigma_{t \text{ in } \mathbb{F}_3} f(t)$$
$$= 2\Sigma_{t \text{ in } \mathbb{F}_3} (t^3 - t^2 + 1)$$
$$= 2.$$

So we have

$$\mathrm{Trace}(\mathrm{Frob}_{\mathbb{F}_3} | H_c^1(\mathbb{A}^1 \otimes_{\mathbb{F}_3} \overline{\mathbb{F}}_3, R^1\pi_{f,\varepsilon g!}\overline{\mathbb{Q}}_\ell)) \equiv 2 \bmod (3). \text{ QED}$$

(11.3) (j^{-1}, twist) families in characteristic 2

(11.3.1) In this section, we consider the characteristic 2 analogue of the (j^{-1}, twist) families we considered in characteristic 3.

(11.3.2) Over a field K of characteristic 2, an elliptic curve with nonzero j is ordinary, and can be written in the form

$$E_{s,t} : y^2 + xy = x^3 + sx^2 + t.$$

Here, $j = 1/t$, so t is determined by j, and the effect of s is to perform a quadratic twist using the Artin-Schreier sheaf $\mathcal{L}\psi$, cf. Twisting Lemma 9.4.1. As s runs over K and t runs over $K^\times$, the curves $E_{s,t}$ exhaust, with infinite repetition, the isomorphism classes of elliptic curves E/K with $j \neq 0$.

(11.3.3) We now specialize to the case when K is a function field in one variable over a finite field of characteristic 2. Thus we work over a finite field k of characteristic 2, in which a prime ℓ is invertible. We fix a projective, smooth, geometrically connected curve C/k, of genus denoted g. We also fix on C two effective divisors D_0 and D_1. We assume that

$$\deg(D_i) \geq 2g + 3,$$

for i=0, 1.

(11.3.4) Given a function f in the Riemann Roch space $L(D_0)$ and a function g in the Riemann Roch space $L(D_1)$, we consider the Weierstrass equation

$$E_{f,g} : y^2 + xy = x^3 + fx^2 + g.$$

Provided that $g \neq 0$, this is an elliptic curve over k(C), with $j = 1/g$. As we allow the efffective divisors D_0 and D_1 to grow, we sweep out (with repetition) all the elliptic curves over $k(C)$ with $j \neq 0$.

(11.3.5) We are interested in the variation with (f, g) in $L(D_0)\times L(D_1)$ of the L-function of $E_{f,g}/k(C)$. Similarly, for each finite extension field k_d/k, we are interested in the variation with (f, g) in $L(D_0)\otimes_k k_d\times L(D_1)\otimes_k k_d$ of the L-function of $E_{f,g}/k_d(C)$. For this to be a reasonable question, we must restrict to a dense open set of $L(D_0)\times L(D_1)$ (viewed as an affine space over k) over which the L-function is a polynomial of constant degree.

(11.3.6) We define

$$V := C - (D_0 \cup D_1).$$

Our next task is to define a dense open set

$$j^{-1}\mathrm{Twist}(D_0, D_1) \subset L(D_0)\times L(D_1)$$

over which the L-function is a polynomial of constant degree. We will do this in two steps. We will first define a preliminary dense open set

$$j^{-1}\mathrm{Twist}(D_0, D_1)_{\mathrm{prelim}} \subset L(D_0)\times L(D_1),$$

and then we will define $j^{-1}\mathrm{Twist}(D_0, D_1)$ as a dense open set

$$j^{-1}\mathrm{Twist}(D_0, D_1) \subset j^{-1}\mathrm{Twist}(D_0, D_1)_{\mathrm{prelim}}.$$

(11.3.7) We define

$$j^{-1}\mathrm{Twist}(D_0, D_1)_{\mathrm{prelim}} \subset L(D_0)\times L(D_1)$$

to be the dense open set whose $\overline{k}$-valued points are the pairs (f, g) with f in $L(D_0)\otimes_k \overline{k}$, g in $L(D_1)\otimes_k \overline{k}$, such that the following five conditions hold:

1) f has divisor of poles D_0,
2) f has Swan-minimal poles, cf. 6.4.6,
3) g has divisor of poles D_1,
4) g has Swan-minimal poles, cf. 6.4.6,
5) g has $\deg(D_1)$ distinct zeroes in $V(\overline{k})$, i.e., g has $\deg(D_1)$ distinct zeroes, which all lie in $V(\overline{k})$.

(11.3.8) For any $d \geq 1$ and for any (f, g) in $j^{-1}\mathrm{Twist}(D_0, D_1)_{\mathrm{prelim}}(k_d)$, $E_{f,g}$ is an elliptic curve over $k_d(C)$. It has good reduction over $(V\otimes_k k_d)[1/g]$, and it has multiplicative reduction at the zeroes of g.

(11.3.9) Denote by $\pi_{f,g} : \mathcal{E}_{f,g}{}^{\mathrm{aff}} \to V\otimes_k k_d$ the relative affine curve of equation

$$y^2 + xy = x^3 + fx^2 + g.$$

Its $R^1\pi_{f,g!}\overline{\mathbb{Q}}_\ell$ on $V\otimes_k k_d$ has a tensor product decomposition as follows. On $\mathbb{A}^1 = \mathrm{Spec}(k[t])$, we have

$$\pi_{0,t} : \mathcal{E}_{0,t}{}^{\mathrm{aff}} \to \mathbb{A}^1,$$

the relative affine curve of equation

$$y^2 + xy = x^3 + t.$$

Its $R^1\pi_{0,t!}\overline{\mathbb{Q}}_\ell$ on $\mathbb{A}^1$ is a middle extension sheaf (by 7.5.5), which is lisse on $\mathbb{G}_m$ and geometrically irreducible (because $j = 1/t$ is nonconstant, cf. the last paragraph of the proof of Theorem 9.3.10). Because the map g is finite etale over 0, the pullback $g^*R^1\pi_{0,t!}\overline{\mathbb{Q}}_\ell$ is a middle extension on $V\otimes_k k_d$. The sheaf $\mathcal{L}_\psi$ on $\mathbb{A}^1$ is lisse, so the pullback $f^*(\mathcal{L}_\psi) := \mathcal{L}_{\psi(f)}$ is lisse on $V\otimes_k k_d$. We have an isomorphism on $V\otimes_k k_d$

$$R^1\pi_{f,g!}\overline{\mathbb{Q}}_\ell \cong \mathcal{L}_{\psi(f)}\otimes g^*R^1\pi_{0,t!}\overline{\mathbb{Q}}_\ell.$$

Here one factor is lisse, so in particular the two tensor factors, $\mathcal{L}_{\psi(f)}$ and $g^*R^1\pi_{0,t!}\overline{\mathbb{Q}}_\ell$, have disjoint ramification. As each is a middle extension, so is their tensor product. Since $\mathcal{E}_{f,g}$ has nonconstant j invariant, the middle extension $R^1\pi_{f,g!}\overline{\mathbb{Q}}_\ell$ is geometrically irreducible. Thus we have

Lemma 11.3.10 For (f, g) in $j^{-1}\mathrm{Twist}(D_0, D_1)_{\mathrm{prelim}}(k_d)$, $R^1\pi_{f,g!}\overline{\mathbb{Q}}_\ell$ on $V\otimes_k k_d$ is a geometrically irreducible middle extension of generic rank two, which is lisse on $(V\otimes_k k_d)[1/g]$, and which has the tensor product structure

$$R^1\pi_{f,g!}\overline{\mathbb{Q}}_\ell \cong \mathcal{L}_{\psi(f)}\otimes g^*R^1\pi_{0,t!}\overline{\mathbb{Q}}_\ell.$$

(11.3.11) Denote by $j : V \to C$ the inclusion. The unitarized L-function of $E_{f,g}/k_d(C)$ is given by

$$L(E_{f,g}/k_d(C), T) = \det(1 - T\mathrm{Frob}_{k_d} \mid H^1(C\otimes_k\overline{k}, j_*R^1\pi_{f,g!}\overline{\mathbb{Q}}_\ell(1))).$$

We can recover $H^1(C\otimes_k\overline{k}, j_*R^1\pi_{f,g!}\overline{\mathbb{Q}}_\ell(1))$ as the weight zero quotient of

$$H_c^1(V\otimes_k\overline{k}, R^1\pi_{f,g!}\overline{\mathbb{Q}}_\ell(1))$$
$$= H_c^1(V\otimes_k\overline{k}, \mathcal{L}_{\psi(f)}\otimes g^*R^1\pi_{0,t!}\overline{\mathbb{Q}}_\ell(1)).$$

[We will see later that if (f, g) lie in $j^{-1}\mathrm{Twist}(D_0, D_1)(k_d)$, then this last cohomology group is in fact pure of weight zero, cf. Lemma 11.3.20, part 2), and Lemma 11.3.21.]

Lemma 11.3.12 For (f, g) in $j^{-1}\mathrm{Twist}(D_0, D_1)_{\mathrm{prelim}}(\overline{k})$, we have

1) $H_c^i(V\otimes_k\overline{k}, R^1\pi_{f,g!}\overline{\mathbb{Q}}_\ell) = 0$ for $i \neq 1$,

2) $\dim H_c^1(V\otimes_k\overline{k}, R^1\pi_{f,g!}\overline{\mathbb{Q}}_\ell) = -\chi_c(V\otimes_k\overline{k}, R^1\pi_{f,g!}\overline{\mathbb{Q}}_\ell)$

$$= -2\chi_c((V\otimes_k\overline{k})[1/g], \overline{\mathbb{Q}}_\ell)$$
$$-\deg(D_1) + \Sigma_{P \text{ in } (D_0 \cup D_1)(\overline{k})} \mathrm{Swan}_P(\mathcal{L}_{\psi(f)}\otimes g^*R^1\pi_{0,t!}\overline{\mathbb{Q}}_\ell).$$

proof Assertion 1) results from the fact that the coefficient sheaf is a geometrically irreducible middle extension which is not geometrically trivial. Assertion 2) is a straightforward application of the Euler Poincaré formula. Indeed, the sheaf on $V\otimes_k\bar{k}$,

$$R^1\pi_{f,g!}\bar{\mathbb{Q}}_\ell \cong \mathcal{L}_{\psi(f)}\otimes g^*R^1\pi_{0,t!}\bar{\mathbb{Q}}_\ell,$$

is lisse where g is invertible. At the $\deg(D_1)$ zeroes of g, it is tame, and local monodromy is a unipotent pseudoreflection. Thus we have

$$\chi_c(V\otimes_k\bar{k},\ R^1\pi_{f,g!}\bar{\mathbb{Q}}_\ell)$$
$$= \chi_c((V\otimes_k\bar{k})[1/g],\ R^1\pi_{f,g!}\bar{\mathbb{Q}}_\ell) + \deg(D_1).$$

The sheaf $R^1\pi_{f,g!}\bar{\mathbb{Q}}_\ell$ on $(V\otimes_k\bar{k})[1/g]$ is lisse of rank two, and tame except possibly at the points of $D_0\cup D_1$, so the Euler Poincaré formula gives

$$\chi_c((V\otimes_k\bar{k})[1/g],\ R^1\pi_{f,g!}\bar{\mathbb{Q}}_\ell)$$
$$= 2\chi_c((V\otimes_k\bar{k})[1/g],\ \bar{\mathbb{Q}}_\ell)$$
$$- \Sigma_{P \text{ in } (D_0\cup D_1)(\bar{k})} \mathrm{Swan}_P(\mathcal{L}_{\psi(f)}\otimes g^*R^1\pi_{0,t!}\bar{\mathbb{Q}}_\ell).$$

Putting this all together, we find assertion 2). QED

(11.3.13) The cohomology groups $H_c^1(V\otimes_k\bar{k},\ R^1\pi_{f,g!}\bar{\mathbb{Q}}_\ell(1))$ can be all captured by a suitable perverse sheaf $M = \mathrm{Twist}(L, K, \mathcal{F}, h)$ attached to suitable "standard input", cf. 1.15.4. We take

the integer $m = 2$,

the perverse sheaf $K = R^1\pi_{s,t!}\bar{\mathbb{Q}}_\ell(3/2)[2]$

$\cong \mathcal{L}_{\psi(s)}(1/2)[1]\boxtimes R^1\pi_{0,t!}\bar{\mathbb{Q}}_\ell(1)[1]$ on $\mathbb{A}^2/k = \mathrm{Spec}(k[s,t])$,

the affine k-scheme V/k given by $V := C - (D_0 \cup D_1)$,

the k-morphism $h : V \to \mathbb{A}^2$ given by $h = 0$,

the perverse sheaf $L = \bar{\mathbb{Q}}_\ell(1/2)[1]$ on V/k,

the integer $d = 1 - 2g + \min(\deg(D_0), \deg(D_1))$,

the space of $\mathbb{A}^2$-valued functions $(\mathcal{F}, \tau)$ on V given by the finite-dimensional k-vector space $\mathcal{F} = L(D_0)\times L(D_1)$ and the k-linear map

$$\tau : \mathcal{F} \to \mathrm{Hom}_{k\text{-schemes}}(V, \mathbb{A}^2),$$
$$(f, g) \mapsto \text{the map } v \mapsto (f(v), g(v)).$$

It results from Theorem 9.4.2 that this is standard input, and, using Kunneth, that $H^*_c((V\times\mathbb{A}^m)\otimes\bar{k},\ pr_1^*L\otimes pr_2^*K) = 0$.

Lemma 11.3.14 On the dense open set $j^{-1}\mathrm{Twist}(D_0, D_1)_{\mathrm{prelim}}$, the perverse sheaf M has only one nonvanishing cohomology sheaf, $\mathcal{H}^{-\dim(\mathcal{F})}(M)$, whose stalks are

$$(f,g) \mapsto H_c^1(V\otimes_k\bar{k},\ R^1\pi_{f,g!}\bar{\mathbb{Q}}_\ell)(2).$$

proof This is just a translation of the previous Lemma 11.3.12. QED

(11.3.15) Thus on the open set $j^{-1}\mathrm{Twist}(D_0, D_1)_{\mathrm{prelim}}$, M is the object $\mathcal{M}(1)[\ell(D_0) + \ell(D_1)]$, with $\mathcal{M} := \mathcal{H}^{-\dim(\mathcal{F})}(M)(-1)$ the sheaf of perverse origin, mixed of weight ≤ 0, given stalkwise by

$$\mathcal{M}_{f,g} := H_c{}^1(V\otimes_k \bar{k}, R^1\pi_{f,g!}\bar{\mathbb{Q}}_\ell)(1).$$

We now invoke the fact [Ka-SCMD, Prop. 12] that for a sheaf of perverse origin $\mathcal{M}$ on a smooth connected k-scheme, here $j^{-1}\mathrm{Twist}(D_0, D_1)_{\mathrm{prelim}}$, the set of points where its stalk has maximum rank is a dense open set,.

$$U_{\max} \subset j^{-1}\mathrm{Twist}(D_0, D_1)_{\mathrm{prelim}},$$

and $\mathcal{M}$ is lisse on $U_{\max}$.

(11.3.16) We now define the dense open set $j^{-1}\mathrm{Twist}(D_0, D_1)$ to be

$$j^{-1}\mathrm{Twist}(D_0, D_1) := U_{\max} \subset j^{-1}\mathrm{Twist}(D_0, D_1)_{\mathrm{prelim}}.$$

Thanks to the dimension formula of Lemma 11.3.12, part 2) above, we can also characterize $U_{\max}$ as the set of points (f, g) in $j^{-1}\mathrm{Twist}(D_0, D_1)_{\mathrm{prelim}}$ at which

$$\Sigma_{P \text{ in } (D_0 \cup D_1)(\bar{k})} \mathrm{Swan}_P(\mathcal{L}_{\psi(f)}\otimes g^*R^1\pi_{0,t!}\bar{\mathbb{Q}}_\ell)$$

attains it maximum value.

Lemma 11.3.17 Write D_1 over $\bar{k}$ as $\Sigma a_P P$. Define integers c_P, one for each P in $D_0 \cup D_1$, by

$$\begin{aligned} c_P &= a_P, \text{ if } a_P \text{ is odd,} \\ &= a_P - 2, \text{ if } a_P \text{ is even and } a_P > 2, \\ &= 1, \text{ if } a_P = 2, \\ &= 0, \text{ if } a_P = 0. \end{aligned}$$

Write D_0 over $\bar{k}$ as $\Sigma \alpha_P P$. Define integers β_P, one for each P in $D_0 \cup D_1$, by

$$\begin{aligned} \beta_P &= \alpha_P, \text{ if } \alpha_P \text{ is odd,} \\ &= \alpha_P - 1, \text{ if } \alpha_P \text{ is even and } \alpha_P \ne 0, \\ &= 0, \text{ if } \alpha_P = 0. \end{aligned}$$

Suppose f in $L(D_0)\otimes_k \bar{k}$ (resp. g in $L(D_1)\otimes_k \bar{k}$) has divisor of poles D_0 (resp. divisor of poles D_1) and has Swan-minimal poles. Suppose also that g has no zeroes in D_0. Fix a point P in $D_0 \cup D_1$. Then we have the inequality

$$\mathrm{Swan}_P(\mathcal{L}_{\psi(f)}\otimes g^*R^1\pi_{0,t!}\bar{\mathbb{Q}}_\ell) \le \mathrm{Max}(2\beta_P, c_P),$$

and, for all but at most four values of λ in $\bar{k}^\times$ we have the equality

$$\mathrm{Swan}_P(\mathcal{L}_{\psi(\lambda f)}\otimes g^*R^1\pi_{0,t!}\bar{\mathbb{Q}}_\ell) = \mathrm{Max}(2\beta_P, c_P).$$

proof This follows from the following more precise Lemma.

Lemma 11.3.17 (bis) Hypotheses and notations as in Lemma 11.3.17 above, we have the following results.

1) At a point P which lies in D_0 but not in D_1, we have

$$\mathrm{Swan}_P(\mathcal{L}_{\psi(f)}\otimes g^*R^1\pi_{0,t!}\overline{\mathbb{Q}}_\ell) = 2\mathrm{Swan}_P(\mathcal{L}_{\psi(f)}).$$

2) At a point P in D_0 with $\alpha_P \neq 2$, we have

$$\mathrm{Swan}_P(\mathcal{L}_{\psi(f)}) = \beta_P.$$

3) At a point P in D_0 with $\alpha_P = 2$, we have

$$\mathrm{Swan}_P(\mathcal{L}_{\psi(f)}) \leq \beta_P = 1,$$

and for all but precisely one value of λ in $\overline{k}^\times$, we have

$$\mathrm{Swan}_P(\mathcal{L}_{\psi(\lambda f)}) = \beta_P.$$

4) At a point P which lies in D_1, we have

$$\mathrm{Swan}_P(g^*R^1\pi_{0,t!}\overline{\mathbb{Q}}_\ell) = c_P,$$

and both I(P)-breaks of $g^*R^1\pi_{0,t!}\overline{\mathbb{Q}}_\ell$ are $c_P/2$.

5) At a point P which lies in D_1 but not in D_0, we have

$$\mathrm{Swan}_P(\mathcal{L}_{\psi(f)}\otimes g^*R^1\pi_{0,t!}\overline{\mathbb{Q}}_\ell) = \mathrm{Swan}_P(g^*R^1\pi_{0,t!}\overline{\mathbb{Q}}_\ell) = c_P.$$

6) At a point P which lies in $D_0\cap D_1$ at which $2\beta_P \neq c_P$, we have

$$\mathrm{Swan}_P(\mathcal{L}_{\psi(f)}\otimes g^*R^1\pi_{0,t!}\overline{\mathbb{Q}}_\ell) = \mathrm{Max}(2\mathrm{Swan}_P(\mathcal{L}_{\psi(f)}), c_P).$$

7) At a point P which lies in $D_0\cap D_1$, and at which $2\beta_P = c_P$, we have the inequality

$$\mathrm{Swan}_P(\mathcal{L}_{\psi(f)}\otimes g^*R^1\pi_{0,t!}\overline{\mathbb{Q}}_\ell) \leq \mathrm{Max}(2\beta_P, c_P),$$

and for all but at most four values of λ in $\overline{k}^\times$ we have the equality

$$\mathrm{Swan}_P(\mathcal{L}_{\psi(\lambda f)}\otimes g^*R^1\pi_{0,t!}\overline{\mathbb{Q}}_\ell) = \mathrm{Max}(2\beta_P, c_P).$$

proof 1) At a point P which lies in D_0 but not in D_1, the function g is invertible, so $g^*R^1\pi_{0,t!}\overline{\mathbb{Q}}_\ell$ is lisse of rank two at P.

2) If α_P is odd, this is standard [De-SomTrig, 3.5.4]. If α_P is even but not 2, say $\alpha_P = 2k$ with $k > 1$, then in terms of a uniformizing parameter z at P, the expansion of f at P can be written in the form

$$A^2/z^{2k} + B/z^{2k-1} + \text{less polar terms}.$$

Because f has a Swan-minimal pole at P, both A and B lie in $\overline{k}^\times$. This expansion in turn is Artin-Schreier equivalent to

$$B/z^{2k-1} + \text{less polar terms}$$

(because A^2/z^{2k} is Artin-Schreier equivalent to A/z^k, which is less polar than B/z^{2k-1} for $k > 1$) and we are reduced to the case when α_P is prime to p.

3) If $\alpha_P = 2$, then the expansion of f at P can be written in the form

$$A^2/z^2 + B/z^1 + \text{holomorphic},$$

which is Artin-Schreier equivalent to

$$(B - A)/z + \text{holomorphic}.$$

For λ in $\overline{k}^\times$, $\lambda^2 f$ then has expansion at P

$$\lambda^2A^2/z^2 + \lambda^2B/z^1 + \text{holomorphic},$$

which is Artin-Schreier equivalent to

$$(\lambda^2 B - \lambda A)/z + \text{holomorphic}.$$

So if λ in $\bar{k}^\times$ is not A/B, then $\mathrm{Swan}_P(\mathcal{L}_{\psi(\lambda^2 f)}) = \beta_P = 1$. Thus we have $\mathrm{Swan}_P(\mathcal{L}_{\psi(\lambda f)}) = \beta_P$ for all λ in $\bar{k}^\times$ save $(B/A)^2$.

4) The Swan conductor is computed in Lemma A6.2.2. That the two I(P)-breaks are equal is Lemma 11.2.16.

5) At a point P which lies in D_1 but not in D_0, $\mathcal{L}_{\psi(f)}$ is lisse of rank one.

6) At a point P which lies in $D_0 \cap D_1$ at which $2\beta_P \neq c_P$, either $\mathcal{L}_{\psi(f)}$ has Swan conductor β_P, or $\beta_P = 1$ and $\mathcal{L}_{\psi(f)}$ is tame. In the first case, both breaks of $g^*R^1\pi_{0,t!}\bar{\mathbb{Q}}_\ell$ are $c_P/2 \neq \beta_P/2$, and hence [Ka-GKM, 1.3] both breaks of $\mathcal{L}_{\psi(f)} \otimes g^*R^1\pi_{0,t!}\bar{\mathbb{Q}}_\ell$ are $\mathrm{Max}(c_P/2, \beta_P)$. In the second case, $\mathcal{L}_{\psi(f)} \otimes g^*R^1\pi_{0,t!}\bar{\mathbb{Q}}_\ell$ has the same Swan_P as $g^*R^1\pi_{0,t!}\bar{\mathbb{Q}}_\ell$, namely c_P, and $\mathcal{L}_{\psi(f)}$ is tame. So in both cases the asserted inequalit holds.

7) At a point P which lies in $D_0 \cap D_1$, and at which $2\beta_P = c_P$, $\mathcal{L}_{\psi(f)}$ has break at most β_P, and all breaks of $g^*R^1\pi_{0,t!}\bar{\mathbb{Q}}_\ell$ are $c_P/2$. So [Ka-GKM, 1.3] all breaks of $\mathcal{L}_{\psi(f)} \otimes g^*R^1\pi_{0,t!}\bar{\mathbb{Q}}_\ell$ are at most $\mathrm{Max}(c_P/2, \beta_P)$. This gives the asserted inequality.

To see that for all but at most four values of λ in $\bar{k}^\times$ we have the equality

$$\mathrm{Swan}_P(\mathcal{L}_{\psi(\lambda f)} \otimes g^*R^1\pi_{0,t!}\bar{\mathbb{Q}}_\ell) = \mathrm{Max}(2\beta_P, c_P),$$

we argue as follows. By its very definition, β_P is odd. So the unique break of $g^*R^1\pi_{0,t!}\bar{\mathbb{Q}}_\ell$, namely $c_P/2 = \beta_P$, is odd. So each of the at most two irreducible constituents M_i of $g^*R^1\pi_{0,t!}\bar{\mathbb{Q}}_\ell$ as I(P)-representation is purely of odd integer slope β_P.

If α_P is odd, then $\beta_P = \alpha_P$ is the order of pole of f at P. Then by the Break Depression Lemma [Ka-GKM, 8.5.7.1], we see that for each of the M_i, for all but one values of λ in $\bar{k}^\times$, $\mathcal{L}_{\psi(\lambda f)} \otimes M_i$ has all its breaks β_P. As there are at most two M_i, there are at most two exceptional λ in $\bar{k}^\times$.

If α_P is even, say $\alpha_P = 2k$, then in terms of a uniformizing parameter z at P, the expansion of f at P can be written in the form

$$A^2/z^{2k} + B/z^{2k-1} + \text{less polar terms},$$

with A and B in $\bar{k}^\times$. If $k > 1$, then $\lambda^2 f$ is Artin-Schreier equivalent to

$$\lambda^2 B/z^{2k-1} + \text{less polar terms},$$

and we repeat the above argument, to get at most two exceptional λ in $\bar{k}^\times$. If $k = 1$, then $\lambda^2 f$ is Artin-Schreier equivalent to

$$(\lambda^2B - \lambda A)/z + \text{holomorphic}.$$

By the Break Depression Lemma, there are at most two nonzero values of $\lambda^2B - \lambda A$ we must avoid, so there are at most four nonzero values of λ we must avoid. QED

Corollary 11.3.18 A point (f, g) in $j^{-1}\mathrm{Twist}(D_0, D_1)_{\mathrm{prelim}}(\bar{k})$ lies in $j^{-1}\mathrm{Twist}(D_0, D_1)$ if and only if for each point P in $(D_0 \cup D_1)(\bar{k})$, we have the equality

$$\mathrm{Swan}_P(\mathcal{L}_{\psi(f)} \otimes g^*R^1\pi_{0,t!}\bar{\mathbb{Q}}_\ell) = \mathrm{Max}(2\beta_P, c_P),$$

or equivalently (cf. Lemma 11.3.10), the equality

$$\mathrm{Swan}_P(R^1\pi_{f,g!}\bar{\mathbb{Q}}_\ell) = \mathrm{Max}(2\beta_P, c_P),$$

Corollary 11.3.19 Given a point (f, g) in $j^{-1}\mathrm{Twist}(D_0, D_1)_{\mathrm{prelim}}(\bar{k})$, for all but finitely many values of λ in $\bar{k}^\times$, $(\lambda f, g)$ lies in $j^{-1}\mathrm{Twist}(D_0, D_1)$, and we have

$$\mathrm{Swan}_P(\mathcal{L}_{\psi(\lambda f)}) = \beta_P.$$

at every point P in $D_0(\bar{k})$.

Lemma 11.3.20 Denote by $j : V \to C$ the inclusion. Let (f, g) be a point of $j^{-1}\mathrm{Twist}(D_0, D_1)(\bar{k})$. Then we have the following results.

1) The canonical map of sheaves on $C\otimes_k\bar{k}$,

$$j_!(R^1\pi_{f,g!}\bar{\mathbb{Q}}_\ell) \to j_*(R^1\pi_{f,g!}\bar{\mathbb{Q}}_\ell)$$

is an isomorphism.

2) For every i, we have

$$H_c^i(V\otimes_k\bar{k}, R^1\pi_{f,g!}\bar{\mathbb{Q}}_\ell) \cong H^i(C\otimes_k\bar{k}, j_*R^1\pi_{f,g!}\bar{\mathbb{Q}}_\ell).$$

3) $H_c^i(V\otimes_k\bar{k}, R^1\pi_{f,g!}\bar{\mathbb{Q}}_\ell) = 0$ for $i \neq 1$,

4) $\dim H_c^1(V\otimes_k\bar{k}, R^1\pi_{f,g!}\bar{\mathbb{Q}}_\ell) = -\chi_c(V\otimes_k\bar{k}, R^1\pi_{f,g!}\bar{\mathbb{Q}}_\ell)$

$$= -2\chi_c((V\otimes_k\bar{k})[1/g], \bar{\mathbb{Q}}_\ell) - \deg(D_1) + \Sigma_{P \text{ in } (D_0 \cup D_1)(\bar{k})} \mathrm{Max}(2\beta_P, c_P)$$

$$= 4g - 4 + \deg(D_1) + \Sigma_{P \text{ in } (D_0 \cup D_1)(\bar{k})} (2 + \mathrm{Max}(2\beta_P, c_P)).$$

proof By Corollary 11.3.18, at each point P in $(D_0 \cup D_1)(\bar{k})$, we have

$$\mathrm{Swan}_P(R^1\pi_{f,g!}\bar{\mathbb{Q}}_\ell) = \mathrm{Max}(2\beta_P, c_P).$$

Since $\mathrm{Max}(2\beta_P, c_P) > 0$, we see from Lemma 11.2.16 that $R^1\pi_{f,g!}\bar{\mathbb{Q}}_\ell)$ is totally wild at P, and hence has no nonzero I(P)-invariants. This proves 1). To prove 2), apply the functor $H^i(C\otimes_k\bar{k}, \cdot)$ to the isomorphism of 1). Assertion 3) was already proven in Lemma 11.3.12, part 1), and 4) is simply Lemma 11.3.12, part 2), combined with the exact value of the Swan conductors given by 11.3.18. QED

Lemma 11.3.21 For (f, g) in $j^{-1}\mathrm{Twist}(D_0, D_1)(k_d)$, the unitarized L-function of $E_{f,g}/k_d(C)$ is given by

$$L(E_{f,g}/k_d(C), T) = \det(1 - T\mathrm{Frob}_{k_d} \mid H_c^1(V\otimes_k\bar{k}, R^1\pi_{f,g!}\bar{\mathbb{Q}}_\ell(1))).$$

It is a polynomial of degree

$$4g - 4 + \deg(D_1) + \Sigma_{P \text{ in } (D_0 \cup D_1)(\bar{k})} (2 + \mathrm{Max}(2\beta_P, c_P)).$$

proof This is immediate from Lemma 11.3.20, part 2) above, and the description of the unitarized L-function as

$$L(E_{f,g}/k_d(C), T) = \det(1 - T\mathrm{Frob}_{k_d} \mid H^1(C\otimes_k\bar{k}, j_*R^1\pi_{f,g!}\bar{\mathbb{Q}}_\ell(1))).$$

QED

Lemma 11.3.22 Denote by N the common dimension

$$N := \dim H_c^1(C\otimes_k\bar{k}, j_*R^1\pi_{f,g!}\bar{\mathbb{Q}}_\ell(1))$$
$$= \dim H_c^1(V\otimes_k\bar{k}, R^1\pi_{f,g!}\bar{\mathbb{Q}}_\ell(1))$$

for (f, g) in $j^{-1}\mathrm{Twist}(D_0, D_1)(\bar{k})$. We have the inequality

$$N \geq 4g - 4 + \deg(D_1) + 2\deg(D_0) + 2\#\{P \text{ in } D_0(\bar{k}) \text{ with } \alpha_P \text{ odd}\}.$$

proof 1) By Lemma 11.3.21 above, we have

$$N = 4g - 4 + \deg(D_1) + \Sigma_{P \text{ in } (D_0 \cup D_1)(\bar{k})} (2 + \mathrm{Max}(2\beta_P, c_P))$$
$$\geq 4g - 4 + \deg(D_1) + \Sigma_{P \text{ in } D_0(\bar{k})} (2 + \mathrm{Max}(2\beta_P, c_P))$$
$$\geq 4g - 4 + \deg(D_1) + \Sigma_{P \text{ in } D_0(\bar{k})} (2 + 2\beta_P).$$

Recall that for P in $D_0(\bar{k})$, β_P is either α_P, if α_P is odd, or is $\alpha_P - 1$, if α_P is even, so we have

$$2 + 2\beta_P = 2\alpha_P + 2\delta_{\alpha_P,\mathrm{odd}}$$

and hence

$$\Sigma_{P \text{ in } D_0(\bar{k})} (2 + 2\beta_P) = \Sigma_{P \text{ in } D_0(\bar{k})} (2\alpha_P + 2\delta_{\alpha_P,\mathrm{odd}})$$
$$= 2\deg(D_0) + 2\#\{P \text{ in } D_0(\bar{k}) \text{ with } \alpha_P \text{ odd}\}.$$

QED

Corollary 11.3.23 1) We always have $N \geq 10g + 7$.
2) We have $N \geq 9$ if any of the following three conditions is satisfied:
a) $g \geq 1$,
b) $\deg(D_0) + \#\{P \text{ in } D_0(\bar{k}) \text{ with } \alpha_P \text{ odd}\} \geq 5$,
c) $\deg(D_1) \geq 5$.

proof 1) Using the fact that $\deg(D_1) \geq 2g + 3$, we get

$$N \geq 4g - 4 + \deg(D_1) + 2\deg(D_0) + 2\#\{P \text{ in } D_0(\bar{k}) \text{ with } \alpha_P \text{ odd}\}$$
$$\geq 4g - 4 + (2g + 3) + 2\deg(D_0) + 2\#\{P \text{ in } D_0(\bar{k}) \text{ with } \alpha_P \text{ odd}\}$$
$$= 6g - 1 + 2\deg(D_0) + 2\#\{P \text{ in } D_0(\bar{k}) \text{ with } \alpha_P \text{ odd}\}.$$

Now use the fact that $\deg(D_0) \geq 2g + 3$. If $\deg(D_0)$ is odd, then some P in D_0 has odd α_P, so in this case we have

$$2\deg(D_0) + 2\#\{P \text{ in } D_0(\bar{k}) \text{ with } \alpha_P \text{ odd}\} \geq 2(2g+3) + 2 = 4g + 8.$$

If $\deg(D_0)$ is even, then we must have $\deg(D_0) \geq 2g + 4$, and we again obtain the same inequality

$$2\deg(D_0) + 2\#\{P \text{ in } D_0(\bar{k}) \text{ with } \alpha_P \text{ odd}\} \geq 4g + 8.$$

Thus we get

$$N \geq 6g - 1 + 4g + 8 = 10g + 7.$$

2), case 2a) Obvious from 1).

2), case 2b) In this case, we have

$$N \geq 4g - 4 + \deg(D_1) + 10$$
$$\geq 4g - 4 + (2g + 3) + 10$$
$$\geq 6g + 9.$$

2), case 2c) In this case we have

$$N \geq 4g + 1 + 2\deg(D_0) + 2\#\{P \text{ in } D_0(\bar{k}) \text{ with } \alpha_P \text{ odd}\}.$$

If $\deg(D_0) = 3$, at least one P in D_0 has α_P odd, so we have

$$N \geq 4g + 1 + 6 + 2 \geq 4g + 9.$$

If $\deg(D_0) \geq 4$, then $2\deg(D_0) \geq 8$, and we get the same inequality

$$N \geq 4g + 9. \text{ QED}$$

(11.3.24) Our situation now is this. On the space $j^{-1}\mathrm{Twist}(D_0, D_1)$, we have the lisse sheaf $\mathcal{M}$, of rank

$$N = 4g - 4 + \deg(D_1) + \Sigma_{P \text{ in } (D_0 \cup D_1)(\bar{k})} (2 + \mathrm{Max}(2\beta_P, c_P)),$$

which is pure of weight zero, orthogonally self dual, and which captures the L-function; for any finite extension k_d/k, and any point (f, g) in $j^{-1}\mathrm{Twist}(D_0, D_1)(k_d)$, we have

$$\mathcal{M}_{(f,g)} = H_c^1(V\otimes_k\bar{k}, R^1\pi_{f,g!}\bar{\mathbb{Q}}_\ell(1)) \cong H_c^1(C\otimes_k\bar{k}, j_\star R^1\pi_{f,g!}\bar{\mathbb{Q}}_\ell(1)),$$

and the unitarized L-function of $E_{f,g}/k_d(C)$ is given by

$$L(E_{f,g}/k_d(C), T) = \det(1 - T\mathrm{Frob}_{k_d} \mid H^1(C\otimes_k\bar{k}, j_\star R^1\pi_{f,g!}\bar{\mathbb{Q}}_\ell(1))).$$

Theorem 11.3.25 Let k be a finite field of characteristic $p = 2$, and ℓ an odd prime. Fix a projective, smooth, geometrically connected curve C/k, of genus denoted g. Fix on C two effective divisors D_0 and D_1, both of degree $\geq 2g + 3$. Then the group G_{geom} for the lisse sheaf $\mathcal{M}$ on the dense open set $j^{-1}\mathrm{Twist}(D_0, D_1)$ is $O(N)$, for $N \geq 10g + 7$ the rank of $\mathrm{Gr}^0(\mathcal{M})$.

proof We know by Corollary 1.20.3 that G_{geom} is a semisimple subgroup of $O(N)$ with fourth moment $M_4 \leq 3$. For $N \geq 2$, $O(N)$ in its standard representation has $M_4 = 3$, cf. [Ka-LAMM, proof of 1.1.6 2)]. Since $N \geq 10g + 7$, G_{geom} has $M_4 = 3$. By Larsen's Alternative Theorem 2.2.2, G_{geom} is either $SO(N)$, or $O(N)$, or finite.

We will first give a diophantine proof that G_{geom} is not finite. We will then give a middle convolution proof to show that G_{geom}

contains a reflection. Then the only remaining possibility for G_{geom} is $O(N)$, and we are done.

To prove that G_{geom} is not finite, we argue as follows. Over the entire space $L(D_0)\times L(D_1)$, we have the perverse sheaf M constructed in 11.3.13. The sheaf of perverse origin on $\mathcal{F} := L(D_0)\times L(D_1)$,

$$\mathcal{M} := \mathcal{H}^{-\dim(\mathcal{F})}(M)(-1),$$

extends the lisse sheaf $\mathcal{M}$ on $j^{-1}\mathrm{Twist}(D_0, D_1)$. Its stalk at any point (f, g) in $(L(D_0)\times L(D_1))(\bar{k})$ is

$$\mathcal{M}_{f,g} := H_c{}^1(V\otimes_k\bar{k}, \mathcal{L}_{\psi(f)}\otimes g^*R^1\pi_{0,t!}\overline{\mathbb{Q}}_\ell)(1).$$

By the Scalarity Corollary 2.8.13, it suffices to exhibit a single finite extension k_d/k, and a single point (f, g) in $(L(D_0)\times L(D_1))(k_d)$ such that no power of Frob_{k_d} acts as a scalar on

$$H_c{}^1(V\otimes_k\bar{k}, \mathcal{L}_{\psi(f)}\otimes g^*R^1\pi_{0,t!}\overline{\mathbb{Q}}_\ell)(1).$$

For this, take for g the constant function 1. Then $g^*R^1\pi_{0,t!}\overline{\mathbb{Q}}_\ell$ is the constant sheaf on V with value

$$g^*R^1\pi_{0,t!}\overline{\mathbb{Q}}_\ell = R^1\pi_{0,1!}\overline{\mathbb{Q}}_\ell = H^1(E_{0,1}\otimes_k\bar{k}, \overline{\mathbb{Q}}_\ell),$$

for $E_{0,1}$ the elliptic curve over k whose affine equation, $\mathcal{E}_{0,1}$, is

$$y^2 + xy = x^3 + 1.$$

So for $g = 1$, we have

$$\mathcal{M}_{f,1} := H_c{}^1(V\otimes_k\bar{k}, \mathcal{L}_{\psi(f)})\otimes_{\overline{\mathbb{Q}}_\ell}H^1(E_{0,1}\otimes_k\bar{k}, \overline{\mathbb{Q}}_\ell)(1).$$

The curve $E_{0,1}$ is ordinary, so no power of Frobenius on $H^1(E_{0,1}\otimes_k\bar{k}, \overline{\mathbb{Q}}_\ell)(1)$ is scalar. Suppose we can choose an f in some $L(D_0)(k_d)$ such that $H_c{}^1(V\otimes_k\bar{k}, \mathcal{L}_{\psi(f)})$ is nonzero. Then no power of Frobenius on $\mathcal{M}_{f,1}$ is scalar. [Indeed, over any field E, given any integers $n \ge 1$ and $m \ge 1$, if A in $GL(n, E)$ and B in $GL(m, E)$ have $A\otimes B$ scalar, then both A and B are scalar, cf. [Ka-TLFM, 1.1.1].]

By Corollary 11.3.17 (bis), parts 1-3), we can choose a function f in $L(D_0)(\bar{k})$ with divisor of poles D_0 such that

$$\mathrm{Swan}_P(\mathcal{L}_{\psi(f)}) = \beta_P$$

at every point P in $D_0(\bar{k})$. We claim $H_c{}^1(V\otimes_k\bar{k}, \mathcal{L}_{\psi(f)})$ is nonzero. The sheaf $\mathcal{L}_{\psi(f)}$ is lisse of rank one on V, so $H_c{}^0(V\otimes_k\bar{k}, \mathcal{L}_{\psi(f)}) = 0$. As $\mathcal{L}_{\psi(f)}$ is wildly, and hence nontrivially, ramified at all points of D_0, we have $H_c{}^2(V\otimes_k\bar{k}, \mathcal{L}_{\psi(f)}) = 0$. So it suffices to show that $\chi_c(V\otimes_k\bar{k}, \mathcal{L}_{\psi(f)})$ is nonzero. By the Euler Poincaré formula, we have

$$\begin{aligned}-\chi_c(V\otimes_k\bar{k}, \mathcal{L}_{\psi(f)}) &= -\chi_c(V\otimes_k\bar{k}, \overline{\mathbb{Q}}_\ell) + \Sigma_{P \text{ in } D_0(\bar{k})} \mathrm{Swan}_P(\mathcal{L}_{\psi(f)})\\ &= 2g - 2 + \#(D_0\cup D_1)(\bar{k}) + \Sigma_{P \text{ in } D_0(\bar{k})} \mathrm{Swan}_P(\mathcal{L}_{\psi(f)})\\ &\ge -2 + \Sigma_{P \text{ in } D_0(\bar{k})} (1 + \mathrm{Swan}_P(\mathcal{L}_{\psi(f)}))\end{aligned}$$

$= -2 + \deg(D_0) + \#\{P \text{ in } D_0(\bar{k}) \text{ with } \alpha_P \text{ odd}\}$

$\geq -2 + \deg(D_0) \geq 1.$

This concludes the proof that G_{geom} is not finite.

(11.4) End of the proof of 11.3.25: Proof that G_{geom} contains a reflection

(11.4.1) The question is geometric, so we may at will replace k by a finite extension of itself. Having done so, we may choose a k-valued point (f, -g) in $L(D_0)\times L(D_1)$ such that all three of the following genericity conditions are satisfied, cf. Corollary 11.3.19 for the first condition, Corollary 6.2.15 for the second condition, and [Ka-TLFM, 2.4.2] for the third condition.

1) (f, -g) lies in the dense open set $j^{-1}\mathrm{Twist}(D_0, D_1)$, and

$$\mathrm{Swan}_P(\mathcal{L}_{\psi(f)}) = \beta_P$$

at every point P in $D_0(\bar{k})$.

2) View g as a finite flat map of degree $d_1 := \deg(D_1)$ from $C - D_1$ to $\mathbb{A}^1$. Then the restriction to $\mathbb{A}^1 - \mathrm{CritValues}(g)$ of $g_*\overline{\mathbb{Q}}_\ell$ is a lisse sheaf, whose geometric monodromy group G_{geom} is the full symmetric group S_{d_1}, in its standard d_1-dimensional permutation representation. Local monodromy at each finite singularity (i.e., at each critical value of g in $\mathbb{A}^1$) is a reflection, the action of a transposition in S_{d_1}. Moreover, the sheaf $g_*\overline{\mathbb{Q}}_\ell$ has at least one finite singularity.

3) The map $g : C - D_1 \to \mathbb{A}^1$ separates the points of $(D_0 - D_0\cap D_1)(\bar{k})$, and for each point P in $(D_0 - D_0\cap D_1)(\bar{k})$, the fibre $g^{-1}(P)$ consists of d_1 distinct points $\bar{k}$-valued points. [This third condition is vacuous if $(D_0 - D_0\cap D_1)(\bar{k})$ is empty, i.e., if $D_0(\bar{k}) \subset D_1(\bar{k})$.]

(11.4.2) The idea is to freeze f and g, and to consider the restriction of $\mathcal{M}$ to the one-parameter family

$$\lambda \mapsto (f, \lambda - g).$$

(11.4.3) Let us denote by $\mathcal{G}$ the sheaf on $\mathbb{A}^1$ which is this restriction:

$$\mathcal{G} := [\lambda \mapsto (f, \lambda - g)]^*\mathcal{M}.$$

Then $\mathcal{G}$ is a sheaf of perverse origin on $\mathbb{A}^1$, being the pullback of the sheaf of perverse origin $\mathcal{M}$ on $L(D_0)\times L(D_1)$, cf. [Ka-SMD, Proposition 7]. This means precisely that $\mathcal{G}$ has no nonzero punctual sections on $\mathbb{A}^1\otimes_k\bar{k}$, and hence that $\mathcal{G}[1]$ is a perverse sheaf on $\mathbb{A}^1$.

(11.4.4) Whatever the value of λ in $\bar{k}$, the point $(f, \lambda - g)$ lies in $j^{-1}\mathrm{Twist}(D_0, D_1)_{prelim}(\bar{k})$, and hence [Lemma 11.3.12, part 1)] we have

$$H_c^i(V\otimes_k\bar{k}, \mathcal{L}_{\psi(f)}\otimes(\lambda - g)^*R^1\pi_{0,t!}\bar{\mathbb{Q}}_\ell)(1) = 0 \text{ for } i \neq 1.$$

(11.4.5) We denote by

$$U \subset \mathbb{A}^1$$

the dense open set

$$U := [\lambda \mapsto (f, \lambda - g)]^{-1}(j^{-1}\mathrm{Twist}(D_0, D_1)).$$

Thus $\mathcal{G}$ is lisse on U. To show that G_{geom} contains a reflection, it suffices to show that some local monodromy of $\mathcal{G}|U$ is a reflection. Thus, to conclude the proof of Theorem 11.3.25, it suffices to prove the following theorem.

Theorem 11.4.6 Notations and hypotheses as above, $\mathbb{A}^1 - U$ is nonempty, and the local monodromy of $\mathcal{G}|U$ at each point of $(\mathbb{A}^1 - U)(\bar{k})$ is a reflection.

proof For technical reasons, it is important to work on the possibly slightly larger affine curve

$$V_1 := C - D_1.$$

Each λ-g is holomorphic on $V_1\otimes_k\bar{k}$, and makes $V_1\otimes_k\bar{k}$ a finite flat covering of $\mathbb{A}^1\otimes_k\bar{k}$ of degree $\deg(D_1)$.

Let us denote by

$$j_1 : V := C - (D_0\cup D_1) \to V_1$$

the inclusion. At any point of $V_1 - V$, i.e., at any point of D_0 which does not lie in D_1, $\lambda - g$ is holomorphic, and hence $(\lambda - g)^*R^1\pi_{0,t!}\bar{\mathbb{Q}}_\ell$ is tame, but $\mathcal{L}_{\psi(f)}$ is wild. So $\mathcal{L}_{\psi(f)}\otimes(\lambda - g)^*R^1\pi_{0,t!}\bar{\mathbb{Q}}_\ell$ is totally wild at each point of $V_1 - V$, and hence we have

$$H_c^i(V\otimes_k\bar{k}, \mathcal{L}_{\psi(f)}\otimes(\lambda - g)^*R^1\pi_{0,t!}\bar{\mathbb{Q}}_\ell)(1)$$
$$\cong H_c^i(V_1\otimes_k\bar{k}, (j_{1*}\mathcal{L}_{\psi(f)})\otimes(\lambda - g)^*R^1\pi_{0,t!}\bar{\mathbb{Q}}_\ell)(1).$$

We now use the projection formula for $\lambda - g$ to rewrite

$$H_c^i(V_1\otimes_k\bar{k}, (j_{1*}\mathcal{L}_{\psi(f)})\otimes(\lambda - g)^*R^1\pi_{0,t!}\bar{\mathbb{Q}}_\ell)(1)$$
$$\cong H_c^i(\mathbb{A}^1\otimes_k\bar{k}, ((\lambda\text{-}g)_*(j_{1*}\mathcal{L}_{\psi(f)}))\otimes R^1\pi_{0,t!}\bar{\mathbb{Q}}_\ell)(1).$$

In terms of the sheaf $\mathcal{G}$ on $\mathbb{A}^1$, we have

$$\mathcal{G}_\lambda := \mathcal{M}_{f,\lambda-g} = H_c^1(\mathbb{A}^1\otimes_k\bar{k}, ((\lambda\text{-}g)_*(j_{1*}\mathcal{L}_{\psi(f)}))\otimes R^1\pi_{0,t!}\bar{\mathbb{Q}}_\ell)(1),$$

and the other H_c^i vanish.

Our next task is to relate the sheaf $\mathcal{G}$, or more precisely the perverse sheaf $\mathcal{G}[1]$, to an additive convolution. This will require a certain amount of preparation.

On $\mathbb{A}^1$, the sheaves

$$\mathcal{K} := g_*(j_{1*}\mathcal{L}_{\psi(f)})$$

and

$$\mathcal{L} := R^1\pi_{0,t!}\bar{\mathbb{Q}}_\ell$$

are both middle extensions. On any dense open set on which it is lisse, $\mathcal{K}$ has finite arithmetic monodromy (being the finite direct image of a character of finite arithmetic order, $\mathcal{L}_{\psi(f)}$) and is hence pure of weight 0. In particular, $\mathcal{K}$ is the middle extension of a sheaf with finite geometric monodromy, so is geometrically the direct sum of geometrically irreducible middle extensions, say

$$\mathcal{K}^{geom} \cong \oplus_i \mathcal{K}_i.$$

Of these $\mathcal{K}_i$, some have $\mathcal{P}$, and some may not. But if we lump together the $\mathcal{K}_i$ acccording as to whether or not they have $\mathcal{P}$, we get an arithmetic direct sum decomposition,

$$\mathcal{K} = \mathcal{K}_{\mathcal{P}} \oplus \mathcal{K}_{not\mathcal{P}},$$

with

$$\mathcal{K}_{\mathcal{P}}{}^{geom} := \oplus_{i \text{ such that } \mathcal{K}_i \text{ has } \mathcal{P}} \mathcal{K}_i,$$

$$\mathcal{K}_{not\mathcal{P}}{}^{geom} := \oplus_{i \text{ such that } \mathcal{K}_i \text{ does not have } \mathcal{P}} \mathcal{K}_i.$$

The sheaf $\mathcal{L}$ is lisse of rank 2 and pure of weight 1 on $\mathbb{G}_m$, its local monodromy at 0 is a unipotent pseudoreflection, and its two $I(\infty)$-slopes are both 1/2.

The local monodromy of $\mathcal{K}$ at finite distance will be of essential importance later.

Lemma 11.4.7 The middle extension sheaf $\mathcal{K} := g_*(j_{1*}\mathcal{L}_{\psi(f)})$ on $\mathbb{A}^1$ has finite singularities. At each finite singularity of $\mathcal{K}$, its local monodromy is a reflection.

proof Consider the map $g : V_1 := C - D_1 \to \mathbb{A}^1$. The local monodromy of $\mathcal{K} := g_*(j_{1*}\mathcal{L}_{\psi(f)})$ at a point s in $\mathbb{A}^1(\bar{k})$, i.e., the attached $I(s)$-representation, is the direct sum of the inductions of the local mondromies of $j_{1*}\mathcal{L}_{\psi(f)})$ at the points in the fibre:

$$g_*(j_{1*}\mathcal{L}_{\psi(f)}) \text{ as } I(s)\text{-representation}$$
$$\cong \oplus_{\text{points t in } g^{-1}(s)} \mathrm{Ind}_{I(t)}{}^{I(s)}(\mathcal{L}_{\psi(f)} \text{ as } I(t)\text{-rep'n.})$$

At a point s in $\mathbb{A}^1(\bar{k})$ whose fibre contains a point P of $(D_0 - D_0 \cap D_1)(\bar{k})$, it contains exactly one such point, and g is finite etale over s. The sheaf $j_{1*}\mathcal{L}_{\psi(f)}$ on $C - D_1$ has local monodromy a (wild) reflection at P, and is lisse of rank one at any other point in the fibre $g^{-1}(s)$. So the local monodromy of $\mathcal{K} := g_*(j_{1*}\mathcal{L}_{\psi(f)})$ at s is a wild reflection. Over a point s in $\mathbb{A}^1(\bar{k})$ whose fibre does not meet $(D_0 - D_0 \cap D_1)(\bar{k})$, the sheaf $j_{1*}\mathcal{L}_{\psi(f)}$ is lisse of rank one at every point in the fibre, and so the local monodromy of $\mathcal{K} := g_*(j_{1*}\mathcal{L}_{\psi(f)})$ is the same as that of $g_*\bar{\mathbb{Q}}_\ell$. This monodromy is trivial, if g is finite etale over s, or it is a reflection, if s is one of the nonzero number of critical values of g. QED for Lemma 11.4.7

We define perverse sheaves K, $K_{\mathcal{P}}$, $K_{not\mathcal{P}}$, and L on $\mathbb{A}^1$ by

$$K := \mathcal{K}(1/2)[1],$$
$$K_{\mathcal{P}} := \mathcal{K}_{\mathcal{P}}(1/2)[1],$$
$$K_{\mathrm{not}\mathcal{P}} := \mathcal{K}_{\mathrm{not}\mathcal{P}}(1/2)[1],$$
$$L := \mathcal{L}(1)[1].$$

The perverse sheaf L is geometrically irreducible, lisse of rank 2 on $\mathbb{G}_m$, and its local monodromy at 0 is a unipotent pseudoreflection. At ∞, both its $I(\infty)$-slopes are 1/2. We remark that L has $\mathcal{P}$, being a geometrically irreducible middle extension of generic rank 2.

Becuause L has $\mathcal{P}$, we can form both the ! additive convolution $K*_!L$ of K and L on $\mathbb{A}^1$, and their middle convolution $K*_{\mathrm{mid}}L$ on $\mathbb{A}^1$, cf. 6.1.1-2.

Lemma 11.4.8 1) The perverse sheaf $K*_!L$ on $\mathbb{A}^1$ is equal to $\mathcal{H}^{-1}(K*_!L)[1]$.

2) On U, the sheaves $\mathcal{H}^{-1}(K*_!L)$ and $\mathcal{G}(1/2)$ are both lisse, pure of weight -1, and have the same trace function.

3) On U, the perverse sheaves $K*_!L$ and $\mathcal{G}(1/2)[1]$ are geometrically isomorphic, i.e., the lisse sheave $\mathcal{H}^{-1}(K*_!L)|U$ and $\mathcal{G}(1/2)|U$ are geometrically isomorphic.

proof 1) We must show that $\mathcal{H}^{-i}(K*_!L)$ vanishes for $i \neq 1$. This can be checked fibre by fibre. At a point λ in $\bar{k}$, the stalk of $\mathcal{H}^{-i}(K*_!L)$ is by definition the cohomology group

$$H_c^{2-i}(\mathbb{A}^1\otimes_k\bar{k}, ((t \mapsto \lambda - t)^*(g_*(j_{1*}\mathcal{L}_{\psi(f)})))\otimes R^1\pi_{0,t!}\bar{\mathbb{Q}}_\ell)(3/2).$$

But for any sheaf $\mathcal{N}$ on V_1, here $j_{1*}\mathcal{L}_{\psi(f)}$, we have

$$(t \mapsto \lambda - t)^*g_*\mathcal{N} \cong (\lambda - g)_*\mathcal{N}$$

on $\mathbb{A}^1$. This is just proper base change for the proper map g, and the base change $t \mapsto \lambda - t$, via the Cartesian diagram

$$\begin{array}{ccc} & \mathrm{id} & \\ V & \rightarrow & V \\ \lambda - g \downarrow & & \downarrow g \\ \mathbb{A}^1 & \rightarrow & \mathbb{A}^1 \\ & t \mapsto \lambda - t. & \end{array}$$

So we can rewrite this stalk as

$$\cong H_c^{2-i}(\mathbb{A}^1\otimes_k\bar{k}, ((\lambda-g)_*(j_{1*}\mathcal{L}_{\psi(f)}))\otimes R^1\pi_{0,t!}\bar{\mathbb{Q}}_\ell)(3/2).$$

We have already noted above (second paragraph of the proof of 11.4.6) that these groups vanish for $i \neq 1$.

2) The above calculation of stalks shows that sheaves $\mathcal{H}^{-1}(K*_!L)$ and $\mathcal{G}(1/2)$ have, fibre by fibre, isomorphic stalks. On the open set U, these stalks have constant rank, so both $\mathcal{H}^{-1}(K*_!L)$ and $\mathcal{G}(1/2)$, being sheaves of perverse origin, are lisse on U. The calculation of stalks

shows that these lisse sheaves on U are both punctually pure, hence pure, of weight -1.

3) As $\mathcal{H}^{-1}(K*_!L)|U$ and $\mathcal{G}(1/2)|U$ have isomorphic stalks, they have the same trace function. By Chebotarev, they have isomorphic semisimplifications as representations of $\pi_1(U)$, and hence a fortiori they have isomorphic semisimplifications as representations of $\pi_1^{geom}(U)$. As these lisse sheaves are both pure of weight -1, they are already semisimple as representations of $\pi_1^{geom}(U)$. Hence they are isomorphic as representations of $\pi_1^{geom}(U)$., i.e, they are geometrically isomorphic on U. QED

Corollary 11.4.9 On U, the perverse sheaf $\mathcal{G}(1/2)[1]$ is geometrically isomorphic to the middle convolution $K*_{mid}L$.

proof By Theorem 6.5.4, with the roles of K and L interchanged, we have an isomorphism of perverse sheaves on $\mathbb{A}^1$

$$Gr^0(K*_!L) \cong K*_{mid}L.$$

On U, $K*_!L$ is itself (lisse and) pure of weight zero, so we get

$$(K*_!L)|U \cong (K*_{mid}L)|U.$$

Combine this with the geometric isomorphism

$$(\mathcal{G}(1/2)[1](|U \cong (K*_!L)|U. \qquad \text{QED}$$

Lemma 11.4.10 The perverse sheaf $K*_{mid}L$ on $\mathbb{A}^1$ is geometrically semisimple and non-punctual, it has finite singularities, and all of its local monodromies at finite singularities are reflections.

proof We have a direct sum decomposition of perverse sheaves

$$K = K_{\mathcal{P}} \oplus K_{not\mathcal{P}},$$

and hence a corresponding decomposition

$$K*_{mid}L = K_{\mathcal{P}}*_{mid}L \oplus K_{not\mathcal{P}}*_{mid}L.$$

We first claim that $K_{not\mathcal{P}}*_{mid}L$ is lisse on $\mathbb{A}^1$. Indeed, the sheaf $K_{not\mathcal{P}}$ is geometrically the direct sum of perverse sheaves of the form $\mathcal{L}_{\psi(\alpha x)}[1]$, with α in $\overline{k}$. For any perverse L which has $\mathcal{P}$, the perverse sheaf on $\mathbb{A}^1\otimes_k\overline{k}$

$$(\mathcal{L}_{\psi(\alpha x)}[1])*_{mid}L$$

is of the form

$$\mathcal{L}_{\psi(\alpha x)}\otimes(\text{a constant sheaf})[1],$$

cf. the proof of Theorem 6.5.4, so in particular is lisse on $\mathbb{A}^1\otimes_k\overline{k}$. Thus $K_{not\mathcal{P}}*_{mid}L$ is of the form

$$(\text{a lisse sheaf on } \mathbb{A}^1)[1].$$

So we are reduced to showing that $K_{\mathcal{P}}*_{mid}L$ is geometrically semisimple and non-punctual, that it has finite singularities, and that all of its local monodromies at finite singularities are reflections.

We have already proven in Lemma 11.4.7 above that the middle extension sheaf $\mathcal{K}$ has finite singularities, and that its local monodromy at each is a reflection. On the other hand, $\mathcal{K}_{\mathrm{not}\,\mathcal{P}}$ is itself lisse on $\mathbb{A}^1$, being geometrically the direct sum of sheaves of the form $\mathcal{L}_{\psi(\alpha x)}$, with α in $\bar{k}$. So we infer from the direct sum decomposition

$$\mathcal{K} = \mathcal{K}_{\mathcal{P}} \oplus \mathcal{K}_{\mathrm{not}\,\mathcal{P}}$$

that $\mathcal{K}_{\mathcal{P}}$ is a middle extension sheaf which has finite singularities, and that its local monodromy at each is a reflection.

We now apply Theorem 6.1.16. Because L has local monodromy at 0 which is a unipotent pseudoreflection, while all the local monodromies of $\mathcal{K}_{\mathcal{P}}$ at finite singularities are reflections, we find that the perverse sheaf $K_{\mathcal{P}}*_{\mathrm{mid}}L$ is geometrically semisimple and non-punctual.

To compute the local monodromies of $K_{\mathcal{P}}*_{\mathrm{mid}}L$ at its finite singularities, we first note that L has only one finite singularity, at 0, so by Theorem 6.1.18 the finite singularities of $K_{\mathcal{P}}*_{\mathrm{mid}}L$ are located at the finite singularities of $K_{\mathcal{P}}$. Since $K_{\mathcal{P}}$ has finite singularities, $K_{\mathcal{P}}*_{\mathrm{mid}}L$ has finite singularities. Because the singularity of L at 0 is a unipotent pseudoreflection, while the local monodromy of $K_{\mathcal{P}}$ at each of its finite singularities is a reflection, it now follows from the Unipotent Pseudoreflection Input Corollary bis 6.1.20 that the local monodromy of $K_{\mathcal{P}}*_{\mathrm{mid}}L$ at each finite singularity is a reflection. QED

(11.4.11) We now obtain the desired Theorem 11.4.6.

Corollary 11.4.12 (= Theorem 11.4.6) Denote by $j : U \subset \mathbb{A}^1$ the inclusion. Then

1) we have a geometric isomorphism of perverse sheaves on $\mathbb{A}^1$

$$(j_*(\mathcal{G}|U))(1/2)[1] \cong K*_{\mathrm{mid}}L,$$

2) $\mathbb{A}^1 - U$ is nonempty,

3) the local monodromy of $\mathcal{G}|U$ at each point of $(\mathbb{A}^1 - U)(\bar{k})$ is a reflection.

proof 1) Since $K*_{\mathrm{mid}}L$ is geometrically semisimple and non-punctual, it is the middle extension of its restriction to any dense open set, here U, where it is geometrically isomorphic to $(\mathcal{G}|U)(1/2)[1]$ by Corollary 11.4.9.

2) We cannot have $U = \mathbb{A}^1$, otherwise $K*_{\mathrm{mid}}L$ would have no finite singularities.

3) This is simply a restatement of the fact that the local monodromy of $K*_{\mathrm{mid}}L$ at each finite singularity is a reflection. QED

This concludes the proof of Theorem 11.3.25. QED

Chapter 12: Uniformity results

(12.1) Fibrewise perversity: basic properties

(12.1.1) In earlier chapters, our main actors were $\overline{\mathbb{Q}}_\ell$-perverse sheaves on separated schemes of finite type over a field k of posititive characteristic $p \neq \ell$. Most of the time, k was a finite field. What happens if instead of working over a finite field k, we work over a base scheme S, which we assume to be separated and of finite type over $\mathrm{Spec}(\mathbb{Z}[1/\ell])$?

(12.1.2) Let $S/\mathbb{Z}[1/\ell]$ be a separated $\mathbb{Z}[1/\ell]$-scheme of finite type, and let $f : X \to S$, or simply X/S if no confusion is possible, be a separated S-scheme of finite type. Suppose given an object M in $D^b{}_c(X, \overline{\mathbb{Q}}_\ell)$. For a field E, and an E-valued point s in S(E), we denote by $X_{s,E}/E$ the E-scheme which is the fibre of X/S over the E-valued point s in S(E), by $i_{s,E} : X_{s,E} \to X$ the "inclusion" of this fibre, and by $M \mid X_{s,E}$ the restriction $(i_{s,E})^*M$ of M to this fibre.

(12.1.3) We say that M is fibrewise perverse on X/S if for every field E, and for every E-valued point s in S(E), $M \mid X_{s,E}$ is perverse on $X_{s,E}$. Since perversity is invariant under field extension, it is equivalent, in this definition, to require "only" that the restriction of M to every geometric fibre of X/S (i.e., to every $X_{s,E}$ whose field E is algebraically closed) be perverse. Or it is equivalent to let s run over the points of the scheme S, and to test on the fibre $X_{s,\kappa(s)}$, where we view s as a $\kappa(s)$-valued point of S.

(12.1.4) Instead of looking at all fields, or at all algebraically closed fields, we can take any property **P** of isomorphism classes of fields, and look only at the restriction of M to all fibres $X_{s,E}$, E any field with property **P**, s any E-valued point of S. We say that M is **P**-fibrewise perverse on X/S if If $M \mid X_{s,E}$ is perverse, whenever E is any field with property **P**, and s is any E-valued point of S.

(12.1.5) Taking for **P** the property of being a finite field, we get the notion of finite fibrewise perversity of M on X/S. Taking for **P** the property of having positive characteristic, we get the notion of positive characteristic fibrewise perversity of M on X/S.

(12.1.6) Happily, for schemes of finite type over $\mathbb{Z}[1/\ell]$, all these notions coincide.

Lemma 12.1.7 Let $S/\mathbb{Z}[1/\ell]$ be a separated $\mathbb{Z}[1/\ell]$-scheme of finite type, and $f : X \to S$ a separated S-scheme of finite type. The following conditions on an object M in $D^b{}_c(X, \overline{\mathbb{Q}}_\ell)$ are equivalent.

1) M is fibrewise perverse on X/S.
2) M is positive characteristic fibrewise perverse on X/S.

3) M is finite fibrewise perverse on X/S.

proof It is trivial that 1) ⇒ 2) ⇒ 3). To show that 3) ⇒ 1), we argue as follows. As noted in 12.1.3 above, it suffices to treat fibres $X_{s,\kappa(s)}$, for s a point of the scheme S. Replacing S by the closure s in S, we reduce to the case when S is reduced and irreducible, and s is its generic point η. Replacing S by a dense open set of itself, we may further assume that the formation of the relative Verdier dual

$$D_{X/S}(M) := \underline{\mathrm{RHom}}(M, f^!\overline{\mathbb{Q}}_{\ell,S}),$$

for $f : X \to S$ the structural morphism, commutes with arbitrary base change on S [Ka-Lau, 1.1.7]. We then pick a nice stratification of X which is adapted to both M and to.$D_{X/S}(M)$ Concretely, we pick a partition of X^{red} into a finite disjoint union of locally closed subschemes Z_i, each of which is a connected, regular scheme, such that for each cohomology sheaf $\mathcal{H}^j(M)$, each cohomology sheaf $\mathcal{H}^j(D_{X/S}(M))$, and for each i, the restrictions $\mathcal{H}^j(M) \mid Z_i$ and $\mathcal{H}^j(D_{X/S}(M)) \mid Z_i$ are lisse on Z_i. Further shrinking on S, we may assume that S is regular and connected. For each i, denote by d(i) the dimension of $(Z_i)_\eta$, the generic fibre of Z_i/S, and denote by

$$\pi_i : Z_i \to S$$

the structural morphism. Further shrinking on S, we may assume that all the sheaves $R^j\pi_{i!}\mathbb{F}_\ell$ on S are lisse. Then we see from looking at the stalks at η that $R^j\pi_{i!}\mathbb{F}_\ell = 0$ for $j > 2d(i)$, and $R^{2d(i)}\pi_{i!}\mathbb{F}_\ell$ is a nonzero lisse sheaf on S. Therefore every fibre of Z_i/S has dimension d(i).

Fix a closed point s of S. Because S is of finite type over $\mathbb{Z}[1/\ell]$, $\kappa(s)$ is a finite field. Now $(Z_i)_{s,\kappa(s)}$ has dimension d(i). The perversity of $M \mid X_{s,\kappa(s)}$ means that for each integer j, we have the two inequalities

$$\mathrm{dimSupp}(\mathcal{H}^j(M) \mid X_{s,\kappa(s)}) \le -j,$$

$$\mathrm{dimSupp}(\mathcal{H}^j(D_{X/S}(M)) \mid X_{s,\kappa(s)}) \le -j.$$

Thus we see that for each pair (an index i, an integer j) such that

$$d(i) > -j,$$

we have

$$\mathcal{H}^j(M) \mid (Z_i)_{s,\kappa(s)} \text{ is somewhere zero on } (Z_i)_{s,\kappa(s)},$$

and

$$\mathcal{H}^j(D_{X/S}(M)) \mid (Z_i)_{s,\kappa(s)} \text{ is somewhere zero on } (Z_i)_{s,\kappa(s)}.$$

But both $\mathcal{H}^j(M) \mid Z_i$ and $\mathcal{H}^j(D_{X/S}(M)) \mid Z_i$ are lisse on the regular connected scheme Z_i, so if they vanish anywhere they are identically zero. Therefore we find that for each pair (an index i, an integer j) such that

$$d(i) > -j,$$

we have

$$\mathcal{H}^j(M) \mid Z_i = 0, \text{ and } \mathcal{H}^j(D_{X/S}(M)) \mid Z_i = 0.$$

Now restrict to the generic fibre $(Z_i)_\eta$ of Z_i/S. This generic fibre is regular and connected, being a limit of open sets of Z_i, and its dimension is $d(i)$. Therefore the vanishings

$$\mathcal{H}^j(M) \mid (Z_i)_\eta = 0, \text{ and } \mathcal{H}^j(D_{X/S}(M)) \mid (Z_i)_\eta = 0,$$

which hold whenever $d(i) > -j$, show that $M \mid X_\eta$ is perverse, as required. QED

Remark 12.1.8 A similar argument gives the following more general result, whose proof is left to the reader.

Lemma 12.1.9 Let S be a noetherian separated $\mathbb{Z}[1/\ell]$-scheme which is excellent, $f : X \to S$ a separated morphism of finite type, and M an object in $D^b{}_c(X, \overline{\mathbb{Q}}_\ell)$. Then the set of points s in S such that $M \mid X_s$ is perverse on X_s is constructible.

(12.2) Uniformity results for monodromy; the basic setting

(12.2.1) Suppose now that S is a separated scheme of finite type over $\mathbb{Z}[1/\ell]$, and that $f : X \to S$ is a separated morphism of finite type which is smooth and surjective, with geometrically connected fibres, all of some common dimension $n \geq 1$. We suppose given
1) an open set $U \subset X$ which meets every geometric fibre of X/S,
2) a field embedding $\iota: \overline{\mathbb{Q}}_\ell \subset \mathbb{C}$,
3) an object M in $D^b{}_c(X, \overline{\mathbb{Q}}_\ell)$,
4) an integer w,
5) an integer $N \geq 1$,
6) a Zariski closed $\overline{\mathbb{Q}}_\ell$-algebraic subgroup $G \subset GL(N)$.

(12.2.2) We make the following hypotheses G1), G2), G3), M1), M2), M3), MG1), and MG2) about this plethora of data.

(12.2.3) We begin with three hypotheses on G.

G1) The (not necessarily connected) algebraic group G is semisimple, in the sense that its identity component is semisimple.

G2) G is an irreducible subgroup of GL(N), i.e., the given N-dimensional $\overline{\mathbb{Q}}_\ell$-representation of G is irreducible.

G3) The normalizer of G in GL(N) is $\mathbb{G}_m \cdot G$.

(12.2.4) We next give three hypotheses on M.

M1) M is fibrewise perverse on X/S, and ι-mixed of weight $\leq$ w.

(12.2.5) For each finite field E and for each point s in S(E), we denote by $M_{s,E}$ the perverse sheaf on $X_{s,E}$ defined by

$$M_{s,E} := M \mid X_{s,E},$$

and by $U_{s,E} \subset X_{s,E}$ the dense open set $U \cap X_{s,E}$.

M2) $M_{s,E}$ has lisse cohomology sheaves on $U_{s,E}$, i.e., $M_{s,E} \mid U_{s.E}$ is of the form $\mathcal{M}_{s,E}[n]$, for a lisse sheaf $\mathcal{M}_{s,E}$ on $U_{s,E}$.

(12.2.6) Recall that M is ι-mixed of weight $\le$ w. Thus $\mathcal{M}_{s,E}$ is mixed of weight $\le$ w-n. Form the lisse sheaf $Gr^{w-n}(\mathcal{M}_{s,E})$:= the weight w-n quotient of $\mathcal{M}_{s,E}$, and then form the lisse sheaf $Gr^{w-n}(\mathcal{M}_{s,E})_{ncst}$:= the nonconstant quotient of $Gr^{w-n}(\mathcal{M}_{s,E}\otimes)$, cf. 1.10.3.

M3) $Gr^{w-n}(\mathcal{M}_{s,E})_{ncst}$ has rank N.

(12.2.7) We now give a hypothesis involving M and G.

MG1) Under the representation $\rho_{s,E} : \pi_1{}^{arith}(U_{s,E}) \to GL(N, \overline{\mathbb{Q}}_\ell)$ which "is" the lisse sheaf $Gr^{w-n}(\mathcal{M}_{s,E})_{ncst}$, the group

$G_{geom,s,E}$:= the Zariski closure of $\rho_{s,E}(\pi_1{}^{geom}(U_{s,E}))$ in GL(N) is conjugate in GL(N) to G.

(12.2.8) As explained in [Ka-Sar-RMFEM, 9.6.3], it follows from these axioms that for each finite field E and for each point s in S(E), there exists an ℓ-adic unit $\alpha_{s,E}$ in $\overline{\mathbb{Q}}_\ell{}^\times$, with

$$|\iota(\alpha_{s,E})|_{\mathbb{C}} = \mathrm{Sqrt}(\#E)^{n-w},$$

such that after replacing $Gr^{w-n}(\mathcal{M}_{s,E})_{ncst}$ by its constant field twist

$$Gr^{w-n}(\mathcal{M}_{s,E})_{ncst}\otimes(\alpha_{s,E})^{deg} = Gr^0(\mathcal{M}_{s,E}\otimes(\alpha_{s,E})^{deg}))_{ncst},$$

the following condition holds:

MG2) Under the representation $\rho_{s,E,\alpha} : \pi_1{}^{arith}(U_{s,E}) \to GL(N, \overline{\mathbb{Q}}_\ell)$ which "is" the lisse sheaf

$$Gr^0(\mathcal{M}_{s,E}\otimes(\alpha_{s,E})^{deg})_{ncst} = Gr^{w-n}(\mathcal{M}_{s,E})_{ncst}\otimes(\alpha_{s,E})^{deg},$$

we have the inclusion

$$\rho_{s,E,\alpha}(\pi_1{}^{arith}(U_{s,E})) \subset G_{geom,s,E}(\overline{\mathbb{Q}}_\ell).$$

[This makes sense, because constant field twists don't alter G_{geom}.]

(12.2.9) So we could add to the initial data 12.2.1.1-6 the

supplementary data 12.2.1.7,

7) for each finite field E and for each point s in S(E), a choice of an ℓ-adic unit $\alpha_{s,E}$ in $\overline{\mathbb{Q}}_\ell^\times$, with $|\iota(\alpha_{s,E})|_{\mathbb{C}} = \mathrm{Sqrt}(\#E)^{n-w}$,

and impose condition MG2). This approach, where we specify the constants $\alpha_{s,E}$ as part of the data, will be used in the next section, where we formulate and prove the Uniformity Theorem 12.3.3. However, in applying the Uniformity Theorem it is more convenient not to choose the constants $\alpha_{s,E}$ in advance.

Remark 12.2.10 In the applications we have in mind, the twist $\mathcal{M}_{s,E}\otimes(\alpha_{s,E})^{\deg})$ turns out to be the (possibly half-integral) Tate twist $\mathcal{M}_{s,E}((w-n)/2)$. When this is the case, and w-n is even, we might as well replace the original M by M((w-n)/2), which has the effect of making all the $\alpha_{s,E}$ become 1, and of slightly simplifying the story. But already when w-n is odd, there may be no global "half a Tate twist", cf. [Ka-Sar-RMFEM, 9.9.6], though there is always "half a Tate twist" up to a character of finite order [Ka-Sar-RMFEM, 9.9.7]. We have taken the more general $\alpha_{s,E}$ formulation, following [Ka-Sar-RMFEM, 9.9.3 and 9.6.10], with an eye to later possible applications.

(12.3) The Uniformity Theorem

(12.3.1) We continue to work in the "basic setting" of the previous section 12.2, with the constants $\alpha_{s,E}$ chosen in advance. Using ι, we form the complex Lie group $G(\mathbb{C})$, and choose in it a maximal compact subgroup K. We denote by $K^\#$ the space of conjugacy classes in K, and by $\mu^\#$ the Borel measure of total mass one on $K^\#$ which is the direct image from K of its Haar measure of total mass one μ_{Haar}. For each finite field E, for each point s in S(E), and for each point x in $U_{s,E}(E)$, we denote by $\mathrm{Frob}_{E,s,x}$ its (geometric, as always) Frobenius conjugacy class in $\pi_1^{\mathrm{arith}}(U_{s,E})$. We denote by

$$\vartheta(E, s, \alpha, x) \text{ in } K^\#$$

the conjugacy class of $\iota(\rho_{s,E,\alpha}(\mathrm{Frob}_{E,s,x}))^{ss}$.

(12.3.2) Recall [Ka-Sar-RMFEM, proof of 9.2.6, 5) and 9.3.3] that there exists a constant A(U/S) such that $U_{s,E}(E)$ is nonempty if $\#E > A(U/S)^2$. Whenever $\#E > A(U/S)^2$, we denote by

(12.3.2.1) $$\mu(E, s, \alpha) := (1/\#U_{s,E}(E))\,\Sigma_{x \text{ in } U_{s,E}(E)}\,\delta_{\vartheta(E, s, \alpha, x)}$$

the probability measure on $K^\#$ obtained by averaging over the Frobenius conjugacy classes, as x runs over $U_{s,E}(E)$.

Uniformity Theorem 12.3.3 Hypotheses and notations as above, suppose we are in the basic setting of the previous section 12.2. Then for any sequence of data (E_i, s_i, α_i) in which each $\#E_i > A(U/S)^2$ and in which $\#E_i$ is strictly increasing, the measures $\mu(E_i, s_i, \alpha_i)$ on $K^\#$ converge weak * to $\mu\#$, i.e., for any continuous $\mathbb{C}$-valued central function f on K, we have

$$\int_K f d\mu_{Haar} = \lim_{i \to \infty} \int_K f d\mu(E_i, s_i, \alpha_i)$$
$$= \lim_{i \to \infty} (1/\#U_{s_i,E_i}(E_i)) \Sigma_{x \text{ in } U_{s_i,E_i}(E_i)} f(\vartheta(E_i, s_i, \alpha_i, x)).$$

More precisely, there exist constants A and C such that for any irreducible nontrivial irreducible unitary representation Λ of K, and for any data (E, s, α) with $\#E > 4A^2$, we have the estimates

$$2(\#E)^n \geq \#U_{s,E}(E) \geq (1/2)(\#E)^n,$$
$$|\Sigma_{x \text{ in } U_{s,E}(E)} \text{Trace}(\Lambda(\vartheta(E, s, \alpha, x)))| \leq C\dim(\Lambda)(\#E)^{n-1/2},$$

and

$$|\int_K \text{Trace}(\Lambda)d\mu(E, s, \alpha)| \leq 2C\dim(\Lambda)/\text{Sqrt}(\#E).$$

proof We will prove the "more precise" form of the theorem, in the form of the next to last inequality

$$|\Sigma_{x \text{ in } U_{s,E}(E)} \text{Trace}(\Lambda(\vartheta(E, s, \alpha, x)))| \leq C\dim(\Lambda)(\#E)^{n-1/2},$$

which trivially (use $\#U_{s,E}(E) \geq (1/2)(\#E)^n$) implies the last one.

We are free to pick any stratification $\{S_i\}_i$ of S, i.e., to write S^{red} as a finite disjoint union of locally closed subschemes, each of finite type over $\mathbb{Z}[1/\ell]$, and then to prove the theorem separately over each S_i, with suitable constants A_i and C_i over S_i. Once we have done this, we simply take $A := \text{Sup}_i(A_i)$, and $C := \text{Sup}_i(C_i)$.

This allows us to reduce to the case when S is normal and connected. Suppose now that S is normal and connected. If there were a single lisse sheaf $\mathcal{G}$ on U whose restriction to each $U_{s,E}$ were a constant field twist of $\text{Gr}^0(\mathcal{M}_{s,E}\otimes(\alpha_{s,E})^{deg}))_{ncst}$, this theorem would be a special case of [Ka-Sar-RMFEM, 9.6.10], applied with its $\mathcal{F}$ = our $\mathcal{G}$, its X = our U, with the constants

$$A := A(U/S),\ C := C(U/S, \mathcal{G})$$

of [Ka-Sar-RMFEM, 9.3.3 and 9.3.4]. The problem is that a priori no such $\mathcal{G}$ need exist. To get around this problem, we argue as follows.

Since S is normal and connected, and X/S is smooth with geometrically connected fibres, X is itself normal and connected. Pick a dense open set $V \subset X$ on which M has lisse cohomology sheaves. After shrinking on S, we may assume that V maps onto S. Then on each geometric fibre of V/S, M has lisse cohomology sheaves. Being perverse on each fibre with lisse cohomology sheaves, only $\mathcal{H}^{-n}(M)$ is possibly nonzero on each fibre. Thus $\mathcal{H}^{-j}(M) \mid V = 0$ for $j \neq n$ (just look fibre by fibre), and $\mathcal{H}^{-n}(M)$ is lisse on V.

The lisse sheaf $\mathcal{M} := \mathcal{H}^{-n}(M)$ on V is ι-mixed, of weight $\leq w - n$, so it is a successive extension of constructible, punctually pure sheaves $\mathcal{M}_i$, each punctually pure of some weight $w_i \leq w - n$. Shrinking on V, we may assume that each $\mathcal{M}_i$ is itself lisse on V. Once again shrinking on S, we once again have V/S surjective.

So it suffices to treat the case when S is normal and connected, and when we have a dense open set $V \subset X$ with V/S surjective and with $M \mid V$ of the form $\mathcal{M}[n]$, with $\mathcal{M}$ a lisse sheaf on V which is a successive extension of lisse pure sheaves. Replacing V by $V \cap U$, we may assume further that $V \subset U$.

Further shrinking S, we may assume that for

$$f : V \to S$$

the structural morphism, all the sheaves

$$R^i f_\star(\text{any irreducible Jordan-Holder constituent of } \mathcal{M})$$

are lisse on S, and of formation compatible with change of base on S.

Now consider the semisimplification $\mathcal{M}^{ss}$ of $\mathcal{M}$ as a lisse sheaf on V. It is a sum of irreducible lisse sheaves, each pure of some weight $\leq w - n$. Denote by

$$\mathcal{G} := \mathcal{M}^{ss}(wt = w-n)$$

the direct summand of $\mathcal{M}^{ss}$ which is the sum of all those irreducible constituents which are pure of weight w-n.

Denote by

$$\mathcal{G}_{ncst}$$

the direct summand of $\mathcal{G}$ which is the sum of those irreducible summands $\mathcal{N}$ of $\mathcal{M}$ for which

$$f_\star(\mathcal{N}) = 0.$$

Denote by

$$\mathcal{G}_{cst}$$

the direct summand of $\mathcal{M}^{ss}(wt = w-n)$ which is the sum of those irreducible summands $\mathcal{N}$ of $\mathcal{M}$ for which

$$f_\star(\mathcal{N}) \neq 0.$$

So we tautologically have a direct sum decomposition

$$\mathcal{G} = \mathcal{G}_{cst} \oplus \mathcal{G}_{ncst}.$$

We will show

Key Lemma 12.3.4 The lisse sheaf

$$\mathcal{G}_{ncst} := \mathcal{M}^{ss}(wt = w-n)_{ncst}$$

on V has the property that its restriction to each arithmetic fibre $V_{s,E}$ of V/S is a constant field twist of $Gr^0(\mathcal{M}_{s,E} \otimes (\alpha_{s,E})^{deg})_{ncst}$.

Let us temporarily admit the truth of the Key Lemma. Then we finish the proof as follows. As observed earlier, we can apply [Ka-Sar-RMFEM, 9.6.10] to the sheaf $\mathcal{G}_{ncst}$ on V/S to conclude that whenever $\#E > 4A(V/S)^2$, we have the estimate

$$|\Sigma_{x \text{ in } V_{s,E}(E)} \operatorname{Trace}(\Lambda(\vartheta(E, s, \alpha, x)))| \le C\dim(\Lambda)(\#E)^{n-1/2},$$

for C the constant $C(V/S, \mathcal{G}_{ncst})$. We now explain how to infer from this estimate an similar estimate when the sum is taken over $U_{s,E}(E)$ instead of "just" over $V_{s,E}(E)$.

Consider the complement U - V of V in U. This is a closed subset of U, which we regard as a closed reduced subscheme of U. Since both U and V meet every fibre of X/S in an open and hence dense set, we see that (U-V)/S has all its fibres of dimension $\le n-1$. Let us denote by

$$\pi : U - V \to S$$

the projection. The sheaves $R^i\pi_!\overline{\mathbb{Q}}_\ell$ on S are constructible, and they vanish for $i > 2n-2$. So we have the a priori inequality

$$0 \le \#U_{s,E}(E) - \#V_{s,E}(E) \le B(\#E)^{n-1},$$

for B the constant

$$B := \operatorname{Sup}_{s \text{ in } S} (\text{rank of } (\oplus_i R^i\pi_!\overline{\mathbb{Q}}_\ell)_s).$$

From the trivial inequality

$$|\operatorname{Trace}(\Lambda(\vartheta(E, s, \alpha, x)))| \le \dim(\Lambda),$$

we get the trivial estimate

$$|\Sigma_{x \text{ in } U_{s,E}(E) - V_{s,E}(E)} \operatorname{Trace}(\Lambda(\vartheta(E, s, \alpha, x)))| \le B\dim(\Lambda)(\#E)^{n-1/2}.$$

Adding this to the above estimate for the sum over V, we get

$$|\Sigma_{x \text{ in } V_{s,E}(E)} \operatorname{Trace}(\Lambda(\vartheta(E, s, \alpha, x)))| \le (B+C)\dim(\Lambda)(\#E)^{n-1/2},$$

as required, with the same constant A, but with C replaced by B+C.

It remains to prove the Key Lemma.

proof of the Key Lemma 12.3.4 For any irreducible consituent $\mathcal{N}$ of $\mathcal{G}$ with $f_*\mathcal{N} \ne 0$, the adjunction map

$$f^*f_*\mathcal{N} \to \mathcal{N}$$

is an isomorphism. Indeed, it is injective (use the base change hypothesis, and look fibre by fibre, where this map is the inclusion of the $\pi_1{}^{\mathrm{geom}}(V_{s,E})$-invariants in the $\pi_1{}^{\mathrm{arith}}(V_{s,E})$-representation which "is" $\mathcal{N} \mid V_{s,E}$), and it is surjective (its source is nonzero, and its target is irreducible).

Thus in the decomposition

$$\mathcal{G} = \mathcal{G}_{cst} \oplus \mathcal{G}_{ncst},$$

we have

$$\mathcal{G}_{cst} = f^*f_*\mathcal{G},$$
$$f_*\mathcal{G}_{ncst} = 0.$$

For each arithmetic fibre $V_{s,E}$ of V/S, with structural map

$$f_{s,E} : V_{s,E} \to \mathrm{Spec}(E),$$

obtain a direct sum decomposition

$$\mathcal{G} \mid V_{s,E} = \mathcal{G}_{cst} \mid V_{s,E} \oplus \mathcal{G}_{ncst} \mid V_{s,E},$$

in which, by the base change hypothesis, we have

$$\mathcal{G}_{cst} \mid V_{s,E} = (f_{s,E})^*(f_{s,E})_*(\mathcal{G} \mid V_{s,E}),$$
$$(f_{s,E})_*(\mathcal{G}_{ncst} \mid V_{s,E}) = 0.$$

To go further, recall that for any lisse pure sheaf $\mathcal{K}$ on $V_{s,E}$, we have a direct sum decomposition

$$\mathcal{K} = \mathcal{K}_{cst} \oplus \mathcal{K}_{ncst}$$

as a direct sum of a geometrically constant lisse sheaf $\mathcal{K}_{cst}$ and of a lisse sheaf $\mathcal{K}_{ncst}$ none of whose irreducible constituents over $\overline{E}$ is geometrically constant, cf. Lemma 1.10.3. In terms of the structural map

$$f_{s,E} : V_{s,E} \to \mathrm{Spec}(E),$$

it is the unique decomposition with

$$\mathcal{K}_{cst} = (f_{s,E})^*(f_{s,E})_*\mathcal{K},$$

and

$$(f_{s,E})_*\mathcal{K}_{ncst} = 0.$$

Since $\mathcal{K}$, being pure, is geometrically semisimple, it follows that the formation of $\mathcal{K}_{cst}$ and of $\mathcal{K}_{ncst}$ commutes with arithmetic semisimplification:

$$(\mathcal{K}^{ss})_{cst} = (\mathcal{K}_{cst})^{ss},$$
$$(\mathcal{K}^{ss})_{ncst} = (\mathcal{K}_{ncst})^{ss}.$$

Applying this decomposition with $\mathcal{K}$ taken to be $\mathcal{G} \mid V_{s,E}$, we see that

$$(\mathcal{G} \mid V_{s,E})_{cst} = \mathcal{G}_{cst} \mid V_{s,E},$$
$$(\mathcal{G} \mid V_{s,E})_{ncst} = \mathcal{G}_{ncst} \mid V_{s,E},$$

and

$$((\mathcal{G} \mid V_{s,E})^{ss})_{cst} = (\mathcal{G}_{cst} \mid V_{s,E})^{ss},$$
$$((\mathcal{G} \mid V_{s,E})^{ss})_{ncst} = (\mathcal{G}_{ncst} \mid V_{s,E})^{ss}.$$

Applying this decomposition with $\mathcal{K}$ taken to be $\mathrm{Gr}^{w-n}(\mathcal{M}_{s,E})$, we get

$$\mathrm{Gr}^{w-n}(\mathcal{M}_{s,E}) = (\mathrm{Gr}^{w-n}(\mathcal{M}_{s,E}))_{cst} \oplus (\mathrm{Gr}^{w-n}(\mathcal{M}_{s,E}))_{ncst}.$$

Passing to arithmetic semisimplification, we obtain

$$((\mathrm{Gr}^{w-n}(\mathcal{M}_{s,E}))^{ss})_{ncst} = ((\mathrm{Gr}^{w-n}(\mathcal{M}_{s,E}))_{ncst})^{ss}$$
$$= (\mathrm{Gr}^{w-n}(\mathcal{M}_{s,E}))_{ncst},$$

the last equality because $(\mathrm{Gr}^{w-n}(\mathcal{M}_{s,E}))_{ncst}$ is geometrically, and hence arithmetically, irreducible.

On the other hand, $\mathcal{G} := \mathcal{M}^{ss}(wt = w-n)$ on each arithmetic fibre $V_{s,E}$ of V/S has the same characteristic polynomials of Frobenius as does $Gr^{w-n}(\mathcal{M}_{s,E})$. Hence by Chebotarev we have

$$(\mathcal{G} \mid V_{s,E})^{ss} \cong (Gr^{w-n}(\mathcal{M}_{s,E}))^{ss}.$$

Passing to ncst parts, we get

$$((\mathcal{G} \mid V_{s,E})^{ss})_{ncst} \cong ((Gr^{w-n}(\mathcal{M}_{s,E}))^{ss})_{ncst}.$$

As we have seen just above, the source is

$$((\mathcal{G} \mid V_{s,E})^{ss})_{ncst} = (\mathcal{G}_{ncst} \mid V_{s,E})^{ss},$$

and the target is

$$((Gr^{w-n}(\mathcal{M}_{s,E}))^{ss})_{ncst} \cong (Gr^{w-n}(\mathcal{M}_{s,E}))_{ncst},$$

Thus we obtain an isomorphism

$$(\mathcal{G}_{ncst} \mid V_{s,E})^{ss} \cong (Gr^{w-n}(\mathcal{M}_{s,E}))_{ncst}.$$

As the target is geometrically and hence arithmetically irreducible, so is the source, and hence $\mathcal{G}_{ncst} \mid V_{s,E}$ is itself arithmetically irreducible. So we end up with an isomorphism

$$\mathcal{G}_{ncst} \mid V_{s,E} \cong (Gr^{w-n}(\mathcal{M}_{s,E}))_{ncst}.$$

But $(Gr^{w-n}(\mathcal{M}_{s,E}))_{ncst}$ is a constant field twist of $(Gr^0(\mathcal{M}_{s,E}\otimes(\alpha_{s,E})^{deg}))_{ncst}$ (because $|\iota(\alpha_{s,E})|_{\mathbb{C}} = \mathrm{Sqrt}(\#E)^{n-w}$). This concludes the proof of the Key Lemma 12.3.4, and with it the proof of the Uniformity Theorem 12.3.3. QED

(12.4) Applications of the Uniformity Theorem to twist sheaves

(12.4.1) In previous chapters, we worked over a finite field k in which ℓ is invertible. We fixed a choice of square root of char(k) in $\overline{\mathbb{Q}}_\ell$, so that we could form Tate twists by half-integers. We worked with "standard input", cf. 1.15.4 Thus we fixed

an integer $m \geq 1$,

a perverse sheaf K on $\mathbb{A}^m/k$,

an affine k-scheme V/k of finite type,

a k-morphism $h : V \to \mathbb{A}^m$,

a perverse sheaf L on V/k,

an integer $d \geq 2$,

a space of functions $(\mathcal{F}, \tau)$ on V, i.e., a finite-dimensional k-vector space $\mathcal{F}$ and a k-linear map

$$\tau : \mathcal{F} \to \mathrm{Hom}_{k\text{-schemes}}(V, \mathbb{A}^m).$$

(12.4.2) We made the following four hypotheses on these data.

1) K is ι-mixed of weight ≤ 0, and $Gr^0(K)$, the weight 0 quotient of K, is geometrically irreducible on $\mathbb{A}^m/k$.

2) L is ι-mixed of weight ≤ 0, and $Gr^0(L)$, the weight 0 quotient of L, is geometrically irreducible on V/k.

3) $(\mathcal{F}, \tau)$ is d-separating, and contains the constants.

4) The $\mathbb{Z}$-graded vector space

$$H^*_c((V\times\mathbb{A}^m)\otimes\overline{k},\ pr_1{}^*L\otimes pr_2{}^*K)$$

is concentrated in degree $\le m$.

(12.4.3) We showed in 1.4.4 and 1.5.11 that the object $M = \mathrm{Twist}(L, K, \mathcal{F}, h)$ on the space $\mathcal{F}$ is perverse, and that after we restricted M to a suitable dense open set U of $\mathcal{F}$, $M \mid U$ became $\mathcal{M}(m/2)[\dim\mathcal{F}]$, for a lisse $\overline{\mathbb{Q}}_\ell$-sheaf $\mathcal{M}$ on U which is ι-mixed of weight ≤ 0.

(12.4.4) Let us say that we have "strong standard input" if in addition we have the following two conditions 5) and 6):

5) $d \ge 4$,

6) $\mathrm{Gr}^0(K)$ is not geometrically constant.

(12.4.5) When we have strong standard input, we showed in 1.15.6 that $\mathrm{Gr}^0(\mathcal{M})_{ncst}$ was geometrically irreducible, and gave in 2.2.3 some general results about its geometric monodromy.

(12.4.6) In this finite field situation, where we have Tate twists by half-integers at our disposal, there is no greater generality attained if instead of the six hypotheses above, we instead pick two integers a and b, and replace the hypotheses 1), 2), and 6) of 12.2.1 above by the following trivial variants 1a), 2b), 6a).

1a) 1) K is ι-mixed of weight $\le a$, and $\mathrm{Gr}^a(K)$, the weight a quotient of K, is geometrically irreducible on $\mathbb{A}^m/k$.

2b) L is ι-mixed of weight $\le b$, and $\mathrm{Gr}^b(L)$, the weight b quotient of L, is geometrically irreducible on V/k.

6a) $\mathrm{Gr}^a(K)$ is not geometrically constant.

(12.4.7) Indeed, K satisfies 1) and 6) if and only if $K(-a/2)$ satisfies 1a) and 6a), and L satisfies 2) if and only if $L(-b)$ satisfies 2b). And we have the trivial relation between the corresponding twist sheaves

$$M := \mathrm{Twist}(L, K, \mathcal{F}, h)$$

and

$$M_{a,b} := \mathrm{Twist}(L(-b/2), K(-a/2), \mathcal{F}, h),$$

namely

$$M_{a,b} = M(-(a+b)/2).$$

(12.4.8) Let us say that when 1a), 2b), 3), 4), 5), and 6a) are satisfied, we have "strong standard input of type (a, b)". Thus our earlier notion of strong standard input now becomes "strong standard input of type (0,0)". The operation

$$(K\ L) \mapsto (K(-a/2), L(-b/2))$$

carries us from strong standard input of type (0,0) to that of type (a, b).

(12.4.9) On a dense open set $U \subset \mathcal{F}$ on which M, and hence $M_{a,b}$ as well, have lisse cohomology sheaves, we have

$$M \mid U = \mathcal{M}(m/2)[\dim\mathcal{F}],$$

with $\mathcal{M}$ lisse on U, and ι-mized of weight ≤ 0, and we have

$$M_{a,b} \mid U = \mathcal{M}_{a,b}(m/2)[\dim\mathcal{F}],$$

with

$$\mathcal{M}_{a,b} = \mathcal{M}(-(a+b)/2)$$

ι-mixed of weight $\le$ a+b. Our results about the geometric monodromy of $\mathrm{Gr}^0(\mathcal{M})_{\mathrm{ncst}}$ now become results about the geometric monodromy of $\mathrm{Gr}^{a+b}(\mathcal{M}_{a+b})_{\mathrm{ncst}}$.

(12.4.10) We now formulate a relative version of the notion of strong standard input of type (a, b). We work over an normal and connected affine $\mathbb{Z}[1/\ell]$-scheme S = Spec(A) which is of finite type over $\mathbb{Z}[1/\ell]$. We fix two integers a and b. We suppose given

an integer $m \ge 1$,

an object K in $D^b_c(\mathbb{A}^m_S, \overline{\mathbb{Q}}_\ell)$ which is ι-mixed of weight $\le$ a and which is fibrewise perverse on $\mathbb{A}^m/S$,

an affine S-scheme V/S of finite type,

an S-morphism $h : V \to \mathbb{A}^m$,

an object L in $D^b_c(V, \overline{\mathbb{Q}}_\ell)$.which is ι-mixed of weight $\le$ b and which is fibrewise perverse on V/S,

an integer $d \ge 2$,

a space of functions $(\mathcal{F}, \tau)$ on V, i.e., a locally free A-module of finite rank $\mathcal{F}$ and an A-linear map

$$\tau : \mathcal{F} \to \mathrm{Hom}_{S\text{-schemes}}(V, \mathbb{A}^m).$$

(12.4.11) We say that this is standard input of type (a, b) [resp. strong standard input of type (a, b)] relative to S if for every finite field E, and for every s in S(E), the pullbacks of these data to the appropriate arithmetic fibres over S, to wit,

$K_{s,E}$ on $(\mathbb{A}^m)_{s,E} = \mathbb{A}^m/E$,

$h_{s,E} : V_{s,E} \to (\mathbb{A}^m)_{s,E} = \mathbb{A}^m/E$,

$L_{s,E}$ on $V_{s,E}$

the integer $d \ge 2$,

$(\mathcal{F}_{s,E}, \tau_{s,E})$ as a space of functions on $V_{s,E}$,

constitute standard input of type (a, b) [resp. strong standard input of type (a, b)].

(12.4.12) Given strong standard input of type (a,b) relative to S, we form the twist sheaf M = Twist(L, K, $\mathcal{F}$, h) on $\mathcal{F}$, defined exactly as in 1.3.3, except that we work over S rather than over k. By proper base change and 1.4.4, part 4), applied to each arithmetic fibre over S, we see that M is fibrewise perverse on $\mathcal{F}$/S, and (by 1.5.11) ι-mixed of weight $\le a + b + \dim\mathcal{F} - m$.

(12.4.13) Let (N, G) be a pair consisting of an integer $N \ge 1$, and a

Zariski closed $\overline{\mathbb{Q}}_\ell$-algebraic subgroup $G \subset GL(N)$ which satisfies the conditions G1), G2), and G3) of 12.2.3. Thus G is an irreducible, semisimple (not necessarily connected) $\overline{\mathbb{Q}}_\ell$-algebraic subgroup of GL(N) whose normalizer in GL(N) is $\mathbb{G}_m \cdot G$.

(12.4.14) Viewing $\mathcal{F}$ as an affine space over S, let $U \subset \mathcal{F}$ be an open set which meets every geometric fibre of $\mathcal{F}/S$. We say that strong standard input of type (a,b) relative to S gives uniform output of type (U, N, G) relative to S if the following two conditions hold.

1) On each arithmetic fibre $U_{s,E}$ of U/S, $M_{s,E} \mid U_{s,E}$ has lisse cohomology sheaves.

2) On each arithmetic fibre $U_{s,E}$ of U/S, write $M_{s,E} \mid U_{s,E} \cong \mathcal{M}_{s,E}[\dim \mathcal{F}]$, with $\mathcal{M}_{s,E}$ a lisse sheaf on $U_{s,E}$ which is ι-mixed of weight $\le a+b-m$. Form the lisse sheaf

$$Gr^{a+b-m}(\mathcal{M}_{s,E})_{ncst}.$$

This lisse sheaf on $U_{s,E}$ has rank N, and its geometric monodromy group $G_{geom,s,E}$ is (conjugate in GL(N) to) G.

(12.4.15) Suppose we start with strong standard input of type (a, b) relative to S, and suppose it gives uniform output of type (U, N, G) relative to S. For each finite field E, and for each s in S(E), choose [Ka-Sar-RMFEM, 9.6.3] an ℓ-adic unit $\alpha_{s,E}$ in $\overline{\mathbb{Q}}_\ell^\times$, with

$$|\iota(\alpha_{s,E})|_{\mathbb{C}} = \mathrm{Sqrt}(\# E)^{m-a-b},$$

such that after replacing $Gr^{a+b-m}(\mathcal{M}_{s,E})_{ncst}$ by its constant field twist

$$Gr^{a+b-m}(\mathcal{M}_{s,E})_{ncst} \otimes (\alpha_{s,E})^{deg} = Gr^0(\mathcal{M}_{s,E} \otimes (\alpha_{s,E})^{deg}))_{ncst},$$

the following condition holds:

MG2) Under the representation $\rho_{s,E,\alpha} : \pi_1{}^{arith}(U_{s,E}) \to GL(N, \overline{\mathbb{Q}}_\ell)$ which "is" the lisse sheaf

$$Gr^0(\mathcal{M}_{s,E} \otimes (\alpha_{s,E})^{deg}))_{ncst} = Gr^{a+b-m}(\mathcal{M}_{s,E})_{ncst} \otimes (\alpha_{s,E})^{deg},$$

we have the inclusion

$$\rho_{s,E,\alpha}(\pi_1{}^{arith}(U_{s,E})) \subset G_{geom,s,E}(\overline{\mathbb{Q}}_\ell).$$

(12.4.16) Pick a maximal compact subgroup K of the complex Lie group $G(\mathbb{C})$. Then the Uniformity Theorem 12.3.3 applies to the situation S = S, X/S = $\mathcal{F}/S$ of relative dimension $n = \dim \mathcal{F}$, U, ι, M, $w = a + b + \dim \mathcal{F} - m$, N, G. We denote by $K^\#$ the space of conjugacy classes in K, and by $\mu^\#$ the Borel measure of total mass one on $K^\#$ which is the direct image from K of its Haar measure of total mass one μ_{Haar}. For each finite field E, for each point s in S(E), and for each point x in $U_{s,E}(E)$, we denote by $Frob_{E,s,x}$ its Frobenius conjugacy class in $\pi_1^{arith}(U_{s,E})$. We denote by

$$\vartheta(E, s, \alpha, x) \text{ in } K^\#$$

the conjugacy class of $\iota(\rho_{s,E,\alpha}(Frob_{E,s,x}))^{ss}$. Whenever $U_{s,E}(E)$ is nonempty (a condition which is automatic is $\#E > A(U/S)^2$), we denote by

$$\mu(E, s, \alpha) := (1/\#U_{s,E}(E)) \Sigma_{x \text{ in } U_{s,E}(E)} \delta_{\vartheta(E, s, \alpha, x)}$$

the probability measure on $K^\#$ obtained by averaging over the Frobenius conjugacy classes, as x runs over $U_{s,E}(E)$.

(12.4.17) Applying the Uniformity Theorem 12.3.3, we obtain the following result.

Uniform Output Theorem 12.4.18 Suppose we start with strong standard input of type (a, b) relative to S, and suppose it gives uniform output of type (U, N, G) relative to S. For each finite field E with $\#E > A(U/S)^2$, and for each s in S(E), form the probability measure on $K^\#$ given by

$$\mu(E, s, \alpha) := (1/\#U_{s,E}(E)) \Sigma_{x \text{ in } U_{s,E}(E)} \delta_{\vartheta(E, s, \alpha, x)}.$$

For any sequence of data (E_i, s_i, α_i) in which each $\#E_i > A(U/S)^2$ and in which $\#E_i$ is strictly increasing, the measures $\mu(E_i, s_i, \alpha_i)$ on $K^\#$ converge weak * to $\mu\#$, i.e., for any continuous $\mathbb{C}$-valued central function f on K, we have

$$\int_K f d\mu_{Haar} = \lim_{i\to\infty}(1/\#U_{s_i,E_i}(E_i)) \Sigma_{x \text{ in } U_{s_i,E_i}(E_i)} f(\vartheta(E_i, s_i, \alpha_i, x)).$$

More precisely, there exist constants A and C such that for any irreducible nontrivial irreducible unitary representation Λ of K, and for any data (E, s, α) with $\#E > 4A^2$, we have the estimates

$$2(\#E)^n \geq \#U_{s,E}(E) \geq (1/2)(\#E)^n,$$

$$|\Sigma_{x \text{ in } U_{s,E}(E)} \text{Trace}(\Lambda(\vartheta(E, s, \alpha, x)))| \leq C\dim(\Lambda)(\#E)^{n-1/2},$$

and

$$|\int_K \text{Trace}(\Lambda) d\mu(E, s, \alpha)| \leq 2C\dim(\Lambda)/\text{Sqrt}(\#E).$$

Corollary 12.4.19 Suppose we are in the special case $S = \text{Spec}(\mathbb{Z}[1/\ell D])$ for some nonzero integer D of the above Theorem 12.4.18. For each prime power q which is prime to ℓD and which satisfies $q > A(U/S)^2$, denote by s_{taut} the unique point in $S(\mathbb{F}_q)$, by $U_{\mathbb{F}_q}$ the fibre $U_{s_{taut},\mathbb{F}_q}$ of U/S above $\mathbb{F}_q$, and by μ_q the probability measure on $K^\#$ given by

$$\mu_q := \mu(\mathbb{F}_q, s_{taut}, \alpha)$$
$$:= (1/\#U_{\mathbb{F}_q}(\mathbb{F}_q)) \Sigma_{x \text{ in } U_{\mathbb{F}_q}(\mathbb{F}_q)} \delta_{\vartheta(\mathbb{F}_q, s_{taut}, \alpha, x)}.$$

Then the measures μ_q on $K^\#$ converge weak * to $\mu\#$, i.e., for any continuous $\mathbb{C}$-valued central function f on K, we have

$$\int\int_K f d\mu_{Haar} = \lim_{q \to \infty} \int_K f d\mu_q$$
$$= \lim_{q \to \infty}(1/\#U\mathbb{F}q(\mathbb{F}q)) \Sigma_{x \text{ in } U_{\mathbb{F}_q}(\mathbb{F}_q)} f(\vartheta(\mathbb{F}_q, s_{taut}, \alpha, x)).$$

More precisely, there exist constants A and C such that for any irreducible nontrivial irreducible unitary representation Λ of K, and for any prime power $q > 4A^2$ which is prime to ℓD, we have the estimates

$$2q^n \geq \#U_{\mathbb{F}_q}(\mathbb{F}_q) \geq (1/2)q^n,$$

$$|\Sigma_{x \text{ in } U_{\mathbb{F}_q}(\mathbb{F}_q)} \mathrm{Trace}(\Lambda(\vartheta(\mathbb{F}_q, s_{taut}, \alpha, x)))| \mathrm{Cdim}(\Lambda)q^{n-1/2}$$

and

$$|\int_K \mathrm{Trace}(\Lambda)d\mu_q\} \leq 2\mathrm{Cdim}(\Lambda)/\mathrm{Sqrt}(q).$$

(12.5) Applications to multiplicative character sums

(12.5.1) We first recall the global incarnation of the Kummer sheaves $\mathcal{L}_\rho$ on $\mathbb{G}_m$ over a finite field E, ρ being a $\overline{\mathbb{Q}}_\ell{}^\times$-valued nontrivial character of $E^\times$ of some given order. Given ρ, denote by b the number of roots of unity in the field $\mathbb{Q}(\rho)$. Thus b is an even integer, equal to the order of ρ if order(ρ) is even, otherwise equal to $2\times$order(ρ).

(12.5.2) To make the global construction, denote by $\Phi_b(X)$ in $\mathbb{Z}[X]$ the b'th cyclotomic polymomial. We write $\mathbb{Z}[\zeta_b]$ for the ring $\mathbb{Z}[X]/(\Phi_b(X))$. Its fraction field is the b'th cyclotomic field $\mathbb{Q}[\zeta_b] = \mathbb{Q}(\zeta_b)$, and $\mathbb{Z}[\zeta_b]$ is the ring of integers in $\mathbb{Q}(\zeta_b)$.

(12.5.3) We pick a prime number ℓ, and work over the ring

$$R := \mathbb{Z}[\zeta_\upsilon][1/\ell b].$$

The group $\mu_b(R)$ of b'th roots of unity in R is the cyclic group of order b generated by ζ_b. [Because b is even, $\mu_b(R)$ is the group of all roots of unity in R.] Over R, we have the group-scheme

$$\mathbb{G}_{m,R} := \mathrm{Spec}(R[t, t^{-1}]).$$

By means of the b'th power map

$$[b] : \mathbb{G}_{m,R} \to \mathbb{G}_{m,R},$$
$$t \mapsto t^b,$$

$\mathbb{G}_{m,R}$ becomes a finite etale galois covering of itself, with group $\mu_b(R)$. This exhibits the group $\mu_b(R)$ as a quotient of $\pi_1(\mathbb{G}_{m,R})$:

$$\mathrm{can}_b : \pi_1(\mathbb{G}_{m,R}) \twoheadrightarrow \mu_\upsilon(R).$$

(12.5.4) Fix a nontrivial $R^\times$-valued character χ of $\mu_b(R)$, i.e., a group homomorphism

$$\chi : \mu_b(R) \to \mu_b(R) \subset R^\times.$$

In the case that will be of interest later, χ will either be faithful, i.e., have order b, or, if b/2 is odd, χ might instead have order b/2. In both of these cases, b is the number of roots of unity in the field $\mathbb{Q}(\chi)$, generated over $\mathbb{Q}$ by the values of χ.

(12.5.5) We denote by χ_2 the quadratic character, i.e., the unique nontrivial character of order two of $\mu_b(R)$. We will later have occassion to consider both χ and $\chi\times\chi_2$.

(12.5.6) Denote by $\overline{\chi} := 1/\chi$ the "conjugate" character. Denote by $\mathcal{L}_\chi$ the $R^\times$-valued character of $\pi_1(\mathbb{G}_{m,R})$ defined as $\overline{\chi}\circ\mathrm{can}_\nu$. This primordial Kummer sheaf $\mathcal{L}_\chi$ is thus a lisse, rank one R-sheaf on $\mathbb{G}_{m,R}$. Its more usual ℓ-adic incarnation, also denoted $\mathcal{L}_\chi$ when no ambiguity is likely, is obtained by choosing a ring embedding of R into $\overline{\mathbb{Q}}_\ell$, which allows us to view $\mathcal{L}_\chi$ as a lisse, rank one $\overline{\mathbb{Q}}_\ell$-sheaf on $\mathbb{G}_{m,R}$.

(12.5.7) Let E be a finite field. Then E admits a structure of R-algebra if and only if ℓb is invertible in E and the group $\mu_b(E)$ is cyclic of order b. If this is the case, each ring homomorphism $R \to E$ induces an isomorphism of groups

$$\mu_b(R) \cong \mu_b(E),$$

and each such isomorphism of groups is induced by a unique ring homomorphism. So a structure of R-algebra on E allows us to view our $R^\times$-valued character χ of $\mu_b(R)$ as an $R^\times$-valued character of $\mu_b(E)$. We then compose with the surjective group homomorphism

$$E^\times \twoheadrightarrow \mu_b(E),$$
$$x \mapsto x^{(\#E - 1)/b},$$

and obtain a character

$$\chi_E : E^\times \to R^\times,$$

whose order is the same as that of χ.

(12.5.8) The fundamental compatibility is this. Suppose given a finite field E in which ℓb is invertible and for which $\mu_b(E)$ is cyclic of order b. If we choose a structure of R-algebra on E, then the pullback of $\mathcal{L}_\chi$ to $\mathbb{G}_{m,E}$ is the "usual" Kummer sheaf $\mathcal{L}_{\chi_E}$, but viewed as a lisse, rank one R-sheaf on $\mathbb{G}_{m,E}$. If we then vary the embedding of R into $\overline{\mathbb{Q}}_\ell$, the $\overline{\mathbb{Q}}_\ell$-sheaves on $\mathbb{G}_{m,E}$ we obtain from $\mathcal{L}_{\chi_E}$ exhaust all the Kummer sheaves $\mathcal{L}_\rho$ on $\mathbb{G}_{m,E}$, for all $\overline{\mathbb{Q}}_\ell^\times$-valued characters ρ of $E^\times$ having the same order as χ.

(12.5.9) If instead we fix an embedding of R into $\overline{\mathbb{Q}}_\ell$, but vary the structure of R-algebra on E, then it is also true that the $\overline{\mathbb{Q}}_\ell$-sheaves on $\mathbb{G}_{m,E}$ we obtain from $\mathcal{L}_{\chi_E}$ exhaust all the Kummer sheaves $\mathcal{L}_\rho$ on $\mathbb{G}_{m,E}$, for all $\overline{\mathbb{Q}}_\ell^\times$-valued characters ρ of $E^\times$ having the same order as χ.

(12.5.10) What this means is this. Fix an embedding of R into $\overline{\mathbb{Q}}_\ell$. Form the lisse $\overline{\mathbb{Q}}_\ell$-sheaf $\mathcal{L}_\chi$ on $\mathbb{G}_{m,R}$. Then for any finite field E in which ℓb is invertible and for which $\mu_b(E)$ is cyclic of order b, we obtain all the Kummer sheaves $\mathcal{L}_\rho$ on $\mathbb{G}_{m,E}$, for all $\overline{\mathbb{Q}}_\ell{}^\times$-valued characters ρ of $E^\times$ of the same order as χ, simply by viewing E as an R-algebra in all different possible ways, and restricting $\mathcal{L}_\chi$ to the corresponding $\mathbb{G}_{m,E}$.

(12.5.11) Having chosen b, ℓ, and χ above, we now choose integers $n \geq 1$ and $e \geq 3$. We suppose that

$$b = \text{the number of roots of unity in } \mathbb{Q}(\chi).$$

Under this hypothesis, a finite-field valued point s of $S = \mathrm{Spec}(\mathbb{Z}[\zeta_b][1/e\ell b])$ is a pair (E, ρ) consisting of a finite field E in which $e\ell b$ is invertible and which contains a primitive b'th root of unity, and of a nontrivial multiplicative character

$$\rho : E^\times \to \overline{\mathbb{Q}}_\ell{}^\times,$$

whose order is the same as that of χ. Indeed, there is a unique group isomorphism $\mu_b(R) \cong \mu_b(E)$ under which χ corresponds to ρ, and this isomorphism is induced by a unique ring homomorphism from $\mathbb{Z}[\zeta_b][1/e\ell b]$ to E.

(12.5.12) We work over the ring $\mathbb{Z}[\zeta_b][1/e\ell b]$. We consider the following "strong standard input of type (a,b) = (1, n)" over

$$S = \mathrm{Spec}(\mathbb{Z}[\zeta_b][1/e\ell b]):$$

the integer $m = 1$,

the object $K = j_!\mathcal{L}_\chi[1]$ in $D^b{}_c(\mathbb{A}^1{}_S, \overline{\mathbb{Q}}_\ell)$, for j the inclusion of $\mathbb{G}_m$ into $\mathbb{A}^1$ (it is ι-mixed of weight ≤ 1 and fibrewise perverse on $\mathbb{A}^1/S$),

the affine S-scheme $V/S := \mathbb{A}^n{}_S/S$ of finite type,

the S-morphism $h = 0 : V \to \mathbb{A}^1$,

the object $L = \overline{\mathbb{Q}}_\ell[n]$ in $D^b{}_c(V, \overline{\mathbb{Q}}_\ell)$, (it is ι-mixed of weight $\leq n$ and fibrewise perverse on V/S),

the integer $d = e+1$,

the space of functions $(\mathcal{F}, \tau) = (\mathcal{P}_e, \mathrm{eval})$ on V, with $\mathcal{P}_e$ the free R[1/e]-module of all polynomials of degree $\leq$ e in n variables.

(12.5.13) We attach to this data the integer N, defined by

$$N := (e-1)^n, \text{ if } \chi^e \neq \mathbb{1},$$

$$N := (1/e)((e-1)^{n+1} - (-1)^{n+1}), \text{ if } \chi^e = \mathbb{1}.$$

We denote by

$$\mathcal{SD}(n, e) \subset \mathcal{P}_e$$

the dense open set consisting of strong Deligne polynomials, cf. 5.1.10

and 6.6.1.

(12.5.14) Theorems 6.7.19 and 6.7.21 then give us the following "uniform output" theorems.

Theorem 12.5.15 Suppose that χ has order 2, and that n is odd. Then the above strong standard input of type (1, n) gives uniform output of type $(U, N, G) = (\mathcal{SD}(n, e), N, Sp(N))$ relative to $S = \mathrm{Spec}(\mathbb{Z}[1/2e\ell])$.

Theorem 12.5.16 Suppose that χ has order 2, n is even, and $N > 8$. Then the above strong standard input of type (1, n) gives uniform output of type $(U, N, G) = (\mathcal{SD}(n, e), N, O(N))$ relative to $S = \mathrm{Spec}(\mathbb{Z}[1/2e\ell])$.

(12.5.17) In the case when χ has order 3 or more, we define

$$a := \text{the order of } \chi(\chi_2)^n.$$

Recall that

$$b := \text{the number or roots of unity in } \mathbb{Q}(\chi) = \mathbb{Q}(\chi(\chi_2)^n).$$

Thus b is even, and either $b = a$ or $b = 2a$.

Theorem 12.5.18 Suppose that χ has order 3 or more, and that at least one of the following conditions 1), 2), or 3) holds:

1)$N > 4$,

2) $N > 2$ and $a > 3$,

3) $a \geq 6$.

Then the above strong standard input of type (1, n) gives uniform output of type $(\mathcal{SD}(n, e), N, G)$ relative to $S = \mathrm{Spec}(\mathbb{Z}[\zeta_b][1/e\ell b])$, where G is the algebraic group

$G = GL_a(N)$, if $b = a$, or if $\chi^e = \mathbb{1}$,

$G = GL_a(N)$, if $b = 2a$, n is odd, $\chi^e \neq \mathbb{1}$, and e is odd,

$G = GL_b(N)$, if $b = 2a$, n is odd, $\chi^e \neq \mathbb{1}$, and e is even,

$G = GL_b(N)$, if $b = 2a$, n is even, and $\chi^e \neq \mathbb{1}$.

(12.5.19) We now wish to make explicit the equidistribution consequences of these theorems. A finite-field valued point s of $S = \mathrm{Spec}(\mathbb{Z}[\zeta_b][1/e\ell b])$ is a pair (E, ρ) consisting of a finite field E in which $e\ell b$ is invertible and which contains a primitive b'th root of unity, and of a nontrivial multiplicative character

$$\rho : E^\times \to \overline{\mathbb{Q}}_\ell^\times,$$

whose order is the same as that of χ. In the "uniform output" theorems above, the lisse sheaf $Gr^{w-n}(\mathcal{M}_{s,E})_{ncst}$ of 12.4.14 induced on $\mathcal{SD}(n, e)\otimes E$ is the lisse sheaf $Gr^0(\mathcal{M}(n,e,\rho))(-n/2)$. Recall that its trace function is the following. For L/E a finite extension, and f in

$\mathcal{SD}(n, e)(L)$, with homogenization F, denoting by ρ_L the multiplicative character $\rho\circ\mathrm{Norm}_{L/E}$ of $L^\times$, we have

$$\mathrm{Trace}(\mathrm{Frob}_{L,f} \mid \mathrm{Gr}^0(\mathcal{M}(n,e,\rho))(-n/2))$$
$$= (-1)^n \Sigma_{x \text{ in } \mathbb{A}^n[1/f](L)} \rho_L(f(x)), \text{ if } \rho^e \neq \mathbb{1},$$
$$= (-1)^n \Sigma_{x \text{ in } \mathbb{P}^n[1/F](L)} \rho_L(F(x)), \text{ if } \rho^e = \mathbb{1}.$$

(12.5.20) If ρ has order 2 and n is even, then $\mathrm{Gr}^0(\mathcal{M}(n,e,\rho))$ is orthogonally self dual and ι-pure of weight 0. There is a unique conjugacy class $\vartheta(E, \rho, f)$ in the compact orthogonal group $O(N, \mathbb{R})$ whose characteristic polynomial is given by

$$\det(1 - T\vartheta(E, \rho, f)) = \iota(\det(1 - T\mathrm{Frob}_{E,f} \mid \mathrm{Gr}^0(\mathcal{M}(n,e,\rho)))).$$

(12.5.21) If ρ has order 2 and n is odd, then $\mathrm{Gr}^0(\mathcal{M}(n,e,\rho))$ is symplectically self dual and ι-pure of weight 0. There is a unique conjugacy class $\vartheta(E, \rho, f)$ in the compact symplectic group $USp(N)$ whose characteristic polynomial is given by the same rule,

$$\det(1 - T\vartheta(E, \rho, f)) = \iota(\det(1 - T\mathrm{Frob}_{E,f} \mid \mathrm{Gr}^0(\mathcal{M}(n,e,\rho)))).$$

(12.5.22) If ρ has order 3 or more, there is a bit more normalization we need to do. Recall from Theorem 6.7.21 that the group G_{geom} for $\mathrm{Gr}^0(\mathcal{M}(n,e,\rho))$ is either $GL_b(N)$, or, if b is 2a with a odd, possibly $GL_a(N)$, according to the following rule:

$$G_{geom} = GL_a(N), \text{ if } b = a, \text{ or if } \chi^e = \mathbb{1}, \text{ or if ne is odd},$$
$$G_{geom} = GL_b(N), \text{ if } b = 2a,\ \chi^e \neq \mathbb{1}, \text{ and ne is even}.$$

Let us write $GL_c(N)$ for whichever it is. Then

$$U_c(N) := \{A \text{ in } U(N) \text{ with } \det(A)^c = 1\}$$

is a compact form of $G_{geom} = GL_c(N)$.

(12.5.23) Denote by $fer_{n,e}$ the Fermat polynomial

$$fer_{n,e} := 1 - (\Sigma_{i=1 \text{ to } n} (x_i)^e) \text{ in } \mathcal{SD}(n, e)(E).$$

Define an ℓ-adic unit $A(E, \rho, n, e)$ by

$$A(E, \rho, n, e) := \det(\mathrm{Frob}_{E,fer_{n,e}} \mid \mathrm{Gr}^0(\mathcal{M}(n,e,\rho))).$$

Choose an N'th root $B(E, \rho, n, e)$ of $A(E, \rho, n, e)^{-1}$:

$$A(E, \rho, n, e)\times B(E, \rho, n, e)^N = 1.$$

Having chosen $B(E, \rho, n, e)$, there is a unique conjugacy class $\vartheta(E, \rho, f)$ in the group $U_c(N)$ whose characteristic polynomial is given by the rule

$$\det(1 - T\vartheta(E, \rho, f))$$
$$= \iota(\det(1 - T\mathrm{Frob}_{E,f} \mid \mathrm{Gr}^0(\mathcal{M}(n,e,\rho))\otimes B^{\deg})).$$

(12.5.24) These conjugacy classes $\vartheta(E, \rho, f)$ in the groups
$O(N, \mathbb{R})$, if ρ has order 2 and n is even,
$USp(N)$, if ρ has order 2 and n is odd,
$U_c(N)$, if ρ has order 3 or more,

are then the subject of the following equidistribution theorem, obtained by applying the general Uniformity Theorem 12.3.3 in the particular contexts of the three "uniform output" Theorems 12.5.15, 12.5.16, and 12.5.18 above.

Theorem 12.5.25 Let $b \geq 2$ be an even integer, and χ a nontrivial $\overline{\mathbb{Q}}_\ell^\times$-valued character of $\mu_b(\mathbb{Z}[\zeta_b])$ such that $\mathbb{Q}(\chi) = \mathbb{Q}(\zeta_b)$. Denote by χ_2 the quadratic character of $\mu_b(\mathbb{Z}[\zeta_b])$. Fix integers $e \geq 3$ and $n \geq 1$. Put

$$N := (e-1)^n, \text{ if } \chi^e \neq \mathbb{1},$$
$$N := (1/e)((e-1)^{n+1} - (-1)^{n+1}), \text{ if } \chi^e = \mathbb{1}.$$

Define the compact group K to be
$O(N, \mathbb{R})$, if χ has order 2 and n is even,
$USp(N)$, if χ has order 2 and n is odd,
$U_c(N)$, if χ has order 3 or more.

[Here $c = a$ if $b = a$, or if $\chi^e = \mathbb{1}$, or if ne is odd, and $c = b$ otherwise, i.e., if $b = 2a$, $\chi^e \neq \mathbb{1}$, and ne is even.]

If χ has order 2 and n is even, suppose that
$$N > 8.$$
If χ has order 3 or more, suppose that one of the following three conditions holds:
$$N > 4,$$
$$N > 2 \text{ and } \chi(\chi_2)^n \text{ has order} > 3,$$
$$\chi(\chi_2)^n \text{ has order} \geq 6.$$
Given the data (b, χ, e, n), there exist constants A and C with the following properties. Fix a pair (E, ρ) consisting of a finite field E with $\#E > 4A^2$ in which ℓeb is invertible and which contains a primitive b'th root of unity, and of a multiplicative character ρ of $E^\times$ of the same order as χ. Then
$$1/2 \leq \#\mathcal{SD}(n, e)(E)/(\#E)^{\dim \mathcal{SD}(n, e)} \leq 2.$$
Define the probablilty measure $\mu(E, \rho, n, e)$ on $K^\#$, the space of conjugacy classes in K, by
$$\mu(E, \rho, n, e) := (1/\#\mathcal{SD}(n, e)(E)) \Sigma_{f \text{ in } \mathcal{SD}(n, e)(E)} \delta_{\vartheta(E, \rho, f)}.$$
For any irreducible nontrivial irreducible unitary representation Λ of K, we have the estimates

$$|\Sigma_{f \text{ in } \mathcal{SD}(n, e)(E)} \operatorname{Trace}(\Lambda(\vartheta(E, \rho, f)))| \leq C\dim(\Lambda)(\#E)^{\dim \mathcal{SD}(n, e) - 1/2},$$
and

$$|\int_K \mathrm{Trace}(\Lambda)d\mu(E, \rho, n, e)| \le 2C\dim(\Lambda)/\mathrm{Sqrt}(\#E).$$

In particular, for any sequence of pairs (E_i, ρ_i) as above, in which $\#E_i$ is strictly increasing, the measures $\mu(E_i, \rho_i, n, e)$ on $K^\#$ converge weak $*$ to to $\mu^\#$, the direct image from K of Haar measure μ_{Haar} on K. For any continuous $\mathbb{C}$-valued central function g on K, we have

$$\int_K f d\mu_{Haar}$$
$$= \lim_{i \to \infty} (1/\#\mathcal{S}\mathcal{D}(n, e)(E_i)) \Sigma_{f \text{ in } \mathcal{S}\mathcal{D}(n, e)(E_i)} f(\vartheta(E_i, \rho_i, f)).$$

(12.6) Non-application (sic!) to additive character sums

(12.6.1) Let k be a finite field, p := char(k), $\ell \ne p$, and ψ a nontrivial additive $\overline{\mathbb{Q}}_\ell^\times$-valued of k. Fix $n \ge 1$, $e \ge 3$, with e prime to p. Denote by $\mathcal{D}(n,e)$ the space of Deligne polynomials, and denote by $\mathcal{M}(n,e,\psi)|\mathcal{D}(n,e)$ the lisse, geometrically irreducible and pure of weight zero $\overline{\mathbb{Q}}_\ell$-sheaf of rank $(e-1)^n$ on $\mathcal{D}(n,e)$ whose trace function is given by

$$\mathrm{Trace}(\mathrm{Frob}_{E,f} \mid \mathcal{M}(n,e,\psi)) = (-1)^n(\#E)^{-n/2}\Sigma_{v \text{ in } \mathbb{A}^n(E)} \psi_E(f(v)).$$

(12.6.2) Denote by $\mathcal{D}(n, e)(\mathrm{odd}) \subset \mathcal{D}(n,e)$ the linear subspace consisting of strongly odd Deligne polynomials, those in which every monomial which occurs has odd degree. Recall that the restriction of $\mathcal{M}(n,e,\psi)$ to $\mathcal{D}(n,e,\mathrm{odd})$ is self dual, symplectically if n is odd and orthogonally if n is even, cf. 3.10.6. Combining Theorem 3.10.7 and Theorem 6.8.35, we have the following result.

Theorem 12.6.3 Let k be a finite field of characteristic p, ℓ a prime with $\ell \ne p$, and ψ a nontrivial additive $\overline{\mathbb{Q}}_\ell^\times$-valued of k. Fix integers

$$n \ge 1, e \ge 3,$$

with e prime to p and odd. Suppose that any of the following six conditions holds:

a) $p \ge 7$,
b) $p \ne 3$ and $n \ge 3$,
c) $p = 5$ and $e \ge 7$,
d) $p = 3$ and $e \ge 7$,
e) p= 2 and $e \ge 7$.

Then we have the following results concerning the group G_{geom} for the lisse sheaf $\mathcal{M}(n,e,\psi)|\mathcal{D}(n,e,\mathrm{odd})$.

1) If n is odd, $G_{geom} = Sp((e-1)^n)$.

2) If n is even, $G_{geom} = O((e-1)^n)$.

(12.6.4) Fix integers $n \ge 1$ and $e \ge 3$, with e odd. Then in each characteristic $p \ge 7$ which is prime to $e\ell$, if we fix a nontrivial additive character ψ_p of $\mathbb{F}_p$, we get a lisse sheaf

$$\mathcal{M}(n,e,\psi_p)|\mathcal{D}(n,e,\mathrm{odd})\otimes\mathbb{F}_p$$

on the space

$$\mathcal{D}(n,e,odd)\otimes\mathbb{F}_p,$$

which always has the same G_{geom}, namely $Sp((e-1)^n)$ for odd n, and $O((e-1)^n)$ for even n.

(12.6.5) With the same same input data (n, e), we can form, in the same characteristics p, the sheaves $\mathcal{M}(n, e, \chi_2)$ on the spaces

$$\mathcal{S}\mathcal{D}(n, e)\otimes\mathbb{F}_p.$$

With the exception of the one case (n=2, e = 3), the sheaf we get in this way always has the same G_{geom} as its additive character analogue $\mathcal{M}(n,e,\psi_p)|\mathcal{D}(n,e,odd)\otimes\mathbb{F}_p$, namely $Sp((e-1)^n)$ for odd n, and $O((e-1)^n)$ for even n.

(12.6.6) As we have explained in the last section 12.5, the fact that the Kummer sheaf $\mathcal{L}_{\chi_2}$ "exists over $\mathbb{Z}$" leads to a uniformity theorem for the constants A and C in the estimates proving equidistribution for the sheaves $\mathcal{M}(n,e,\psi_p)|\mathcal{D}(n,e,odd)\otimes\mathbb{F}_p$, as p varies. [The uniformity for A is essentially trivial, it is the uniformity for C which is deep.] There is no analogous "existence over $\mathbb{Z}$" of the Artin-Schreier sheaves $\mathcal{L}_{\psi_p}$ as p varies, and as a result we cannot prove a uniformity for the analogous constant C in the in the estimates proving equidistribution for the sheaves $\mathcal{M}(n,e,\psi_p)|\mathcal{D}(n,e,odd)\otimes\mathbb{F}_p$ as p varies. Nonetheless, it seems quite likely that such a uniformity does in fact hold in this case.

(12.7) Application to generalized Weierstrass families of elliptic curves

(12.7.1) Let S be an arbitrary scheme, g ≥ 0, and

$$\rho : C \to S,$$

or simply C/S, a projective smooth curve with geometrically connected fibres, all of genus g. Let d be an integer with

$$d \geq \mathrm{Max}(2g-1, 0),$$

and D an effective Cartier divisor D in C which is finite and flat over S of degree d (with the convention that D is empty if d = 0). We recall from [Ka-TLFM, 6.1] some basic facts.

(12.7.2) Because d ≥ 2g-1, the functor on S-schemes Y/S given by

$$Y \mapsto H^0(C_Y, I^{-1}(D)_Y)$$

is representable by the vector bundle L(D) over S, of rank d + 1 - g. Inside L(D), the subfunctor which attaches to Y/S those global sections f in $H^0(C_Y, I^{-1}(D)_Y)$ which are invertible near D, and whose zero locus (as a section of $I^{-1}(D)_Y$) is finite etale over Y of degree d, is representable by an open subscheme

$$\mathit{Fct}(C, d, D, \emptyset) \subset L(D).$$

(12.7.3) Suppose now that d ≥ 2g + 3. The fibre product over S of the spaces L(kD) for k = 1, 2, 3, 4, and 6,

$$L(D)\times_S L(2D)\times_S L(3D)\times_S L(4D)\times_S L(6D)$$

represents the functor on S-schemes Y/S given by

$Y/S \mapsto$ all tuples $(f_1, f_2, f_3, f_4, f_6)$, with f_k in $H^0(C_Y, I^{-1}(kD)_Y)$.
The discriminant Δ(f's) is a universal $\mathbb{Z}$-polynomial in the f_i, isobaric of weight 12, which defines an S-morphism

$$\Delta : L(D)\times_S L(2D)\times_S L(3D)\times_S L(4D)\times_S L(6D) \to L(12D).$$

Inside L(12D), we have the open set

$$Fct(C, 12d, 12D, \emptyset) \subset L(D).$$

We denote by

$$GW_{I_1 fd}(C, D) \subset L(D)\times_S L(2D)\times_S L(3D)\times_S L(4D)\times_S L(6D)$$

the open subscheme which is the inverse image of this open set:

$$GW_{I_1 fd}(C, D) := \Delta^{-1}(Fct(C, 12d, 12D, \emptyset)).$$

(12.7.4) Suppose now that $S = \mathrm{Spec}(A)$ is an affine $\mathbb{Z}[1/\ell]$-scheme of finite type. Over $\mathbb{A}^5{}_S$, with coordinates $(a_1, a_2, a_3, a_4, a_6)$, we have the affine curve in $\mathbb{A}^2$ over this base given by the universal generalized Weierstrass equation

$$E_{a's} : y^2 + a_1xy + a_3y = x^3 + a_2x^2 + a_4x + a_6,$$

with structural morphism

$$\pi_{a's} : E_{a's} \to \mathbb{A}^5{}_S.$$

(12.7.4.1) We obtain "strong standard input of type (0, 1)" relative to S, cf. 12.4.4, as follows. We take

the integer $m = 5$,

the fibrewise perverse object $K = R^1\pi_{a's!}\overline{\mathbb{Q}}_\ell(3)[5]$ on $\mathbb{A}^5{}_S$,

the affine S-scheme V/S given by $V := C - D$,

the S-morphism $h : V \to \mathbb{A}^5{}_S$ given by $h = 0$,

the fibrewise perverse object$L = \overline{\mathbb{Q}}_\ell[1]$ on V,

the integer $d = 1 - 2g + \deg(D)$,

the space of $\mathbb{A}^5$-valued functions $(\mathcal{F}, \tau)$ on V given by

$$\mathcal{F} := L(D)\times_S L(2D)\times_S L(3D)\times_S L(4D)\times_S L(6D),$$

with τ the obvious evaluation map.

Theorem 12.7.5 The strong standard input 12.7.4.1 gives uniform output of type (U, N, G), with

$$U := GW_{I_1 fd}(C, D),$$

$$N := 4g - 4 + 12\deg(D),$$

$$G := O(N).$$

proof This is simply a restatement of Lemma 10.1.12 and Theorem 10.1.14. QED

(12.7.6) Let us make explicit the general Uniform Output Theorem 12.4.18 in this case. For each finite field k, for each k-valued point s in S(k), and for each $f = (f_1, f_2, f_3, f_4, f_6)$ in $GW_{I_1 fd}(C, D)_{s,k}(k)$, we have the elliptic curve $E_{f's}$ over $k(C_{s,k})$, with equation

$$y^2 + f_1xy + f_3y = x^3 + f_2x^2 + f_4x + f_6,$$

and its unitarized L-function

$$L(E_{f's}/k(C_{s,k}),T)$$
$$= \det(1 - TFrob_{s,k,f} \mid H_c^1(C_{s,k}\otimes_k\overline{k}, j_*R^1\pi_{f's!}\overline{\mathbb{Q}}_\ell)(1)).$$

There is a unique conjugacy class $\vartheta(k, s, f)$ in $O(N, \mathbb{R})$ such that

$$L(E_{f's}/k(C_{s,k}),T) = \det(1 - T\vartheta(k, s, f)).$$

For k a large finite field, and any s in S(k), these conjugacy classes are nearly equidistributed.

Theorem12.7.7 For each finite field k with $\#k > A(GW_{I_1 fd}(C, D)/S)^2$, and for each s in S(k), form the probability measure on $O(N, \mathbb{R})^\#$ given by

$$\mu(k, s)$$
$$:= \text{average over the classes } \vartheta(k, s, f), \text{ f in } GW_{I_1 fd}(C, D)_{s,k}(k)$$
$$:= (1/\#GW_{I_1 fd}(C, D)_{s,k}(k)) \textstyle\sum_{f \text{ in } GW_{I_1,fd}(C,D)_{s,k}(k)} \delta_{\vartheta(k, s, f)}.$$

For any sequence of data (k_i, s_i) in which each $\#k_i > A(U/S)^2$ and in which $\#k_i$ is strictly increasing, the measures $\mu(k_i, s_i)$ on $O(N, \mathbb{R})^\#$ converge weak $*$ to $\mu\#$. More precisely, there exist constants A and C such that for any irreducible nontrivial irreducible unitary representation Λ of $O(N, \mathbb{R})$, and for any data (k, s) with $\#k > 4A^2$, we have the estimate

$$\left|\int_{O(N, \mathbb{R})} \text{Trace}(\Lambda)d\mu(k, s)\right| \le 2C\dim(\Lambda)/\text{Sqrt}(\#k).$$

(12.8) Application to usual Weierstrass families of elliptic curves

(12.8.1) We work over an affine $\mathbb{Z}[1/6\ell]$-scheme S = Spec(A) of finite type. We fix a genus $g \ge 0$, and two integers

$$d_2 \ge 2g + 3,\ d_3 \ge 2g + 3.$$

We are given

$$\rho : C \to S,$$

or simply C/S, a projective smooth curve with geometrically connected fibres, all of genus g, and two effective Cartier divisors D_2 and D_3 in C, both finite and flat over S, of degrees d_2 and d_3 respectively. We are also given a finite collection of pairwise disjoint sections

$$P_i \text{ in } C(S),\ i \text{ in } I,$$

and collections of integers $a_i \ge 0$ and $b_i \ge 0$, I in I, such that

$$D_2 = \textstyle\sum_i a_iP_i,\ D_3 = \sum_i b_iP_i.$$

We define integers $c_i \ge 0$ by

$$c_i := \text{Max}(3a_i, 2b_i),\ i \text{ in } I.$$

We define the divisor

$$D_{max} = \mathrm{Max}(3D_2, 2D_3)$$

to be

$$D_{max} := \Sigma_i c_i P_i.$$

(12.8.2) Formation of the discriminant

$$\Delta(f_2, f_3) := (f_2)^3 - 27(f_3)^2$$

defines an S-morphism

$$\Delta : L(D_2)\times_S L(D_3) \to L(D_{max}).$$

In $L(D_{max})$, we have the open set

$$Fct(C, \deg(D_{max}), D_{max}, \emptyset) \subset L(D_{max}).$$

We denote by

$$W_{I_1 fd}(C, D_2, D_3) \subset L(D_2)\times_S L(D_3)$$

the open subscheme which is the inverse image of this open set:

$$W_{I_1 fd}(C, D_2, D_3) := \Delta^{-1}(Fct(C, \deg(D_{max}), D_{max}, \emptyset)).$$

(12.8.3) Over $\mathbb{A}^2{}_S$, with coordinates (g_2, g_3), we have the affine curve in $\mathbb{A}^2$ over this base given by the universal (usual) Weierstrass equation

$$E_{g's} : y^2 = 4x^3 - g_2x - g_3,$$

with structural morphism

$$\pi_{g's} : E_{g's} \to \mathbb{A}^2{}_S.$$

(12.8.4) We obtain "strong standard input of type (1, 1)" relative to S, cf. 12.4.8, as follows. We take

the integer $m = 2$,

the fibrewise perverse sheaf object $= R^1\pi_{g's!}\overline{\mathbb{Q}}_\ell(1)[2]$ on $\mathbb{A}^2{}_S$,

the affine S-scheme V/S given by $V := C - D_{max}$,

the S-morphism $h : V \to \mathbb{A}^2{}_S$ given by $h = 0$,

the fibrewise perverse object $L = \overline{\mathbb{Q}}_\ell[1]$ on V/k,

the integer $d = 1 - 2g + \mathrm{Min}(\deg(D_2), \deg(D_3))$,

the space of $\mathbb{A}^2$-valued functions $(\mathcal{F}, \tau)$ on V given by

$$\mathcal{F} = L(D_2)\times_S L(D_3)$$

with τ the obvious evaluation map.

Theorem 12.8.5 Define the integer N by

$$N := 4g - 4 + \deg(D_{max}) + 2\#\{i \text{ in } I \text{ with } c_i \not\equiv 0 \bmod 12\}.$$

Suppose that

$$N \geq 9.$$

Suppose in addition that one of the following conditions a), b), or c) holds:

a) $D_{max} = 3D_2$,

b) $D_{max} = 2D_3$,

c) N is odd, i.e., $\deg(D_{max})$ is odd, and at least one of the following two conditions holds:

c1) $4g - 4 + 3\deg(D_2) \geq 10$,

c2) $4g - 4 + 2\deg(D_3) \geq 9$.

The above strong standard input gives uniform output of type (U, N, G), with

$U := W_{I_1 fd}(C, D_2, D_3)$,

N as above,

$G := O(N)$.

proof This is just a restatement of Lemma 10.2.11, Theorem 10.2.13, and Theorem 10.2.15 QED

(12.8.6) We now spell out the general Uniform Output Theorem 12.4.18 in this case. The statement is essentially identical to that given in Theorem 12.7.7 above for generalized Weierstrass families, except that $GW_{I_1 fd}(C, D)$ is replaced by $W_{I_1 fd}(C, D_2, D_3)$.

(12.8.7) For each finite field k, each k-valued point s in S(k), and each $f = (f_2, f_3)$ in $W_{I_1 fd}(C, D_2, D_3)_{s,k}(k)$, we have the elliptic curve $E_{f's}$ over $k(C_{s,k})$, with equation

$$y^2 = 4x^3 - f_2x - f_3,$$

and its unitarized L-function

$$L(E_{f's}/k(C_{s,k}),T)$$
$$= \det(1 - T\mathrm{Frob}_{s,k,f} \mid H_c^1(C_{s,k}\otimes_k \bar{k}, j_\star R^1\pi_{f's!}\bar{\mathbb{Q}}_\ell)(1)).$$

There is a unique conjugacy class $\vartheta(k, s, f)$ in $O(N, \mathbb{R})$ such that

$$L(E_{f's}/k(C_{s,k}),T) = \det(1 - T\vartheta(k, s, f)).$$

For k a large finite field, and any s in S(k), these conjugacy classes are nearly equidistributed.

Theorem 12.8.8 For $A_0 := A(W_{I_1 fd}(C, D_2, D_3)/S)$, for each finite field k with $\#k > A_0{}^2$, and for each s in S(k), form the probability measure on $O(N, \mathbb{R})^\#$ given by

$\mu(k, s)$

$:=$ average over the classes $\vartheta(k, s, f)$, f in $W_{I_1 fd}(C, D_2, D_3)_{s,k}(k)$.

For any sequence of data (k_i, s_i) in which each $\#k_i > A_0{}^2$ and in which $\#k_i$ is strictly increasing, the measures $\mu(k_i, s_i)$ on $O(N, \mathbb{R})^\#$ converge weak $\star$ to $\mu^\#$. More precisely, there exist constants A and C such that for any irreducible nontrivial irreducible unitary representation Λ of $O(N, \mathbb{R})$, and for any data (k, s) with $\#k > 4A^2$, we have the estimate

$$\left|\int_{O(N,\, \mathbb{R})} \mathrm{Trace}(\Lambda)d\mu(k, s)\right| \leq 2C\dim(\Lambda)/\mathrm{Sqrt}(\#k).$$

(12.9) Application to FJTwist families of elliptic curves

(12.9.1) Again in this section we work over an affine $\mathbb{Z}[1/6\ell]$-scheme S = Spec(A) of finite type. We fix a genus $g \geq 0$, and two integers

$$d_0 \geq 2g + 3,\ d_1 \geq 2g + 3.$$

We are given

$$\rho : C \to S,$$

or simply C/S, a projective smooth curve with geometrically connected fibres, all of genus g, and two effective Cartier divisors D_0 and D_1 in C, both finite and flat over S, of degrees d_0 and d_1 respectively. We are also given a finite collection of pairwise disjoint sections

$$P_i \text{ in } C(S),\ i \text{ in } I,$$

and collections of integers $a_i \geq 0$ and $b_i \geq 0$, i in I, such that

$$D_0 = \Sigma_i a_i P_i,\ D_1 = \Sigma_i b_i P_i.$$

We define integers $c_i \geq 0$ by

$$c_i := 2a_i + b_i,\ i \text{ in } I.$$

(12.9.2) We define an open set

$$\mathrm{FJTwist}(D_0, D_1) \subset L(D_0)\times_S L(D_1)$$

as follows. We have an S-morphism

$$\varphi : L(D_0)\times_S L(D_1) \to L(D_0 + 2D_1),$$
$$(f, g) \mapsto fg(g-1),$$

and inside $L(D_0 + 2D_1)$ we have the open set

$$\mathit{Fct}(C, d_0 + 2d_1, D_0 + 2D_1, \varnothing) \subset L(D_0 + 2D_1).$$

Then we define

$$\mathrm{FJTwist}(D_0, D_1) := \varphi^{-1}\mathit{Fct}(C, d_0 + 2d_1, D_0 + 2D_1, \varnothing).$$

(12.9.3) Over $\mathbb{A}^2{}_S$, with coordinates (s, t), we have the affine curve in $\mathbb{A}^2$ over this base given by the FJTwist Weierstrass equation

$$E_{s,t} : y^2 = 4x^3 - 3s^2tx - s^3t,$$

with structural morphism

$$\pi_{s,t} : E_{s,t} \to \mathbb{A}^2{}_S.$$

(12.9.4) We obtain "strong standard input of type (1, 1)" relative to S, cf. 12.4.8, as follows. We take

the integer m = 2,

the fibrewise perverse object = $R^1\pi_{s,t!}\overline{\mathbb{Q}}_\ell(1)[2]$ on $\mathbb{A}^2{}_S$,

the affine k-scheme V/S given by $V := C - (D_0 \cup D_1)$,

the S-morphism $h : V \to \mathbb{A}^2$ given by h = 0,

the fibrewise perverse object $L = \overline{\mathbb{Q}}_\ell[1]$ on V/S,

the integer $d = 1 - 2g + \min(\deg(D_0), \deg(D_1))$,

the space of $\mathbb{A}^2$-valued functions $(\mathcal{F}, \tau)$ on V given by

$$\mathcal{F} = L(D_0)\times_S L(D_1)$$

with τ the obvious evaluation map.

Theorem 12.9.5 Define the integer N by

$$N = 4g - 4 + 2\deg(D_0) + 3\deg(D_1) + 2\#\{i \text{ in } I \text{ with } c_i \not\equiv 0 \bmod 4\}.$$

Then the above strong standard input gives uniform output of type (U, N, G), with

U := FJTwist(D_0, D_1),

N as above,

G := O(N).

proof This is just a restatement of Lemma 11.1.21 and Theorem 11.1.23. QED

(12.9.6) We now spell out the general Uniform Output Theorem 12.4.18 in this case. The statement is essentially identical to that given in 12.8.8 above for usual Weierstrass families, except that $W_{I_1 fd}(C, D_2, D_3)$ is replaced by FJTwist(D_0, D_1).

(12.9.7) For each finite field k, each k-valued point s in S(k), and each = (f, g) in FJTwist$(D_0, D_1)_{s,k}(k)$, we have the elliptic curve $E_{f,g}$ over $k(C_{s,k})$, with equation

$$y^2 = 4x^3 - 3f^2gx - f^3g$$

and its unitarized L-function

$$L(E_{f,g}/k(C_{s,k}),T)$$
$$= \det(1 - T\mathrm{Frob}_{s,k,f,g} \mid H_c^1(C_{s,k}\otimes_k \bar{k}, j_\star R^1\pi_{f,g}\bar{\mathbb{Q}}_\ell)(1)).$$

There is a unique conjugacy class $\vartheta(k, s, f, g)$ in O(N, $\mathbb{R}$) such that

$$L(E_{f,g}/k(C_{s,k}),T) = \det(1 - T\vartheta(k, s, f, g)).$$

For k a large finite field, and any s in S(k), these conjugacy classes are nearly equidistributed.

Theorem 12.9.8 For A_0 := A(FJTwist(D_0, D_1)/S), for each finite field k with $\#k > A_0^2$, and for each s in S(k), form the probability measure on O(N, $\mathbb{R}$)$^\#$ given by

$\mu(k, s)$

:= average over the classes $\vartheta(k, s, f, g)$, (f, g) in FJTwist$(D_0, D_1)_{s,k}(k)$

For any sequence of data (k_i, s_i) in which each $\#k_i > A^2$ and in which $\#k_i$ is strictly increasing, the measures $\mu(k_i, s_i)$ on O(N, $\mathbb{R}$)$^\#$ converge weak $\star$ to $\mu\#$. More precisely, there exist constants A and C such that for any irreducible nontrivial irreducible unitary representation Λ of O(N, $\mathbb{R}$), and for any data (k, s) with $\#k > 4A^2$, we have the estimate

$$\left|\int_{O(N,\, \mathbb{R})} \mathrm{Trace}(\Lambda)d\mu(k, s)\right| \le 2C\dim(\Lambda)/\mathrm{Sqrt}(\#k).$$

(12.10) Applications to pullback families of elliptic curves

(12.10.1) In this section we work over a normal connected affine $\mathbb{Z}[1/6\ell]$-scheme $S = \mathrm{Spec}(A)$ of finite type. We begin with the affine line $\mathbb{A}^1_S$ over S. We suppose given an integer $t \geq 1$ and an effective Cartier divisor T in $\mathbb{A}^1_S$, which is finite etale over S of degree t. We further suppose that T is defined by the vanishing of a monic polynomial T(x) in A[x] which is monic of degree t, and whose discriminant is a unit in A. We suppose given a relative elliptic curve

$$\pi : \mathcal{E} \to \mathbb{A}^1_S - T.$$

We suppose that, on each geometric fibre of $\mathbb{A}^1_S/S$, $\mathcal{E}$ has multiplicative reduction at some point of T. This $\mathcal{E}$ over $\mathbb{A}^1_S - T$ is our input elliptic curve. On $\mathbb{A}^1_S - T$, we form the lisse sheaf

$$\mathcal{G} := R^1\pi_*\overline{\mathbb{Q}}_\ell.$$

Because 6 is invertible on S, the restriction of $\mathcal{G}$ to every geometric fibre of $(\mathbb{A}^1_S - T)/S$ is everywhere tame (i.e., tame at ∞ and at each geometric point of T).

(12.10.2) The pullback families arise as follows. We fix a genus $g \geq 0$, and an integer $d \geq 2g + 3$. We are given

$$\rho : C \to S,$$

or simply C/S, a projective smooth curve with geometrically connected fibres, all of genus g, and an effective Cartier divisors D in C, finite and flat over S, of degree d. We are also given a finite collection of pairwise disjoint sections

$$P_i \text{ in } C(S),\ i \text{ in } I,$$

and a collection of integers $a_i \geq 1$, i in I, such that

$$D = \Sigma_i\, a_iP_i.$$

Thus

$$D^{red} = \Sigma_i\, P_i$$

is finite etale over the base S.

(12.10.3) With this plethora of hypotheses stated, we now define an open set

$$U_{D,T} \subset L(D).$$

For this, we use the degree t polynomial T(x) defining $T \subset \mathbb{A}^1_S$ to construct an S-morphism

$$T : L(D) \to L(tD),$$
$$f \mapsto T(f).$$

We define $U_{D,T}$ to be the inverse image of the open set

$$Fct(C, td, tD, \varnothing) \subset L(D)$$

under the map T:

$$U_{D,T} := T^{-1}Fct(C, td, tD, \varnothing).$$

(12.10.4) Each f in $U_{D,T}$ is a morphism

$$f : C - f^{-1}T - D \to \mathbb{A}^1 - T.$$

Over $U_{D,T}$, the $f^{-1}T$ form a divisor $f_{univ}{}^{-1}T$ in $C\times U_{D,T}$, which is finite etale over $U_{D,T}$ of degree td, and disjoint from D. Over $U_{D,T}$, the f's form a $U_{D,T}$-morphism

$$f_{univ} : (C - D)\times U_{D,T} - f_{univ}{}^{-1}T \to (\mathbb{A}^1{}_S - T)\times U_{D,T}.$$

Denote by

$$j : \mathbb{A}^1{}_S - T \to \mathbb{P}^1{}_S,$$

$$j_C : (C - D)\times U_{D,T} - f_{univ}{}^{-1}T \to C\times U_{D,T}$$

the inclusions. In both cases, we are dealing with the inclusion into a proper smooth curve of the complement of a divisor (either $T \amalg \infty$ or $f_{univ}{}^{-1}T \amalg D^{red}$) which is finite etale over the base. On the sources, we have the lisse sheaves

$$\mathcal{G} \text{ on } \mathbb{A}^1 - T,$$

$$f_{univ}{}^*(\mathcal{G}\boxtimes\overline{\mathbb{Q}}_\ell) \text{ on } (C - D)\times U_{D,T} - f_{univ}{}^{-1}T,$$

which are tamely ramified along the "missing" divisors. It follows [Ka-SE, 4.7.1] that the formation of

$$j_*(\mathcal{G}) \text{ on } \mathbb{P}^1{}_S,$$

and the formation of

$$j_{C*}(f_{univ}{}^*(\mathcal{G}\boxtimes\overline{\mathbb{Q}}_\ell)) \text{ on } C\times U_{D,T}$$

are compatible with arbitrary change of base on S and on $U_{D,T}$ respectively and that the cohomology sheaves

$$R^i(pr_2)_*(j_*\mathcal{G}), \text{ for } pr_2 : \mathbb{P}^1{}_S \to S,$$

and

$$R^i(pr_2)_*(j_{C*}(f_{univ}{}^*(\mathcal{G}\boxtimes\overline{\mathbb{Q}}_\ell))), \text{ for } pr_2 : C\times U_{D,T} \to U_{D,T},$$

are lisse on S and on $U_{D,T}$ respectively.

(12.10.5) Looking fibre by fibre, we see that these R^i vanish for $i \neq 1$. We define integers

$$N_{down} := \text{rank of } R^1(pr_2)_*(j_*(\mathcal{G})),$$

$$N_{up} := \text{rank of } R^1(pr_2)_*(j_{C*}(f_{univ}{}^*(\mathcal{G}\boxtimes\overline{\mathbb{Q}}_\ell))),$$

$$N := N_{up} - N_{down}.$$

(12.10.6) We obtain "strong input of type (0,1)" relative to S, cf. 12.4.8, as follows. We take

the integer $m = 1$,

the fibrewise perverse object $j_*(\mathcal{G})(1)[1]|\mathbb{A}^1{}_S$ on $\mathbb{A}^1{}_S$,

the affine S-scheme $V := C-D$,

the S-morphism $h : V \to \mathbb{A}^1$ given by $h = 0$,

the fibrewise perverse object $L := \overline{\mathbb{Q}}_\ell[1]$ on C-D,

the integer $d := \deg(D) - (2g-1)$,

the space of $\mathbb{A}^1$valued functions $(\mathcal{F}, \tau)$ on C-D given by

$\mathcal{F} = L(D)$

with τ the obvious evaluation map.

Theorem 12.10.7 Suppose $N \geq 9$. Then the above strong standard input gives uniform output of type (U, N, G), with

$U := U_{D,T}$,

N as above,

$G := O(N)$.

proof This is just a restatement of Lemma 7.1.7 and Theorem 7.2.3. QED

(12.10.8) We now spell out the general Uniform Output Theorem 12.4.18 in this case.

(12.10.9) For each finite field k, each k-valued point s in S(k), and each f in $U_{D,T,s,k}(k)$, we have the elliptic curve $f^*\mathcal{E}$ over $k(C_{s,k})$, and its unitarized L-function

$$L(f^*\mathcal{E}_{s,k}/k(C_{s,k}),T)$$
$$= \det(1 - T\mathrm{Frob}_{s,k,f} \mid H_c^1(C_{s,k}\otimes_k \bar{k}, j_{C_{s,k}*}\mathcal{G}(1))).$$

This L-function is a polynomial of degree N_{up}. It is always divisible, as a polynomial, by the unitarized L-function

$$L(\mathcal{E}_{s,k}/k(\mathbb{P}^1),T)$$

of the original elliptic curve $\mathcal{E}_{s,k}/k(\mathbb{P}^1)$, which is a polynomial of degree N_{down}. There is a unique conjugacy class $\vartheta(k, s, f)$ in $O(N, \mathbb{R})$ such that

$$L(f^*\mathcal{E}_{s,k}/k(C_{s,k}),T)/L(\mathcal{E}_{s,k}/k(\mathbb{P}^1), T) = \det(1 - T\vartheta(k, s, f)).$$

For k a large finite field, and any s in S(k), these conjugacy classes are nearly equidistributed.

Theorem 12.10.10 For $A_0 := A(U_{D,T}/S)$, for each finite field k with $\#k > A_0^2$, and for each s in S(k), form the probability measure on $O(N, \mathbb{R})^\#$ given by

$\mu(k, s)$

$:=$ average over the classes $\vartheta(k, s, f)$, f in $U_{D,T,s,k}(k)$.

For any sequence of data (k_i, s_i) in which each $\#k_i > A^2$ and in which $\#k_i$ is strictly increasing, the measures $\mu(k_i, s_i)$ on $O(N, \mathbb{R})^\#$ converge weak $*$ to $\mu\#$. More precisely, there exist constants A and C such that for any irreducible nontrivial irreducible unitary representation Λ of $O(N, \mathbb{R})$, and for any data (k, s) with $\#k > 4A^2$, we have the estimate

$$|\textstyle\int_{O(N,\ \mathbb{R})} \mathrm{Trace}(\Lambda)d\mu(k, s)| \leq 2C\dim(\Lambda)/\mathrm{Sqrt}(\#k).$$

(12.10.11) Three examples Here are three standard examples, where we give N_{down} and N_{up}.

(12.10.11.1) **FJ family** Over $S = \mathrm{Spec}(\mathbb{Z}[1/6\ell])$, take $T := \{0, 1\}$, and over $\mathbb{A}^1 - T$ with coordinate u take the relative elliptic curve $\mathcal{E}$ given by the FJ equation

$$y^2 = 4x^3 - 3ux - u.$$

This curve has multiplicative reduction at u=1, additive reduction at u=0 which becomes good after taking the sixth root of u, and additive reduction at ∞ which becomes good after taking the fourth root of u. We have

$$N_{down} = 1.$$

For $D = \Sigma_{i \text{ in } I}\, a_iP_i$ on C of degree d, and f in $U_{D,T}$, $f^*\mathcal{E}$ has additive reduction at the d geometric points over u=0, multiplicative reduction at the d geometric points over u=1, additive reduction at those points P_i with $a_i \not\equiv 0 \bmod 4$, and good reduction at those points P_i with $a_i \equiv 0 \bmod 4$. We have

$$N_{up} = 4g - 4 + 3d + 2\#\{i \text{ in } I \text{ with } a_i \not\equiv 0 \bmod 4\}.$$

(12.10.11.2) **Legendre family** Over $S = \mathrm{Spec}(\mathbb{Z}[1/2\ell])$, take $T := \{0, 1\}$, and over $\mathbb{A}^1 - T$ with coordinate λ take the relative elliptic curve $\mathcal{E}$ given by the Legendre equation

$$y^2 = x(x-1)(x-\lambda).$$

This curve has multiplicative reduction at u-0 and at u=1, and additive reduction at ∞ which after a quadratic twist becomes multiplicative. We have

$$N_{down} = 0.$$

For $D = \Sigma_{i \text{ in } I}\, a_iP_i$ on C of degree d, and f in $U_{D,T}$, $f^*\mathcal{E}$ has multiplicative reduction at the d geometric points over u=0, multiplicative reduction at the d geometric points over u=1, additive reduction at those points P_i with a_i odd, and multiplicative reduction at those points P_i with a_i even. We have

$$N_{up} = 4g - 4 + 2d + + \#I + \#\{i \text{ in } I \text{ with } a_i \text{ odd}\}.$$

(12.10.11.3) **Level 3 family** Over $S = \mathrm{Spec}(\mathbb{Z}[1/3\ell])$, take $T = \mu_3$, and over $\mathbb{A}^1 - T$ with coordinate μ take the relative elliptic curve $\mathcal{E}$ given by the "level 3" projective equation

$$X^3 + Y^3 + Z^3 = 3\mu XYZ,$$

with (0, 1, -1) taken as origin. This curve has multiplicative reduction at each cube root of unity, and at ∞. We have

$$N_{down} = 0.$$

For $D = \Sigma_{i \text{ in } I}\, a_iP_i$ on C of degree d, and f in $U_{D,T}$, $f^*\mathcal{E}$ has multiplicative reduction at the 3d geometric points over μ_3, and at each P_i.We have

$$N_{up} = 4g - 4 + 3d + + \#I.$$

(12.11) Application to quadratic twist families of elliptic curves

(12.11.1). We work over an affine $\mathbb{Z}[1/2\ell]$-scheme $S = \mathrm{Spec}(A)$ of finite type. We fix a genus $g \geq 0$, and integers $d \geq 2g + 3$, and $t \geq 1$. We are given

$$\rho : C \to S,$$

or simply C/S, a projective smooth curve with geometrically connected fibres, all of genus g, and effective Cartier divisors D and T in C, both finite and flat over S, of degrees d and t respectively. We assume that D and T are scheme-theoretically disjoint, and that T is finite etale over S. We are also given a finite collection of pairwise disjoint sections

$$P_i \text{ in } C(S),\ i \text{ in } I,$$

and a collection of integers $a_i \geq 1$, i in I, such that

$$D = \Sigma_i\ a_i P_i.$$

(12.11.2) Over $C - D - T$, we are given a relative elliptic curve

$$\pi : \mathcal{E} \to C - D - T.$$

We assume that on each geometric fibre of $(C - D - T)/S$, $\mathcal{E}$ has multiplicative reduction at some geometric point of T, and is everywhere tamely ramified (this last condition is automatic if 6 is invertible).

(12.11.3) We obtain "strong standard input of type (1, 1)" relative to S, cf. 12.4.8, as follows. We take

the integer $m = 1$,

the fibrewise perverse object $K := j_*\mathcal{L}_{\chi_2}[1]$ on $\mathbb{A}^1{}_S$,

the affine S-scheme $V := C - D - T$,

the S-morphism $h : V \to \mathbb{A}^1$ given by $h = 0$,

the fibrewise perverse object $L := R^1\pi_*\overline{\mathbb{Q}}_\ell[1]$ on $C - D - T$,

the integer $d := \deg(D) - (2g-1)$,

the space of $\mathbb{A}^1$-valued functions $(\mathcal{F}, \tau)$ on V given by

$$\mathcal{F} = L(D),$$

with τ the obvious evaluation map.

(12.11.4) Recall from [Ka-TLFM, 6.1] that inside L(D), we have the open subscheme

$$Fct(C\ d, D, T) \subset L(D),$$

which represents the subfunctor of L(D) which attaches to Y/S those global sections f in $H^0(C_Y, I^{-1}(D)_Y)$ which are invertible near D, invertible near T, and whose zero locus (as a section of $I^{-1}(D)_Y$) is finite etale over Y of degree d.

(12.11.5) On $(C - D - T)\times Fct(C\ d, D, T)$, we have the lisse sheaf

$$R^1\pi_*\overline{\mathbb{Q}}_\ell\otimes\mathcal{L}_{\chi_2(f)},$$

which is everywhere tamely ramified. If we denote by j the inclusion

$$j : (C - D - T)\times Fct(C\ d, D, T) \to C\times Fct(C\ d, D, T),$$

then the formation of

$$j_*(R^1\pi_*\overline{\mathbb{Q}}_\ell \otimes \mathcal{L}_{\chi_2(f)})$$

commutes with arbitrary change of base on *Fct*(C d, D, T), and the cohomology sheaves

$$R^i(pr_2)_*(j_*(R^1\pi_*\overline{\mathbb{Q}}_\ell \otimes \mathcal{L}_{\chi_2(f)}))$$

are lisse on *Fct*(C d, D, T). Looking fibre by fibre, we see that these R^i vanish for $i \neq 1$. We define the integer N by

$$N := \text{rank of } R^1(pr_2)_*(j_*(R^1\pi_*\overline{\mathbb{Q}}_\ell \otimes \mathcal{L}_{\chi_2(f)})).$$

Theorem 12.11.6 Suppose $N \geq 9$. Then the above strong standard input gives uniform output of type (U, N, G), with

U := *Fct*(C d, D, T),
N as above,
G := O(N).

proof This is just a restatement of Lemma 8.1.11 and Corollary 8.2.3. QED

(12.11.7) For any finite field k, for any k-valued point s in S(k), and for any f in $Fct(C\ d, D, T)_{s,k}(k)$, we have an elliptic curve $\mathcal{E}_{s,k}/k(C_{s,k})$, and its quadratic twist $\mathcal{E}_{s,k}\otimes\chi_2(f)/k(C_{s,k})$ by f. Its unitarized L-function

$$L(\mathcal{E}_{s,k}\otimes\chi_2(f)/k(C_{s,k}), T)$$
$$= \det(1 - T\mathrm{Frob}_{s,k,f} \mid H_c^1(C_{s,k}\otimes_k\overline{k}, j_*(R^1\pi_*\overline{\mathbb{Q}}_\ell\otimes\mathcal{L}_{\chi_2(f)})(1))).$$

is a polynomial of degree N. There is a unique conjugacy class $\vartheta(k, s, f)$ in $O(N, \mathbb{R})$ such that

$$L(\mathcal{E}_{s,k}\otimes\chi_2(f)/k(C_{s,k}), T) = \det(1 - T\vartheta(k, s, f)).$$

(12.11.8) For k a large finite field, and any s in S(k), these conjugacy classes are nearly equidistributed.

Theorem 12.11.9 For U := *Fct*(C d, D, T), A_0 := A(U/S), for each finite field k with $\#k > A_0^2$, and for each s in S(k), form the probability measure on $O(N, \mathbb{R})^\#$ given by

$\mu(k, s)$
:= average over the classes $\vartheta(k, s, f)$, f in $U_{s,k}(k)$.

For any sequence of data (k_i, s_i) in which each $\#k_i > A^2$ and in which $\#k_i$ is strictly increasing, the measures $\mu(k_i, s_i)$ on $O(N, \mathbb{R})^\#$ converge weak $*$ to $\mu\#$. More precisely, there exist constants A and C such that for any irreducible nontrivial irreducible unitary representation Λ of $O(N, \mathbb{R})$, and for any data (k, s) with $\#k > 4A^2$, we have the estimate

$$\left|\int_{O(N,\ \mathbb{R})} \mathrm{Trace}(\Lambda)d\mu(k, s)\right| \leq 2C\dim(\Lambda)/\mathrm{Sqrt}(\#k).$$

(12.11.10) Three examples Here are the same three standard examples, where we now give N for various kinds of twist families.
(12.11.10.1) **FJ family** Over S = Spec($\mathbb{Z}[1/6\ell]$), take the relative elliptic curve $\mathcal{E}$ over $\mathbb{P}^1 - \{0, 1, \infty\}$ given by the FJ equation

$$y^2 = 4x^3 - 3ux - u.$$

This curve has multiplicative reduction at u=1, additive reduction at u=0 which becomes good after taking the sixth root of u, and additive reduction at ∞ which becomes good after taking the fourth root of u.

Take T = {0, 1}, D = $a_\infty\infty + \Sigma_{i \text{ in } I}\, a_iP_i$. For f in *Fct*($\mathbb{A}^1$, d, D, T), $\mathcal{E}\otimes\chi_2(f)$ has additive reduction at u=0, at the d geometric zeroes of f, at ∞ and at all the points P_i, i in I, with a_i odd. It has multiplicative reduction at u=1. We have

$$N = 2d + 2\#\{i \text{ in } I \text{ with } a_i \text{ odd}\} - 1.$$

If we take T = {0, 1, ∞} and D = $\Sigma_{i \text{ in } I}\, a_iP_i$, then for f in *Fct*($\mathbb{A}^1$, d, D, T), $\mathcal{E}\otimes\chi_2(f)$ has exactly the same reduction types, and the same N.

Finally, take T = {1} and D = $a_0 0 + a_\infty\infty + \Sigma_{i \text{ in } I}\, a_iP_i$. Again for f in *Fct*($\mathbb{A}^1$, d, D, T), $\mathcal{E}\otimes\chi_2(f)$ has exactly the same reduction types, and the same N.
(12.11.10.2) **Legendre family** Over S = Spec($\mathbb{Z}[1/2\ell]$), take the relative elliptic curve $\mathcal{E}$ over $\mathbb{P}^1 - \{0, 1, \infty\}$ given by the Legendre equation

$$y^2 = x(x-1)(x-\lambda).$$

This curve has multiplicative reduction at λ=0 and at λ=1, and additive reduction at ∞ which after a quadratic twist becomes multiplicative.

Take T = {0, 1}, D = $a_\infty\infty + \Sigma_{i \text{ in } I}\, a_iP_i$. For f in *Fct*($\mathbb{A}^1$, d, D, T), $\mathcal{E}\otimes\chi_2(f)$ has multiplicative reduction at λ=0 and at λ=1. It has multiplicative reduction at ∞ if a_∞ is odd, otherwise it has additive reduction at ∞. At the d geometric zeroes of f, and at all the points P_i, i in I, with a_i odd, it has additive reduction. We have

$$N = 2d + 2\#\{i \text{ in } I \text{ with } a_i \text{ odd}\} - \varepsilon_\infty,$$

with

$$\varepsilon_\infty = 1 \text{ if } a_\infty \text{ is odd, } 0 \text{ if } a_\infty \text{ is even.}$$

If instead we take T = {0, 1, ∞} and D = $\Sigma_{i \text{ in } I}\, a_iP_i$, then for f in *Fct*($\mathbb{A}^1$, d, D, T), $\mathcal{E}\otimes\chi_2(f)$ has the same reduction types as in the a_∞ even case, and we have

$$N = 2d + 2\#\{i \text{ in } I \text{ with } a_i \text{ odd}\}.$$

Take T = {1}, and D = $a_0 0 + a_\infty\infty + \Sigma_{i \text{ in } I}\, a_iP_i$. Then for f in

$Fct(\mathbb{A}^1, d, D, T)$, $\mathcal{E}\otimes\chi_2(f)$ has multiplicative reduction at $\lambda=1$. It has multiplicative reduction at $\lambda=0$ if a_0 is even, otherwise it has additive reduction at $\lambda=0$. It has multiplicative reduction at ∞ if a_∞ is odd, otherwise it has additive reduction at ∞. At the d geometric zeroes of f, and at all the points P_i, i in I, with a_i odd, it has additive reduction. So we get

$$N = 2d + 2\#\{i \text{ in } I \text{ with } a_i \text{ odd}\} + 1 - \varepsilon_0 - \varepsilon_\infty,$$

with

$$\varepsilon_0 = 1 \text{ if } a_0 \text{ is even, } 0 \text{ if } a_0 \text{ is odd,}$$
$$\varepsilon_\infty = 1 \text{ if } a_\infty \text{ is odd, } 0 \text{ if } a_\infty \text{ is even.}$$

Similarly for $T = \{0\}$, and $D = a_1 1 + a_\infty\infty + \Sigma_{i \text{ in } I}\, a_iP_i$, we find

$$N = 2d + 2\#\{i \text{ in } I \text{ with } a_i \text{ odd}\} + 1 - \varepsilon_1 - \varepsilon_\infty,$$

with

$$\varepsilon_1 = 1 \text{ if } a_1 \text{ is even, } 0 \text{ if } a_1 \text{ is odd,}$$
$$\varepsilon_\infty = 1 \text{ if } a_\infty \text{ is odd, } 0 \text{ if } a_\infty \text{ is even.}$$

(12.11.10.3) **Level 3 family** Over $S = \mathrm{Spec}(\mathbb{Z}[1/3\ell])$, take the relative elliptic curve $\mathcal{E}$ $\mathbb{P}^1 - \{\mu_3, \infty\}$ given by the "level 3" projective equation

$$X^3 + Y^3 + Z^3 = 3\mu XYZ,$$

with $(0, 1, -1)$ taken as origin. This curve has multiplicative reduction at each cube root of unity, and at ∞.

Take $D = \Sigma_{\zeta \text{ in } \mu_3}\, a_\zeta\zeta + a_\infty\infty + \Sigma_{i \text{ in } I}\, a_iP_i$, such that either a_∞ or some a_ζ vanishes. Take for T those points in $\{\mu_3, \infty\}$ which are absent from D. For f in $Fct(\mathbb{A}^1, d, D, T)$, $\mathcal{E}\otimes\chi_2(f)$ has multiplicative reduction at those points in $\{\mu_3, \infty\}$ which are either absent from D, or occur in D with even multiplicity. It has additive reduction at the points which occur in D with odd multiplicity, and at the d geometric zeroes of f. We have

$$N = 2d + 2\#\{i \text{ in } I \text{ with } a_i \text{ odd}\} + 4 - \varepsilon_\infty - \Sigma_{\zeta \text{ in } \mu_3}\, \varepsilon_\zeta,$$

with

$$\varepsilon_\infty = 0 \text{ if } \infty \text{ occurs in D with odd multiplicity, 1 if not,}$$
$$\varepsilon_\zeta = 0 \text{ if } \zeta \text{ occurs in D with odd multiplicity, 1 if not.}$$

Chapter 13: Average analytic rank and large N limits

(13.1) The basic setting

(13.1.1) In this chapter, we will make explicit the application of our results to L-functions of elliptic curves over function fields. We first give applications to average rank, along the lines of [deJ-Ka, 9.7] and [Ka-TLFM, 10.3]. We then pass to the large N limit, and give applications to the distribution of low lying zeroes, along the lines of [Ka-Sar-RMFEM, Chapters 12 and 13] and [Ka-TLFM, 10.5 and 10.6].

(13.1.2) To begin, we formulate a version, Theorem 13.1.5 below, of Uniform Output Theorem 12.4.18 in the case when G is the orthogonal group O(N), which pays attention to the "sign in the functional equation" ε. For ε either choice of ±1, let us denote by

$$O_{\mathrm{sign}\varepsilon}(N, \mathbb{R}) := \{A \text{ in } O(N, \mathbb{R}) \text{ with } \det(-A) = \varepsilon\}.$$

Thus for A in $O_{\mathrm{sign}\varepsilon}(N, \mathbb{R})$, the polynomial

$$P(T) = \det(1 - TA)$$

has the functional equation

$$T^N P(1/T) = \varepsilon P(T).$$

(13.1.3) For ε either choice of ±1, we denote by $\mu_{\mathrm{sign}\varepsilon}$ the restriction to $O_{\mathrm{sign}\varepsilon}(N, \mathbb{R})$ of twice (sic) the normalized Haar measure μ on $O(N, \mathbb{R})$, i.e., $\mu_{\mathrm{sign}\varepsilon}$ is normalized to give $O_{\mathrm{sign}\varepsilon}(N, \mathbb{R})$ total mass one. A function on $O_{\mathrm{sign}\varepsilon}(N, \mathbb{R})$ is said to be "central" if it is invariant under $O(N, \mathbb{R})$-conjugation. We denote by $O_{\mathrm{sign}\varepsilon}(N, \mathbb{R})^\#$ the quotient of $O_{\mathrm{sign}\varepsilon}(N, \mathbb{R})$ by the equivalence relation of $O(N, \mathbb{R})$-conjugation, and we denote by $\mu_{\mathrm{sign}\varepsilon}{}^\#$ the direct image of $\mu_{\mathrm{sign}\varepsilon}$ on $O_{\mathrm{sign}\varepsilon}(N, \mathbb{R})^\#$. A continuous $\mathbb{C}$-valued function f on $O_{\mathrm{sign}\varepsilon}(N, \mathbb{R})^\#$ is none other than a continuous central $\mathbb{C}$-valued function $\tilde{f}$ on $O_{\mathrm{sign}\varepsilon}(N, \mathbb{R})$, and

$$\int_{O_{\mathrm{sign}\varepsilon}(N, \mathbb{R})} f d\mu_{\mathrm{sign}\varepsilon} = \int_{O_{\mathrm{sign}\varepsilon}(N, \mathbb{R})^\#} \tilde{f} d\mu_{\mathrm{sign}\varepsilon}{}^\#$$
$$= 2\int_{O(N, \mathbb{R})} \{\tilde{f} \text{ extended by } 0\} d\mu.$$

Lemma 13.1.4 Suppose that we are in the situation of the Uniform Output Theorem 12.4.8, with constants A and C. Suppose in addition that G is the full orthogonal group O(N). For each choice of ε = ±1, each choice of finite field E with $\#E > \mathrm{Max}(4A^2, 4C^2)$, and for each s in S(E), the set

$$U_{s,E,\mathrm{sign}\varepsilon}(E) := \{x \text{ in } U_{s,E}(E) \text{ with } \vartheta(E, s, \alpha, x) \text{ in } O_{\mathrm{sign}\varepsilon}(N, \mathbb{R})\}$$

is nonempty. More precisely, for $\#E > 4A^2$, we have the estimate

$$|\#U_{s,E,\mathrm{sign}\varepsilon}(E)/\#U_{s,E}(E) - 1/2| \le C/E^{1/2}.$$

proof Recall from 12.4.18 that for $\#E > 4A^2$, we have

$$2(\#E)^n \ge \#U_{s,E}(E) \ge (1/2)(\#E)^n,$$

and for any irreducible nontrivial representation Λ of $O(N, \mathbb{R})$, we have

$$|\Sigma_{x \text{ in } U_{s,E}(E)} \mathrm{Trace}(\Lambda(\vartheta(E, s, \alpha, x))| \le C\dim(\Lambda)(\#E)^{n-1/2}.$$

We apply this with Λ the one-dimensional representation det. It is tautologous that we have

$$2\#U_{s,E,\mathrm{sign}\varepsilon}(E) = \Sigma_{x \text{ in } U_{s,E}(E)} (1 + \varepsilon\det(-(\vartheta(E, s, \alpha, x))))$$
$$= \#U_{s,E}(E) + \varepsilon\Sigma_{x \text{ in } U_{s,E}(E)} \det(-(\vartheta(E, s, \alpha, x))).$$

Thus we find

$$|2\#U_{s,E,\mathrm{sign}\varepsilon}(E) - \#U_{s,E}(E)| \le C(\#E)^{n-1/2}.$$

Dividing through by $2\#U_{s,E}(E) \ge (\#E)^n$, we find the desired estimate. QED

Theorem 13.1.5 Suppose that we are in the situation of the Uniform Output Theorem 12.4.8, with constants A and C. Suppose in addition that G is the full orthogonal group O(N). For each choice of $\varepsilon = \pm 1$, each choice of finite field E with $\#E > \mathrm{Max}(4A^2, 4C^2)$, and for each s in S(E), form the probability measure $\mu_{\mathrm{sign}\varepsilon}(E, s, \alpha)$ on $O_{\mathrm{sign}\varepsilon}(N, \mathbb{R})^\#$ defined by

$$\mu_{\mathrm{sign}\varepsilon}(E, s, \alpha)$$
$$:= (1/\#U_{s,E,\mathrm{sign}\varepsilon}(E))\ \Sigma_{x \text{ in } U_{s,E,\mathrm{sign}\varepsilon}(E)} \delta_{\vartheta(E, s, \alpha, x)}.$$

For either choice of $\varepsilon = \pm 1$, and for any sequence of data (E_i, s_i, α_i) with each $\#E_i > \mathrm{Max}(4A^2, 4C^2)$ and with $\#E_i$ strictly increasing, the measures $\mu_{\mathrm{sign}\varepsilon}(E_i, s_i, \alpha_i)$ on $O_{\mathrm{sign}\varepsilon}(N, \mathbb{R})^\#$ converge weak $*$ to $\mu_{\mathrm{sign}\varepsilon}{}^\#$, i.e., for any continuous $\mathbb{C}$-valued central function f on $O_{\mathrm{sign}\varepsilon}(N, \mathbb{R})$, we have

$$\int_{O_{\mathrm{sign}\varepsilon}(N, \mathbb{R})} f d\mu_{\mathrm{sign}\varepsilon}$$
$$= \lim_{i\to\infty}(1/\#U_{s_i,E_i,\mathrm{sign}\varepsilon}(E_i))\ \Sigma_{x \text{ in } U_{s_i,E_i,\mathrm{sign}\varepsilon}(E_i)} f(\vartheta(E_i, s_i, \alpha_i, x)).$$

proof Thanks to Lemma 13.1.4, this results from applying 12.4.18 to continuous central functions on $O(N, \mathbb{R})$ which are supported in one of the sets $O_{\mathrm{sign}\varepsilon}(N, \mathbb{R})$. QED

(13.1.6) We define three integer-valued functions on $O(N, \mathbb{R})$, as follows. Each has values in the closed interval [0, N]. The first is the "analytic rank",

$$r_{an}(A) := \mathrm{ord}_{T=1}(\det(1 - TA)).$$

The second is the "quadratic analytic rank", defined as

$r_{quad, an}(A) := r_{an}(A^2) = \Sigma_{\varepsilon \text{ in } \pm 1} \mathrm{ord}_{T=\varepsilon}(\det(1 - TA))$.

The third is the "geometric analytic rank"

$r_{geom, an}(A) := \Sigma_{\zeta \text{ in } \mu_\infty(\mathbb{C}) \text{ of deg} \leq N \text{ over } \mathbb{Q}} \mathrm{ord}_{T=\zeta}(\det(1 - TA))$.

[The definition of geometric analytic rank we have taken above is only the "correct" one when we are dealing with a situation where all the classes $\vartheta(E, s, \alpha, x)$ have characteristic polynomials in $\mathbb{Q}[T]$, as is the case, e.g, when we are dealing with L-functions of elliptic curves. In general, one must pick a finite extension $E_\lambda/\mathbb{Q}_\ell$ such that all the characteristic polynomials lie in $E_\lambda[T]$, and replace "N" by the maximum degree over $\mathbb{Q}$ of any of the finitely many roots of unity in $\overline{\mathbb{Q}}_\ell$ whose degree over E_λ is at most N. With any such definition, the Analytic Rank Theorem 13.1.7 below remains valid.]

Analytic Rank Theorem 13.1.7 Suppose that we are in the situation of the Uniform Output Theorem 12.4.8, with constants A and C. Suppose in addition that G is the full orthogonal group O(N). For each choice of $\varepsilon = \pm 1$, each choice of finite field E with $\#E > \mathrm{Max}(4A^2, 4C^2)$, and for each s in S(E), we denote by

$U_{s,E,\mathrm{sign}\varepsilon}(E) := \{x \text{ in } U_{s,E}(E) \text{ with } \vartheta(E, s, \alpha, x) \text{ in } O_{\mathrm{sign}\varepsilon}(N, \mathbb{R})\}$.

For any sequence of data (E_i, s_i, α_i) with each $\#E_i > \mathrm{Max}(4A^2, 4C^2)$ and with $\#E_i$ strictly increasing, we have the following tables of limit formulas for various average analytic ranks. In the tables, the first column is the quantity being averaged, the second is the set over which it is being averaged, and the third column is the limit, as $i \to \infty$, of the average.

$r_{an}(\vartheta(E_i, s_i, \alpha_i, x))$	$U_{s_i,E_i}(E_i)$	1/2,
$r_{an}(\vartheta(E_i, s_i, \alpha_i, x))$	$U_{s_i,E_i,\mathrm{sign}+}(E_i)$	0,
$r_{an}(\vartheta(E_i, s_i, \alpha_i, x))$	$U_{s_i,E_i,\mathrm{sign}-}(E_i)$	1,
$r_{quad,an}(\vartheta(E_i, s_i, \alpha_i, x))$	$U_{s_i,E_i}(E_i)$	1,
$r_{geom,an}(\vartheta(E_i, s_i, \alpha_i, x))$	$U_{s_i,E_i}(E_i)$	1,

Supplementary table, when N is odd

$r_{quad,an}(\vartheta(E_i, s_i, \alpha_i, x))$	$U_{s_i,E_i,\mathrm{sign}+}(E_i)$	1,
$r_{quad,an}(\vartheta(E_i, s_i, \alpha_i, x))$	$U_{s_i,E_i,\mathrm{sign}-}(E_i)$	1,
$r_{geom,an}(\vartheta(E_i, s_i, \alpha_i, x))$	$U_{s_i,E_i,\mathrm{sign}+}(E_i)$	1,
$r_{geom,an}(\vartheta(E_i, s_i, \alpha_i, x))$	$U_{s_i,E_i,\mathrm{sign}-}(E_i)$	1,

Supplementary table, when N is even

$r_{\text{quad,an}}(\vartheta(E_i, s_i, \alpha_i, x))$	$U_{s_i,E_i,\text{sign}+}(E_i)$	0,
$r_{\text{quad,an}}(\vartheta(E_i, s_i, \alpha_i, x))$	$U_{s_i,E_i,\text{sign}-}(E_i)$	2,
$r_{\text{geom,an}}(\vartheta(E_i, s_i, \alpha_i, x))$	$U_{s_i,E_i,\text{sign}+}(E_i)$	0,
$r_{\text{geom,an}}(\vartheta(E_i, s_i, \alpha_i, x))$	$U_{s_i,E_i,\text{sign}-}(E_i)$	2,

proof Straightforward application of equidistribution as incarnated in Theorem 13.1.5, cf. the proof of [deJ-Ka, 6.11], or of [Ka-TLMF, 8.3.3]. QED

(13.1.8) Let us now recall the definition of the eigenvalue location measures $\nu(c, O_{\text{sign}\varepsilon}(N, \mathbb{R}))$ of [Ka-Sar-RMFEM, 6.20.1]. Here $r \geq 1$ is an integer, $c = (c(1),..., c(r))$ in $\mathbb{Z}^r$ is an "offset vector", i.e.,

$$0 < c(1) < c(2) < ... < c(r),$$

and

$$N > 2c(r) + 2.$$

We must proceed through cases, depending on the the parity of N. and the value of $\varepsilon = \pm 1$.

(13.1.9) For N even = 2d, the eigenvalues of an element A in $O_{\text{sign}+}(N, \mathbb{R}) = SO(2d, \mathbb{R})$ are of the form

$$e^{\pm i\varphi(1)}, e^{i\varphi(2)},..., e^{\pm i\varphi(d)},$$

for a unique sequence of angles

$$0 \leq \varphi(1) \leq \varphi(2) \leq ... \leq \varphi(d) \leq \pi.$$

Formation of any given $\varphi(i)$ defines a continuous central function on $O_{\text{sign}+}(N, \mathbb{R}) = SO(2d, \mathbb{R})$,

$$A \mapsto \varphi(i)(A).$$

We rescale this function, and call it $\vartheta(i)$:

$$\vartheta(i)(A) := N\varphi(i)(A)/2\pi.$$

Given the offset vector c, we define the continuous central function

$$F_c : O_{\text{sign}+}(N, \mathbb{R}) \to \mathbb{R}^r,$$
$$F_c(A) := (\vartheta(c(1))(A),..., \vartheta(c(r))(A)).$$

We then define the probability measure $\nu(c, O_{\text{sign}+}(N, \mathbb{R}))$ on $\mathbb{R}^r$ to be

$$\nu(c, O_{\text{sign}+}(N, \mathbb{R}))$$
$$:= (F_c)_*(\text{normalized Haar measure } \mu_{\text{sign}+} \text{ on } O_{\text{sign}+}(N, \mathbb{R})).$$

(13.1.10) For N even = 2d, the eigenvalues of an element A in $O_{\text{sign}-}(N, \mathbb{R}) = O_-(2d, \mathbb{R})$ are of the form

$$\pm 1, e^{\pm i\varphi(1)}, e^{i\varphi(2)},..., e^{\pm i\varphi(d-1)},$$

for a unique sequence of angles

$$0 \leq \varphi(1) \leq \varphi(2) \leq ... \leq \varphi(d-1) \leq \pi.$$

Formation of any given $\varphi(i)$ defines a continuous central function on $O_{\text{sign}-}(N, \mathbb{R}) = O_-(2d, \mathbb{R})$,

$$A \mapsto \varphi(i)(A).$$

We rescale this function, and call it $\vartheta(i)$:

$$\vartheta(i)(A) := N\varphi(i)(A)/2\pi.$$

Given the offset vector c, we define the continuous central function

$$F_c : O_{sign-}(N, \mathbb{R}) \to \mathbb{R}^r,$$
$$F_c(A) := (\vartheta(c(1))(A),..., \vartheta(c(r))(A)).$$

We then define the probability measure $\nu(c, O_{sign-}(N, \mathbb{R}))$ on $\mathbb{R}^r$ to be

$$\nu(c, O_{sign-}(N, \mathbb{R}))$$
$$:= (F_c)_*(\text{normalized Haar measure } \mu_{sign-} \text{ on } O_{sign-}(N, \mathbb{R})).$$

(13.1.11) For N odd = 2d+1, the eigenvalues of an element A in $O_{sign\varepsilon}(N, \mathbb{R})$ are of the form

$$-\varepsilon,\ e^{\pm i\varphi(1)},\ e^{i\varphi(2)},...,\ e^{\pm i\varphi(d)},$$

for a unique sequence of angles

$$0 \le \varphi(1) \le \varphi(2) \le ... \le \varphi(d) \le \pi.$$

Formation of any given $\varphi(i)$ defines a continuous central function on $O_{sign\varepsilon}(N, \mathbb{R})$,

$$A \mapsto \varphi(i)(A).$$

We rescale this function, and call it $\vartheta(i)$:

$$\vartheta(i)(A) := N\varphi(i)(A)/2\pi.$$

Given the offset vector c, we define the continuous central function

$$F_c : O_{sign\varepsilon}(N, \mathbb{R}) \to \mathbb{R}^r,$$
$$F_c(A) := (\vartheta(c(1))(A),..., \vartheta(c(r))(A)).$$

We then define the probability measure $\nu(c, O_{sign\varepsilon}(N, \mathbb{R}))$ on $\mathbb{R}^r$ to be

$$\nu(c, O_{sign\varepsilon}(N, \mathbb{R}))$$
$$:= (F_c)_*(\text{normalized Haar measure } \mu_{sign\varepsilon} \text{ on } O_{sign\varepsilon}(N, \mathbb{R})).$$

Eigenvalue Location Theorem 13.1.12 Fix an integer $r \ge 1$, and an offset vector c in $\mathbb{Z}^r$. Suppose that we are in the situation of the Uniform Output Theorem 12.4.8, with constants A and C. Suppose in addition that G is the full orthogonal group O(N), with $N > 2c(r) + 2$. For each choice of $\varepsilon = \pm 1$, each choice of finite field E with $\#E > \text{Max}(4A^2, 4C^2)$, and for each s in S(E), the $\#U_{s,E,sign\varepsilon}(E)$ points $F_c(\vartheta(E, s, \alpha, x))$ in $\mathbb{R}^r$ are approximately equidistributed for the measure $\nu(c, O_{sign\varepsilon}(N, \mathbb{R}))$. More precisely, define a probability measure $\nu_{sign\varepsilon}(c, E, s, \alpha)$ on $\mathbb{R}^r$ by

$$\nu_{sign\varepsilon}(c, E, s, \alpha) := (F_c)_*\mu_{sign\varepsilon}(E, s, \alpha)$$
$$:= (1/\#U_{s,E,sign\varepsilon}(E))\ \Sigma_{x \text{ in } U_{s,E,sign\varepsilon}(E)}\ \delta_{F_c(\vartheta(E,\ s,\ \alpha,x))}.$$

In any sequence of data (E_i, s_i, α_i) with each $\#E_i > \text{Max}(4A^2, 4C^2)$ and with $\#E_i$ strictly increasing, the measures $\nu_{sign\varepsilon}(c, E, s, \alpha)$ tend weak * to the measure $\nu(c, O_{sign\varepsilon}(N, \mathbb{R}))$ on $\mathbb{R}^r$. In fact, for any continuous $\mathbb{C}$-valued function g on $\mathbb{R}^r$ of polynomial growth (and not just of compact support!), we have the limit formula

$$\int_{\mathbb{R}^r} g d\nu(c, O_{\mathrm{sign}\varepsilon}(N, \mathbb{R})) = \lim_{i\to\infty} \int_{\mathbb{R}^r} g d\nu_{\mathrm{sign}\varepsilon}(c, E_i, s_i, \alpha_i).$$

proof That the measures $\nu_{\mathrm{sign}\varepsilon}(c, E, s, \alpha) := (F_c)_* \mu_{\mathrm{sign}\varepsilon}(E, s, \alpha)$ converge weak $*$ to $(F_c)_*(\mu_{\mathrm{sign}\varepsilon}$ on $O_{\mathrm{sign}\varepsilon}(N, \mathbb{R}))$ is the direct image by F_c of Theorem 13.1.5. That we have convergence for continuous functions of polynomial growth results from the tail estimates of [Ka-Sar-RMFEM, 7.11.2]. QED

(13.2) Passage to the large N limit: general results

(13.2.1) Fix a prime ℓ. Suppose we are given a sequence, indexed by integers $j \geq 1$, of data as in 12.4.10, i.e.,

S_j, a normal connected affine $\mathbb{Z}[1/\ell]$-scheme $\mathrm{Spec}(A_j)$ of finite type,

a_j and b_j, integers,

$m_j > 1$ an integer,

V_j/S_j an affine S_j-scheme of finite type,

an S_j-morphism $h_j : V_j \to \mathbb{A}^{m_j}$,

an object L_j in $D^b{}_c(V_j, \overline{\mathbb{Q}}_\ell)$.which is ι-mixed of weight $\leq b_j$ and which is fibrewise perverse on V_j/S_j,

an integer $d_j \geq 2$,

a space of functions $(\mathcal{F}_j, \tau_j)$ on V_j, i.e., a locally free A_j-module of finite rank $\mathcal{F}_j$ and an A_j-linear map

$$\tau_j : \mathcal{F}_j \to \mathrm{Hom}_{S_j\text{-schemes}}(V_j, \mathbb{A}^{m_j}).$$

(13.2.2) We suppose that for each j, this data is strong standard input of type (a_j, b_j), in the sense of 12.4.11 We further suppose that for each j, we are given a dense open set $U_j \subset \mathcal{F}_j$ which meets every geometric fibre of $\mathcal{F}_j/S_j$, and an integer $N_j \geq 1$, such that for every j, our strong standard input data produces uniform output of type $(U_j, N_j, O(N_j))$ relative to S_j, in the sense of 12.4.14. Assume further that the sequence of integers N_j is strictly increasing. Denote by (A_j, C_j) the constants (A, C) occuring in Uniform Output Theorem 12.4.18 for the j'th input data.

Theorem 13.2.3 Suppose the hypotheses of 13.2.1 and 13.2.2 hold. Fix an integer $r \geq 1$, and an offset vector c in $\mathbb{Z}^r$. Then we have the following double limit formulas for the Katz-Sarnak measures $\nu(\varepsilon, c)$ on $\mathbb{R}^r$. Fix a continuous $\mathbb{C}$-valued function g of polynomial growth on $\mathbb{R}^r$. For each j large enough that $N_j > 2c(r) + 2$, pick a sequence of pairs $(E_{i,j}, s_{i,j})$ consisting of

a finite field $E_{i,j}$ with $\#E_{i,j} > \mathrm{Max}(4A_j{}^2, 4C_j{}^2)$, a point $s_{i,j}$ in $U_j(E_{i,j})$,

such that $\#E_{i,j}$ is a strictly increasing function of i. Form the measures

$$\mu_{\text{sign}\varepsilon}(E_{i,j}, s_{i,j}, \alpha_{i.j}) \text{ on } O_{\text{sign}\varepsilon}(N, \mathbb{R})^{\#},$$

as in 13.1.5, and their direct images

$$\nu_{\text{sign}\varepsilon}(c, E, s, \alpha) := (F_c)_{*}\mu_{\text{sign}\varepsilon}(E, s, \alpha) \text{ on } \mathbb{R}^r,$$

as in 13.1.9-11. Then we have the double limit formula

$$\int_{\mathbb{R}^r} g d\nu(,\varepsilon,c) = \lim_{j\to\infty} \lim_{i\to\infty} \int_{\mathbb{R}^r} g d\nu_{\text{sign}\varepsilon}(c, E_{i,j}, s_{i,j}, \alpha_{i,j}).$$

Remark 13.2.4 In the next section, we will regard this double limit formula as a statement about the totality of conjugacy classes $\vartheta(E_{i,j}, s_{i,j}, \alpha_{i,j}, x)$.

(13.3) Application to generalized Weierstrass families of elliptic curves

(13.3.1) Here we are given a sequence, indexed by integers $j \geq 1$, of data ($S_j/\mathbb{Z}[1/\ell]$, C_j/S_j of genus g_j, D_j on C_j of degree $d_j \geq 2g_j + 3$), as in 12.7.1-5. We define the integer N_j by

$$N_j := 4g_j - 4 + 12\deg(D_j).$$

We suppose that N_j is a strictly increasing function of j. For each j, we have the smooth S_j-scheme $GW_{Ifd}(C_j, D_j)$, which carries the corresponding family of elliptic curves in generalized Weierstrass form

$$y^2 + f_1xy + f_3y = x^3 + f_2x^2 + f_4x + f_6,$$

with f_ν in $L(\nu D_j)$. For each finite field k_j with $\#k_j > \text{Max}(A_j^2, C_j^2)$, for each point s_j in $S_j(k_j)$, and for each point f's in $GW_{Ifd}(C_j, D_j)_{s_j}$, we have an elliptic curve $E_{f's,s_j,k_j}$ over the function field of $C_{j,s_j}/k_j$, whose unitarized L-function $L(E_{f's,s_j,k_j}/k_j(C_{j,s_j}), T)$ is given as follows, cf. 12.7.8. There is a unique conjugacy class $\vartheta(k_j, s_j, \alpha_j, f's)$ in $O(N, \mathbb{R})^{\#}$ such that

$$L(E_{f's,s_j,k_j}/k_j(C_{j,s_j}), T) = \det(1 - T\vartheta(k_j, s_j, \alpha_j, f's)).$$

[Here $\alpha_j = (\#k_j)^2$. In 12.7.8, α_j is omitted in the notation.]

(13.3.2) For each j, the input data gives uniform output of type $(U, N, G) = (GW_{Ifd}(C_j, D_j), N_j, O(N_j))$, cf. 12.7.5. So we obtain the following two theorems. [Notice that not only the divisors D_j, but also the bases S_j and the curves C_j/S_j are allowed to vary with j. We expect that in most applications, S and C will be fixed, and that only D_j on C/S will vary. But by prudence we state the general version.]

Theorem 13.3.3 Hypotheses and notations as in 13.3.1 above, fix any $j \geq 1$. Consider all the conjugacy classes $\vartheta(k_j, s_j, \alpha_j, f's)$, which define the L-functions of the elliptic curves in the corresponding generalized Weierstrass family. The Analytic Rank Theorem 13.1.7 and the Eigenvalue Location Theorem 13.1.12 apply to these conjugacy classes.

Theorem 13.3.4 Fix an integer $r \geq 1$, and an offset vector c in $\mathbb{Z}^r$. Hypotheses and notations as in 13.3.1 above, for j with $N_j > 2c(r) + 2$, consider all the conjugacy classes $\vartheta(k_j, s_j, \alpha_j, f\text{'s})$. Then the double limit formulas of Theorem 13.2.3 hold for them, cf. 13.2.4.

(13.4) Application to usual Weierstrass families of elliptic curves

(13.4.1) Here we are given a sequence, indexed by integers $j \geq 1$, of data ($S_j/\mathbb{Z}[1/6\ell]$, C_j/S_j of genus g_j, pairwise disjoint sections $\{P_{i,j}\}_i$ in $C_j(S_j)$, effective divisors $D_{2,j} = \Sigma_i a_{i,j}P_{i,j}$ and $D_{3,j} = \Sigma_i b_{i,j}P_{i,j}$ on C_j of degrees $d_{2,j}$ and $d_{3,j}$, both $\geq 2g_j + 3$), as in 12.8.1-5. [In particular, all the hypotheses of Theorem 12.8.5 are assumed to hold hold.] We define the integers $c_{i,j}$ by

$$c_{i,j} := \mathrm{Max}(3a_{i,j}, 2b_{i,j}),$$

and the divisor $D_{max,j}$ by

$$D_{max,j} := \Sigma_i c_{i,j}P_{i,j}.$$

We then define the integer N_j by

$$N_j := 4g_j - 4 + \deg(D_{max,j}) + 2\#\{i \text{ with } c_{i,j} \not\equiv 0 \bmod 12\}.$$

We suppose that N_j is a strictly increasing function of j. For each j, we have the smooth S_j-scheme $W_{Ifd}(C_j, D_{2,j}, D_{3,j})$, which carries the corresponding family of elliptic curves in Weierstrass form

$$y^2 = 4x^3 - f_2x - f_3,$$

with f_2 in $L(D_{2,j})$ and f_3 in $L(D_{3,j})$. For each finite field k_j with $\#k_j > \mathrm{Max}(A_j^2, C_j^2)$, for each point s_j in $S_j(k_j)$, and for each point (f_2, f_3) in $W_{Ifd}(C_j, D_j)_{s_j}(k_j)$, we have an elliptic curve $E_{f\text{'s},s_j,k_j}$ over the function field of $C_{j,s_j}/k_j$, whose unitarized L-function $L(E_{f\text{'s},s_j,k_j}/k_j(C_{j,s_j}), T)$ is given as follows, cf. 12.8.7. There is a unique conjugacy class $\vartheta(k_j, s_j, \alpha_j, f\text{'s})$ in $O(N_j, \mathbb{R})^\#$ such that

$$L(E_{f\text{'s},s_j,k_j}/k_j(C_{j,s_j}), T) = \det(1 - T\vartheta(k_j, s_j, \alpha_j, f\text{'s})).$$

[Here $\alpha_j = 1$. In 12.8.7, α_j is omitted in the notation.]

(13.4.2) For each j, the input data gives uniform output of type $(U, N, G) = (W_{Ifd}(C_j, D_j), N_j, O(N_j))$, cf. 12.8.5. So we obtain the following two theorems. [Once again, notice that not only the divisors $D_{2,j}$ and $D_{3,j}$, but also the bases S_j and the curves C_j/S_j are allowed to vary with j.]

Theorem 13.4.3 Hypotheses and notations as in 13.4.1 above, fix any $j \geq 1$. Consider all the conjugacy classes $\vartheta(k_j, s_j, \alpha_j, f\text{'s})$, which define the L-functions of the elliptic curves in the corresponding usual Weierstrass family. The Analytic Rank Theorem 13.1.7 and the Eigenvalue Location Theorem 13.1.12 apply to these conjugacy classes.

Theorem 13.4.4 Fix an integer $r \geq 1$, and an offset vector c in $\mathbb{Z}^r$. Hypotheses and notations as in 13.4.1 above, for j with $N_j > 2c(r) + 2$, consider all the conjugacy classes $\vartheta(k_j, s_j, \alpha_j, f\text{'s})$. Then the double limit formulas of Theorem 13.2.3 hold for them, cf. 13.2.4.

(13.5) Applications to FJTwist families of elliptic curves

(13.5.1) Here we are given a sequence, indexed by integers $j \geq 1$, of data ($S_j/\mathbb{Z}[1/6\ell]$, C_j/S_j of genus g_j, pairwise disjoint sections $\{P_{i,j}\}_i$ in $C_j(S_j)$, effective divisors $D_{0,j} = \Sigma_i a_{i,j}P_{i,j}$ and $D_{1,j} = \Sigma_i b_{i,j}P_{i,j}$ on C_j of degrees $d_{0,j}$ and $d_{1,j}$, both $\geq 2g_j + 3$), as in 12.9.1-5. We define the integers $c_{i,j}$ by

$$c_{i,j} := 2a_{i,j} + b_{i,j},$$

and the integer N_j by

$$N_j := 4g_j - 4 + 2\deg(D_{0,j}) + 3\deg(D_{1,j}) + 2\#\{i \text{ with } c_{i,j} \not\equiv 0 \bmod 4\}.$$

We suppose that N_j is a strictly increasing function of j. For each j, we have the smooth S_j-scheme $\mathrm{FJTwist}(D_{0,j}, D_{1,j})$, which carries the corresponding family of elliptic curves in FJTwist form

$$y^2 = 4x^3 - 3f^2gx - f^3g,$$

with f in $L(D_{0,j})$ and g in $L(D_{1,j})$. For each finite field k_j with $\#k_j > \mathrm{Max}(A_j^2, C_j^2)$, for each point s_j in $S_j(k_j)$, and for each point (f, g) in $\mathrm{FJTwist}(D_{0,j}, D_{1,j})_{s_j}(k_j)$, we have an elliptic curve E_{f,g,s_j,k_j} over the function field of $C_{j,s_j}/k_j$, whose unitarized L-function $L(E_{f,g,s_j,k_j}/k_j(C_{j,s_j}), T)$ is given as follows, cf. 12.9.7. There is a unique conjugacy class $\vartheta(k_j, s_j, \alpha_j, f,g)$ in $O(N_j, \mathbb{R})^\#$ such that

$$L(E_{f,g,s_j,k_j}/k_j(C_{j,s_j}), T) = \det(1 - T\vartheta(k_j, s_j, \alpha_j, f,g)).$$

[Here $\alpha_j = 1$. In 12.9.7, α_j is omitted in the notation.]

(13.5.2) For each j, the input data gives uniform output of type $(U, N, G) = (\mathrm{FJTwist}(D_{0,j}, D_{1,j}), N_j, O(N_j))$, cf. 12.9.5. So we obtain the following two theorems. [Once again, notice that not only the divisors $D_{0,j}$ and $D_{1,j}$, but also the bases S_j and the curves C_j/S_j are allowed to vary with j.]

Theorem 13.5.3 Hypotheses and notations as in 13.5.1 above, fix any $j \geq 1$. Consider all the conjugacy classes $\vartheta(k_j, s_j, \alpha_j, f,g)$, which define the L-functions of the elliptic curves in the corresponding FJTwist family. The Analytic Rank Theorem 13.1.7 and the Eigenvalue Location Theorem 13.1.12 apply to these conjugacy classes.

Theorem 13.5.4 Fix an integer $r \geq 1$, and an offset vector c in $\mathbb{Z}^r$. Hypotheses and notations as in 13.5.1 above, for j with $N_j > 2c(r) + 2$, consider all the conjugacy classes $\vartheta(k_j, s_j, \alpha_j, f, g)$. Then the double limit formulas of Theorem 13.2.3 hold for them, cf. 13.2.4.

(13.6) Applications to pullback families of elliptic curves
(13.6.1) Here we fix a normal connected affine $\mathbb{Z}[1/6\ell]$-scheme $S = \mathrm{Spec}(A)$ of finite type, an integer $t \geq 1$, and a Cartier divisor T in $\mathbb{A}^1{}_S$ which is finite etale over S of degree t. We suppose further that T is defined by the vanishing of a monic polynomial T(x) in A[x] which is monic of degree t, and whose discriminant is a unit in A. We suppose given a relative elliptic curve

$$\pi : \mathcal{E} \to \mathbb{A}^1{}_S - T,$$

which, on each geometric fibre of $\mathbb{A}^1{}_S/S$, has multiplicative reduction at some point of T.
(13.6.2) Having fixed this initial input data as in 12.10.1, we suppose given a sequence, indexed by integers $j \geq 1$, of data (C_j/S of genus g_j, pairwise disjoint sections $\{P_{i,j}\}_i$ in $C_j(S)$, an effective divisors $D_j = \Sigma_i a_{i,j}P_{i,j}$ in C_j), as in 12.10.2. For each j, we form the integers N_{down}, $N_{up,j}$, and $N_j := N_{up,j} - N_{down}$, as in 12.10.5. We suppose that each $N_j \geq 9$, and that N_j is a strictly increasing function of j. For each j, we have the smooth S-scheme $U_{D_j,T}$ of 12.10.3, which carries the family of pullback elliptic curves

$$f^*\mathcal{E} \text{ over } C_j - D_j - f^{-1}T,$$

f in $U_{D_j,T}$. For each finite field k_j with $\#k_j > \mathrm{Max}(A_j{}^2, C_j{}^2)$, for each point s_j in $S(k_j)$, and for each point f in $(U_{D_j,T})_{s_j}(k_j)$, we have an elliptic curve $f^*\mathcal{E}_{s_j,k_j}$ over the function field of $C_{j,s_j}/k_j$. The unitarized L-function $L(f^*\mathcal{E}_{s_j,k_j}/k_j(C_{j,s_j}), T)$ is a polynomial of degree $N_{up,j}$, which is always divisible as a polynomial by the unitarized L-function $L(\mathcal{E}_{s_j,k_j}/k_j(\mathbb{P}^1), T)$ of the original elliptic curve $\mathcal{E}_{s_j,k_j}/k_j(\mathbb{P}^1)$. This latter L-function is a polynomial of degree N_{down}. The quotient of these L-functions, the "new part", is given as follows, cf. 12.10.9. There is a unique conjugacy class $\vartheta(k_j, s_j, \alpha_j, f)$ in $O(N_j, \mathbb{R})^\#$ such that

$$L(f^*\mathcal{E}_{s_j,k_j}/k_j(C_{j,s_j}), T)/L(\mathcal{E}_{s_j,k_j}/k_j(\mathbb{P}^1), T)$$
$$= \det(1 - T\vartheta(k_j, s_j, \alpha_j, f)).$$

[Here $\alpha_j = 1$. In 12.10.9, α_j is omitted in the notation.]

(13.6.3) For each j, the input data gives uniform output of type $(U, N, G) = (U_{D_j,T}, N_j, O(N_j))$, cf. 12.10.7. So we obtain the following two theorems.

Theorem 13.6.4 Hypotheses and notations as in 13.6.1 and 13.6.2 above, fix any $j \ge 1$. Consider the conjugacy classes $\vartheta(k_j, s_j, \alpha_j, f)$, which define the "new parts" of the L-functions of the elliptic curves in the corresponding pullback families. The Analytic Rank Theorem 13.1.7 and the Eigenvalue Location Theorem 13.1.12 apply to these conjugacy classes.

Theorem 13.6.5 Fix an integer $r \ge 1$, and an offset vector c in $\mathbb{Z}^r$. Hypotheses and notations as in 13.6.1 and 13.6.2 above, for j with $N_j > 2c(r) + 2$, consider all the conjugacy classes $\vartheta(k_j, s_j, \alpha_j, f)$. Then the double limit formulas of Theorem 13.2.3 hold for them, cf. 13.2.4.

(13.7) Applications to quadratic twist families of elliptic curves

(13.7.1) We work over an affine $\mathbb{Z}[1/2\ell]$-scheme $S = \mathrm{Spec}(A)$ of finite type. We fix a genus $g \ge 0$ and a projective smooth curve C/S with geometrically connected fibres of genus g. We suppose given a sequence, indexed by integers $j \ge 1$, of data (pairwise disjoint sections $\{P_{i,j}\}_i$ in C(S), disjoint effective Cartier divisors T_j and $D_j = \Sigma_i a_{i,j}P_{i,j}$ in C, with T_j/S finite etale of degree t_j, D_j/S finite flat of degree $d_j \ge 2g+3$, an elliptic curve $\mathcal{E}_j/C - T_j - D_j$ which, on every geometric fibre of $(C - T_j - D_j)/S$ has multiplicative reduction at some point of T_j, and is everywhere tamely ramified), as in 12.11.1 and 12.11.2. We form the integer N_j as in 12.11.5. We suppose that each $N_j \ge 9$, and that N_j is a strictly increasing function of j. For each j, we have the smooth S-scheme $\mathcal{F}ct(C\ d_j, D_j, T_j)$, cf. 12.11.4, which carries the quadratic twist family

$$\mathcal{E}_j \otimes \chi_2(f),$$

f in $\mathcal{F}ct(C\ d_j, D_j, T_j)$. For each finite field k_j with $\#k_j > \mathrm{Max}(A_j^2, C_j^2)$, for each point s_j in $S(k_j)$, and for each point f in $\mathcal{F}ct(C\ d_j, D_j, T_j)_{s_j}(k_j)$, we have an elliptic curve $\mathcal{E}_{j,s_j,k_j} \otimes \chi_2(f)$ over the function field of C_{s_j}/k_j. The unitarized L-function $L(\mathcal{E}_{j,s_j,k_j} \otimes \chi_2(f)/k_j(C_{s_j}), T)$ is given as follows, cf. 12.11.7. There is a unique conjugacy class $\vartheta(k_j, s_j, \alpha_j, f)$ in $O(N_j, \mathbb{R})^\#$ such that

$$L(\mathcal{E}_{j,s_j,k_j} \otimes \chi_2(f)/k_j(C_{s_j}), T) = \det(1 - T\vartheta(k_j, s_j, \alpha_j, f,g)).$$

[Here α_j is a choice of $\mathrm{Sqrt}(\#k_j)$. In 12.11.7, α_j is omitted in the notation.]

(13.7.2) For each j, the input data gives uniform output of type $(U, N, G) = (Fct(C\ d_j, D_j, T_j), N_j, O(N_j))$, cf. 12.11.6. So we obtain the following two theorems.

Theorem 13.7.3 Hypotheses and notations as in 13.1.1 above, fix any $j \geq 1$. Consider the conjugacy classes $\vartheta(k_j, s_j, \alpha_j, f)$, which define the L-functions of the elliptic curves in the corresponding quadratic twist families. The Analytic Rank Theorem 13.1.7 and the Eigenvalue Location Theorem 13.1.12 apply to these conjugacy classes.

Theorem 13.7.4 Fix an integer $r \geq 1$, and an offset vector c in $\mathbb{Z}^r$. Hypotheses and notations as in 13.7 above, for j with $N_j > 2c(r) + 2$, consider all the conjugacy classes $\vartheta(k_j, s_j, \alpha_j, f)$. Then the double limit formulas of Theorem 13.2.3 hold for them, cf. 13.2.4.

References

[ATLAS] Conway, J. H., Curtis, R. T., Norton, S. P., Parker, R. A., and Wilson, R. A., *Atlas of finite groups. Maximal subgroups and ordinary characters for simple groups. With computational assistance from J. G. Thackray*, Oxford University Press, Oxford, 1985.

[Ax] Ax, J., Zeroes of polynomials over finite fields, Amer. J. Math. 86 (1964), 255-261.

[BBD] Beilinson, A. A., Bernstein, I. N., and Deligne, P., Faisceaux Pervers, entire contents of *Analyse et Topologie sur les espaces singuliers I*, Conference de Luminy, Asterisque 100, 1982.

[Bbki-Lie VI] Bourbaki, N., *Groupes et Algèbres de Lie*, Chapitres 4, 5, et 6, Masson, Paris, 1981.

[Blich] Blichfeldt, H. F., *Finite Collineation Groups*, U. Chicago Press, Chicago, 1917.

[Chev] Chevalley, C., *Théorie des Groupes de Lie Vol. III*, Hermann, Paris, 1954.

[Dav-Has] Davenport, H., and Hasse, H., Die Nullstellen der Kongruenzzetafunktionen in gewissen zylkischen Fallen, J. Reine Angew. Math. 172 (1934), 151-162.

[De-Const] Deligne, P., Les constantes des équations fonctionelles des fonctions L, pp. 168-232 in *Modular Functions of One Variable III*, Springer Lecture Notes in Mathematics 349, 1973.

[De-Fin] Deligne, P., Théorèmes de finitude en cohomologie ℓ-adique, pp. 233-251 in *Cohomologie Etale (SGA 4 1/2)*, Springer Lecture Notes in Mathematics 569, 1977.

[De-SomTrig] Deligne, P., Application de la formule des traces aux sommes trigonométriques, pp. 168-232 in *Cohomologie Etale (SGA 4 1/2)*, Springer Lecture Notes in Mathematics 569, 1977.

[De-Weil I] Deligne, P., La conjecture de Weil I, Pub. Math. I.H.E.S. 48 (1974), 273-308.

[De-Weil II] Deligne, P., La conjecture de Weil II, Pub. Math. I.H.E.S. 52 (1981), 313-428.

[deJ-Ka] de Jong, A. J., and Katz, N., Monodromy and the Tate conjecture: Picard numbers and Mordell-Weil ranks in families, Israel J. Math. 120 (2000), part A, 47-79.

[Dorn] Dornhoff, L., *Group representation theory. Part A: Ordinary representation theory*, Pure and Applied Mathematics 7, Marcel Dekker, New York, 1971.

[EGA] Grothendieck, A., rédigé avec la collaboration de Dieudonné, J., *Eléments de Géométrie Algébrique*, Pub. Math. I.H.E.S. 4, 8, 11, 17, 20, 24, 28, 32, 1960-1967.

[Erne] Erné, R., Construction of a del Pezzo surface with maximal Galois action on its Picard group, J. Pure Appl. Algebra 97 (1994), no. 1, 15-27.

[Fuj-Indep] Fujiwara, K., Independence of ℓ for intersection cohomology (after Gabber), in *Algebraic Geometry 2000, Azumino*, Advanced Studies in Pure Mathematics 36 (2002), 145-151.

[GT] Guralnick, R., and Tiep, P., Decomposition of small tensor powers, 2003 preprint available at http://math.usc.edu/~guralnic/Preprints/.

[Hasse] Hasse, H., Zur Theorie der abstrakten elliptischen Funktionenkörper I, II, III, J. Reine Angew. Math. 175 (1936), 55-62, 69-88, and 193-208.

[Huff-Eig] Huffman, W. C., Linear groups containing an element with an eigenspace of codimension two, J. Algebra 34 (1975), 260-287.

[Huff-Wales-Eig] Huffman, W. C., and Wales, D. B., Linear groups containing an element with an eigenspace of codimension two, pp. 425-429 in *Proceedings of the Conference on Finite Groups* (Univ. Utah, Park City, Utah, 1975), Academic Press, New York, 1976.

[Huff-Wales-Equal] Huffman, W. C., and Wales, D. B., Linear graphs of degree n containing an element with exactly n-2 equal eigenvalues, Collection of articles dedicated to Olga Taussky-Todd, Linear and Multilinear Algebra 3 (1975/76), no. 1/2, 53-59.

[Huff-Wales-Inv] Huffman, W. C. , and Wales, D. B., Linear groups containing an involution with two eigenvalues -1, J. Algebra 45 (1977), no. 2, 465-515.

[Ka-ACT] Katz, N., Affine cohomological transforms, perversity and monodromy, JAMS 6 (1993), no. 1, 149-222.

[Ka-ENSMCS] Katz, N., Estimates for nonsingular multiplicative character sums, Int. Math. Res. Not. (2002), no. 7, 333-349.

[Ka-ESDE] Katz, N., *Exponential sums and differential equations*, Annals of Math. Study 124, Princeton Univ. Press, Princeton, N.J., 1990.

[Ka-GKM] Katz, N., *Gauss sums, Kloosterman sums, and monodromy groups*, Annals of Math. Study 116, Princeton Univ. Press, Princeton, N.J., 1988.

[Ka-LAMM] Katz, N., Larsen's Alternative, moments, and the monodromy of Lefschetz Pencils, pp. 521-560 in *Contributions to automorphic forms, geometry, and number theory*, Johns Hopkins Univ. Press, Baltimore, MD, 2004.

[Ka-Lau] Katz, N., and Laumon, G., Transformation de Fourier et majoration de sommes exponentielles, Pub. Math. I.H.E.S. 62 (1986), 361-418.

[Ka-LFM] Katz, N., L-functions and monodromy: four lectures on Weil II, Adv. Math. 160 (2001), no. 1, 81-132

[Ka-LG] Katz, N., Local to global extensions of representations of fundamental groups, Ann. Inst. Fourier 36 (1986), no 4, 59-106.

[Ka-MG] Katz, N., On the monodromy groups attached to certain families of exponential sums, Duke Math. J. 54 (1987), no. 1, 41-56.

[Ka-Pan] Katz, N., and Pandharipande, R., Inequalities related to Lefschetz pencils and integrals of Chern classes, pp. 805-819 in *Geometric Aspects of Dwork Theory, Vol. 2,* de Gruyter, Berlin, 2004.

[Ka-RLS] Katz, N., *Rigid Local Systems,* Annals of Math. Study 138, Princeton Univ. Press, Princeton, N.J., 1995.

[Ka-Sar-RMFEM] Katz, N., and Sarnak, P., *Random Matrices, Frobenius Eigenvalues, and Monodromy,* A.M.S. Colloquium Publications 45, 1999.

[Ka-SE] Katz, N., *Sommes Exponentielles,* rédigé par G. Laumon, Astérisque 79, 1980.

[Ka-SMD] Katz, N., A semicontinuity result for monodromy under degeneration, Forum Math. 15 (2003), no. 2, 191-200.

[Ka-TLFM] Katz, N., *Twisted L-Functions and Monodromy,* Annals of Math. Study 150, Princeton Univ. Press, Princeton, N.J., 2002.

[Lar-Char] Larsen, Michael, A characterization of classical groups by invariant theory, preprint, middle 1990's.

[Lar-Normal], Larsen, Michael, The normal distribution as a limit of generalized Sato-Tate measures, preprint, early 1990's.

[Lau-TF] Laumon, G., Transformation de Fourier, constantes d'équations fonctionelles et conjecture de Weil, Pub. Math. I.H.E.S. 65 (1987), 131-210.

[Lau-SC] Laumon, G., Semi-continuité du conducteur de Swan (d'après P. Deligne), in *Caractéristique d'Euler-Poincaré, Seminaire E.N.S. 1978-79*, Astérisque 82-83 (1981), 173-219.

[LW] Lang, S., and Weil, A., Number of points of varieties in finite fields, Amer. J. Math. 76 (1954), 819-827.

[Manin] Manin, Y. I., *Cubic Forms: Algebra, Geometry, Arithmetic*, North-Holland, Amsterdam, 1974.

[Mit] Mitchell, H., Determination of all primitive collineation groups in more than four variables which contain homologies, Amer. J. Math. 36 (1914), 1-12.

[Rains] Rains, E., Increasing subsequences and the classical groups, Electron. J. Combin. 5 (1998), no. 1, Research Paper 12, 9 pp. (electronic).

[Ray] Raynaud, M., Caractéristique d'Euler-Poincaré d'un faisceau et cohomologie des variétés abéliennes, Exposé 286 in *Séminaire Bourbaki 1964/65*, W.A. Benjamin, New York, 1966.

[Serre-CL] Serre, J.-P., *Corps locaux*, Troisième édition, corrigée, Publications de l'Université de Nancago, No. VIII. Hermann, Paris, 1968.

[Serre-TGT] Serre, J.-P., *Topics in Galois theory* (Lecture notes prepared by Henri Darmon, with a foreword by Darmon and the author), Research Notes in Mathematics 1, Jones and Bartlett Publishers, Boston, MA, 1992.

[SGA] Grothendieck, A., et al, *Séminaire de Géométrie Algébrique du Bois-Marie*, SGA 1, SGA 4 Parts I, II, and III, SGA 4½, SGA 5, SGA 7 Parts I and II, Springer Lecture Notes in Mathematics, volumes 224, 269, 270, 305, 569, 589, 288, 340, 1971-1977.

[Shioda-MWL] Shioda, T., On the Mordell-Weil lattices, Comment. Math. Univ. St. Paul., 39 (1990), 211-240.

[Sil-ATEC] Silverman, J., *Advanced topics in the arithmetic of elliptic curves*, Graduate Texts in Mathematics 151, Springer-Verlag, New York, 1994.

[Stan-ECI] Stanley, R. P., *Enumerative combinatorics Vol. 1.* With a foreword by Gian-Carlo Rota. Corrected reprint of the 1986 original. Cambridge Studies in Advanced Mathematics 49, Cambridge University Press, Cambridge, 1997.

[vdG-vdV-RM] van der Geer, G., and van der Vlugt, M., Reed-Muller codes and supersingular curves I, Compositio Math. 84 (1992), no. 3, 333-367.

[Wales] Wales, David, Quasiprimitive linear groups with quadratic elements, J. Algebra 245 (2001), no. 2, 584-606.

[Wales-Inv] Wales, D. B., Linear groups of degree n containing an involution with two eigenvalues -1 II, J. Algebra 53 (1978), no. 1, 58-67.

[Weyl] Weyl, H., *Classical Groups*, Princeton Univ. Press, Princeton, N.J., 1946.

[Zal] Zalesskii , A. E., Linear groups, Russian Math Surveys 36 (1981), no. 5, 63-128.

Notation Index

Subject Index

Boldface page numbers indicate definitions or statements of theorems.